Editorial Office
ASM International
Materials Park, OH 44073-0002
USA
Tel: 440/338-5151, ext. 5240
Fax: 440/338-4634

Office of Publication
Springer, 233 Spring Street, New York, NY 10013, USA
Telephone: 212-460-1500
E-mail: journals-ny@springer.com
Website: www.springer.com

Subscription Information
Journal of Materials Engineering and Performance (coden: JMEPEG) is published monthly except January, May, and September for a total of 9 times per year by Springer Science+Business Media, LLC (Springer), Volume 18 (9 issues) will be published in 2009.

ISSN: 1059-9495 (print version)
ISSN: 1544-1024 (electronic version)

Subscription Rates, Orders, Inquiries:
Please contact our Customer Service department for the latest rates and pricing information: *The Americas (North, South, Central America, and the Caribbean)*:
MAIL: Journals Customer Service
P.O. Box 2485
Secaucus, NJ 07096, USA
TEL: 800-777-4643; 212-460-1500
(outside North America)
FAX: 201-348-4505
E-MAIL: journals-ny@springer.com; servicio-ny@springer.com (Central and South America)

Outside of the Americas:
MAIL: Journals Customer Service
Springer Customer Service Center
Haberstrasse 7
69126 Heidelberg, GERMANY
TEL: 49-6221-345-4303
FAX: 49-6221-345-4229
E-MAIL: subscriptions@springer.com

The *Journal of Materials Engineering and Performance®* publishes contributions on all aspects of materials selection, design, processing, characterization, and evaluation. The scope includes all materials used in engineering applications: those that typically result in components for larger systems.

Relationships between material type, secondary processing, and component performance are emphasized. Improvement of materials properties through processes such as casting, forming, heat treating, surface modification and coating, fabrication, and the control of these processes is presented. Testing and characterization are demonstrated through mechanical and physical tests, NDE, metallography, failure analysis, corrosion resistance, chemical analysis, surface characterization, and microanalysis of surfaces, features and fractures.

Case studies, examples, and illustrations using these technologies to achieve innovative applications of engineered materials are highlighted.

The *Journal of Materials Engineering and Performance®* (ISSN: 1059-9495; CODEN: JMEPEG) is published monthly except January, May, and September for a total of 9 times per year by Springer Science+Business Media, LLC (Springer). Periodicals postage paid at New York, New York and additional mailing offices.

Postmaster: Send address changes to: *Journal of Materials Engineering and Performance*, Springer, 233 Spring Street, New York, NY 10013, USA.

Printed in the United States of America on acid-free paper.

The Materials Information Society

Vol. 18, Nos. 5–6
August 2009
ISSN: 1059-9495
JMEPEG 18(5–6) 447–848 (2009)

Journal of Materials Engineering and Performance®

Contents

Special Issue Editorial

Editorial
Ausonio Tuissi, M.R. Mitchell, and José San Juan 447

Special Issue: SMST 2008

Characterization of Non-Metallic Inclusions in Superelastic NiTi Tubes
Alejandro Toro, Fei Zhou, Ming H. Wu, William Van Geertruyden, and Wojciech Z. Misiolek 448

Cu-Al-Ni-SMA-Based High-Damping Composites
Gabriel A. López, Mariano Barrado, Jose San Juan, and María Luisa Nó 459

Use of Circular Foldable Nitinol Blades for Resecting Calcified Aortic Heart Valves
Florian Hauck, Daniel Wendt, Sebastian Stühle, Emilia Kawa, Hermann Wendt, Wiebke Müller, Matthias Thielmann, Brigitte Kipfmüller, Bernd Vogel, and Heinz Jakob 463

Nitinol Surfaces for Implantation
Svetlana Shabalovskaya, Gianni Rondelli, and Markus Rettenmayr 470

Influence of Quenching Rates on Equiatomic NiTi Ribbons Fabricated by Melt-Spinning
K. Mehrabi, M. Bruncko, B.J. McKay, and A.C. Kneissl 475

A Comparison of Chemistry and Inclusion Distribution and Morphology Versus Melting Method of NiTi Alloys
George M. Kramer 479

Processing of Shape Memory CuZnAl Open-cell Foam by Molten Metal Infiltration
E.M. Castrodeza, C. Mapelli, M. Vedani, S. Arnaboldi, P. Bassani, and A. Tuissi 484

Design and Characterization of New Ti-Nb-Hf Alloys
M. González, J. Peña, J.M. Manero, M. Arciniegas, and F.J. Gil 490

High-Temperature Transformation Processes in Cu-13.4Al-5Ni Shape Memory Alloy Single Crystals
M. Benke, V. Mertinger, and L. Daróczi 496

Factors Influencing the Reversion of Stress-induced Martensite to Austenite in a Fe-Mn-Si-Cr-Ni Shape Memory Alloy
Leandru G. Bujoreanu, Sergiu Stanciu, Radu I. Comaneci, Markus Meyer, Vasile Dia, and Ciprian Lohan 500

Optimization of the Ti-16.2Hf-24.8Nb-1Zr Alloy by Cold Working
M. González, J. Peña, J.M. Manero, M. Arciniegas, and F.J. Gil 506

Properties and Heat Treatment of High Transition Temperature Ni-Ti-Hf Alloys
C. Craig Wojcik 511

Mechanical Analysis of Hybrid Textile Composites with NiTi Wires
Elena Villa, Sergio Arnaboldi, Ausonio Tuissi, Marco Giacomelli, and Elena Turco 517

Embedding of Superelastic SMA Wires into Composite Structures: Evaluation of Impact Properties
Silvio Pappadà, Rocco Rametta, Luca Toia, Alberto Coda, Luca Fumagalli, and Alfonso Maffezzoli 522

Mechanical and Vibration Characteristics of Laminated Composite Plates Embedding Shape Memory Alloy Superelastic Wires
Silvio Pappadà, Per Gren, Kourosh Tatar, Tord Gustafson, Rocco Rametta, Ettore Rossini, and Alfonso Maffezzoli 531

Surface Form Memory in NiTi: Energy Density of Constrained Recovery During Indent Replication
Xueling Fei, Corey J. O'Connell, David S. Grummon, and Yang-Tse Cheng 538

Texture Development and Phase Transformation Behavior of Sputtered Ni-Ti Films
R.M.S. Martins, N. Schell, K.K. Mahesh, L. Pereira, R.J.C. Silva, and F.M. Braz Fernandes 543

Properties of Co-alloyed Ni-Fe-Ga Ferromagnetic Shape Memory Alloys
V.A. Chernenko, K. Oikawa, M. Chmielus, S. Besseghini, E. Villa, F. Albertini, L. Righi, A. Paoluzi, P. Müllner, R. Kainuma, and K. Ishida 548

Analysis of Variant Orientation Before and After Compression in Polycrystalline $Ni_{50}Mn_{29}Ga_{21}$ MSMA
C. Hürrich, H. Wendrock, M. Pötschke, U. Gaitzsch, S. Roth, B. Rellinghaus, and L. Schultz 554

A Study of the Effect of Diameter on the Fatigue Properties of NiTi Wire
Dennis W. Norwich and Audrey Fasching 558

Fatigue Performance of Nitinol Round Wire with Varying Cold Work Reductions
Jeremy E. Schaffer and David L. Plumley 563

Fracture Behaviour of Nickel-Titanium Laser Welded Joints
C. Maletta, A. Falvo, F. Furgiuele, G. Barbieri, and M. Brandizzi 569

Influence of Carbon on the Microstructure of a Fe-Mn-Si-Cr-Ni Alloy
Khaled M. Mostafa, J. De Baerdemaeker, N. Van Caenegem, D. Segers, and Y. Houbaert 575

Editorial Committee

The *Journal of Materials Engineering and Performance*® is covered in the following Institute for Scientific Information® products: Current Contents—Engineering, Computing & Technology, Research Alert®, and the Materials Science Citation Index®.

The contents of this journal are also indexed or abstracted by: Cambridge Scientific Abstracts (CSA), Compendex, Engineering Index, Institut de L'Information Scientifique et Technique (INIST), Urethane Abstracts (UA) and Polymer Blends, Alloys & Interpenetrating Polymer Networks Abstracts (PBAI), Technomic Publishing Company, Inc., the American Institute of Aeronautics and Astronautics, Inc., Applied Science & Technology Index, and Applied Science & Technology Abstracts.

Journal of Materials Engineering and Performance®

Editor
Jeffrey A. Hawk

Associate Editors
Rajiv Asthana
Narendra Dahotre
Omar S. Es-Said

Published by
Springer Science+Business Media, LLC

ASM Staff

Manager, Technical Journals
Mary Anne Fleming

Manager, Production
Bonnie Sanders

Production Coordinator
Diane Whitelaw

Change of Address: Allow six weeks for all changes to become effective. All communications should include both old and new addresses (with zip codes) and should be accompanied by a mailing label from a recent issue.
Back Volumes: Prices for back volumes are available on request.
Microform Editions: Available from: University Microfilms International, 300 N. Zeeb Road, Ann Arbor, MI 48106, USA.
Reprints: Copies of the *Journal of Materials Engineering and Performance*® articles are available as reprints from Sheridan Press, Reprint Service, 450 Fame Avenue, Hanover, PA 17331, USA, www.sheridan.com/springer/eoc or Phone 800-352-2110, Ext. 8134

Submitting a manuscript?
See Author Instructions,
inside the front and back covers.

Structure-Property Relationships in Conventional and Nanocrystalline NiTi Intermetallic Alloy Wire
Jeremy E. Schaffer 582

Development of Texture and Microstructure During Cold Rolling and Annealing of a Fe-Based Shape Memory Alloy
Bikas C. Maji, Madangopal Krishnan, Vijay Hiwarkar, Indradev Samajdar, and R.K. Ray 588

Thermomechanical Properties of Porous NiTi Alloy Produced by SHS
P. Bassani, P. Giuliani, A. Tuissi, and C. Zanotti 594

Effect of Impact-Induced Strain on the Stress-Induced Martensitic Transformation of Superelastic NiTi Shape-Memory Alloy Wires
J. Zurbitu, G. Castillo, I. Urrutibeascoa, and J. Aurrekoetxea 600

Factors Controlling Superelastic Damping Capacity of SMAs
L. Heller, P. Šittner, J. Pilch, and M. Landa 603

Application of SMA Composites in the Collectors of the Railway Pantograph for the Italian High-Speed Train
A. Tuissi, P. Bassani, R. Casati, M. Bocciolone, A. Collina, M. Carnevale, A. Lo Conte, and B. Previtali 612

Characterization of Transformation Temperatures with the Bend and Free Recovery Technique: Parameters and Effects
Masao Drexel, Jim Proft, and Scott Russell 620

Implementation of the Müller-Achenbach-Seelecke Model for Shape Memory Alloys in ABAQUS
Frank Richter, Oliver Kastner, and Gunther Eggeler 626

Numerical Study on the Influence of Material Characteristics on Ni-Ti Endodontic Instrument Performance
Lorenza Petrini, Silvia Necchi, Silvio Taschieri, and Francesco Migliavacca 631

Conceptual Design and Simulation of a Compact Shape Memory Actuator for Rotary Motion
I. Spinella, G. Scirè Mammano, and E. Dragoni 638

SMA Numerical Modeling Versus Experimental Results: Parameter Identification and Model Prediction Capabilities
Ferdinando Auricchio, Alberto Coda, Alessandro Reali, and Marco Urbano 649

Study of a Nitinol Stent Deployed into Anatomically Accurate Artery Geometry and Subjected to Realistic Service Loading
Nuno Rebelo, Rachel Fu, and Michael Lawrenchuk 655

Carbon Fiber Reinforced Smart Laminates with Embedded SMA Actuators—Part I: Embedding Techniques and Interface Analysis
P. Bettini, M. Riva, G. Sala, L. Di Landro, A. Airoldi, and J. Cucco 664

Carbon Fiber-Reinforced Smart Laminates with Embedded SMA Actuators—Part II: Numerical Models and Empirical Correlations
M. Riva, P. Bettini, L. Di Landro, G. Sala, and A. Airoldi 672

Stress-Induced Martensitic Transformation in the Crack Tip Region of a NiTi Alloy
A. Falvo, F. Furgiuele, A. Leonardi, and C. Maletta 679

Simulation-Based Design of a Rotatory SMA Drive
Sascha Dilthey and Horst Meier 686

SmartFlex® NiTi Wires for Shape Memory Actuators
L. Fumagalli, F. Butera, and A. Coda 691

A Novel SMA-based Concept for Airfoil Structural Morphing
S. Barbarino, R. Pecora, L. Lecce, A. Concilio, S. Ameduri, and E. Calvi 696

Shape Memory Hooks Employed in Fasteners
D. Vokoun, D. Majtás, M. Frost, P. Sedlák, and P. Šittner 706

"One-Module"-Actuators Based on Partial Activation of Shape Memory Components
S. Langbein and E.G. Welp 711

Electrically Actuated Antiglare Rear-View Mirror Based on a Shape Memory Alloy Actuator
T. Luchetti, A. Zanella, M. Biasiotto, and A. Saccagno 717

Investigation on the Hysteretic Behavior of NiTi Shape Memory Wires Actuated Under Quasi-Equilibrium and Dynamic Conditions
A. Coda, M. Urbano, L. Fumagalli, and F. Butera 725

Feasibility Investigation of Superelastic Effect Devices for Seismic Isolation Applications
G. Attanasi, F. Auricchio, and G.L. Fenves 729

Damping in Civil Engineering Using SMA. The Fatigue Behavior and Stability of CuAlBe and NiTi Alloys
V. Torra, A. Isalgue, C. Auguet, G. Carreras, F.C. Lovey, H. Soul, and P. Terriault 738

Shape Memory Alloy Tension/Compression Device for Seismic Retrofit of Buildings
Matthew Speicher, Darel E. Hodgson, Reginald DesRoches, and Roberto T. Leon 746

The Effect of Exposure to Simulated Body Fluids on Breakdown Potentials
Clifford P. Warner 754

A Comparative Biocompatibility Analysis of Ternary Nitinol Alloys
Waseem Haider, Norman Munroe, Chandan Pulletikurthi, Puneet K. Singh Gill, and Sushma Amruthaluri 760

Enhanced Biocompatibility of Porous Nitinol
Norman Munroe, Chandan Pulletikurthi, and Waseem Haider 765

Acceptance Criteria for Corrosion Resistance of Medical Devices: Statistical Analysis of Nitinol Pitting in In Vivo Environments
Lawrence E. Eiselstein, Duane Steffey, Andrew Nissan, Nigel Corlett, Roberto Dugnani, Esra Kus, and Sarah G. Stewart 768

Analysis and Evaluation of the Dynamic Performance of SMA Actuators for Prosthetic Hand Design
Kevin T. O'Toole, Mark M. McGrath, and Eugene Coyle 781

Nitinol Embolic Protection Filters: Design Investigation by Finite Element Analysis
Michele Conti, Matthieu De Beule, Peter Mortier, Denis Van Loo, Pascal Verdonck, Frank Vermassen, Patrick Segers, Ferdinando Auricchio, and Benedict Verhegghe 787

Static and Cyclic Load-Deflection Characteristics of NiTi Orthodontic Archwires Using Modified Bending Tests
Mahmoud Nili Ahmadabadi, Tahereh Shahhoseini, Mohamad Habibi-Parsa, Maryam Haj-Fathalian, Tahereh Hoseinzadeh-Nik, and Hananeh Ghadirian 793

Stress-Shielding Effect of Nitinol Swan-Like Memory Compressive Connector on Fracture Healing of Upper Limb
Q.G. Fu, X.W. Liu, S.G. Xu, M. Li, and C.C. Zhang 797

Pseudoelastic Nitinol-Based Device for Relaxation of Spastic Elbow in Stroke Patients
S. Viscuso, S. Pittaccio, M. Caimmi, G. Gasperini, S. Pirovano, E. Villa, S. Besseghini, and F. Molteni 805

Determination of Transformation Properties of Thin Medical Grade Ni-Ti Wire by High-Resolution Bend and Free Recovery Testing
A. Undisz, M. Fink, and M. Rettenmayr 814

Superelasticity of NiTi Ring-Shaped Springs Induced by Aging for Cranioplasty Applications
Henryk Z. Morawiec, Zdzisław H. Lekston, Kazimierz F. Kobus, Marek C. Węgrzyn, and Jan T. Drugacz 818

SHADE: A Shape-Memory-Activated Device Promoting Ankle Dorsiflexion
S. Pittaccio, S. Viscuso, M. Rossini, L. Magoni, S. Pirovano, E. Villa, S. Besseghini, and F. Molteni 824

Design Considerations for Nitinol Bone Staples
Scott M. Russell 831

Tensile Testing of Nitinol Tubes and Wires with Higher Strain Rates
Rainer Johe 836

The Mechanical and Thermal Behaviors of Heat-Treated Ni-Rich NiTi Orthodontic Archwires
S.M. Seyyed Aghamiri, M. Nili Ahmadabadi, Sh. Raygan, I. Haririan, and M.S. Ahmad Akhondi 843

JMEPEG (2009) 18:447
DOI: 10.1007/s11665-009-9475-x

1059-9495/$19.00

EDITORIAL

Shape Memory and Superelastic Technologies 2008

Ausonio Tuissi

M.R. Mitchell

José San Juan

The International Conference on Shape Memory and Superelastic Technologies—SMST 2008—took place in Stresa (Maggiore Lake), Italy, 21-25 September 2008. It was sponsored by the SMST International Organization, an affiliate society of ASM International, and locally endorsed by CNR IENI (Lecco) an Institute of the National Research Council of Italy.

As was the case for previous SMST conference editions, SMST 2008 was dedicated to the technical/scientific aspects and applications of Shape Memory Alloys. The specific scopes of SMST 2008 were directed toward a better, deeper, and shared understanding of the interplay between the most recent developments in SMA research and the positioning of applications of SMA on the market place. The conference program included approximately 160 presentations arranged in 19 oral and 2 poster sessions during which lively discussions of overall participation was appreciated. A positive atmosphere was also created in the "Product Exhibition" of SMA companies, "R&D exhibition" for research centers and "Workshop on SMA" conference events. The strong and widespread interaction between academic, government and industrial enterprises operating worldwide in the field of SMA and sharing their knowledge and experience represented the main result of that event.

We would like to acknowledge the members of the organizing and advisory committees, the sponsors, the session chairs, the event chairs and particularly CNR IENI and ASM staff members. The success of SMST 2008 was only possible through their generosity and hard work.

This special issue of the *Journal of Materials Engineering and Performance* (JMEP) features a selection of contributions presented at SMST 2008. Authors of the selected articles were invited to submit their manuscripts to JMEP where they passed through the full peer review process required for publishing refereed papers. These 63 selected papers represent the broad cross-section of the topics covered at the conference and provide the latest scientific and engineering developments in the exciting field of applied shape memory materials.

Ausonio Tuissi
SMST 2008 Chairman
National Research Council of Italy
CNR IENI, Lecco, Italy

M.R. Mitchell
SMST Organization President
Northern University of Arizona
Mechanical Engineering Department

José San Juan
SMST 2008 Technical Program Chair
University of the Basque Country
Dpt. Physics of Condensed Matter (Bilbao), Spain

JMEPEG (2009) 18:448–458
DOI: 10.1007/s11665-009-9410-1

Characterization of Non-Metallic Inclusions in Superelastic NiTi Tubes

Alejandro Toro, Fei Zhou, Ming H. Wu, William Van Geertruyden, and Wojciech Z. Misiolek

(Submitted October 20, 2008; in revised form February 10, 2009)

Non-metallic inclusions have been shown frequently to lead to crack initiation in superelastic Nitinol fatigue specimens. While prior studies suggested that both carbide (TiC) and oxide ($Ti_4Ni_2O_x$) inclusions can develop in superelastic Nitinol alloys, questions remain on whether or how the chemistry and morphology of these non-metallic inclusions are affected by the melting and subsequent tube manufacturing process. In the present study, samples of Ti-55.8wt.%Ni alloy were taken from tubes fabricated from materials of various VIM and VAR melt processes. Additional samples were taken from various stages of the tube drawing process for studying the stringer formation. Our results suggest that both carbide and oxide inclusions are present in VIM/VAR materials and the oxide break-down during tube drawing appears to be the primary mechanism for stringer formation. Carbides in VIM materials generally remain as isolated particles during tube fabrication while the primary inclusion of oxide in the VAR material explains its higher stringer density. In addition, the carbide inclusion has been confirmed to contain a noticeable amount of oxygen; hence, we suggest the "Ti(C, O)" nomenclature. Further study on the role of oxygen on carbide and oxide formation in VIM/VAR materials may be beneficial for improving the future melt quality of NiTi alloys.

Keywords advanced characterization, biomaterials, electron Microscopy, fabricated metal, metallography

1. Introduction

Superelasticity is a key characteristic for selecting NiTi alloys for medical implant devices because it offers the possibility of designing a self-expanding device with a small delivery profile and compliance similar to that of the natural anatomy. For example, treating diseases such as stenosis in superficial femoral, iliac and popliteal arteries where the devices are subjected to complex deformations, self-expanding NiTi stents have been proven to perform better than do stainless steel stents in clinical outcomes owing to the more compliant mechanical properties and reduced thrombogenicity (Ref 1). However, these difficult anatomies inevitably impose severe dynamic deformations on the material, which can lead to device fracture and negative clinical outcomes (Ref 2). Hence, fatigue lifetime is a critical consideration for device durability in self-expanding NiTi implant designs.

This article is an invited paper selected from presentations at Shape Memory and Superelastic Technologies 2008, held September 21-25, 2008, in Stresa, Italy, and has been expanded from the original presentation.

Alejandro Toro, National University of Colombia, Medellin, Colombia; **Fei Zhou** and **Ming H. Wu**, Edwards LifeSciences LLC, Irvine, CA; **William Van Geertruyden**, EMV Technologies, LLC, Bethlehem, PA; and **Wojciech Z. Misiolek**, Institute for Metal Forming, Lehigh University, Bethlehem, PA. Contact e-mail: Ming_Wu@Edwards.com.

Fatigue fractures in NiTi test specimens have often been observed to initiate from non-metallic inclusions (Ref 3, 4). It is well known in the metal industry that non-metallic inclusions are influenced primarily by the melting technique. At the present time, commercial NiTi alloys are produced by vacuum induction melting (VIM), vacuum arc melting (VAR), or a hybrid process where the primary VIM ingot is further refined using a VAR process (VIM/VAR) (Ref 5). Because the VIM process uses graphite crucibles, carbon pickup during melting leads to a greater level of carbon impurity of 300-700 ppm in VIM ingots as compared to less than 100 ppm in VAR ingots (Ref 5). Reinoehl et al. (Ref 3) studied the effects of carbon content on inclusions and rotating-beam fatigue of re-draw wires of significantly different carbon contents from two different suppliers. They reported that the material of lower carbon content had less inclusion count but the inclusion size was much larger than those of material having higher carbon content. The authors however did not observe any significant difference in the fatigue lifetimes between the two materials.

It has been recognized that both TiC and $Ti_4Ni_2O_x$ non-metallic inclusions can develop in NiTi alloys (Ref 6). Studying the inclusions in Ti-50.7at.%Ni alloys prepared by a laboratory scale VIM furnace using a graphite crucible, Zhang et al. (Ref 7) detected eutectic TiC in an as-cast ingot with 200 ppm carbon. But for ingot having 710 ppm carbon, both eutectic TiC and primary TiC were present in the as-cast structure. Habu et al. (Ref 4) in their study on VIM alloys observed only TiC in the as-cast ingot. While TiC remained present at various stages of the tube manufacturing process, $Ti_4Ni_2O_x$ was occasionally detected in finished tubes.

Graham et al. (Ref 8) studied the inclusions in VAR Ti-50.8at.%Ni alloys in details. By controlling the melt process and input stocks, the authors produced sample ingots of 250-280 ppm oxygen and less than 30 ppm carbon as well as an ultra-clean variant with oxygen and carbon contents of less than 50 and 20 ppm, respectively. The non-metallic chemistry was not analyzed but assumed to be $Ti_4Ni_2O_x$. The study demonstrated that the ingot size, and hence the solidification rate, can influence the inclusion size. In the ultra-clean variant, both the area fraction and size of inclusion were significantly reduced. However, they reported inferior fatigue lifetimes of ultra-clean material when compared to those of the typical VAR melt variants attributing the anomaly to the dominant influence of surface finish. A further study by Morgan et al. (Ref 9) suggested that ultra-low carbon and oxygen impurity levels do improve melt cleanliness and reduce inclusion. Fatigue study of electro-polished wires demonstrated slightly improved fatigue lifetimes in the long life regime with the low inclusion material compared with those of standard VAR and VIM/VAR materials.

In our experience of fatigue testing superelastic NiTi tube specimens, we have encountered significant variations in inclusion morphology and fatigue durability of materials processed using various melt processes. These inconsistencies were observed even in materials fabricated by the same melt technique. It is therefore not our intent to compare the fatigue durability of materials prepared by various melting techniques based on limited sample population. However, quite often crack initiation in prematurely failed specimens can be linked to stringers, an elongated cluster of fractured inclusions as shown in Fig. 1(a) while crack initiation in specimens failing at "normal" durability cycles is more typically linked to a small isolated inclusion particle; an example is shown in Fig. 1(b). As we continue to develop knowledge of the prejudicial effect of non-metallic inclusions on the fatigue resistance of Ni-Ti alloys, it is reasonable to anticipate some correlations that must exist among their chemical composition, size, shape, and distribution in the microstructure. Many of these factors are influenced by not only the melting technique but also the subsequent thermal and mechanical conversion process for making semi-finished products such as wires and tubes. In the present study, chemistries of non-metallic inclusions in commercial Ti-50.8at.%Ni alloys manufactured by VIM, VIM/VAR, and VAR processes were analyzed using various techniques of electron microscopy. The inclusion morphologies at various stages of the tube fabrication process of a VIM/VAR alloy were examined in detail. The aim is to provide further understanding on the effect of melting process on inclusion chemistry as well as those on stringer formation during the subsequent tube manufacturing process.

2. Experimental Procedure

Tubes of 2.2 mm outside diameter fabricated from materials of VIM/VAR, VIM, and VAR melt processes were used in the present study. Samples of a VIM/VAR alloy taken at different stages of the tube drawing process were also analyzed. The specimen identifications and related material information are tabulated in Table 1.

All the samples were prepared according to standard metallographic specimen preparation procedure and finished by automatic vibratory polishing in cloth with 0.05 μm SiO_2 particles. Specimens for optical microscopy were examined in

Table 1 Description of specimens used in the study

ID#	Melt process	Ingot Ap temperature, °C	Carbon, ppm	Oxygen, ppm	Tube fabrication stage
A1	VIM/VAR	−8	324	262	Hot rolled bar 31.0 mm OD
A2	VIM/VAR	−8	324	262	Intermediate 6.7 mm OD
A3	VIM/VAR	−8	324	262	Finished tube 2.2 mm OD
B	VIM	+4	240	300	Finished tube 2.2 mm OD
C	Standard VAR	+4	34	310	Finished tube 2.2 mm OD
D	HP VAR	−19	<20	<50	Finished tube 2.2 mm OD

OD, outside diameter; HP, high purity

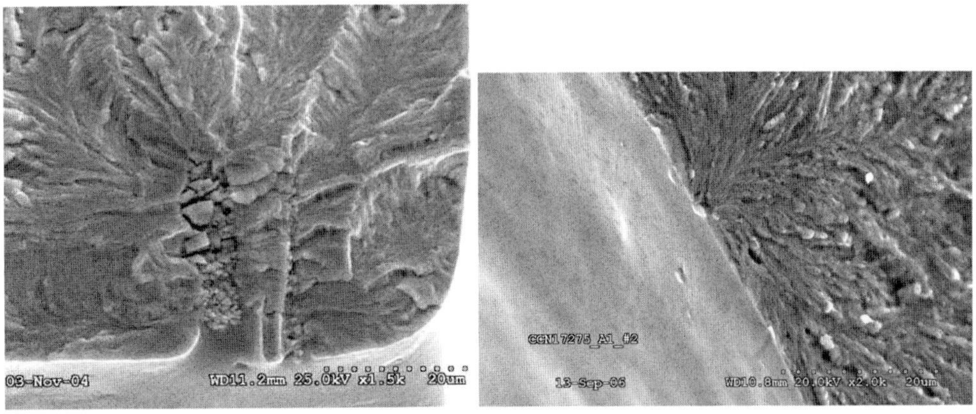

(a) (b)

Fig. 1 SEM micrographs of fatigue fractured surface showing the crack initiation site of (a) a stringer of inclusions in a prematurely failed specimen and (b) an isolated inclusion particle in a specimen failed at a typical durability cycle

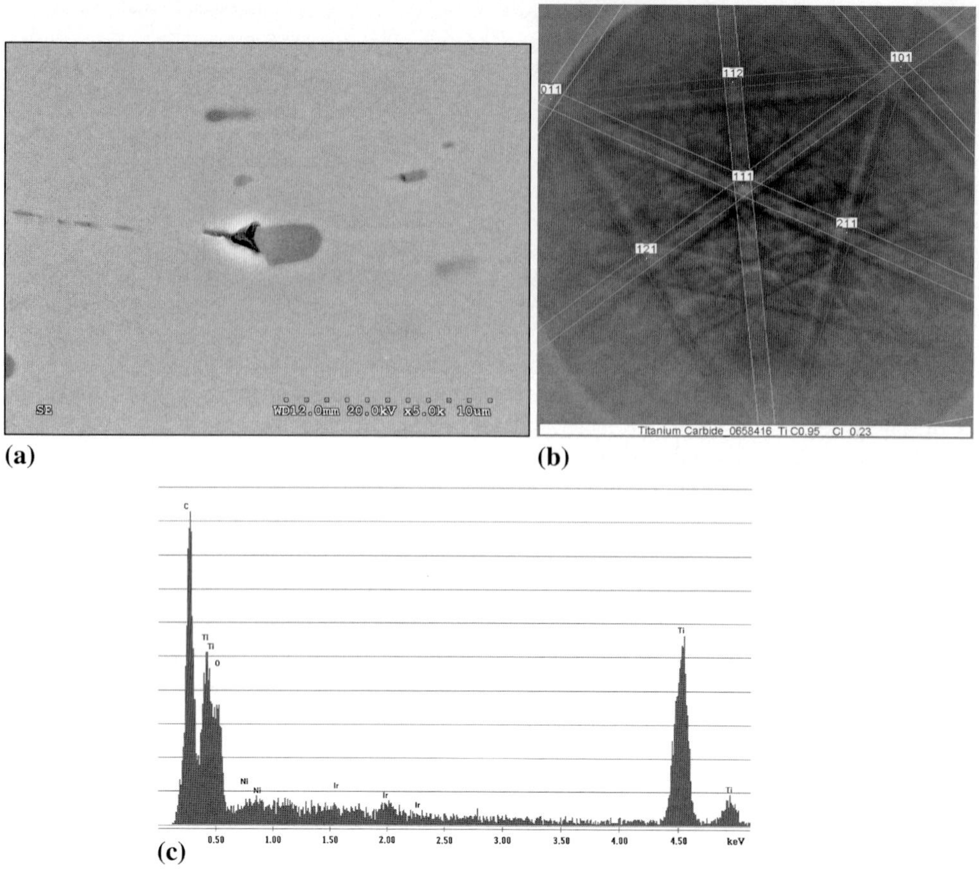

Fig. 2 (a) Secondary electron SEM image of a typical carbide in specimen A1, (b) corresponding indexed EBSD pattern, and (c) EDS spectrum which determined the most likely phase to be Ti(C, O) carbide

lightly etched condition. Etchant of $HF/HNO_3/H_2O$, 3:10:87 in volumetric ratio, was used to bring up the inclusion contrast. Preferential etching was found to often remove certain inclusions from the surface. Therefore, for the SEM study, all samples were polished and coated with a thin layer of amorphous Ir by using a sputtering current of 20 nA and sputtering time of 3 s. The analyses were made using an Olympus BH2 light optical microscope (LOM), a Hitachi 4300 field emission gun scanning electron microscope (FEG-SEM) in conjunction with TSL/EDAX electron backscatter diffraction (EBSD) and energy dispersive spectroscopy (EDS) systems, and Auger electron spectroscopy (AES) using a PHI Smart-200 FE-Nanoprobe instrument. An acceleration voltage of 10 kV was used for all the EDS measurements to improve spatial resolution, while 20 kV were applied for all the EBSD and AES measurements. The AES specimen was sputter-cleaned using Ar^+ ion prior to the analysis.

3. Results and Discussion

3.1 Basic Identification of Inclusions' Type

3.1.1 Carbide-Type Inclusions. Figure 2(a) shows a secondary electron image of a carbide inclusion in the A1 specimen. Some plastic flow of the matrix around these particles was observed, namely the "tail" to the left of the inclusion. The EBSD pattern obtained closely matches the cubic structure of TiC (Fig. 2b) and the EDS spectra shows mostly C, O, and Ti peaks with only a trace amount of Ni (Fig. 2c). Figure 3(a) shows an SEM micrograph of both carbide and oxide inclusions in specimen A3. The carbide appears darker than oxide due to the Z-contrast of lighter elements in the composition. The EDS line scan of Fig. 3(b) indicates that the carbide is more enriched in Ti, C, and O, and depleted in Ni than the matrix.

The reason for using the "Ti(C, O)" nomenclature is that after a significant number of EDS and line scan analyses some amount of O was consistently detected together with Ti and C. The presence of O in carbide was further confirmed by AES; a typical spectrum of the carbide inclusion shown in Fig. 4(b) undoubtedly demonstrates the presence of O in Auger electron signal. Since the crystal structures of both TiC and TiO are cubic and the mean difference between their lattice parameters is only 6% (Ref 10), these results may indicate that part of the C atoms in the TiC structure could have been replaced by O atoms, configuring a substitutional crystal. It is worth noticing that these particles are not complex oxide-carbide compounds, given that their structure is always cubic and compatible with that of TiC.

3.1.2 Oxide-Type Inclusions. Analyses of an oxide inclusion in specimen A1 are shown in Fig. 5. These particles are, in general, larger than the carbides and their shape is more irregular (Fig. 5a). The EBSD patterns closely matched the cubic structure of the $Ti_4Ni_2O_x$ (Fig. 5b) (Ref 10) and the EDS spectra revealed the presence of less Ti and C, about the same amount of O but much higher Ni content (Fig. 5c) than in the

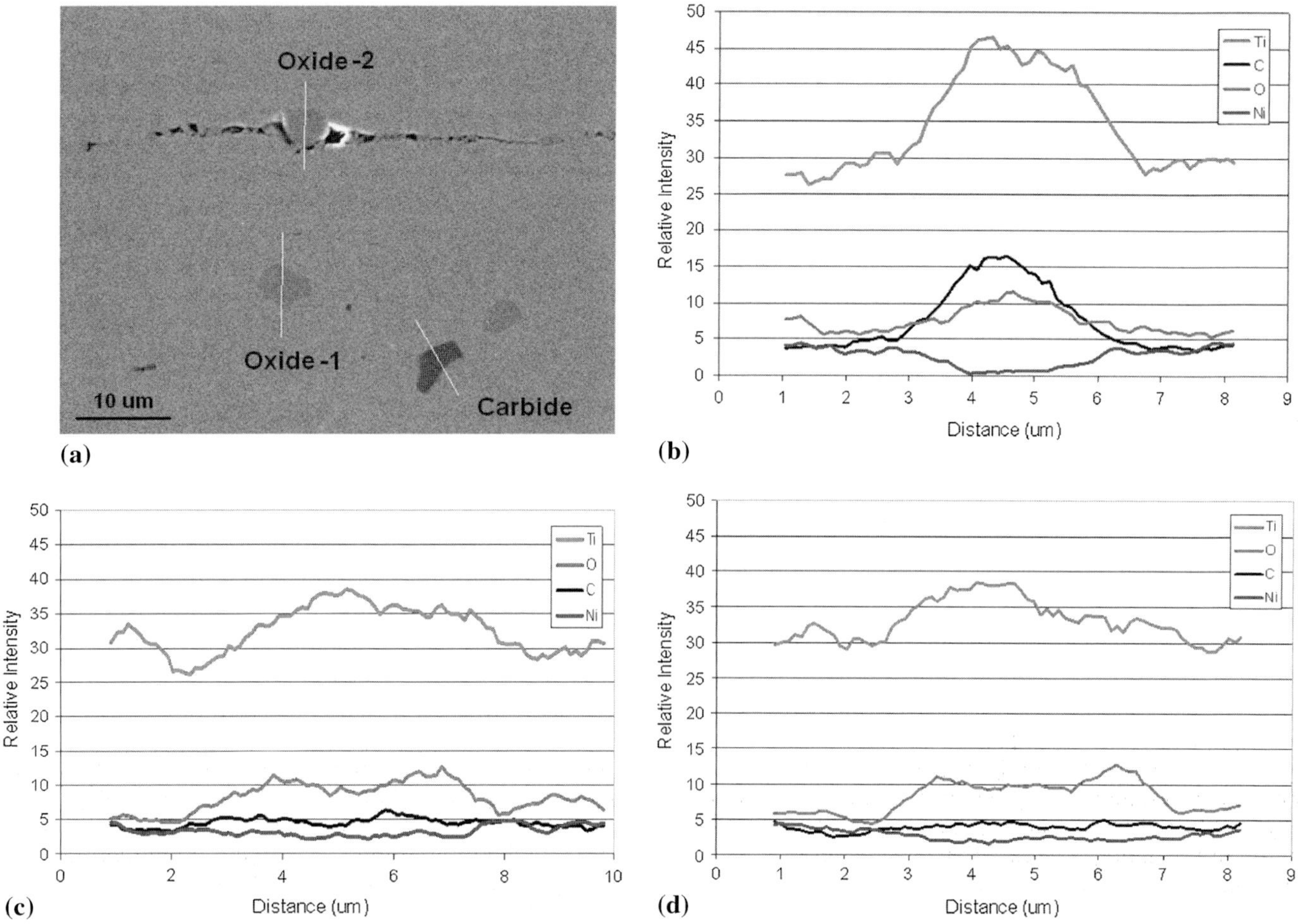

Fig. 3 (a) SEM image of carbide and oxides in specimen A3, and EDS line scans of (b) carbide, (c) isolated oxide (oxide-1), and (d) oxide in stringer (oxide-2)

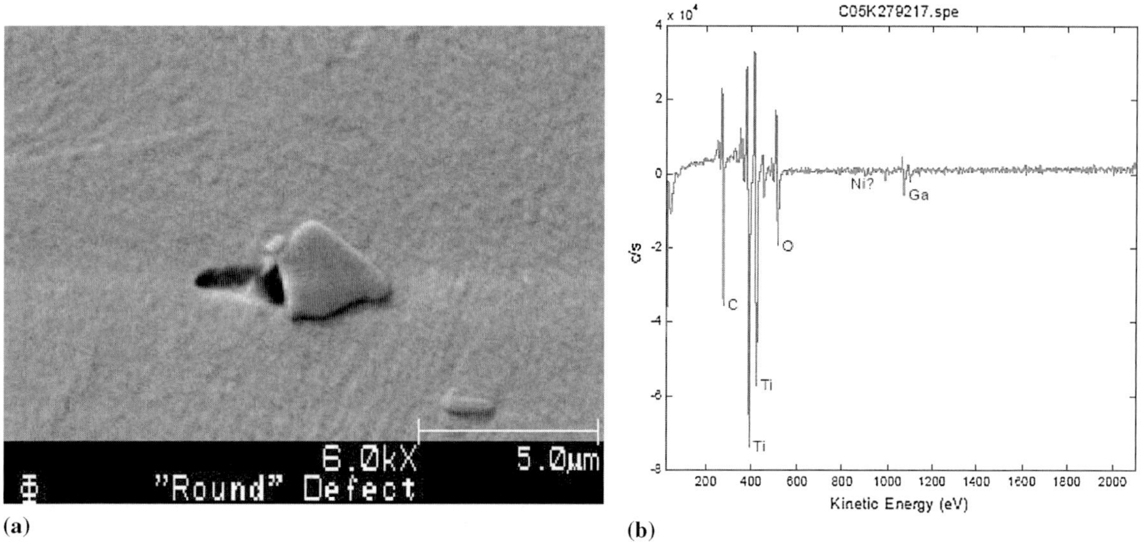

Fig. 4 (a) An SEM image and (b) the corresponding AES spectrum of a Ti(C, O) carbide inclusion

carbides. In the analysis of A3 finished tube specimen, oxide inclusions were found as isolated particle as well as in elongated stringer. Figures 3(c) and (d) show the EDS line scans of an isolated oxide (oxide-1) and an oxide in a stringer (oxide-2), respectively. The results suggest that both oxides are identical in chemical composition that is rich in Ti and O but the Ni and C contents are similar to those of the matrix. The lack of a noticeable difference in the C content between oxide and matrix suggests that the presence of C in the EDS spectrum of Fig. 5(c) must come from surface contamination. EDS mapping provides further verification on the elemental distribution within carbide and oxide inclusions as described above. Figure 6(a) displays an SEM micrograph of an A3 specimen containing both carbide and a fractured oxide. C, O, Ti, and Ni mapping in Fig. 6(b) confirms that the carbide contains higher levels of C, O, and Ti, but less Ni while the oxide is rich in O, Ti, and its C and Ni concentrations are comparable to those of the matrix.

A file in the International Centre for Diffraction Data (ICDD) suggests the O content in the oxide inclusion to be $Ti_4Ni_2O_{0.3}$ (Ref 10) but the stoichiometry would support the "Ti_4Ni_2O" atomic ratio. Since present analytical techniques are limited in providing quantitative data to resolve this issue, we continue to use the "$Ti_4Ni_2O_x$" nomenclature.

3.2 Effects of Melting and Tube Drawing Processes

As mentioned earlier, current supplies of NiTi materials exhibit noticeable lot-to-lot variations in inclusion size and distribution. For the sake of discussion, we present in Fig. 7 exemplary optical micrographs in the longitudinal view of 2.2 mm diameter tube specimens of A3, B, C, and D fabricated of VIM/VAR, VIM, standard VAR, and HP VAR materials, respectively, that we consider as typically representative. Two forms of inclusion morphology were observed; one of elongated stringer along the tube drawing direction and the other as isolated particles. The isolated inclusion count is quite high in VIM/VAR, VIM, and standard VAR materials, but appears to be reduced in HP VAR specimen. For the stringer type of inclusion, the tube specimen of VIM material (Fig. 7b), in general, has the lowest while the standard VAR material (Fig. 7c) has the highest stringer density.

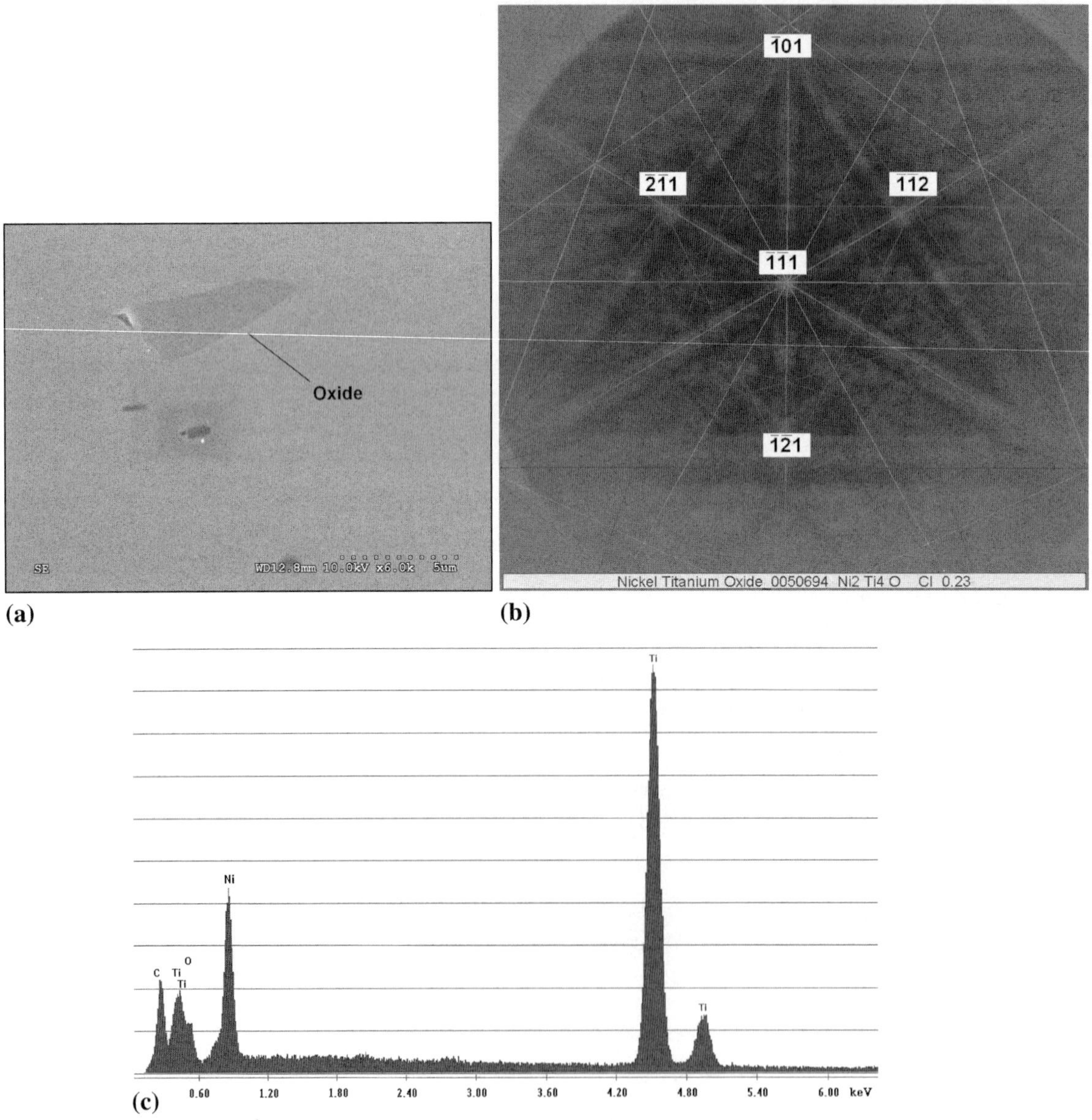

Fig. 5 (a) SEM image, (b) corresponding indexed EBSD pattern, and (c) EDS spectrum of an oxide-type inclusion in sample A1, which determined the most likely phase to be $Ti_4Ni_2O_x$ oxide

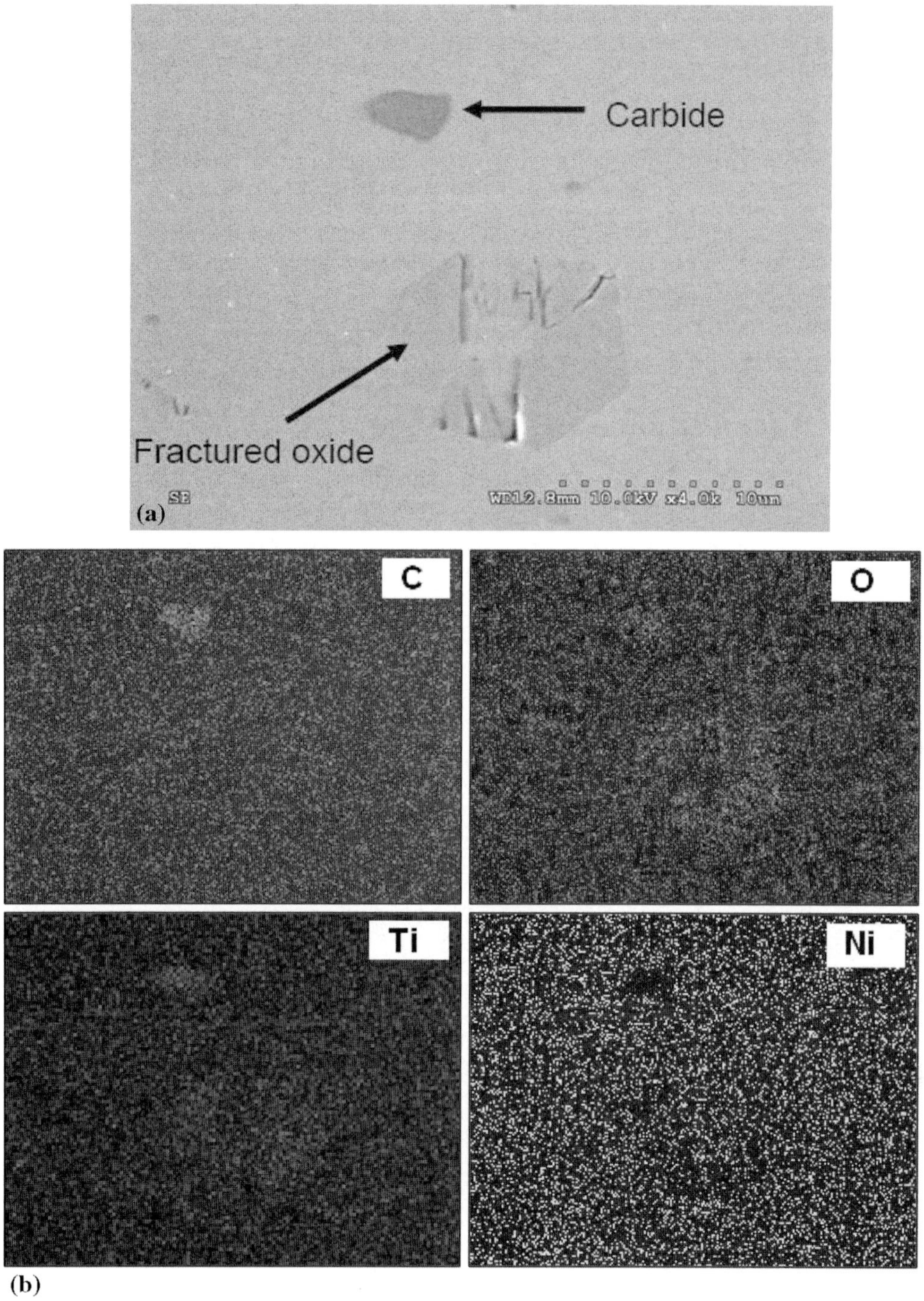

Fig. 6 EDS mapping of sample A3 showing the C, O, Ni, and Ti distribution in carbide and oxide inclusions

3.3 VIM/VAR Specimens

Regarding the type of inclusion in these materials of various melt processes, it is clear from Fig. 3 that both carbide and oxide are present in specimen A3, the tube of VIM/VAR material. The VIM/VAR material is therefore selected for studying the effect of inclusion chemistry on stringer formation. Light optical micrographs in Fig. 8 reveal the microstructure as the material goes through the tube drawing process from 31.0 mm diameter hot-rolled bar to 2.2 mm diameter tube. All inclusions in the starting hot-rolled bar (Fig. 8a) appear as isolated particles and no stringer was observed. In the micrograph of Fig. 8(a), carbides exhibit a rosy hue while the oxide has a similar contrast to the matrix. The color contrast presents a good way to differentiate carbides from oxide inclusions in optical metallography. Linear arrays of stringer along the drawing direction emerge at an intermediate size of 6.7 mm OD (Fig. 8b) and become more densely populated in the microstructure of the finished tube (Fig. 8c).

The SEM and EDS analyses on the stringers provided more detailed information about the effects of carbide and oxide on

Fig. 7 Exemplary optical micrographs of finished 2.2 mm diameter tube specimens: (a) A3 (VIM/VAR), (b) B (VIM), (c) C (standard VAR), and (d) D (HP VAR)

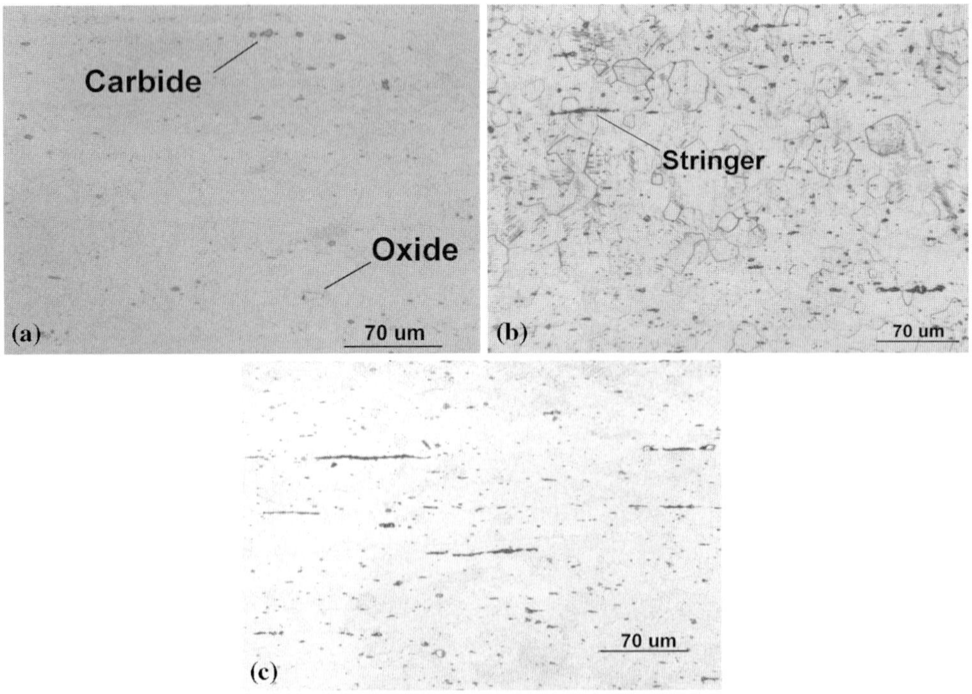

Fig. 8 Light optical micrographs of VIM/VAR specimens at various stages of tube drawing process: (a) A1 (31.0 mm OD hot rolled bar), (b) A2 (6.7 mm OD intermediate), and (c) A3 (2.2 mm finished tube)

their formation. The SEM micrographs in Fig. 9 display several examples of stringer morphology in specimen A3. The most frequently observed stringer is shown in Fig. 9(a) consisting of broken oxide particles clustered in an elongated array along the tube drawing direction while carbides which are generally smaller than oxide in dimension usually remain as isolated particles. Occasionally, fractured oxides not quite developed into stringers were observed (Fig. 9b). In some cases, particles of the carbide type were also found, preferentially in the area surrounding the end of the stringers (Fig. 9c). The observation of a number of A3 and other lots of tube specimens indicated that breaking-up of large oxides during cold drawing could be the most likely mechanism for stringers' formation, since in all cases the particles inside the stringers seemed to be small, fractured parts of a larger original particle (Fig. 9a-c). Occasionally, carbide particles were found to line up along the drawing direction forming a similar morphology to that of a stringer (Fig. 9d). But it is less apparent if the carbide particles inside this morphology evolved from breaking-up of a larger particle of carbide inclusion.

3.4 VIM Specimens

The majority of inclusion particles are small (less than 5 μm in diameter) and remained as isolated particles in tube specimen B as shown in the SEM micrograph of Fig. 10(a). The corresponding EDS line scan in Fig. 10(b) confirms that the inclusion is of the carbide type. No oxide inclusions were found in the specimen. Much less frequently than in the VIM/VAR material, stringers were found in the VIM tube specimen, an example of which is shown in the SEM micrograph of Fig. 11. Resembling the oxide stringer morphology in Fig. 9, a cluster of small carbide particles are closely packed in an elongated array. Comparing the scale to that of a nearby isolated carbide particle, the string was likely developed from a large carbide particle (larger than 10 μm in diameter) which broke up during tube drawing.

3.5 VAR Specimens

Other than the lower inclusion density in the HP VAR material, the inclusion's type and morphology are fundamentally similar between the standard VAR (specimen C) and HP VAR (specimen D) tube specimens. All inclusions were found to be of the $Ti_4Ni_2O_x$ type as no Ti(C, O) type was observed. An SEM image and the corresponding line scan of an inclusion in specimen C are shown in Fig. 12 confirming the oxide chemistry. The morphology represents an example of small oxide particles, less than 2 μm in diameter, which were aligned along the drawing direction during tube drawing. A more typical stringer in specimen C is shown in the SEM micrograph of Fig. 13(a) where broken-up oxide particles are apparent. Oxide break-up can be more clearly seen in Fig. 13(b), an SEM micrograph showing fractured oxide inclusions before actually forming a stringer.

A typical example of oxide stringer in HP VAR tube specimen (specimen D) is exhibited in Fig. 14(a) where an enlarged SEM micrograph of the stringer is shown in

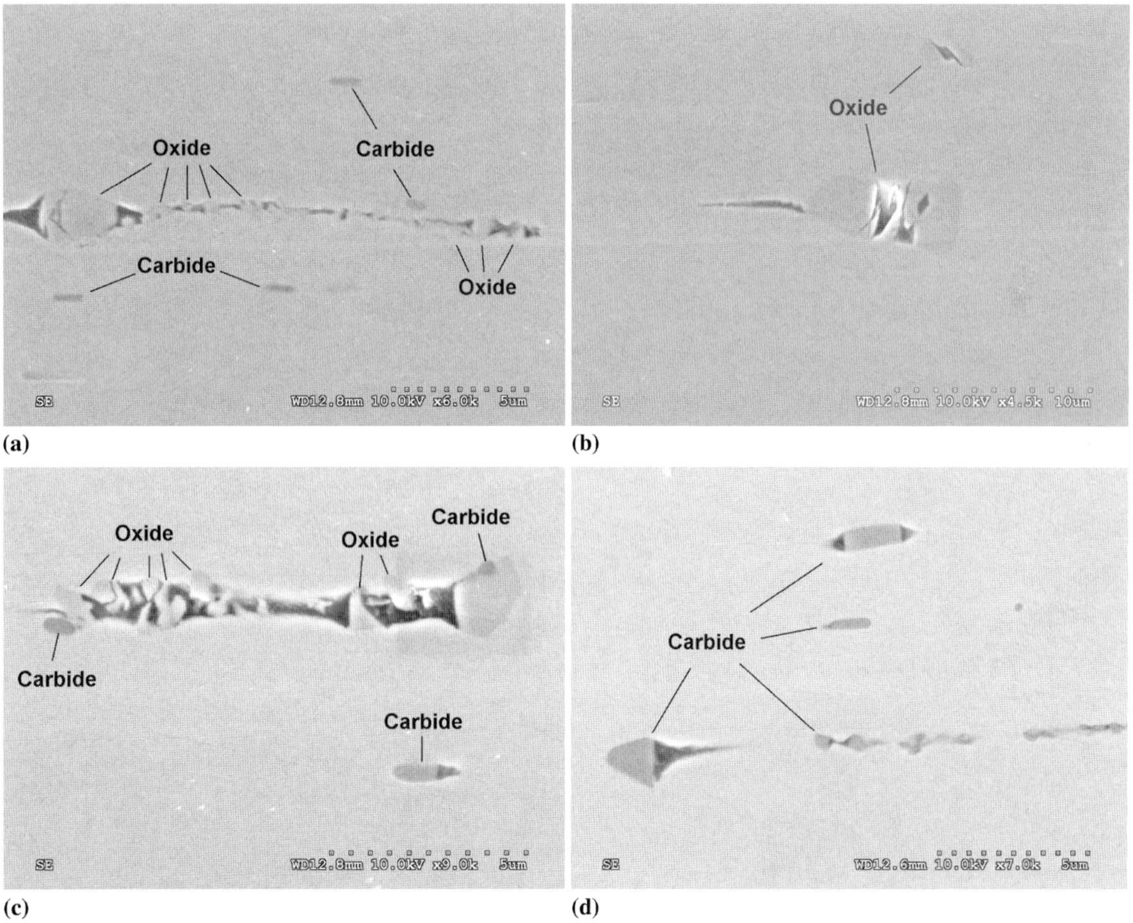

(a) (b) (c) (d)

Fig. 9 SEM micrographs showing examples of stringer morphology in the 2.2 mm OD tube specimen A3

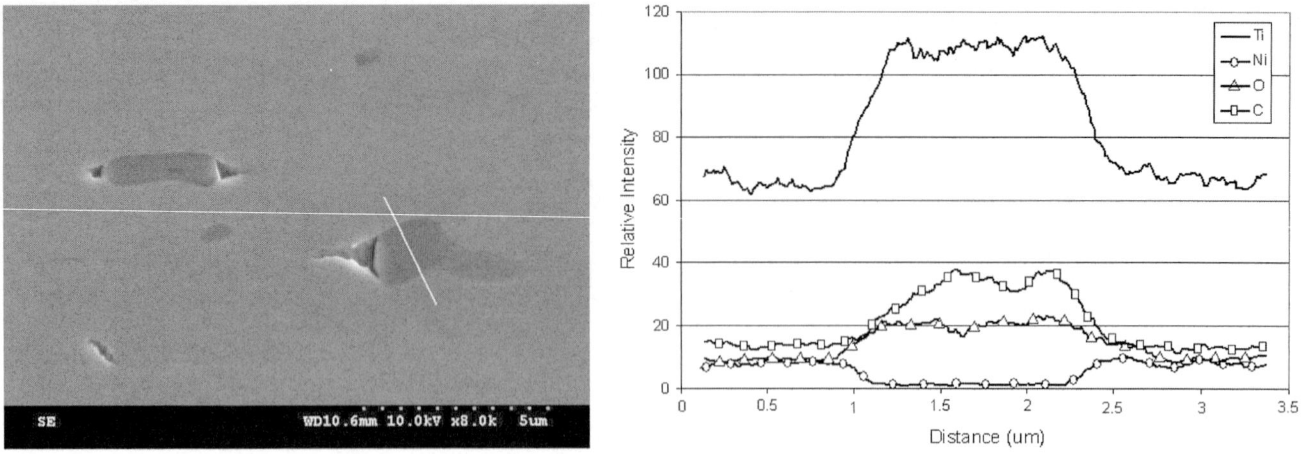

(a) **(b)**

Fig. 10 (a) An SEM micrograph and (b) corresponding EDS line scan of Ti(C, O) carbide inclusion in specimen B

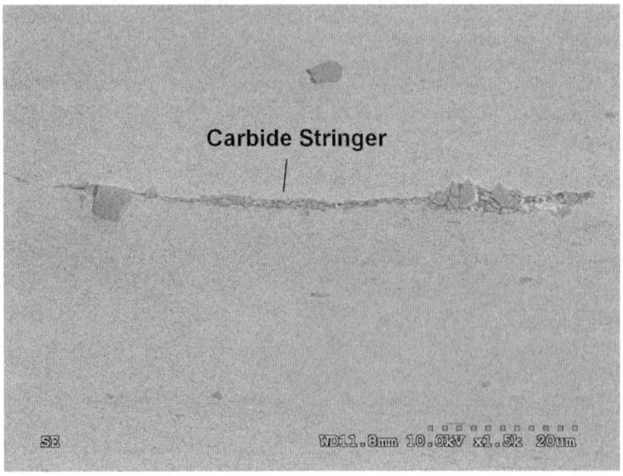

Fig. 11 SEM micrograph of a carbide stringer in specimen B

Fig. 14(b). Both micrographs exhibit the same characteristics of fractured oxide particles as of the stringers in the standard VAR tube material (specimen C).

4. Conclusions

1. Combined use of SEM, EBSD, EDS, and AES demonstrated that oxygen is present in the carbide inclusion, hence the Ti(C, O) nomenclature, while the $Ti_4Ni_2O_x$ oxide inclusions have been shown to be rich in Ti and O relative to the matrix chemistry. However, present analyses are limited in providing quantitative data to determine the oxygen content in the oxide inclusion.
2. Both carbide and oxide inclusions are present in the VIM/VAR material. These inclusions appear as isolated particles in the hot-rolled bar but the oxide, generally larger in size than carbide, exhibits a greater propensity to

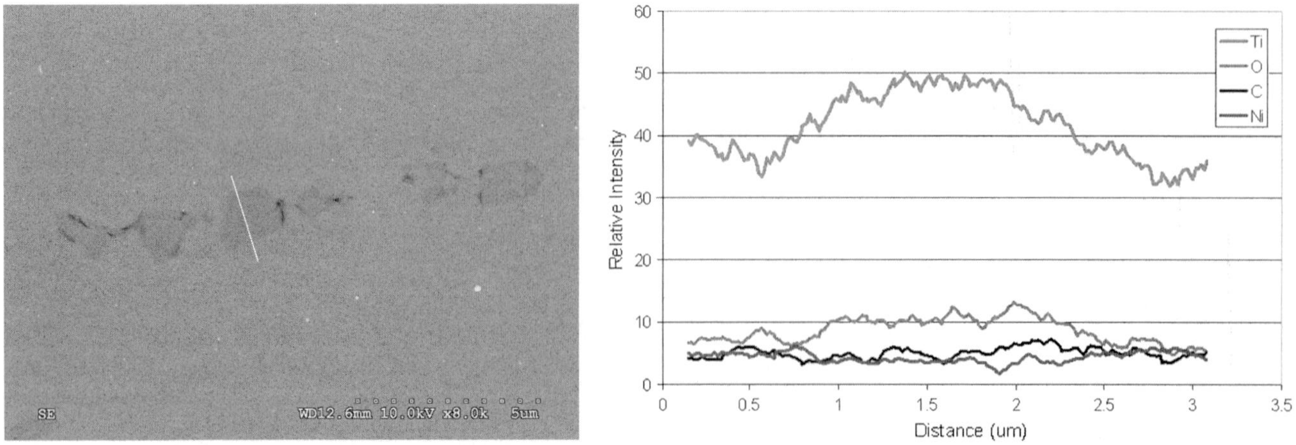

(a) **(b)**

Fig. 12 (a) An SEM image of $Ti_4Ni_2O_x$ inclusions in the tube specimen of standard VAR material (specimen C) and (b) the corresponding EDS line scans

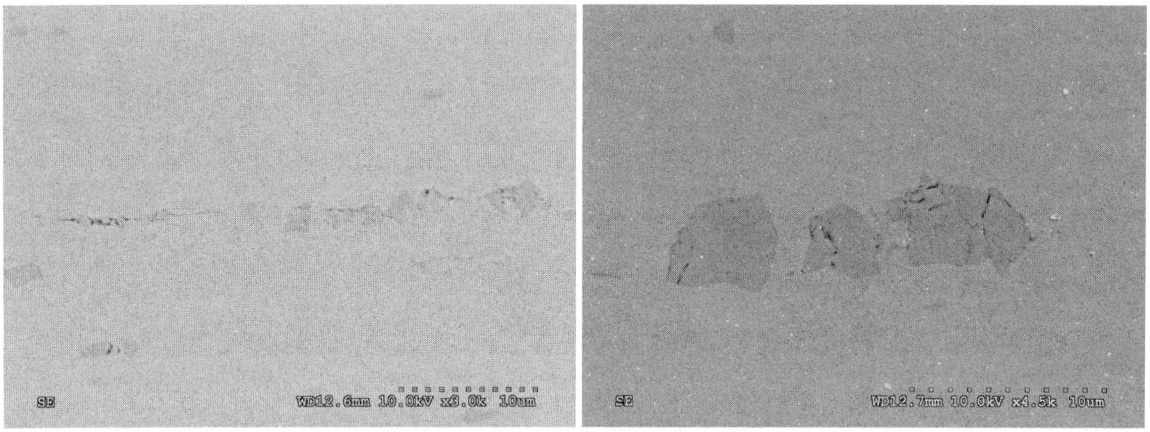

(a) (b)

Fig. 13 SEM micrographs of (a) a typical oxide stringer and (b) an early stage of stringer formation showing partially broke-up oxide particles in standard VAR tube specimen C

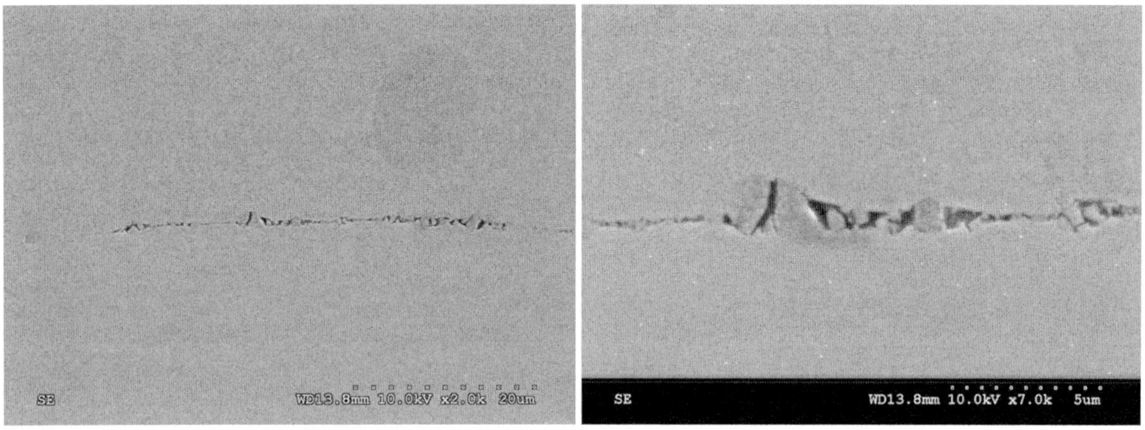

(a) (b)

Fig. 14 SEM micrographs of (a) a typical oxide stringer and (b) an enlarged area of (a) in HP VAR specimen D

fracture during tube drawing, breaking-up and forming elongated stringer along the drawing direction.

3. Inclusions in VIM material are essentially of the Ti(C, O) type and are generally small, less than 5 μm in diameter. The majority of these carbide inclusions remain as isolated particles during tube drawing, but occasionally, stringers of fractured carbide can develop from large carbide inclusions of greater than 10 μm in diameter.
4. Inclusions in both standard and HP VAR materials are of the $Ti_4Ni_2O_x$ type. While HP VAR material exhibits less inclusion and stringer densities than the standard VAR material, which has the highest stringer density among the specimens studied, the characteristics of stringers in both standard and HP VAR materials are fundamentally the same consisting of fractured oxide particles that broke-up during the tube drawing process.

Acknowledgments

The authors thank the Loewy Family Foundation for their support of Alejandro Toro as the Loewy Visiting Professor and partial support of Wojciech Z. Misiolek through Loewy Professorship at Lehigh University.

References

1. S. Sabeti, M. Schillinger, J. Amighi, C. Sherif, W. Mlekusch, R. Ahmadi, and E. Minar, Primary Patency of Femoropopliteal Arteries Treated with Nitinol Versus Stainless Steel Self-Expanding Stents: Propensity Score-Adjusted Analysis, *Radiology*, 2004, **232**(2), p 516–521
2. D. Schneinert, S. Schneinert, J. Sax, C. Piorkowski, S. Bräunlich, M. Ulrich, G. Biamino, and A. Schmit, Prevalence and Clinical Impact of Stent Fractures After Femoropopliteal Stenting, *J. Am. Coll. Cardiol.*, 2005, **45**(2), p 312–315
3. M. Reinoehl, D. Bradley, R. Bouthot, and J. Proft, The Influence of Melt Practice on Final Fatigue Properties of Superelastic NiTi Wires, *Proceedings of The International Conference on Shape Memory and Superelastic Technologies*, S.M. Russell and A.R. Pelton, Eds., April 30–May 4, 2000 (Pacific Grove, CA), p 397–403
4. T. Habu, I. Kato, K. Mikajiri, and F. Yamashita, Study of Inclusion in Ni-Ti Alloys, *Proceedings of the International Conference on Shape Memory and Superelastic Technologies*, December 3–6, 2007 (Tsukuba, Japan), in press

5. S.M. Russell, Nitinol Melting and Fabrication, *Proceedings of the International Conference on Shape Memory and Superelastic Technologies*, S.M. Russell and A.R. Pelton, Eds., April 30–May 4, 2000 (Pacific Grove, CA), p 1–9
6. Standard Specification for Wrought Nickel-Titanium Shape Memory Alloys for Medical Devices and Surgical Implants, ASTM F2063-05, ASTM, 2005
7. Z. Zhang, J. Frenzel, C. Somsen, J. Pesicka, K. Neuking, and G. Eggler, Orientation Relationship Between TiC Carbides and B2 Phase in As-Cast and Heat Treated NiTi Shape Memory Alloys, *Mater. Sci. Eng. A*, 2006, **438–440**, p 879–882
8. R. Graham, B. Van Doren, R. Henson, and J. DiCello, Characteristics of High Purity Nitinol, *Proceedings of the International Conference on Shape Memory and Superelastic Technologies*, A.R. Pelton and T. Duerig, Eds., May 5–May 8, 2003 (Pacific Grove, CA), p 7–14
9. N. Morgan, A. Wick, J. DiCello, and R. Graham, Carbon and Oxygen Levels in Nitinol Alloys and the Implications for Medical Device Manufacture and Durability, *Proceedings of the International Conference on Shape Memory and Superelastic Technologies*, May 7–11, 2006 (Pacific Grove, CA), p 821–828
10. International Centre for Diffraction Data (www.icdd.com), Newtown Square, Pennsylvania, Release 2005

JMEPEG (2009) 18:459–462
DOI: 10.1007/s11665-008-9339-9

©ASM International
1059-9495/$19.00

Cu-Al-Ni-SMA-Based High-Damping Composites

Gabriel A. López, Mariano Barrado, Jose San Juan, and María Luisa Nó

(Submitted October 11, 2008)

Recently, absorption of vibration energy by mechanical damping has attracted much attention in several fields such as vibration reduction in aircraft and automotive industries, nanoscale vibration isolations in high-precision electronics, building protection in civil engineering, etc. Typically, the most used high-damping materials are based on polymers due to their viscoelastic behavior. However, polymeric materials usually show a low elastic modulus and are not stable at relatively low temperatures (≈323 K). Therefore, alternative materials for damping applications are needed. In particular, shape memory alloys (SMAs), which intrinsically present high-damping capacity thanks to the dissipative hysteretic movement of interfaces under external stresses, are very good candidates for high-damping applications. A completely new approach was applied to produce high-damping composites with relatively high stiffness. Cu-Al-Ni shape memory alloy powders were embedded with metallic matrices of pure In, a In-10wt.%Sn alloy and In-Sn eutectic alloy. The production methodology is described. The composite microstructures and damping properties were characterized. A good particle distribution of the Cu-Al-Ni particles in the matrices was observed. The composites exhibit very high damping capacities in relatively wide temperature ranges. The methodology introduced provides versatility to control the temperature of maximum damping by adjusting the shape memory alloy composition.

Keywords Cu-Al-Ni shape memory alloys, high damping, In-Sn, metal matrix composites

1. Introduction

In the last decade, absorption of vibration energy by mechanical damping has attracted much attention in several fields such as vibration reduction in aircraft and automotive industries, nanoscale vibration isolations in high-precision electronics, building protection in civil engineering, etc. (Ref 1-5). For the design of structural applications, materials combining both a high-damping capacity and high stiffness at moderate temperatures are required, but, unfortunately, to find materials with these characteristics is not frequent (Ref 6-9). Indeed, the merit index for damping materials is a combination of the elastic modulus (E) and the damping coefficient tan ϕ, where ϕ is the loss angle between the strain experienced by the sample and the stress applied (Ref 6, 7). Typically, polymers were used as high-damping materials due to their viscoelastic behavior (Ref 7, 8). However, polymeric materials usually show a low elastic modulus and are not stable at relatively low temperatures (≈323 K). These drawbacks have encouraged the scientific community to look for alternate materials for damping applications.

High damping metallic materials (HIDAMETS), which, in principle, exhibit higher modulus and thermal stability than polymers, with similar damping properties were proposed (Ref 9). Another approach was the development of metal matrix composites, in which each component or phase has a specific role: damping or stiffness (Ref 10). Among HIDAMETS, shape memory alloys (SMA) intrinsically show very high damping properties thanks to the dissipative hysteretic movement of interfaces under external stresses (Ref 11-13) and have found some practical applications (Ref 4, 5, 14-17).

Making use of both the high-damping capacity of Cu-Al-Ni SMAs and the flexibility of composite production, high-damping metal matrix composites were developed (Ref 18, 19). Cu-based SMAs were used because they definitively exhibit higher damping capacity and are not so expensive as the commercially widespread Ni-Ti ones. Phenomena like interdiffusion, intermetallic formation, or grain boundary penetration were observed in the different composites studied (Ref 19, 20).

The processing method to elaborate these materials as well as their microstructure and damping properties will be reviewed in the present work. Results on the damping behavior will be discussed as well.

This article is an invited paper selected from presentations at Shape Memory and Superelastic Technologies 2008, held September 21-25, 2008, in Stressa, Italy, and has been expanded from the original presentation.

Gabriel A. López and **María Luisa Nó**, Dpto. de Física Aplicada II, Facultad de Ciencia y Tecnología, Universidad del País Vasco, Apdo. 644, Bilbao 48080, Spain; and **Mariano Barrado** and **Jose San Juan**, Dpto. de Física Materia Condensada, Facultad de Ciencia y Tecnología, Universidad del País Vasco, Apdo. 644, 48080 Bilbao, Spain. Contact e-mail: gabrielalejandro.lopez@ehu.es.

2. Experimental Procedures

Cu-Al-Ni SMA powders produced by Ar atomization (Ref 21) with two different compositions and a particle diameter of 25-50 μm were embedded with metallic matrices by infiltration (Ref 18, 19). Pure In, a In-10wt.%Sn alloy, and a eutectic In-Sn alloy were employed as matrices. To produce the composites, firstly, a weighted amount of powders of the corresponding Cu-Al-Ni SMA was placed at the bottom of a

Teflon mould under vacuum ($p \approx 13.3$ Pa). Pieces of the matrix material were also suspended inside the tube. The assembly was heated up to a temperature below 500 K, depending on the particular matrix, and subsequently a moderate pressure (0.3-0.4 MPa) for a short time ($\approx$10 min) was applied. Finally, the assembly was cooled down in water. More details can be found elsewhere (Ref 18, 19). Relevant data for the SMA powders, matrices, and composites are summarized in Table 1. The transformation temperatures, i.e. martensite start (M_s) and martensite finish (M_f) and the analogous ones for austenite (A_s and A_f), of the as-atomized powders determined by differential scanning calorimetry are also given. The volume fraction of the embedded SMA powders was determined by image analysis.

Specimens were prepared for metallographic analysis by standard procedures. The final polishing step was performed

Table 1 Data of the SMA powders, matrices, and composites produced

	Composite		
	A	B	C
Powder composition, wt.%	Cu-13.1 Al-3.1Ni (P1)	Cu-14.15 Al-3.3Ni (P2)	Cu-14.15 Al-3.3Ni (P2)
Transformation temperature of the as-atomized powders, K			
M_s	338	253	253
M_f	300	206	206
A_s	324	220	220
A_f	368	289	289
Matrix	Pure In	In-10wt.%Sn	In-49.1wt.%Sn (eutectic)
SMA, vol.%	61 ± 1.5	57 ± 2	60 ± 2

using colloidal silica suspension. Optical microscopy (OM) was performed in a Leica DMRXA equipped with a Nomarski prism. Differential scanning calorimetry (DSC) measurements were performed using a Perkin–Elmer DSC7 equipment.

The damping behavior of the composites was investigated by mechanical spectroscopy using a forced inverted torsion pendulum (Ref 22, 23).

3. Experimental Results and Discussion

3.1 Composite Microstructure

Images of the composites microstructures acquired using polarized light in the optical microscope at room temperature are presented in Fig. 1. In the overview of composite A (Fig. 1a), it is noticeable how the Cu-Al-Ni SMA particles are well surrounded by the In matrix. All the Cu-Al-Ni particles have undergone the martensitic transformation after the composite production and are in martensite state. This was expected, because the transformation temperatures of the as-atomized powders used in this composite are above room temperature (see Table 1). According to the image analysis applied to several images of this composite, there is a 61 vol.% of SMA particles.

A similar microstructure was observed for composite B (Fig. 1b), where Cu-Al-Ni SMA particles, with composition P2, were embedded by a In-10wt.%Sn matrix in a uniform way too. Moreover, irregularly shaped particles are very well impregnated with the matrix even in intricate zones. The polycrystalline character of the SMA particles in the austenitic state can be appreciated clearly. The amount of SMA particles was determined to be 57 ± 2 vol.%. The occurrence of

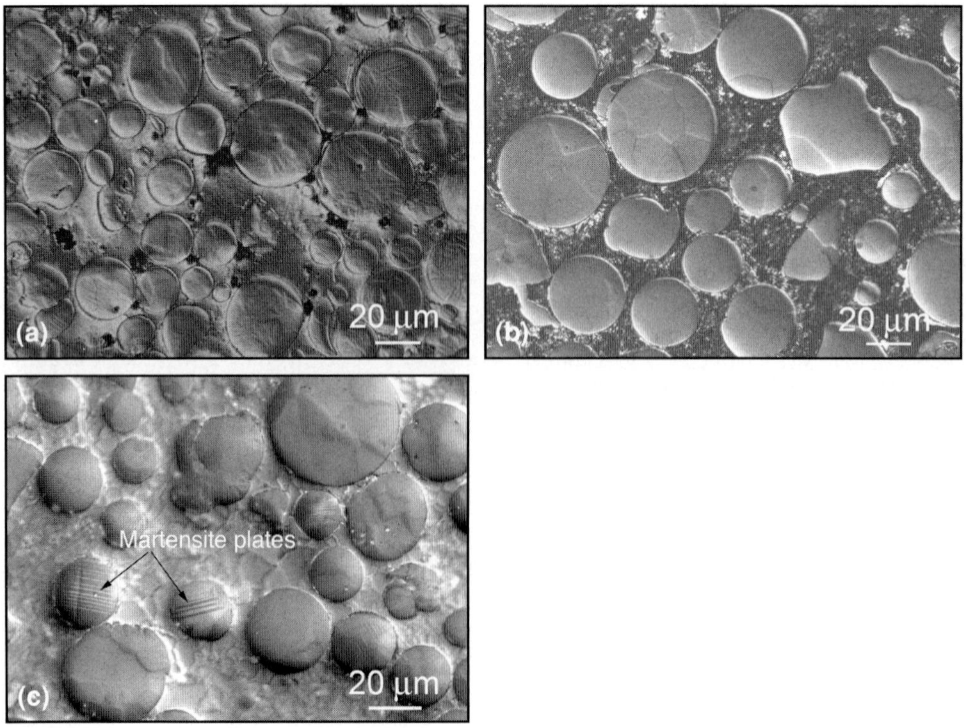

Fig. 1 Optical micrographs of composite A (a), B (b), and C (c)

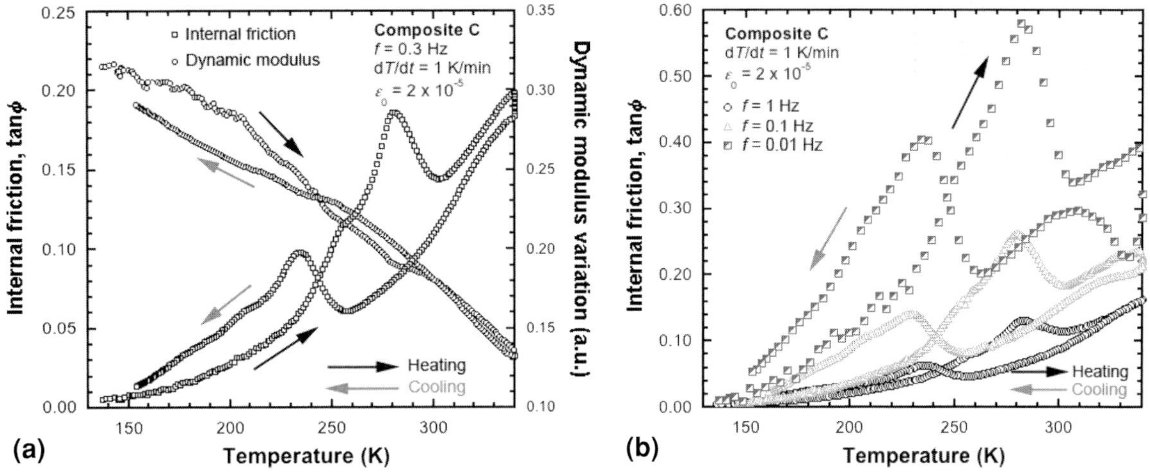

Fig. 2 (a) Internal friction spectrum and dynamic modulus variation of composite C. (b) Internal friction measurements of composite C at three different frequencies

martensite plates in the SMA particles at room temperature was not observed in composite B. This was expected because the martensitic transformation for these SMA powders takes place below room temperature (see Table 1).

Figure 1(c) shows the microstructure of composite C, where the same type of SMA particles as that used in composite B, P2, was imbedded by a eutectic In-49.1wt.%Sn matrix. The amount of SMA particles was about 60 vol.%. Even though both composites B and C were produced with the same type of SMA powders (same composition) and applying the same fabrication procedure, some differences were observed in the microstructure. As mentioned above, no particles that have undergone the martensitic transformation could be observed in composite B at room temperature; however, particles that have experienced this transformation at room temperature were clearly visible in composite C. Evidently the higher Sn content of the In-Sn eutectic matrix used in composite C has played a role in the interaction between the Cu-Al-Ni SMA particles and the molten matrix during the composite fabrication.

3.2 Damping Behavior

Figure 2(a) shows the internal friction spectra, as well as the dynamic modulus variation, of composite C for a heating/cooling cycle at a frequency of 0.3 Hz with an initial oscillation amplitude of 2×10^{-5}. It is clearly observed that there exists an internal friction peak during heating, between approximately 210 and 300 K, due to the reverse thermo-elastic martensitic transformation of the Cu-Al-Ni SMA particles of the composite, and another relatively broad peak during cooling, due to the direct transformation, between 170 and 250 K. Here, it is worth to note the very high damping observed, which reaches almost 0.2. The peaks have an associated decrease in the dynamic modulus. In addition to the peaks related to the martensitic transformation, an increasing background linked to the In-Sn eutectic matrix can also be observed when increasing temperature. The temperatures, at which the internal friction peaks are observed, are appreciably shifted toward higher temperatures with respect to the transformation temperatures of the as-atomized powders (see Table 1). This result is consistent with the microstructural observation of transformed particles at room temperature mentioned in the previous section. In addition, a broadening in the internal friction peaks was observed.

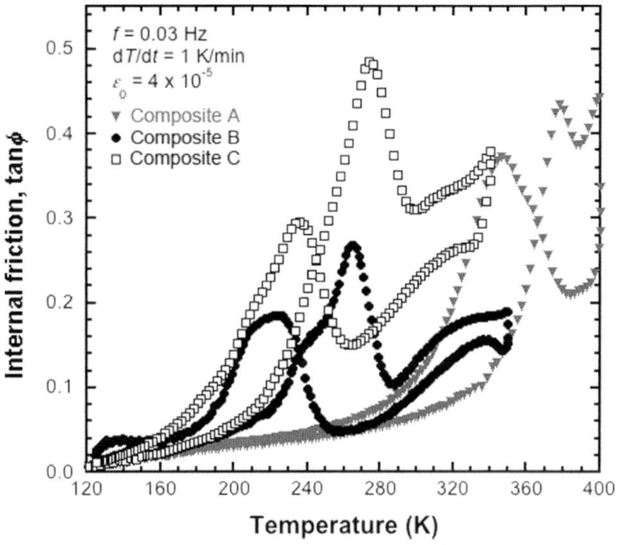

Fig. 3 Comparison of the internal friction spectra of composites A, B, and C at 0.03 Hz for an initial oscillation amplitude of 4×10^{-5}

Qualitatively, similar spectra were obtained for composites A and B (see Fig. 3 and Ref 22, 23). For composite A, the internal friction peak during heating exists between approximately 335 and 385 K, and upon cooling between 290 and 380 K. There is only a slight shift toward higher temperature, as well as a slight broadening of the transformation cycle in this composite, with respect to that of the as-atomized powders. In the case of composite B, the internal friction peaks appeared between approximately 210 and 290 K during heating, and between 170 and 250 K during cooling. These temperatures are close to the transformation temperatures of the as-atomized powders (see Table 1), although once more a certain broadening can be appreciated.

Slight temperature shifts as those observed for composites A and B are expected, because the transformation cycle in Cu-Al-Ni SMA evolves in this way upon aging (Ref 24); indeed, the powders of the composite have been aged during the elaboration process. However, even though composite C has been produced with the same powders (P2) as those used for composite B, the temperature shift in the transformation cycle was more significant. It seems that the interaction of the Cu-Al-Ni SMA particles with the higher Sn content matrix during the composite fabrication is stronger than in the other cases.

The damping of composite C for three different frequencies is shown in Fig. 2(b). As expected from the theory for the internal friction during martensitic transformations (Ref 25, 26), the lower the frequency is the higher is the internal friction exhibited by the composite. The peak position was not affected by the frequency change, which confirms that it is related to the martensitic transformation. Here, the remarkably high internal friction, reaching values higher than 0.5, is to be emphasized. Obviously, from the well-known dependence of the transitory internal friction on $\dot{T}/\omega$ (Ref 25, 26), it is expected that the damping due to the martensitic transformation of the SMA particles should be much lower in isothermal conditions, and this fact is, actually, a drawback from a practical point of view. This aspect, however, was out of the scope of the present work and only data at 1 K/min were taken.

A comparison among the internal friction spectra obtained with the three produced composites at 0.03 Hz is shown in Fig. 3. As common characteristics it can be mentioned that all composites show high damping values over a relatively broad range of temperatures. The highest damping, in a wider temperature range, has been measured for the composite produced with the In-Sn eutectic matrix, i.e. for composite C. It is evident that the higher intrinsic damping of the eutectic matrix introduces a larger background in the internal friction curves and is more attractive from a practical point of view. From this figure it can be clearly observed that a slight change in the composition leads to significant modifications in the transformation temperatures, as expected from the literature data (Ref 27). Finally, we emphasize the fact that the applied methodology enables one to match the temperature of the maximum of damping with that of work by adjusting the SMA composition. This opens new possibilities for designing high-damping materials for specific industrial or technological applications.

4. Conclusions

Composites based on Cu-Al-Ni SMAs with very high damping capacities were produced. The composite microstructures and damping properties were characterized. A good particle distribution of the Cu-Al-Ni particles in the matrices was observed. The composites exhibit very high damping capacities in relatively wide temperature ranges, values higher than 0.5 were measured. From the composites studied, the one produced with the In-Sn eutectic matrix showed the highest damping capacity in a wider temperature range. The methodology introduced provides versatility to control the temperature of maximum damping by adjusting the shape memory alloy composition.

Acknowledgments

The financial support from the Basque Government through the ACTIMAT-2005 project under the Strategic Programme ETORTEK and from the Spanish Ministry of Education through the Materials National Plan (Contract No. MAT2004-03166) is also acknowledged.

References

1. N. Igata, K. Nishiyama, K. Ota, Y. Yin, W. Wuttig, I.S. Golovin, J. Van Humbeeck, and J. San Juan, Panel Discussion on the Application of HDM, *J. Alloys Compd.*, 2003, **355**, p 230–240
2. D.D.L. Chung, Materials for Vibration Damping, *J. Mater. Sci.*, 2001, **36**, p 5733–5737
3. M. Kireitseu, D. Hui, and G. Tomlinson, Advanced Shock-Resistant and Vibration Damping of Nanoparticle-Reinforced Composite Material, *Compos. B*, 2008, **39**, p 128–138
4. Q. Pan and C. Cho, The Investigation of a Shape Memory Alloy Micro-Damper for MEMS Applications, *Sensors*, 2007, **7**, p 1887–1900
5. L. Janke, C. Czaderski, M. Motavalli, and J. Ruth, Applications of Shape Memory Alloys in Civil Engineering Structures—Overview Limits and New Ideas, *Mater. Struct.*, 2005, **38**, p 578–592
6. M.F. Ashby, *Materials Selection in Mechanical Design*, Elsevier, Amsterdam, 2005
7. R.S. Lakes, *Viscoelastic Solids*, CRC Press, Boca Raton, FL, 1999
8. Y.C. Wang, M. Ludwigson, and R.S. Lakes, Deformation of Extreme Viscoelastic Metals and Composites, *Mater. Sci. Eng. A*, 2004, **370**, p 41–49
9. R. Schaller, High Damping Materials, *Mater. Sci. Forum*, 2001, **366–368**, p 621 ()
10. R. Schaller, Metal Matrix Composites, a Smart Choice for High Damping Materials, *J. Alloys Compd.*, 2001, **355**, p 131–135
11. K. Otsuka and C.M. Wayman, *Shape Memory Materials*, Cambridge University Press, Cambridge, UK, 1998
12. J. San Juan and M.L. No, Damping Behavior During Martensitic Transformation in Shape Memory Alloys, *J. Alloys Compd.*, 2003, **355**, p 65–71
13. J. Van Humbeeck, Damping Capacity of Thermo-Elastic Martensite in Shape Memory Alloys, *J. Alloys Compd.*, 2003, **355**, p 58–64
14. J. Van Humbeeck, Non-Medical Applications of Shape Memory Alloys, *Mater. Sci. Eng. A*, 1999, **273–275**, p 134–148
15. J. Van Humbeeck and S. Kustov, Active and Passive Damping of Noise and Vibrations Through Shape Memory Alloys: Applications and Mechanisms, *Smart Mater. Struct.*, 2005, **14**, p S171–S185
16. B. Andrawes and R. DesRoches, Comparison Between Shape Memory Alloy Seismic Restrainers and Other Bridge Retrofit Devices, *J. Bridge Eng.*, 2007, **12**, p 700–709
17. G. Song, N. Ma, and H.-N. Li, Applications of Shape Memory Alloys in Civil Structures, *Eng. Struct.*, 2006, **28**, p 1266–1274
18. J. San Juan and M.L. No, Internal Friction in a New Kind of Metal Matrix Composites, *Mater. Sci. Eng. A*, 2006, **442**, p 429–432
19. G.A. López, M. Barrado, E.H. Bocanegra, J.M. San Juan, and M.L. Nó, Influence of the Matrix and of the Thermal Treatment on the Martensitic Transformation in Metal Matrix Composites, *Mater. Sci. Eng. A*, 2008, **481–482**, p 546–550
20. G.A. López, M. Barrado, J.M. San Juan, and M.L. Nó, Interaction of Cu-Al-Ni Shape Memory Alloys Particles with Molten In and In+Sn Matrices, *Mater. Sci. Eng. A*, 2008, **495**, p 304–309
21. R.B. Pérez-Sáez, V. Recarte, M.L. Nó, O.A. Ruano, and J. San Juan, Advanced Shape Memory Alloys Processed by Powder Metallurgy, *Adv. Eng. Mater.*, 2000, **2**, p 49–53
22. G.A. López, M. Barrado, J.M. San Juan, and M.L. Nó, Mechanical Spectroscopy Measurements on SMA High-Damping Composites, *Mater. Sci. Eng. A*, 2008 (in press)
23. M. Barrado, G.A. López, M.L. Nó, and J.M. San Juan, Composites with Ultra High Damping Capacity Based on Powder Metallurgy Shape Memory Alloys, *Mater. Sci. Eng. A*, 2008 (in press)
24. V. Recarte, R.B. Pérez-Sáez, M.L. Nó, and J. San Juan, Evolution of Martensitic Transformation in Cu-Al-Ni Shape Memory Alloys During Low-Temperature Aging, *J. Mater. Res.*, 1999, **14**, p 2806–2813
25. R.B. Perez-Saez, V. Recarte, M.L. No, and J. San Juan, Anelastic Contributions and Transformed Volume Fraction During Thermoelastic Martensitic Transformations, *Phys. Rev. B*, 1998, **54**, p 5684–5692
26. J. Van Humbeeck, The Martensitic Transformation, *Mater. Sci. Forum*, 2001, **366–368**, p 382–415
27. V. Recarte, R.B. Pérez-Sáez, J. San Juan, E.H. Bocanegra, and M.L. Nó, Influence of Al and Ni Concentration on the Martensitic Transformation in Cu-Al-Ni Shape-Memory Alloys, *Metall. Mater. Trans. A*, 2002, **33**, p 2581–2591

JMEPEG (2009) 18:463–469
DOI: 10.1007/s11665-009-9436-4

1059-9495/$19.00

Use of Circular Foldable Nitinol Blades for Resecting Calcified Aortic Heart Valves

Florian Hauck, Daniel Wendt, Sebastian Stühle, Emilia Kawa, Hermann Wendt, Wiebke Müller, Matthias Thielmann, Brigitte Kipfmüller, Bernd Vogel, and Heinz Jakob

(Submitted September 19, 2008; in revised form March 16, 2009)

The use of percutaneous aortic valve implantation is limited, as the native calcified valve is left in situ. A new device has been developed for resecting calcified aortic valves, using collapsible nickel-titanium blades: laser-cut T-structures of Nitinol sheet-material (Ni51Ti49 at.%) have been grinded on a high-speed milling cutter to produce cutting edges which have been given the shape of half-circles afterwards. These have been connected to each other and to struts by using rivets which also serve as articulating axes for the cutting ring. The blades are folded around these axes and retreated into a tube to be inserted in the heart through the calcified valve leaflets. Once released, the cutting edges regain their ring-shape. By combining rotation of the ring with a translating movement against a second ring of slightly greater diameter on the instrument, a punching process is created which cuts the calcified valve leaflets and leaves a circular annulus, where a prosthesis can be fixed. In vitro cutting of artificially calcified valves (n = 6) resulted in a resection time of $t = 22 \pm 6.29$ s with a maximum turning moment of $M = 2.4 \pm 1.27$ Nm, proving the function and the feasibility of the concept.

Keywords aortic valve, biomaterials, blade design, Nitinol, percutaneous valve implantation, resection tool, superalloys

1. Introduction

The standardized surgical procedure for treating aortic stenosis is aortic valve replacement (AVR) using open heart surgery (Ref 1-3): This "Goldstandard" procedure is performed "on-pump" using extracorporeal circulation after median sternotomy. The aorta is opened and the calcified native aortic valve is resected subsequently. Afterward, the new prosthesis is implanted either by a running suture line or pledget-armed single stitches. However, this procedure is rather time-consuming (20-60 min), and may therefore lead to trauma, especially in the case of additional comorbidities (e.g., poor left ventricular function or renal insufficiency). These comorbidities are currently becoming more frequent because of the aging population and so, morbidity and mortality following this surgery will increase (Ref 4-7).

One possibility to reduce the time on cardiopulmonary bypass is the use of sutureless valves. Therefore, after resection of the native diseased calcified tissues, the prosthesis—mostly an equine or porcine pericardial valve on a metal stent structure—is placed within the aortic annulus. It is fixed either by self-expanding (Nitinol frame) or by dilatation with a balloon catheter. For additional safety, up to three stay sutures are placed. However, experience shows that there is the problem of severe aortic insufficiency and paravalvular leakage caused mainly by the irregular geometries of the native tissues wherein the prosthesis is placed (Ref 8).

Percutaneous techniques for AVR via transfemoral or transapical access have become an alternative treatment option for high risk patients during the recent years (Ref 9-12). These methods minimize concomitant trauma and can be used in beating-heart conditions: The native valve is pressed against the aortic wall via balloon valvuloplasty and represents the landing zone, where a stented biological prosthesis is implanted afterward. However, as this calcified layer may be non-circular, often paravalvular leakage and regurgitation can be observed; even distortion and geometry-change of the prosthesis and of the aortic annulus may result (Ref 8, 13, 14). These factors may limit the function and long-term durability of the implanted valve.

As a result, a fast resection of the degenerated valve leaving an exact pathologic geometry would be advantageous not only for improving catheter-based implantation (Ref 15, 16) (if done endovascularly), but also for reducing the time of extracorporeal circulation by accelerating resection and thereby enabling the use of sutureless valves in open heart surgery.

In this report, we describe a resection device developed for this purpose by Endosmart GmbH in collaboration with the West German Heart Centre, University Hospital Essen and the Department of Physical Engineering, Fachhochschule Gelsenkirchen (patents pending). Production and function of the instrument which is intended for minimal-invasive surgery,

This article is an invited paper selected from presentations at Shape Memory and Superelastic Technologies 2008, held September 21-25, 2008, in Stresa, Italy, and has been expanded from the original presentation.

Florian Hauck and **Bernd Vogel**, Endosmart GmbH, Stutensee, Germany; **Daniel Wendt, Sebastian Stühle, Hermann Wendt, Wiebke Müller, Matthias Thielmann,** and **Heinz Jakob**, Department of Thoracic and Cardiovascular Surgery, University Hospital Essen, Essen, Germany; and **Emilia Kawa** and **Brigitte Kipfmüller**, Department of Physical Engineering, Fachhochschule Gelsenkirchen, Gelsenkirchen, Germany. Contact e-mail: florian.hauck@endosmart.de.

using collapsible Nitinol cutting edges for the resection of calcified aortic valves, are described, as well as the results of several tests which were carried out under different configurations of the device.

2. Materials and Methods

2.1 Requirements

The basic idea was to design an easy-handling instrument, which allows a quick resection of the native aortic valve (less than 30 s), thus enabling reduction of ECC (Extracorporeal Circulation) times in open heart surgery and with the potential to be used in surgery on beating heart in a future application. As the tool was intended to be inserted through the strongly calcified leaflets of a native stenotic aortic valve with an opening area of only 0.75 cm^2 to 1.0 cm^2, the maximum outer diameter of the instrument tip which crosses the valve should be less than 10 mm (30 French). Furthermore, the cutting of different diameters according to the size of the calcified valve should be possible, leaving a circular rim of 1-2 mm, which functions as a "neo-annulus" for the insertion of the prosthesis. The valve leaflets should be resected mechanically, using a blade-expansion mechanism to create a circular cutting edge of up to 27 mm diameter in the expanded configuration (Fig. 1).

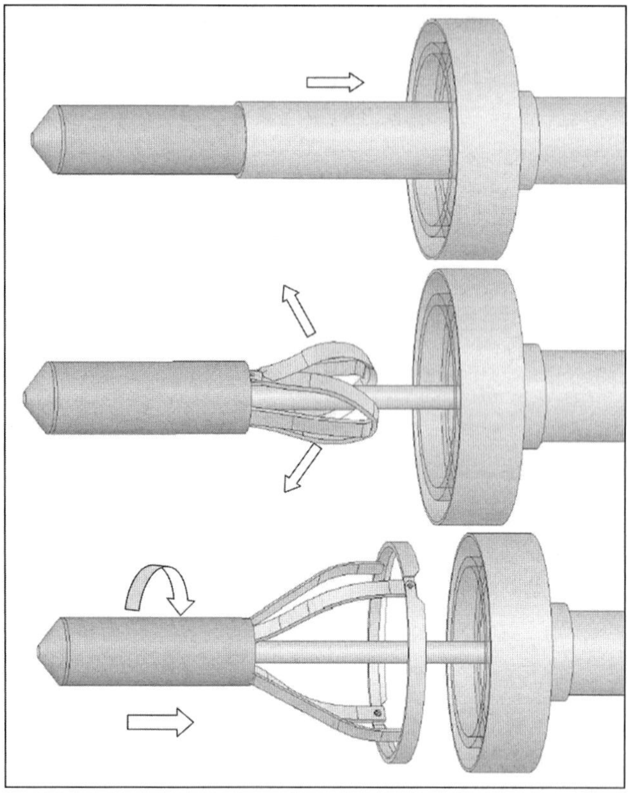

Fig. 1 Concept of resection

2.2 Cutting Edges

The shape of the circular cutting edges should ensure great stability and hardness to allow the creation of precise geometries during the cutting process. On the other hand, they should be collapsible into an insertion tube with an outer diameter of less than 10 mm. Thus, high flexibility was required. In view of the required properties, Nitinol has become the material of choice because of its superelastic behavior: The material can undergo extreme elastic deformation (up to 8%) while it regains its predefined shape when it is released and unloaded.

The blades were made from Nitinol (Ni51Ti49 at.%) sheet (thickness 0.35 mm), cut with a Nd:YAG-laser into T-structures, and shaped by heating at 530 °C while fixed on a 3D-structure. The resulting shapes were precise semi-circles with defined radius, created by flexing the T cross member. The T long member was bent as well to serve as support and connection to the instrument via the cross-structure at the 'foot' end. Furthermore, supporting struts were laser-cut and shaped, possessing a cross-shaped 'foot' end, too. In order to create the blade circle/cutting ring, two T-structures were connected to each other and to two of the additional struts by special rivets, which are perpendicular to the circle axis and are serving as articulating joints for the cutting edges. The different assembling steps are displayed in Fig. 2.

2.3 Sharpening

Before shaping, the blades underwent multi-stage grinding on a high-speed milling cutter (Aciera F1, SA Muller Machines Brügg, Switzerland, 10.000 rpm) using cera-gloss (Cera Gloss diamond, Edenta AG, AU/SG, Switzerland). After dressing of the rims (SRD Finisher 200, Service Precision Grinding Inc., Port Byron, NY, USA, 260 rpm), the T-structures were fixed in a custom-made holder, enabling a precise infeed and feed motion during grinding. The parameters of the three different steps of finish-milling are shown in Table 1 (Ref 17).

2.4 Resection Technique

For resection, a rotating punching technique was chosen. Therefore, two cutting rings are needed with a slight gap

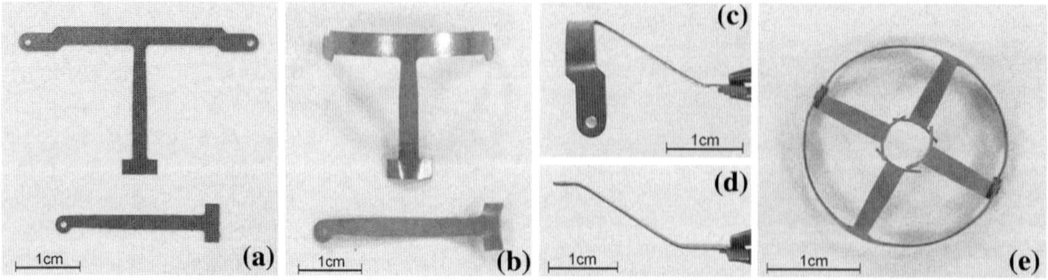

Fig. 2 Assembly of Nitinol cutting ring: (a) laser-cut blade and supporting strut; (b-d) blade and strut after shaping; (e) assembled cutting ring

(in this case 0.15 mm) between the outer surface of the inner ring and the inner surface of the outer ring. The blades of the two rings face each other with the material to be cut placed in-between. While one ring stays fixed, the other one approaches with a combination of translation and rotation movements until the final punching, when the mobile ring penetrates the fixed one.

2.5 Resection Tool

The resection tool used in this study was a prototype for open heart surgery, possessing one foldable and expandable blade, which has to pass the calcified valve, and a stable non-foldable ring structure as its counter part. Assembling of the prototype is shown in Fig. 3 and 4. The foldable Nitinol blade (outer diameter $d = 16.0$ mm, sheet thickness 0.35 mm) is mounted on the distal end of the instrument (A), with its cutting edge pointing at the main body. The "feet" of the blades are fixed on the distal sheath whose chamfered cap facilitates the instrument insertion through the center of the valve leaflets and protects the activating mechanism of the blades. Inside the distal sheath, the supporting struts are fixed on the body which can be moved along the longitudinal axis. The distal body (A) is connected to the main body via a tube, which transmits the turning and translation motion and also houses the activating struts of the cutting edge. Opposite to the collapsible blade, a massive, non-foldable ring, made of stainless steel (inner diameter 16.3 mm, outer diameter 18 mm, height 9 mm) acts as the counter-blade. This ring is chamfered on its outer rim for insertion in the aorta, but only burred at the inner rim to facilitate the punching process. This ring is connected to the "activating body" via a shaft of acrylic plastic (B), which by being translucent, enables visibility during resection. The activating body (C) houses two mechanisms: By operating lever (1), a contact protecting sheath (outer diameter 9 mm) in the acrylic shaft can be translated along the longitudinal axis to enclose/release the collapsed cutting edges; lever (2) is used to push the supporting struts in proximal direction and to expand the cutting ring to its circular shape. The body (D) of the mechanism houses an epicyclic gear (planetary gear 30/1, Faulhaber GmbH, Schönaich, Germany, 3.71:1), which increases the turning speed of the blade, and a screw thread (lead length 0.35 mm), which in turn transforms the turning motion of the handle (E) into a combination of translation and rotation movements. Handling is facilitated by using a ratchet.

Table 1 Parameters for finish-milling

Setting	Pre-grinding	Fine-milling #1	Fine-milling #2
Infeed, mm	0.1	0.04	0.01
Feed motion, mm/s	2.0	2.0	2.0
Rotation speed, rpm	10.000	10.000	10.000
Grinding tool	Cera Gloss HP green	Cera Gloss HP yellow	Cera Gloss HP yellow

2.6 Resection Process

The collapsible blades are folded, and the proximal contact protecting sheath is pushed over them until it makes contact with the distal insertion cap. From the aortic side, the so closed distal head is advanced through the center of the aortic valve until the massive counter ring touches the valve. The protective sheath is pulled back and the cutting edges are expanded. By turning the handle, the distal member (A) with the blade is moved toward the counter ring, thus punching out the tissues entrapped in-between. When the valve leaflets are cut, the instrument as a whole is pulled back out of the created "neo-annulus".

2.7 Test Specimens

Commercially available ingredients were mixed in order to replace native contents of calcifications (Table 2) and obtain material with comparable properties to native calcifications. Collagen was mixed with water (1:7) and after dissolution, 0.5 g hydroxyl-apatite and 3.5 g calcium carbonate were added. The mixture was dried for 48 h at room temperature and then fragmented to form particles. Moderately calcified valves were simulated by gluing (Loctite 4011, Zyanakrylat, Henkel®, Germany) one calcified fragment on each of the three leaflets of the commercially available (worn out/expired) stented aortic valve bioprostheses with geometries similar to native valves. After removing parts of the rigid stent structure, the prostheses were fixed in a steel tube to allow for stable conditions during resection (Fig. 5, 6).

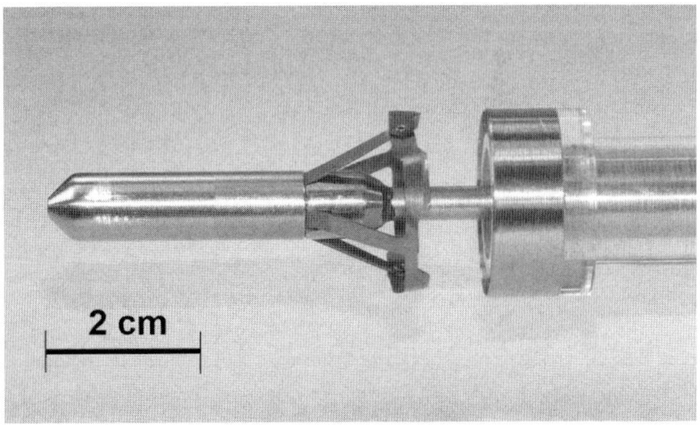

Fig. 4 Zoom on distal end of resection tool for open heart surgery

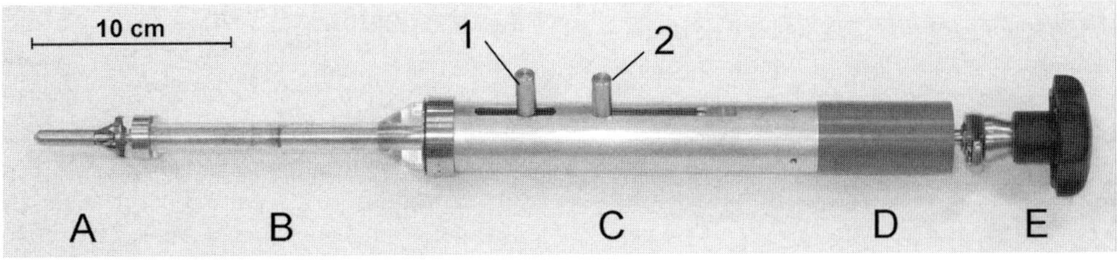

Fig. 3 Resection tool for open heart surgery

Table 2 Components of calcifications

Component	Equivalent	
Calcium	Calciumcarbonate	"Weinkalk," Art.Nr.: 35500, Schlag GmbH®, Aalen, Germany
Collagen	Bone-glue	Perlform, MH + P®, Berlin, Germany
Hydroxylapatite	Hydroxylapatite	"Ostim" < 250 μm, Ch: 7A001, AAP Biomaterialien®, Dieburg, Germany

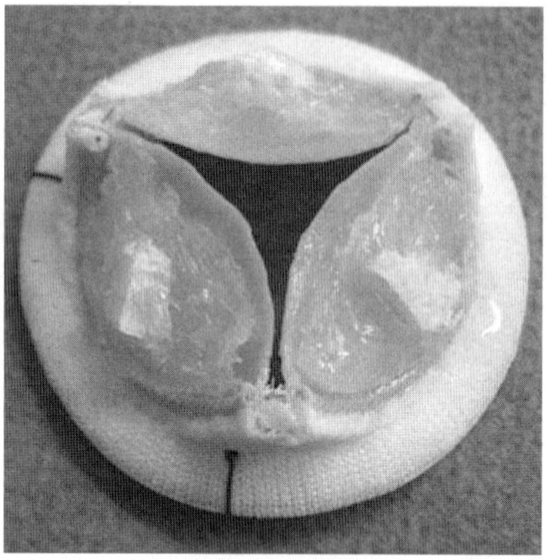

Fig. 5 Test specimen: Artificial particles glued on a bioprosthesis to simulate a moderately calcified aortic valve

Fig. 6 Test specimen fixed in stainless steel tube

2.8 Measurements

This study was carried out to obtain: (1) The overall function and handling of the instrument, (2) the time needed for resection, and (3) an idea of the turning moments during resection. Turning moments were recorded by inserting a torque sensor (DR-2477, Lorenz Messtechnik GmbH, Altdorf, Germany) between the handle and the main body of the instrument and processing the information using a digital connector (CB-68LP, National Instruments, Austin, Texas) and a PCA-card (PCI-MIO-16E-4, National Instruments, Austin, Texas). The measurements were saved and evaluated after the experiments using LabVIEW (version 6.1, National Instruments, Austin, Texas) for numerical and graphical analyses. Resection time and the maximum turning moment during this testing period were registered.

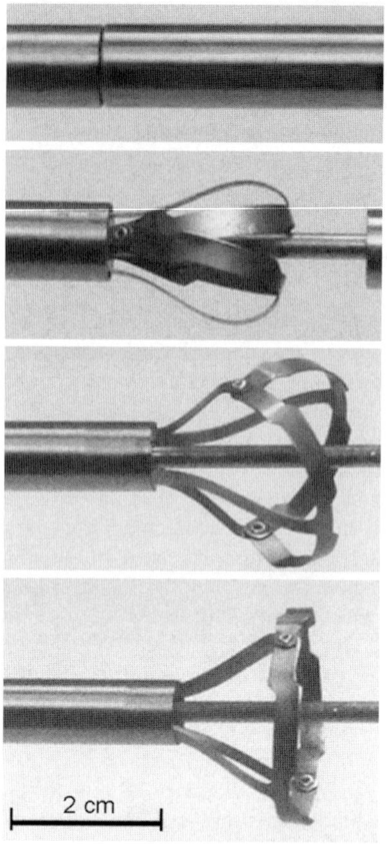

Fig. 7 Expansion of the Nitinol cutting blades

3. Results

Six artificially calcified aortic bioprostheses were used for the resection procedure with the prototype for open access. The cutting of the leaflets was successful in all the specimens without any complication during resection. The resection tool was easy to handle, even when elevated forces had to be applied; the expansion of the cutting ring worked well (Fig. 7).

Resection time was 22 ± 6.29 s (Mean $\pm$ SD); maximum turning moment of the instrument was 2.4 ± 1.27 Nm (Mean $\pm$ SD). The surface of the specimens was quite smooth after resection (Fig. 8). The results are listed in Table 3.

4. Discussion

The major findings of this study include (1) the function of the resection tool prototype using Nitinol blades was confirmed and (2) the realization of an aortic valve resection in less than 30 s using this tool seems to be possible regarding the results obtained for the resection of stented valves fixed in a steel tube.

Sutureless valves are rather a new application in the field of aortic valve replacement, but they will certainly become more important and more widely available when the problems of insufficiency and paravalvular leakage of the prostheses are solved. In our opinion, it is therefore necessary to create a defined circular geometry in the native tissue before inserting a new valve. As of date, to our knowledge no instrument exists for this purpose. Resection methods which can be found in several patents, especially on endovascular resection of aortic valves (e.g., Ref 18, 19), are neither feasible nor clinically reasonable. Most of the concepts cannot produce a geometrically circular cut—being one of the main requirements—as they only "chip off" small parts of the leaflets with each cutting movement of the device. Other methods—mostly non-mechanical techniques such as laser or water jet cutting—are very difficult to handle and time-consuming, and thus do not allow a very fast resection. Therefore, we decided to develop an instrument which cuts the native valve by a rotational punching process between two circular cutting edges.

The configuration of this prototype is a further development of a previous study, in which we investigated grinding possibilities, stability, and fundamental design of the cutting edges (Ref 17). The use of flat cutting edges (instead of waved ones or saw-tooth blades) and the rotating punching process showed the best results. The sheet thickness of 0.35 mm for the Nitinol blades was chosen as tests revealed that it offers the best combination of shape stability and flexibility: Using thinner sheets facilitates the folding of the cutting edges and their insertion into the protecting sheaths. On the other hand, stability gets poor with turning moments ranging from 2.5 to 3.0 Nm as they were obtained in this study, thus causing elastic deformation during cutting. Contrarily, the use of thicker sheets decreases flexibility, i.e., the blades can be folded only with difficulty, causing deformation and ultimately leading to non-circular/oval cutting rings in the expanded state. Further

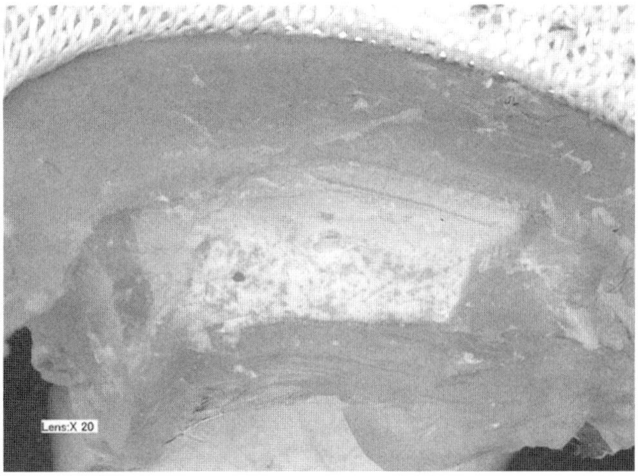

Fig. 8 Cutting result of moderately calcified bioprosthesis: Zoom (×20) on artificial calcification

Table 3 Resection results

	I	II	III	IV	V	VI	Mean	SD
Resection time, s	32	26	18	23	18	15	**22**	**6.29**
Maximum turning moment, Nm	4.6	3	1.9	2.4	1.2	1.3	**2.4**	**1.27**

experiments will show whether the right parameters were chosen or whether the alloy grade or the thermo-mechanical treatment needs to be changed.

In this study, the shaping of the cutting edges was performed on a complex 3D structure, using 25 min of heat treatment at 530 °C in normal atmosphere, followed by quenching in water. The temperature of 530 °C was chosen as a compromise, as we were not able to create a non-oxidising environment in our furnace. On the one hand, lower temperatures would have increased the heating time while not totally preventing the creation of an oxide layer. On the other hand, higher temperatures would have reduced time, however, resulting in a much thicker oxide layer on the blades. The rather long time span was determined by tests: We increased heating times at fixed temperature until the sheet material did not bend back elastically when the loading was removed. In future studies, the effect of the temperature curve of the blades during the nominal 25 min of heating should be dealt with to learn more about its influence on the thermodynamic, micro-structural, and mechanical properties and about its effects on stability and elasticity of the blades.

There was no special treatment of the Nitinol structures before aging, but dressing of the rims and sharpening of the cutting edges. We performed sharpening before heat treatment mainly for production reasons. Flat sheeting was much easier to fix and to grind on a high-speed milling cutter than complex structures. Thus, the sharpening procedure could be optimized and nearly standardized, resulting in reproducible good sharpness of cutting edges, comparable to razor blades (Fig. 9) (Ref 17). Heat treatment following this process created a thin oxide layer on the blades which reduced the initial sharpness, but as the cutting edges still passed the sharpness test according to DIN 58298, we decided to use them for this study. For future studies, we will have to think about possibilities to remove the oxide layer and to restore the initial state, for example, using electro-polishing. We also thought about DLC (Diamond-like carbon) coating to improve the durability of the cutting edges, but due to the high stresses induced by the folding of the blades during application (radius of up to 14 mm got reduced to radius of about 3 mm), there was the risk of damage to and detachment of the thin layer; therefore, this method was not used.

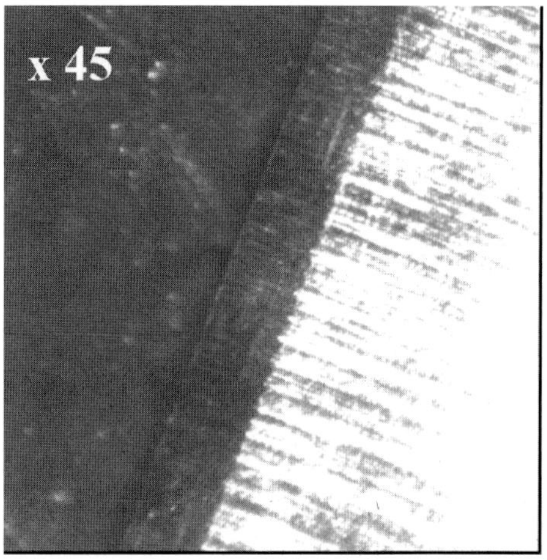

Fig. 9 Microscopic view (×45) of a sharpened blade after multi-stage grinding

As this study is meant to be a first step in testing the instrument, the test-setup was important for analyzing the results. Therefore, we chose to use commercially available worn-out or expired stented valves, for example, the "Edwards 25 mm Aorta S.A.V. 2650"-bioprosthesis (Edwards Lifesciences, Irwine, CA) or the "Medtronic 25 mm Aorta Mosaic Ultra 25-A 305"-bioprosthesis (Medtronic Inc., Minneapolis, MN). All the specimens had in common, that the valve consisted of natural tissue, being attached to a stented frame in a semilunar fashion similar to geometries of native valves. As the rigidity of the stent was not comparable to native tissues, it would have prevented taking the desired blade diameter. Therefore, the upper tips of the semilunar structures were cut before we added the particles for simulating calcifications. Finally, the suture cuff of each valve was fixed in a steel tube to guarantee stable conditions during resection. For comparable results, the instrument was attached to a holding fixture, thus eliminating differences due to the instrument's angle to the valve or due to a positioning out of the valve's center.

Another important point to note relates to choosing the right size for the cutting edges. Before normal surgery, the diameter of the diseased valve can be measured using TEE guidance (transoesophageal echocardiography), CT (computer tomography), or MRT (magnetic resonance tomography). The nominal diameter of the blade set is chosen to be smaller to generate a circular rim of 1-2 mm, which will act as "neo-annulus" for the prostheses. As we wanted to cut prostheses of 25 mm diameter, we had to subtract additional 5 mm for the supporting stent frame and suture cuff, thus performing resection with a nominal blade diameter of 16 mm.

All these abovementioned points have to be considered when analyzing the results.

Operating the instrument was easy. There was neither a problem to insert the instrument head through the leaflets of the prostheses nor to open the foldable blades which regained their circular shape and enabled it to perform the cutting process. Furthermore, the rotational punching was easy to perform as a ratchet was included in the handle. On the other hand, quite relevant forces had to be applied on the handle because of the elevated turning moments ranging from 1.2 to 4.6 Nm. Without the use of the holding fixture for the instrument, stable cutting and accurate positioning of the instrument would have been very difficult. It is not really clear, why torques had to be that high, especially as cutting seemed quite uniform. Regarding the turning moments, a much more discontinuous movement would have been expected. Presuming that the forces are not only produced by the cutting itself and that the contact-free torque sensor is no source for torques, we concluded that at least one of the components between blades and handle had to be responsible. With an efficiency of 88%, the transmission gear contributed to the increase of forces, but probably the main cause was the thread that was used for the transformation of the turning motion into a combination of translation and rotation. In order to investigate the influence of the thread, this part of the instrument should be redesigned to separate these two movements.

In this study, resection of all the three valve leaflets could be completed within a time range of 15 to 32 s, validating our concept of a fast resection using a simple mechanical procedure. Other concepts which are described in the literature required much more time: from 2.3 ± 0.3 min to 12.2 ± 0.8 min per leaflet using a water stream scalpel (Ref 15), and to 15.5 ± 3 min per leaflet using a YAG laser (Ref 16). Thus, an explicit acceleration of the procedure was obtained when cutting the moderately calcified valves with our new device. Further tests will show whether these results remain the same for organic tissues which are more calcified.

It is also important to perform additional tests to identify the risks of using the resection tool in its real working environment, when cutting conditions are not as optimized as in this study. As 40 mm of the distal end of the instrument is pushed through the valve inside the left ventricle, there is the risk of the blades damaging ventricular structures during operation. Before experimenting it on living patients, surgeons have to be sure that neither an unintended lesion is created in the organic tissues, such as the ventricular wall or the mitral valve annulus, nor parts of the instrument stick to ventricular structures as the mitral cordae. Risk is reduced by the concept of not excising the valve and all the calcified tissues completely, but of leaving a rim of 1-2 mm, serving as "neo-annulus". Thus, a certain distance between rotating blades and ventricular tissues is already installed. Nonetheless, it would be appropriate to control the cutting procedure by endoscopic means, at least to enable a right positioning. There is less risk of damages to aortic structures, such as the coronary ostia, as the outer edge of the counter ring can be chamfered, and a rotating movement is absent.

During our tests, we observed the creation of fragments and debris. The resected valve leaflets, as well as small- and middle-sized particles generated by the cutting process were either enclosed in the counter ring or fell through the struts of the foldable blade to where there would be the left ventricle. The protection against potential embolism caused by these particles will be of utmost importance during surgery to prevent obstruction of the coronary arteries or stroke. Therefore, first of all, a filter structure should be incorporated in the instrument, being placed between the struts of the distal cutting ring, and thereby collecting most of the particles which are not already enclosed within the proximal counter ring. In addition, in the open heart situation, there is the possibility to block the coronary ostia (e.g., with probes) and to use a suction device for aspiration of particles which escape from the blades and fall into the ventricle.

Further important questions to be noted are the influence of working environment on the results and the limitations of this study. We are aware of the complexity of cardiac anatomy, and that we created optimized conditions for our tests. Replacing the steel frame by the native, non-circular aortic root could pose a problem, but possibly the circular counter ring, being nearly as big as the native valve, will spread the aorta to circular shape. Further studies will have to show, whether this is sufficient for stabilization of the instrument or other means have to be designed and integrated. In addition, cadaver studies would be necessary to investigate the influence of comorbidities or congenital defects on the function of the resection tool, as well as to cut native diseased aortic valves with blade diameters ranging from 21 to 27 mm. Regarding the Nitinol cutting edges, this would allow us to learn about forces, turning moments, and abrasion of the blades. As a result, optimizing of geometry and sizes of the blades will be possible, for example, using the Finite Element Method (FEM), especially for the dimensioning of blades with a diameter greater than 21 mm, as with increasing diameter, achieving stability during collapsing/expanding becomes more difficult. In addition, as changing of blades is rather time-consuming for the prototype of this study (approx. 5 min), a new design should allow changing of blades more easily and quickly even during surgery, in the event of their malfunction.

However, in conclusion, the first results with this prototype show that a rapid resection of an aortic valve seems to be possible. If the use of this device can be combined with the use of a sutureless valve, then a duration for cardiopulmonary bypass below 10 min should become possible, meaning a reduction of 50-80%. Thus, morbidity and mortality rates of AVR could be reduced, and therefore this novel method of an accelerated aortic valve resection in combination with the insertion of a sutureless aortic valve will probably make conventional surgery more attractive. Further, it would be the first step toward the use of these techniques (resection and sutureless valve) in surgery on beating heart. Performing the AVR within 1 min after resection of the pathology with much better results than till now (e.g., without regurgitation) seems possible owing to the precisely shaped annulus remaining *in situ*. Further optimization of the instrument is necessary, as well as *in vitro* tests, cadaver studies, and perhaps animal trials, to continue this "development" and to use this minimally invasive instrument during open heart surgery.

Acknowledgment

This study was supported by the "Arbeitsgemeinschaft industrieller Forschung" (AIF), Berlin, Germany (Grant PROINNO KF0373601PK6).

References

1. P. Kvidal, R. Bergstrom, L.G. Horte, and E. Stahle, Observed and Relative Survival After Aortic Valve Replacement, *J. Am. Coll. Cardiol.*, 2000, **35**, p 747–756
2. B.A. Carabello, Clinical Practice. Aortic Stenosis, *N. Engl. J. Med.*, 2002, **346**, p 677–682
3. C.M. Otto, Valvular Aortic Stenosis: Disease Severity and Timing of Intervention, *J. Am. Coll. Cardiol.*, 2006, **47**, p 2141–2151
4. L.Z. Bloomstein, I. Gielchinsky, A.D. Bernstein, V. Parsonnet, C. Saunders, R. Karanam, and B. Graves, Aortic Valve Replacement in Geriatric Patients: Determinants of In-Hospital Mortality, *Ann. Thorac. Surg.*, 2001, **71**, p 597–600
5. K.P. Alexander, K.J. Anstrom, L.H. Muhlbaier, R.D. Grosswald, P.K. Smith, R.H. Jones, and E.D. Peterson, Outcomes of Cardiac Surgery in Patients > or = 80 Years: Results from the National Cardiovascular Network, *J. Am. Coll. Cardiol.*, 2000, **35**, p 731–738
6. R. Sharony, E.A. Grossi, P.C. Saunders, C.F. Schwartz, G.B. Ciuffo, F.G. Baumann, J. Delianides, R.M. Applebaum, G.H. Ribakove, A.T. Culliford, A.C. Galloway, and S.B. Colvin, Aortic Valve Replacement in Patients with Impaired Ventricular Function, *Ann. Thorac. Surg.*, 2003, **75**, p 1808–1814
7. K.K. Yeo and R.I. Low, Aortic Stenosis: Assessment of the Patient at Risk, *J. Interv. Cardiol.*, 2007, **20**, p 509–516
8. D. Wendt, M. Thielmann, T. Buck, R.-A. Jánosi, T. Bossert, N. Pizanis, M. Kamler, and H. Jakob, First Clinical Experience and 1-year Follow-up with the Sutureless 3F-Enable Aortic Valve Prosthesis, *Eur. J. Cardiothorac. Surg.*, 2008, **33**, p 542–547
9. E. Grube, U. Gerckens, and L. Buellesfeld, Perkutaner Aortenklappenersatz (Percutaneous Aortic Valve Replacement), *Herz*, 2006, **31**, p 697–697 (in German)
10. A. Vahanian, O. Alfieri, N. Al-Attar, M. Antunes, J. Bax, B. Cormier, A. Cribier, J.P. De, G. Fournial, A.P. Kappetein, J. Kovac, S. Ludgate, F. Maisano, N. Moat, F. Mohr, P. Nataf, L. Pierard, J.L. Pomar, J. Schofer, P. Tornos, M. Tuzcu, B. van Hout, L.K. von Segesser, and T. Walther, Transcatheter Valve Implantation for Patients with Aortic Stenosis: A Position Statement from the European Association of Cardio-Thoracic Surgery (EACTS) and the European Society of Cardiology (ESC), in Collaboration with the European Association of Percutaneous Cardiovascular Interventions (EAPCI), *Eur. Heart. J.*, 2008, **29**, p 1463–1470
11. T. Walther, V. Falk, M.A. Borger, T. Dewey, G. Wimmer-Greinecker, G. Schuler, M. Mack, and F.W. Mohr, Minimally Invasive Transapical Beating Heart Aortic Valve Implantation—Proof of Concept, *Eur. J. Cardiothorac. Surg.*, 2007, **31**, p 9–15
12. J.G. Webb, S. Pasupati, K. Humphries, C. Thompson, L. Altwegg, R. Moss, A. Sinhal, R.G. Carere, B. Munt, D. Ricci, J. Ye, A. Cheung, and S.V. Lichtenstein, Percutaneous Transarterial Aortic Valve Replacement in Selected High-risk Patients with Aortic Stenosis, *Circulation*, 2007, **116**, p 755–763
13. T.M. Joudinaud, E.M. Flecher, J.W. Curry, C.L. Kegel, P.A. Weber, and C.M. Duran, Sutureless Stented Aortic Valve Implantation Under Direct Vision: Lessons from a Negative Experience in Sheep, *J. Card. Surg.*, 2007, **22**, p 13–17
14. J. Ye, A. Cheung, S.V. Lichtenstein, S. Pasupati, R.G. Carere, C.R. Thompson, A. Sinhal, and J.G. Webb, Six-month Outcome of Transapical Transcatheter Aortic Valve Implantation in the Initial Seven Patients, *Eur. J. Cardiothorac. Surg.*, 2007, **31**, p 16–21
15. R. Quaden, T. Attmann, A. Boening, J. Cremer, and G. Lutter, Percutaneous Aortic Valve Replacement: Resection Before Implantation, *Eur. J. Cardiothorac. Surg.*, 2005, **27**, p 836–840
16. R. Quaden, T. Attmann, M. Schunke, D. Theisen-Kunde, J. Cremer, and G. Lutter, Percutaneous Aortic Valve Replacement: Endovascular Resection of Human Aortic Valves In Situ, *J. Thorac. Cardiovasc. Surg.*, 2008, **135**, p 1081–1086
17. D. Wendt, S. Stühle, E. Kawa, M. Thielmann, B. Kipfmüller, H. Wendt, F. Hauck, H. Fischer, and H. Jakob, NiTinol-based Cutting Edges for Endovascular Heart Valve Resection: First In Vitro Cutting Results, *Minim. Invasive Ther. Allied. Technol.*, 2009, **18**(1), p 1–7
18. Patent WO 2004089250 A2: Apparatus and Methods for Valve Repair
19. Patent WO 2003088809: Method and Apparatus for Resecting and Replacing an Aortic Valve

JMEPEG (2009) 18:470–474
DOI: 10.1007/s11665-009-9498-3

Nitinol Surfaces for Implantation

Svetlana Shabalovskaya, Gianni Rondelli, and Markus Rettenmayr

(Submitted September 12, 2008; in revised form April 23, 2009)

Nitinol, a group of nearly equiatomic Ni-Ti alloys, steadily conquers new areas of application. Because of the need to keep a low profile of miniature implant devices, and considering the lack of compatibility between Nitinol superelasticity and the mechanical properties of traditional coatings, bare surfaces are of interest. In this article, an overview of our studies of bare Nitinol surfaces is presented, and the performance of coated surfaces is outlined. Together dense and porous Nitinol offer a wide array of surface topographies, suitable for attachment and migration of biological cells and tissue ingrowth. Native Nitinol surface oxides vary from amorphous to crystalline and exhibit semiconducting properties associated with better blood compatibility. Nitinol surfaces are analyzed with regard to high and lasting nickel release in vitro. Surface oxide thickness and Nitinol intermetallic particulates are discussed in relation to corrosion resistance and mechanical performance of the material.

Keywords advanced characterization, biomaterials, coatings, corrosion testing, failure analysis, intermetallics, titanium

1. Introduction

Owing to mechanical compatibility with living tissues (Ref 1), Nitinol is suitable for a diverse field of implantation from cardiovascular to orthopedics. The biocompatibility of Nitinol, however, is constantly challenged because of high content (up to ~56% by weight) of Ni, known for its toxic and allergenic effects (Ref 1). High Ni content of these alloys does not imply high Ni release in the body (Ref 2). However, the knowledge on Ni release from Nitinol evolves. Earlier studies on the alloys prepared under laboratory conditions showed a reduction of Ni release to almost undetectable levels after 10-14 days of exposure to physiological solutions. In contrast, recent results obtained for certain commercial materials used in as-received wire and sheet forms (Ref 3, 4) demonstrated that Ni release may significantly increase with time up to six months (more references in (Ref 2)). This correlates well with the complications after implantation of Nitinol ASD (Atrium Septal Defect) closure devices attributed to Ni release and sensitivity reaction (Ref 5). What are the sources of Ni release? Nitinol as an intermetallic compound is stable, and Ni is not readily soluble as in the case of Ni-rich alloys or pure Ni (Ref 1). Furthermore, due to a gain in the formation energy, Ti oxides prevail over Ni oxides on Nitinol surface providing good biocompatibility associated with their inertness. Chemically homogeneous surfaces with an optimal thickness would protect Nitinol against corrosion and Ni release. However, this is not always the case. Importantly, with the miniaturization of Nitinol devices, the role of the surface in the implant performance is increased. In this article, we overview our studies of Nitinol surfaces related to their safe biological performances. It is not our goal, however, to provide here a detailed review of the literature on this topic. Our corresponding chronological reviews focused on various Nitinol's aspects and all the references can be found elsewhere (Ref 1, 2, 6, 7).

This article is an invited paper selected from presentations at Shape Memory and Superelastic Technologies 2008, held September 21-25, 2008, in Stresa, Italy, and has been expanded from the original presentation.

Svetlana Shabalovskaya and **Markus Rettenmayr**, Institute of Materials Science and Technology, Metallic Materials, University of Jena, Loebdergraben 32, 07743 Jena, Germany; and **Gianni Rondelli**, Institute of Energy and Interphases, Center of National Research, Milan, Italy. Contact e-mail: svetinol@yahoo.com.

2. Performance of Bare Nitinol Surfaces

Comparative analysis of the corrosion resistance of strained Nitinol demonstrated that bare surfaces outperformed coated ones and those modified with energy sources (Ref 7). Bare surfaces also offer low profile for the miniature devices and variable thrombogenicity (Ref 2), important for cardiovascular applications. At present, the chemistry of Nitinol surfaces can be varied in a wide range providing Ni content from zero to about 30% (Ref 1). Although in the latter case Ni is bound into an oxide, these surfaces should not be used for medical applications since they proved to be as toxic as pure Ni (Ref 1). Low Ni concentrations are easily achieved by means of simple surface treatments, such as electropolishing, chemical etching with subsequent boiling in water, anodizing, and brief heat treatments (Ref 8).

Nitinol surface topographies can also be largely modulated (Fig. 1) rendering desirable cell responses. For example, electropolishing of NiTi in different states, austenite versus martensite, results either in smooth or structured surfaces (Fig. 1a, b). Elimination of inclusions and precipitates using chemical etching results in surfaces more suitable for cell attachment and migration (Fig. 1c, d). The structured surfaces associated with the R phase (Fig. 2a), as opposed to the B19' surface martensite (Fig. 1b), may be more appropriate for certain applications. As one can see from the AFM images

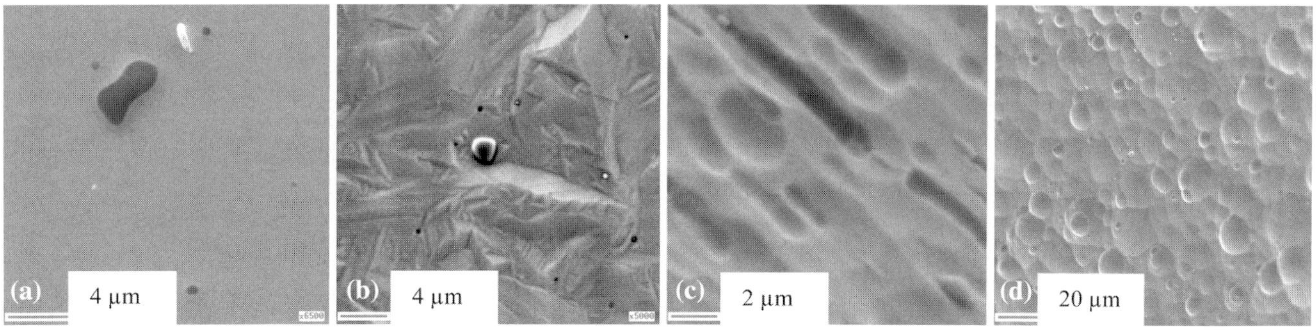

Fig. 1 SEM images of Nitinol surfaces electropolished in austenitic (a) and martensitic (b) states; wire etched in HF + HNO_3 solutions—lateral side (c) and cross section (d). Adopted and reprinted from Ref 7

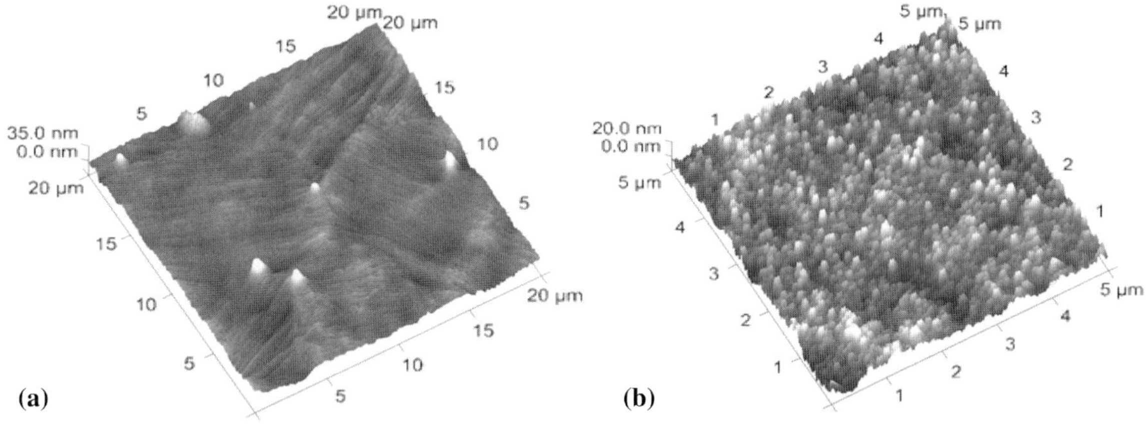

Fig. 2 AFM images of mechanically polished surface (a), and chemically etched plus heat treated in air for 20 min at 500 °C (b). Peaks observed on the image (a) correspond to the hard particulates protruding through the surface. Ti dioxide crystallites of a 20-30-nm size were formed during heat treatment (b). Adopted and reprinted from Ref 7

(Fig. 2), Nitinol surfaces differ significantly in their topography on a nano scale level. Heat treatments of the material promote growth of Ti dioxide crystals that may reach 20-30 nm in size after a brief exposure to 500 °C (Fig. 2b) (Ref 2). According to recent studies (Ref 8), chemically etched Nitinol surfaces are amorphous, and the surfaces heat treated at 500 °C are crystalline, and both are semiconductive with the band gaps of 3.4 and 3.0 eV, respectively. Electropolished surfaces are also semiconductive with a band gap of 3.4 eV (Ref 9). Anodizing of Nitinol to generate new surfaces is also being explored (Ref 8) (see other Ref in 7); the increase of polarization resistance up to 67 MΩ cm^2 can be achieved by anodizing.

Porous Nitinol with permeable structural network and desirable degree of porosity is another material suitable for bone substitute (Ref 1, 2, 6). The living tissues grow directly into the body of porous implants transforming it into a new material. It has been shown that after three months, a newly formed implant-bone composite acquires the properties superior to those of the original alloy and the bone. New bone in the porous Nitinol is detected as early as 2-3 weeks after implantation. The performance of porous Nitinol has always been found to be similar or superior to pure porous Ti or its alloys (Ref 2, 6).

3. Nickel Enrichment of Surface and Interfaces

High and lasting ion release observed sometimes in vitro is an indication of Ni accumulation in Nitinol surface. There are quite different avenues that can lead to this outcome. Thus, high temperature treatments of the material to improve its mechanical properties or to anneal lattice defects after plastic deformation are accompanied by alteration of the phase composition of the alloy. Various binary Ni_xTi_y phases as well as their oxides can be formed and, as a result, the composition of the adjacent matrix alloy is altered. This affects not only the mechanical properties of Nitinol but also the phase transformation temperatures, corrosion resistance, and Ni release. During the production of wires, Nitinol is subjected to numerous drawings with intermediate annealing at the temperatures up to 700 °C and higher. This material has a heavily oxidized lustrous surface, which is sometimes considered for medical applications. Indeed, the external surface layers may have relatively low Ni content. However, a seemingly low Ni content may not ensure an inert surface behavior. Thus in Fig. 3, the TEM images of NiTi wires with original surface are shown (Ref 10). Wire (a) had less than 1% Ni and wire (b) approx. 7% Ni on the surface. Both wires showed the increase

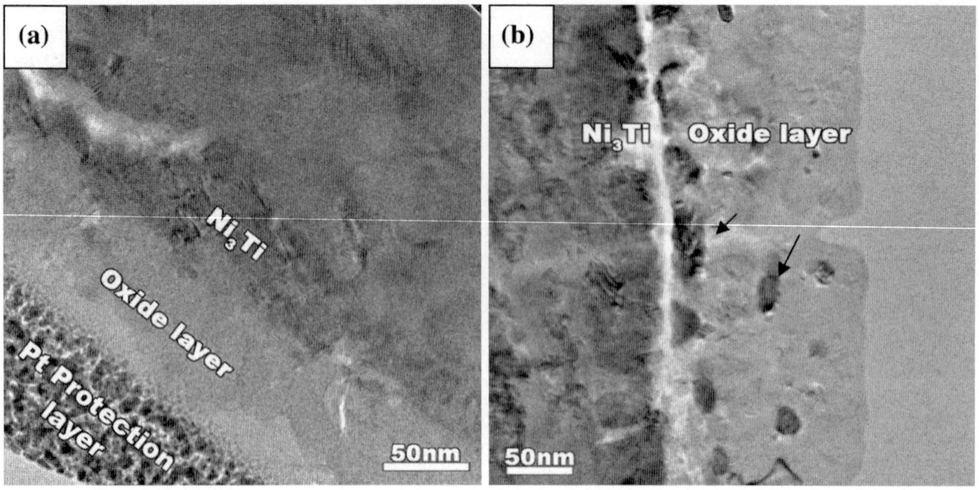

Fig. 3 TEM images of the surfaces of two Nitinol wires with original oxides. Formation of a thick Ti-based surface oxide layer resulted in the enrichment with Ni of the interface between the surface and the bulk of the alloy: Ni_3Ti layer on both images (a and b) and Ni particles marked by arrows (b). Adopted and reprinted from Ref 10

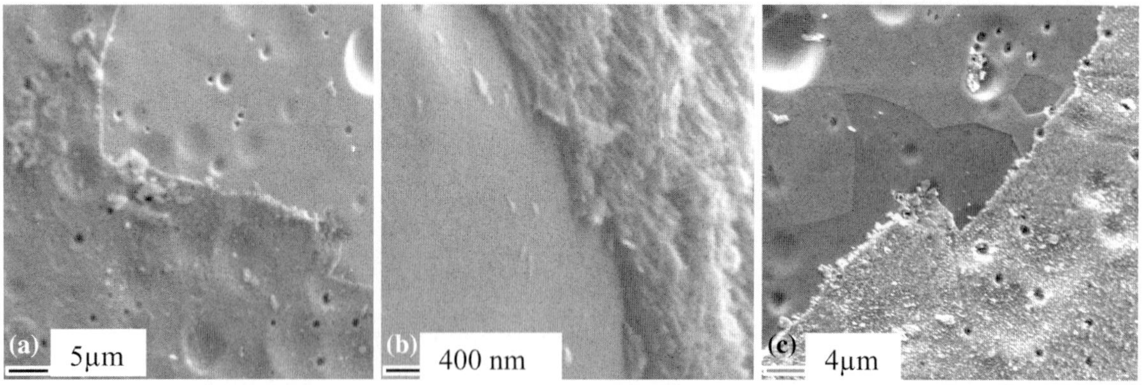

Fig. 4 SEM images of Nitinol surface pre-treated using a routine protocol developed for pure Ti for deposition of a CaP coating (boiling in 30% hydrogen peroxide with subsequent etching in 4 M KOH solution at 120 °C for 30 min). Adopted and reprinted from Ref 7

of Ni release with immersion time, but the Ni release from wire (b) increased to 1.8 mg/L after six months and exceeded that from wire (a) by as much as 100 times (Ref 10). For comparison, human blood serum contains 1-6 μg/L of Ni (Ref 2). In addition to differences in the surface topography, these wires also revealed differences in Ni chemical state and its distribution in the surface. In wire (a) there was only an intermediate Ni_3Ti phase formed between the bulk of the alloy and the Ti surface oxides. In wire (b), however, there was also segregated pure Ni in agreement with the earlier studies of Nitinol oxidation behavior (Ref 11-13). The Ni particulates of 10-100 nm diameters adjacent to the Ni_3Ti layer and located as close as 20 nm from the surface were observed. This Ni reservoir is easily activated in corrosive environments and can be a cause of associated allergenic reactions observed after Nitinol implantation (Ref 5). These surfaces showed low breakdown potentials (200 mV) (Ref 10) implying the need for their optimization before implantation.

Surprisingly, another avenue, which may lead to Ni accumulation in the surface, is associated with the low temperature protocols developed for pure Ti, aiming at the formation of thick dioxide layers. Soaking or boiling of Nitinol in hydrogen peroxide with a subsequent long exposure to alkaline KOH or $Ca(OH)_2$ solutions at temperatures from 60 to 160 °C is employed as a pre-treatment for CaP deposition. The resulting Nitinol surfaces induced Ni release (up to 9 mg/L) of as much as a few thousands times higher than that observed with bare surfaces (Ref 7, 14). The Auger study (Ref 7) revealed two layers formed during this pre-treatment (Fig. 4): an external flake-like layer less than 1 μm thick ($C_{35}Ti_{20}Ni_{12}O_{33}$) and a smooth internal layer with high Ni content ($C_8Ti_{11}Ni_{58}O_{22}$). A precise TiO_2 stoichiometry of this buried sub-layer ($Ti_{11}O_{22}$) indicates that the goal of the pre-treatment has been in fact achieved, but this also implies that the detected Ni (~58 at.%) is in elemental state and can be readily released.

The temperatures mentioned above are not high, and one does not expect a significant atomic diffusion. The situation with the atomic diffusion in surface oxides is not trivial, however. The Ti oxide layers are forming spontaneously even at room temperature, and Ni atoms are liberated from the Ni-Ti atomic bonds. These Ni atoms are defects in the Ti oxide layer. Due to the smaller size of the Ni atoms compared to Ti and oxygen atoms, Ni can easily diffuse through an interstitial path. On the other hand, because of oxygen shortage, the Ti surface oxides are commonly non-stoichiometric. The sites of missing oxygen atoms present atomic vacancies, which can also be used by Ni to

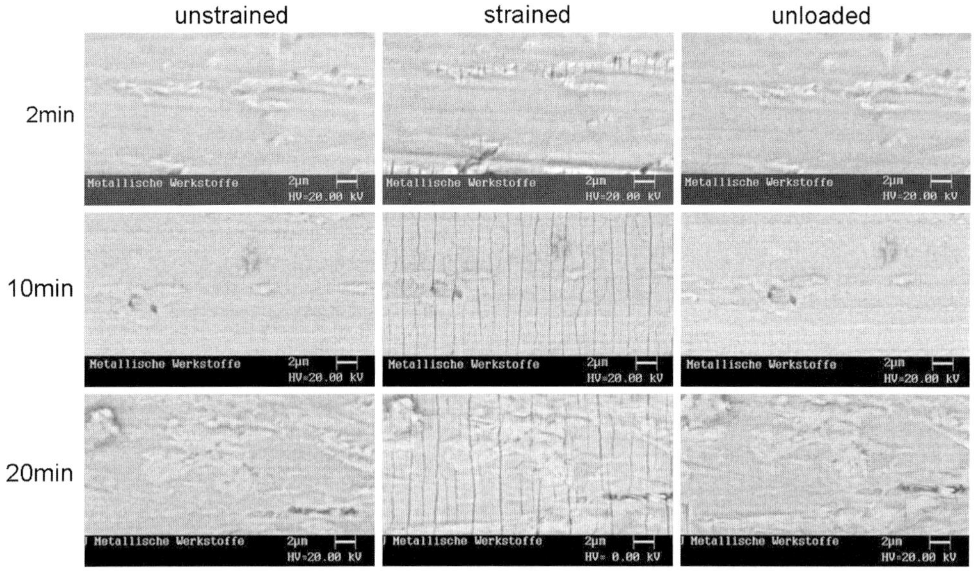

Fig. 5 SEM images of the surfaces of heat-treated, as-received Nitinol wire in strain-free, strained to 8%, and unloaded states. Adopted and reprinted from Ref 17

migrate through Ti oxides. Thus, in the presence of ready-to-use lattice vacancies, the diffusion of Ni atoms within Ti oxides may occur effectively already at low temperatures, and Ni-rich surface sublayers can be generated during long exposures.

4. The Effect of Surface Oxide Thickness

Heat treatments of Nitinol are continuously explored for Nitinol surface oxidation. There is a belief that "the ceramic-like nature of Ti dioxide must resist cracking in the pulsative oscillation" (Ref 15). Unfortunately, the ceramic nature of TiO_2 acts against both Nitinol superelasticity and cyclic pulsation regime. Ceramic materials are held together by ionic and covalent bonds, which tend to fracture before any plastic deformation occurs; superelastic deformation experienced by Nitinol is in plastic limits for TiO_2.

Being strained, thick surface oxides, which were formed during prolonged heat treatments crack (Ref 8, 10, 16, 17). There should be an optimal oxide thickness that would retain surface integrity despite its ceramic-like nature under strain and provide a barrier for Ni release. Figure 5 indicates that the surface of Nitinol wires heated for 2 min at 540 °C (30-nm oxide thickness determined using Rutherford spectroscopy) cracks at 8% strain, but only at surface inhomogeneity locations; after a 10-min heat treatment (80 nm), the whole surface is uniformly cracked in both perpendicular and longitudinal directions; further increase in oxide thickness to 100 nm results in surface flaking (Ref 17). According to the studies of the effects of heat treatments on Nitinol breakdown potentials (Ref 13, 16), this oxide thickness is on the border between high and very low breakdown potentials.

5. Surface Particulates

Nitinol is a heterogeneous material, which acquires inclusions (Ti-based oxides and carbides) during melting, and Ti_xNi_y precipitate particles through the solid state reactions during annealing. The latter phases are easily oxidized into $Ti_xNi_yO_z$. All these particulates are hard and brittle and can vary in size substantially (Ref 18, 19). While inclusions are distributed quite uniformly in Nitinol, the precipitate particles can form clusters or chains of 50-300-μm length. The type of the particle distribution in the material is essential for its mechanical strength. Because the stent wall thickness is being reduced to 30-50 μm to obtain better biocompatibility and the diameters of wires employed are around 100 μm, and because, in two-thirds of stent graft failures, were metal fractures involved (Ref 20), a special attention should be given to material optimization, which, in turn, will improve Nitinol surface.

The areas adjacent to the intermetallic particles are more vulnerable to localized corrosion because of altered chemical composition. The localized corrosion around particles was observed at potentials about 1 V (Ref 19). These potentials are much higher of those expected in the body ($\leq$500 mV (Ref 6)). However, our analysis (Ref 2) of the images of the failed stents with very thick surface oxides 0.2-1 μm (Ref 15) revealed that localized corrosion occurs along the chains of the particulates. Although "the next-generation Nitinol stents were improved and only minimal evidences of corrosion reported, the mechanical instability is still a problem" (Ref 20), which may also be related to the surface. The analysis of elute from the grafts recovered a granular material of 0.5-2.5 μm identified as 'nickel-free titanium oxide' (Ref 20). The accumulation of intermetallic particles of the size comparable to the cell's size ($\sim$2 μm) due to corrosion in the areas adjacent to the implants is a possibility that needs to be considered. Fortunately, the thickness of the fibrous membranes formed around Nitinol by the eighth week of implantation were only 21 versus 24 μm for stainless steel and 425 μm pure Ni (Ref 21). This implies low Nitinol toxicity involved.

6. Surface Coatings and Modifications

Coatings and modifications of native Nitinol surfaces aimed at the design of a barrier for Ni release in the human body have been widely explored (Ref 7). The situation with the Nitinol implant devices, which are expected to perform in the superelastic regime, is not trivial. It was observed in situ that electropolished NiTi surface withstood 8% of tension without cracking, but coated or ion-implanted surfaces lost their integrity at lower strains (Ref 22). Surface cracking was observed at 1% strain for the nitrogen implanted and at 3% for DLC coated and Ti-sputtered surfaces. The coated or modified NiTi surfaces already in a strain-free state might not have passivity, and their corrosion resistance deteriorated under strain. Alteration of the surface topography and the heterogeneous distribution of local strains, both associated with the nature of the martensitic transformation in Nitinol, and the presence of particulates undermine coating adhesion and cohesion. Certain polymer coatings could be better choices as far as the mechanical compatibility is concerned. However, damaged polymer does not repassivate and causes crevice resulting in enhanced material corrosion. The corrosion rates for polymer coated Nitinol of the order of 13 μm/year reported in Ref 23 are significantly higher than those known for bare surfaces (0.1-2.26 μm/year) (Ref 7). In addition, in strained states, the coatings may not be effective against Ni release (Ref 7, 24). Considering the specifics of Nitinol, we believe that native surfaces appropriately passivated with optimized oxide thickness are a better choice for implantation.

References

1. S. Shabalovskaya, On the Nature of Biocompatibility and Medical Applications of NiTi Shape Memory Alloys, *Bio. Med. Mater. Eng.*, 1996, **6**, p 267–290
2. S. Shabalovskaya and J. Van Humbeeck, Biocompatibility of Nitinol for Medical Applications, *Shape Memory Alloys for Medical Applications*, Chap. 9, 1st ed., T. Yoneyama and S. Miyazaki, Ed., Woodhead Publishing Limited, Cambridge, 2008
3. O. Cisse, O. Savagodo, M. Wu, and L. Yahia, Effect of Surface Treatment of NiTi Alloy on its Corrosion Behavior in Hank's Solution, *J. Biomed. Mater. Res.*, 2002, **61**, p 339–345
4. B. Clarke, W. Carroll, Y. Rochev, M. Hynes, D. Bradley, and D. Plumley, Influence of Nitinol Wire Surface Treatment on Oxide Thickness and Composition and its Subsequent Effect on Corrosion Resistance and Nickel Ion Release, *J. Biomed. Mater. Res.*, 2006, **79A**, p 61–70
5. B. Wertman, B. Azabal, M. Riedl, and J. Tobis, Adverse Events Associated with Nickel Allergy in Patients Undergoing Percutaneous Atrial Seoptal Defect of Paten Foramen Ovale Closure, *J. Am. Coll. Cardiol.*, 2006, **47**, p 1226–1227. doi:10.1016/j.jacc.2005.12.017
6. S. Shabalovskaya, Surface, Corrosion and Biocompatibility Aspects of Nitinol as an Implant Material, *Bio. Med. Mater. Eng.*, 2002, **12**, p 69–109
7. S. Shabalovskaya, J. Anderegg, and J. Van Humbeeck, Critical Overview of Nitinol Surfaces and their Modifications for Medical Applications, *Acta Biomater.*, 2008, **4**, p 447–467
8. S. Shabalovskaya, G. Rondelli, A. Undisz, J. Anderegg, D. Burleigh, and M. Rettenmayr, The Electrochemical Characteristics of Native Nitinol Surfaces, *Biomaterials*, 2009, **30**, p 3662–3671. doi:10.1016/j.biomaterials.2009.03.034
9. S. Barison, S. Cattarin, S. Daolio, M. Musiani, and A. Tuissi, Characterization of Surface Oxidation of Nickel-Titanium Alloy by Ion-beam and Electrochemical Techniques, *Electrochem. Acta*, 2004, **50**, p 11–18
10. S. Shabalovskaya, H. Tian, J. Anderegg, D. Schryvers, W. Carroll, and J. Van Humbeeck, The Influence of Surface Oxides on the Distribution and Release of Nickel from Nitinol Wires, *Biomaterials*, 2009, **30**, p 468–477
11. V. Chuprina, Examination of the Process of Oxidation of Titanium Nickelide, *Poroshkovaya Metall. (Eng.)*, 1989, **316**(4), p 310–313 and 1989, **318**(6), p 57–61
12. C. Chan, C. Tringwell, and T. Duerig, Oxidation of NiTi Alloy, *Surf. Interface Anal.*, 1990, **15**, p 349–354
13. L. Zhu, J. Fino, and A. Pelton, Oxidation of Nitinol, *Proceedings of the Conference on Shape Memory and Superelastic Technologies (SMST), Asilomar*, A. Pelton and T. Duerig, Ed., 2003, p 357–365
14. D. Bogdanski, "Untersuchungen zur Biocompatibilität and Biofunktionalität von Implantatmaterialien am Beispiel von Nickel-Titan-Formgedächtnislegierungen," Ph.D. Thesis, Ruhr-Universität Bochum, Germany, 2005
15. C. Heintz, G. Riepe, L. Birken, E. Kaiser, N. Chakfe, M. Morlock, G. Delling, and H. Imig, Corroded Nitinol Wires in Explanted Aortic Endografts: An Important Mechanism or Failure?, *J. Endovasc. Ther.*, 2001, **8**, p 248–253
16. C. Trepanier, L. Zhu, J. Fino, and A. Pelton, Corrosion Resistance of Oxidized Nitinol, Oxidation of Nitinol, *Proceedings of the SMST, Asilomar*, A. Pelton and T. Duerig, Ed., 2003, p 367–375
17. A. Undisz, F. Schrempel, W. Wesch, and M. Rettenmayr, In Situ Observation of Surface Oxide Layers on Medical NiTi Alloy During Straining, *J. Biomed. Mater. Res.*, 2009, **88A**, p 1000–1009
18. S. Shabalovskaya, J. Anderegg, G. Rondelli, and J. Xiong, The Effect of Surface Particulates on the Corrosion Resistance of Nitinol Wire, *Proceedings of the SMST, Asilomar*, A. Pelton and T. Duerig, Ed., 2003, p 399–408
19. S. Shabalovskaya, J. Anderegg, and J. Van Humbeeck, Recent Observations of Particulates in Nitinol, *Mater. Sci. Eng. A*, 2008, **481-482**, p 431–436
20. T. Jacobs, J. Won, C. Gravereaux, P. Faries, N. Morryssey, V. Teodorescu, L. Hollier, and M. Marin, Mechanical Failure of Prosthetic Human Implants: A 10-year Experience with Aortic Stent Graft Devices, *J. Vasc. Surg.*, 2003, **37**, p 16–26
21. M. Morita, T. Hashimoto, K. Yamauchi, Y. Suto, T. Homma, and Y. Kimura, Evaluation of Biocompatibility for Titanium-Nickel Shape Memory Alloys In Vivo and In Vitro Environments, *Mater. Trans.*, 2007, **48**, p 352–360
22. C. Heßing, J. Frenzel, M. Pohl, and S. Shabalovskaya, Effect of Martensitic Transformation on the Performance of Coated NiTi Surfaces, *Mater. Sci. Eng. A*, 2008, **486**, p 461–469
23. G. Tepe, J. Schmehl, H. Wendel, S. Schaffner, S. Heller, M. Gianotti, C. Claussen, and S. Duda, Reduced Thrombogenicity of Nitinol Stents—In Vitro Evaluation of Different Surface Modifications and Coatings, *Biomaterials*, 2006, **27**, p 643–650
24. T. Peitsch, A. Klocke, B. Kahl-Nieke, O. Prymak, and M. Epple, The Release of Nickel from Orthodontic NiTi Wires is Increased by Dynamic Mechanical Loading but not Constrained by Surface Nitridation, *J. Biomed. Mater. Res.*, 2007, **82A**, p 731–739

JMEPEG (2009) 18:475–478
DOI: 10.1007/s11665-009-9396-8

1059-9495/$19.00

Influence of Quenching Rates on Equiatomic NiTi Ribbons Fabricated by Melt-Spinning

K. Mehrabi, M. Bruncko, B.J. McKay, and A.C. Kneissl

(Submitted September 18, 2008; in revised form December 16, 2008)

Quenching rates, in terms of circumferential wheel speed in melt-spinning, play an important role in the resultant characteristics of the ribbons. In this study, the influence of various wheel speeds on ribbon dimensions, cross-sectional microstructures, and crystallographic phases within samples were investigated. The Ni-50.3at.%Ti melt-spun ribbons were produced under 200 mbar He atmosphere at wheel speeds of 5-30 ms^{-1} using a quartz crucible coated internally with Y_2O_3. The different wheel speeds led to different sample dimensions and cooling rates. The microstructures of the samples were observed using optical microscopy and TEM, while the shape memory effect of samples and the existing phases were studied by differential scanning calorimetry, x-ray diffraction, and tensile tests.

Keywords melt-spinning, NiTi, shape memory alloys, wheel speed

1. Introduction

NiTi shape memory alloys (SMAs) have received considerable research attention and are widely used because they combine special functional properties with high mechanical strength (Ref 1). These characteristics are due to the martensitic transformation and its reversion, which can be activated by thermal or mechanical loads. SMAs show good corrosion resistance, wear resistance, and high specific electric resistance (which is beneficial for direct current heating) and can be exploited many times before functional or structural fatigue limit service life (Ref 2).

A very promising field for high-volume applications of SMAs in the near future lies in actuator technology, since with a shape memory element, a pre-determined response can be obtained very easily by thermal or electric stimulus. Especially the possibility to realize even complicated movements with an element of simple design and compact size makes shape memory actuators very attractive. There is also a trend to very small-dimensioned shape memory elements which could be used as microactuators. The application of rapid solidification can drastically change the microstructure, improving the ductility and shape memory characteristics, leading to small-dimensioned samples (Ref 3).

Generally there are several advantages of rapid solidification (quenching rate varies from 10^3 up to 10^7 K/s (Ref 4)) over the slower conventional solidification techniques. These are an ability to form metastable phases, increasing the solubility above the equilibrium solubility, decreasing the segregation of additions, and refining the microstructure. It is considered that all of these effects have been attempted to improve shape memory effects in the rapidly solidified ribbons.

Many physical properties of melt-spun ribbons, as well as their microstructures, sensitively depend on the values of the processing parameters such as the wheel speed, gas pressure, melt temperature, nozzle-wheel gap, etc. (Ref 5). As known from the literature (e.g. Ref 6), cooling rates are proportional to the wheel speed. In the present study, the microstructure and shape memory behavior of NiTi alloy ribbons, fabricated at different cooling rates by melt-spinning, were investigated.

2. Experimental Procedure

In this study, an investigation of the microstructures and shape memory effect of Ni-50.3at.%Ti melt-spun ribbons was carried out. First of all, the alloy was prepared in ingot form by vacuum arc-melting technique on a water-cooled copper crucible in a reduced Ar atmosphere. To ensure homogeneous SMAs, arc-melting was repeated three times for each alloy.

The melt-spun ribbons were produced under a 200 mbar He atmosphere using quartz-glass crucibles with a nozzle diameter of 0.9 mm and coated internally with Y_2O_3. By applying an Ar overpressure of 90 mbar within the crucible, the melt was ejected onto the surface of a polished Cu wheel (200 mm diameter) having a circumferential wheel speed of 5-30 m/s. The distance between the nozzle and the wheel surface was 2 mm.

Specimens for metallographic investigation were cut from the longitudinal cross sections of melt-spun ribbons. They were

This article is an invited paper selected from presentations at Shape Memory and Superelastic Technologies 2008, held September 21-25, 2008, in Stresa, Italy, and has been expanded from the original presentation.

K. Mehrabi, M. Bruncko, and **A.C. Kneissl**, Department of Physical Metallurgy and Materials Testing, University of Leoben, Leoben, Austria; **M. Bruncko**, Faculty of Mechanical Engineering, University of Maribor, Maribor, Slovenia; and **B.J. McKay**, Department of Metallurgy, University of Leoben, Leoben, Austria. Contact e-mail: kambiz.mehrabi@unileoben.ac.at.

ground on SiC paper to a final mesh size of 2400 and polished. The samples were etched with conventional and wipe-etching in a solution containing $HF:HNO_3:H_2O$ = 1:4:5 and studied by optical microscopy using interference contrast. For TEM experiments, thin discs (3 mm in diameter) were mechanically ground and dimpled to a thickness of 25-30 μm. Ion-milling was then performed on the samples until electron transparency was achieved. The energy of the Ar+ ions was 3.6 keV, the angle to the surface of the specimen was ±4°, and the specimen was kept at a rotation rate of 3 rpm during ion-milling. Conventional TEM studies were carried out with an acceleration voltage of 120 kV.

The phase transformation temperatures were measured using a Mettler DSC 821e device (Mettler Toledo GmbH, Schwerzenbach, CH). Thermograms were recorded under static air from −40 to 180 °C at a heating/cooling rate of 10 °C/min. A circular sample disc was cut with a punch (5 mg), put in a 20 μL aluminum pan, and closed with a perforated lid. The crystallographic structure was determined using Bragg-Brentano x-ray diffraction with CuKα radiation at room temperature. Tensile specimens with gauge size of 55 mm in length, 2 mm in width, and 80 μm in thickness were cut from the melt-spun ribbons and tested on a universal testing machine with a crosshead speed of 0.1 mm/min at room temperature.

3. Results and Discussion

Figure 1(a) shows an optical micrograph of a conventionally etched ribbon (wheel speed 15 m/s). The microstructure is primarily single phase, with small amounts of secondary phase distributed in the matrix. Since the stoichiometric range of NiTi is very narrow at low temperatures, it is quite usual that the material contains precipitates of a second phase. By EDX analysis these particles were found to be Ti_2Ni precipitates. In this picture because of the very fine and sophisticated microstructure, it is not easy to characterize the martensitic microstructure with conventional etching method. Figure 1(b) shows the martensitic structure of the same ribbon after wipe-etching using interference contrast. From Fig. 1(a) it seems that the grain size is about 5-10 μm but it is hard to relate this to the martensitic structure seen in Fig. 1(b), therefore we cannot be sure that this is the real grain size. The same observations have also been found by Undisz et al. (Ref 7).

Furthermore, TEM was employed to investigate the martensitic structures of the NiTi ribbons, which were too fine to be sufficiently resolved with optical microscopy, especially in the ribbons fabricated at the higher wheel speed. Figure 2 displays a fine martensitic structure (B19′ phase) of a melt-spun ribbon (wheel speed 30 m/s) with nano-sized twin boundaries.

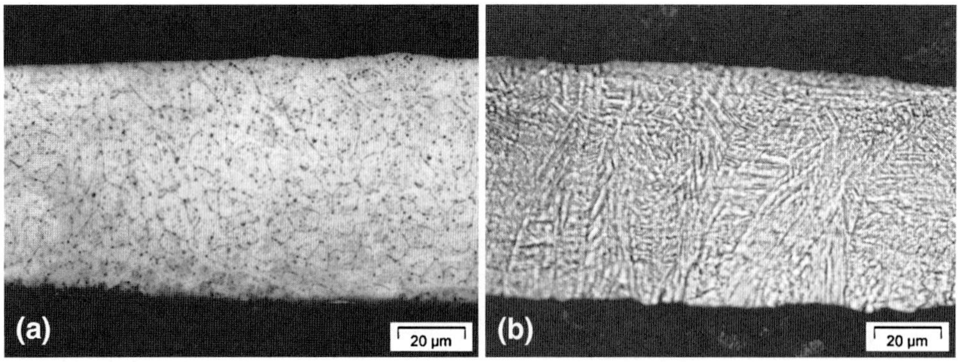

Fig. 1 Optical microstructures of melt-spun ribbon: (a) conventionally etched surface and (b) martensite structure after wipe-etching using interference contrast

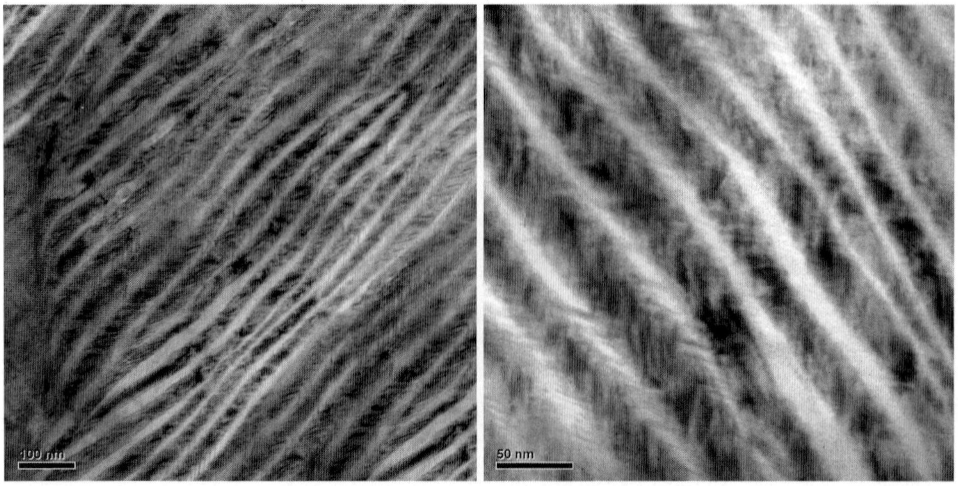

Fig. 2 TEM images of the B19′ martensite of a melt-spun shape memory alloy

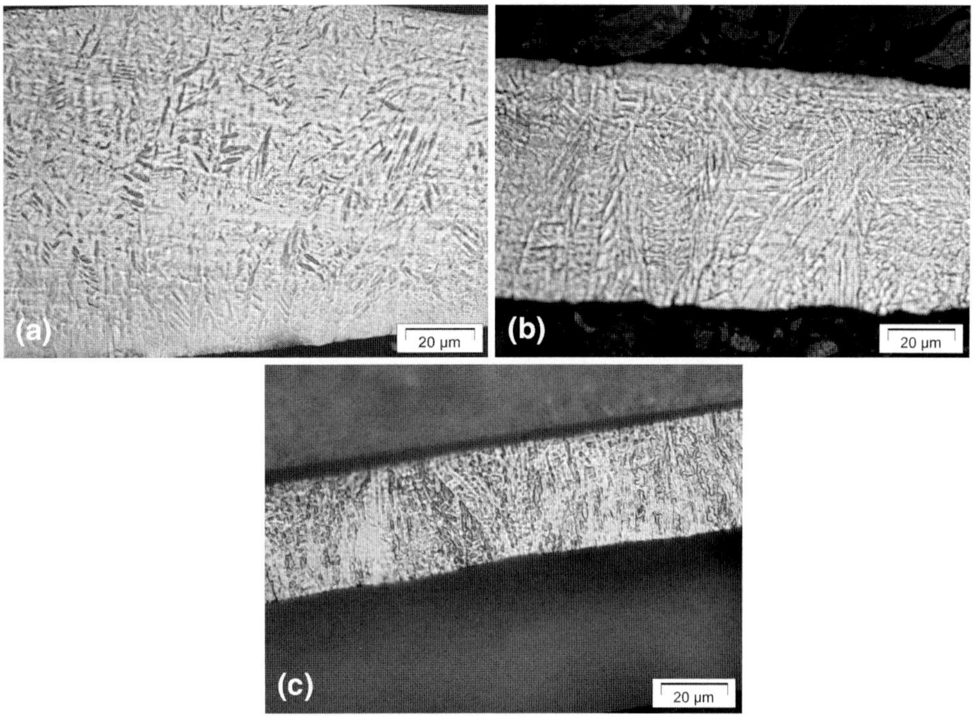

Fig. 3 Optical cross-sectional view of NiTi ribbons at various wheel speeds, wipe-etching using interference contrast: (a) 5 m/s, (b) 15 m/s, and (c) 30 m/s

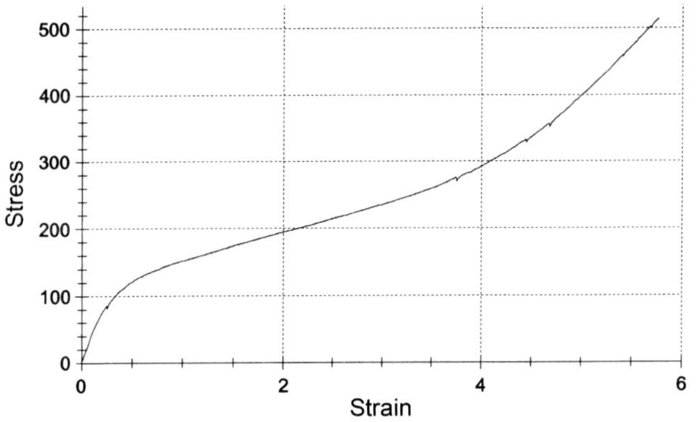

Fig. 4 Stress-strain curve of the melt-spun ribbon; martensite plateau

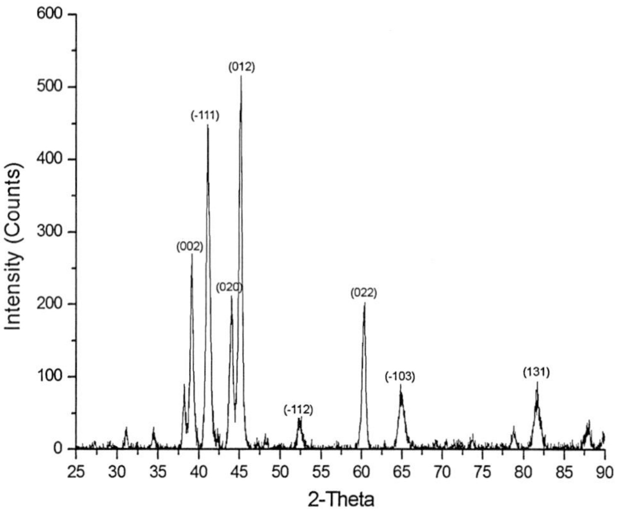

Fig. 5 XRD patterns of the ribbons, B19′ phase

The variation of cooling rates was achieved by the speed of the Cu wheel that was changed from 5 to 30 m/s while the melt-spinning temperature was fixed at 1350 °C. Figure 3 shows the cross-sectional view of ribbons at various wheel speeds. An increase of the wheel speed from 5 to 30 m/s results in a decrease of the ribbon thickness from 100 to 20 μm. As the increase of the wheel speed leads to a reduced ribbon thickness, the cooling rate increases.

Tensile tests of melt-spun ribbons fabricated at various wheel speeds show clearly the martensitic plateau even without subsequent heat treatment (e.g. Fig. 4). The shape memory one-way effect is about 4% in these samples.

The results of XRD experiments were obtained from the free side of the ribbons. The results also show that all of the ribbons manufactured at the different wheel speeds are crystalline and martensitic (e.g. Fig. 5). No transition from crystalline phase to amorphous phase with increasing cooling rate even at the wheel speed of 30 m/s could be found.

Figure 6 displays the DSC curves for the melt-spun ribbons at speeds of 5, 15, and 30 m/s. Figure 6(a) shows the phase transformation for the lowest wheel speed and therefore the lowest cooling rate while Fig. 6(b) and (c) represents the higher wheel speeds and cooling rates. There is not much difference in the transformation temperatures but a slight shift to higher martensite transformation temperatures which is not in accordance with other observations, e.g. Ref 8, 9. These papers deal

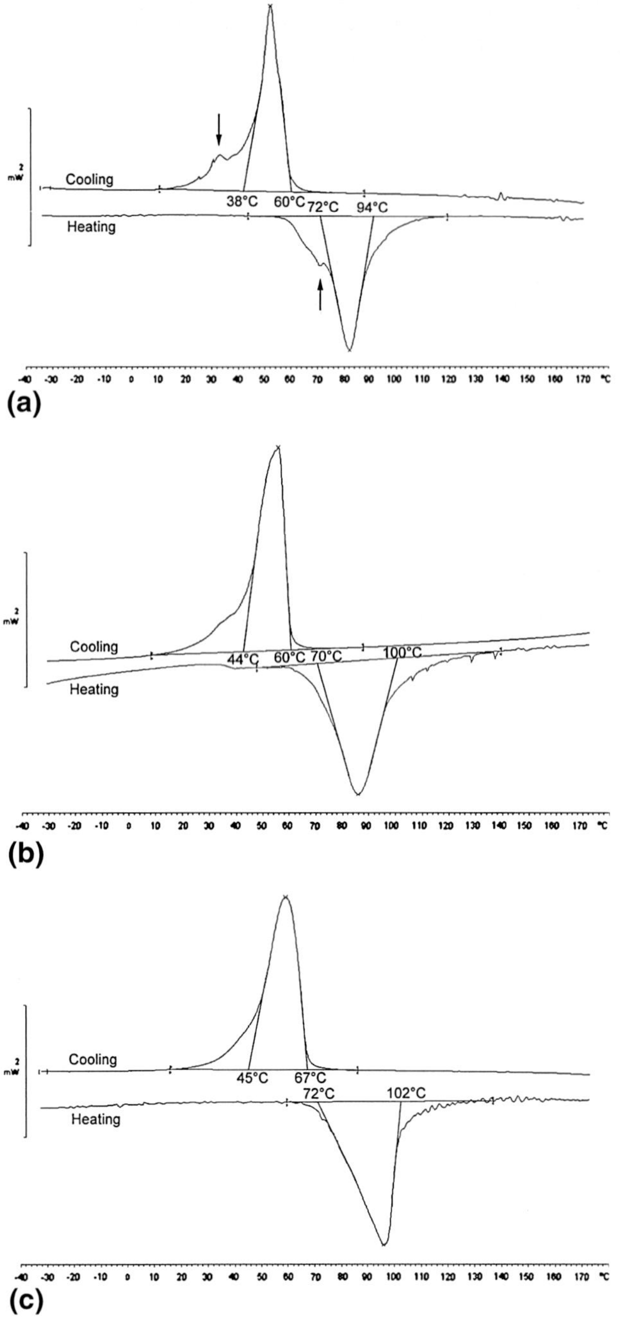

Fig. 6 DSC curves of NiTi ribbons at various wheel speeds: (a) 5 m/s, (b) 15 m/s, and (c) 30 m/s

with Cu-based SMAs and obtain lower martensite transformation temperatures with decreasing grain size. In this study, it was not possible to make quantitative comparisons of the grain size. However, the martensitic microstructure became finer with increasing wheel speed. The very slight increase of the martensite transformation temperatures could be caused probably by higher internal stresses because of higher cooling rates. It also seems that there are additional peaks at the lower cooling rates, especially in the final stage of the martensitic transformation (evident in Fig. 6a). Perhaps these additional peaks could be attributed to the fact that transformation temperatures of the chilled side are different from those of the free side of the ribbons.

4. Conclusions

The work of this paper has shown that the melt-spin technique is able to produce samples with good shape memory properties immediately after cooling even without subsequent heat treatment. The higher the wheel speed, the smaller the thickness of the resulting ribbon and therefore the higher the cooling rate. With increasing cooling rate the microstructure gets finer and finer and therefore metallographic preparation was not easy at all. Using wipe-etching and interference contrast, it was possible to reveal the fine martensitic structure; by TEM investigations also the twin boundaries of the martensitic structure could be shown. XRD and DSC investigations revealed that all samples were fully martensitic at room temperature. At the higher cooling rates, the martensite transformation temperatures were very slightly raised, probably due to higher internal stresses. The material with very fine microstructure has also good mechanical properties (strength and ductility) and therefore exhibits good potential for future microactuator applications.

References

1. J. Frenzel, Z. Zhang, K. Neuking, and G. Eggeler, High Quality Vacuum Induction Melting of Small Quantities of NiTi Shape Memory Alloys in Graphite Crucibles, *J. Alloys Compd.*, 2004, **385**, p 214–223
2. S. Miyazaki, T.W. Duerig, K.N. Melton, D. Stockel, and C.M. Wayman, eds., *Engineering Applications of Shape Memory Alloys*, Butterworth-Heinemann Applications, London, 1990, p 394
3. P. Ochin, V. Kolomytsev, A. Pasko, P. Vermaut, F. Prima, and R. Portier, Phase Transformations in Rapidly Solidified (Ti-Zr)50(Ni-Cu-Sn)50 Alloys, *Mater. Sci. Eng. A*, 2006, **438–440**, p 630–633
4. S.H. Whang, Ductile, Single Phase-Continuous Super-Conducting Oxide Conductors, U.S. Patent 4,968,663
5. Y.-w. Kim, Y.-m. Yun, and T.-h. Nam, The Effect of the Melt Spinning Processing Parameters on the Solidification Structures in Ti–30 at.% Ni–20 at.% Cu Shape Memory Alloys, *Mater. Sci. Eng. A*, 2006, **438–440**, p 545–548
6. M.J. Kramer, H. Mecco, K.W. Dennis, E. Vargonova, R.W. McCallum, and R.E. Napolitano, Rapid Solidification and Metallic Glass Formation—Experimental and Theoretical Limits, *J. Non-Cryst. Solids*, 2007, **353**, p 3633–3639
7. A. Undisz, S. Flauder, and M. Rettenmayr, Metallographische Präparation und Gefüge von Ni-Ti Legierungen mit Formgedächtnis/Pseudoelastizität, *Prakt. Met. Sonderband*, 2008, **40**, p 331–335 (in German)
8. J. Dutkiewicz, T. Czeppe, and J. Morgiel, Effect of Titanium on Structure and Martensic Transformation in Rapidly Solidified Cu–Al–Ni–Mn–Ti Alloys, *Mater. Sci. Eng. A*, 1999, **273–275**, p 703–707
9. G.A. Lara-Rodriguez, G. Gonzalez, H. Flores-Zúñiga, and J. Cortés-Pérez, The Effect of Rapid Solidification and Grain Size on the Transformation Temperatures of Cu–Al–Be Melt Spun Alloys, *Mater. Charact.*, 2006, **57**, p 154–159

JMEPEG (2009) 18:479–483
DOI: 10.1007/s11665-009-9438-2

1059-9495/$19.00

A Comparison of Chemistry and Inclusion Distribution and Morphology Versus Melting Method of NiTi Alloys

George M. Kramer

(*Submitted September 16, 2008; in revised form March 20, 2009*)

NiTi alloys are produced by three melting methods. The first method requires compaction of nickel and titanium raw material into sections that can be joined together for melting in a Vacuum Arc Remelt unit (VAR). This ingot is melted two or more times in a VAR. The second method utilizes a Vacuum Induction Melting (VIM) unit to alloy the nickel and titanium, with the use of a graphite crucible. The resulting ingot is prepared and remelted in a VAR. The third method begins with primary melting in a vacuum Induction Skull Melter (ISM). The ISM produces ingots that are assembled into an electrode for VAR melting. For each of the melting methods, the final product depends on the quality and handling of the raw materials, the control of the process at each unit, and the preparation of the intermediate ingots for further processing. The melting method influences the final chemistry as well as the type and number of inclusions present in the final product. This study compares the chemistry and microcleanliness of product manufactured by each method to determine the appropriate melting technique that produces NiTi with the lowest residual elements, such as carbon, as well as the lowest size, and number of inclusions.

Keywords advanced characterization, biomaterials, intermetallics, primary metals, titanium

1. Introduction

Nitinol possesses unique mechanical properties that have sparked an explosion of product innovation (Ref 1, 2). The superelasticity and shape-memory properties impart characteristics that are engineered for use in human bodies as well as auto bodies.

The superelastic and shape-memory effects result from an atomic arrangement of approximately 50 atm.% Ni and 50 atm.% Ti. The alloy undergoes a martensitic/austenitic shift at a temperature determined by the alloy chemistry (Ref 3). Either the Differential Scanning Calorimetry (DSC) method or the active A_f method is used to measure the transformation temperature. The transformation temperature is sensitive to small chemical changes and is not accurately predicted by a chemistry analysis. Chemical analysis is not precise enough for transformation temperature prediction, but chemical homogeneity along the final ingot is important to the end-user of the ingot for consistent properties (Ref 4).

Inclusion size, composition, and morphology are the subjects of investigation into their effect on the physical properties of nitinol. Multiple studies have attempted to determine the effects of inclusion composition and inclusion size versus mechanical properties (Ref 5, 6, 7).

This investigation compares three melt methods (VIM/VAR, multiple VAR, and ISM/VAR) and characterizes the chemical and inclusion results.

2. Melt Method

Vacuum Induction Melting (VIM) followed by Vacuum Arc Remelting (VAR), or multiple Vacuum Arc Remelting (MVAR) are the common methods of industrial melting of NiTi (Ref 2).

When the primary melting is performed in the VIM, the use of graphite crucible results in carbon pickup during the melt. The charge material is melted quickly at a controlled temperature to minimize carbon pickup and is thoroughly mixed before pouring. The VIM ingot displays a chemical homogeneity that results in minimal transformation temperature variation from top to bottom (as good as ±3 °C). The VIM ingot is subsequently processed through the VAR (Ref 1).

When all the melt cycles are performed in the VAR, raw materials are compacted and assembled into a primary melt electrode. The VAR remelts the ingot multiple times to homogenize the chemistry. This method maintains low carbon levels by melting in a copper as against versus graphite crucible. The carbon and oxygen levels in the ingot are then only dependent on raw material control and vacuum management. The ingot varies in chemistry from top to bottom since a relatively small melt pool is mixing with the material at any time. Consequently, the transformation temperatures vary up to ±10 °C along the ingot as well (Ref 2).

3. Alternate Melting

Multiple alternate methods of melting nitinol have been attempted (Ref 2, 8, 9). As in the VAR, a method known as

This article is an invited paper selected from presentations at Shape Memory and Superelastic Technologies 2008, held September 21-25, 2008, in Stresa, Italy, and has been expanded from the original presentation.

George M. Kramer, Metalwerks PMD, Inc., 401 Steel Street, Aliquippa, PA 15001. Contact e-mail: george.kramer@metalwerks.com.

Induction Skull Melting (ISM) maintains low carbon and oxygen levels through proper raw material and vacuum management. The ISM is similar to the VIM in that the induction field mixes the metal well prior to pouring. The ISM ingots are processed singly or tested and matched with other ingots for VAR processing.

The ISM process utilizes a water-cooled copper crucible that is loaded with charge material. An induction coil surrounds the copper crucible. The copper crucible is constructed of copper "fingers" that are separated by thin layers of di-electric. This allows the field produced by the induction coil to couple with the charge materials. The coupling of the charge material with the induction field is inefficient when compared to other melt methods, and therefore, ISM units must use high power (up to 1150 kW) for melting (Ref 10).

A melt in the ISM progresses as follows. The melter cleans the crucible and charges the raw material into the crucible. After evacuation of the chamber and the application of power, the charge melts rapidly, undergoing vigorous stirring and mixing. The liquid metal solidifies, and when in contact with the copper crucible effectively results in the melting of the alloy within a solid skin of the same composition, i.e., without any contaminant pick-up. When the first melt is poured from the ISM, a layer of solidified metal coats the surface of the crucible, and subsequent melting of the same alloy is conducted without removing the skull. Some authors have suggested that the ISM method be studied for high-purity nitinol production (Ref 1).

4. Experimental Setup

The Metalwerks melting facility has produced common NiTi compositions for some time. The experimental procedure followed proprietary standard operating procedures. Procedures that did not previously exist for nitinol on VIM and VAR furnaces were developed and standardized for the experiment. Every effort is adopted to maintain similar raw material and melting conditions for each process path.

High-purity electro-refined titanium crystal from the same lot and electrolytic nickel from the same lot were obtained for use in the melting of ingots during all the three processes (the nickel was minimum 99.97% pure and the electro-refined titanium was minimum 99.99% pure). Three melts of 45.36 kg each were formulated for the VIM and for the ISM. A 136.08-kg ingot was formulated for the first VAR melt.

The VIM utilized a new graphite crucible to produce three ingots from three individual melts. The melts were sequential and poured into molds sized for assembly to melt in the VAR.

The first ISM ingot poured was not equal in weight to the charge makeup, since the melt formed the skull for the subsequent two melts. The ISM melts were sequential and poured into the same size molds as were the VIM ingots for VAR processing.

The preparation of VAR compacts was considerably more complicated. The titanium crystal and the nickel were weighed on the same scale as the material for the other units. Packets of mixed Ti and Ni were weighed for the target composition. Each packet weighed 2.27 kg. The material was placed in a titanium-lined tube and compressed into cylinders of the same diameter as the primary VIM and ISM ingots. The compressed piece was removed from the tube, and the compacts were prepared for assembly.

Table 1 Ingot chemistries

Ingot	C%	O%	Ni%	Ti%
ISM/VAR—Top—V1608683	0.013	0.024	55.50	44.42
ISM/VAR—Bottom—V1608683	0.013	0.026	55.07	44.85
VIM/VAR—Top—V1608684	0.028	0.021	55.75	44.15
VIM/VAR—Bottom—V1608684	0.038	0.025	55.29	45.59
Twice VAR—Top—V1608690	0.015	0.031	45.73	54.15
Triple VAR—Center—V1608693	0.013	0.029	46.96	52.98

The VAR electrode of welded NiTi compacts did not possess sufficient strength to be loaded into the VAR as a single piece. The electrode was cut into two pieces and melted. The resulting two ingots were joined and remelted. This ingot was cut into sections and reassembled to further mix the ingot chemistry by melting a third time.

5. Experimental Results

The ingots from the VIM and the ISM were analyzed (as detailed in Table 1), and the chemistries conformed to ASTM F-2063-05. The ingot chemistries were consistent with formulation aims. Conversely, the ingot produced by triple-VAR of compacted raw materials did not meet ASTM F-2063-05 for Ni and Ti. This is likely due to inhomogeneous raw material arrangement of the Ti and Ni in the primary electrode. Second, inadequate mixing of the raw materials in the smaller melt-pool of the VAR caused segregation of the Ti and Ni (which is visually apparent in Fig. 7).

The ISM/VAR and VIM/VAR ingots were heat-treated, then forged on an in-house GFM. The ingots were forged to a bar size of 34.29 mm (1.35″) and cut to obtain samples for microcleanliness analysis. Because of multiple remelting, the size of the triple-VAR ingot was not compatible with the GFM, and it was sent to an outside conversion source. The ingot broke apart during forging. Sections from this ingot were cut into sections to attempt lab-scale processing for limited data analysis. The ingot sections were not formable, therefore processing was stopped. The ingot sections were sent for microstructural analysis to determine the phases present (Fig. 7).

The microcleanliness analyses for all the samples were performed on 50 fields of view at 500×, with methodology adhering to ASTM F-2063-05. Backscattered Electron Imaging (BEI) was used to perform the inclusion analysis. This method can distinguish between lower and higher atomic weight elements that contribute to the inclusions' composition. Lower atomic weight compositions appear lighter. However, through this technique, the voids adjacent to the inclusions were highlighted as part of the microcleanliness analysis. The inclusion plus void total were compared with the requirements of ASTM-F-2063-05. This was conducted in this manner to eliminate completely the defects found in the material.

As mentioned previously, the triple VAR ingot was not usable for microcleanliness analysis. Table 2 summarizes the results for the inclusion analysis for the other two ingots.

Table 3 further details the distribution of inclusion sizes for each ingot.

Table 4 details the dimensions of the largest inclusions.

Table 2 Inclusion maximums by ingot

Ingot	Diameter	Maximum dimension of porosity + non-metallic inclusions	Maximum %area of inclusions and porosity in one field of view	Total number of inclusions + voids	Total number of voids
(ISM/VAR) V1608683	34.29 mm	40.0 μm	1.4%	4,817	135
(VIM/VAR) V1608684	34.29 mm	28.3 μm	1.0%	12,236	54
(triple VAR) V1608693	~20 mm rectangular	NA	NA	NA	NA

Table 3 Inclusion distribution by ingot

	(ISM/VAR) V1608683		(VIM/VAR) V1608684	
Size range, μm	Void	Void + inclusion	Void	Void + inclusion
1.0-3.9	124	4,466	50	11,152
4.0-7.9	10	7	4	967
8.0-11.9	0	157	0	96
12.0-19.9	1	47	0	20
20.0-29.9	0	2	0	1
30.0-39.0	0	2	0	0
>39.0	0	1	0	0
Total	135	4,682	54	12,236

Table 4 Further dimensional breakdown of inclusions

Ingot	Length, largest dark feature, μm	Breadth, longest dark feature, μm	Area, longest dark feature, μm^2
ISM/VAR—V1608683	40.0	5.1	107.0
VIM/VAR—V1608684	28.3	3.1	32.8

6. VIM/VAR Analysis

The microcleanliness of the VIM/VAR ingot revealed primarily carbon-containing inclusions. The inclusions were generally globular and small. The inclusions were mostly carbide inclusions with some mixed carbide/intermetallic inclusions. The percent of voids versus the total number of inclusions was 0.44%.

The image of the longest inclusion in the VIM/VAR ingot is seen in Fig. 1.

Another typical carbide/void inclusion is seen in Fig. 2 at a higher magnification.

The EDS analysis of a typical inclusion in this forged ingot section is shown in Fig. 3.

7. ISM/VAR Analysis

The microcleanliness evaluation of the ISM/VAR ingot revealed primarily NiTi inclusions. There were two classes of these inclusions: smaller inclusions that were globular in nature, and larger inclusions that were angular. The ISM/VAR sample contained one inclusion/void 1 μ longer than the permissible limit by the ASTM standard. The percent of voids versus the total number of inclusions was 2.88%, compared to the 0.44% in the VIM/VAR ingot. Figure 4 shows the longest inclusion observed in the ISM/VAR ingot.

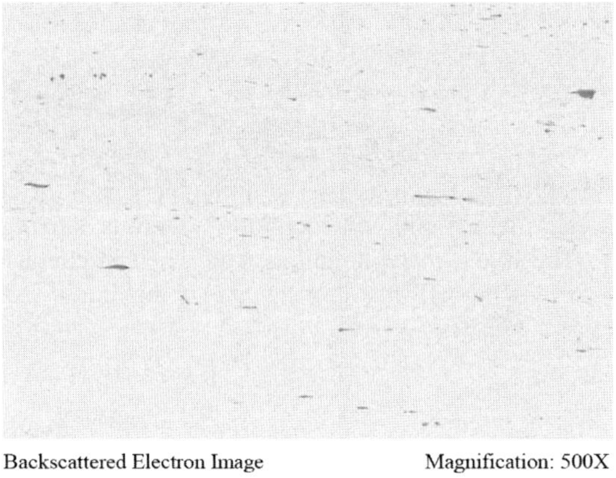

Fig. 1 Longest inclusion in Vim/Var ingot

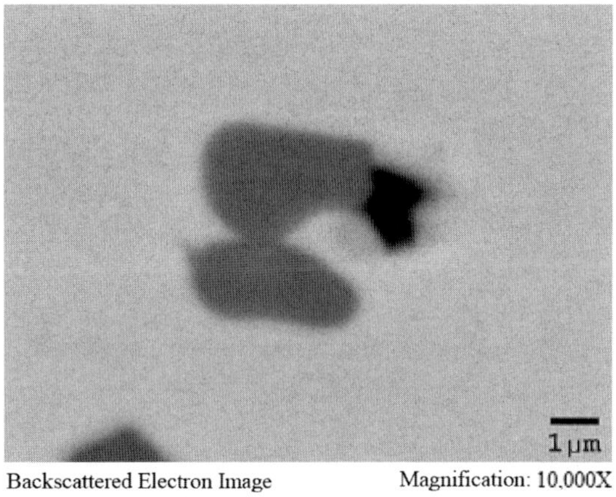

Fig. 2 High magnification VIM/VAR inclusion

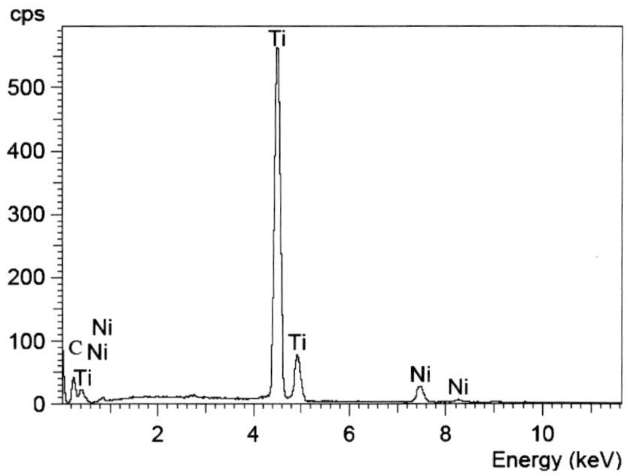

Fig. 3 EDS analysis VIM/VAR ingot

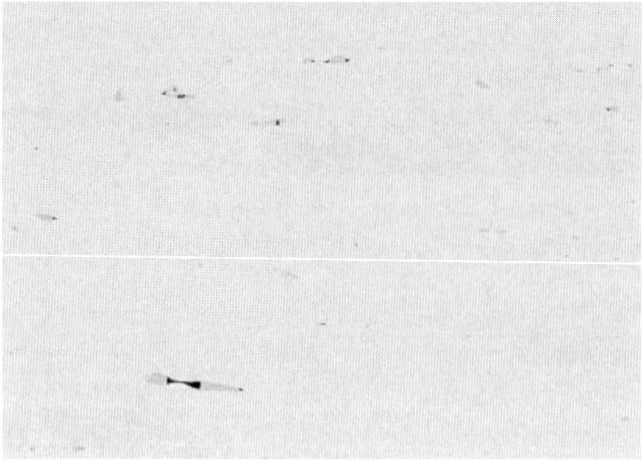

Fig. 4 ISM/VAR inclusions—500×

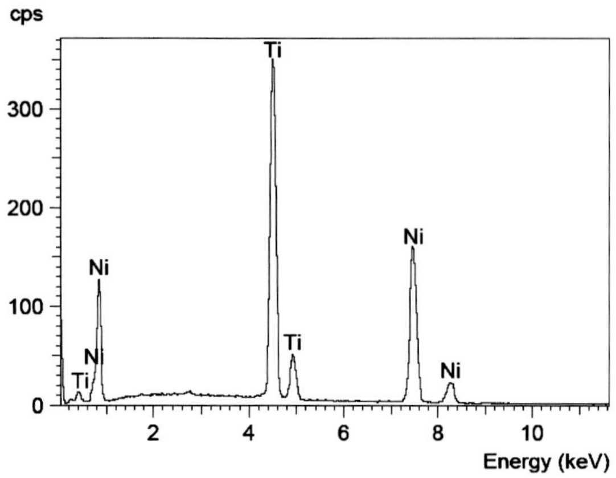

Fig. 5 EDS analysis of ISM/VAR inclusion

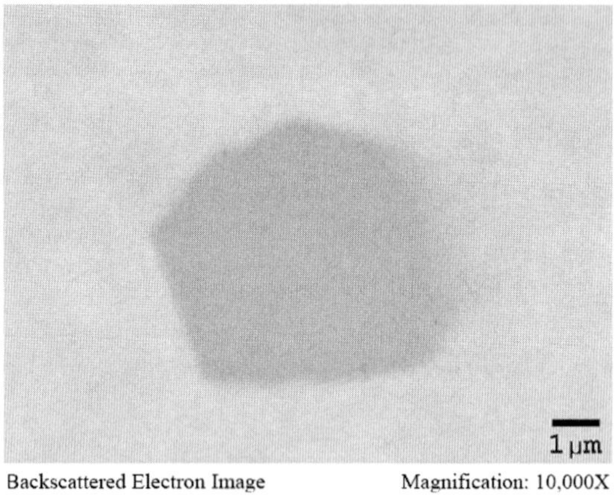

Fig. 6 A globular inclusion in ISM/VAR ingot

Figure 5 shows the EDS analysis for the typical inclusion in this ingot. The analysis reveals no carbon signature.

A second image of typical inclusion morphology is shown, minus an adjacent void (Fig. 6).

8. Triple VAR Ingot

The microstructural/microcleanliness analysis for the triple-VAR ingot could not be completed, because of a dual-phase microstructure that possessed a matrix with a dark appearance. This dark phase is likely composed of NiTi intermetallic compounds with islands of the lighter NiTi alloy. The dark phase propagated cracks through the ingot structure (Fig. 7).

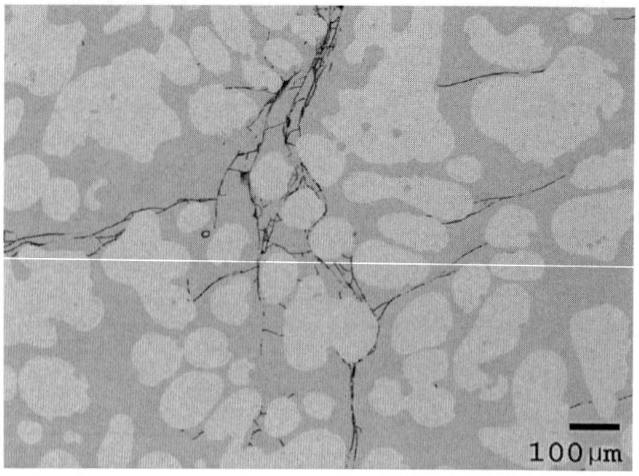

Fig. 7 Triple VAR ingot structure

9. Summary/Conclusion

The production of Nitinol ingots adopting three melt paths, VIM/VAR, VAR/VAR/VAR, and ISM/VAR, was undertaken to compare the effect of each method on the chemistry and inclusion content of the final material. Single lots of Ti and Ni were divided equally for use in each process. The nickel purity was minimum 99.97% and that of electro-refined titanium minimum 99.99%.

The ingot produced by compaction and multiple VAR melting was not usable. The ingot was melted three times from NiTi compacts, but the composition and physical properties were poor. The resulting electrode was extremely brittle and exhibited a segregated microstructure (Fig. 7). The microstructure manifests evidence of inadequate molten mixing and homogenization of the ingot, possibly because of the compaction method. Even though this process path is widely used in the production of nitinol, the processing of this ingot in this experiment did not result in a usable ingot. Further development of materials, electrode compaction methods, electrode assembly, and melting parameters is required for successful melting of NiTi by multiple-VAR in this facility.

The composition of the ingots produced by the ISM/VAR versus the VIM/VAR process demonstrates that the VIM ingots possess higher carbon than ingots produced by ISM, keeping the raw materials the same. The compositions varied from top to bottom on both the ISM/VAR and VIM/VAR ingot. This resulted from using input ingots regardless of composition for the purpose of this experiment. The ingots were welded together end-to-end to form the VAR melt electrode. The variation of the chemistry from each set of three ingots resulted from the first ingot melted by each unit. Further study is needed to determine improved methods to produce a "first ingot" from the ISM and VIM that is close to the target chemistry.

A comparison of the inclusion characteristics for the VIM/VAR versus the ISM/VAR product shows that the ISM/VAR process produces nitinol with low carbon content and inclusions with undetectable C content in this experiment (Fig. 5).

The VIM/VAR product contained higher carbon and the inclusions were composed mostly of Ti and C. While the VIM/VAR ingot possessed 2.6 times the total number of inclusions and relative to the ISM/VAR ingot (Table 3), the ISM/VAR ingot displayed larger inclusions composed of hard NiTi intermetallic compositions. The larger intermetallic inclusions were angular, and these inclusions resulted in a larger amount of voids in the adjacent matrix. The total reduction of the ingots from VAR through the GFM was 77.5%, which may not have been sufficient to break up the NiTi inclusions and heal voids.

The fraction of voids was 2.88% in the ISM/VAR ingot versus 0.44% in the VIM/VAR ingot. The voids were adjacent to the intermetallic inclusions, primarily the larger inclusions. The largest defect was a combination of an inclusion and a void. As pointed out above, the number of voids is proportional to the number and distribution of "hard" intermetallic NiTi inclusions.

It is hypothesized that the number, distribution, and size of the inclusions in the ISM/VAR were negatively affected by thermal conditions prior to forging. Future investigations will focus on the relationship between ingot melting and forging parameters and the resulting inclusion morphology and composition.

Acknowledgments

Recognition must be given to Mr. Michael Popper, who committed the resources of Metalwerks PMD, Inc. to this project. Metalwerks offers a special thanks to the entire workforce for their dedication in making this project come together, from ordering and receiving the raw material through sampling of the final bar. Finally, Metalwerks recognizes the employees at Materials Evaluation and Engineering, Inc. for their response, effort, and evaluation of the materials analyzed.

References

1. L.McD. Schetky and M.H. Wu, Issues in the Further Development of Nitinol Properties and Processing for Medical Device Applications. *Memry Corporation*, 2003 [Online], [Cited: January 24, 2008], Memry Corporation, http://www.memry.com
2. S. Russell, Nitinol Melting and Fabrication, Pacific Grove, California: International Organization on SMST, 2000, 2001, *SMST-2000: Proceedings of the International Conference on Shape Memory and Superelastic Technologies*, Article sourced from the website http://www.nitinol.info/pdf_files/russell.pdf
3. C.M. Jackson, H.J. Wagner, and R.J. Wasilewski, *Nitinol—The Alloy With a Memory: Its Physical Metallurgy, Properties, and Applications*, Technology Utilization Office, NASA, Washington, DC, 1972
4. A.R. Pelton, J. DiCello, and S. Miyazaki, Optimisation of Processing and Properties of Medical Grade Nitinol Wire, A. Pelton and S. Russell, Eds., *Proceedings of the International Conference on Shape Memory and Superelastic Technologies*, 2000, p 361–374, http://www.nitinol.info/pdf_files/027.pdf
5. M. Reinoehl, D. Bradley, R. Bouthot, and J. Proft, *The Influence of Melt Practice on Final Fatigue Properties of Superelastic NiTi Wires*, International Organization on SMST, Pacific Grove, California, 2001
6. M. Patel, D. Plumley, and R. Bouthot, The effects of Varying Active Af Temperatures on the Fatigue Properties of Nitinol Wire, *Presentation Presented at ASM Material and Processes for Medical Devices Conference and Exposition*, Boston, MA, November 2005, http://www.fortwaynemetals.com/resources_whitepapers/MMP_ASM_MPMD.pdf
7. M. Patel and R. Gordon, An Investigation of Diverse Surface Finishes on Fatigue Properties of Superelastic Nitinol Wire, *Presented at the International Conference on Shape Memory and Superelastic Technologies (SMST)*, Pacific Grove CA, May 2006, http://www.fwmetals.com/resources_whitepapers/surface_finish_comparison.pdf
8. J. Otubo, O. Rigo, C. Neto, M. Kaufman, and P. Mei, Scale Up of NiTi Shape Memory Alloy Production by EBM, *J. Phys. IV France*, 2003, **112**, p 873–876, EDP Sciences, Les Uliis
9. M. Wu, Fabrication of Nitinol Materials and Components, *Proceedings of the International Conference on Shape Memory and Superelastic Technologies*, Kunming, China, 2001
10. A. Muhlbauer, *Innovative Induction Melting Technologies: A Historical Review*, International Scientific Colloquium, Modelling for Material Processing, Riga, 2006

JMEPEG (2009) 18:484–489
DOI: 10.1007/s11665-009-9398-6

Processing of Shape Memory CuZnAl Open-cell Foam by Molten Metal Infiltration

E.M. Castrodeza, C. Mapelli, M. Vedani, S. Arnaboldi, P. Bassani, and A Tuissi

(Submitted October 15, 2008; in revised form January 26, 2009)

Foams and other highly porous materials with a cellular structure are known to have many interesting combinations of physical and mechanical properties. In addition, foaming of shape memory alloys (SMA) greatly improves the set of application possibilities. In this work an open-cell metal foam of a CuZnAl SMA is presented. This foam was produced through a recently proposed process, which consists mainly in molten metal infiltration of a bed of silica-gel particles and the subsequent use of hydrofluoric acid (HF) as solvent. The results showed that very regular open-cell foams could be obtained, having an almost spherical cell morphology and relative densities of approximately 0.3. Microstructural and compositional analyses on foamed specimens showed uniform microstructure of ligaments and the absence of SiO_2 interaction with the metal. In this way the possibility of foaming CuZnAl system through the proposed low-cost process was clearly demonstrated.

Keywords cellular materials, CuZnAl, metal foam, shape memory alloys

1. Introduction

1.1 Metal Foams

The unique combination of morphology and material in cellular metals offers an interesting set of physical-chemical and mechanical properties that makes these materials very attractive both for structural and functional applications. Cellular metals can be produced by several processes that include molten metal infiltration in a particular leachable bed of solid particles (Ref 1-4).

This work introduces the use of an amorphous SiO_2 (silica-gel) bed as space holder for open-cell CuZnAl shape memory alloy (SMA) foam production through molten metal infiltration. Once infiltrated, SiO_2 particles are dissolved by a wet solution of hydrofluoric acid (HF). By this route, highly homogeneous foams of CuZn alloys, having almost spherical open-cell morphologies, can be produced (Ref 5, 6). The proposed methodology was based on cheap commercial consumables and a simple technology, focusing on intermediate-density low-cost foams with interesting cost/benefit ratios.

Metal foams obtained by liquid infiltration of leachable space holder are usually Al-based. In this case, NaCl particles leached by pure water are used (Ref 1, 2, 4, 6-8). More recently, open-cell amorphous metal foams were processed using refractory fluoride salts (Ref 9). No matter the mineral salt used for metal foaming, all the reported materials have melting point lower than SiO_2, being this a limitation for the application of the route to high melting point metals and alloys. Besides, due to the irregular morphology of mineral salt particles, the final products have mainly irregular cell morphologies, even if good cell distribution and size are obtained (Ref 4, 7, 9).

The use of SiO_2 beads as space holder results in cellular metals with a very homogeneous arrangement of almost spherical cavities. Additionally, because of the high chemical stability of SiO_2, reactions between space holder and molten metal tend to be minimal or inexistent.

1.2 CuZnAl Shape Memory Alloys

Among the large number of SMAs which have been discovered, most are copper-based. The most known NiTi SMAs exhibit better properties in strength, ductility and resistance to corrosion, in addition to excellent SME characteristics, but they are very expensive compared with Cu-based SMAs. Hence, two groups of cheap Cu-based alloys are currently being developed: Cu-Al based and Cu-Zn-X ternary alloys. Among ternary alloys, the Cu-Zn-Al system has often been the subject of research (Ref 10, 11).

The studies on this system have been mainly focused on shape memory and superelastic effect, aging phenomena and fatigue properties (Ref 12-18). Corrosion resistance was seldom examined (Ref 19). Due to the relative low melting temperature, these SMAs are easy to manufacture and have excellent electric performance, heat conductivity and a strong memory effect. Cu-based SMAs are in general also prone to

This article is an invited paper selected from presentations at Shape Memory and Superelastic Technologies 2008, held September 21-25, 2008, in Stresa, Italy, and has been expanded from the original presentation.

E.M. Castrodeza, C. Mapelli, and **M. Vedani**, Dipartimento di Meccanica, Politecnico di Milano, Via G. La Masa 34, 20156 Milan, Italy; **E.M. Castrodeza**, Departamento de Ingeniería de Materiales, Universidad de Concepción, Edmundo Larenas 270, 4070409 Concepcion, Chile; and **S. Arnaboldi, P. Bassani**, and **A Tuissi**, CNR-I-ENI, Unità di Lecco, Corso Promessi Sposi 29, 23900 Lecco, Italy. Contact e-mails: sergio.arnaboldi@ieni.cnr.it, and tuissi@ieni.cnr.it.

aging effects and show excessive grain growth when heat-treated (Ref 20, 21). These phenomena cause the well-known and unwanted problems, such as martensite stabilization or intergranular cracking, that hindered a wider diffusion of these alloys in the past decades. However, all the studies carried out in recent years seem to confirm that innovative applications by the use of these materials are feasible and well promising.

2. Experimental

2.1 Shape Memory Alloy Foaming

In this work the tested material was a typical CuZnAl SMA. The chemical composition and some physical properties of the alloy are reported in Table 1. The SMA was cast in an Aseg Galloni VCMIII induction melt casting machine, under pure Ar flow at room pressure. The maximum electrical power of this machine is 6 kW and it is equipped with a cylindrical graphite crucible of approximately 250 ml. A small cylinder of SMA having almost 30% of the intended foam volume was placed at the bottom of a vertical graphite crucible (80 mm in depth, 60-mm internal diameter). Then a bed of silica-gel beads was placed on the top of the SMA cylinder, also inside the crucible, and gently shaken for improving piling. Finally, a short cylindrical piece of graphite was positioned on top as a cap, closing the crucible for better temperature distribution over the silica beads. This initial configuration can be seen on Fig. 1(a). At this point the inductor was turned on and after some minutes, the metal was molten. During the melting, the metal temperature was continually measured by the thermocouple positioned in a blind hole in the solid SMA cylinder. After complete melting, when approximately 100 °C overheating was reached, the crucible cap and thermocouple were retired, the induction main power was turned off and silica-gel bed was pressed down by a steel cylinder having 85 N in weight (Fig. 1b and c). The infiltrated metal was at this point able to solidify between the silica particles (Fig. 1d). After complete solidification and cooling, the remaining solid (silica-gel beads plus infiltrated SMA) was unmolded and lathe machined to its final shape. Finally, the piece was submerged in an aqueous HF bath (25 vol.%) till complete SiO_2 dissolution was achieved.

The material was heat treated in a resistance furnace and quenched in water at room temperature (betatization treatment). In order to investigate the best condition for betatization, heat treatments with different temperatures and times were carried out.

The relative density of the final foam, the ratio of foamed to bulk material density, was theoretically calculated based on the bulk SMA density (Table 1), on weight measurements of the samples and on careful measurements of physical dimensions of the final (post-machined) cellular solid.

During the different steps of the whole process, material samples were collected and characterized. Calorimetric analyses were performed in a Seiko SCC/5200 differential scanning calorimeter at a rate of 10 °C/min in a range −100/80 °C which includes the direct and reverse transformations. The specimens were metallographically polished and etched with an alcoholic solution of $FeCl_3$. The material microstructure and its composition were investigated using a conventional optical microscope and a scanning electron microscope, equipped with energy dispersive spectroscopy (EDS) microanalyses system.

2.2 Amorphous SiO_2 (Silica Gel) Beads

Commercial non-indicating Sigma S7500 silica-gel Type II beads were used. The chemical composition of the particles and some of their physical properties are summarized in Table 2. Synthetic amorphous silica is virtually inert and has no known

Table 1 Chemical composition and physical properties of the investigated alloy

Chemical composition, wt.%			Physical properties	
Cu	Zn	Al	Melting temperature, °C	Density, g/cm^3
72.5	21.5	6	920	8.16

Table 2 Typical properties of the used silica-gel beads

Composition	Average diameter, mm	SiO_2 density, g/cm^3	Bead density (bulk), g/cm^3	Melting point, °C	Boiling point, °C
SiO_2 > 99.7%	3.175	2.2	0.7	~1600	2230

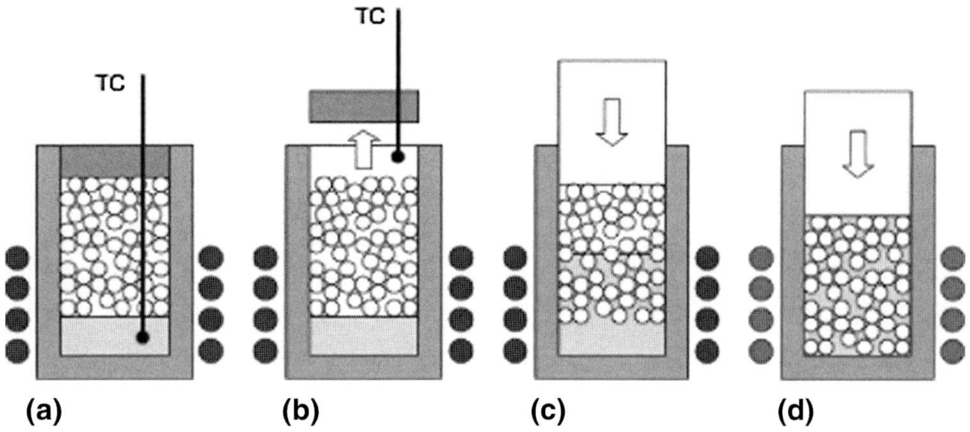

Fig. 1 Schematic of the infiltration process

adverse effects on the environment. This material is normally used as desiccant and it is commercially available worldwide as a cheap commodity.

The silica-gel beads were treated before infiltration to remove humidity content at 200 °C for 1 h and then heat treated in air at 900 °C for approximately 45 min.

2.3 Wet Chemical Etching of Silica-gel

Silica-gel was dissolved by wet chemical etching in aqueous solution of HF (25 vol.%). In literature it is possible to find several proposed reactions for HF etching of SiO_2. The stoichiometric reaction is defined as (Ref 22):

$$4HF + SiO_2 \leftrightarrow SiF_4 + 2H_2O.$$

Having aqueous HF in excess (as for our case) the main chemical reactions seems to be (Ref 23):

$$SiO_2 + 6HF \rightarrow H_2SiF_6 + 2H_2O,$$

and

$$SiO_2 + 3HF_2^- + H^+ \rightarrow SiF_6^{2-} + 2H_2O.$$

It must be specified that the highly porous morphology of silica-gel particles and the stirring action of the HF solution make the etching of beads rather easy.

3. Results and Discussion

3.1 Process

A little batch of CuZnAl foams were produced in order to evaluate feasibility and reproducibility of the process. EDS microanalyses show an excellent mean homogeneity of the chemical composition. A resulting cylindrical piece of metal and silica-gel beads after solidification and external lathe machining is shown in Fig. 2. The irregular face on the right of Fig. 2 corresponds to the upper side of the crucible. Figure 3 shows a SEM image of a heat treated and polished specimen surface. In both images the infiltrated silica-gel beads can be clearly seen. Figure 4 shows the final sample of the CuZnAl foam after silica-gel dissolution. Figure 5 shows a SEM image of the polished specimen surface. It is possible to see the uniform pore distribution of almost spherical cells, as well as many interconnecting windows.

As can be seen from Fig. 4 and 5, silica-gel beads were completely dissolved by HF, without optical evidence of any significant metal corrosion. From these images it is possible to see that the silica-gel beads remained in the initial shape and size during metal infiltration. SEM and EDS analyses show no reactions products or inter-diffusive phase between metal and space holder spheres and no traces of Si and O on metal ligaments after SiO_2 leaching. This behavior was expected and it is in accordance to the mechanical and chemical stability of SiO_2 beads at the processing temperature selected (maximum processing temperature 1050 °C). The inner surface of the open cells shows the "negative image" of the surface morphology of pre-heated silica-gel beads.

Density calculations for the final foam specimen of Fig. 4 led to a relative density of 0.31 (corresponding to 69% of empty space). Empty space proportion is slightly higher than the reported 63-64% that corresponds to random packing of shacking hard spheres of the same size (Ref 24).

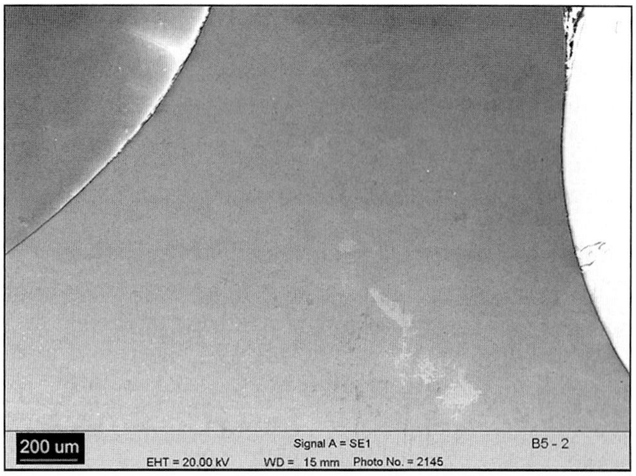

Fig. 3 SEM image of a heat-treated and polished specimen surface

Fig. 2 Infiltrated SiO_2 beads after cooling and external lathe machining

Fig. 4 Final sample of foam after machining and SiO_2 dissolution

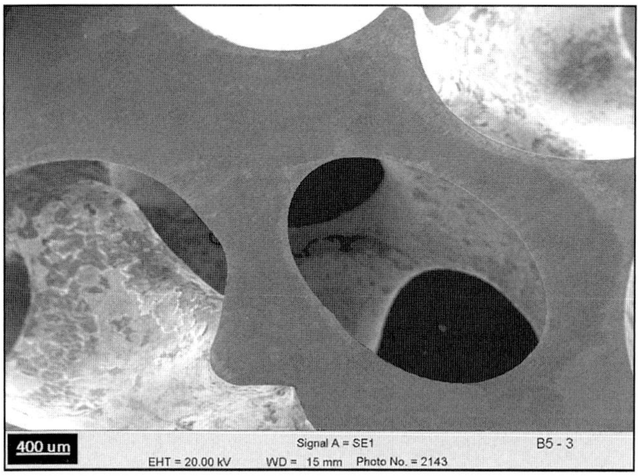

Fig. 5 SEM image of a polished specimen surface

Fig. 6 Representative images of the betatized bulk microstructure

This result could be partially explained as a result of the non-complete wettability of silica-gel beads by molten brass that leaves part of the interstitial space unfilled, counterbalancing some piling faults and also permitting the existence of these small interconnection windows between silica-gel beads (Ref 4). Other reasons for this positive deviation of the theoretical space holder volume are (Ref 24, 25): (a) silica-gel beads are not perfectly spherical, (b) their size is not uniform, and (c) there is a localized contraction of the metal during the solidification.

No matter the contribution of each mechanism, the wettability of silica-gel beads by molten metal plays an important role in the process (Ref 4). If silica-gel beads were completely wetted by molten brass the interconnection windows would be very small or inexistent and dissolution of the space holder could be harder or even not possible. Although metal infiltration of porous space holders depends on many other variables (i.e., metal pressure, infiltration velocity, etc.) (Ref 1-4) the study of those influences was beyond the scope of the work.

As for other production processes of metal foams, the proposed route has some dimensional limitations and is non-continuous, batch method. The achievable relative density of the metal sponges seems to be in the range 0.25-0.35, depending mainly on silica-gel morphology (i.e., irregular and/or almost spherical particles) and granulometric distribution. These features are common to processes that use leachable particles as space holder (Ref 1, 2). The main difference of the proposed process is the suitability of the space holder for higher temperatures. Non-indicating silica-gel (very stable silicon oxide SiO_2) is non-toxic and commonly used in drugs and food industries. On the other hand, the use of HF in the proposed process seems to have the same limitations and regulations that in the semiconductor industry, which also uses HF for SiO_2 leaching.

3.2 Microstructure and Transformation Behavior

In as cast condition the alloy microstructure is made up by a mixture of alpha and beta-phases. After heat treatment at 830 °C for 15 min and water quenching (Fig. 6 and 7), a single-phase microstructure is evident both in bulk and foamed specimens. The grain size grew up to 2 mm in the bulk material during the betatization. It is also possible to notice that the grain growth in the foamed materials is less evident. This phenomenon could be due to the different cooling conditions during the quench in the foamed and in the bulk material. Moreover, the geometrical limitation in the foamed state could slow down the grain growth rate. No alpha precipitates can be observed at grain boundaries and the microstructure shows only localized martensite plates.

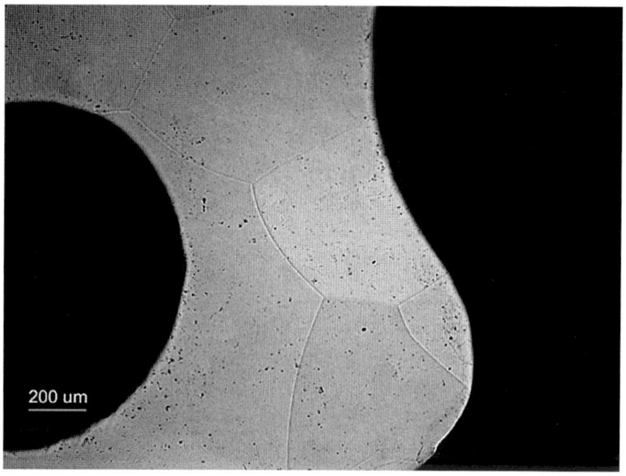

Fig. 7 Representative images of the betatized foam microstructure

Table 3 Bulk material DSC results after betatization treatment

M_s, °C	M_f, °C	ΔH, mJ/mg	A_s, °C	A_f, °C	ΔH, mJ/mg
−7.6	−12.5	−5.6	−5.7	−1	5.9

The transformation behavior was investigated by DSC measurements. The values reported in Table 3 and 4, show transformation temperatures lower than room temperature with comparable values in both specimens. In the reference bulk betatized state, the investigated alloy has a sharp transformation peak, which is depicted in Fig. 8. In this condition Ms, ΔH, and hysteresis are derived (Table 3). DSC scans of the foamed alloy in as cast condition, after betatization and after dissolution of silica-gel beads, are reported in Fig. 9. It is possible to observe

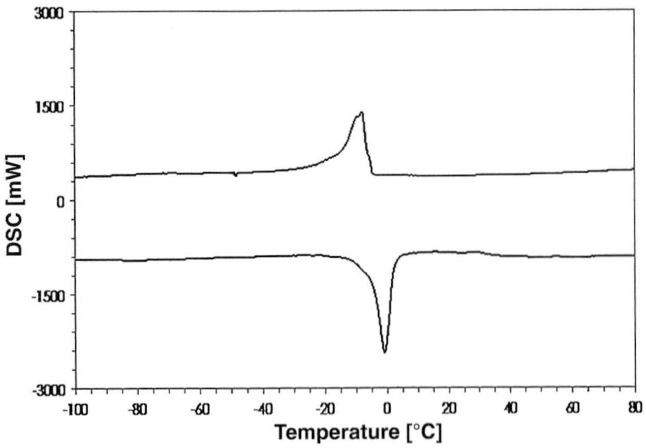

Fig. 8 DSC profiles of betatized bulk specimens

Table 4 Foamed material DSC results

State	M_s, °C	M_f, °C	ΔH, mJ/mg	A_s, °C	A_f, °C	ΔH, mJ/mg
As cast	−19.2	−86	−3.5	−40.7	−7	3.9
Betatized (before dissolution)	−9.2	−20	−5.8	−7.9	1.4	6.2
Betatized (after dissolution)	−2	−27.3	−6	−10.2	7.9	6

that betatization treatment promotes the transformation behavior and DSC peak in the same temperature range of the bulk material is detected. After dissolution a broader double peak is observed in the temperature range reported in Table 4. The changes in the transformation behavior, is likely to be connected to intrinsic inhomogeneities of the DSC specimens, taken from different positions in the foamed SMA. In fact, possible variation in the microstructure or morphology of the analyzed DSC samples may affect more peak intensity and shape than onset and offset transformation temperatures. Moreover, residual internal mechanical stresses, that could araise in the foamed SMA during its water quench and change after the dissolution of the silica-gel beads, may vary the peak intensity and subsequently the transformation enthalpy.

4. Conclusions

Based on experimental results it is possible to draw the following conclusions.

1. The production of metal foams of CuZnAl SMAs by molten metal infiltration of silica-gel beads, subsequently leached by aqueous HF was experimentally demonstrated.
2. Silica-gel remains stable in shape and dimensions during liquid-brass infiltration (maximum processing temperature 1050 °C). Due to the high chemical stability of SiO_2 particles, no interaction with the metallic system during liquid infiltration was found.
3. Silica-gel beads are completely dissolved by an aqueous solution (25 vol.%) of HF, without macroscopic evidences of chemical attack of the metal.

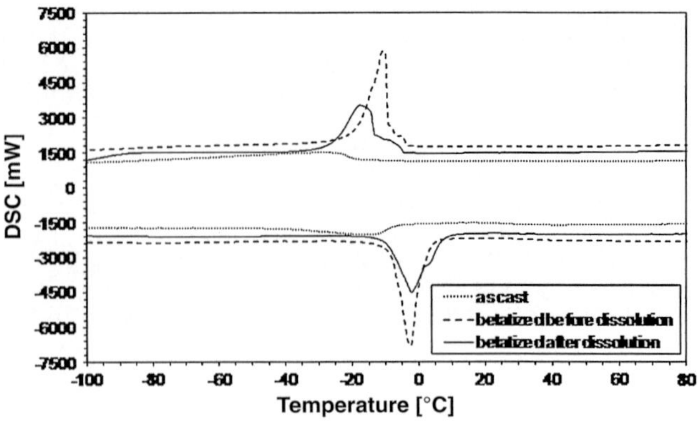

Fig. 9 DSC profiles of betatized foam specimens

4. The resulting metal foam features a relative density of 0.31 and all the observed cells are open and interconnected.
5. A thermoelastic martensitic transformation on foamed material was observed.
6. The foaming process is practically irrelevant on the transformation temperature of the SMA. In order to clarify possible effects of the chemical etching necessary to dissolve the silica-gel beads, on the functional properties of the material, further study are in progress.

Acknowledgments

The authors would like to acknowledge Fondazione CARIPLO and the Department of Mechanics of Politecnico di Milano for the financial support, the Department of Engineering Materials of University of Concepción, Prof. W. Nicodemi for his encouragement to the activity of E.M Castrodeza in Italy and Mr. M. Pini for his irreplaceable work during the experimental activities.

References

1. M.F. Ashby, A. Evans, N.A. Fleck, L. Gibson, J.W. Hutchinson, and H. Wadley, *Metal Foams: A Design Guide*, Butterworth-Heinemann, Boston, 2000
2. J. Banhart, Manufacture Characterization and Application of Cellular Metals and Metal Foams, *Prog. Mater. Sci.*, 2001, **46**(6), p 559–632
3. H.N.G. Wadley, Cellular Metals Manufacturing, *Adv. Eng. Mater.*, 2002, **4**(12), p 726–733
4. J.F. Despois, A. Marmottant, L. Salvo, and A. Mortensen, Spherical Pore Replicated Microcellular Aluminium: Processing and Influence on Properties, *Mater. Sci. Eng. A*, 2007, **462**(1–2), p 68–75
5. E.M. Castrodeza and C. Mapelli, *Italian Industrial Patent Process MO 2008 A 166* (in press)
6. E.M. Castrodeza and C. Mapelli, Processing of Brass Open-cell Foam by Silica-gel Replication, *J. Mater. Process. Technol.*, 2009 (in press)
7. A. Pollien, Y. Conde, L. Pambaguian, and A. Mortensen, Graded Open-cell Aluminium Foam Core Sandwich Beams, *Mater. Sci. Eng. A*, 2005, **404**(1–2), p 9–18
8. C. Gaillard, J.F. Despois, and A. Mortensen, Processing of NaCl Powders of Controlled Size and Shape for the Microstructural Tailoring of Aluminium Foams, *Mater. Sci. Eng. A*, 2004, **374**(1–2), p 250–262
9. A.H. Brothers, R. Scheunemann, J.D. DeFouw, and D. Dunand, Processing and Structure of Open-celled Amorphous Metal Foams, *Scr. Mater.*, 2005, **52**(4), p 335–339
10. H. Funakubo, *Shape Memory Alloys*, Gordon and Breach Science Publishers, Tokyo, 1984
11. K. Otsuka and C.M. Wayman, *Shape Memory Materials*, Cambridge University Press, London, 1998

12. M. Stipcich and R. Romero, The Effect of Post-quench Aging on Stabilization of Martensite in Cu–Zn–Al and Cu–Zn–Al–Ti–B Shape Memory Alloys, *Mater. Sci. Eng. A*, 1999, **273–275**, p 581–585
13. J.L. Pelegrina and M. Ahlers, Stabilization and Ferroelasticity in Cu–Zn Based Martensites, *Scr. Mater.*, 2004, **50**(2), p 213–218
14. M. Ahlers, Stability of Martensite in Noble Metal Alloys, *Mater. Sci. Eng. A*, 2003, **349**(1–2), p 120–131
15. K. Otsuka and X. Ren, Mechanism of Martensite Aging Effect, *Scr. Mater.*, 2004, **50**(2), p 207–212
16. S. Kustov, J. Pons, E. Cesari, and J. Van Humbeeck, Pinning-induced Stabilization of Martensite: Part II. Kinetic Stabilization in Cu–Zn–Al Alloy Due to Pinning of Moving Interfaces, *Acta Mater.*, 2004, **52**(10), p 3083–3096
17. J.A. Giampaoli, J.L. Pelegrina, and M. Ahlers, Ferroelasticity or Rubber Like Behaviour in Stabilized Cu-Zn-Al Single Crystals, *Acta Mater.*, 1998, **46**(10), p 3333–3342
18. R.J. Garcia, Stabilization of Martensite in Cu-Zn-Al Shape Memory Alloys: Effects of γ Precipitates and Thermal Cycling, *Scr. Mater.*, 2000, **42**(6), p 531–536
19. B. Chen, C. Liang, D. Fu, and D. Ren, Corrosion Behavior of Cu and Cu-Zn-Al Shape Memory Alloy in Simulated Uterine Fluid, *Contraception*, 2005, **72**(3), p 221–224
20. F.J. Gil, J.M. Guilemany, and J. Fernandez, Kinetic Grain Growth in Beta-copper Shape Memory Alloys, *Mater. Sci. Eng. A*, 1998, **241**(1–2), p 114–121
21. F.J. Gil and J.M. Guilemany, Effect of Cobalt Addition on Grain Growth Kinetics in Cu-Zn-Al Shape Memory Alloy, *Intermetallics*, 1997, **6**(5), p 445–450
22. S.A. Lermontov and A.N. Malkova, A New Method for Dissolution of Silica Gel in a Pentafluoropropionyl Fluoridetertiary Amine System, *Russian Chem. Bull.*, 2007, **56**(3), p 435–437
23. G. Spierings, Wet Chemical Etching of Silicate Glasses in Hydrofluoric Acid Based Solutions, *J. Mater. Sci.*, 1993, **28**(23), p 6261–6273
24. A.J. Matheson, Computation of a Random Packing of Hard Spheres, *J. Phys. C Solid State Phys.*, 1974, **7**(15), p 2569–2576
25. A. Donev, S. Torquato, F.H. Stillnger, and R. Cornelly, Jamming in Hard Sphere and Disk Packings, *J. Appl. Phys.*, 2004, **95**(3), p 989–999

JMEPEG (2009) 18:490–495
DOI: 10.1007/s11665-009-9381-2

1059-9495/$19.00

Design and Characterization of New Ti-Nb-Hf Alloys

M. González, J. Peña, J.M. Manero, M. Arciniegas, and F.J. Gil

(Submitted September 15, 2008; in revised form January 26, 2009)

Three new Ni-free Ti alloys Ti-16.2Hf-24.8Nb-1Zr, Ti-5.2Hf-31.2Nb-0.4Zr, and Ti-16Hf-36.2Nb-1Zr (wt.%), were designed and produced in order to obtain shape memory and/or low elastic modulus materials for the use in the load transfer implant field. For that, a method based on the molecular orbital theory was implemented to design the three new Ti-Nb-Hf system alloys. A vacuum arc-melted button of each alloy was treated at 1100 °C for 1.5 h and quenched in a mixture of ethanol/water at 0 °C. Finally, the alloys were microstructurally and mechanically characterized. Special attention on studying the elastic modulus and the thermoelastic martensitic transformation was given by means of nanoindentation tests using a Berkovich and a spherical tip, respectively. X-ray diffraction results showed the presence of β-phase in the three studied alloys. Moreover, one of the alloys exhibited reversible phase transformation due to the presence of thermoelastic martensitic α″-plates inside the β-grains observed by transmission electron microscopy. Results showed a low elastic modulus in all the studied alloys with values between 70 and 90 GPa, which are lower than those of the commercial alloys used in load transfer bone implants.

Keywords low elastic modulus, nanoindentation, Ni-free Ti alloys, shape memory effect

1. Introduction

Shape memory effect (SME) refers to the ability of certain materials to recover a predetermined shape when heated or stressed (Ref 1, 2). This ability is due to a reversible solid-state phase transformation from austenite (β-phase) to martensite (α″-phase). The shape memory alloys exhibit a number of remarkable properties such as shape memory effect, superelasticity, high damping capacity, low elastic modulus, and wear resistance (Ref 1-3). These properties make them promising candidates for biomedical applications such as prosthesis, osteosynthesis plates and screws, and orthodontic wires. Furthermore, such alloys present excellent corrosion resistance and biocompatibility (Ref 4). NiTi alloys are the only shape memory alloys which have been used as implantable materials in the human body (Ref 5).

However, Ni ions released from the implant to the surrounding tissue produces adverse reactions such as allergy, which affects up to 20% of women (Ref 6, 7). There are two main ways to solve the problem: the surface modification treatments such as oxidizing to reduce Ni surface concentration (Ref 8) and the design of new Ni-free Ti alloys (Ref 9-11).

The most commercial alloys used as implant materials present elastic modulus in the range of 110 to 220 GPa from Ti and Co-Cr-Mo alloys. A lower elastic modulus, near to that of cortical bone (25 GPa) (Ref 12), is required in order to decrease the stress-shielding effect and to enhance bone remodeling (Ref 13). Current research is specially focusing on new β-type Ti alloys, using β-stabilizing alloying elements such as Nb, Ta, Zr, and Hf, which exhibit lower elastic modulus and possible martensitic transformation.

In 1988, Morinaga and co-workers (Ref 14) introduced a theoretical method based on molecular orbitals for designing Ti-alloys with a significant reduction in terms of both time and cost. The purpose of this method is to consider the effect of alloying elements for predicting the stable phase of Ti-alloys based on phase stability maps as functions of two quantum parameters: the bond order (Bo) and the d-orbital energy level (Md). Bo represents the covalent bond strength between Ti and the alloying element, and Md is correlated with the electronegativity and metallic radius of the element.

The aim of this study is to achieve a material with shape memory properties and/or low elastic modulus to be used in load transfer implantology. For this purpose, three new Ni-free Ti alloys were designed and produced using the new properties map based on the abovementioned theoretical method for decreasing time and experimentation cost. Finally, the alloys were microstructurally and mechanically characterized.

2. Experimental

A theoretical method based on molecular orbitals was used to predict the stable phase and design three new Ti alloys (Ref 15). The method consists on the creation of a new map based on two quantum parameters, Bo and Md, calculated by the implementation of Khon-Sham from the Density Functional

This article is an invited paper selected from presentations at Shape Memory and Superelastic Technologies 2008, held September 21-25, 2008, in Stressa, Italy, and has been expanded from the original presentation.

M. González, J. Peña, J.M. Manero, M. Arciniegas, and **F.J. Gil**, Department of Materials Science and Metallurgy, Universitat Politècnica de Catalunya, Avda. Diagonal 647, Barcelona 08028, Spain; and **J. Peña**, Materials Science, Elisava Escola Superior de Disseny, C/Ample 11-13, Barcelona 08002, Spain. Contact e-mails: Marta.Gonzalez.Colominas@upc.edu, jpenya@elisava.es, Jose.Maria.Manero@upc.edu, Milena.Arcienagas@upc.edu, Francesc.Xavier.Gil@upc.edu.

Theory (DFT) (Ref 16) and based on a database of 132 alloys. The application of the resulting parameters to the data base permitted to build a new phase stability map in which two zones can be distinguished, delimited by the dot and dashed line corresponding to the probability to obtain a low elastic modulus and the shape memory effect, respectively. Three pairs of $(\overline{Bo}, \overline{Md})$, were selected on the β-zone, in the frontier between the low elastic modulus and the shape memory effect, in order to study a new system and a new zone of the design map (Fig. 1). The compositions were determined by using Nb, Hf, and Zr as alloying elements and are listed in Table 1.

The determination of β-transus temperature and Mo equivalent was carried out by using Eq 1 and 2, respectively, in order to design the heat treatment (Ref 17). The β-transus for pure Ti is reported as 882 °C and the addition of β-stabilizing solutes reduces this value.

$$T_{\beta\text{transusalloy}} = T_{\beta\text{transusTi}} - 7.22\,\text{wt.\%}\,\text{Nb} \qquad \text{(Eq 1)}$$

$$[\text{Mo}]_{\text{eq}} = 0.28\,\text{wt.\%}\,\text{Nb} \qquad \text{(Eq 2)}$$

A 40 g button of each alloy was made by vacuum arc melting, using high purity (99.99%) starting elemental materials. The buttons were sealed under vacuum into a quartz capsule, homogenized at 1100 °C for 12 h, solution treated at 1100 °C for 1.5 h and then quenched in ice water. The samples were mechanically polished and finished with colloidal silica to give a surface roughness with a Ra < 100 nm. The chemical composition was verified by x-ray fluorescence using Philips PW2400 equipment. Optical microstructural analysis was carried out after Keller reactive attack (2 mL HF, 3 mL HCl, 5 mL HNO_3, and 190 mL distilled H_2O). Thin disks of 3-mm diameter were electropolished for transmission electron microscopy (TEM) in an electrolyte consisting of 400 mL butoxy-ethanol, 400 mL methanol, and 100 mL perchloric acid.

Differential scanning calorimetry studies were carried out on a DSC Q1000 TA Instrument. Two entire heating-cooling cycles from −90 to 200 °C at a heating and cooling rate of 10 °C/min were used to investigate the presence of martensitic transformation and that of in thermoelastic martensitic transformation.

Finally, nanoindentation tests were conducted using a MTS Nano Indenter XP with Continuous Stiffness Measurement (CSM) module. Elastic modulus was measured from monotonic load tests to a depth of 2000 nm using a Berkovich tip with a measured radius of 750 nm. Data have been analyzed using the Oliver and Pharr method (Ref 18, 19). Furthermore, cyclic nanoindentation tests to a depth of 2000 nm were developed

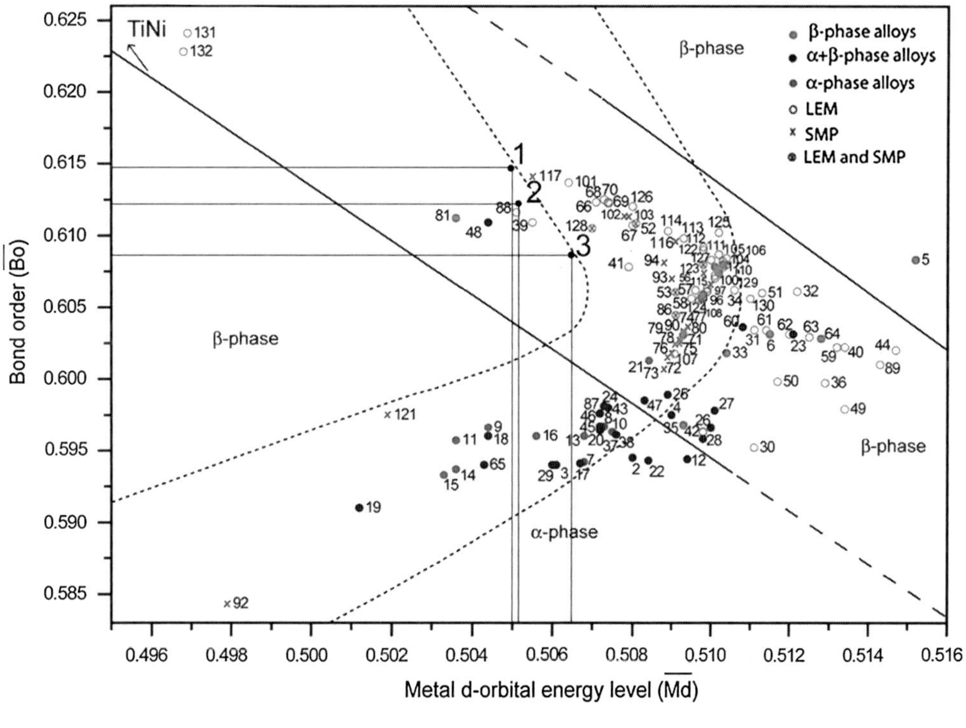

Fig. 1 Design map based on two quantum parameters, Bo and Md

Table 1 Relevant properties of the three new alloys: chemical composition (wt.%), microstructure, grain-size, β-transus temperature, Mo equivalent, thermoelastic phase transformation (TPT), and elastic modulus calculated by using a Berkovich tip (GPa)

Chemical composition, wt.%	No.	Microstructure	Grain-∅, μm	β-transus, °C	Mo eq.	TPT	*E*, Gpa
Ti-16.2Hf-24.8Nb-1Zr	1	$\beta + \alpha''$	900 ± 482	709	6.7	Yes	74 ± 1.5
Ti-5.2Hf-31.2Nb-0.4Zr	2	$\beta + \alpha'$	850 ± 492	658	8.7	No	76 ± 4
Ti-16Hf-36.2Nb-1Zr	3	β	525 ± 178	629	9.8	No	91 ± 2

using a spherical tip of radius 25 μm for evidencing the martensitic thermoelastic phase transformation from the load-displacement curves.

3. Results and Discussion

The relevant properties of the microstructural and mechanical characterization of the three new alloys are shown in Table 1.

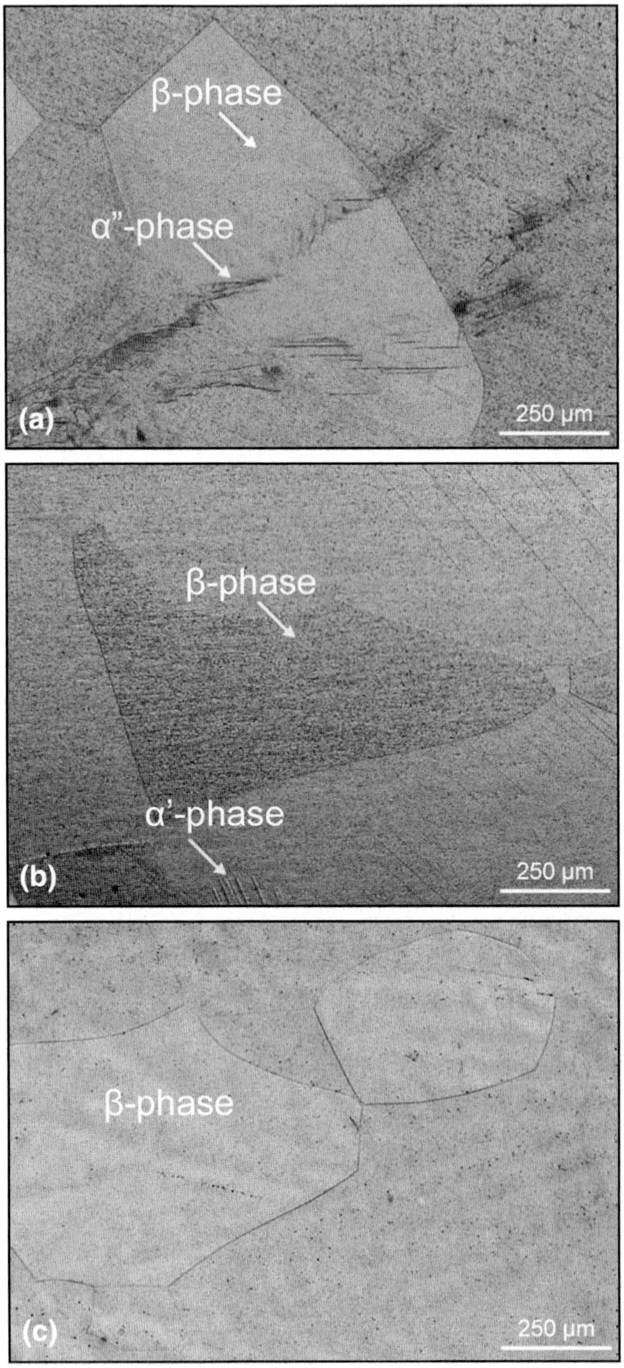

Fig. 2 Optical microscopy images of the alloys numbered as 1 (a), 2 (b), and 3 (c)

A notable decrease of β-transus temperature as a function of Mo equivalent was observed in the studied alloys. Heat treatment could be done at lower temperature which would reduce the average grain size.

As it was expected, because of the high content of β-stabilizing elements mainly β-phase was observed in the new alloys with a size range of 525 to 900 μm (Fig. 2). Moreover, small areas with fine orthorhombic martensitic plates nucleated from grain boundaries were observed in the alloys number 2 and 3.

X-ray diffraction results confirmed the presence of β-phase in the three alloys associated with (110), and (200) diffraction peaks and two low peaks associated with (013) and (131), indexed as α″-phase. In alloys numbered as 2, α′ peaks were detected associated with the (110) and (211) diffraction planes. In alloy number 3, with the higher Mo equivalent, only β-phase was observed (Fig. 3). The results confirmed that the presence of martensitic phase is conditioned by the β-stabilizer content.

The DSC patterns obtained from the three alloys did not evidence any martensitic transformation. This fact can be due to the low enthalpy values reported for the Ni-free Ti alloys (Ref 20).

However, the images obtained by TEM showed the presence of α″ martensite plates which got re-oriented and disappeared when increasing the intensity of the electron current beam (Fig. 4). This means that the heat produced by the electron beam induced the martensitic transformation from the austenitic phase, justifying the reversibility of the transformation (thermoelastic transformation).

In order to obtain more details about the martensitic transformation of alloy 1, TEM analysis was repeated by using an electron microscope with temperature control unit. The test started at 0 kV corresponding to 0 °C temperature. In this condition, the alloy presented a microstructure formed by martensitic plates inside the β-matrix (Fig. 5a). At 0.40 kV, corresponding to 60 °C, the complete transformation from martensite (α″) to austenite (β) was observed (Fig. 5b).

The elastic modulus of the three alloys (between 74 and 91 GPa) was lower than those presented by the commercial alloys used as load transfer implant materials. Moreover, the

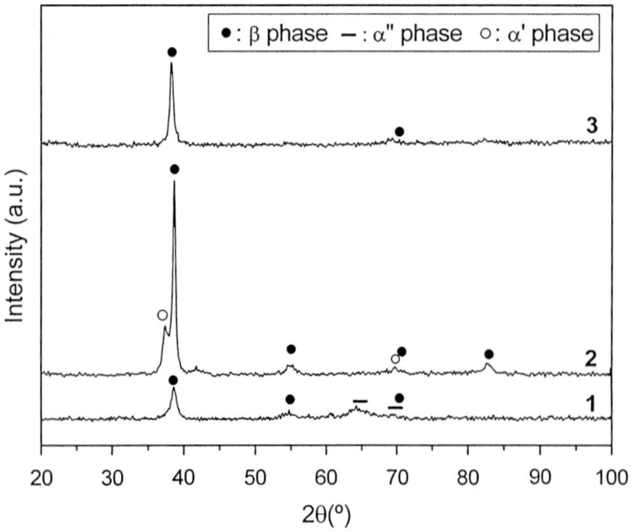

Fig. 3 X-ray diffraction patterns of the three new alloys

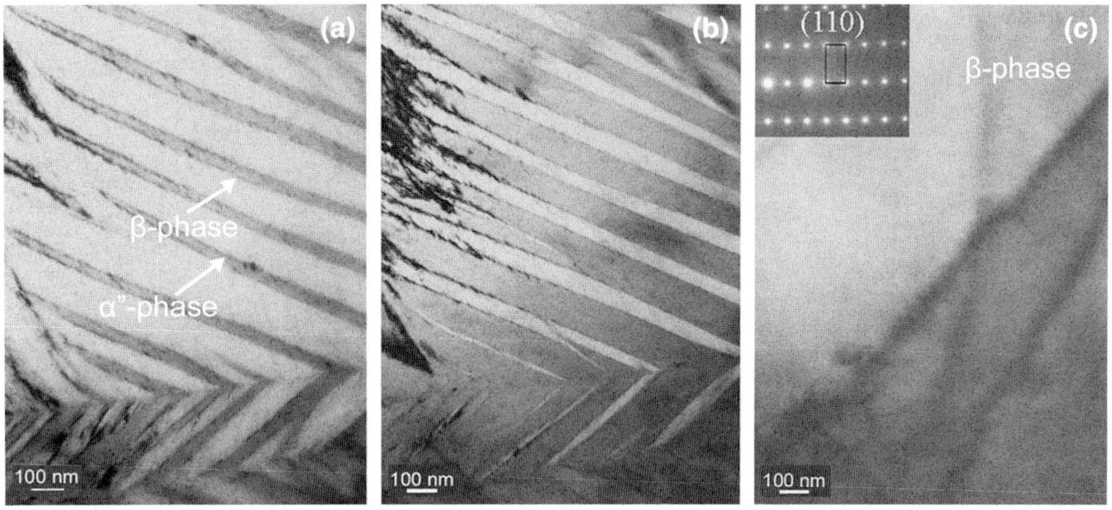

Fig. 4 Bright-field TEM image of the alloy number 1 at low (a), medium (b), and high (c) electron current beam intensity

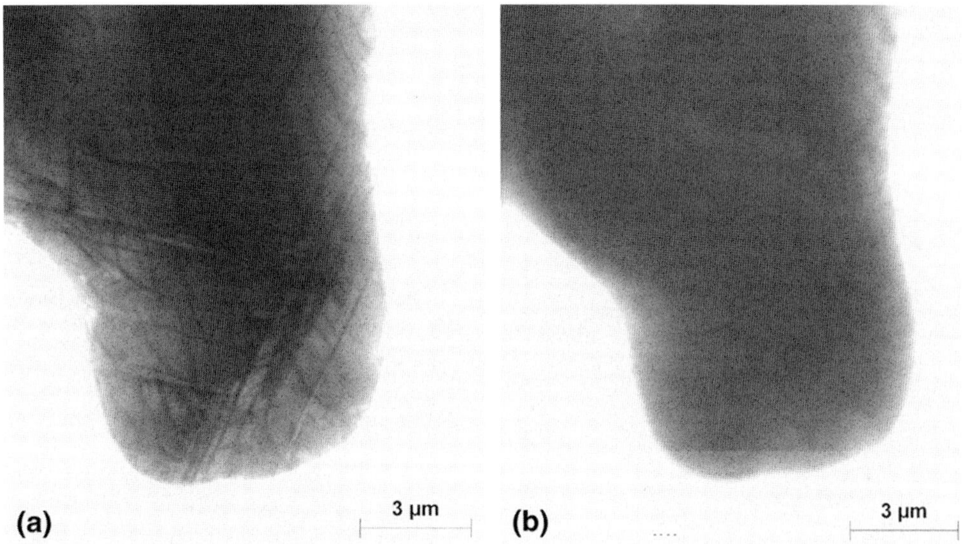

Fig. 5 Bright-field TEM image of the alloy number 1 at 0 kV (a) and 0.40 kV (b)

lowest elastic modulus corresponded to alloy number 1, which also presented thermoelastic martensitic transformation observed by TEM.

Finally, Fig. 6 shows one representative load-displacement curve of each alloy obtained by cyclic nanoindentation tests to a depth of 2000 nm. Load-displacement curve of alloy number 1 (Fig. 6a) evidenced a discontinuity on the loading portion reported as pop-in from the fourth loading-unloading cycle. A pop-in is characterized by the increase of the applied load without the consequent increase in the contact stress as a mechanism of deformation accommodation (Ref 21). Hysteresis loops generated between the unloading and the re-loading curves are especially observed after the appearance of the first pop-in in the fourth cycle. In transforming materials, when load is released, the elastic response and the reverse phase transformation contribute to the recovery, causing the observed loops. Both events pop-in and hysteresis loops of the alloy numbered as 1, are associated with the martensitic transformation. On the other hand, as it was expected, alloys number 2 and 3 did not evidence changes on the loading curve and neither hysteresis loops, indicating that thermoelastic martensitic transformation did not take place in that alloys (Fig. 6b, c).

Alloy numbered as 2, also showed the presence of martensitic plates, which remained stable when increasing the current electron beam intensity (Fig. 7).

4. Conclusions

The following conclusions can be drawn from this study:

Three new Ni-free Ti alloys were designed by implementing a new design method. From the obtained results it was verified the utility of the method in the studied zone.

The results showed a low elastic modulus in the three new alloys, lower than those presented by the commercial alloys used as load transfer implant materials.

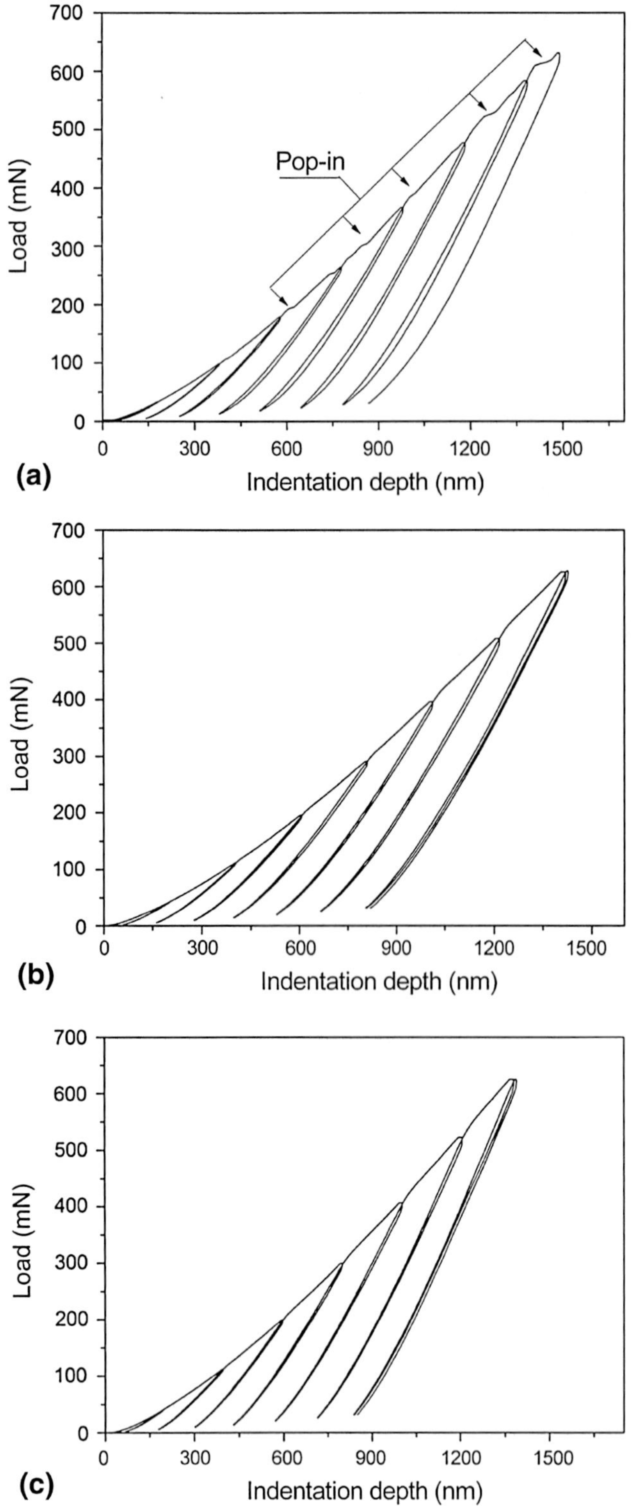

Fig. 6 Cyclic nanoindentation load-displacement curves of the alloys numbered as 1 (a), 2 (b), and 3 (c)

TEM images and also pop-in event and hysteresis loops observed on the alloy numbered as 1 were indicative that the alloy presented thermoelastic martensitic transformation. It is worth highlighting that the low elastic modulus of alloy number 1, with 74 GPa, also presented thermoelastic martensitic transformation observed by TEM and cyclic nanoindentation using a spherical tip.

Fig. 7 Bright-field TEM image of the alloy number 2

However, the results of alloy number 1 should be optimized by thermomechanical treatments.

Acknowledgments

With the support of the Generalitat de Catalunya Commission for the Universities and Research of the Department of Innovation, Universities and Companies and the European Social Fund.

References

1. T.W. Duerig, K.N. Melton, D. Stoeckel, and C.M. Wayman, *Engineering Aspects of Shape Memory Alloys*, Butterworth-Heinemann Ltd., London, 1990
2. T.W. Duering and A.R. Pelton, Ti-Ni Shape Memory Alloys, in *Materials Properties Handbook Titanium Alloys*, R. Boyer, G. Welsch, and E.W. Collings, Eds., ASM International, Materials Park, OH, 1994, p 1035
3. K. Otsuka and C.M. Wayman, Eds., *Shape Memory Materials*, University of Illinois, Urbana-Champaign, 1999
4. G. Rondelli, Corrosion Resistance Tests on NiTi Shape Memory Alloy, *Biomaterials*, 1996, **17**(20), p 2003–2008
5. C.D.J. Barras and K.A. Myers, Nitinol—Its Use in Vascular Surgery and Other Applications, *Eur. J. Vasc. Endovasc. Surg.*, 2000, **19**(6), p 564–569
6. G.C. McKay, R. Macnair, C. MacDonald, and M.H. Grant, Interactions of Orthopaedic Metals with an Immortalized Rat Osteoblast Cell Line, *Biomaterials*, 1996, **17**(13), p 1339–1344
7. H. Kerosuo, A. Kullaa, E. Kerosuo, L. Kanerva, and A. Hensten-Pettersen, Nickel Allergy in Adolescents in Relation to Orthodontic Treatment and Piercing of Ears, *Am. J. Orthod. Dentofacial Orthop.*, 1996, **109**(2), p 148–154
8. A. Michiardi, C. Aparicio, J.A. Planell, and F.J. Gil, Electrochemical Behaviour of Oxidized NiTi Shape Memory Alloys for Biomedical Applications, *Surf. Coat. Technol.*, 2007, **201**(14), p 6484–6488
9. K. Nitta, S. Watanabe, N. Masahashi, H. Hosada, and S. Hanawa, *Symposium on Structural Biomaterials for the 21st Century*, M. Niinomi, T. Okabe, E.M. Taleff, D.R. Lesuer, and H.F. Lippard, Eds., TMS, 2001, p 25–34
10. D. Kuroda, M. Niinomi, M. Morinaga, Y. Kato, and T. Yashiro, Design and Mechanical Properties of New β Type Titanium Alloys for Implant Materials, *Mater. Sci. Eng. A*, 1998, **243**(1–2), p 244–249

11. E. Takahashi, S. Watanabe, and S. Hanada, *Proceedings of the International Conference on Shape Memory and Superelastic Technologies*, USA, 2003, p 91–99
12. J.-Y. Rho, T.Y. Tsui, and G.M. Pharr, Elastic Properties of Human Cortical and Trabecular Lamellar Bone Measured by Nanoindentation, *Biomaterials*, 1997, **18**(20), p 1325–1330
13. M. Niinomi, Recent Research and Development in Titanium Alloys for Biomedical Applications and Healthcare Goods, *Sci. Technol. Adv. Mater.*, 2003, **4**(5), p 445–454
14. M. Morinaga, M. Kato, T. Kamimura, M. Fukumoto, I. Harada, and K. Kubo, *Sixth World Conference on Titanium*, France, 1, 1988, p 1601
15. M. Arciniegas, J.M. Manero, J. Peña, F.J. Gil, and J.A. Planell, Study of New Multifunctional Shape Memory and Low Elastic Modulus Ni-Free Ti Alloys, *Metall. Mater. Trans. A*, 2008, **39**(4), p 742–751
16. P.J. Hay and W.R. Wadt, Ab Initio Effective Core Potentials for Molecular Calculations Potentials for the Transition Metal Atoms Sc to Hg, *J. Chem. Phys.*, 1985, **82**, p 270
17. R. Boyer, G. Welsch, and E.W. Collings, *Materials Properties Handbook: Titanium Alloys*, ASM International, Materials Park, OH, 1998, p 5–11
18. A.C. Fischer-Cripps, *Nanoindentation*, 2nd ed., Springer, New York, 2004, p 16–45
19. G.M. Pharr, Measurement of Mechanical Properties by Ultra-low Load Indentation, *Mater. Sci. Eng. A*, 1998, **253**(1–2), p 151–159
20. Y.L. Hao, S.J. Li, S.Y. Sun, and R. Yang, Effect of Zr and Sn on Young's Modulus and Superelasticity of Ti-Nb-based Alloys, *Mater. Sci. Eng. A*, 2006, **441**(1–2), p 112–118
21. S. Bhagavat and I. Kao, Nanoindentation of Lithium Niobate: Hardness Anisotropy and Pop-in Phenomenon, *Mater. Sci. Eng. A*, 2005, **393**(1–2), p 327–331

JMEPEG (2009) 18:496–499
DOI: 10.1007/s11665-009-9397-7

High-Temperature Transformation Processes in Cu-13.4Al-5Ni Shape Memory Alloy Single Crystals

M. Benke, V. Mertinger, and L. Daróczi

(Submitted November 12, 2008; in revised form January 29, 2009)

Cu-13.4Al-5Ni alloy single crystals were investigated after performing the corresponding heat treatments to reveal the formation of equilibrium α and γ_2 phases and to examine transformation processes occurring in the alloy during heating-cooling cycles between room temperature and 970 K. Transformation processes were examined using heat flux DSC, structure examinations were performed using optical microscope and SEM devices, and phase identifications were carried out by XRD method. Identification of the equilibrium α phase using XRD method was not possible on the single crystal samples. According to the results, the equilibrium phases (α and γ_2) form only after long-term heat treatments. Because of that, aging processes do not affect the thermoelastic martensitic transformation of the CuAlNi alloy. This phenomenon is probably caused by the high Ni content that decreases the diffusion rate in CuAlNi alloys.

Keywords advanced characterization, metallography, nonferrous metals

1. Introduction

Cu-based shape memory alloys (SMAs) may be suitable in high-temperature applications because of their high transformation temperatures (320 to 420 K) compared to other SMAs (Ref 1, 2). However, after exposing to higher temperatures, aging process can advance and the metastable parent (austenite) phase can decompose into equilibrium phases. Examinations of the aging process were reported in many papers. These examinations were carried out on alloys containing 3 to 4% w/w Ni. V. Recarte et al. (Ref 2) examined the Cu-13.15Al-3.25Ni alloy from room temperature to 970 K. In their experiments, the samples, processed by powder metallurgy, were homogenized and quenched before the examination. During quenching, an ordering process started in the body centered cubic (bcc) β (parent) phase solid solution which could not be finished. On further cooling, partially ordered β phase transformed to β' martensite. The martensitic sample was heated at 10 K/min up to 970 K. In situ neutron diffraction was carried out at temperatures located before and after the DSC peaks to examine the evolved phases. The first endothermic peak, connecting to the β' (martensite) $\rightarrow$ β (austenite) transformation was followed by the first exothermic peak which could be connected to the continued ordering of the β phase. The second exothermic peak was ascribed to the dissolution of the β phase to equilibrium α and γ_2 (eutectoid) phases. The second endothermic peak was associated to the transformation of α and γ_2 to β phase (Ref 2). The processes can be traced on the section of the Cu-Al phase diagram (Fig. 1).

In this article, a Cu-13.4Al-5Ni with higher Ni content was examined. The aim of our research was not to characterize the aging process quantitatively but to examine the processes of the alloy when heated far above its transformation temperature.

2. Experimental

The CuAlNi single crystals were produced by Bridgeman method at the Catholic University of Leuven, Belgium. Samples were spark cut from the 3 mm diameter rods. According to optical microscopic examinations, the samples were not actually single crystals, since they contained three to six crystals. Inductive coupled plasma (ICP) examinations (Metalcontrol Kft) revealed that the Al content of the alloy is 12.6 to 12.9% w/w and the Ni content is 4.4 to 4.6% w/w. The initial samples were homogenized at 1170 K for 1 h in furnace then quenched in room temperature water. Samples were etched in HCl-$FeCl_3$ solution. DSC examinations were carried out by a NETZSCH 404 heat flux device using 10 K/min scanning rates under Ar atmosphere. SEM examinations were performed with an AMRAY 1830 (with installed EDS microprobe) instrument. Four-point electrical resistance measurements were carried out in vacuum using AGILENT 34970A data acquisition unit by means of the Department of Solid State Physics, University of Debrecen.

3. Results

The as-quenched Cu-13.4Al-5Ni sample was identified as NiK5. The DSC curves of the alloy are seen in Fig. 2.

This article is an invited paper selected from presentations at Shape Memory and Superelastic Technologies 2008, held September 21-25, 2008, in Stressa, Italy, and has been expanded from the original presentation.

M. Benke and **V. Mertinger**, Institute of Physical Material Science, University of Miskolc, H-3515 Miskolc-Egyetemváros, Hungary; and **L. Daróczi**, Department of Solid State Physics, University of Debrecen, Bem József u. 18/b, H-4010 Debrecen, Hungary. Contact e-mail: marton_benke@hotmail.com.

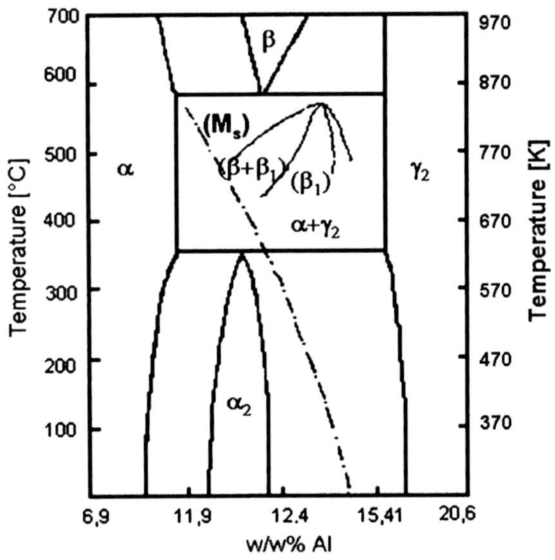

Fig. 1 The specified Cu-Al phase diagram

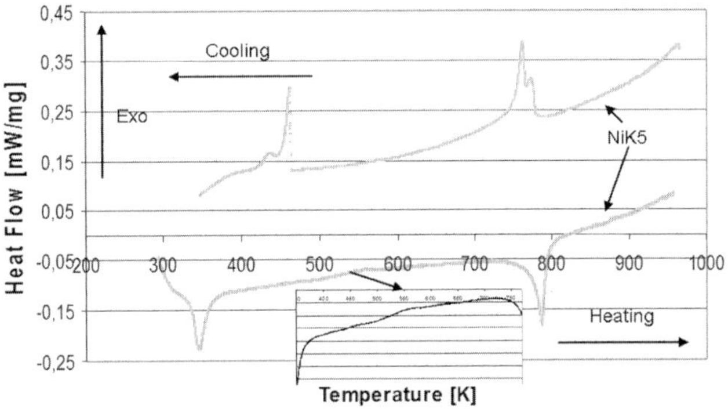

Fig. 2 DSC curves of the as-quenched Cu-13.4Al-5Ni alloy (NiK5)

During heating, the first endothermic peak would correspond to the β′ (martensite) → β (austenite) transformation at around 320 K. This is followed by a small exothermic peak at around 570 K which could be associated to the continued ordering of the β (Ref 2) phase. The next transformation is illustrated by the second endothermic peak at around 750 K in contrast to the results from the literature (Ref 2, 3). After that no peak can be seen during heating. During cooling there are two exothermic peaks which consist both of larger and smaller maxima. These peaks seem to be the pairs of the endothermic ones. The one on higher temperature will be identified later with its endothermic pair. The one on lower temperature indicates the formation of martensite. This is confirmed by the fact that after DSC cooling the sample was in martensitic state (Fig. 3).

The endothermic martensite → austenite transformation is on lower temperature (starts at around 320 K) than the exothermic peak of the austenite → martensite transformation (starts at around 470 K). This shift of the martensite → austenite transformation is often observed during the first heating cycle of the samples.

To reveal what process causes the second endothermic peak during heating, a series of examinations were performed. First,

Fig. 3 Optical microscopic image of the martensitic state of NiK5 sample after DSC examination

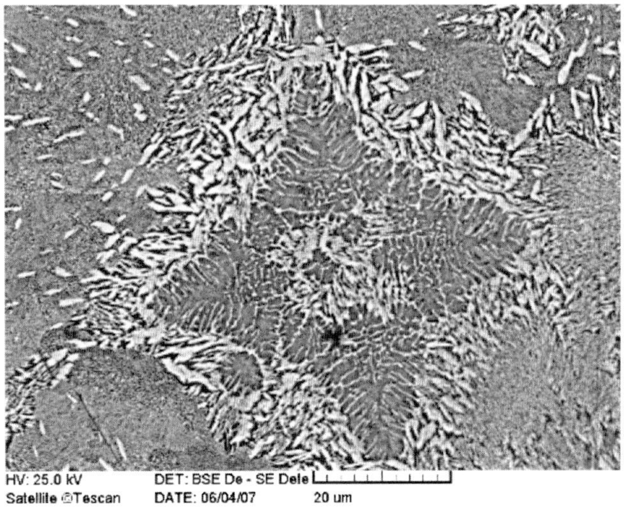

Fig. 4 SEM image of a γ_2 precipitate in the slowly cooled NiK8 sample

samples were produced with equilibrium phases. For that a sample was heat treated in the furnace at 1170 K for 1 h, and then the furnace was turned off and let to slowly cool down with the sample in it. The estimated average cooling rate was 0.9 K/min. The SEM images of the sample cooled in the furnace (identified as NiK8) can be seen in Fig. 4.

Figure 4 shows a single, cross-shaped γ_2 precipitate with α crystals in its vicinity as it evolved. The identification of the phases was carried out by EDS method. The NiK8 sample contained purely eutectoid phases in both single precipitate and lamellar forms. The small cooling rate enabled the formation of the equilibrium phases. However, the sample was not purely eutectic, it also contained some martensite.

The DSC curves of the NiK8 sample (dark gray) together with as-quenched NiK5 sample (light gray) are seen in Fig. 5.

The β′ (martensite) → β (austenite) transformation (first endothermic peak) shifted to higher temperature (to around 650 K) and became less intense in the NiK8 sample. This is because the precipitation of the eutectoid phases changed the composition of the matrix. The γ_2 phase has high Al content

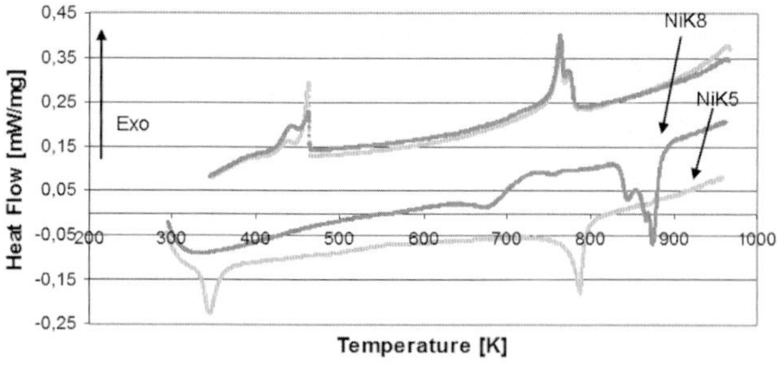

Fig. 5 DSC curves of the NiK5 and NiK8 samples

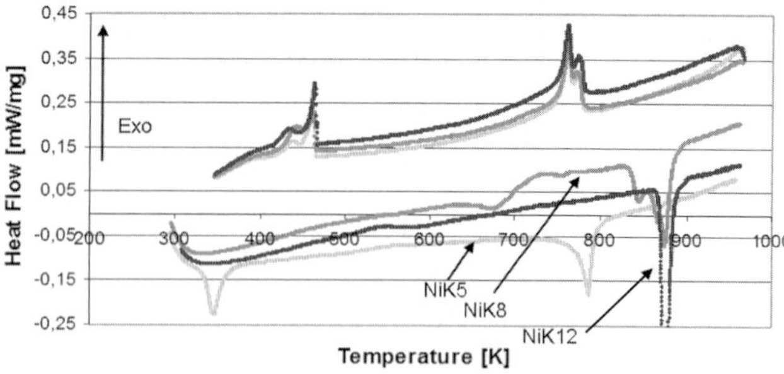

Fig. 7 DSC curves of the long-term heat-treated NiK12 sample together with the NiK8 and NiK5

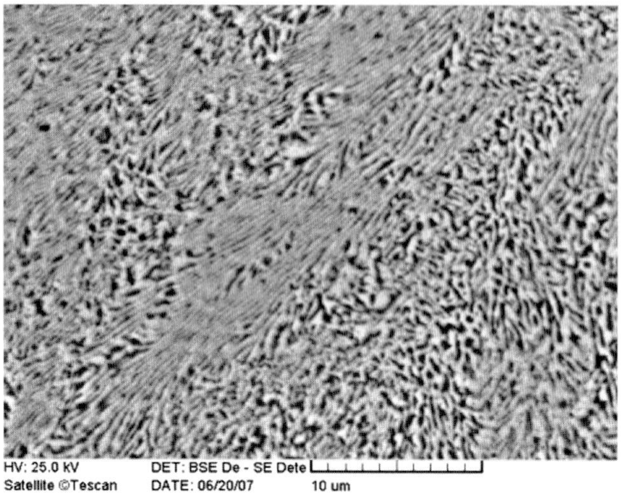

Fig. 6 SEM image of the long-term heat-treated NiK12 sample

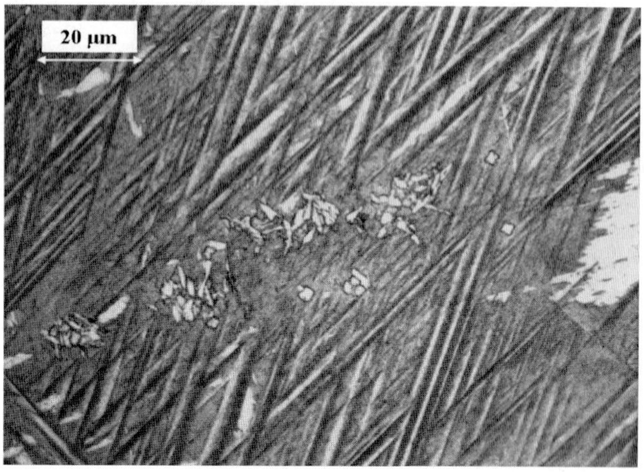

Fig. 8 Optical microscopic image of the sample heat treated at 670 K for 2 h

compared to the overall composition of the alloy (see Fig. 1). The remaining matrix is depleted in Al, which causes the martensite transformation to shift to higher temperature (Ref 4). The sample has less martensite than the NiK5; therefore, NiK8 has lower martensite $\rightarrow$ austenite peak. There is a second endothermic process with three peaks at around 830, 860, and at around 870 K. One of these peaks correspond to the $\alpha + \gamma_2 \rightarrow \beta$ transformation. We assume that the peak at 870 K corresponds to the eutectoid reaction, being the sharpest peak. We assume that the other two peaks correspond to the same process as in the case of NiK5 sample. This process is also shifted because the composition of the matrix was changed. The peaks during cooling are the same as in the previous NiK5 sample. They will be discussed later.

To produce a purely equilibrium α and γ_2 phase structure, a sample was heat treated in furnace at 690 K for 20 h. The sample was identified as NiK12. SEM image of the NiK12 sample can be seen in Fig. 6.

The sample was in a purely lamellar eutectoid state. The DSC examination of this sample will clearly indicate the $\alpha + \gamma_2 \rightarrow \beta$ transformation. The DSC curves of the NiK12 sample (heat treated for 20 h at 690 K) is seen in Fig. 7 (black curves).

There is one sharp endothermic peak during the heating of the NiK12 sample at around 870 K. This peak is the $\alpha + \gamma_2 \rightarrow \beta$ transformation. The transformation occurred in one step due to the fine lamellar eutectoid structure. The cooling stage is the same as the previous ones. This shows that the processes occurring during cooling are independent of the structure of the sample before the examinations. The same processes occur during cooling in the sample that was in purely martensitic state before the DSC examination (NiK5) as in the one that contained both martensite and eutectoid (NiK8) and as in the one that was in purely eutectoid state (NiK12). This shows that, at the end of the heating, the same structure evolved in all the samples.

The next step was to make samples with the structures before and after the second endothermic peak of the martensitic NiK5 sample. For this, a sample was heat treated before the endothermic peak at 690 K (Fig. 8) and another one after the endothermic peak at 870 K (Fig. 9) for 2 h with subsequent quenching.

The sample heat treated at 670 K (before the endothermic peak) contained martensite and some γ_2 precipitates with α crystals in their vicinity.

The sample heat treated at 870 K (after the endothermic peak) contained martensite matrix with small γ_2 precipitates in it. The α phase was not found in this sample. This is intelligible because the $\alpha \rightarrow \beta$ transformation starts at 830 K (Fig. 5) which means that at this temperature no α phase can be formed. If the starting or the finishing structure of the endothermic process was an equilibrium structure, it would have remained during quenching and the whole sample would have an equilibrium structure. However, the samples heat treated at 670 and 870 K both had predominantly martensitic structure. This means that the samples before and after the endothermic

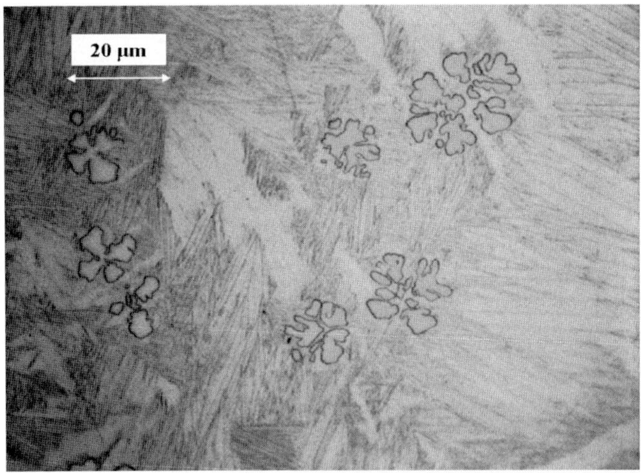

Fig. 9 Optical microscopic image of the sample heat treated at 870 K for 2 h

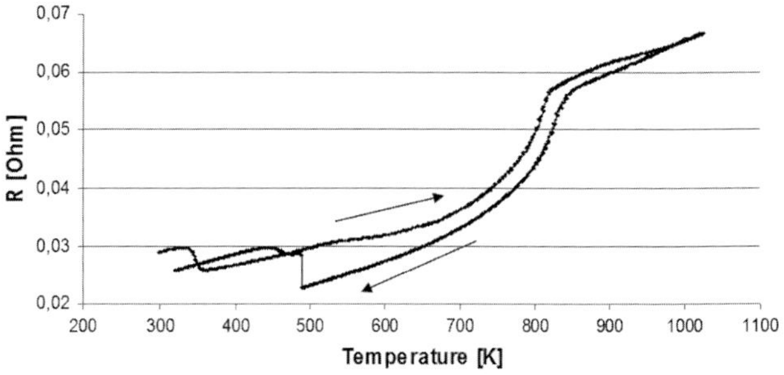

Fig. 10 Electrical resistance curve of the as-quenched NiK5 CuAl-Ni sample

process both had β structures. The β phase transformed to martensite during the subsequent quenching.

The second endothermic process of the martensitic NiK5 sample and its exothermic pair must be the heat effects of the order-disorder processes of the β phase. This explains why the samples had β structure both before and after the endothermic process. These processes occur at the temperatures where the peaks appeared (Fig. 2). To demonstrate the order-disorder process, electrical resistance measurement was performed on the as-quenched NiK5 CuAlNi sample (Fig. 10).

There are two shifts in the resistance. The first shift indicates the martensite ↔ austenite transformations at 330 K during heating and 470 K during cooling. The resistivity of the sample drops as it transforms from martensite to austenite. Note that the martensite → austenite transformation during heating is on lower temperature than the austenite → martensite transformation during cooling. This is the abovementioned first cycle effect. A small decrease in the resistance can be seen at around 570 K. This decrease could be caused by the continued ordering process, previously observed on DSC curves, as well-ordered solid solutions are more electrically conductive than disordered ones. The second shift indicates the order ↔ disorder transition starting at around 750 K. The resistivity of the sample shifts to higher value as the ordered structure becomes disordered. This transition occurs about between 730 and 790 K at 12.6 to 12.9 w/w Al content (Fig. 1). The latent heat of the transition was about 22 kJ/kg, while of the martensitic transformation was about 13 kJ/kg. The continued ordering of the β phase at 570 K during the first heating of the as-quenched sample occurred with only a small peak because most of the process took place during the quenching of the sample. In contrast to the as-quenched Cu-13.15Al-3.25Ni alloy, the equilibrium $\beta \rightarrow \alpha + \gamma_2$ transformation did not take place in the as-quenched Cu-13.4Al-5Ni alloy during the DSC examinations. This phenomenon could be caused by the larger Ni content since Ni decreases the diffusion velocity in the CuAlNi alloys (Ref 5). Since formation of the equilibrium phases is retarded by the increased Ni content, the resistance to aging is believed to be increased.

4. Summary

DSC heating-cooling scans of high (nominal 5% w/w) Ni content CuAlNi alloy single crystals in various heat-treated states were performed between room temperature and 970 K. The equilibrium $\beta \rightarrow \alpha + \gamma_2$ transformation did not occur during DSC cooling in as-quenched sample. The formation of the equilibrium phases is retarded by the larger Ni content which hereby makes the alloy more resistant against aging. The order ↔ disorder transition of the β phase was detected with larger amount of latent heat as compared to the martensitic transformation.

Acknowledgments

The authors would like to thank to Jan Van Humbeeck and the University of Leuven for providing the samples. The research was carried out with the support of OTKA T049513 and OTKA NI 68129.

References

1. V. Pelosin and A. Riviére, Structural and Mechanical Spectroscopy Study of the β′ Martensite Decomposition in Cu-12Al-3Ni (wt%) Alloy, *J. Alloys Compd.*, 1998, **268**, p 166–172
2. V. Recarte, J.I. Pérez-Landazábal, A. Ibarra, M.L. Nó, and J. San Juan, High Temperature β Phase Decomposition Process in a Cu-Al-Ni Shape Memory Alloy, *Mater. Sci. Eng. A*, 2004, **378**, p 238–242
3. J.I. Perez-Landazabal, V. Recarte, V. Sanchez-Alarcos, M.L. No, and J. San Juan, Study of the Stability and Decomposition Processes of the β Phase in Cu-Al-Ni Shape Memory Alloys, *Mater. Sci. Eng. A*, 2006, **438–440**, p 734–737
4. V. Recarte, R.B. Pérez-Sáez, E.H. Bocanegra, M.L. Nó, and J. San Juan, Dependence of the Martensitic Transformation Characteristics on Concentration in Cu-Al-Ni Shape Memory Alloys, *Mater. Sci. Eng. A*, 1999, **273–275**, p 380–384
5. E.G. West, *Copper and its Alloys*, Ellis Harwood Ltd., Chichester, 1982, p 110–116

JMEPEG (2009) 18:500–505
DOI: 10.1007/s11665-009-9499-2

Factors Influencing the Reversion of Stress-induced Martensite to Austenite in a Fe-Mn-Si-Cr-Ni Shape Memory Alloy

Leandru G. Bujoreanu, Sergiu Stanciu, Radu I. Comaneci, Markus Meyer, Vasile Dia, and Ciprian Lohan

(Submitted September 2, 2008; in revised form April 16, 2009)

It is well known that one way shape memory effect (SME) in Fe-Mn-Si-based shape memory alloys (SMAs) is related to the thermally induced reversion of ε (hexagonal close packed, hcp) stress-induced martensite (SIM) to γ (face centered cubic, fcc) austenite. In the case of a Fe-Mn-Si-Cr-Ni SMA, this reverse martensitic transformation was analyzed in regard to the critical temperature for the beginning of austenite formation (A_s) in different states characterized by quenching temperature and permanent tensile strain. For this purpose, dynamic mechanical analysis (DMA), dilatometry (DIL), differential thermal analysis (DSC), and optical microscopy (OM) were employed to determine the influence of quenching temperature and permanent tensile straining on SIM reversion to austenite during heating.

Keywords critical transformation temperature, Fe-Mn-Si base shape memory alloys, martensite reversion to austenite, stress-induced martensite, thermomechanical and thermodynamic response

1. Introduction

The development of most successful ferrous shape memory alloys (SMAs), based on the Fe-Mn-Si system discovered in Japan at the beginning of 1980's (Ref 1), started with the obtainment of one-way shape memory effect (SME) with recovery degree larger than 97% in Fe-30 Mn-1 Si (hereafter, all the chemical compositions are in mass%) single crystals (Ref 2, 3). Afterward, Murakami et al. first reported that almost 100% SME can be obtained in polycrystalline Fe-30 Mn-6 Si SMAs (Ref 4), then extended composition range to Fe-(28-34) Mn-(4-6) Si (Ref 5) and observed an increased recoverable strain to 4%, by employing thermomechanical training comprising pre-straining and heat treatment (Ref 6). By maintaining silicon content in the range of 5-6% to obtain a good SME, and by adding Cr and Ni to increase corrosion resistance (Ref 7), two potential candidates for low-cost (as compared to NiTi base alloys) SMA applications were developed: Fe-28 Mn-6 Si-5 Cr and Fe-14 Mn-5 Si-9 Cr-5 Ni (Ref 8).

For a comprehensive characterization of these alloys that undergo a γ (face-centered cubic: fcc) ⇔ ε (hexagonal close-packed: hcp) martensitic transformation (Ref 9), several shape memory phenomena were investigated such as pseudoelastic effect (PSE) (Ref 10, 11) and "apparent" two-way shape memory effect (TWSME) which was revealed (Ref 12) after suitable training (Ref 13). A mechanism of reversible stress-induced transformation, from tension-induced to compression-induced ε martensite, was proposed (Ref 14).

However, the SME, related to the thermally induced reversion of ε stress-induced martensite (SIM) to γ austenite, accompanied by recoverable strains of about 5%, has presently remained the most prominent characteristic of the Fe-Mn-Si SMAs (Ref 15) which are able to develop recovery stresses as high as 500 MPa (Ref 16) by tensile-constrained recovery (Ref 17).

In the particular case of corrosion-resistant Fe-Mn-Si-Cr-Ni SMAs, intensive research was carried out on the ε SIM regarding its growth (Ref 18) and preferred formation due to the occurrence (Ref 19) and alignment (Ref 20) of second-phase precipitates, as well as on the influence of variant intersections on the transformation behavior (Ref 21) and on the reversibility-assistance by back-stress developed at martensite-band tips (Ref 22). The basic conclusion that has been drawn is that ε SIM plates should be as narrow as possible, with a single variant orientation and should interact neither with each other nor with thermally induced martensite (Ref 23).

Among the factors that might influence ε SIM reversion on heating, the annealing temperature and the amount of pre-strain were analyzed in the case of Fe-13 Mn-5 Si-(9-10) Cr-(6-7) Ni SMAs. The results showed that in Fe-13 Mn-5 Si-10 Cr-6 Ni alloy (Ref 24) the start temperature, A_s, for martensite reversion, on heating, tends to decrease with increasing annealing temperature while it remains almost constant with increasing pre-strain. On the other hand, in Fe-13 Mn-5 Si-9

This article is an invited paper selected from presentations at Shape Memory and Superelastic Technologies 2008, held September 21-25, 2008, in Stresa, Italy, and has been expanded from the original presentation.

Leandru G. Bujoreanu, Sergiu Stanciu, Radu I. Comaneci, and **Ciprian Lohan**, Faculty of Materials Science and Engineering, The "Gheorghe Asachi" Technical University from Iasi, Bd. D. Mangeron 61A, Iaşi 700050, Romania; **Markus Meyer**, BU Analyzing & Testing, NETZSCH, Gerätebau GmbH, Wittelsbacherstrasse 42, 95100 Selb, Bavaria, Germany; and **Vasile Dia**, Laboratory Department, ARCELOR MITTAL Tubular Products Iaşi, Calea Chişinăului 132, Iaşi 700180, Romania. Contact e-mail: lgbujor@tuiasi.ro.

Cr-7 Ni alloy (Ref 25), the magnitude of shape recovery augmented with increasing annealing temperature from 870 K to 1070 K and cold rolling up to 15%.

The presence of carbon, limited to a maximum value of 0.3%, has been considered to be beneficial for Fe-Mn-Si-Cr-Ni SMAs since it contributes to the reinforcement of the austenite (Ref 26).

This article aims to bring additional evidence on the influence of both the temperatures of the heat treatment and the pre-straining on the start temperature A_s for ε SIM reversion, on heating, in the case of a Fe-Mn-Si-Cr-Ni SMA with high carbon content.

2. Experimental Procedure

From a Fe-Mn-Si-Cr-Ni SMA which was cast, homogenized (1270 K, 5×3.6 ks, water), and hot rolled (1270 K), specimens were cut with gauge dimensions $1 \times 5 \times 10 \times 10^{-3}$ m. The chemical composition was determined by spectrogravimetry as 13.9 Mn, 6.3 Si, 3.34 Cr, 1.61 Ni, 0.6 C, 0.12 Cu, 0.1 P, 0.05 S, and balance Fe. The specimens were analyzed in 12 different states characterized by three different heat treatments ($Q_{1\text{-}3}$) in each of which the formation of ε SIM, as compared to initial state E_0, was expected to occur as an effect of applying three different nominal permanent elongations ($E_{1\text{-}3}$), by means of an INSTRON 3382 tensile testing machine at room temperature (*RT*). Owing to uncontrollable spring back during unloading, a tolerance of ±0.05% was considered as compared to the nominal values of permanent strain listed in Table 1. Accordingly, the 12 states were designated as Q_iE_j, where $i = 1, 2, 3$ and $j = 0, 1, 2, 3$; for instance, state Q_2E_3 refers to specimens quenched from 1320 K with permanent pre-strains ranging from 2.05% to 2.15%. In each of the above 12 states, three specimens were prepared for dynamic mechanical analyzer (DMA) and three for dilatometer (DIL).

In order to determine the influence of quenching temperature and permanent elongation on thermally induced reversion of martensite to austenite, both thermomechanical and thermodynamic responses of Fe-Mn-Si-Cr-Ni SMA were evaluated during heating.

Thermomechanical response was examined by means of a DMA and a DIL. Three DMA experiments were conducted, on specimens belonging to each of the above 12 states, on a NETZSCH DMA 242 device, functioning in 3 point bending mode, also called forced 3 point beam mode (Ref 27), at a frequency of 1 Hz and an amplitude of 20×10^{-6} m. Heating was performed at a rate of 5×10^{-2} K/s from RT to 620 K. Based on specimen dimensions, and on the force and displacement signals, both internal friction (Tan Φ) and elastic modulus (*E*) were evaluated during heating. DIL measurements were performed on a NETZSCH DIL 402 CD dilatometer, at a heating rate of 8.33×10^{-2} K/s up to 870 K, under He atmosphere. Fused silica and alumina were used as sample holder material and standard calibration material, respectively. In this case, as well, three measurements were performed on each state of the specimens. For DIL evaluation, the variations with temperature of the relative elongation (dL/L_0), expansion rate $d(dL/L_0)/dt$, and thermal expansion coefficient (α) were analyzed.

Table 1 Designation of the 12 heat-treated specimens of Fe-Mn-Si-Cr-Ni SMA analyzed in initial and permanently elongated states

State	Quenching temperature, K	Q_1 1270	Q_2 1320	Q_3 1370
E_0	Nominal permanent strain, %	0	0	0
E_1		1.1	1.4	1.9
E_2		1.5	1.7	2.15
E_3		1.9	2.1	2.5

Thermodynamic response was recorded by a NETZSCH STA 449 F3 device during a heating-cooling cycle between RT and 870 K, under Ar atmosphere, at a rate of 1.66×10^{-1} K/s.

In order to reveal the formation of ε SIM under an applied nominal permanent elongation, optical micrographs were recorded on fragments cut from the deformed gauge of the elongated specimens, which were embedded into resin, electro polished, and etched with 10% Nital solution before being analyzed on a NEOPHOT 32 optical microscope.

3. Results and Discussions

Two typical examples of DMA results recorded on specimens quenched from 1320 K are illustrated in Fig. 1. It is known that, at low frequency such as 1 Hz, a high internal friction peak (Tan Φ) (Ref 28) and a step-like modulus (*E*) variation (Ref 29) are associated with the hcp → fcc transition, in the case of low-manganese Fe-Mn-Si-based SMAs (Ref 30). This evolution is observed in Fig. 1(a) corresponding to an unstrained specimen which was quenched from 1320 K. Tan Φ peak exceeds 0.02 and *E* increases with approx. 3.7 GPa during heating, which suggests that a solid state transition, such as reverse martensitic transformation, could be responsible for these changes. As an effect of pre-straining, multiple peaks were noticeable in most of the cases, in the variation of Tan Φ with temperature, such as that shown in Fig. 1(b), which corresponds to a specimen pre-strained with 2.07%. This rather erratic behavior, in the variation of Tan Φ with temperature could be an effect of specimen shrinkage, due to SME triggering, since the specimens tend to recover their unstrained shape on heating, or could be an effect of multiple stages of the martensite reversion. On the other hand, it has been shown that the change in the temperature coefficient, occurring on the modulus curve, is related to shape recovery of the specimen during heating since "modulus softening" is associated with the reverse ε → γ transformation (Ref 31). Therefore, in those cases where it was possible, A_s was evaluated as the point where modulus increase on heating started for the first time (Ref 32). The determined values of A_s are summarized in Table 2. As known from Table 1, permanent elongation increases from Q_{i1} to Q_{i3} which suggests that the SIM amount also increases in the same way. On the other hand, it is known that Tan Φ is increased by SIM formation (Ref 33) and, for this reason, the highest value, of 0.033, was reached in state Q_3E_3, which corresponds to the largest permanent elongation of 2.53%. From an analysis of the A_s values listed in Table 2, we can notice that A_s tends to decrease with increasing permanent applied strain, within the quenched states, Q_2 and Q_3. Therefore, from the point of view of thermomechanical response on heating, as determined by DMA, it may be estimated that the critical start temperature for SIM reversion

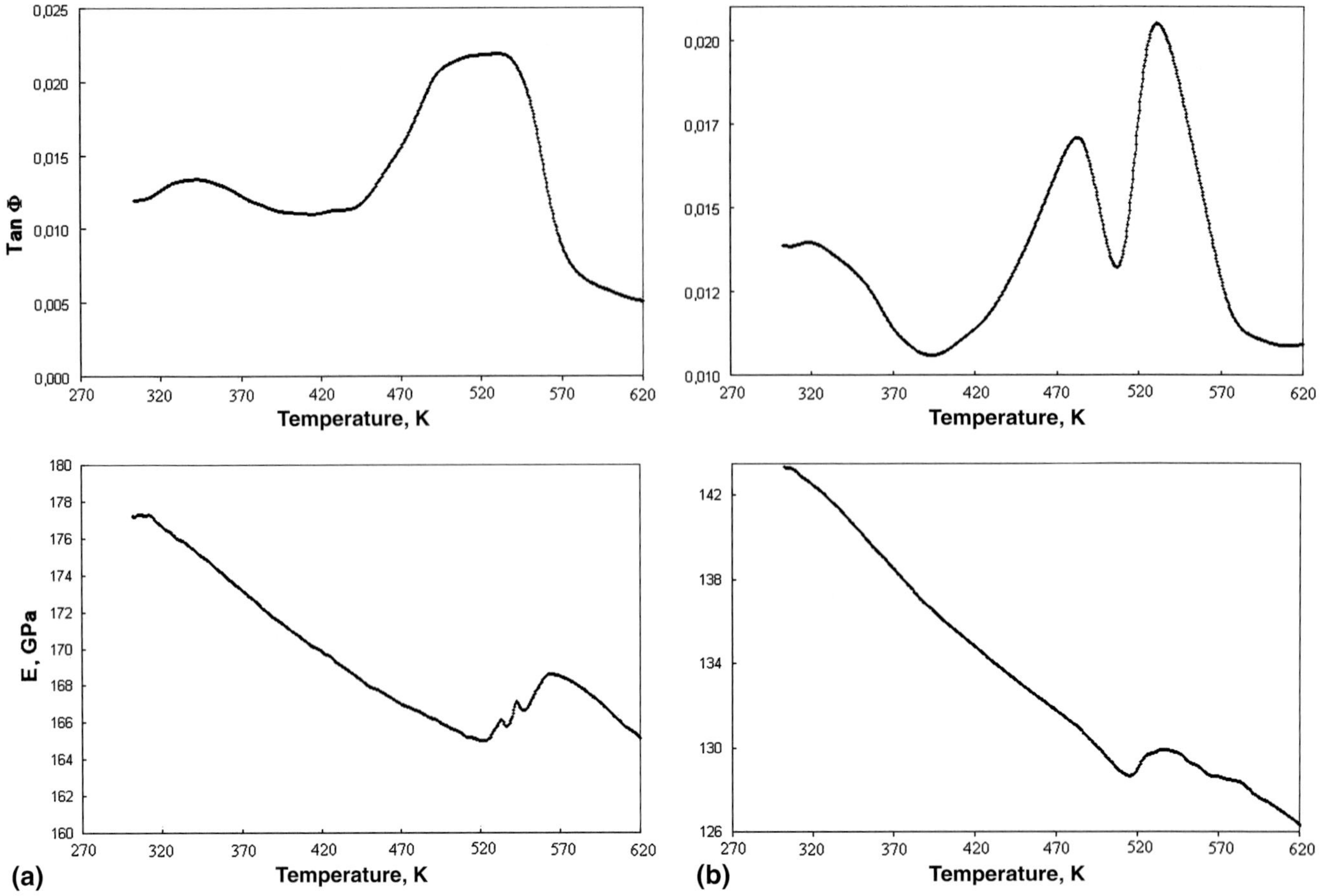

Fig. 1 Variations of internal friction (Tan Φ) and elastic modulus (E) determined by DMA measurements in 3 point bending mode, with a frequency of 1 Hz and an amplitude of 20×10^{-6} m, during heating to 620 K of the specimens of Fe-Mn-Si-Cr-Ni SMA under study, quenched from 1320 K (Table 1): (a) Q_2E_0; (b) Q_2E_3

Table 2 Values of critical transformation temperatures for martensite reversion start on heating, estimated from DMA experiments

	Heat treatment state											
	Q_1				Q_2				Q_3			
Permanent elongation state:	E_0	E_1	E_2	E_3	E_0	E_1	E_2	E_3	E_0	E_1	E_2	E_3
A_s, K	…	553	…	…	522	510	502	506	…	498	485	477

to austenite tends to decrease with increasing permanent elongation.

The second evaluation, aiming to reveal the effects of both quenching temperature and pre-straining on thermomechanical response of Fe-Mn-Si-Cr-Ni SMA under study, was performed by dilatometry which enables accurate monitoring of relative elongation (dL/L_0), expansion rate $d(dL/L_0)/dt$, and thermal expansion coefficient (α) due to temperature variation on heating. The effect of quenching temperature on unstrained specimens is illustrated in Fig. 2. Since the specimens were not deformed, they should not contain any SIM. Consequently no obvious inflections were noticeable on the dilatograms, in spite of the amount of thermally induced martensite which is expected to exist, according to DMA results shown in Fig. 1(a). Moreover, we can notice that both expansion rate $d(dL/L_0)/dt$ and thermal expansion coefficient (α) varied in the same way in the three quenching states, during heating, although α seemed to decrease with increasing quenching temperature. On the other hand, with increasing quenching temperature, an increase of the amount of thermally induced martensite is expected to occur as indicated by Fig. 1(a) and by the variations of tensile mechanical characteristics listed in Table 3. As a matter of fact, the ultimate tensile stress decreased and the ultimate strain exhibited a tendency to increase, so it may be assumed that the specimens became more ductile with increasing quenching temperature. The only transformation (confirmed by the sharp decrease of $d(dL/L_0)/dt$ is located at around 830 K in Fig. 2(b), corresponding to Q_2E_0 state (1320 K) and probably also to Q_3E_0 (although heating was stopped at 770 K). It is assumed that this inflection could correspond to the transformation of stabilized thermally induced martensite to bainite (Ref 34). The effect of permanent elongation on the thermomechanical

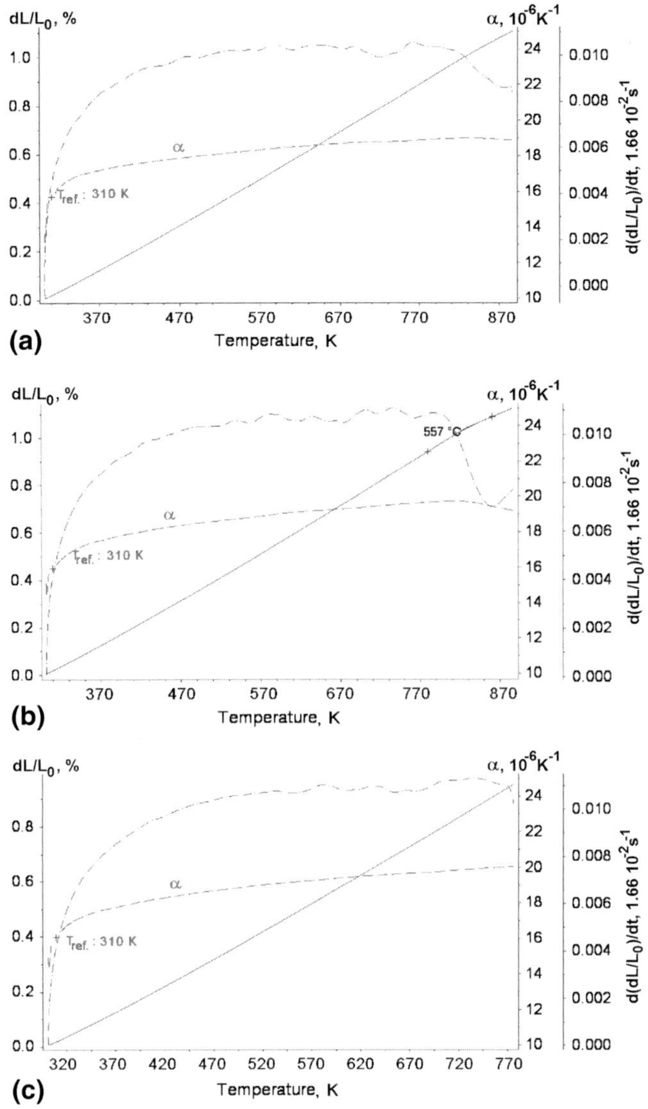

Fig. 2 Variations with temperature of relative elongation (dL/L_0 with solid line), expansion rate $d(dL/L_0)/dt$, and thermal expansion coefficient (α), on the dilatograms recorded during the heating of unstrained specimens of Fe-Mn-Si-Cr-Ni SMA under study, in different heat-treated states (Table 1): (a) Q_1E_0; (b) Q_2E_0; (c) Q_3E_0

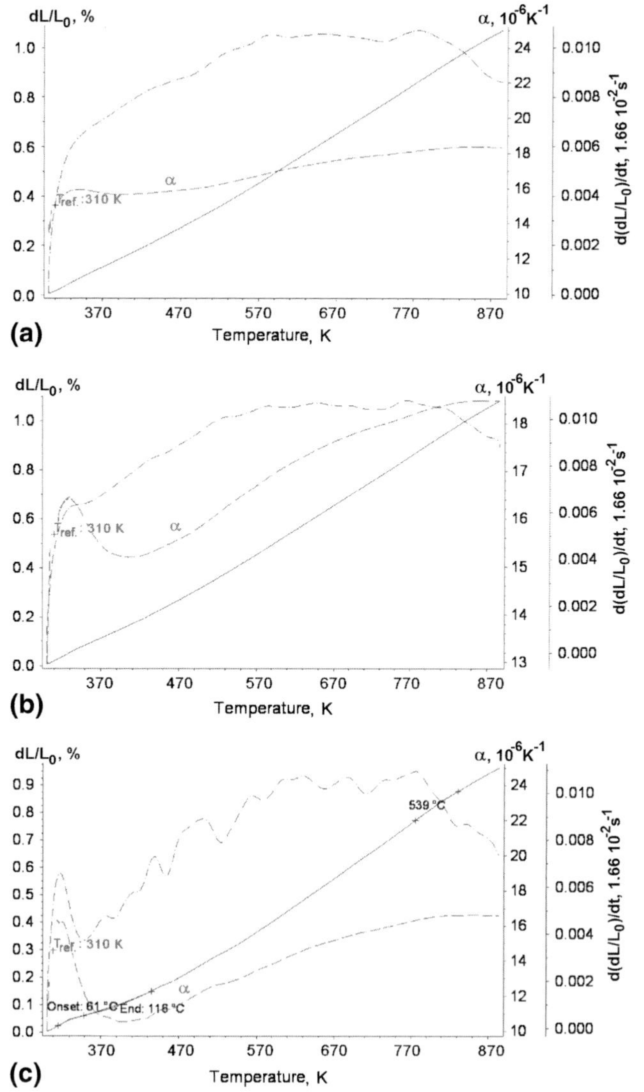

Fig. 3 Variations with temperature of relative elongation (dL/L_0 with solid line), expansion rate $d(dL/L_0)/dt$, and thermal expansion coefficient (α), on the dilatograms recorded during the heating of specimens of Fe-Mn-Si-Cr-Ni SMA under study, in heat-treated state Q_1 in different permanently elongated states (Table 1): (a) Q_1E_1; (b) Q_1E_2; (c) Q_1E_3

Table 3 Values of mechanical parameters determined by tensile tests

Heat treatment state	Q_1	Q_2	Q_3
Ultimate stress, MPa	523	395	373
Ultimate strain, %	4.68	14.88	11.48

response of the specimens quenched from 1270 K is illustrated in Fig. 3. It is obvious that, with increasing nominal permanent elongation from 1.1 to 1.9, a solid-state transition becomes more and more evident. Therefore, from the point of view of thermomechanical response determined by dilatometry during heating, it may be assumed that the amount of SIM has been very low so that it is only at high permanent elongations, such as in specimen Q_1E_3, where SIM reversion to austenite could be detected. In Fig. 3(c), corresponding to state Q_1E_3, the start temperature of the transition is about 331 K. This transition is accompanied by marked fluctuations of $d(dL/L_0)/dt$ and (α), and the entire aspect of the dilatogram is very similar to that found in literature (Ref 35).

The thermodynamic response is illustrated in Fig. 4 by the DSC versus temperature diagram of a fragment cut from the elongated gauge of Q_1E_3 specimen. It is assumed that the first shallow endothermic peak could fairly correspond to SIM reversion to austenite. This is suggested for two reasons: (i) it starts at around 370 K which is close to the value A_s = 331 K determined by dilatometry studies and (ii) it has a flat and not a sharp minimum since it was generated by a small fragment of the elongated gauge which could incorporate very small amounts of SIM. The second endothermic peak shows that an obvious transition occurred during heating, between 663 and 822 K, which is close to 830 K, the value presumed to

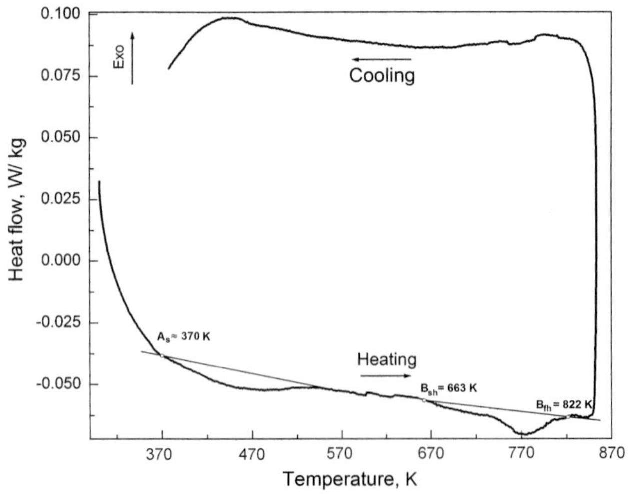

Fig. 4 DSC thermogram, recorded during a heating-cooling cycle between room temperature and 870 K, indicating a reversible martensitic transformation in a fragment cut from the elongated gauge of a specimen of Fe-Mn-Si-Cr-Ni SMA under study in state Q_1E_3

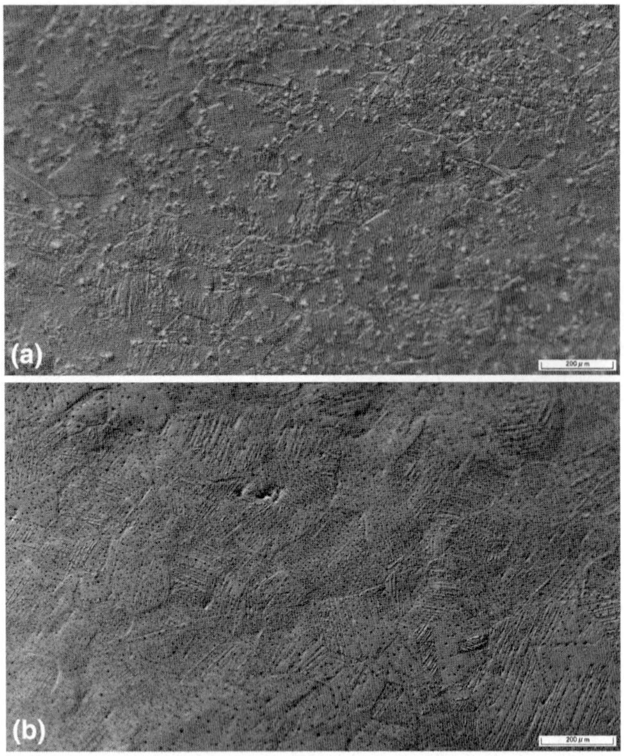

Fig. 5 OM micrographs of specimens of Fe-Mn-Si-Cr-Ni SMA under study, quenched from 1270 K: (a) unstrained, Q_1E_0; (b) prestrained 1.9%, Q_1E_3

correspond to the transformation of stabilized thermally induced martensite to bainite, as determined by dilatometry.

Finally, two optical micrographs are shown in Fig. 5 for the states Q_1E_0 and Q_1E_3, respectively. Figure 5(a) reveals that after 1270 K-quenching, scarce thermally induced ε martensite plates are noticeable on some of the grains, while Fig. 5(b) shows that, as an effect of applying 1.9% permanent elongation, the plates became significantly denser and finer.

4. Summary and Conclusions

A Fe-13.9 Mn-6.3 Si-3.34 Cr-1.61 Ni-0.6 C SMA was subjected to three different quenching temperatures (1270, 1320, and 1370 K, respectively) and three permanent elongations (between 1.1% and 2.5%), for each of the quenched states, to stress induce ε martensite. Martensite reversion to austenite during heating was evaluated by means of thermomechanical and thermodynamic responses. The former response was determined by DMA and dilatometry and the latter by STA.

Owing to the low amount of stress-induced martensite and to the different sensitivities of the three techniques, different aspects were revealed by each experimental investigation:

(i) DMA recorded multiple Tan Φ peaks, reaching values as high as 0.033, suggesting a multiple-stage reversion of stress-induced martensite to austenite during heating. A_s values, estimated by DMA and corresponding to the point where elastic modulus increase occurred on heating, for the first time, tend to slightly increase with increasing quenching temperature and to decrease with increasing permanent elongation;

(ii) Dilatometry revealed reversion of the stress-induced martensite to austenite and the formation of increasingly larger amounts of thermally induced martensite, which was found to transform to recovery bainite at around 830 K;

(iii) STA data showed that the reversion of the stress-induced martensite to austenite absorbs much lower specific energy as compared to the transition of stabilized thermally induced martensite to recovery bainite;

(iv) The existence of thermally induced martensite in quenched state and of stress-induced martensite in prestrained state was ascertained by optical microscopy.

Acknowledgments

This study was financially supported by UEFISCU by means of the research Grant PN II-ID 301-PCE-2007-1, Contract no. 279/01.10.2007. Special thanks are due to Dr. Takahiro Sawaguchi, NIMS Tsukuba, for fruitful scientific discussions.

References

1. T. Maki, Ferrous Shape Memory Alloys, *Shape Memory Materials*, K. Otsuka and C.M. Wayman, Ed., Cambridge University Press, Cambridge, 1998, p 117–132
2. A. Sato, E. Chishima, K. Soma, and T. Mori, Shape Memory Effect in $\gamma \leftrightarrow \varepsilon$ Transformation in Fe-30Mn-1Si Alloy Single Crystals, *Acta Metall.*, 1982, **30**, p 1177–1183
3. A. Sato, E. Chishima, Y. Yamaji, and T. Mori, Orientation and Composition Dependencies of Shape Memory Effect in Fe-Mn-Si Alloys, *Acta Metall.*, 1984, **32**, p 539–547
4. M. Murakami, H. Suzuki, and Y. Nakamura, Effect of Si on the Shape Memory Effect of Polycrystalline Fe-Mn-Si Alloys, *Trans. ISIJ*, 1987, **27**, p B-87
5. M. Murakami, H. Otsuka, H. Suzuki, and S. Matsuda, Effect of Alloying Content, Phase and Magnetic Transformation on the Shape Memory Effect of Fe-Mn-Si Alloys, *Trans. ISIJ*, 1987, **27**, p B-88
6. M. Murakami, H. Otsuka, and S. Matsuda, Improvement of Shape Memory Effect of Fe-Mn-Si Alloys, *Trans. ISIJ*, 1987, **27**, p B-89

7. H. Otsuka, H. Yamada, T. Maruyama, H. Tanahashi, S. Matsuda, and M. Murakami, Effects of Alloying Additions on Fe-Mn-Si Shape Memory Alloys, *ISIJ Int.*, 1990, **30**, p 674–679
8. S. Kajiwara, Characteristic Features of Shape Memory Effect and Related Transformation Behavior in Fe-Based Alloys, *Mater. Sci. Eng. A*, 1999, **273-275**, p 67–88
9. T.Y. Hsu and X. Zuyao, Martensitic Transformation in Fe-Mn-Si Based Alloys, *Mater. Sci. Eng. A*, 1999, **273-275**, p 494–497
10. O. Matsumura, T. Sumi, N. Tamura, K. Sakao, T. Furukawa, and H. Otsuka, Pseudoelasticity in an Fe-28Mn-6Si-5Cr Shape Memory Alloy, *Mater. Sci. Eng. A*, 2000, **279**, p 201–206
11. T. Sawaguchi, T. Kikuchi, and S. Kajiwara, The Pseudoelastic Behavior of Fe-Mn-Si-based Shape Memory Alloys, Containing Nb and C, *Smart Mater. Struct.*, 2005, **14**, p S317–S322
12. J.H. Yang, H. Chen, and C.M. Wayman, Development of Fe-Based Shape Memory Alloys Associated with Face-Centered Cubic-Hexagonal Close-Packed Martensitic Transformations: Part I. Shape Memory Behavior, *Metall. Trans. A*, 1992, **23A**, p 1431–1437
13. T. Shiming, L. Jinhai, and Y. Shiwei, Two-Way Shape Memory Effect of an Fe-Mn-Si Alloy, *Scripta Metall. Mater.*, 1991, **25**, p 1119–1121
14. T. Sawaguchi, L.G. Bujoreanu, T. Kikuchi, K. Ogawa, M. Koyama, and M. Murakami, Mechanism of Reversible Transformation-Induced Plasticity of Fe-Mn-Si Shape Memory Alloys, *Scripta Mater.*, 2008, **59**, p 826–829
15. N. Stanford, D.P. Dunne, and H. Li, Re-examination of the Effect of NbC Precipitation on Shape Memory in Fe-Mn-Si-Based Alloys, *Scripta Mater.*, 2008, **58**, p 583–586
16. Y. Wen, N. Li, and M. Tu, Effect of Quenching Temperature on Recovery Stress of Fe-18Mn-5Si-8Cr-4Ni Alloy, *Scripta Mater.*, 2001, **44**, p 1113–1116
17. L.G. Bujoreanu, V. Dia, S. Stanciu, M. Susan, and C. Baciu, Study of the Tensile Constrained-Recovery Behavior of a Fe-Mn-Si Shape Memory Alloy, *Eur. Phys. J. Special Top.*, 2008, **158**, p 15–20
18. J.H. Yang and C.M. Wayman, Self-Accommodation and Shape Memory Mechanism of ε-Martensite—I. Experimental Observations, *Mater. Charact.*, 1992, **28**, p 23–35
19. S. Kajiwara, D. Liu, T. Kikuchi, and N. Shinya, Remarkable Improvement of Shape Memory Effect in Fe-Mn-Si Based Shape Memory Alloys by Producing NbC Precipitates, *Scripta Mater.*, 2001, **44**, p 2809–2814
20. Y.H. Wen, W. Zhang, N. Li, H.B. Peng, and L.R. Xiong, Principle and Realization of Improving Shape Memory Effect in Fe-Mn-Si-Cr-Ni Alloy through Aligned Precipitations of Second-Phase Particles, *Acta Mater.*, 2007, **55**, p 6526–6534
21. J.H. Yang, H. Chen, and C.M. Wayman, Development of Fe-Based Shape Memory Alloys Associated with Face-Centered Cubic-Hexagonal Close-Packed Martensitic Transformations: Part II. Transformation Behavior, *Metall. Trans. A*, 1992, **23A**, p 1439–1444
22. N. Bergeon, G. Guenin, and C. Esnouf, Microstructural Analysis of the Stress-induced ε Martensite in a Fe-Mn-Si-Cr-Ni Shape Memory Alloy. Part II: Transformation Reversibility, *Mater. Sci. Eng. A*, 1998, **242**, p 87–95
23. Y.H. Wen, N. Li, and L.R. Xiong, Composition Design Principles for Fe-Mn-Si-Cr-Ni Based Alloys with Better Shape Memory Effect and Higher Recovery Stress, *Mater. Sci. Eng. A*, 2005, **407**, p 31–35
24. H. Li, D. Dunne, and N. Kennon, Factors Influencing Shape Memory Effect and Phase Transformation Behaviour of Fe-Mn-Si Based Shape Memory Alloys, *Mater. Sci. Eng. A*, 1999, **273-275**, p 517–523
25. N. Stanford and D.P. Dunne, Thermo-Mechanical Processing and the Shape Memory Effect in an Fe-Mn-Si-Based Shape Memory Alloy, *Mater. Sci. Eng. A*, 2006, **422**, p 352–359
26. L. Chengxin, W. Guixin, W. Yandong, L. Qingsuo, and Z. Jianjun, Effect of Addition of V and C on Strain Recovery Characteristics in Fe-Mn-Si Alloy, *Mater. Sci. Eng. A*, 2006, **438-440**, p 808–811
27. T. Sawaguchi, L.G. Bujoreanu, T. Kikuchi, K. Ogawa, and F. Yin, Effects of Nb and C in Solution and in NbC Form on the Transformation-related Internal Friction of Fe-17 Mn (mass. %) Alloys, *ISIJ Int.*, 2008, **48**(1), p 99–106
28. J.E. Bidaux, R. Schaller, and W. Benoit, Study of the h.c.p-f.c.c. Phase Transition in Cobalt by Acoustic Measurements, *Acta Metall.*, 1989, **37**, p 803–811
29. J. Van Humbeeck, J. Stoiber, L. Delaey, and Rolf. Gotthardt, The High Damping Capacity of Shape Memory Alloys, *Z. Metallkd.*, 1995, **86**, p 176–183
30. A.K. De, N. Cabanas, and B.C. De Cooman, Fcc-hcp Transformation-Related Internal Friction in Fe-Mn Alloys, *Z. Metallkd.*, 2002, **93**, p 228–235
31. J.F. Wan, S.P. Chen, T.Y. Hsu, and Y.N. Huang, Modulus Softening during the $\gamma \rightarrow \varepsilon$ Martensitic Transformation in Fe-25 Mn-6 Si-5 Cr-0.14 N Alloy, *Mater. Sci. Eng. A*, 2006, **438–440**, p 887–890
32. A. Sato, K. Ozaki, Y. Watanabe, and T. Mori, Internal Friction due to $\varepsilon \rightarrow \gamma$ Reverse Transformation in an Fe-Mn-Si-Cr Shape Memory Alloy, *Mater. Sci. Eng. A*, 1988, **101**, p 25–30
33. T. Sawaguchi, T. Kikuchi, F. Yin, and S. Kajiwara, Internal Friction of an Fe-28Mn-6Si-5Cr-0.5NbC Shape Memory Alloy, *Mater. Sci. Eng. A*, 2006, **438–440**, p 796–799
34. H.K.D.H. Bhadeshia, *Bainite in Steels*, 2nd ed., IOM Communications Ltd, London, 2001
35. N. Van Caenegem, L. Duperez, K. Verbeken, D. Segers, and Y. Houbaert, Stress Related to the Shape Memory Effect in Fe-Mn-Si-based Shape Memory Alloys, *Mater. Sci. Eng. A*, 2008, **481–482**, p 183–189

JMEPEG (2009) 18:506–510
DOI: 10.1007/s11665-009-9393-y

1059-9495/$19.00

Optimization of the Ti-16.2Hf-24.8Nb-1Zr Alloy by Cold Working

M. González, J. Peña, J.M. Manero, M. Arciniegas, and F.J. Gil

(Submitted September 15, 2008; in revised form November 21, 2008)

The objective of the present work is to characterize the influence of cold work on the thermoelastic martensitic transformation and on the apparent elastic modulus of the Ni-free Ti-21.6Hf-23.7Nb-1Zr alloy in order to determine the key factor that promotes the desired shape memory properties and/or low apparent elastic modulus. A vacuum arc melted button of each alloy was heat treated at 1100 °C during 1.5 h and quenched with a mixture of ethanol/water at 0 °C. Samples of the alloy were cold rolled from 5% up to 95% and, finally, microstructurally and mechanically characterized. The apparent elastic modulus for each condition as well as the reversibility percentages were evaluated by instrumented nanoindentation using a Berkovich tip and a spherical tip, respectively. A higher proportion of martensite was found in the low cold work percentages compared to the untreated material as it was observed by optical and TEM microscopy. A decrease in the apparent elastic modulus was observed when increasing the cold work percentage. The lowest value was found in the 99% cold work condition with 44 GPa, value closer to that of cortical bone. Cyclic nanoindentation tests show an increase in the reversibility percentages in the cold worked condition compared to the untreated material.

Keywords	low elastic modulus, nanoindentation, Ni-free Ti alloys, shape memory effect

1. Introduction

In the biomaterials field, NiTi alloys have been widely used as implantable materials in the human body due to their shape memory effect, superelasticity, high damping capacity, and corrosion resistance. Their ability to recover a predetermined shape when heated or stressed is due to a reversible solid-state phase transformation from austenite to martensite (Ref 1). It has been recently reported that the Ni ions released produce adverse hypersensitivity reactions, which has stimulated the development of new Ni-free Ti alloys. Due to the lower elastic modulus exhibited by the β-phase of Ti and the possible transformation from β to α'' phase, the research is focused on β-type Ti alloys. The parameters average valence electron number (e/a) and bond order (Bo) values are parameters that relate to β-phase stability and depend on the alloy composition. Saito et al. (Ref 2) reported that the magic numbers to obtain the characteristics of Gum Metal are e/a ~4.24 and Bo = 2.87 eV. However, Hao et al. (Ref 3) reported that the relation e/a ~4.15 corresponds to a low value of elastic shear modulus of the body-centered cubic structure crystal and because of that to a low elastic modulus. This property reduces the shielding effect and enhances bone remodeling (Ref 4). β-Ti alloys offer also good coldworkability which permits the modification of the apparent elastic modulus and mechanical strength through microstructural changes (Ref 3-5).

Fabrication of new Ni-free Ti alloys has to be produced in high-temperature vacuum and controlled atmosphere furnaces due to the affinity for interstitial elements and the high melting point of the alloying elements. This limitation increases the cost of fabrication and reduces the amount of material produced. The lack of material justifies the election of the nanoindentation technique which is appropriated to evaluate the mechanical properties of small size samples. It is also non-destructive and permits the indentation at the martensitic plate level (Ref 6). The three-sided sharp Berkovich tips are useful to determine elastic modulus from the first unloading segment of the load-displacement curve. Spherical tips can also be used to determine elastic modulus by using the Hertz contact theory. The elastoplastic response can be studied by using a spherical tip (Ref 6, 7), which will determine the presence of discontinuities in the load-displacement curve, such as pop-in and the appearance of hysteresis loops between unloading and re-loading curves. Both events are associated to the martensitic transformation under the indenter tip (Ref 8, 9).

Although the contribution of second deformation mechanisms (stress-induced martensitic transformation or martensite reorientation) causes errors in the determination of the true elastic modulus, the measurement of the apparent modulus of elasticity is obviously of engineering importance (Ref 10).

This article is an invited paper selected from presentations at Shape Memory and Superelastic Technologies 2008, held September 21-25, 2008, in Stresa, Italy, and has been expanded from the original presentation.

M. González, J.M. Manero, M. Arciniegas, and **F.J. Gil**, Department of Materials Science and Metallurgy, Universitat Politècnica de Catalunya, Avda. Diagonal 647, Barcelona 08028, Spain; and **J. Peña**, Department of Materials Science and Metallurgy, Universitat Politècnica de Catalunya, Avda. Diagonal 647, Barcelona 08028, Spain and Materials Science, Elisava Escola Superior de Disseny, C/Ample 11-13, Barcelona 08002, Spain. Contact e-mails: Marta.Gonzalez.Colominas@upc.edu, jpenya@elisava.es, Jose.Maria.Manero@upc.edu, Milena.Arcienagas@upc.edu, and Francesc.Xavier.Gil@upc.edu.

The aim of the work is to achieve a material with shape memory properties and/or low elastic modulus to be used in load transfer implantology. For this purpose, the influence of cold work on the thermoelastic martensitic transformation and elastic modulus of the Ti-16.2Hf-24.8Nb-1Zr alloy has been investigated to determine optimal conditions.

2. Experimental

A theoretical method based on molecular orbitals was used to predict the stable phase and design a new Ti alloy with a significant reduction in terms of both time and cost (Ref 11). One pair of $(\overline{Bo}, \overline{Md})$, was selected from the β-zone, in the frontier between the low elastic modulus and the shape memory effect. The chemical composition was determined by using Nb, Hf, and Zr as alloying elements and the result was the Ti-21.6Hf-23.7Nb-1Zr (wt.%) alloy .

A 40 g button of the alloy was made by arc vacuum melting from bars of the starting elements of 99.99% purity and non-consumable W-electrode. A sample of the alloy was encapsulated under vacuum in a quartz tube and homogenized at 1100 °C for 12 h, solution treated at 1100 °C for 1.5 h, and quenched with a mixture of ethanol/water at 0 °C. Samples of the alloy were cold-rolled with a reduction of 5 up to 99% in thickness. Specimens from the quenched and the cold-rolled alloy were mechanically polished and finished with colloidal silica to give a surface roughness with a Ra < 100 nm. The chemical composition was verified by x-ray fluorescence using Philips PW2400 equipment. The samples were etched with Keller reactive (2 mL HF, 3 mL HCl, 5 mL HNO_3, and 190 mL distilled H_2O) for optical microstructural analysis. Electropolishing technique using an electrolyte consisting of 400 mL butoxyethanol, 400 mL methanol, and 100 mL perchloric acid was carried out to prepare transmission electron microscopy (TEM) samples from thin disks of 3 mm diameter.

Nanoindentation tests were conducted using a MTS Nano Indenter XP with continuous stiffness measurement (CSM) module. Apparent elastic modulus and hardness were measured using the Oliver & Pharr method (Ref 6, 12) from loading-unloading tests to a depth of 2000 nm using a Berkovich tip with a measured radius of 750 nm. In addition, cyclic nanoindentation tests to a depth of 300 and 2000 nm were developed using a spherical tip of radius 25 μm to characterize the thermoelastic martensitic transformation. For each test condition, 9 indentations were performed for each sample. The reduced elastic modulus was calculated from spherical nanoindentation by using the Hertz contact theory (Ref 12) in the elastic deformation range through Eq 1.

$$F = \frac{4}{3} E^* R^{1/2} h^{3/2} \quad \text{(Eq 1)}$$

where E^* is the reduced elastic modulus, R is the tip radius, and h the indentation depth.

The apparent elastic modulus was calculated from the reduced modulus by using Eq 2.

$$\frac{1}{E^*} = \frac{1 - \nu_m^2}{E_m} + \frac{1 - \nu_i^2}{E_i} \quad \text{(Eq 2)}$$

where E^* is the reduced modulus, E_m, ν_m and E_i, ν_i are the elastic modulus and Poisson coefficient of the material, and of the indenter material (diamond), respectively (Ref 13).

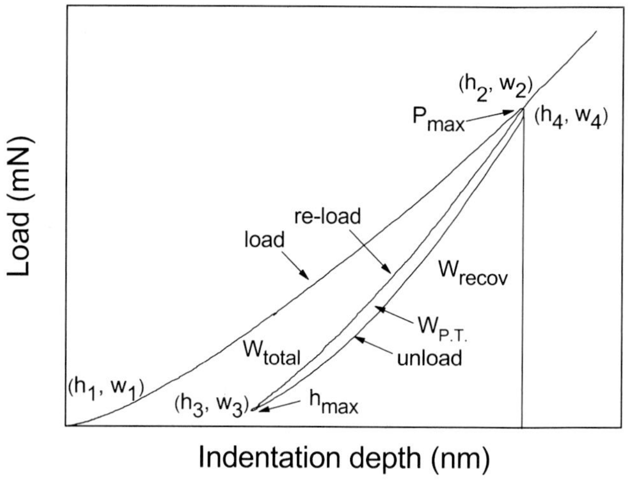

Fig. 1 P-h curve model of a nanoindentation test

The authors examined work hardening behavior by calculating the work hardening ratio, which is the rate of change of hardness between quenched and cold-rolled alloy. Figure 6 shows the change tendency of the work hardening ratio by cold work percentage.

The total recovery work, η_w, was calculated using Eq 3 (Ref 14).

$$\eta_w = \frac{W_{recov}}{W_{total}} \cdot 100 = \frac{\int_{h_3}^{h_4} P \cdot dh}{\int_{h_1}^{h_2} P \cdot dh} \cdot 100 \quad \text{(Eq 3)}$$

which correlates the work necessary to indent the material (activation of the deformation mechanisms), W_{total} $(W_2 - W_1)$, to the total work recovered during the unloading segment, W_{recov} $(W_4 - W_3)$, indicated in Fig. 1. The W values correspond to the load-displacement points where the work has been evaluated. Such points change in each cycle, W_1 of the second cycle is equal to W_3 of the first cycle. The W_3 value corresponds to a 95% unload of the maximum applied load during each cycle in order to avoid the lateral displacement of the indenter.

The calculated recovery work corresponds to elasticity and reversible phase transformation. The work recovery produced by the phase transformation, η_{PT}, has been evaluated through Eq 4.

$$\eta_{TF} = \frac{W_{loop}}{W_{recup}} \cdot 100 = \frac{W_4 - W_2}{W_4 - W_3} \cdot 100 \quad \text{(Eq 4)}$$

where the energy of the hysteresis loop between the unloading and the re-loading curve has been correlated with the recovery total work. The effect of thermal drift was corrected by using Ti-6Al-4V, which does not present phase transformation in the tested conditions.

Both total recovery work, η_w, and the recovery produced by the phase transformation, η_{PT}, have been calculated for the 1st and the 7th load-unload cycle of the cyclic nanoindentation tests to a depth of 2000 nm, corresponding to the material without and with mechanical history, respectively.

3. Results and Discussion

The chemical composition of the new alloy was the Ti-21.6Hf-23.7Nb-1Zr (wt.%) alloy, with an average valence

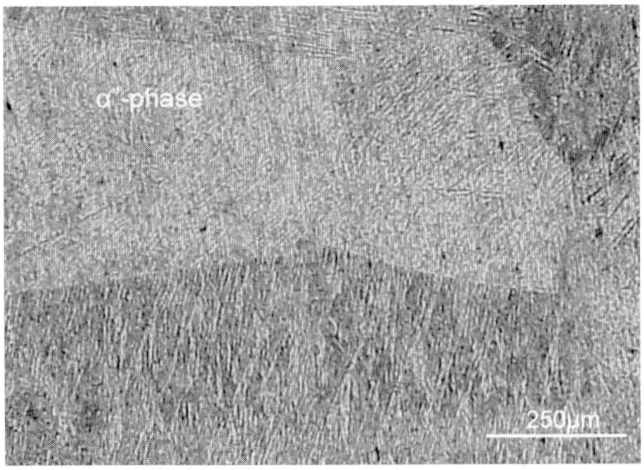

Fig. 2 Optical microscopy images of the quenched alloy with 5% cold work

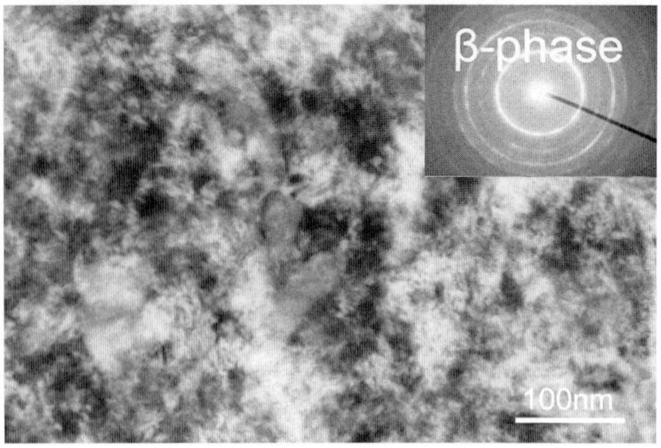

Fig. 4 Bright-field TEM image of the 90% cold worked alloy and electron diffraction pattern of the selected area with continual diffraction rings

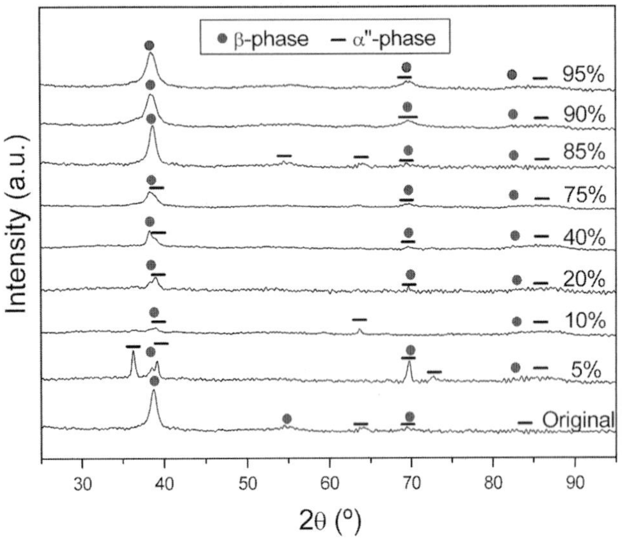

Fig. 3 X-ray diffraction patterns of the quenched alloy and after different cold-roll percentages

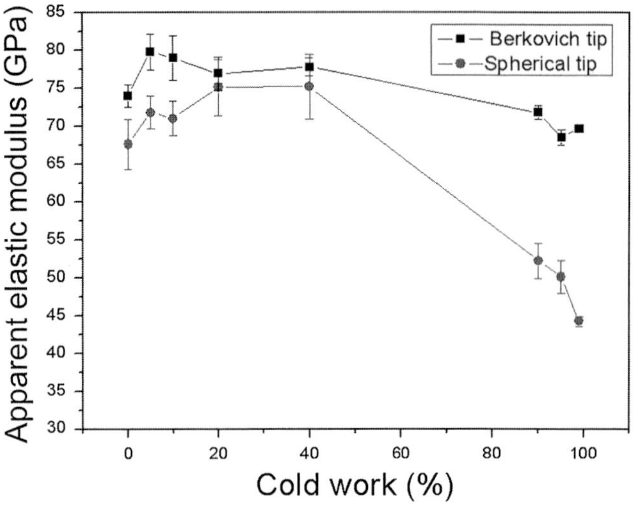

Fig. 5 Apparent elastic modulus vs. cold work percentages calculating by using a Berkovich and a spherical tip

electron number (e/a) of 4.17, near to the value reported by Hao et al. (Ref 5) as indicated to obtain low elastic modulus and a bond order (Bo) of 2.86 eV near to the 2.87 eV of the Gum Metal alloy investigated by Saito et al. (Ref 2).

Optical microscopy images showed a higher proportion of martensitic plates in the samples with 5-40% reduction in thickness, especially in 5%, than in the quenched alloy (Fig. 2). As Hwang et al. reported (Ref 15), as Bo is lower than 2.87, athermal martensite α'' by quenching and stress-induced martensite α'' by cold working, were generated.

Above 75% cold work, the recrystallization process started and it was difficult to observe the microstructure by optical microscopy, due to the formation of a nanocrystalline structure, recently reported in metastable Ti alloys with reduction in thickness above 50% (Ref 5).

X-ray diffraction analysis (Fig. 3) confirmed the presence of β-phase in all the conditions. Stress-induced martensitic transformation was formed at 5% of cold work and started to disappear after the 10% of cold work. The stress level generated by cold working produced the disappearance of the stress-induced martensite until the recrystallization of the austenitic parent phase took place after the 75% of cold work.

Figure 4 shows a bright field image obtained by TEM from the 90% cold-rolled alloy. The picture shows the formation of a nanocrystalline structure with grain size less than 50 nm. Inset in Fig. 3 shows corresponding selected area electron diffraction pattern with continual diffraction rings, which confirms the presence of many crystalline grains. The indexation of the rings corresponds to the reticular parameters of β-phase.

The nanoindentation results associated with the Berkovich tip showed a decrease in elastic modulus with increasing cold rolling reduction, from 74 to 68 GPa, corresponding to the quenched alloy and the 95% cold work condition, respectively (Fig. 5). The results associated with the spherical tip showed a more pronounced decrease in apparent elastic modulus, from 67 to 44 GPa, corresponding to the quenched alloy and after 99% reduction in thickness (Fig. 5). The plastic deformation induced by cold-roll refined coarse grains to tens of nanometers, dislocation activity is reduced and grain boundary mediated deformation becomes dominant. As a result, strengthening in

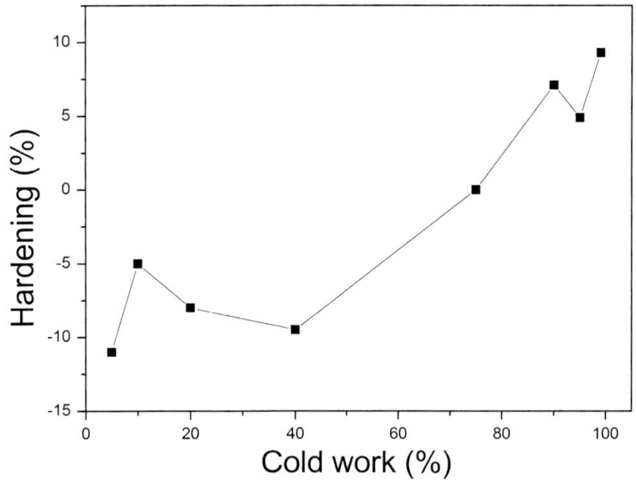

Fig. 6 Work-hardening ratio vs. cold work percentage

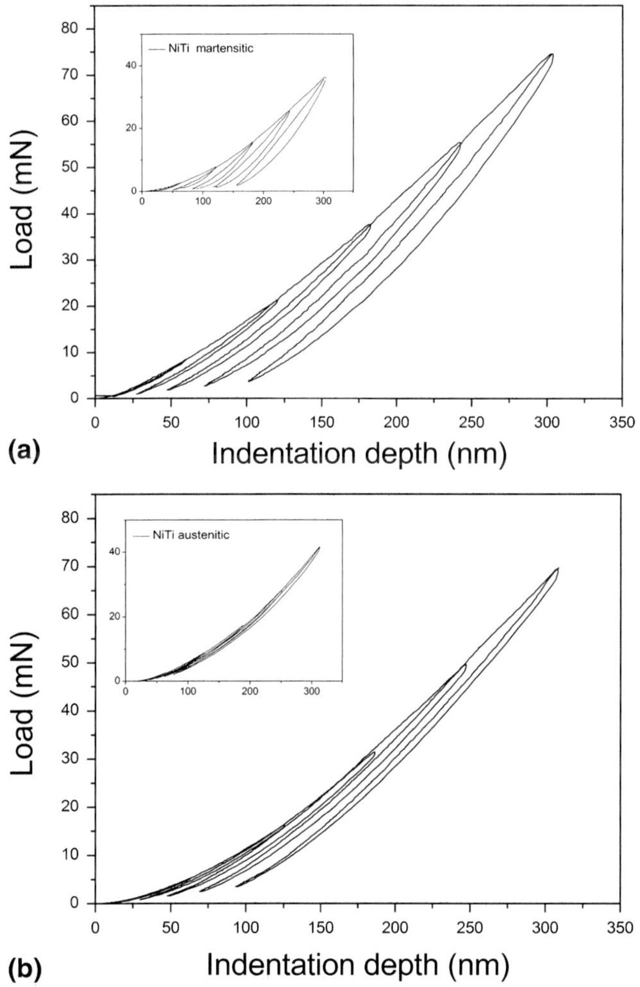

Fig. 7 Nanoindentation load-displacement curves of the alloy with 5% (a) and with 95% reduction in thickness (b) obtained under cyclic increasing-load control conditions

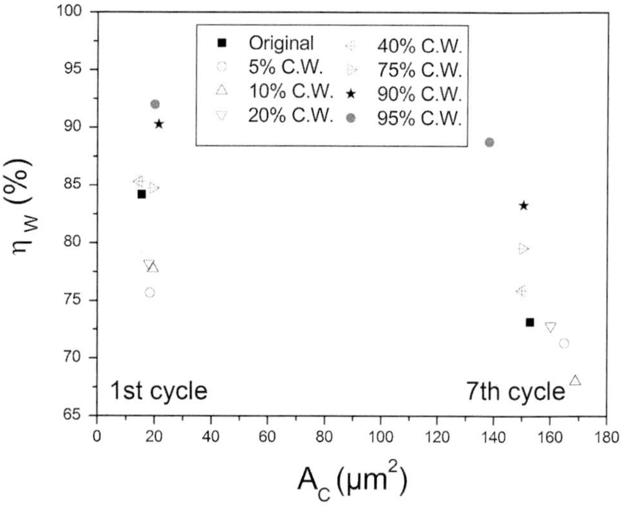

Fig. 8 Total recovery work vs. contact area calculated for the different cold work conditions

accordance with Hall-Petch relation is less significant, the material has high cold workability and apparent elastic modulus decreases, caused by the large fraction of atoms in the grain boundaries having a lower elastic modulus (Ref 16, 17). Apparent elastic modulus values obtained by spherical nanoindentation are lower than the ones associated with the Berkovich tip. The higher contact area obtained with a spherical tip generates lower stress levels and activates different deformation mechanisms around the indenter tip (Ref 14).

Figure 6 shows the changed tendency of the work hardening ratio by cold work percentage, where two different zones were observed. A first softening zone, until the 40% of cold work, due to the stress-induced martensitic transformation. After 40% cold work, the recrystallization process started and the typical β-phase work hardening was observed in the second zone. It is interesting to note that the work hardening ratio for the 95% cold-rolled alloy was as low as 5, which means that the alloy was hard to work-harden even after high reductions in thickness.

Finally, Fig. 7 shows one representative load-displacement curve of the alloy with 5% (Fig. 7a) and 95% reduction in thickness (Fig. 7b) obtained by cyclic nanoindentation tests to a depth of 300 nm. Hysteresis loops generated between the unloading and the re-loading curves are especially observed after the 5% cold work, which are associated to the martensitic transformation. In transforming materials, when load is released, the elastic response and the reverse phase transformation contributes to the recovery causing the observed loops (Ref 8). The elastic response of the 5% cold worked alloy is similar to that of a martensitic NiTi alloy, as is shown in the inset of Fig. 7(a). In the alloy with 95% reduction in thickness, a high deformation recovery is observed in the load-displacement curve similar in shape to that of an austenitic NiTi alloy (inset Fig. 7b) although with inferior recovery values as it is reported for this kind of materials. The lower recovery values demonstrated the minimal effect of the phase transformation effect in the work recovery.

The total recovery work results, η_w, confirmed that the higher values corresponded to the cold worked condition compared to the quenched alloy, especially to the 95% reduction in thickness (Fig. 8). A similar tendency was found in the work hardening ratio and the inverse behavior was observed at 40% of cold work. After the 40% cold work the η_w increased due to the recrystallization process and decreased for

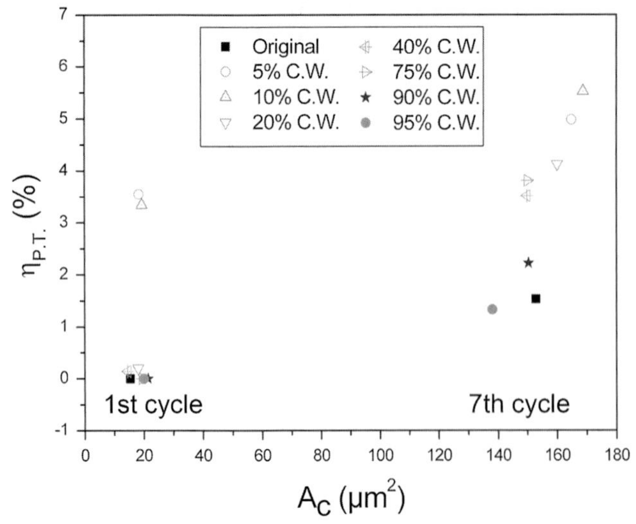

Fig. 9 Recovery produced by the phase transformation vs. contact area calculated for the different cold work conditions

the lower cold work percentages due to the stabilization of the stress-induced martensitic plates.

The results were concordant with η_{PT} values shown in Fig. 9 where η_{PT} were maximum for the lower cold work percentages due to the ability of the material to transform by stress. The η_{PT} values decreased for the higher cold work percentages, confirming the results observed in Fig. 7, showing a practically complete elastic behavior of the material.

4. Conclusions

The following conclusions can be drawn from this work.

As Hwang et al. (Ref 15) reported for alloys with Bo < 2.87 eV, stress-induced martensitic transformation was generated in the Ti-21.6Hf-23.7Nb-1Zr (wt.%) alloy by cold working from 5 up to 40% reduction in thickness and athermal martensite α'' was generated by quenching. The highest proportion of martensitic plates was found in the 5% cold worked alloy. The highest recovery produced by the phase transformation, η_{PT}, corresponded to the alloy with 5% reduction in thickness.

After 75% cold work, the recrystallization process started and a nanocrystalline structure was formed. The plastic deformation mechanism changed from martensitic transformation to grain boundary mediated deformation. As a consequence, the alloy was superplastic permitting cold plastic working to 99% with no work hardening at room temperature. The higher deformation recovery was found after the higher reduction in thickness as well as the higher total recovery work results, η_w.

The influence of cold work in the apparent elastic modulus was investigated and the lowest value was found after the 99% reduction in thickness, with 44 GPa, value close to that of cortical bone. This suggests the viability of employing them in load transfer bone implants in order to reduce the shielding effect and enhance bone remodelling.

Acknowledgments

This work was carried out with the support of the Generalitat de Catalunya Commission for the Universities and Research of the Department of Innovation, Universities and Companies and the European Social Fund.

References

1. T.W. Duering, K.N. Melton, D. Stoeckel, and C.M. Wayman, *Engineering Aspects of Shape Memory Alloys*, Butterworth-Heinemann Ltd., London, 1990
2. T. Saito, T. Furuta, J.H. Hwang, S. Kuramoto, K. Nishino, N. Suzuki, R. Cheng, A. Yamada, K. Ito, Y. Seno, T. Nonaka, H. Ikehata, N. Nagasako, C. Iwamoto, Y. Ikuhara, and T. Sakuma, Multifunctional Alloys Obtained via a Dislocation-Free Plastic Deformation Mechanism, *Science*, 2003, **300**, p 464–467
3. Y.L. Hao, Titanium Alloy with Extra-Low Modulus and Superelasticity and its Producing Method and Processing Thereof, US 2007/0137742 A1, 2007
4. M. Niinomi, Recent Research and Development in Titanium Alloys for Biomedical Applications and Healthcare Goods, *Sci. Technol. Adv. Mater.*, 2003, **4**(5), p 445–454
5. Y.L. Hao, S.J. Li, S.Y. Sun, C.Y. Zheng, and R. Yang, Elastic Deformation Behaviour of Ti-24Nb-4Zr-7.9Sn for Biomedical Applications, *Acta Biomater.*, 2007, **3**(2), p 277–286
6. A.C. Fischer-Cripps, *Nanoindentation*, 2nd ed., Springer, New York, 2004
7. W. Yan, Q. Sun, and H.Y. Liu, Spherical Indentation Hardness of Shape Memory Alloys, *Mater. Sci. Eng. A*, 2006, **425**(1–2), p 278–285
8. C.P. Frick, T.W. Lang, K. Spark, and K. Gall, Stress-Induced Martensitic Transformations and Shape Memory at Nanometer Scales, *Acta Mater.*, 2006, **54**(8), p 2223–2234
9. T. Juliano, Y. Gogotsi, and V. Domnich, Effect of Indentation Unloading Conditions on Phase Transformation Induced Events in Silicon, *J. Mater. Res.*, 2003, **18**(5), p 1192–1201
10. Y. Liu and H. Xiang, Apparent Modulus of Elasticity of Near-Equiatomic NiTi, *J. Alloys Compd.*, 1998, **270**(1–2), p 154–159
11. M. Arciniegas, J.M. Manero, J. Peña, F.J. Gil, and J.A. Planell, Study of New Multifunctional Shape Memory and Low Elastic Modulus Ni-Free Ti Alloys, *Metall. Mater. Trans. A*, 2008, **39**(4), p 742–751
12. G.M. Pharr, Measurement of Mechanical Properties by Ultra-Low Load Indentation, *Mater. Sci. Eng. A*, 1998, **253**(1–2), p 151–159
13. W. Yan, Q. Sun, F. Xi-Qiao, and L. Qian, Analysis of Spherical Indentation of Superelastic Shape Memory Alloys, *Int. J. Solids Struct.*, 2007, **44**(1), p 1–17
14. A.C. Fischer-Cripps, A Review of Analysis Methods for Sub-Micron Indentation Testing, *Vacuum*, 2000, **58**(4), p 569–585
15. J. Hwang, S. Kuramoto, T. Furuta, K. Nishino, and T. Saito, Phase-Stability Dependence of Plastic Deformation Behavior in Ti-Nb-Ta-Zr-O Alloys, *J. Mater. Eng. Perform.*, 2005, **14**(6), p 747–754
16. Y.L. Hao, S.J. Li, Y. Sun, C.Y. Zheng, Q.M. Hu, and R. Yang, Super-Elastic Titanium Alloy with Unstable Plastic Deformation, *Appl. Phys. Lett.*, 2005, p 87
17. J. Schiøtz, F.D. Di Tolla, and K.W. Jacobsen, Softening of Nanocrystalline Metals at Very Small Grain Sizes, *Nature*, 1998, **391**, p 561–563

JMEPEG (2009) 18:511–516
DOI: 10.1007/s11665-009-9357-2

1059-9495/$19.00

Properties and Heat Treatment of High Transition Temperature Ni-Ti-Hf Alloys

C. Craig Wojcik

(Submitted September 11, 2008; in revised form November 21, 2008)

The high transition temperature Ni-Ti (Hf, Zr) alloys have long been of interest for actuators and other applications requiring transition temperatures greater than 100 °C. Unfortunately, the high hardness and poor fabricability of these alloys have prohibited the scale up to commercial production. Some of these alloys are so "hot short" that even modest size ingots cannot be cast without internal cracks formed by solidification shrinkage stresses. Hot rolling methods have recently been demonstrated that can produce crack free Ni-Ti-(6-10 at.%)Hf thin sheets having austenite transition temperatures up to approximately 170 °C. Since these alloys are soft martensite phase at room temperature, they can easily be formed and bent at ambient temperature but cold rolling can only be performed to a limited extent due to high work hardening rates which are typical for Ni-Ti alloys. Progress is now underway to scale up these methods to produce 500-600 mm wide sheets. The effects of composition variations, heat treatment and cold working on transition temperatures are discussed. Microstructural features unique to these ternary alloys and impurity effects are also discussed. The effects of stress on transition temperature have been determined. Austenite transition temperatures, as measured by DSC and bend-free recovery testing, can be controlled within 100-170 °C for these alloys.

Keywords aerospace, biomaterials, intermetallics, refractory metals, titanium

1. Introduction

Binary NiTi alloys have a long history of use as superelastic and shape memory components. The martensite to austenite transformation temperature range possible with these alloys is very useful since it can be tailored from well below room temperature to about 100 °C. Combinations of very controlled Ni/Ti ratio and thermo-mechanical treatments are routinely used to control the exact transformation temperature within just a few degrees. Additionally, these alloys can be made slightly nickel rich to allow precipitation of Ni_4Ti_3 type particles to further enhance strength and shape stability. Unfortunately, these materials cannot be used in applications as a shape memory actuator where operating environments are not cold enough to induce a return back to the martensite phase. An example of this would be in or near an internal combustion or turbine engine compartment. In other potential applications, it may take too long for an actuator to cool to the maximum martensite transformation temperature obtainable with conventional NiTi alloys. This would result in the actuator having too long of a response time. For these reasons, many researchers have tried to develop substantially higher temperature shape memory alloys. The most alluring of these possible alloys have been Ni-Ti-X alloys. The third X component of which can be Hf, Zr, Pd, or Pt. Austenite transformations temperatures of 300-400 °C have been well documented for these alloys yet none have ever approached commercial status (Ref 1-3). Aside from the cost considerations of Hf, Pd, and Pt, all these alloys suffer from very poor hot fabricability, and deficient mechanical properties compared to normal NiTi alloys. The properties of these alloys have often been reported on very small pieces of cast material with little or no hot and cold working to homogenize and refine the grain structure. Conventional NiTi binary alloys must always be hot worked to obtain acceptable shape memory properties. For this reason, it is believed that some data reported on these ternary alloys may be far from optimum data obtainable on a well-refined wrought type structure.

From the group of previously mentioned ternary additions, Hf, Zr, Pt, and Pd, researchers at ATI Wah Chang have always been interested in Ni-Ti-Hf system because it is much more economical than Pt or Pd and it affects the transformation temperature more strongly than Zr for a given increase in alloy substitution. Both Hf and Zr substitute directly for the Ti atoms in the NiTi lattice structure. In 1994, Tuominen (Ref 4) reported on attempts to fabricate Ni-Ti alloys containing 8-16 at.% Hf. Even the relatively modest 10% Hf alloy proved very difficult to fabricate. Even more disappointing was a reported very low percentage of recovered strain, only 0.44% at a test stress of 138 MPa. Later attempts to fabricate Ni-40Ti-10Hf material by high temperature extrusion proved that fabricability was still a major problem that could not be easily resolved. The resultant extrusion from this work (Ref 5) is

This article is an invited paper selected from presentations at Shape Memory and Superelastic Technologies 2008, held September 21-25, 2008, in Stressa, Italy, and has been expanded from the original presentation.

C. Craig Wojcik, ATI Wah Chang, an Allegheny Technologies Company, Albany, OR 97321. Contact e-mail: craig.wojcik@wahchang.com.

shown in Fig. 1. Enthusiasm for that alloy system waned and further work was discontinued. More recently, that alloy system was revisited in 2006. By 2006, additional processing knowledge had been gained from learning how to make higher purity Ni-Ti alloys and learning how to manufacture a much more challenging Ni-40Ti alloy that previously could not be fabricated. That work, which led to a lab scale process capable of making crack free Ni-40Ti-10Hf sheets, formed the basis of this study (Ref 6). Some of the early unsuccessful attempts and later successfully rolled buttons are shown in Fig. 2.

Fig. 1 Ni-40Ti-10Hf Extrusion attempt, unpublished research (Ref 5) at ATI Wah Chang

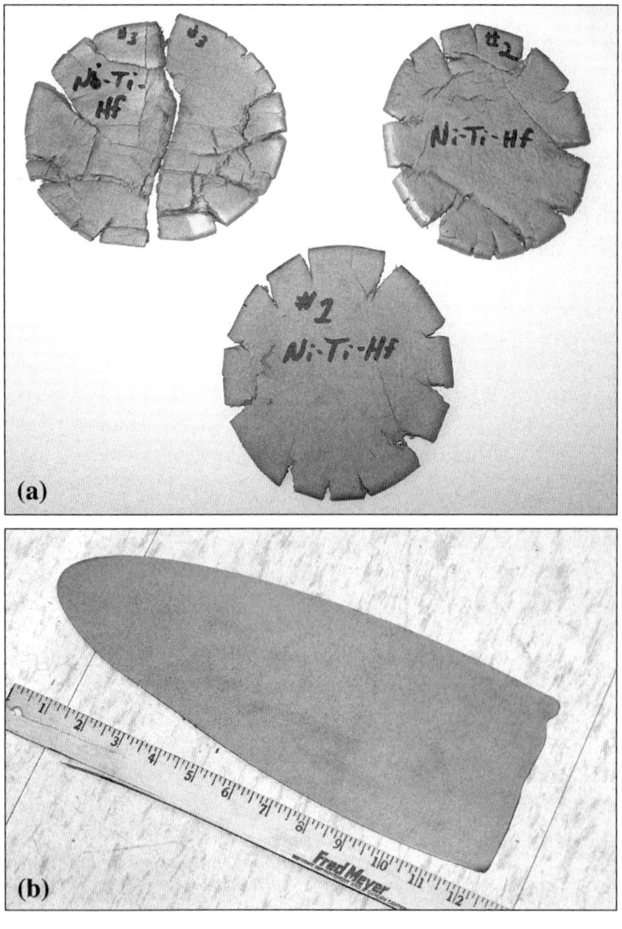

Fig. 2 Preliminary attempts to hot roll Ni-40Ti-10Hf arc melted buttons encased in steel cans (a), and successfully hot-rolled button 0.6 mm thick (b) (Ref 6)

2. Material Preparation

A total of five alloys were made into thin test sheets to investigate the effects of oxygen content and hafnium concentration. All alloys were melted into buttons approximately 10 mm thick × 70 mm in diameter using the nonconsumable vacuum arc melting method with a water-cooled copper mold. Each alloy was melted at least five times to obtain sufficient mixing of the alloy ingredients. The raw materials used included high-purity electrolytic nickel plate, iodide crystal titanium, and commercial purity hafnium sheet. Initially, it was expected that these high-purity ingredients would be essential for obtaining sufficient hot workability. After hot rolling success was achieved, then commercial purity Grade I titanium sheet was used for the lower purity buttons. The faces of the buttons were machined to remove any laps and then welded in steel cans for hot rolling. The cans provided a greater thermal mass to keep the buttons uniformly hot during rolling and allow hot rolling to a thinner cross section.

After rolling, the sheets were grit blasted, and then chemically cleaned in a mixture of 50% HNO_3 and 25% HF, balance water. Samples for aging heat treatment were cut to ~1 cm wide × 8 cm long. Samples for constant load dilatometry (CLD) were sheared with a shaped punch and die directly into the tensile testing configuration with a gage section measuring 6.35 mm wide × 76.2 mm long. One of the hot-rolled sheets is shown in Fig. 3 with one of the CLD test specimens. The tensile axis of these samples corresponded to the rolling direction. Prior to any heat treatment, the samples were wrapped in titanium foil and sealed in helium backfilled quartz capsules. All heat-treated samples were cooled immediately by water quenching the capsules. The 1 × 8 cm sample strips were used for bend-free recovery (BFR) testing, metallographic samples, and differential scanning calorimetry (DSC). Austenite start and finish temperatures were measured with the BFR method using a bending mandrel that produced a surface strain of 3% after cooling in liquid nitrogen. The 5 mm diameter samples for DSC testing were punched directly out of the heat-treated sheets and then pickled in acid to remove 20-40 μm of surface metal. A scan rate of 10 °C per min was used for all DSC tests.

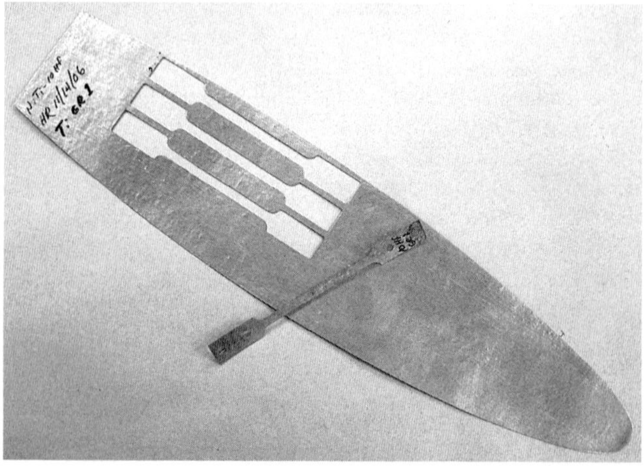

Fig. 3 Hot-rolled Ni-40Ti-10Hf sheet 0.6 mm thick showing CLD transformation test samples punched out at room temperature. Overall sample length is 16 × 1.7 cm wide at grip section

Austenite and martensite transition temperatures measured with the CLD method were determined while under a constant tensile stress of 138 and 276 MPa (20 and 40 ksi). In this test, the specimen was initially heated above the austenite finish temperature (250-300 °C) and then a static tensile load was applied with weights. A metal pipe wrapped with heat tape surrounded the specimen and full length of the grips to maintain temperature uniformity. The temperature was lowered past the martensite finish temperature and then raised well above the austenite finish temperature while the strain was continuously measured. The CLD tests were conducted as a series of three consecutive tests on the same sample to see how reproducibly the strain was recovered after a second cycle at 138 MPa. Then on the third cycle, the stress was increased to 276 MPa to see how much the transformation temperatures, and strain would increase at a higher stress.

3. Material Evaluation and Discussion

Chemical analysis of the alloys was obtained on samples of the sheet after hot rolling and acid cleaning. These results summarized in Table 1 show good agreement between the target and measured compositions. Buttons #1 and #2 had the lowest oxygen content compared with the other three alloys that were made with lower purity Grade 1 titanium sheet.

Microstructural samples were examined after hot rolling, quenching, and aging heat treatments. Using light microscopy, the effects of heat treatment could not be seen in the structure. These structures are shown in Fig. 4. The martensite platelets and second-phase particles are clearly visible in all the alloys. A greater concentration of second-phase particles was seen in the alloys with higher oxygen concentration. The alloy with only 6% Hf had a noticeably finer martensite structure. That alloy also had the highest level of carbon and oxygen impurities.

Second-phase particles in conventional binary Ni-Ti alloys made by arc melting contain a complex Ti_4-Ni_2-O_x lower oxide phase (Ref 7) so it is assumed that a similar oxygen rich phase is present in these alloys however, it may be even more stabilized by the hafnium content. These particles must be substantially harder than the matrix at rolling temperatures since they do not exhibit the long aspect ratio expected from the degree of rolling deformation performed on the buttons. We attempted to perform a quantitative SEM analysis on that phase; however, the small size of the particles limited our precision. Images of these particles in the backscattered SEM detection mode are shown in Fig. 5, along with the spectrum obtained in Fig. 6. There is a slight indication that the particles have a higher concentration of titanium and a lower concentration of hafnium and nickel than the matrix.

After hot rolling and cleaning, the samples were heat treated by either a high temperature anneal or low temperature aging treatment. The annealing heat treatments of 900 and 950 °C were carried out for a time of 60 min. Samples of the alloys were aged at 500-600 °C for several time periods up to 120 h. These samples were then tested with the DSC to see what effects aging had on the austenite and martensite transition temperatures. These results for the 10% hafnium alloy after aging 120 h are shown in Fig. 7 and after 72 h for the 6%

Table 1 Analysis of Ni-Ti-Hf alloy sheets hot rolled to 0.6 mm thick

Alloy element	$Ni_{50}Ti_{40}Hf_{10}$ (44.2Ni-28.9Ti-26.9Hf wt.%), Button #1	$Ni_{50}Ti_{40}Hf_{10}$ (44.2Ni-28.9Ti-26.9Hf wt.%), Button #2	$Ni_{50}Ti_{40}Hf_{10}$ (44.2Ni-28.9Ti-26.9Hf wt.%), Button #3	$Ni_{50}Ti_{42}Hf_8$ (46.0Ni-31.6Ti-22.4Hf wt.%), Button #4	$Ni_{50}Ti_{44}Hf_6$ (48.0Ni-34.5Ti-17.5Hf wt.%), Button #5
Hafnium	26.70 wt.%	26.92	26.86	21.80	17.0
Nickel	44.27 wt.%	44.21	44.12	46.70	49.77
Oxygen	110 wppm	150	310	240	410
Nitrogen	<20 wppm	<20	25	20	31
Hydrogen	50 wppm	24	22	20	23
Carbon	45 wppm	32	43	50	76

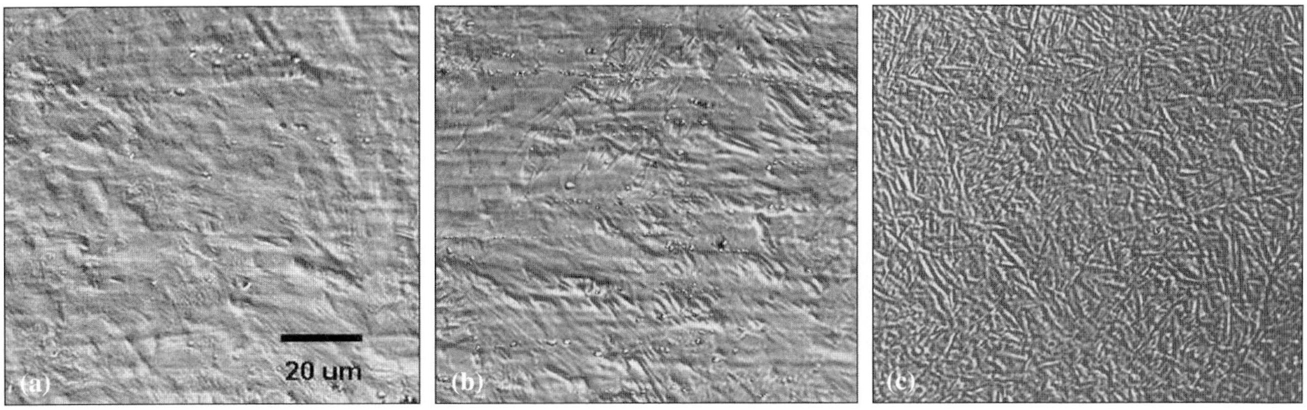

Fig. 4 Microstructure of alloy sheets after hot rolling and quenching. (a) 10% Hf 150 ppm oxygen, (b) 10% Hf 310 ppm oxygen, and (c) 6% Hf 410 ppm oxygen, longitudinal sections

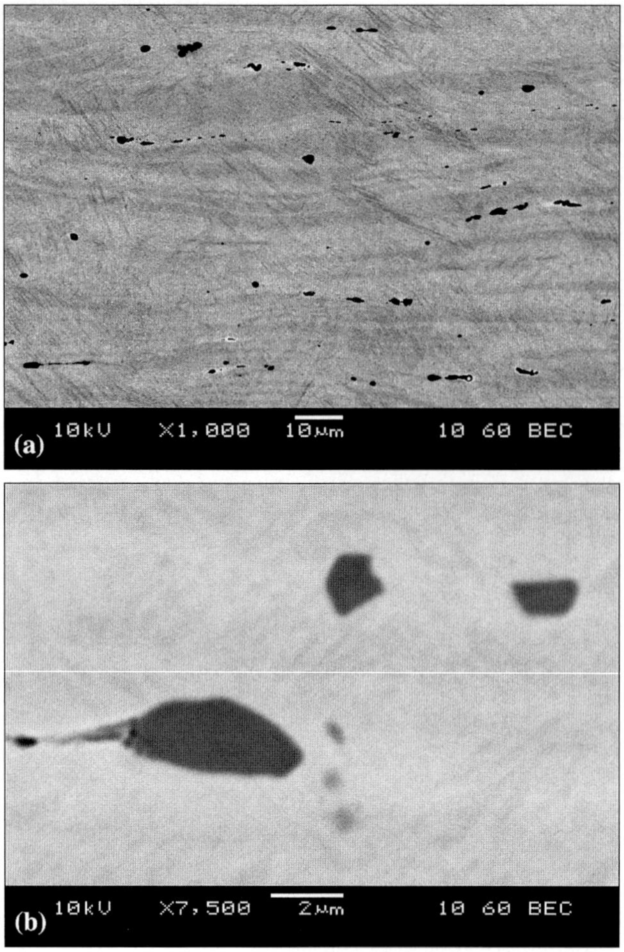

Fig. 5 Backscattered images of second-phase particles observed in the 10% Hf alloy sheet containing 310 ppm oxygen, longitudinal section

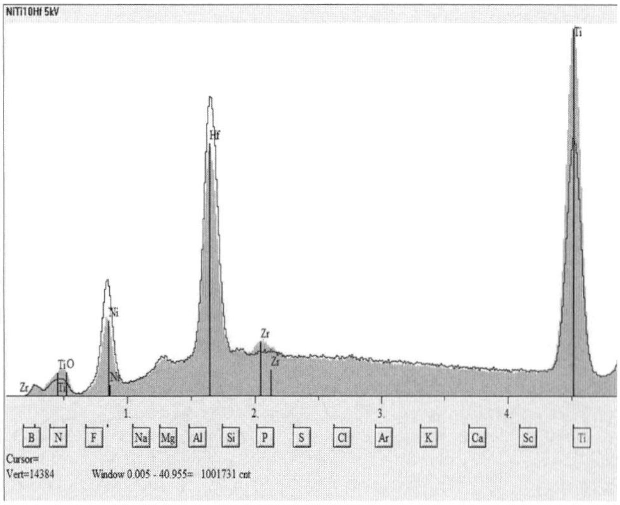

Fig. 6 SEM qualitative composition analyses of second-phase and matrix phase. Line is matrix, gray-shaded area is second-phase particle

hafnium alloy in Fig. 8. Although we observed some minor temperature shifts and reshaping of phase transition peaks, there was no major shift in the transition temperatures. This was

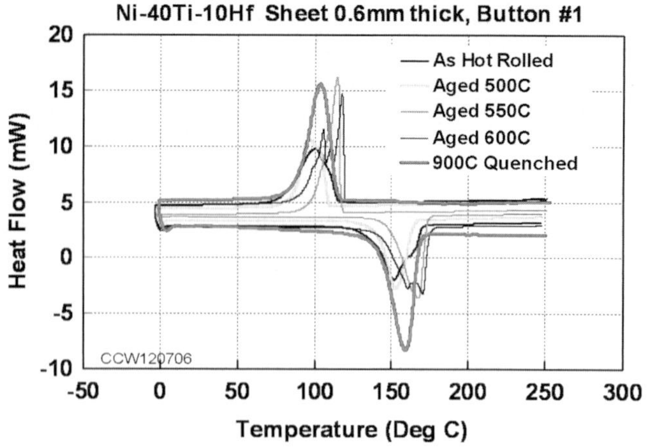

Fig. 7 DSC test results after various aging and quenching heat treatments of the 10% Hf alloy

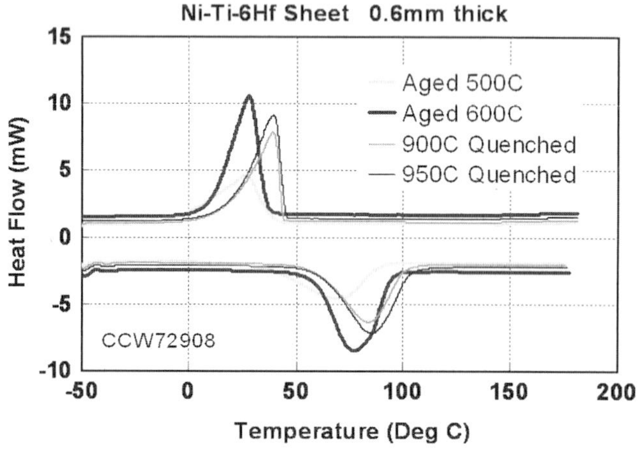

Fig. 8 DSC test results after various aging and quenching heat treatments of the 6% Hf alloy

a good indication that these alloys did not contain a significant excess of nickel. When excess (i.e., supersaturated) nickel is present, it typically will precipitate out as Ni_4Ti_3 type lenticular-shaped precipitates during aging (Ref 8-10). This lowers the nickel concentration in the matrix which results in increases in the transition temperatures after aging. Since aging heat treatments appeared to have only a minor affect on these alloys an anneal/quench heat treatment of 60 min at 900 °C was used on all the subsequent test samples.

The effects of oxygen on the 10% Hf alloy are shown in Fig. 9 with all the alloys in the quenched heat treatment. It should be noted that the martensite peak for the alloy containing 150 ppm oxygen has an initial spike that should not be confused with an R phase peak. This is a testing artifact that is usually attributable to poor thermal coupling to the DSC test pan. The transition temperatures are shown for the various oxygen levels after being aged for 120 h at 650 °C in Table 2. A temperature decrease of about 20° was noted for the 310 ppm oxygen level with both heat treatments. Since it is not precisely known what the composition ratio of Ni to Ti and Hf is in the second-phase particles, this shift in transition temperature could be simply a result of oxygen rich second-phase particles shifting the metal alloy ratio rather than an intrinsic effect of lower oxygen dissolved in the matrix phase.

The effect of hafnium concentration on transition temperatures is shown in Fig. 10 for buttons 2, 4, and 5. All these alloys display sharp and symmetrical transition peaks with approximately 60 °C hysteresis between the austenite and martensite peaks.

Results of CLD testing the 6% Hf alloy are shown in Fig. 11. This alloy displays very consistent recoverable strain for both the first and second cycle at 138 MPa. At 276 MPa, the total strain increased only about 1% which indicates a high rate of strain hardening in the martensite. Approximately 5.5 of the total 6% strain was recovered in that alloy when heated back to the austenite phase. It is expected that after more cycles at 276 MPa the total strain would decrease and the percentage of strain recovery would increase. Both the martensite and austenite transformation temperatures increase with stress, approximately 20 °C for 276 MPa compared to 138 MPa.

CLD test results for the 10% Hf alloy with low (150 ppm) and high (310 ppm) oxygen are shown in Fig. 12 and 13. These

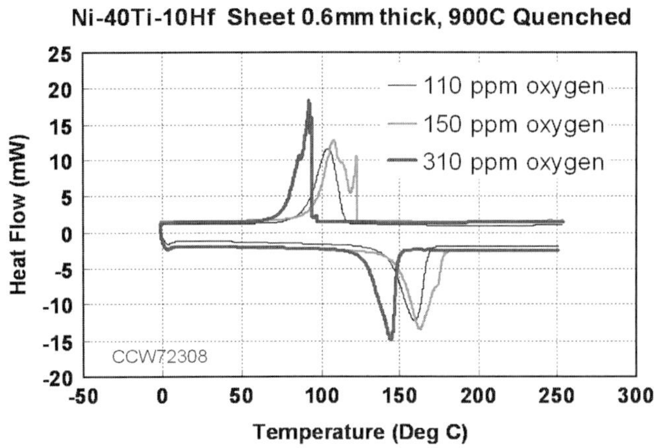

Fig. 9 DSC test results for various levels of oxygen in 10% Hf alloy sheet

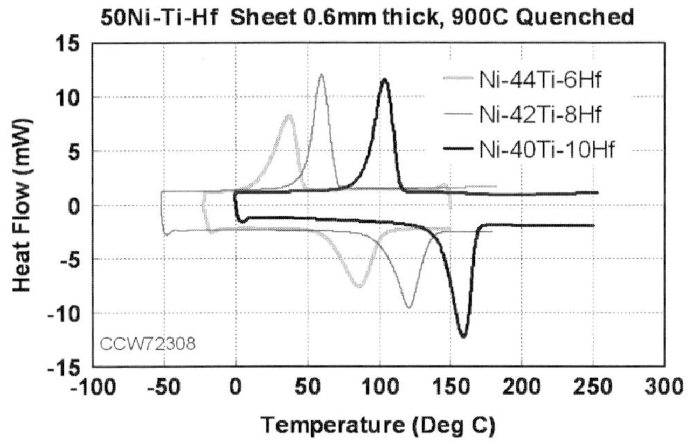

Fig. 10 DSC scans for various concentrations of hafnium

Table 2 Transformation temperatures as measured by bend-free recovery method for various oxygen levels, quenched, and aged heat treatments

$Ni_{50}Ti_{40}Hf_{10}$ bend-free recovery austenite transition temperature				
	A_s, °C		A_f, °C	
Oxygen concentration, wppm	He quenched 900 °C, 60 min	Aged 600 °C, 120 h	He quenched 900 °C, 60 min	Aged 600 °C, 120 h
110	188	184	196	195
150	188	178	197	193
310	164	157	174	169

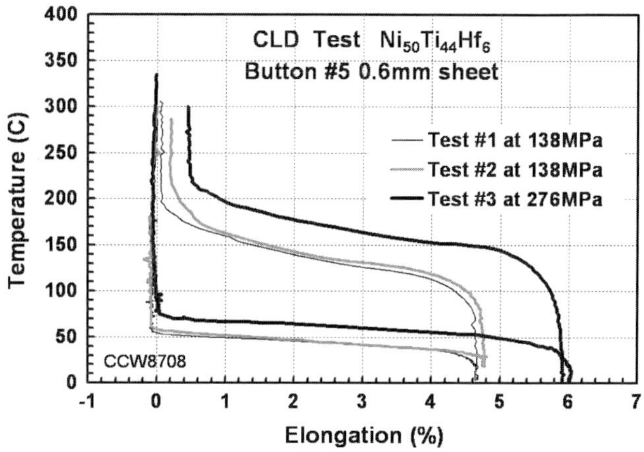

Fig. 11 CLD test results for 6% Hf alloy sheet

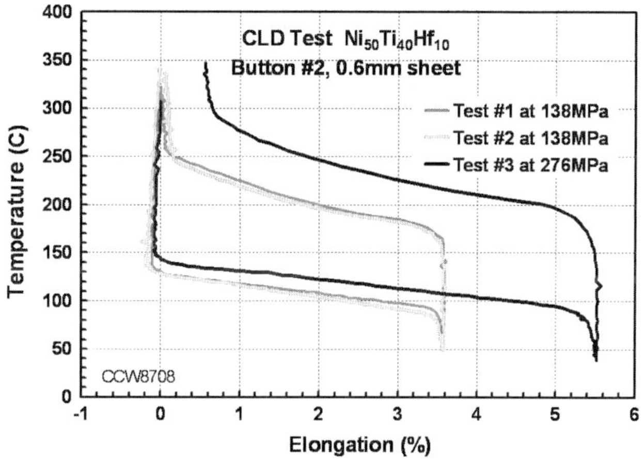

Fig. 12 CLD test results for 10% Hf, 150 ppm oxygen sheet

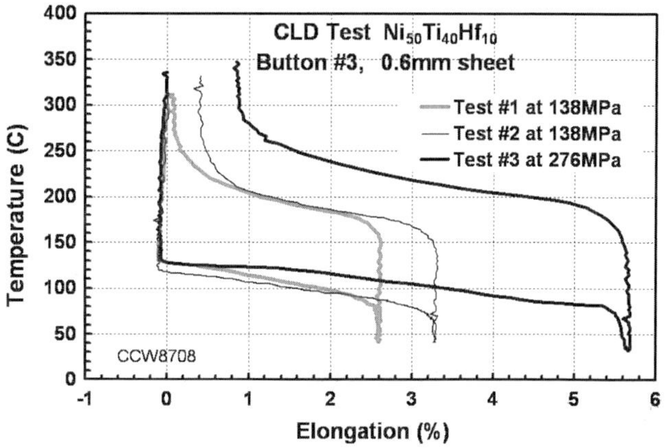

Fig. 13 CLD test results for 10% Hf, 310 ppm oxygen sheet

Table 3 Summary of CLD test transition temperatures and recovered energy

Alloy button	Summary of CLD transition temperatures					Recovered energy, J/cm^3
	Stress, MPa	A_s, °C	A_f, °C	M_s, °C	M_f, °C	
Ni-44Ti-6Hf, Button #5	138 test #1	95	175	58	33	6.2
	138 test #2	97	172	53	36	6.2
	276 test #3	124	199	74	47	15.2
Ni-40Ti-10Hf, Button #2	138 test #1	159	250	129	91	4.8
	138 test #2	158	239	130	86	4.8
	276 test #3	174	275	141	91	13.8
Ni-40Ti-10Hf, Button #3	138 test #1	169	224	134	87	3.5
	138 test #2	161	216	121	77	4.1
	276 test #3	175	256	140	74	13.5

alloys show similar results to the 6% Hf alloy; however, the transitions temperatures are all shifted higher and the total strains are lower as would be expected for a martensite phase with a higher yield strength. The first cycle at 138 MPa on the specimen from button #3 had an unusually low strain. This may be a result of bending damage during fixture assembly. At room temperature, the 0.6-mm-thick martensite samples can very easily be bent during handling.

All of the CLD data are summarized in Table 3. Additionally a calculation was made to determine how much recoverable energy was observed for each test condition. At lower stress, 138 MPa, the 6% Hf alloy had the highest energy recovery due to the increased strain compared to the 10% Hf alloys. The highest recovered energy for all alloys was at the highest stress, 276 MPa, which produced more recoverable strain. The maximum energy value of 15.2 J/cm^3 for the 6% Hf alloy at 276 MPa is a very high value for any shape memory metal. It compares favorably even with cold worked conventional NiTi alloy sheet as measured in the same CLD test in a previous study (Ref 8), and cold worked and aged commercially available SM495 wire (Ref 11).

4. Conclusions

In summary, we have shown that the Ni-Ti-Hf system can be processed to yield useful shape memory properties with high recovery strains and high recovery energy. Hafnium additions up to 10% can be used to enhance the austenite transition temperature almost 100 °C higher than conventional Ni-Ti alloys. The effects of changes in oxygen level and hafnium content have been further documented at zero stress and two levels of constant tensile stress. Additional testing needs to be completed to determine the shape memory properties after many cycles, as well as the superelastic properties of these alloys which should occur at elevated temperatures of 100 °C or more. Ongoing work to scale up the processing size capability for this family of alloys however will determine if these compositions can become a commercial success.

References

1. D.N. AbuJudom, P.E. Thoma, Ming-Yuan Kao, and D.R. Angst, U.S. Patent 5,114,504
2. P.G. Lindquist and C.M. Wayman, Shape Memory and Transformation Behavior of Martensitic Ti-Pd-Ni and Ti-Pt-Ni Alloys, in *Engineering Aspects of Shape Memory Alloys*, T.W. Durig, K.N. Melton, D. Stockel, and C.M. Wayman, Eds., Butterworth-Heinemann, Boston, 1990, p 58–68
3. D. Goldberg, Y. Xu, Y. Murakami, S. Morito, K. Otsuka, T. Ueki, and H. Horikawa, Improvement of a $Ti_{50}Pd_{30}Ni_{20}$ High Temperature Shape Memory Alloy by Thermomechanical Treatments, *Scripta Metall. Mater.*, 1994, **30**, p 1349–1354
4. S.M. Tuominen, High Temperature Ni-Ti-Hf Alloys, *SMST-1994*, A.R. Pelton, D. Hodgson, and T. Durig, Eds., MIAS, Monterey, 1995, p 49–54
5. S.M. Tuominen, Unpublished Research, ATI Wah Chang, 1994
6. C.C. Wojcik, Unpublished Research, ATI Wah Chang, 2006
7. *Standard Specification for Wrought Nickel-Titanium Shape Memory Alloys for Medical Devices and Surgical Implants*, ASTM Standard F 2063-05, ASTM International, West Conshohocken, 1995
8. C.C. Wojcik, Aging Effects in Nickel Rich Ni-Ti Alloys, *SMST-2004*, M. Mertmann, Ed., ASM International, Materials Park, 2006, p 229–236
9. R.J. Wasilewski, S.R. Butler, J.E. Hanlon, and D. Worden, *Met. Trans.*, 1971, **2**, p 229–238
10. M. Nishida, C.M. Wayman, and T. Honma, *Met. Trans.*, 1986, **17A**, p 1505–1515
11. R. Noebe, D. Gaydosh, S. Padula II, A. Garg, T. Biles, and M. Nathal, Properties and Potential of Two (Ni,Pt)Ti Alloys for use as High-Temperature Actuator Materials, *Smart Structures and Materials 2005, Proceedings of SPIE 5761*, W.D. Armstrong, Ed., International Society for Optical Engineering, Bellingham, 2005, p 364–375

JMEPEG (2009) 18:517–521
DOI: 10.1007/s11665-009-9382-1

1059-9495/$19.00

Mechanical Analysis of Hybrid Textile Composites with NiTi Wires

Elena Villa, Sergio Arnaboldi, Ausonio Tuissi, Marco Giacomelli, and Elena Turco

(Submitted September 15, 2008; in revised form December 10, 2008)

Hybrid textile structures composed by polyamide (PA), Lycra (or Elastane EL), and NiTi thin wires were manufactured. The fabrics were realized by knitting Lycra (EL) as weft and warp filaments and a coupled PA/NiTi yarn through in-laying technique. Superelastic NiTi wire with diameter of 50 μm in both straight annealed and snake-like form and shape memory with a snake-like memorized form were used. The polyamide filament used coupled with the NiTi wire has a diameter around 140 μm, while the textile structure is composed by various Lycra filaments, characterized by different diameters. The textile structures were realized by a circular machine appropriately equipped and used in order to minimize problems related to the use of the thin NiTi alloy filament. To study the influence of NiTi filaments on mechanical properties, specimens taken from textile fabric were analyzed by using a DMA Q800 TA Instruments, equipped with a tension film clamp, in static and dynamic conditions. Force/strain measurements up to 150% in elongation and Tandelta versus frequency were carried out on fabrics with and without NiTi filaments. Finally, some tests strain recovery under load versus temperature are conducted.

Keywords DMA analysis, hybrid fabric, NiTi thin wires

1. Introduction

The opportunity to use the NiTi shape memory or pseudo-elastic wires in textile field is today a real interesting application which opens different alternative for new devices (Ref 1).

Smart fabrics with new particular damping, elastic properties, or shape memory effect have chances to be applied in different fields (Ref 2, 3) (biomedical, sport, or other). For example, the articulation rehabilitation after a physical shock, i.e. elbow dislocation, has become of quite high importance in medical and health care area, since it is often particularly difficult to allow the various connection parts in the arm, and the body generally speaking, to return to the initial functionality conditions and to completely repair the occurred damage. Grado Zero Espace has achieved the idea of creating an orthopedic textile support which should not take the place of the classic steel or plastic tutor, but which should become an intermediate step between the rigid tutoring structures and the complete removal of the same, in order to provide a more efficient rehabilitation therapy and to avoid the body part to displace again. Starting from its previous experience in working with NiTi wires, Grado Zero went forward in the development of shape setting, processing, and utilization of the thinnest superelastic or shape memory wires in complex fabric structures which can include the high added value of this innovative material. Physical and mechanical properties of the NiTi wires allow the construction of hybrid semi-rigid textile structures characterized by lightness and complete bio-compatibility. The characterization of mechanical properties is then the most important study to clarify the effective role of NiTi wires in the whole fabric, and to measure the force developed and the strain raised. To study the influence of NiTi filaments on mechanical properties, specimens taken from different textile fabrics were analyzed by using a DMA equipped with a tension film clamp. We used several kinds of mechanical tests to investigate static and dynamic properties of the new material. This system is able to characterize article sample limited in size, but it is possible to obtain stable test conditions in terms of temperature and applied load. These are necessary conditions to obtain reliable data.

The textile samples tested in this work are hybrid structures polymer-NiTi thin wire with different design.

2. Materials and Characterizations

2.1 Materials

The new smart materials are hybrid textile structures composed by polyamide (PA), Lycra (or Elastane EL), and three kinds of NiTi thin wires. The fabrics were realized using an industrial circular knitting machine by knitting EL as weft and warp filaments and inserting a coupled PA/NiTi yarn through in-laying technique. The polyamide filament used coupled with the NiTi wire has a diameter around 140 μm, while the textile structure is composed by various Lycra (EL)

This article is an invited paper selected from presentations at Shape Memory and Superelastic Technologies 2008, held September 21-25, 2008, in Stresa, Italy, and has been expanded from the original presentation.

Elena Villa, Sergio Arnaboldi, and **Ausonio Tuissi**, CNR IENI, Lecco, Italy; and **Marco Giacomelli** and **Elena Turco**, Grado Zero Espace srl, Empoli, Italy. Contact e-mail: e.villa@ieni.cnr.it.

filaments, characterized by different diameters. The employed machinery had been properly equipped, set up, and used in order to minimize problems related to the use of Ni-Ti filaments: these wires have mechanical and electrical characteristics which are completely different from those usually utilized for textile applications, and that might create processing problems and/or damages to various machine parts. We studied three kinds of fabric:

(1) *Fabric A*: manufactured by using one PolyAmide yarn (with a diameter around 140 μm), two different Lycra (EL) yarns (with diameters around 65 and 130 μm), and a superelastic commercial Ni-Ti wire (with a diameter of 50 μm, which has been properly thermo-mechanically treated by Grado Zero Espace through a treatment process which is currently under patenting) in order to obtain a snake-like (with serpentine shape) set filament.

(2) *Fabric B*: manufactured by using one PolyAmide yarn (with a diameter around 140 μm), two different Lycra yarns (with diameters around 65 and 130 μm), and a superelastic commercial straight annealed Ni-Ti wire (with a diameter of 50 μm).

  Figure 3(a) reports the scheme of these two fabrics, where the knitting structure which has been manufactured consists in the insertion, through in-laying technique, of the Ni-Ti wire coupled with the PolyAmide yarn, closing subsequent turns composed by Lycra (EL) yarns (inserted both in weft and warp directions).

(3) *Fabric C*: for a useful comparison we tested also the fabric without NiTi wires, named in the following like Fabric C.

(4) *Fabric D*: manufactured by using one PolyAmide yarn (with a diameter around 140 μm), one Lycra yarn (with diameter around 65 μm), and a shape memory commercial Ni-Ti wire (with a $T_{Af} \approx 87$ °C and a diameter of 50 μm), which has been properly thermo-mechanically treated (by Grado Zero Espace through a treatment process which is currently under patenting) in order to obtain a snake-like re-memorized filament.

  In this case, the knitting structure (reported in Fig. 3b) which has been manufactured is different from the previous ones and consists, as shown in the scheme, in the insertion of the Ni-Ti wire, coupled with a Lycra (EL)/PA yarn, every two turns.

2.2 Characterization

The Dynamic Mechanical Analyzer Q800 TA Instruments, with a GCA cooling accessory, is a suitable equipment to study all the mechanical properties of thin NiTi or NiTi-based thin wires or ribbons (Ref 4). It is possible to use many test configurations, in static or dynamic experimental mode. Force/strain measurements were carried out on fabrics with and without NiTi filaments, up to 150% in elongation, and mechanical degradation is studied by cycling test. Also, damping properties at room temperature are obtained by isothermal frequency scanning measurements. Moreover, the fabric prepared with NiTi shape memory thin wire were tested in strain recovery configurations. For a suitable comparison, previous tests on the single NiTi superelastic and shape memory wire in snake-like form have been carried out.

3. Results and Discussion

At first some tests on a single snake-form NiTi wire have been realized. For this kind of sample it is necessary, for a correct comparison among the different measurements, to use the same number of loops for the different samples, according to the displacement range of the system (0-25 mm).

For both the NiTi wires (superelastic and shape memory), the snake form is impressed after a first straight annealing process.

We measured the NiTi superelastic in stress strain configuration at 35 and 60 °C in order to assess the temperature effect on the mechanical behavior. Moreover, the aim of these new tests is to study the mechanical behavior of the material from the initial serpentine shape to a linear shape, before reaching the superelastic plateau. Figure 1 shows the comparison of the two curves.

On the shape memory NiTi snake, we made preliminary tests to find information about the recovery capacity of the wire under different loads, from martensite to austenite field. In Fig. 2 it is possible to observe that the wire is able to recover completely the snake-form re-memorized shape only under very low forces, due to the low stress developed by the wire in the bending configuration during heating and also due to the softening of the material after two thermal treatments.

Based on these first results we tested the hybrid fabrics with the structure showed in Fig. 3(a) and (b). For all the specimens, small samples (nominally 20 mm in length and 6 mm in width) have been taken from the fabric. In every article test it is possible to find 5 NiTi wires and the results have been normalized to a single NiTi wire.

The fabric A was composed by PolyAmide, Lycra (EL), and a superelastic NiTi filament, 50 μm of diameter, in the snake-like form. Static tests, static force versus strain to 150% in strain, showed an hysteretic behavior and we can observe an important (about 20%) residual strain after the first cycle which is reduced to zero in the next cycles, showing a good stabilization of the recovery. The curves of this sample are reported in Fig. 4.

To investigate also the damping properties of this new hybrid materials we carried out dynamic tests, in particular we

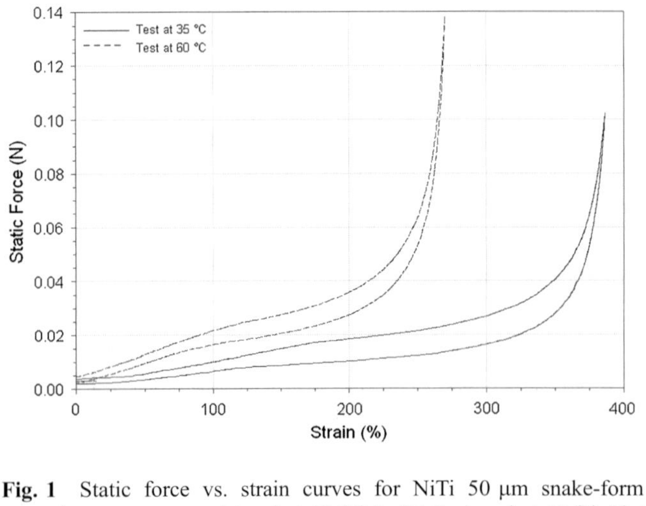

Fig. 1 Static force vs. strain curves for NiTi 50 μm snake-form wire; the curves are registered at 35 °C (solid line) and at 60 °C (dot line)

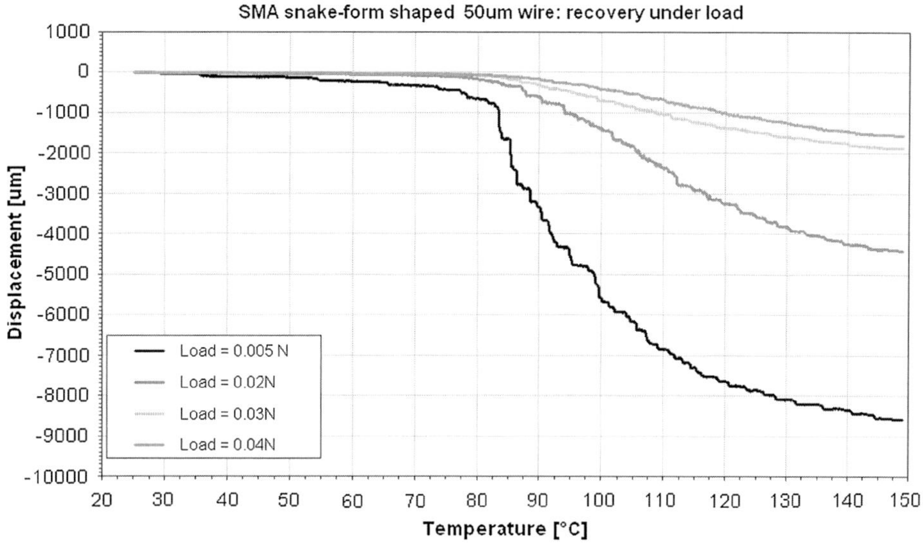

Fig. 2 Recovery under load vs. temperature (°C) of NiTi SMA snake form 50 μm wire

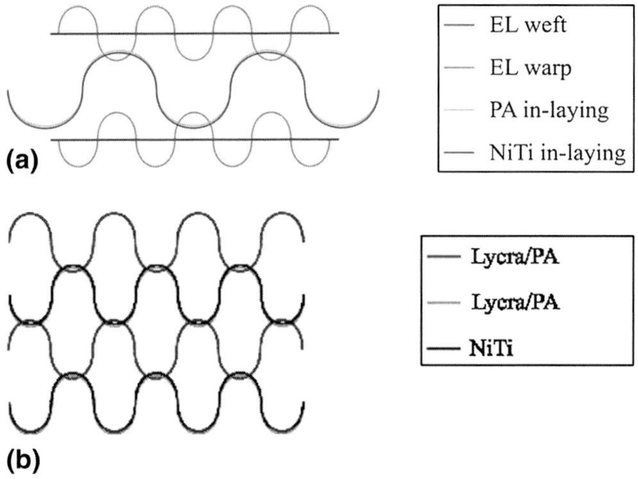

Fig. 3 (a) Fabrics structure for fabric A (PA + EL + NiTi superelastic snake-form wires) and fabric B (PA + EL + NiTi superelastic straight annealed wires). The height of the NiTi loops is 2 mm. (b) Fabrics structure of fabric D (PA + EL + Shape Memory Ni-Ti wires). The height of the NiTi loops is 2 mm

measured the internal friction coefficient, Tandelta, which gives a measure of the mechanical energy dissipation properties, in isothermal condition at different frequencies, ranging from 1 to 100 Hz. Moreover, the pattern obtained in this way can represent a characteristic pattern to use like product quality assessment of the different kind of fabrics.

The frequency scanning on fabric A (reported in the complete comparison of Fig. 7) shows two peaks in the Tandelta curve at 44 and 56 Hz.

The fabric B, composed by PolyAmide, Lycra (EL), and a superelastic NiTi filament, 50 μm of diameter, in the straight annealed form, was tested following the same protocol of measurements. The static tests, reported in Fig. 5, showed a hysteretic behavior with a lower hysteresis than the fabric A and lower residual strain after the first cycle (about 17%). Also,

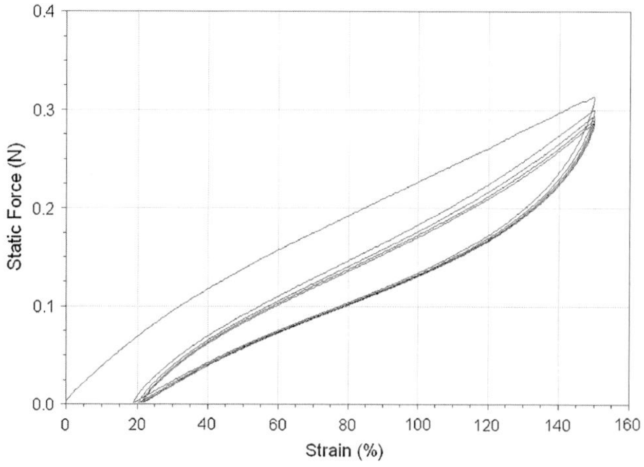

Fig. 4 Static force vs. strain cycling curves for fabric A (PA + EL + NiTi superelastic snake-form wires)

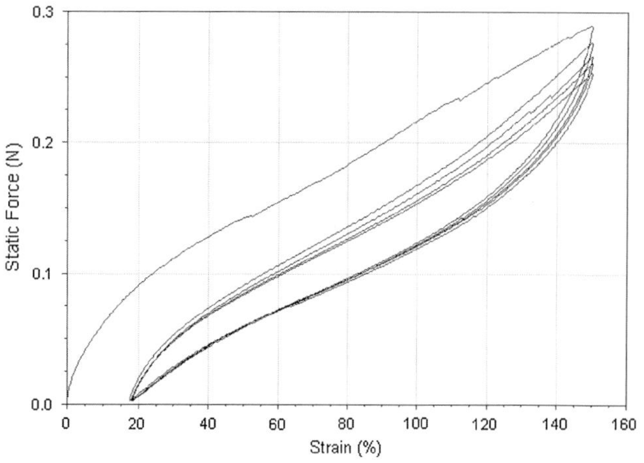

Fig. 5 Static force vs. strain cycling curves for fabric B (PA + EL + NiTi superelastic straight annealed wires)

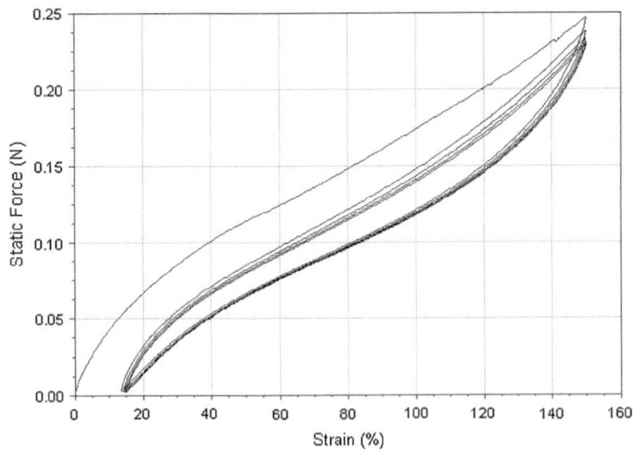

Fig. 6 Static force vs. strain cycling curves for fabric C (PA + EL, no NiTi wires)

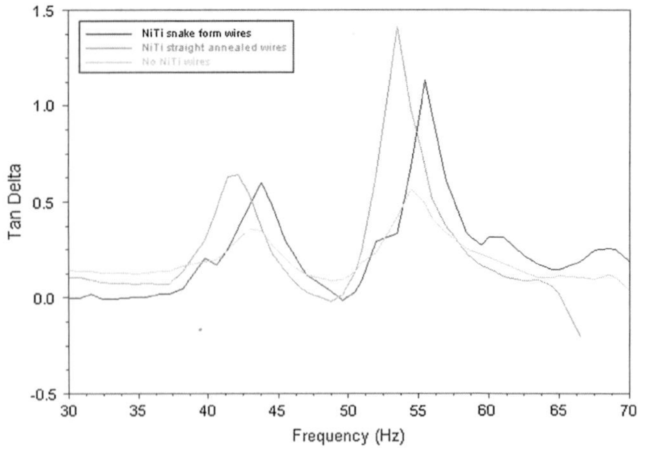

Fig. 7 Tandelta vs. frequency curves comparison for fabrics A, B, and C

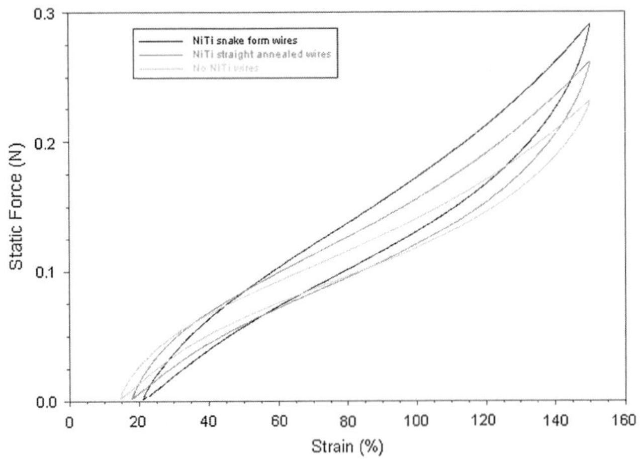

Fig. 8 Static force vs. strain curves comparison for fabrics A, B, and C

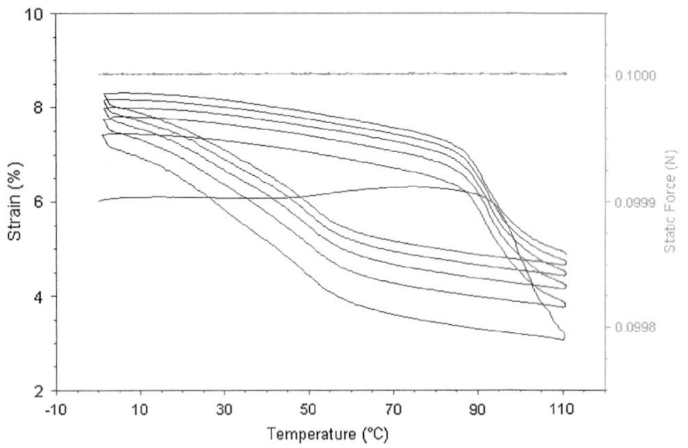

Fig. 9 Strain recovery under load cycles vs. temperature for fabric D and load = 0.1 N

for this sample, residual strain after the first cycle is reduced to zero in the next cycles. In the frequency scanning measurements (Fig. 7), the peaks are different in frequencies and intensities (41 and 53.5 Hz).

In order to evaluate the influence of the presence of the NiTi wires, further test on the fabric C has been conducted. Small samples without NiTi wires have been taken from the fabric too and tested in the same way. In Fig. 6, the static tests to 150% in strain present the smallest hysteresis and the best residual strain after the first cycle. The pattern in dynamic conditions (Fig. 7) shows shift in frequencies and intensities. At the end in Fig. 8 we report a suitable comparison among the curves force versus strain of the three fabrics for the last cycle.

Dynamic tests (Tandelta vs. frequency at room temperature) on the fabric with and without NiTi showed the contribution of the NiTi wires in terms of Tandelta. It seems that the damping properties increase with the presence of the NiTi wires and the different shape of the NiTi wires changes the frequency of the peaks. The shift in the peaks position and the change in the intensities associated to each kind of fabric give us a characteristic Tandelta pattern for fabrics A, B, C. Then we can use this pattern like a control test of the good result of the fabric production.

In the force-strain tests, the presence of NiTi wires generally increases the rigidity of the fabric. Moreover, superelastic NiTi filament enhances the textile stiffness without reducing the elastic performance of the hybrid structures. The plastic deformation seems to be more probable with the snake-like form NiTi wires, according to softer microstructure due to the two thermal treatments.

Finally we tested the fabric D sample, obtained from a textile structure, composed by a Lycra (EL)/PolyAmide yarn and a snake-form re-memorized shape memory NiTi wire ($T_{Af} \approx 87$ °C).

The results on a single snake-form shape memory wire provided information about the recovery capacity of the wire under different loads, from martensite to austenite field. After these measurements we chose the applied load to investigate the recovery strain behavior of this last fabric in 5 cycles of heating-cooling. In Fig. 9-11 it is possible to observe the recovery of the yarn in terms of strain recovered versus temperature.

It is possible to observe that the higher the load, the larger the recovery, due to the larger initial deformation of the yarn. For all the measurements it is possible to note a residual strain after each cycle, this residual strain is bigger when bigger load

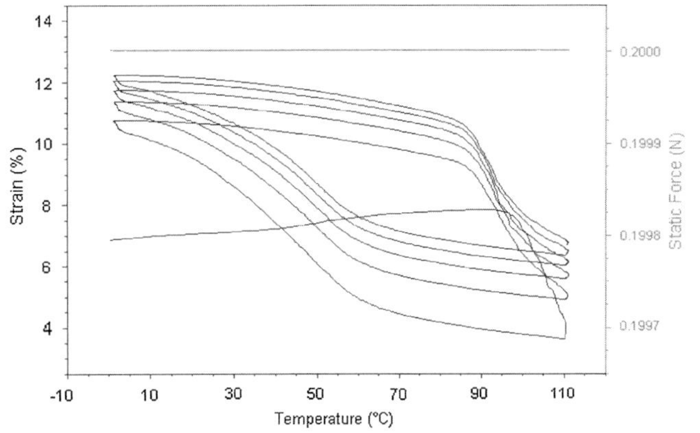

Fig. 10 Strain recovery under load cycles vs. temperature for fabric D and load = 0.2 N

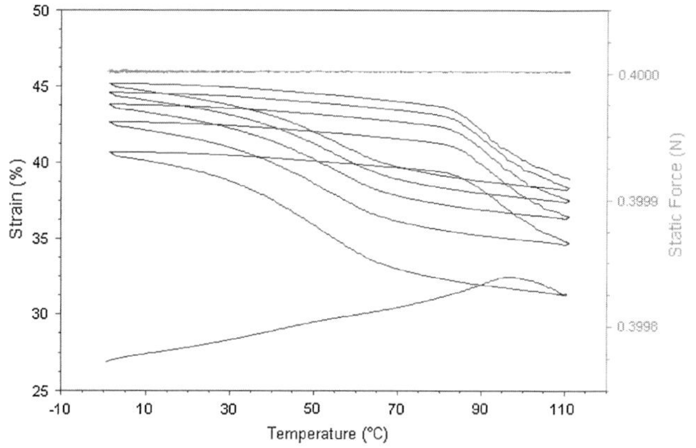

Fig. 11 Strain recovery under load cycles vs. temperature for fabric D and load = 0.4 N

is applied. In any case, the stabilization of the recovery is visible in the reduction of the residual strain with the number of cycles for the load 0.1 and 0.2 N.

The collected data for all the tests should be considered reliable, because at the end of the tests, all samples were intact.

4. Conclusions

The DMA equipment proved to be a suitable instrument to investigate the mechanical properties of non-conventional samples with hybrid and composite structures. The data obtained are reproducible and significative, or well correlated with the physical and mechanical properties of the hybrid material.

The fabric NiTi wires-PA-Lycra(EL) with different kinds of NiTi wires showed good elastic properties. The presence of NiTi wires increases the stiffness of the materials, giving the particular intermediate rigidity useful in articular rehabilitation. It will be interesting to study in depth the influence of NiTi wires in other microstructural conditions (cold worked or different thermal treatment of annealing) and also the tests in bending or torsion will be important for a complete characterization. The dynamic tests versus frequency showed first remarkable damping properties which can also be better studied in future to obtain an exhaustive description of the material from various points of view.

In the strain recovery under constant load tests, the fabrics showed good properties only for low applied loads. For the future activities it is possible to improve this aspect changing the thermal treatment of the NiTi wires; in this way we can obtain a more hard microstructure which is able to give better performance in terms of recovery under applied load.

Acknowledgment

These studies were carried out in the framework of the European IP AVALON EC Contract NMP2 CT 2005-515813.

References

1. S. Arnaboldi, P. Bassani, R. Casati, F. Passaretti, E. Villa, and A. Tuissi, Wire Drawing in Functional Characterization of Thin NiTi Wires, International Conference on Shape memory and Superelastic Technologies, SMST 2008
2. S. Pittaccio, A. Nespoli, M. Pini, E. Villa, S. Besseghini, F. Molteni, and A.C. Turconi, Mechanics of Dropfoot Rehabilitation with a Shape Memory Alloy Active Split, V World Congress of Biomechanics, München, Luglio-Agosto 2006
3. S. Pittaccio, A. Nespoli, S. Viscuso, E. Villa, and S. Besseghini, On the Use of Smart Materials in the Development of Dynamic Orthoses, 4th World Congress of the ISPRM, Seoul, Giugno, 2007
4. S. Besseghini, E. Villa, and J. Portman, DMA Characterization of a Ni50.5at%Ti Shape Memory Alloys, *Int. J. Appl. Electromagn. Mech.*, 2006, **23**(1/2), p 33–38

JMEPEG (2009) 18:522–530
DOI: 10.1007/s11665-009-9366-1

Embedding of Superelastic SMA Wires into Composite Structures: Evaluation of Impact Properties

Silvio Pappadà, Rocco Rametta, Luca Toia, Alberto Coda, Luca Fumagalli, and Alfonso Maffezzoli

(Submitted September 11, 2008; in revised form December 5, 2008)

Shape memory alloy (SMA) represents the most versatile way to realize smart materials with sensing, controlling, and actuating functions. Due to their unique mechanical and thermodynamic properties and to the possibility to obtain SMA wires with very small diameters, they are used as smart components embedded into the conventional resins or composites, obtaining active abilities, tunable properties, self-healing properties, and damping capacity. Moreover, superelastic SMAs are used to increase the impact resistance properties of composite materials. In this study, the influence of the integration of thin superelastic wires to suppress propagating damage of composite structures has been investigated. Superelastic SMAs have very high strain to failure and recoverable elastic strain, due to a stress-induced martensitic phase transition creating a plateau region in the stress-strain curve. NiTi superelastic wires ($A_f = -15$ °C fully annealed) of 0.10 mm in diameter have been produced and characterized by SAES Getters. The straight annealed wire shows the typical flag stress-strain behavior. The measured loading plateau is about 450 MPa at ambient temperature with a recoverable elastic strain of more than 6%. For these reasons superelastic SMA fibers can absorb much more strain energy than other fibers before their failure, partly with a constant stress level. In this paper, the improvement of composite laminates impact properties by embedding SMA wires is evaluated and indications for design and manufacturing of SMA composites with high-impact properties are also given.

Keywords advanced characterization, aerospace, mechanical testing, structural plastics

1. Introduction

Composite structures often present small damages and defects resulting either from the manufacturing process or from accidental impacts occurring during their use. These small impacts may result in initial deformations and reduce the strength and the lifetime of composite material. In fields where the event of an impact needs to be considered, such as aerospace and transportation, it is of fundamental importance to predict the impact behavior and to collect data on impact resistance of candidate materials. In these cases, the composite material has to resist to the static and fatigue loads and to random impact loads which can occur during its lifetime. For these reasons there is a need of improving impact resistance behavior of composite structures, without increasing their weights and compromising other mechanical properties.

Shape memory alloys (SMA) represent the most versatile way to fabricate smart materials with sensing, controlling, and actuating functions (Ref 1-4). Due to their unique mechanical and thermodynamic properties, and to the possibility to obtain SMA wires with very small diameters (down to 20-30 μm diameter), they are used as smart components embedded into conventional composites, obtaining active abilities, tunable properties, self-healing properties, and damping capacity. Moreover, SMA superelastic properties can be exploited to increase the impact properties of composite materials. SMA presents a very high strain to failure and a large recoverable elastic strain, thanks to a stress-induced martensitic phase transition creating a plateau region in the stress-strain curve. For these reasons superelastic SMA fibers can absorb much more strain energy than other fibers before their failure, with a nearly constant stress level.

In this study, the increasing of impact properties of fiber reinforced polymer composites by the integration of thin superelastic SMA wires has been investigated. Other authors have already investigated the impact damage behavior of SMA composites, showing that for low-velocity impacts, embedding SMA wires increases the impact properties (Ref 5-10). For this reason, SMA wires of small diameter (100 μm) are embedded in glass and carbon reinforced composite laminates, in order to understand how the impact properties of composite materials change with this hybridization, and to find out some indications for the design of SMA/carbon or SMA/glass hybrid composites with improved impact properties. These results will be applied to new energy-absorbing structures for the aerospace and transportation industries and for the individual protection industry.

This article is an invited paper selected from presentations at Shape Memory and Superelastic Technologies 2008, held September 21-25, 2008, in Stressa, Italy, and has been expanded from the original presentation.

Silvio Pappadà and **Rocco Rametta**, Consorzio CETMA, Departments of Materials and Structures Engineering, Technologies and Processes Area, Cittadella della Ricerca, SS7-Km706+300, 72100 Brindisi, Italy; **Luca Toia, Alberto Coda,** and **Luca Fumagalli**, SAES Getters, Viale Italia 77, 20020 Lainate (Milan), Italy; and **Alfonso Maffezzoli**, Department of Engineering for Innovation, University of Salento, Via per Monteroni, Edificio "La Stecca", 73100 Lecce, Italy. Contact e-mails: silvio.pappada@cetma.it, Luca_Toia@saes-group.com, alfonso.maffezzoli@unile.it.

In order to discriminate the influence of the martensitic transformation from the simple introduction of metallic wires into the polymeric matrix, preliminary tests were carried out on composite panels embedding steel wires of the same diameter of the SMA ones.

The hybrid laminates were tested by Charpy pendulum, using a hammer equipped with a piezoelectric sensor for the measurement of force-time and force-displacement curves during the impact test. Moreover, glass and carbon reinforced samples were subjected to repeated low-velocity impacts up to fracture, in order to evaluate the influence of this hybridization on the damage tolerance of the composite. Results of repeated impact loads were plotted as peak load (F_{max}) versus repeated impact number, as in a traditional fatigue plot, thus providing preliminary design rules. In addition, a quasi-static analysis was performed, according to ASTM D790, to compare the failure mechanisms at different strain rates.

2. Materials and Methods

2.1 Materials

A SAES Getters SMA wire made of superelastic 56.00 wt.% Ni balance Ti alloy straight annealed was used. The superelastic wires have 100 μm diameter, transformation temperature $A_f = -15$ °C as measured on finished wires, upper plateau load (at 4% strain) ≅ 400 MPa, lower plateau load (at 4% strain) ≅ 100 MPa. In Fig. 1 the results of tensile test on these wires are reported. Moreover, steel wires made of AISI 304L alloy with 100 μm diameter, 700 MPa break load and 5% break deformation, provided by COMAR S.r.l., were embedded in the composite laminates to make the comparison with equivalent SMA composites.

To embed SMA wires into the composite layers, special frames were designed for a proper wire spacing into the matrix, obtaining a 1 mm distance between the wires and alignment in the principal direction of the sample. Four layers of wires were embedded into the composites at different distances from the neutral axis, near the external surfaces of the specimen. The volume percentage of the SMA and steel wires embedded into the composites is about 1%. The laminates were realized by vacuum-assisted resin infusion process.

In Table 1 materials and stacking sequences of the composite laminates in which SMA and steel wires were embedded are reported, and in Table 2 their main mechanical characteristics are represented.

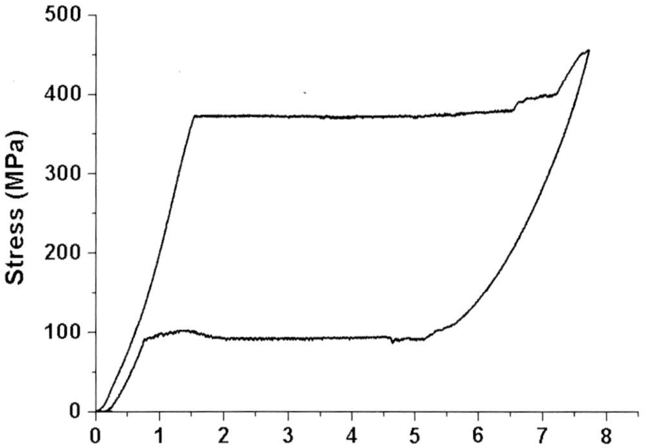

Fig. 1 Tensile stress-strain curve for a SAES Getter shape memory superelastic wire

2.2 Experimental Methods

Three kinds of specimens were obtained from each laminates for impact characterization: specimens without any metallic wires and hybrid specimens embedding steel and SMA wires.

Impact tests were performed according to the Charpy test method (ASTM D256 (Ref 11)) by an instrumented impact tester. A schematization of sample geometry, positioning, and testing procedure is represented in Fig. 2. The pendulum hammer was equipped with an ICP® Dynamic Force Sensor provided by PCB Piezoelectronics for the measurement of the force as a function of the time during the impact test.

The maximum energy used to impact the specimens was 25 J: with this energy level the specimens break with a single hit. Moreover, the samples were hit with less energy, regulating the pendulum holding and releasing mechanism so to start the swing from a minor vertical height. Using such energy levels the specimens do not break with a single hit and so they were repeatedly impacted up to fracture. Results of repeated impact loads were represented as peak load (F_{max}) versus repeated impacts number, as in a traditional fatigue plot.

Static flexural tests were performed by three-point bend procedure (ASTM D790 (Ref 12)): the results are an average of 15 specimens of each type.

3. Approach

A typical damage sequence in Charpy laminate specimens under increasing low-velocity impacts is represented in Table 3, where F_{max} is the maximum in the force-time curve, F_i the force at the onset of delamination, and F_{ult} is the breaking load for the composite.

For low-energy impacts (Table 3a), localized matrix cracking occurred in the contact area before the onset of delaminations, but it was insufficient to have a noticeable effect on the impact response curve that remains symmetric. When the onset of delamination took place, there was an initial drop as indicated in Table 3b. For higher energy impacts, this load drop increased for the ply shear-out. By further increasing the energy impacts, ply shear continued through the thickness, resulting in the growth of a truncated pyramid of shear-out material. Ultimately this pyramid forced the remaining plies, to fracture with an explosive bang as illustrated in Table 3e.

In Fig. 3 a schematic representation of the development of damage during the fatigue life of a composite laminate is reported. Composite laminate was assumed to experience a flexural load history and to include plies with different orientations with respect to a principle load direction (0° direction). It can be observed that the changes in F_{max} values take place in three regions (Ref 13-16). At first region, F_{max} values decrease gradually and this diminution is associated to primary crack formation and multiplication. In the second region, there is a plateau (or crack saturation) region, where crack coupling, interfacial debonding, and fiber breaking take

Table 1 Materials and stacking sequences of the composite laminates in which SMA superelastic wires have been embedded

Laminate	Material	Stacking sequences	Side views
TS-glass	Vinylester resin Glass fabric EE425 Plain 5.5/6.3	$[0°,W,90°,W,0°,90°,0°]_S$	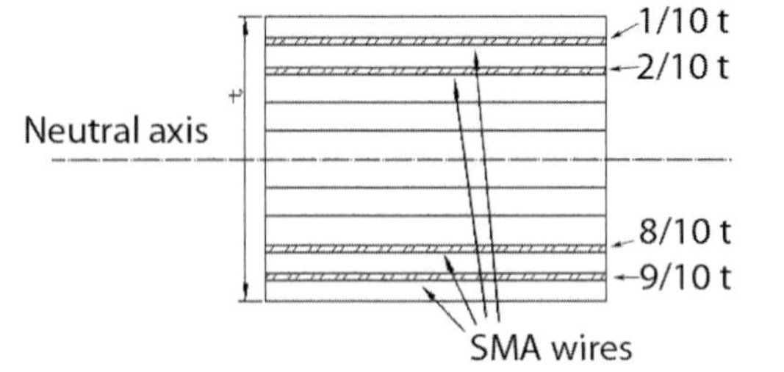
TS-carb	Vinylester resin Carbon fabric CC420 Twill 2/2	$[0°,W,90°,W,0°,90°]_S$	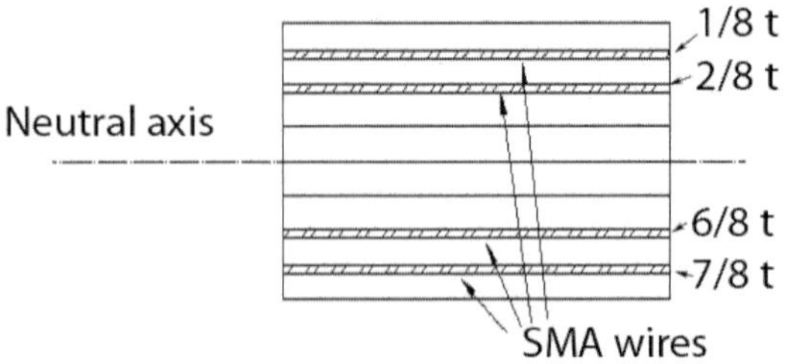

W indicates the metallic wires

Table 2 Main mechanical characteristics of the composite laminates in which SMA wires have been embedded

Laminate code	Fiber content, wt.%	Thickness, mm	E_1, GPa	E_2, GPa	ε_1, %	ε_2, %	σ_1, MPa	σ_2, MPa
TS-glass	70	3	16.8	16.8	4.2	4.2	650	650
TS-carbon	70	3.2	32.5	32.5	2	2	530	530

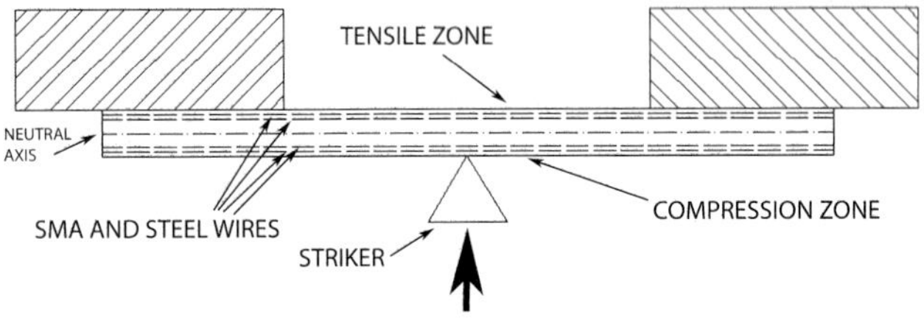

Fig. 2 Schematization of sample geometry, positioning, and testing procedure

place. At last in the third region, an acceleration of the damage takes place and F_{max} values decrease sharply due to delamination growth till the sample breaks.

Another important aspect to be considered is the increase of absorbed energy as a consequence of the possible stiffness increase associated with addition of the SMA (or steel) wires. An important parameter that can be used to measure the effect of wire embedding is the ductility index D (Ref 13), which is the ratio between the propagation energy E_p and the initiation energy E_i:

$$D = \frac{E_p}{E_i}$$

where E_i is the area up to the maximum load under the force-displacement curve and E_p is the area from the force peak until the end of the test. The hybridization with SMA wires is more effective if it increases the parameter D, otherwise it can be considered as a brittle reinforcement of the material behaving like the other reinforcement. The effect of hybridization can be also observed from the stress-strain curves obtained in a static flexural experiment.

4. Experimental Results

A synthesis of the Charpy and static flexural results is reported in Table 4 and 5. The average force and displacements impact curves for the three types of specimens for each composite are shown in Fig. 4 and 5. As visible in Table 4, SMA hybridization has a little positive effect on the impact energy for TS-glass (+1%), while for TS-carbon it is more considerable (+9.1%): this behavior is due to the smaller break deformation of carbon reinforced composite, which is more brittle than glass reinforced one. Moreover, for steel hybridization, impact energy of TS-glass composite decreases (−4.6%), while for TS-carbon there is an increase of impact energy smaller than for SMA hybridization (+5%). It is interesting to observe the behavior of ductility index D: it sensibly increases for TS-glass hybrid laminate. This is principally due to the decrease of mechanical properties of glass reinforced laminate as a consequence of hybridization: this is responsible for a sensitive decrease of E_i, reflected by an increase of D (see results of static flexural characterization in Table 5).

Table 3 Damage sequence in Charpy laminate specimens under increasing low-velocity impacts

Stage I $F_{max} < F_i$	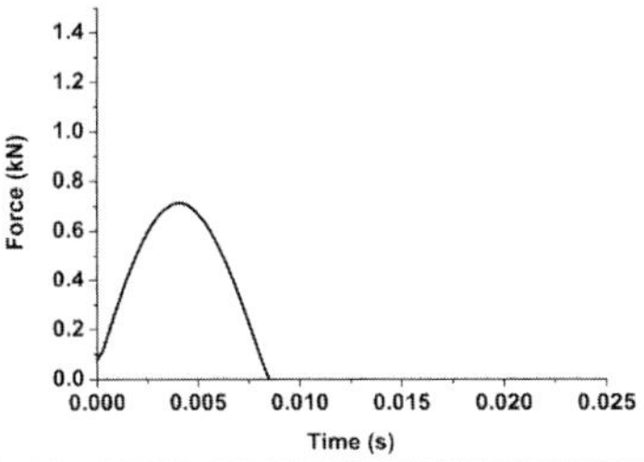	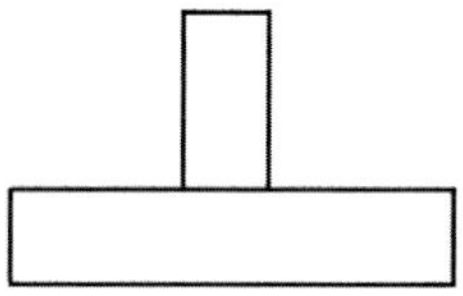 **a**
Stage II $F_{max} = F_i$	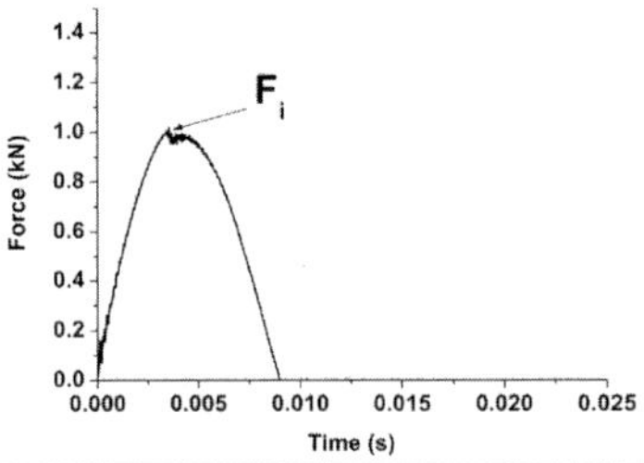	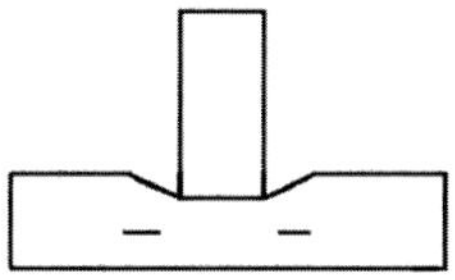 **b**
Stage III $F_{ult} > F_{max} > F_i$	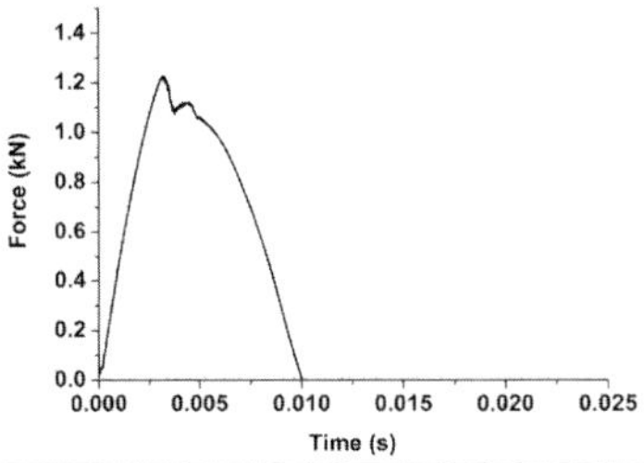	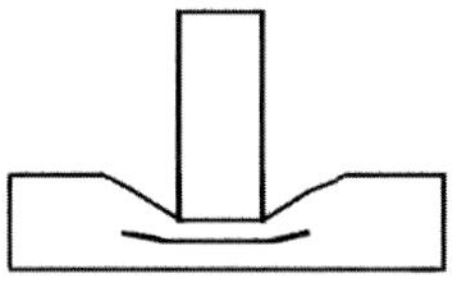 **c**
Stage IV $F_{ult} = F_{max} > F_i$	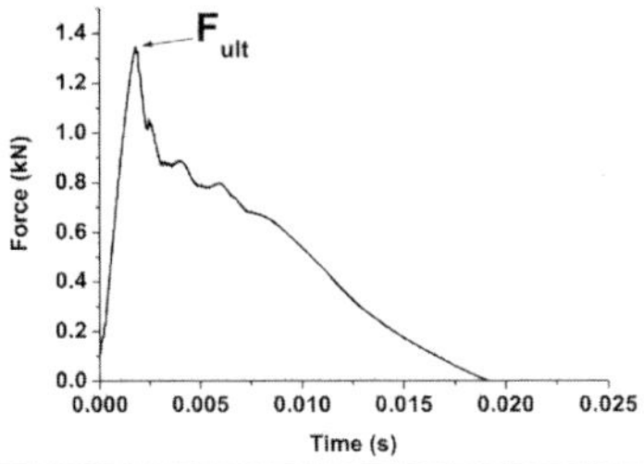	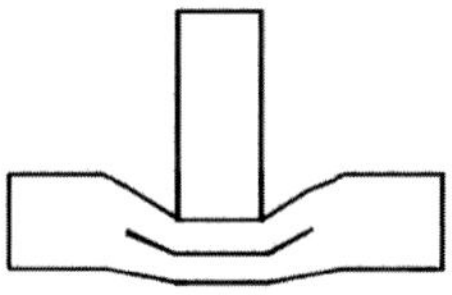 **d**
Stage V $F_{ult} < F_{max}$	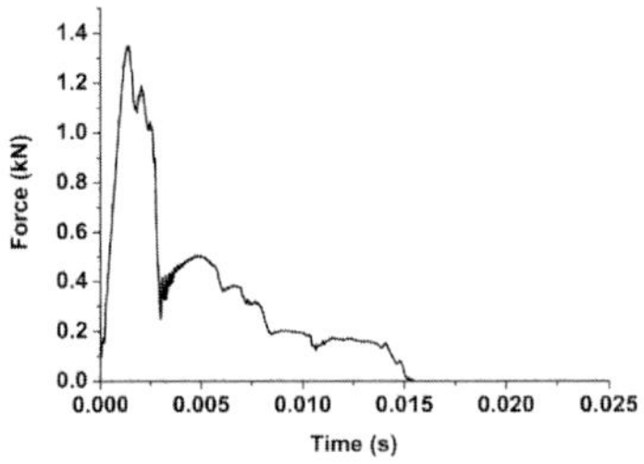	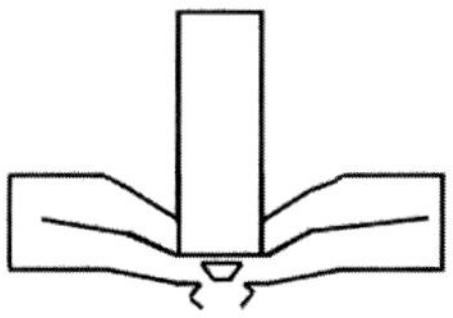 **e**

We can conclude that the impact energy is marginally affected by SMA wires in the lay-up; however, to have a more sensitive increase of impact energy, more wires have to be embedded into the composite structures, even if this is not technologically easy.

For this reason the samples were hit with less energy, for a more general investigation of the effect of SMA hybridization on composite damage tolerance. So the specimens did not break with a single hit and they were repeatedly impacted up to fracture.

In Fig. 6 the force-time (F-t) curves of impacts at energy levels lower than 25 J are illustrated for non-hybrid (a-d), SMA-hybrid (e-h), and steel-hybrid (i-l) TS-glass samples. In this case, the hammer struck each sample only once and the samples were preserved from additional strikes. As Fig. 6 shows, for SMA-hybrid samples up to 3.94 J, total elastic deformations were observed (symmetrical F-t curves), while for non-hybrid samples the energy value for the onset of visible damages was lower (equal to 1.8 J). Even for steel-hybrid samples the onset of visible damages was 3.94 J, but in this case a higher sudden drop in F value was measured. Moreover, hybrid samples reach higher F_{max} and F_{ult} values than non-hybrid ones.

Starting from these first results, impact fatigue tests on TS-glass samples have been carried out. The composite samples were positioned on the sample holder and hit by Charpy hammer repeatedly. The composite samples were subjected to single impacts repeatedly up to fracture at each

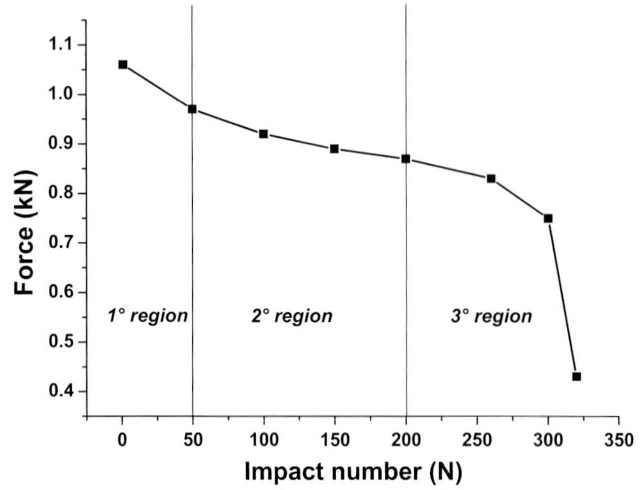

Fig. 3 Peak load (F_{max}) vs. repeated impact number curves during the impact fatigue experiments

impact energy level. In Fig. 7 the F_{max} values during the fatigue impact experiments on TS-glass specimens are reported for 0.8 J impact energy, and in Fig. 8 the impact fatigue curves are shown. As Fig. 7 shows, embedding SMA wires increases the damage tolerance of the composite laminate. In fact, SMA samples fractured after nearly 1200 impacts, so much later than

Table 5 Synthesis of static flexural characterization results and evaluation of the changes in mechanical properties of the composite due to hybridization

Laminate	E_c, GPa	ε_r, %	σ_r, MPa
TS-glass			
No wires	16.8	4.2	650
Steel wires	17.8	4.2	541
	+6%		−17%
SMA wires	15.5	4	513
	−8%	−5%	−21%
TS-carbon			
No wires	32.5	1.99	528
Steel wires	35	1.66	471
	+7.7%	−16.6%	−10.8%
SMA wires	32.5	1.89	451
	//	−5%	−15%

The increases are referred to the absorbed energy of the samples without metallic wires

E_c = longitudinal flexural modulus, ε_r = longitudinal flexural strain, σ_r = longitudinal flexural strength

Table 4 Summary of the average impact energies to fracture not hybridized and steel and SMA hybridized samples and ductility index values measured in Charpy tests

Laminate	Energy, J No wires	Energy, J Steel wires	Energy, J SMA wires	*D* No wires	*D* Steel wires	*D* SMA wires
TS-glass	8.7	8.3	8.8	1.1	1.4	1.5
		−4.6%	+1%		+27%	+37%
TS-carbon	12	12.6	13.1	8.9	9.5	9.13
		+5%	+9.2%		+2.5%	+2.5%

The increases are referred to the absorbed energy of the samples without metallic wires

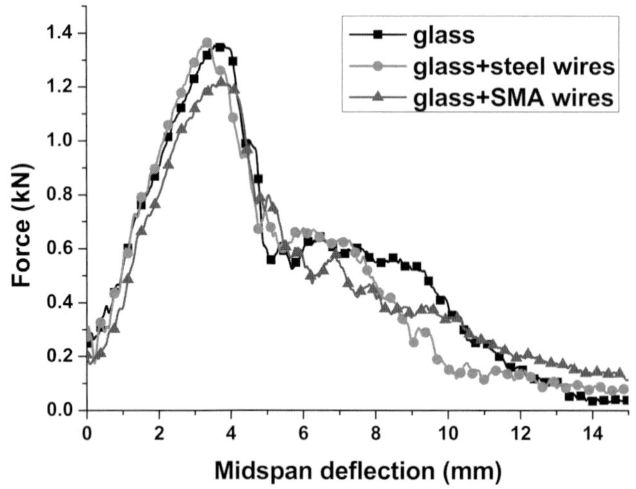

Fig. 4 Average force and displacements impact curves for the three types of specimens for TS-glass composite

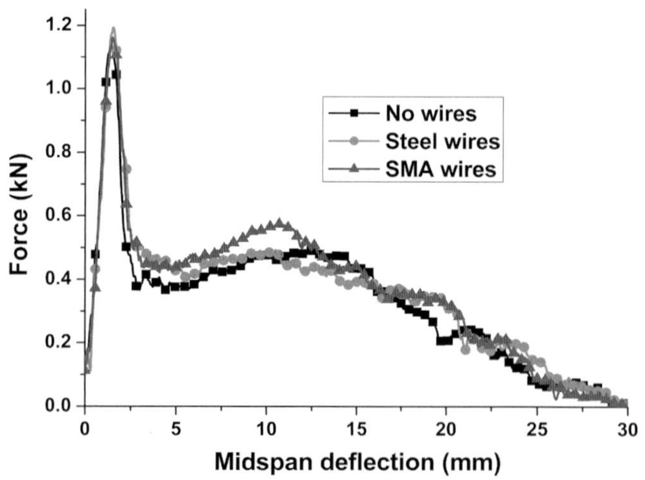

Fig. 5 Average force and displacements impact curves for the three types of specimens for TS-carbon composite

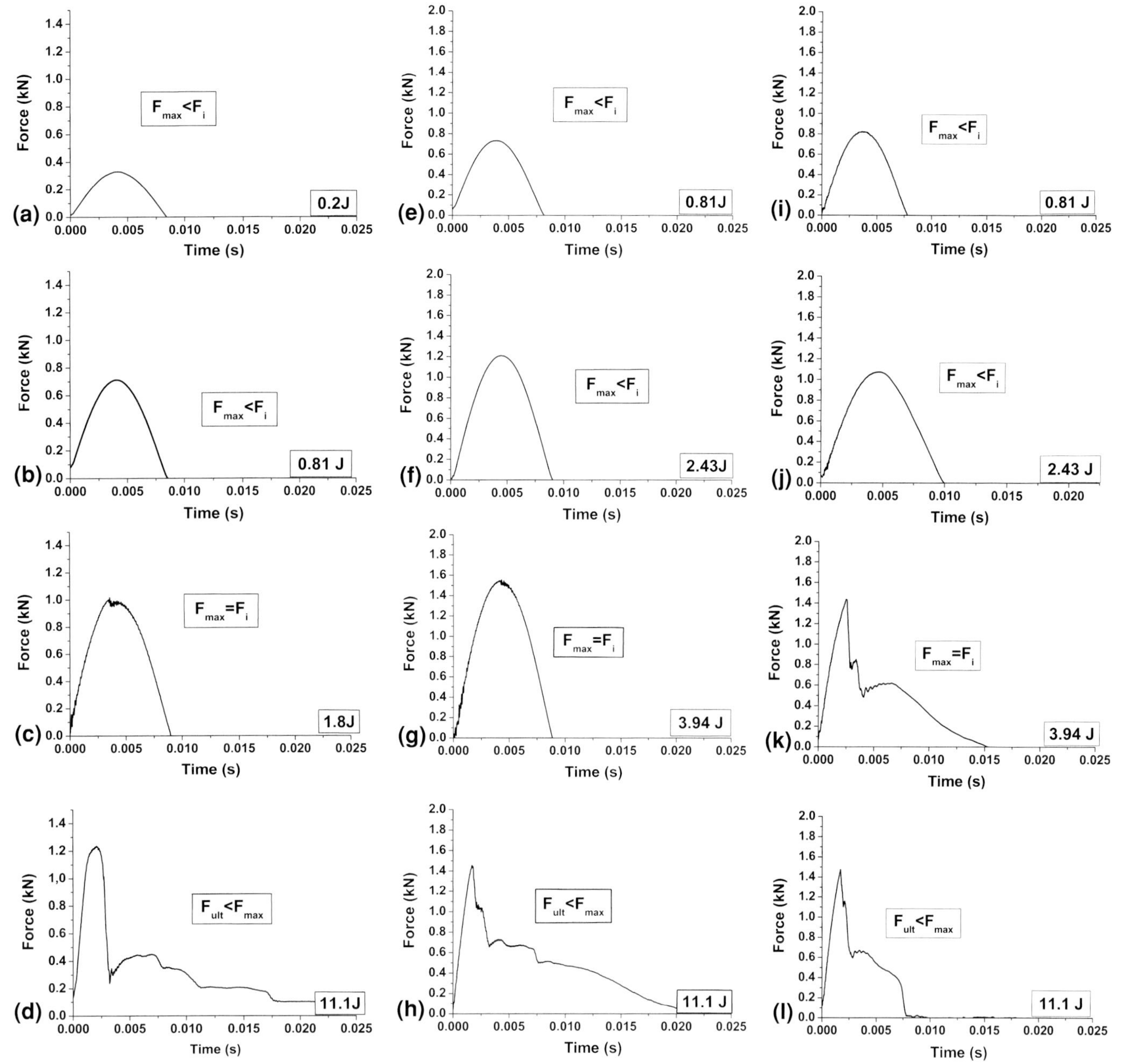

Fig. 6 Force-time (F-t) curves of impacts at energy levels lower than 25 J for non-hybrid (a-d), SMA-hybrid (e-h) and steel-hybrid (i-l) TS-glass samples

non-hybrid samples, which fracture after about 500 impacts. Instead, for steel hybrid samples, there is a worsening of fatigue impact properties (fracture after about 170 impacts). Moreover, for SMA-hybrid samples a lower "slope" in F_{max}-N curve can be observed.

This effect was observed for different impact energies, even if it becomes less sensitive for higher energy levels. Moreover, as for low energy levels, steel hybridization shows a negative effect on the damage tolerance of TS-glass composite for higher energy levels too. In Fig. 9 the results of fatigue impact experiments have been reported for TS-glass and TS-glass SMA specimens for 1.8, 2.43, and 3.14 J energy levels.

Finally, impact fatigue tests were performed on specimens obtained from the carbon reinforced laminate: in Fig. 10 the F_{max} values during the fatigue impact experiments are reported for 0.46 and 0.8 J impact energies. In this case, SMA influence on the damage tolerance of the composite is negative: in fact, for hybrid samples, there is no plateau region and fracture takes place before non-hybrid sample. Instead, steel wires have a little positive effect both on the slope curve and impacts number; however, this effect is not so sensitive as for SMA wires for TS-glass laminate.

5. Conclusions

In this work, the influence of hybridization by means of SMA and steel wires on the damage tolerance of glass and carbon fiber reinforced composites was investigated. Initially,

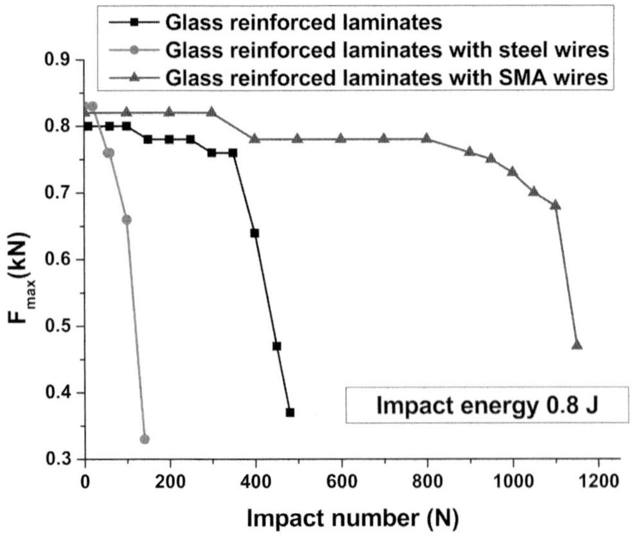

Fig. 7 Variations of F_{max} vs. repeated impact number for non-hybridized and steel and SMA hybridized samples during the impact fatigue experiments (TS-glass laminate)

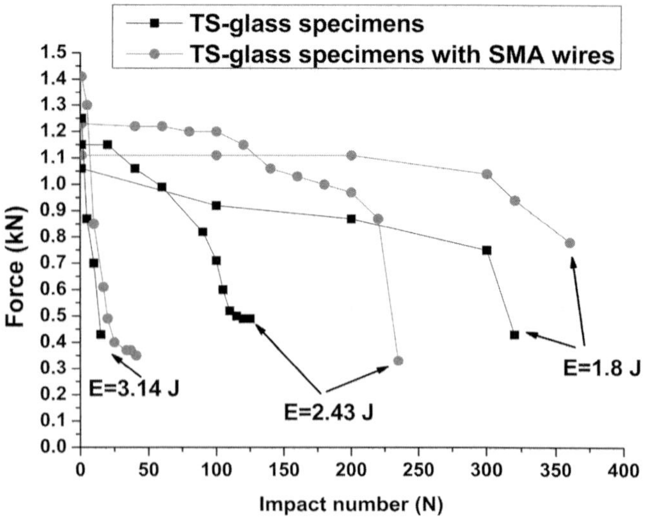

Fig. 9 Variations of F_{max} vs. repeated impact number during the impact fatigue experiments for 1.8, 2.43, and 3.14 J impact energies for steel and SMA hybridized samples (TS-glass laminate)

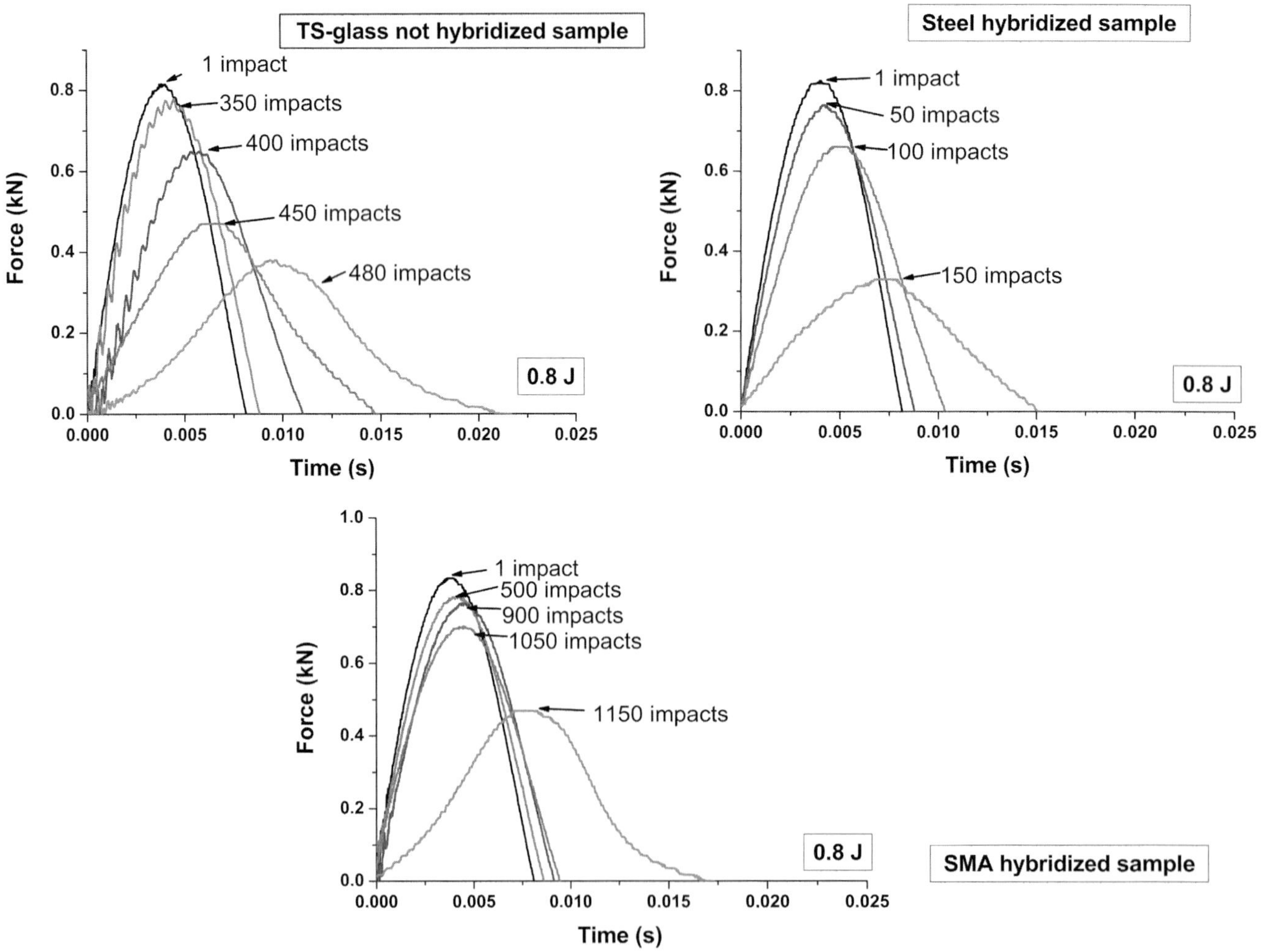

Fig. 8 Force-time (F-t) curves of impacts for 0.8 J energy level for non-hybridized and steel and SMA hybridized samples (TS-glass laminate)

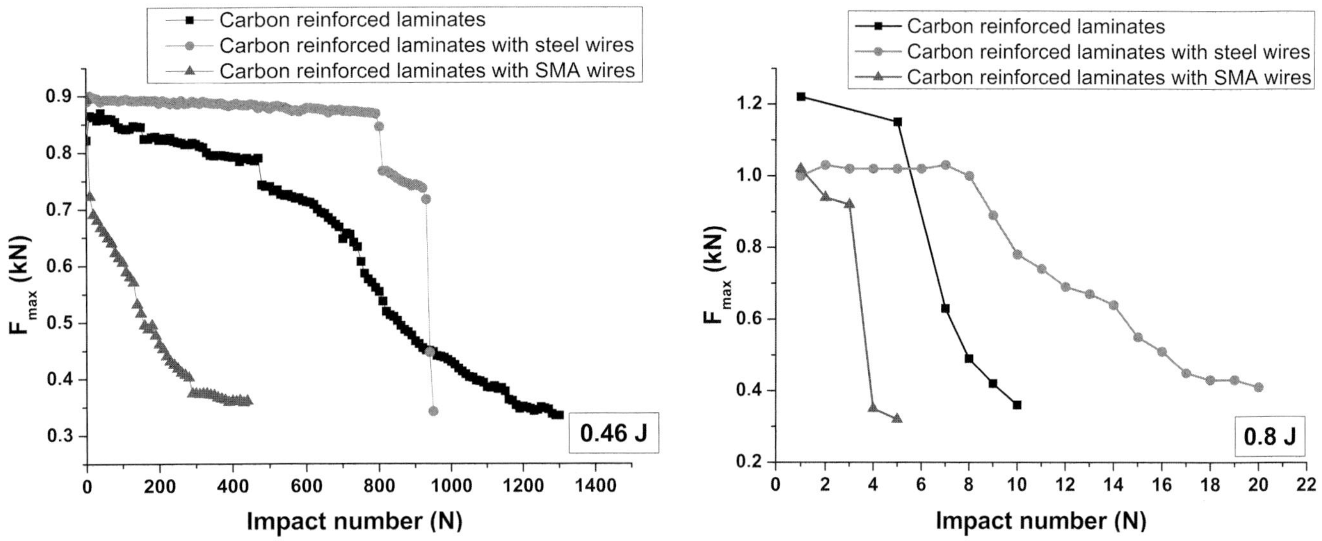

Fig. 10 Variations of F_{max} vs. repeated impact number during the impact fatigue experiments for 0.46 and 0.8 J impact energies for non-hybridized and steel and SMA hybridized samples (TS-carbon laminate)

the possibility to increase the energy to break standard Charpy samples by SMA and steel hybridization was studied. By the use of SMA hybridization a little increase of impact energy was observed, while with steel wires the increase is absent or less sensitive. This is probably due to the small amount of metallic wires embedded into the composite structures. Further tests are to be performed trying to increase the total amount of SMA wires to be embedded.

Subsequently, the samples were repeatedly impacted with less energy up to fracture, to investigate more generally the damage tolerance of hybrid composite. As reported before, SMA wires have a sensitive effect on the improving of the damage tolerance of glass reinforced laminates, while they have a worsening effect for carbon reinforced laminate. Steel wires influence is bad for glass reinforced laminate, while they have a little positive effect for carbon reinforced ones.

To understand these effects, it is to be considered that embedding metallic wires into a composite structure implies the introduction of an inhomogeneity. It is well known that inhomogeneities have a great influence on laminate response to cyclic loading, and that from behavioral standpoint, they present a dichotomy (Ref 13). In many respects, they are the most important factors contributing to the generally superior resistance of laminated composite materials to fatigue damage development, because they obstruct the crack growing. At the same time, they are almost certainly the greatest contributing factors in the initiation of damage at the micro-level. For these reasons, SMA wires make glass reinforced laminates more resistant to the growth of the cracks, thanks to their very good effect on obstructing this growth. Instead, in carbon reinforced laminates, they cause inhomogeneous deformations in the material, since their modulus is very lower than carbon fiber modulus and so they promote the cracks initiation. For the same reason steel wires decrease the damage tolerance of glass reinforced composites, because their higher modulus promote the cracks initiation. This effect is inverted for carbon reinforced laminates: the modulus of carbon fibers is higher than the elastic modulus of SMA wires, while it is closer to the steel modulus. However, in this case, the positive effect of steel wires is less sensitive, because carbon laminates are generally more resistant to the crack growth during cyclic loading.

Acknowledgments

This research has been completed in the framework of the AVALON project, a European Commission Community Research 6th Framework Programme, Priority 3 NMP Project. The authors would like to acknowledge Polynt for the supply of resins.

References

1. R. Stalmans, V. Michaud, J.E. Bidaux, R. Gotthardt, and J.A.E. Manson, Adaptive Composites with Embedded Shape Memory Alloy Wires, *Proceedings of the 4th European Conference on Smart Structures and Materials*, G.R. Tomlinsonen and W.A. Bullough, Eds. (Bristol, UK), Institute of Physics Publishing, 1998, p 801–804
2. C. Boller, Shape Memory Alloys – Their Challenge to Contribute to Smart Structures, Annual Meeting of the Materials Research Society, December 1999, Boston, USA (invited paper)
3. R. Stalmans, Adaptive Hybrid Composites with a Focus on the Integration of Shape Memory Elements, *Advances in Science and Technology, 25, Smart Materials Systems*, P. Vincenzini, Ed. (Faenza, Italy), Techna, 1999, p 83–94 (invited lecture)
4. R. Stalmans, K. Tsoi, and J. Schrooten, The Transformational Behaviour of Shape Memory Wires Embedded in a Composite Matrix, *Fifth European Conference on Smart Structures and Materials, Proceedings of SPIE*, Vol 4073, P.F. Gobin and C.M. Friend, Eds., 2000, p 88–96
5. K. Tsoi, et al., Impact Damage Behavior of Shape Memory Alloy Composites, *Mater. Sci. Eng.*, 2003, **342**(1–2), p 207–215
6. K.-t. Lau, et al., Low Velocity Impacts on Shape Memory Alloy Stitched Composite Plates, *Smart Mater. Struct.*, 2004, **13**, p 364–370
7. S. John, and M. Hariri, Effect of Shape Memory Alloy Actuation on the Dynamic Response of Polymeric Composite Plates, *Compos. A: Appl. Sci. Manuf.*, 2008, **39**(5), p 769–776
8. S.M.R. Khalili, et al., Low-Velocity Impact Response of Active Thin Walled Hybrid Composite Structures Embedded with SMA Wires, *Thin-Walled Struct.*, 2007, **45**(9), p 799–808
9. S.M.R. Khalili, et al., Effect of Smart Stiffening Procedure on Low-Velocity Impact Response of Smart Structures, *J. Mater. Process. Technol.*, 2007, **190**(1–3), p 142–152

10. R.-x. Zhang, et al., Mechanical Properties of Composites Filled with SMA Particles and Short Fibers, *Compos. Struct.*, 2007, **79**(1), p 90–96
11. ASTM D256-06ae1 Standard Test Methods for Determining the Izod Pendulum Impact Resistance of Plastics, 2006
12. ASTM D790 07e1 Standard Test Methods for Flexural Properties of Unreinforced and Reinforced Plastics and Insulating Materials, 2007
13. S.R. Reid and G. Zhou, *Impact Behaviour of Fibre-Reinforced Composite Materials and Structures*, Woodhead Publishing Ltd., 2000
14. K.L. Reifsnider, *Fatigue of Composite Materials*, Elsevier Science Publishers B. V., 1991
15. M.O. Bora, and O. Coban, et al., On the Life Time Prediction of Repeatedly Impacted Thermoplastic Matrix Composites, *Mater. Des.*, 2009, **30**(1), p 145–153
16. O.S. David-West, et al., An Experimental Study of Damage Accumulation in Balanced CFRP Laminates Due to Repeated Impact, *Compos. Struct.*, 2008, **83**(3), p 247–258

JMEPEG (2009) 18:531–537
DOI: 10.1007/s11665-009-9403-0

1059-9495/$19.00

Mechanical and Vibration Characteristics of Laminated Composite Plates Embedding Shape Memory Alloy Superelastic Wires

Silvio Pappadà, Per Gren, Kourosh Tatar, Tord Gustafson, Rocco Rametta, Ettore Rossini, and Alfonso Maffezzoli

(Submitted September 12, 2008; in revised form November 18, 2008)

Currently, there is a great interest in the study of shape memory alloy (SMA) composites, since SMA wires with a small diameter have become commercially available. Many potential uses have been found for SMA composites in shape control, vibration control, and for the realization of structures with improved damage tolerance. In this work, two types of SMA-hybridized composites are presented for investigating the mechanical and vibration characteristics. The first one contains unidirectional superelastic SMA wires, while the other has been realized with embedded knitted SMA layers. The samples from these laminates have been tested according to "Charpy method" (ASTM D256) and static flexural test method (ASTM D790) to evaluate the influence of the integration of thin superelastic SMA wires on the impact behavior and the mechanical properties of the hybrid composites. Moreover, since the SMA wires are expected to give damping capacity, by measuring the vibration mode of a clamped cantilever using laser vibrometry, the influence of both SMA arrangements on the vibration characteristics has been investigated. Finally, further tests have been carried out on composite panels realized by embedding unidirectional steel wires to distinguish the influence of the martensitic transformation from the pure introduction of a metallic wire into the polymeric matrix.

Keywords advanced characterization, aerospace, automotive, nondestructive testing, polymer matrix composites, structural plastics

1. Introduction

Composite materials are commonly used in fields in which high performances, such as high resistance to different types of loads and lightweight, are requested. In many cases, these materials have to resist to static and fatigue loads or to random impacts which can occur during their lifetime. For these reasons, it is fundamental to improve the resistance behavior of composite structures, without increasing their weights and compromising other mechanical properties.

This article is an invited paper selected from presentations at Shape Memory and Superelastic Technologies 2008, held September 21-25, 2008, in Stresa, Italy, and has been expanded from the original presentation.

Silvio Pappadà and **Rocco Rametta**, Departments of Materials and Structures Engineering, Technologies and Processes Area, Consorzio CETMA, Cittadella della Ricerca, SS7-Km706+300, 72100 Brindisi, Italy; **Per Gren** and **Kourosh Tatar**, Division of Experimental Mechanics, Lulea University of Technology, SE-97187, Luleå, Sweden; **Tord Gustafson**, APC Composite, Karlsviksvagen 81, S-97594 Luleå, Sweden; **Ettore Rossini**, Extreme Materials S.r.l., Via dei Cipressi 2, Costa Masnaga, LC, Italy; and **Alfonso Maffezzoli**, Department of Engineering for Innovation, University of Salento, Via per Monteroni, Edificio "La Stecca", 73100 Lecce, Italy. Contact e-mails: silvio.pappada@cetma.it, rocco.rametta@cetma.it, ettore_rossini@virgilio.it, and alfonso.maffezzoli@unile.it.

In these years, there is an increasing interest on hybrid composite materials in which shape memory alloy (SMA) wires are embedded with the aim to increase their static and dynamic properties or for vibration and shape control (Ref 1-10). In fact, SMA materials, presenting unique mechanical and thermodynamic properties such as shape memory effect and pseudoelasticity, are being used as "smart" materials embedded into conventional resins or composites to obtain tunable properties and shapes, damping capacity, and self-healing capabilities. Moreover, SMA wires have been obtained with very small diameters (down to 20 to 30 μm diameter), and hybrid textile with SMA wires have been realized and used as reinforcements of hybrid composite materials for different applications. For example, SMA wires or yarns can be used as "smart fibers" combined with conventional glass/carbon fibers to realize new hybrid textile preforms, even if the fabrication process of these SMA-based preforms is still complicate and nonstandardized.

In this study, the possibility of realizing new hybrid composite with enhanced damping properties has been investigated by means of laser vibrometry (Ref 7-10). The primary advantage of using a laser vibrometer is the noncontact nature of the transducer, which eliminates mass loading of structure due to response measurements transducer. It is well known that there are two types of damping: material damping and system damping. Material damping is the damping inherent in the material, while system or structural damping includes the damping at the supports, boundaries, joints, interface, etc. Since utilizing damping materials is the most common way to reduce resonance responses, accurate measurements of damping are crucial to proper design, optimization, and modeling of systems from a vibration reduction standpoint.

Three types of hybrid composites have been realized:

- A glass-reinforced composite plate with a vinylester matrix containing unidirectional superelastic wires;
- Composite plates embedding steel wires of the same diameter of the SMA ones to discriminate the influence of the martensitic transformation of SMA from the simple introduction of metallic wires into the polymeric matrix; and
- Hybrid composites by means of glass hybrid preforms (HEP), in which SMA wires are knitted together with glass bundles to evaluate the possibility of enhancing the damping effect of superelastic wires by introducing predeformed regions into the wires.

By measuring the vibration mode of a clamped cantilever, the influences of both SMA arrangements on the vibration characteristics of the laminated plates have been evaluated. Furthermore, the possibility of utilizing SMA composite in applications where the event of an impact need to be considered is of fundamental importance to collect data on impact resistance of candidate materials and to predict the impact resistance. For this reason, the hybrid samples were tested in a Charpy geometry (Ref 11) using a hammer equipped with a piezoelectric sensor providing force-time and force-displacement curves during the impact test. Finally, a quasi-static analysis was performed, according to ASTM D790 (Ref 12), to evaluate the effect of hybridization on the static flexural properties of the laminates.

2. Materials and Methods

2.1 Materials

A SMA wire made of superelastic 56.00 wt.% Ni balance Ti alloy, straight annealed, and provided by SAES GETTERS was used. The superelastic wires are of 100 μm diameter in the straight configuration and 200 μm in the "zigzag shape" configuration. They include Austenite Finish temperature $A_f = -15$ °C as measured on finished wires, upper plateau stress (at 4% strain) ≅ 400 MPa, and lower plateau stress (at 4% strain) ≅ 100 MPa. Moreover, steel wires made of AISI 304L alloy with 100 μm diameter, 700 MPa break stress, and 5% break deformation, provided by COMAR S.r.l., were embedded in the straight configuration to make the comparison with equivalent SMA composites.

To embed SMA wires into the composite layers, special frames were designed to allow embedding the wires into the matrix equally spaced (with a distance between the wires equal to 1 mm) and aligned in the principal direction of the sample. Four layers of wires have been embedded in the composites at different distances from the neutral axis, near the external surfaces of the specimen. Instead, in the second configuration, special hybrid textile preforms (HEP) were realized by Extreme Materials, a company involved with the development of innovative hybrid textiles for technical uses. In these preforms, SMA wires are knitted within the fabric structure together with reinforcing glass bundles with a zigzag shape (patent pending). The volume percentage of metallic wires embedded into the two laminates is about the same and equal to 1%.

The laminates have been realized by vacuum-assisted resin infusion process. In the straight configuration, a vinylester resin Distitron VE370SC provided by Polynt was poured into a vacuum bag in which the frame necessary for the embedding of wires was positioned. Instead, in the second configuration, the wires are positioned by means of the textile preforms in which SMA wires are knitted with the glass fibers. These hybrid preforms have been laminated far from the neutral axis of the laminated to enhance the influence of SMA. So five different laminates have been realized and tested to evaluate the influence of hybridization:

- Traditional glass-reinforced laminates with SMA wires in a straight configuration by infusion of EE425 preforms (laminate 1);
- Hybrid glass-reinforced laminates with SMA wires in a straight configuration by infusion of EE425 preforms (laminate 1-SMA);
- Hybrid glass-reinforced laminates with steel wires in straight configuration by infusion of EE425 preforms (laminate 1-steel);
- Glass-reinforced laminates with HEP preforms (laminate 2-SMA); and
- Glass-reinforced laminates with Extreme Materials glass (EP) preforms (laminate 2).

The dimensions of the samples for damping characterization were 250 × 40 mm, whereas for impact and static flexural characterization the dimensions were 80 × 12.7 mm.

In Table 1, materials and stacking sequences of the composite laminates in which SMA and steel wires have been embedded are reported.

2.2 Experimental Methods

From the realized laminates, two types of specimens were obtained for damping and mechanical characterization: specimens without any metallic wires and hybrid specimens embedding SMA wires. Moreover, for the unidirectional configuration, specimens with embedded steel wires have also been fabricated. The wires have been embedded along the principal direction of the samples (see Fig. 1). In this figure, it is possible to see the predeformed regions of these wires in the knitted configuration.

2.2.1 Laser Vibrometry Measurement Setup. The schematic of the vibration testing setup is shown in Fig. 2. The samples (target) were clamped on one end by a vice and were excited in a controlled and repeatable manner by impacts with a pendulum. The active length of the samples was 212 mm and the impact point was positioned 110 mm from the free end. The release of the pendulum was controlled by an electromagnet. A Brüel & Kjaer type 4393 accelerometer with a mass of 5.6 g (including mount) and a sensitivity of 0.316 pC/(m/s^2) was attached to the pendulum to register the impact start. The impact velocity was 0.87 m/s.

The laser vibrometer system used is a Polytec PSV 300 scanning system. It is an interferometric device that uses a continuous wave He-Ne laser with a wavelength of 632.8 nm. The system measures the velocity of the vibrating target without contact.

The time responses of the composite plates were recorded after each impact. The sensitivity of the laser vibrometer was set to 10 mm/s/V with a range of 10 V. The measurements were trigged by the sharp slope of the accelerometer signal caused by the impact. Both signals were sampled at 1.28 kHz by the laser vibrometer system. The frequency response of each

Table 1 Materials and stacking sequences of the realized hybrid composites

SMA arrangements	Materials	Stacking sequences	Side views
Straight Laminate 1	SMA wires 100 μm provided by SAES Getters Steel wires 100 μm Glass Fabric EE425 Vinylester resin Distritron VE370SC by Polynt Thickness 3 mm	$[0^\circ, \mathrm{W}, 0^\circ, \mathrm{W}, (0^\circ, 90^\circ, 0^\circ)]_\mathrm{S}$	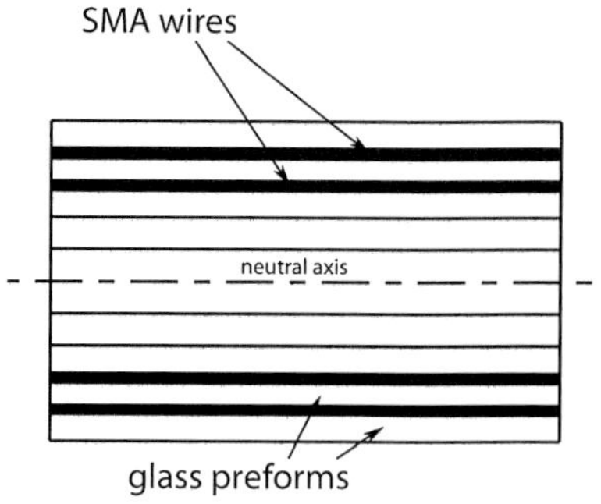
Knitted Laminate 2	SMA wires 200 μm Extreme hybrid preforms Extreme Materials' glass preforms Vinylester resin Distritron VE370SC by Polynt Thickness 4.2 mm	$[\mathrm{HEP}, \mathrm{HEP}, \mathrm{EP}_2, \overline{\mathrm{EP}}]_\mathrm{S}$	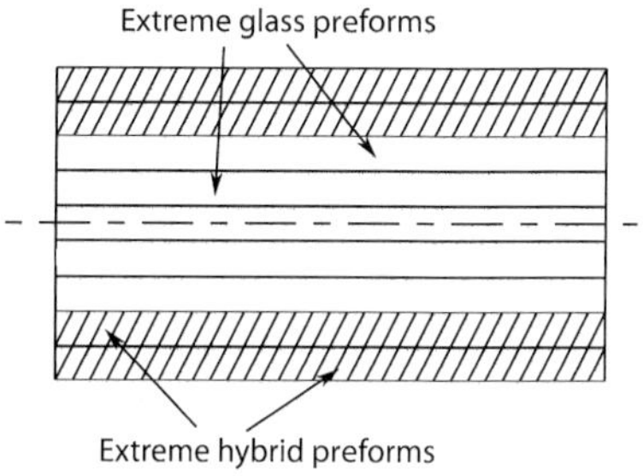

W indicates the metallic wires in the straight configuration, HEP indicates Hybrid Extreme Materials preforms, and EP indicates Extreme Materials preforms

measurement was calculated where natural frequencies and damping of each resonant mode were obtained.

2.2.2 Mechanical Characterization. Impact tests have been performed according to the Charpy test method (ASTM D256, Ref 11) by an instrumented impact tester. In Fig. 3, a schematization of sample geometry, positioning, and testing procedure is reported. The pendulum hammer was equipped with an ICP® Dynamic Force Sensor provided by PCB Piezoelectronics providing the force as a function of the time during the impact test. The impact height, H, was 0.74, while E was determined using $E = mgh$, where E is the energy, m is the mass of the impact head (equal to 3.48 kg), and g is the acceleration due to gravity. Non-notched samples of dimensions 12.7 × 80 mm were rested freely against two anvils and struck in the center by a pendulum. The results shown are an average between four and eight specimens of each type.

Static flexural tests have been performed by 3-point bend procedure (ASTM D790, Ref 12): the results are an average of 15 specimens of each type, and their dimensions were the same of Charpy tests.

2.3 Approach

2.3.1 Laser Vibrometry Analysis. Damping or loss factor measurements are rarely straightforward due to the complexity of the dynamic interaction of system joints, trim, and geometry (Ref 13-16). Furthermore, a variety of nomenclature exists to denote damping. These include: damping ratio (ζ), log decrement (δ), loss factor (η), loss angle (Φ), etc. Consequently, equal to the number of different descriptors for damping levels there are different test methods.

In this work, we have chosen to report the measure of the damping system by damping ratio values, and therefore how long the system will take to reach its final value. It is defined as the ratio of the damping constant to the critical damping constant:

$$\zeta = \frac{c}{c_\mathrm{c}}$$

If the damping ratio is low, the system can oscillate for longer period (underdamping), whereas if it is high, it may not oscillate at all (overdamping). For this reason, an increase of damping ratio of the system leads to an enhancing of its damping properties and a reduction of the resonant effects.

A possible way to determine the amount of damping in a system is to measure the rate at which the oscillation decays. The logarithmic decrement method described in Ref 13 was used in this work to calculate the damping ratio.

In this experiment, not only the material and support causes damping but also the surrounding air. But, since the geometry and the eigenfrequencies are nearly the same for the samples (e.g., Laminate 1 or 2), the air damping will be the same.

2.4 Impact Analysis

In Fig. 4, a typical force-displacement curve resulting from a Charpy test on a composite sample is reported (Ref 17). The energy needed to break the sample is the area obtained integrating this curve. One method of expressing impact energy, for a plastic or a composite, is in terms of the formula: $U = E/bd$, where U is the impact energy, E is the energy registered in the test, b is the width of the specimen, and d is its thickness. Moreover, by integrating the area under the load-displacement curve and noting the value up to the maximum load or the load at first failure, it is possible to deduce the energy required to start or initiate damage and relate it to the energy of propagate damage. The ratio of the propagation energy E_p to the initiation energy E_i is known as the ductility

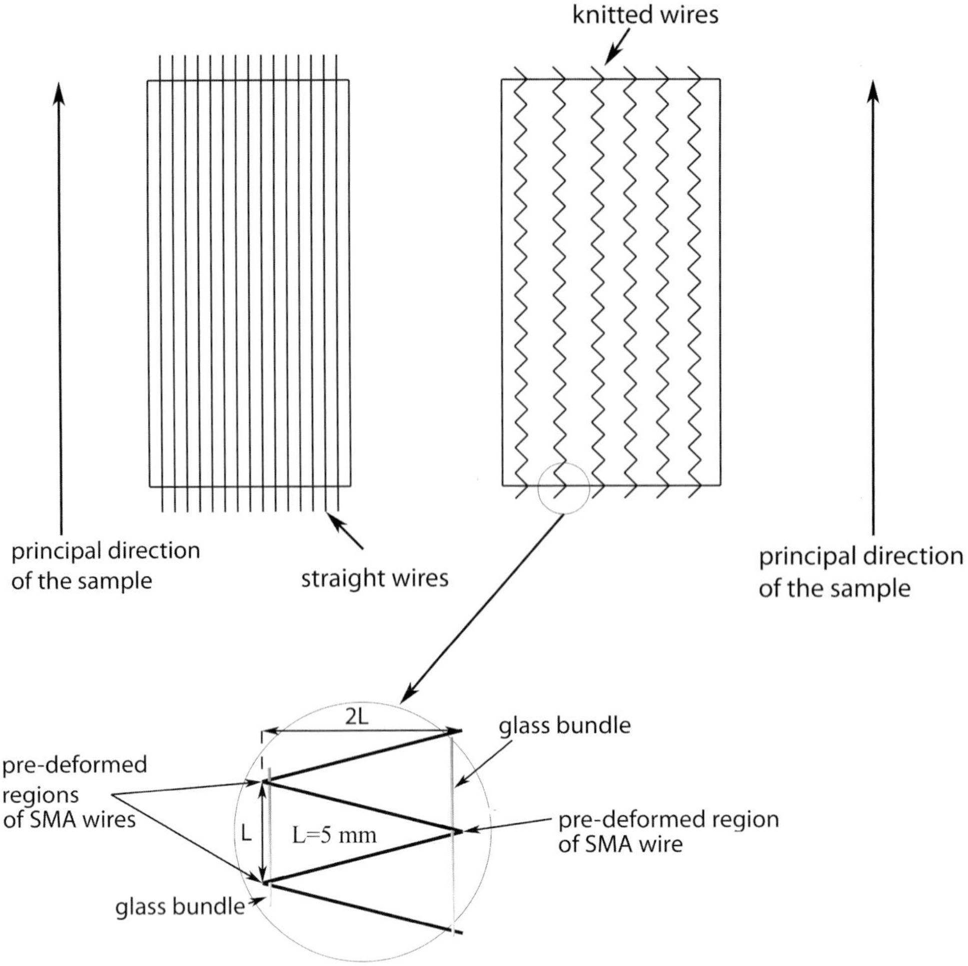

Fig. 1 Schematic diagram of the two configurations of the wires in the samples

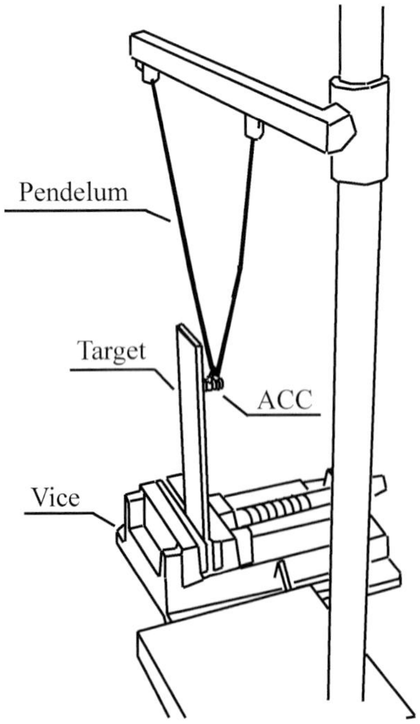

Fig. 2 The schematic of the vibration testing setup; ACC, accelerometer

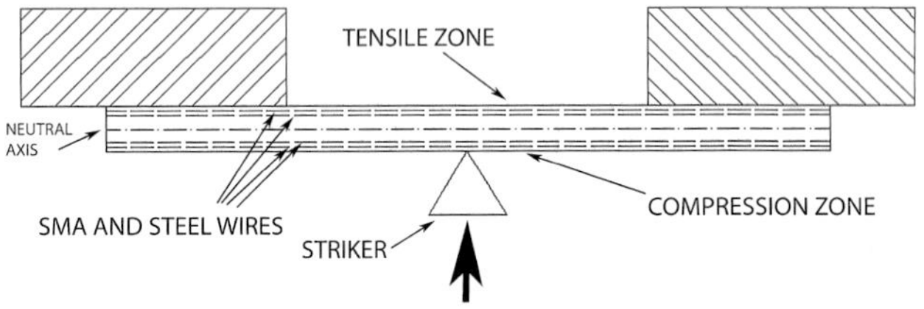

Fig. 3 Schematization of sample geometry, positioning, and testing procedure

index D, which is calculated by the formula $D = E_p/E_i$. For these reasons, the ductility of the material is enhanced only if both U and D increase, otherwise if only U increases it can be considered as a brittle reinforcement of the material behaving like the other reinforcement.

3. Experimental Results

3.1 Laser Vibrometry Analysis

Figure 5 shows the calculated frequency responses of the laminate 1 sample obtained from laser vibrometry measurements. In Fig. 5(a), the first bending frequencies of the samples

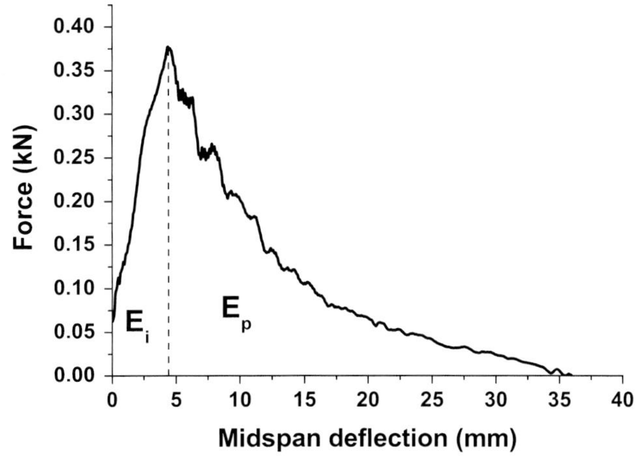

Fig. 4 Typical force-displacement curve resulting from a Charpy test on a composite sample

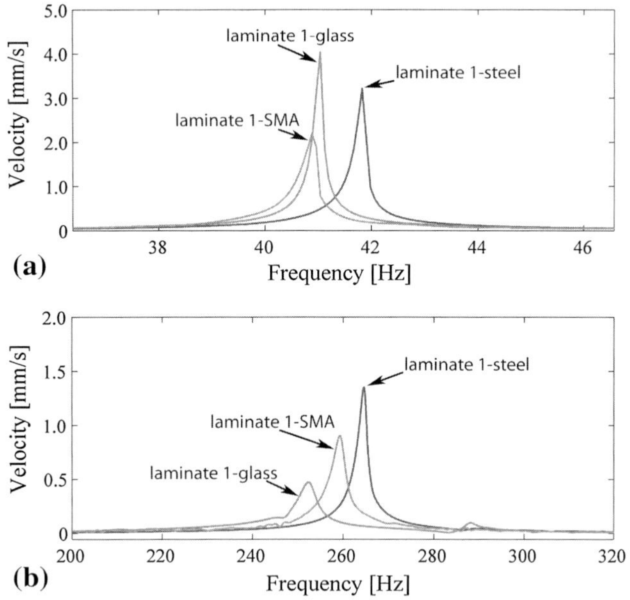

Fig. 5 (a) Plots of the first bending frequency response for the laminate 1 samples and (b) Plots of the second bending frequency response for the laminate 1 samples

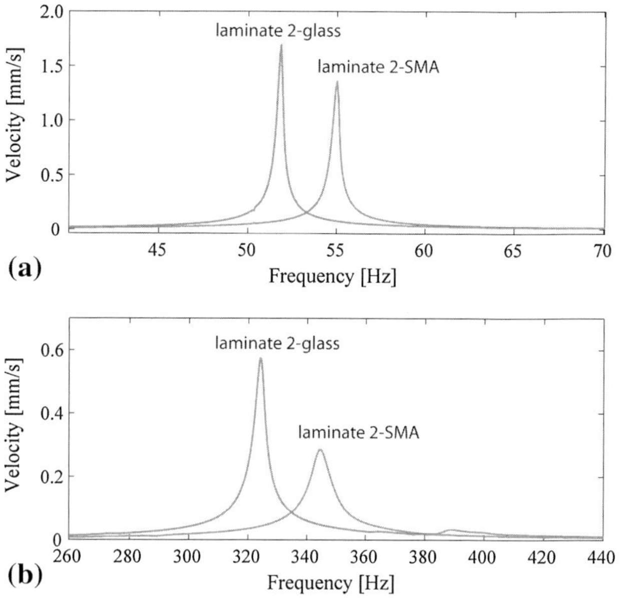

Fig. 6 (a) Plots of the first bending frequency response for the laminate 2 samples and (b) Plots of the second bending frequency response for the laminate 2 samples

are shown. Clearly, the SMA-fiber-reinforced composite has the smallest vibration velocity amplitude at this frequency. The damping ratios are: laminate 1-SMA $\zeta = 0.25\%$, laminate 1-steel $\zeta = 0.19\%$, and laminate 1-glass $\zeta = 0.18\%$. In Fig. 5(b), the second bending frequencies are shown. The corresponding damping ratios are: laminate 1-SMA $\zeta = 0.50\%$, laminate 1-steel $\zeta = 0.32\%$, and laminate 1-glass $\zeta = 0.73\%$. The laminate 1-glass sample has a small peak at 288 Hz which is its first torsional frequency.

So, embedding straight SMA wires into traditional composites has increased the damping ratio for the first bending ratio, while the influence of steel wires is lower. However, the positive effect provided by SMA wires cannot be related to the hysteresis loss which characterize these materials; in fact, the maximum strain at the beginning of vibration at the fixed end is about 50 microstrain, not big enough for the onset of the reverse martensitic transformation (that is about 10,000 microstrain). Moreover, for the second bending frequency, SMA hybridization causes a decrease of damping ratio (even if less considerable than for steel wires).

The calculated frequency responses of the laminate 2 samples obtained from laser vibrometry measurements are shown in Fig. 6. In Fig. 6(a), the first bending frequencies of the samples are shown. The natural frequency of the laminate 2-SMA sample is slightly higher compared to the laminate 2-glass sample. The damping ratios are also increased: laminate 2-SMA $\zeta = 0.32\%$ and laminate 2-glass $\zeta = 0.29\%$. In Fig. 6(b), the second bending frequencies are shown, where the damping ratios are laminate 2-SMA $\zeta = 0.93\%$ and laminate 2-glass $\zeta = 0.51\%$.

Clearly, the wrinkled SMA-fiber configuration induces a significant increase of the damping ratio for both the first and the second bending frequencies. In this case, this positive result is probably due to the initial deformation applied to the wires during knitting process. Further studies are necessary for concluding if this difference has a considerable effect in practical situation.

Finally, field measurements were obtained by scanning the laser beam across the surface where a grid of 36 points was measured. The mode shapes of the samples are shown in Fig. 7 where the lower ends are the fixed positions and the upper ends are the free ends. The color coding represents velocities, where the blue and the red areas are vibrating out of phase relative to each other. In Fig. 7(a), the first bending mode of the laminate 1-SMA sample is shown where the farther the fixed end the larger the velocities are. In Fig. 7(b), the second bending mode of the laminate 1-SMA sample is shown where the surface is split by a horizontal nodal line in the upper part of the sample. The first twisting mode of the laminate 1-glass sample is shown in Fig. 7(c) where the surface is split by a vertical nodal line. This mode is, however, only slightly generated since the impact point is positioned in the middle of cantilever.

Fig. 7 Mode shapes of the laminate 1-samples: (a) laminate 1-SMA first bending mode, (b) laminate 1-SMA second bending mode, and (c) laminate 1-glass first twisting mode. Color coding: light blue, zero velocity and light red, maximum velocity

Table 2 Synthesis of Charpy test results

	E/A, J/mm^2			*D*		
Laminate	**No wires**	**Steel wires**	**SMA wires**	**No wires**	**Steel wires**	**SMA wires**
Laminate 1	0.215	0.205 −4.7%	0.211 −1.86%	1.14	1.37 +20.2%	1.47 +28.9%
Laminate 2	0.112	···	0.153 +36.6%	1.03	···	1.19 +15.5%

The increases are referred to the absorbed energy or *D* index of the samples without metallic wires

3.2 Impact Analysis

A synthesis of the Charpy and static flexural results is reported in Table 2 and 3. The average *F*/*A* (where *F* is the load measured during Charpy tests and *A* is the area of the cross section of the samples) and displacements impact curves for laminates 1 and 2 are shown in Fig. 8 and 9. The results are reported in terms of *E*/*A* (where *E* is the energy obtained by integrating the force-displacements curves) to take into account little increases of the thickness of the hybrid samples due to the introduction of the metallic wires.

As it is evident from tables and figures, SMA hybridization has a positive effect only for laminate 2, in which impact energy increases by +36.6%, compared with impact energy of the analogous nonhybrid samples. So, embedding SMA wires with the knitted configuration has increased the ductility of the laminate, as the increase of ductility index *D* (+15.5%) confirms. Instead, laminate 1 is only marginally affected by the hybridization, both for steel and SMA wires, since impact energy remains almost unchanged compared with analogous

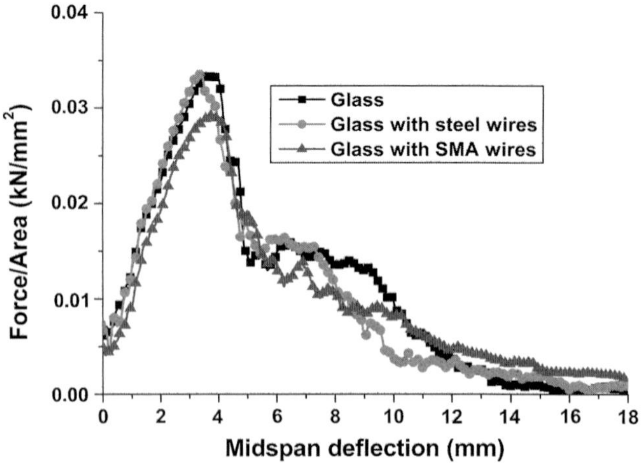

Fig. 8 Average force/area and displacements impact curves for laminate 1

Table 3 Synthesis of static flexural characterization results

	E_c, GPa	ε_r, %	σ_r, MPa
Laminate 1			
No wires	16.8	4.2	650
Steel wires	17.8 +6%	4.2 ···	541 −17%
SMA wires	15.5 −8%	4 −5%	513 −21%
Laminate 2			
No wires	7.4	6.5	232.5
SMA wires	5.5 −25.7%	7.1 +9.2%	171 −26.5%

The increases are referred to the absorbed energy of the samples without metallic wires
E_c, longitudinal flexural modulus, ε_r, longitudinal flexural strain, and σ_r, longitudinal flexural strength

nonhybrid sample. Moreover, the increase of *D* for hybrid laminate 1 (+28.9%) has to be related to the decrease of E_i caused by the decreases of the stiffness and ultimate loads of the laminate due to hybridization (see blue curve in Fig. 8).

This different influence of hybridization on the impact energy for the two laminates is probably due to the higher elastic modulus of laminate 1 which is twice the modulus of laminate 2. In fact, for laminate 1, the small amount of SMA wires (1% by volume) cannot influence the impact behavior of the composite. Moreover, laminate 1 has a higher break deformation, and so it is more ductile than laminate 2. On the other hand, laminate 2 presents lower modulus and lower deformation to break, which coupled with the knitted configuration of the embedded wires is probably responsible for the positive effect of hybridization. However, in both cases, hybridization causes a decrease of the static flexural properties that are affected by the external layers in each lay up configuration. Factors such as fiber crimping and fiber matrix adhesion can be responsible for the lower ultimate properties. For these reasons, tensile properties should be less affected by the presence of SMA wires.

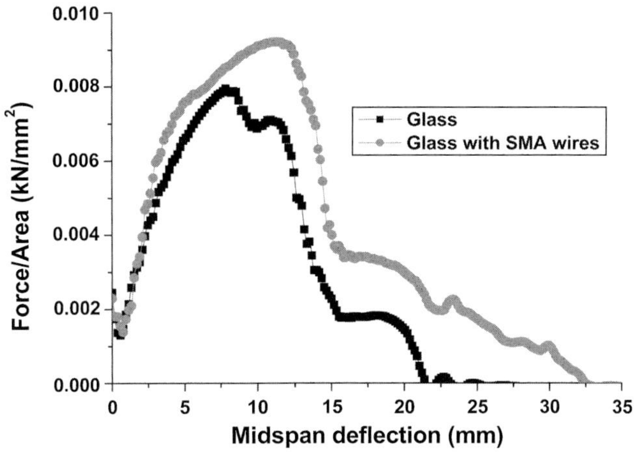

Fig. 9 Average force/area and displacements impact curves for laminate 2

4. Conclusions

In this work, the possibility of enhancing the damping properties of traditional glass-reinforced composites by means of SMA wires has been evaluated. Results show embedding SMA wires with a knitted configuration; thus, providing an initial deformation to the wires increases the positive effect of the hybridization. However, even with straight configuration there is an increase of the damping ratio for the first bending frequency, even if this effect cannot be related to the hysteretic behavior of superelastic alloys. Probably, this positive effect is due to the austenitic structure of SMA material in the laminate 1 and additionally to the martensitic structure of the predeformed regions in the knitted configuration. Further tests are needed to well investigate this phenomenon, for example, embedding SMA wires of smaller diameter with knitted configuration or embedding SMA predeformed wires in the straight configuration. Instead, it is clear that steel wires cannot increase damping properties of the composite.

Then, impact properties of hybrid composites indicated that SMA wires in the knitted configuration cause a considerable increase of impact strength, while a decrease of flexural mechanical properties due to hybridization was observed.

Acknowledgments

This research was completed in the framework of the AVALON project, a European Commission Community Research 6th Framework Programme, Priority 3 NMP Project. The authors would like to acknowledge Polynt for the resins' supply.

References

1. R. Stalmans, V. Michaud, J. E. Bidaux, R. Gotthardt, and J. A. E. Manson, Adaptive Composites with Embedded Shape Memory Alloy Wires, in *Proceedings of the 4th European Conference on Smart Structures and Materials*, G. R. Tomlinsonen and W. A. Bullough, Eds., Institute of Physics Publishing, Bristol, England, 1998, p 801–804
2. C. Boller, Shape Memory Alloys – Their Challenge to Contribute to Smart Structures, *Annual Meeting of the Materials Research Society*, Boston, MA (Invited paper)
3. R. Stalmans, Invited Lecture—Adaptive Hybrid Composites with a Focus on the Integration of Shape Memory Elements, *Adv. Sci. Technol.*, 1999, **25**, p 83–94, *Smart Materials Systems*, P. Vincenzini, Ed., June 14–19, 1998 (Florence, Italy), Techna, Faenza, Italy
4. R. Stalmans, K. Tsoi, and J. Schrooten, The Transformational Behaviour of Shape Memory Wires Embedded in a Composite Matrix, *Proceedings of SPIE Fifth European Conference on Smart Structures and Materials*, P. F. Gobin and C. M. Friend, Eds., May 22-25, 2000 (Glasgow, Scotland), 2000, **4073**, p 88–96
5. K. Tsoi, et al., Impact Damage Behavior of Shape Memory Alloy Composites, *Mater. Sci. Eng.*, 2003, **342**(1–2), p 207–215
6. K.-T. Lau, et al., Low Velocity Impacts on Shape Memory Alloy Stitched Composite Plates, *Smart Mater. Struct.*, 2004, **13**, p 364–370
7. R.-X. Zhang, and Q.-Q. Ni, et al., Vibration Characteristics of Laminated Composite Plates with Embedded Shape Memory Alloy, *Compos. Struct.*, 2006, **74**, p 389–398
8. Q.-Q. Ni, and R.-X. Zhang, et al., Stiffness and Vibration Characteristics of SMA/ER3 Composites with Shape Memory Alloy Short Fibers, *Compos. Struct.*, 2007, **79**, p 501–507
9. W. M. Ostachowicz, and S. Kaczmrczyk, Vibration of Composite Plates with SMA Fibers in a Gas Stream with Defects of the Type of Delamination, *Compos. Struct.*, 2001, **54**, p 305–311
10. T. L. Turner, Structural Acoustic Response of a Shape Memory Alloy Hybrid Composite Panel, *Proc. SPIE*, 2002, **4701**, p 592–603
11. ASTM D256-04, *Standard Test Methods for Determining the Izod Pendulum Impact Resistance of Plastics*, ASTM International, 2004, p 1–20
12. ASTM D790-03 *Standard Test Methods for Flexural Properties of Unreinforced and Reinforced Plastics and Electrical Insulating Materials*, ASTM International, 2003, p 1–11
13. D.J. Inman, *Engineering Vibration*, 3rd ed., Prentice-Hall International Editions, Upper Saddle River, NJ, 2008
14. W.T. Thomson and M. D. Dahleh, *Theory of Vibration with Applications*, 5th ed., Prentice-Hall, Upper Saddle River, NJ, 1998
15. A.A. Shabana, "Theory of Vibration," Drugaizdaja, University of Illinois, Chicago, 1995
16. L.E. Drain, 1980, *The Laser Doppler Technique*, Wiley, London
17. S.R. Reid and G. Zhou, *Impact Behaviour of Fibre-Reinforced Composite Materials and Structures*, Woodhead Publishing Ltd., Cambridge, UK, 2000

JMEPEG (2009) 18:538–542
DOI: 10.1007/s11665-009-9494-7

1059-9495/$19.00

Surface Form Memory in NiTi: Energy Density of Constrained Recovery During Indent Replication

Xueling Fei, Corey J. O'Connell, David S. Grummon, and Yang-Tse Cheng

(Submitted October 7, 2008; in revised form February 9, 2009)

Spherical indentation of NiTi shape memory alloys (SMA) to depths greater than about 3% of the indenter radius results in two-way shape-memory training in a deformation zone beneath the indent. If deep spherical or cylindrical indents are subsequently machined away just sufficiently to remove traces of the original indent (in the martensitic condition), a thermally induced and cyclically reversible flat-to-protruded surface topography is enabled. We term the phenomenon surface form memory. The amplitude of cyclic protrusions, or 'exdents', is related to the existence of a subsurface deformation zone in which indentation has resulted in plastic strains beyond that which can be accomplished by martensite detwinning reactions. Dislocation generation in this zone is thought to underlie the observed two-way shape-memory (TWSME) training effect. In this article, we show that these cyclic exdents can perform appreciable mechanical work when displacing under load against a base-metal substrate (constrained recovery). This "non-Hertzian" indentation, which appears to be able to exert the full energy density of SMA actuation, may have use for assembly of micromachines, bond-release, microforging, microjoining, electrical switching, microconnectors, and variable heat transfer devices, among many other potential applications.

Keywords energy density, NiTi, shape memory, surfaces

1. Introduction

Several groups have investigated shape memory and superelastic effects using micro- and nanoindentation techniques (Ref 1-5). Recently, Zhang et al. (Ref 6, 7) reported thermally reversible flat-to-bumpy surface transitions using an indentation-planarization method applied to both wrought and thin film NiTi. The planarization step, which removes visible traces of the indent, converts an indented surface having two-way indent depth change to one with thermally controllable *form*.

The essential features of the technique are schematically illustrated in Fig. 1. Here, r is the initial indenter radius, a is the indent contact radius, and h is the initial indent depth. The ratio a/r can be used to characterize the representative strain, ε_r, produced by the indent, which in turn correlates with the plastic component of total strain in tensile tests (Ref 8, 9). For spherical indenters, $\varepsilon_r = 0.2\ a/r$. Previous work has shown that two-way indent recovery effects are observed only when a/r exceeds 0.25, equivalent to $\varepsilon_r = 0.05$, i.e., the strain corresponding to the end of the detwinning plateau in tensile deformation of martensitic NiTi.

Both the amplitude of reversible indent depth changes (δ_1 in Fig. 1) and the height δ_2 of the surface protrusions (exdents) can be related to the depth of a subsurface zone in which indentation has induced strains beyond those that can be mediated by deformation twinning reactions in the martensite. Dislocation production in this zone probably accounts for the observed two-way effect. The depth, D^*, of this 'active' zone has been determined by experiment (as described elsewhere, Ref 10) for both spherical and cylindrical indentation. As can be seen in Fig. 2, which plots the exdent height, δ_2, as a function of measured values of D^*, indicates that an average reversible strain within the active zone of ~2.4% can account for the observed displacements.

For surface form memory (SFM) to be useful as an actuation strategy, or for application to variable-friction surfaces, mold-release, or lithographic devices, it is necessary to demonstrate that useful work can be extracted during austenitization, i.e., that SFM can operate robustly in a constrained-recovery mode.

In this article, we present results on experiments designed to estimate the mechanical work that can be exerted by spherical protrusions when they displace against a base-metal substrate, and thereby produce a 'replicant' indent. It is shown that, by conservative estimate, the energy density associated with constrained recovery in SFM is on the order of the maximum generally available from NiTi actuators.

2. Experiments

A 3 mm thick 50.26 at.% Ni wrought NiTi alloy sheet acquired from Special Metals Corporation had transformation temperatures, M_f, M_s, A_s, and A_f, of 22, 58, 70, and 110 °C

This article is an invited paper selected from presentations at Shape Memory and Superelastic Technologies 2008, held September 21-25, 2008, in Stresa, Italy, and has been expanded from the original presentation.

Xueling Fei, Corey J. O'Connell, and **David S. Grummon**, Department of Chemical Engineering and Materials Science, Michigan State University, East Lansing, MI 48824; and **Yang-Tse Cheng**, Department of Chemical and Materials Engineering, University of Kentucky, Lexington, KY 40506. Contact e-mail: grummon@egr.msu.edu.

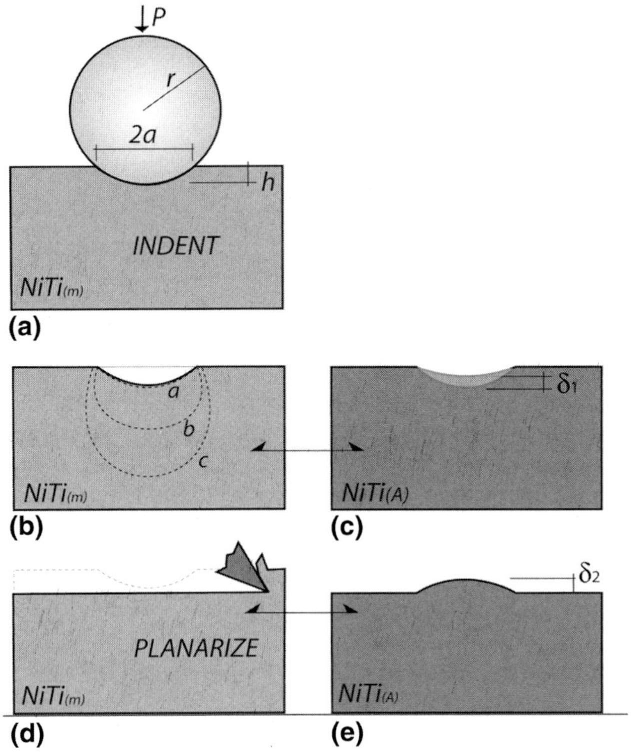

Fig. 1 Indent training (a-c) and planarization (d) to obtain surface form memory (SFM) 'exdents' in NiTi. Thermally cycling between (b) ↔ (c) and (d) ↔ (e) is indefinitely repeatable

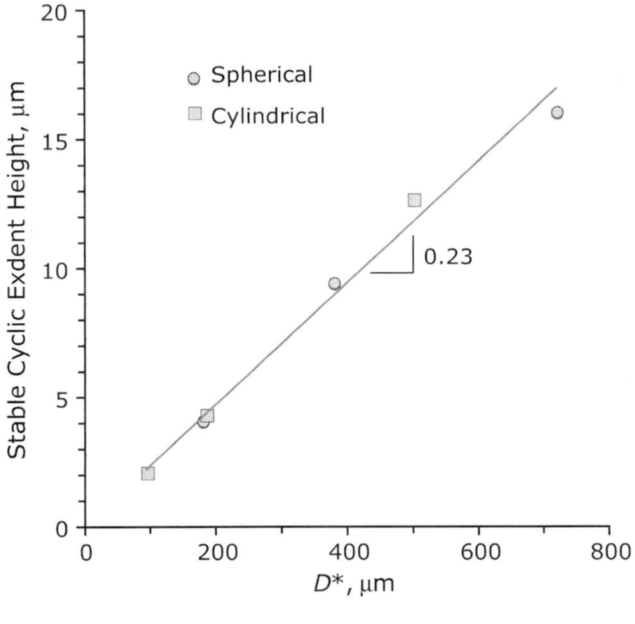

Fig. 2 Stable cyclic protrusion height for spherical and cylindrical indents as a function of the depth of the subsurface active zone, D^*. The slope of the linear curve fit gives an average strain in the volume defined by D^* of ~2.4%

respectively, as determined by differential scanning calorimetry. Specimens were first electrodischarge cut into 12.7 mm diameter discs and manually ground with 2000-grit SiC paper. They were then machine polished at room temperature with 0.05 µm colloidal silica. Before indentation, the specimens were chilled to approximately −50 °C to ensure a fully martensitic condition before later indentation at room temperature.

Table 1 Initial spherical indents into martensitic NiTi

Load, N	Depth, µm	Contact radius (a), µm	a/r ratio
200	37	167	0.42
700	107	271	0.68
3000	380	397	1.00

Spherical indents were made with an 0.794 mm diameter tungsten carbide ball using loads of 200, 700, and 3000 N. Each sample was indented at the same location twice since this had been found in previous work to give a stronger training effect. Surface profiles of the as-indented surfaces were acquired with a WYKO NT 1000 optical surface profilometer. The loads used gave a/r ratios of 0.42, 0.68, and 1.0, respectively, as summarized in Table 1. After indentation, and while still in the martensitic condition, the surfaces were ground and polished just enough to remove visible traces of the indent.

Constrained-recovery experiments were designed to obtain a rough estimate of the work output of the protrusions during constrained recovery. After planarization, the optically flat NiTi samples were placed in contact with coupons of 3 mm thick 6061-T6 aluminum and positioned between compression platens in a screw-driven load frame. The NiTi-aluminum sandwich was preloaded to a compressive stress of 100 MPa to insure positive contact. The assembly was then slowly heated by convection to 150 °C to austenize the NiTi, thus allowing the NiTi exdent to indent the aluminum coupon. (Thermal expansion in the assembly caused the stress to rise to about three times the initial preload.) Dimensional profiles of the indented aluminum surfaces were subsequently acquired using the WYKO profilometer. Additionally, profiles were obtained for the residual exdents in the NiTi samples, both directly following the indent replication experiment (at room temperature), and also in the austenitic condition, with samples heated to 150 °C, in situ in the profilometer stage, using a thermoelectric module.

To provide a basis for estimation of the work output of the NiTi exdent, conventional spherical indents were made in the same 6061-T6 aluminum plate using a 6.3 mm diameter tungsten ball under various loads. The diameter of this ball was chosen to match the radius of curvature of the replicant indents. Indents that were found to have a/r ratios comparable to those of the replicants were chosen to provide force-displacement data to give a rough measure of the work of conventional (Hertzian) indentation in 6061-T6 aluminum.

3. Results and Discussion

3.1 Spherical Replicant Indents in Aluminum

Surface profiles of replicant indents are shown in Fig. 3, together with profiles of the corresponding NiTi exdents taken at room temperature after the replicants were made. Measurements derived from these profiles are summarized in Table 2, which additionally includes measurement of the austenitic

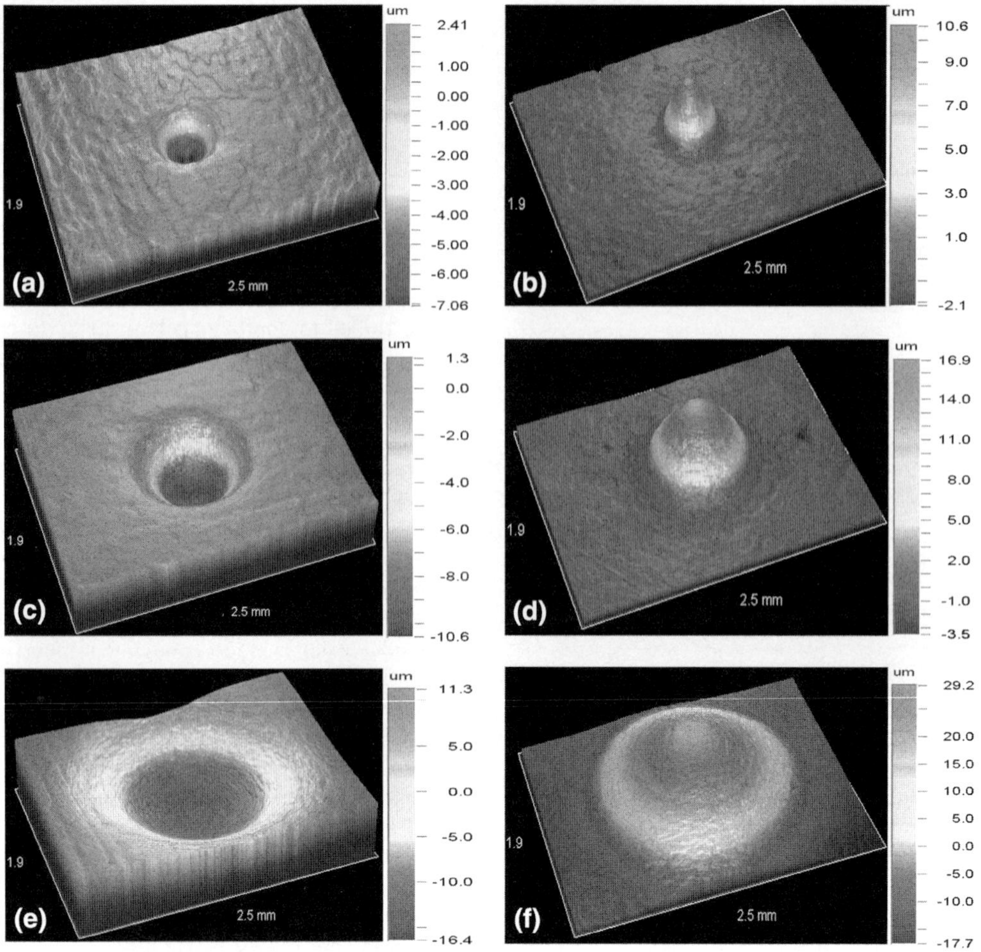

Fig. 3 Replicant indents in aluminum (left) and residual exdents in NiTi at 20 °C made directly following the replication experiment. Images (a, b), (c, d) and (e, f) correspond to initial indent *a/r* ratios of 0.42, 0.68, and 1.00, respectively. In each image, the field size is 1.9 mm × 2.5 mm. Vertical scales vary, as shown

Table 2 Replicant indent and residual exdent dimensions

Initial indent *a/r* ratio	Replicant indent depth (d_{repl}), μm	Replicant radius of curvature, μm	Replicant indent *a/r* ratio	NiTi exdent height, μm	
				20 °C	150 °C
0.42	7.1	3,808	0.061	10.7	16.4
0.68	11.4	8,812	0.051	20	27.6
1.00	25.8	15,443	0.058	42	51.2

exdent height after the replication experiment with NiTi samples re-heated to 150 °C in the profilometer. The latter measurement may be construed as the free-recovery amplitude of the exdents.

The replicant indent depth (normalized to the original indenter radius) is plotted as a function of the initial indentation *a/r* ratio in Fig. 4. The curve-fit shown is a second-order polynomial (constrained to pass through the origin) which achieved an R^2 value of 0.977. From this it may be concluded that, for the maximum initial indent *a/r* ratio of 1.0 (i.e., indentation of NiTi to a depth equal to the indenter radius), a

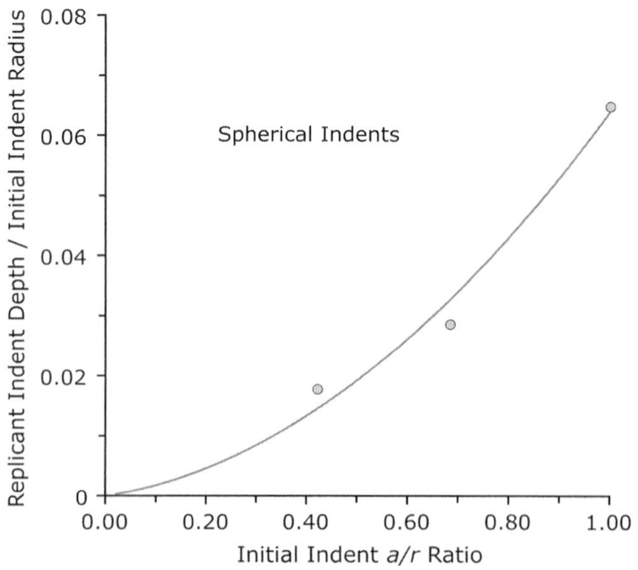

Fig. 4 Depth of replicant indents (normalized to the initial NiTi indenter radius) plotted as a function of the initial indent *a/r* ratio

replicant indent can be produced in 6061-T6 aluminum that has a depth that is approximately 6.5% of the initial NiTi indenter radius. This value declines to less than 2% for an initial *a/r* ratio of 0.42.

It is also of interest to compare the replicant indent depth to the height of the corresponding NiTi exdent in the austenitic condition at 150 °C (see leftmost column in Table 2). This is shown in Fig. 5 which plots these measurements, together with results of similar experiments conducted using cylindrical indenters to be described in detail elsewhere. For the present case of indent replication in 6061-T6 aluminum, a simple relationship between these two quantities emerges: The replicant indent depth, for both spherical and cylindrical indentation, is approximately half (47%) of the austenitic exdent height. The difference may be presumed to be composed of elastic springback in the aluminum and limitation of the SFM displacement under the imposed constraint conditions.

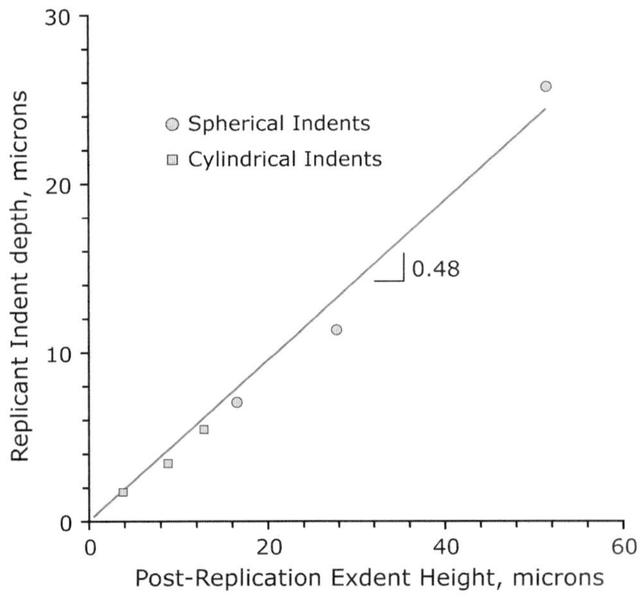

Fig. 5 The depth of replicant indents plotted as a function of the postreplication height of corresponding NiTi exdents when re-warmed to 150 °C (austenitic condition)

Experiments are under way on various different replication-substrate alloys to further study these relationships.

3.2 Calibration Indents in Aluminum

We have made a very rough approximation of the apparent energy density associated with exdent formation during constrained recovery in SFM. This has been accomplished by making conventional spherical indents in 6061-T6 aluminum using a ball with a radius close to that found for replicant indents in the study, and indenting to *a/r* ratios that are also close to those for of the replicants. Results for these calibration indentations are given in Table 3. For *a/r* ratios between 0.51 and 0.68, these experiments determined the spherical indentation hardness (H_{Al}, defined as the load divided by the projected contact area) to be ~1.3 GPa.

The projected contact area during replicant indentation is then calculated as $A_{repl} = \pi a_{repl}^2$ allowing us to estimate the peak replication force (assumed to be an increment above the preload force) as $P_{repl} = A_{repl} H_{Al}$. The energy of indentation then becomes $U_i = 1/2 P_{repl} d_{repl}$.

To obtain the effective energy density, it is necessary to make an estimate of the volume of material in which shape-memory recovery occurs, i.e., the active subsurface volume contributing to formation of the exdent. Previous experiments to determine the active volume that drives exdent formation (producing the results shown in Fig. 2) were made on specimens which had been planarized after having been thermally cycled several times between M_f and A_f, such that exdent formation was cyclically stable. In these cases, only the two-way component of shape memory is retained, and the depth of the active zone, D^*, was found to be on the order of 1.5 times the contact radius for the initial indent. (This is a depth that can be well correlated with the 5 to 7% strain contour in the subsurface deformation zone.)

However, in the present case, exdent formation without prior thermal cycling occurs by one-shot shape memory, and the depth of the active zone should extend roughly to the elastic-plastic boundary for the initial indent. This depth is not known directly in the present case, but experimental work on conventional spherical indentation in work-hardening materials (Ref 11) suggests that it is on the order of four times the contact radius. To arrive at a conservative estimate, we have adopted the latter approximation for D^*.

Table 3 Calibration indentation data for 6061-T6 aluminum

Indent depth, μm	Load, N	Indenter radius, m	*a/r* ratio	Contact radius, m	Contact area, m^2	Hardness (H_{Al}), Pa
4	100	3.08E − 03	0.051	1.57E − 04	7.75E − 08	1.29E + 09
5.2	135	3.08E − 03	0.058	1.79E − 04	1.00E − 07	1.35E + 09
5.7	150	3.08E − 03	0.061	1.88E − 04	1.11E − 07	1.35E + 09

Table 4 Quantities used in the calculation of energy density

Indent depth in Al, μm	Replicant contact radius (a_{repl}), μm	Replicant *a/r* ratio	Replicant contact area (A_{repl}), m^2	Equivalent replicant load (P_{repl}), N	Indentation energy (U), J	Estimated D^*, m	Estimated active volume (V^*), m^3	Estimated energy density, J/m^3
7.1	232.29	0.061	1.70E − 07	2.19E + 02	7.76E − 04	6.68E − 04	6.24292E − 10	1.24E + 06
11.4	449.41	0.051	6.35E − 07	8.54E + 02	4.87E − 03	1.08E − 03	2.66776E − 09	1.83E + 06
25.8	895.69	0.058	2.52E − 06	3.41E + 03	4.40E − 02	1.59E − 03	8.36174E − 09	5.26E + 06

With the further simplifying assumption that D^* represents the radius of a hemisphere centered on the indent, we estimate the volume of the active zone as $V^* = (1/2)[4/3\pi D^{*2}]$. Dividing the energy of indentation, U_i, by this quantity then provides an estimate of the energy density associated with exdent formation in NiTi during an SFM event. Results of this calculation are summarized in Table 4.

Energy densities estimated in this way ranged from 1.2 to 5.3 MJ/m^3. This may be compared to the energy density exerted by an NiTi actuator that undergoes a shape memory strain of 4% under constraint conditions that cause a reaction stress to rise linearly from 0 to 400 MPa during the actuation stroke. This energy is just $U = 1/2\sigma\varepsilon$, or about 8 MJ/m^3. Thus, the formation of exdents in NiTi under the present conditions of constrained recovery appears to carry a very appreciable fraction of the maximum energy density associated with NiTi actuator materials.

4. Conclusion

An indentation-planarization method has been demonstrated for NiTi shape memory alloys that results in two-way shape memory-driven transformation of surface form between flat and bumpy ('exdented') states during thermal cycling. We have termed this phenomenon 'surface form memory,' or SFM. Exdents formed by the SFM effect in NiTi are able to produce replicant indents in a base metal when the latter and the SFM alloy are brought into contact under a suitable preload and then heated.

Consideration of the hardness of the replicant alloy and the depth of the replicant indents leads to the conclusion that SFM involves shape memory energy density greater than 1 MJ/m^3, an appreciable fraction of the theoretical maximum for NiTi SMAs.

The present experiments used single spherical indenters and 'one-shot' shape-memory recovery during exdent formation. Preliminary experiments with cyclically stable exdents indicate that similar indent replication results can be obtained repeatedly from a single NiTi SFM 'tool.' Furthermore, the indentation-planarization technique is even more pronounced when cylindrical indenters are used, and in general, can be readily extended to close-packed indent arrays. In view of the fact that the process operates robustly at the nanoscale, the new technique may be applicable to a wide range of engineering problems from MEMS microassembly to nanolithography to thermally variable friction surfaces.

Acknowledgments

The authors gratefully acknowledge funding from the National Science Foundation under grants CMS0336810 and CMS0510294, and from General Motors Corporation.

References

1. R. Liu, D.Y. Li, Y.S. Xie, R. Llewellyn, and H.M. Hawthorne, Indentation Behavior of Pseudoelastic TiNi Alloy, *Scripta Mater.*, 1999, **41**, p 691
2. F.T. Cheng, P. Shi, and H.C. Man, Correlation of Cavitation Erosion Resistance with Indentation-Derived Properties for a NiTi Alloy, *Scripta Mater.*, 2001, **45**, p 1083
3. K. Gall, K. Juntunen, H.J. Maier, H. Sehitoglu, and Y.I. Chumlyakov, Instrumented Micro-Indentation of NiTi Shape-Memory Alloys, *Acta Mater.*, 2001, **49**, p 3205
4. G.A. Shaw, D.S. Stone, A.D. Johnson, A.B. Ellis, and W.C. Crone, Shape Memory Effect in Nanoindentation of Nickel-Titanium Thin Films, *Appl. Phys. Lett.*, 2003, **83**, p 257
5. X.G. Ma and K. Komvopoulos, Nanoscale Pseudoelastic Behavior of Indented Titanium-Nickel Films, *Appl. Phys. Lett.*, 2003, **83**, p 3773
6. Y.J. Zhang, Y.T. Cheng, and D.S. Grummon, Two-Way Indent Depth Recovery in a NiTi Shape Memory Alloy, *Appl. Phys. Lett.*, 2006, **88**, p 1904
7. Y.J. Zhang, Y.T. Cheng, and D.S. Grummon, Shape Memory Surfaces, *Appl. Phys. Lett.*, 2006, **89**, p 1912
8. D. Tabor, Indentation Hardness: Fifty Years On—A Personal View, *Philos. Mag. A*, 1996, **74**, p 1207
9. D. Tabor, A Simple Theory of Static and Dynamic Hardness, *Proc. R. Soc. Lond. A*, 1948, **192**, p 247
10. X. Fei, Y. Zhang, D.S. Grummon, and Y.-T. Cheng, Indentation-Induced Two-Way Shape Memory Surfaces, *J. Mater. Res.*, Focus issue on Indentation Methods in Advanced Materials Research, in review July, 2008
11. L.E. Samules and T.O. Mulhearn, An Experimental Investigation of the Deformed Zone Associated with Indentation Hardness Impressions, *J. Mech. Phys. Solids*, 1957, **5**, p 125

JMEPEG (2009) 18:543–547
DOI: 10.1007/s11665-009-9484-9

1059-9495/$19.00

Texture Development and Phase Transformation Behavior of Sputtered Ni-Ti Films

R.M.S. Martins, N. Schell, K.K. Mahesh, L. Pereira, R.J.C. Silva, and F.M. Braz Fernandes

(Submitted September 15, 2008; in revised form February 28, 2009)

It is essential to identify and control the preferential orientation of Ni-Ti shape memory alloy (SMA) films since it is a crucial factor in determining the shape memory behavior. In the present work, in situ studies by synchrotron radiation scattering enabled to identify the different steps of the structural evolution of Ni-Ti films during co-sputtering deposition. For micro-electromechanical systems (MEMS) integration, there is a need for an electrically and thermally insulating or sacrificial layer. Widening the scope of previous experiments concerning the influence of the deposition parameters on the structure of the Ni-Ti films, the incorporation of a TiN buffer layer has been tested. Here, it is established a relationship between the TiN substrates and Ni-Ti texture development (B2 phase). Ni-Ti films mainly containing grains with (110) or (211) planes of the B2 phase parallel to the film surface could be produced using TiN buffer layers with distinct thickness values. The electrical resistivity measurements performed during temperature cycling have shown that the crystallographic orientations of the Ni-Ti films influence their phase transformation characteristics. The resistivity increase during R-phase transformation, especially visible on cooling, is higher for Ni-Ti films with a higher fraction of grains of the B2 phase with (211) parallel to the film surface.

Keywords deposition by sputtering, in situ x-ray diffraction, Ni-Ti, phase transformation behavior, shape memory alloy, texture development

1. Introduction

In physical vapor deposition (PVD) processes the atoms are deposited from a gas phase onto a substrate and the deposition parameters play an important role in determining the microstructure and physical properties of the grown films (Ref 1). There are some published structure-zone models in the literature to describe the observed morphology of PVD grown films (Ref 2). However, Hultman and Sundgren (Ref 3) referred that though the structure zone models seem to be adequate in giving a qualitative classification of film microstructure formation, they do not include effects from the substrate material itself on the film growth mode. It has been found that the observed microstructure of hard coatings is not simply dependent on process parameters such as temperature and energetic particle bombardment during growth, but also significantly dependent on the substrate crystal structure, morphology, composition and/or surface contamination, all of which have important influences on the nucleation. For example, substrate surface roughness often exists at the same time on several size scales. This roughness can be reproduced in exaggerated form on the coating surfaces in *zone I* structures, due to superimposed shadow growth. It is also essential to choose the substrate material carefully since preferred crystallographic orientation in coatings has been associated with effects such as minimization in surface energies and a variation in condensation coefficients (Ref 4).

The composition and structure of sputtered Ni-Ti shape memory alloy (SMA) films are significantly affected by the sputtering conditions: target power, gas pressure, target to substrate distance, deposition temperature, substrate bias voltage etc. (Ref 5-10). Additionally, important issues like formation of film texture and its control are not yet understood. It is fundamental to identify and control their preferential orientation, as it is a crucial factor in determining the extent of the strain recovery (Ref 11). For micro-electro-mechanical systems (MEMS) integration, there is a need for an electrically and thermally insulating or sacrificial layer (Ref 12, 13). Therefore, the study of Ni-Ti film deposition on different types of substrates is an important topic of research.

In the present work, naturally oxidized Si(100) wafers as well as TiN buffer layers have been used as substrates for Ni-Ti deposition. Crystallized near equiatomic Ni-Ti films have been obtained at the end of the deposition, exhibiting distinct crystallographic orientations depending of the substrate type. It has been established a relationship between the substrates and texture development.

In order to achieve the goals proposed for this work, a two-magnetron sputter deposition chamber has been used allowing to heat and to apply a bias voltage to the substrate. This equipment could be mounted into the six-circle diffractometer

This article is an invited paper selected from presentations at Shape Memory and Superelastic Technologies 2008, held September 21-25, 2008, in Stresa, Italy, and has been expanded from the original presentation.

R.M.S. Martins, Institute of Ion Beam Physics and Materials Research, Forschungszentrum Dresden-Rossendorf, P.O. Box 510119, 01314 Dresden, Germany; **R.M.S. Martins, K.K. Mahesh, L. Pereira, R.J.C. Silva,** and **F.M. Braz Fernandes**, CENIMAT/I3N, Campus da FCT/UNL, 2829-516 Monte de Caparica, Portugal; and **N. Schell**, GKSS Research Center Geesthacht, Max-Planck-Str. 1, 21502 Geesthacht, Germany. Contact e-mail: rmsm@fct.unl.pt.

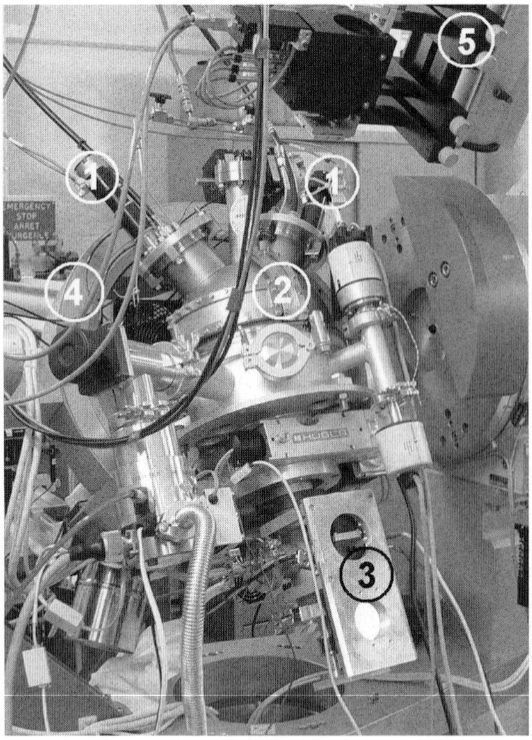

Fig. 1 Deposition chamber inserted into the six-circle diffractometer of ROBL: (1) magnetrons, (2) semicircular windows for diffracted beams in horizontal and vertical planes, (3) drive for height adjustment, (4) Eulerian cradle, and (5) detector arm

of the Rossendorf Beamline (ROBL) (Ref 14) at the European Synchrotron Radiation Facility (ESRF), enabling an in situ characterization by x-ray diffraction (XRD) of the films during their growth. In addition to the in situ studies, complementary ex situ characterization techniques such as electrical resistivity (ER) measurements during temperature cycling have been used to characterize the transformations characteristics of the Ni-Ti samples. Wang et al. (Ref 15), based on a previous work (Ref 16), reported that the greater the extent of sub micro twinning, the higher the increase in the resistivity. They also mentioned the finding of Cross et al. (Ref 17), i.e., there is a one-to-one correspondence between the mechanical "memory" effect and the ER curve; the larger the electrical resistivity "peak" [between austenite start temperature (A_s) and martensite finish temperature (M_s)], the more pronounced is the "memory" effect (i.e., the shape recovery performance).

2. Experimental

The sputtering experiments were carried out in the D.C. magnetron sputtering chamber that has been described elsewhere (Ref 18). Two unbalanced magnetrons, equipped with a 25.4 mm Ni-Ti target (49at.%Ni-51at.%Ti) and a 25.4 mm Ti target (purity 99.99%), respectively, were positioned at a distance of 100 mm from the substrate. They are each tilted 30° away from the substrate normal, including their cylindrical chimneys. Figure 1 shows the actual mounting of the deposition chamber in the Huber goniometer at ROBL.

Pieces of size 15×15 mm^2, cut from naturally oxidized Si(100) wafers (oxide layer of ≈2 nm) and thermally oxidized

Table 1 Selected deposition parameters for the various samples investigated

Substrate	Buffer layer	Deposition TiN: Time, min	Deposition TiN: Thickness, nm	Deposition Ni-Ti: Time, min	Deposition Ni-Ti: Thickness, nm
Si(100)	...	...	...	143	860
SiO_2/Si(100) (a)	TiN	3	15	120	720
SiO_2/Si(100) (a)	TiN	15	80	120	720
SiO_2/Si(100) (a)	TiN	40	215	120	720

(a) 140 nm amorphous SiO_2 capping layer

Si(100) wafers (with a 140 nm amorphous SiO_2 capping layer), have been used as substrates. The base pressure at the deposition temperature of ≈470 °C was 2×10^{-5} Pa and the working pressure during deposition was 0.42 Pa. For the deposition of the TiN buffer layers, the Ti target was run at a constant power of 80 W with an Ar/N_2 gas flow of 2/0.5 sccm and for the Ni-Ti films the Ni-Ti and Ti magnetrons were driven at a power of 40 and 20 W, respectively. The Ni-Ti films exhibited final compositions values of approximately 50at.%Ni-50at.%Ti (compositional results obtained from the combination of Rutherford backscattering spectroscopy and Auger electron spectroscopy techniques—results not shown here). Selected processing conditions of the samples studied are presented in Table 1.

The x-rays were monochromatized to 18.367 keV (0.0675 nm) for the in situ studies at ROBL. Two different scattering geometries were employed: (1) off-plane (vertical) large-angle XRD in Bragg-Brentano geometry. This method provides a measure for the off-plane texture when comparing the peak areas. (2) Low-angle specular x-ray reflectivity (XRR) has been performed for the study of the thinner TiN layers. Information on film density and surface roughness has been obtained, and from the Kiessig fringes the film thickness has been determined. This was achieved using the Bede REFS computer code to fit the data (Ref 19).

In order to investigate the phase transformations, the temperature dependence of the ER of the samples was measured using the four-probe van der Pauw geometry (BIO-RAD HL 5550). The thermal cycles comprised (i) heating from room temperature (RT) up to 110 °C, followed by (ii) cooling down to −110 °C, and finishing by (iii) heating up to 110 °C. The analysis of the transformation characteristics was based on the cycles (ii) and (iii). The measurements were performed with the Ni-Ti films attached to the substrate.

3. Results and Discussion

Experimentally, the growth of four Ni-Ti samples was studied. The varying deposition parameter was the substrate type, i.e., Ni-Ti films were deposited directly on naturally oxidized Si(100) and on TiN grown on thermally oxidized Si(100) substrates (Table 1). TiN buffer layers of thickness of approximately 15, 80 and 215 nm were chosen (corresponding to 3, 15 and 40 min deposition, respectively).

Figure 2 shows the variation of the net areas of the diffraction peaks of the Ni-Ti B2 phase as a function of deposition time (peaks observed during growth):

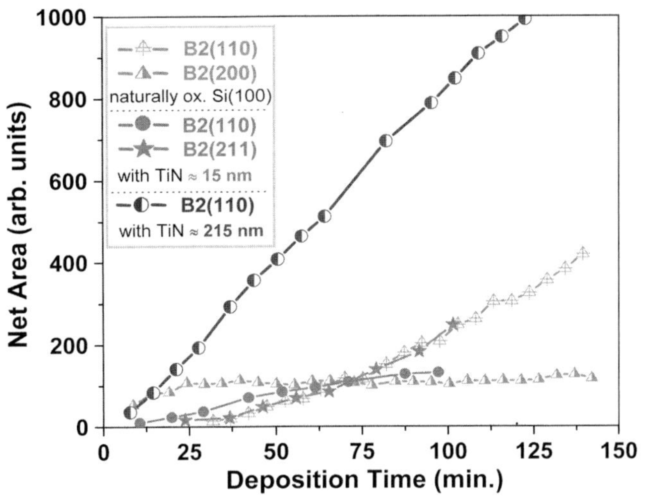

Fig. 2 In situ XRD results for Ni-Ti films deposited directly on a naturally oxidized Si(100) substrate and on TiN buffer layers with thickness values of approximately 15 and 215 nm. Net areas of the diffraction peaks of the B2 phase of Ni-Ti recorded as a function of time

(i) B2(110) and B2(200) peaks for the Ni-Ti film deposited directly on naturally oxidized Si(100) substrate;
(ii) B2(110) and B2(211) peaks for the Ni-Ti film deposited on a TiN buffer layer with a thickness of ≈15 nm;
(iii) B2(110) peak for the Ni-Ti film deposited on a TiN buffer layer with a thickness of ≈215 nm.

In the case of the film deposited directly on the naturally oxidized substrate, after the growth of an initial layer with the {200} planes parallel to the surface during the first 30 min (≈180 nm thick, taking into account that the growth rate was estimated as ≈0.1 nm/s), the B2(200) peak intensity saturated and stayed constant. The B2(110) peak intensity was linearly growing, starting from the B2(200) peak intensity stabilization, until the end of the deposition. The experiments on the growth of Ni-Ti on TiN buffer layers have shown that the B2(200) diffraction peak is hardly detected (in contrast to the depositions on oxidized Si substrates) and, thus, not considered for this representation.

A key role is attributed to the amorphous Si oxide layer on the initial preferential stacking of the B2 phase with (*h*00) planes parallel to the film surface during the initial growth stage. This has been demonstrated by the use of thermally oxidized Si(100) substrates (Ref 20). Kim et al. (Ref 21) proposed that a strong interfacial adsorption on the heated substrate promotes the preferential coverage by a first layer of Ti on top of thermally oxidized Si substrates. This leads to the preferential formation of the (*h*00) oriented Ni-Ti film since, in the B2 cubic structure, the (*h*00) planes are alternately occupied by Ti and Ni atoms. For depositions of Ni-Ti on TiN buffer layers there is no Ni-Ti/SiO_2 interface and, therefore, the development of the (100) orientation of B2 phase parallel to the substrate surface is not observed.

TiN films grown by vapor phase deposition techniques usually have a preferred growth orientation that varies according to the growth conditions. The orientations most often observed are <100> and <111> (<*hkl*> grains are here defined as grains with a plane from the {*hkl*} family parallel to the film surface). Primarily <100> and <111> oriented grains nucleate during the initial stages of growth but, with increasing film thickness, the development of a preferred layer orientation either <100> or <111> is observed. The deposition parameters selected for the growth of the TiN buffer layers used in the present study led to the nucleation and growth of <100> and <111> oriented grains during the initial stages of film growth and <111> grains taking over at larger thickness values. TiN(111) and TiN(200) diffraction peaks could be observed for thickness values of approximately 80 and 215 nm (no peaks could be detected for the thinner TiN layer of ≈15 nm). However, only for a TiN layer with ≈215 nm it is possible to conclude that <111> grains dominate (results presented in Ref 22).

Figure 2 shows that in the case of the deposition of Ni-Ti on the TiN film with a thickness of ≈15 nm, the 110 diffraction peak of the B2 phase of Ni-Ti (in Bragg-Brentano geometry) dominated at the beginning of the deposition, while the B2(211) diffraction peak assumed higher values at a later stage of the deposition. It suggested that for this thinner TiN layer (≈15 nm) a rough or granular surface with different crystal facets plays an important role on the growth direction of the columnar crystals of Ni-Ti, together with the fact that the targets are tilted 30° away from the substrate normal. During the initial Ni-Ti deposition period, there is a (110) stacking and a columnar growth such that <110> lies close to the substrate normal. This type of stacking is due to the fact that (110) is the more densely packed crystallographic plane for the B2 structure (Ref 23). However, due to the surface morphology of the TiN layer (without a dominating crystallographic orientation—most likely a granular uneven surface) and the geometrical shadowing effects, as growth proceeds the growth direction is more influenced by the direction of incident particles. Columnar grains stacking on (110) will then grow tilted to the surface normal (toward the magnetron with the Ni-Ti target running at 40 W) due to the shadowing effects. This tilting leads to (211) planes more closely parallel to the substrate surface and a higher intensity of the corresponding diffraction peak. Thus, it is suggested that this crystallographic orientation development (for the Ni-Ti sample with a TiN buffer layer with a thickness of ≈15 nm) is associated to the combined effects of low surface mobility and shadowing. Although the results are not shown in Fig. 2 this behavior could be expected for the deposition of Ni-Ti on a TiN buffer layer with a thickness of ≈80 nm. It has been mentioned above that in both cases a dominating orientation could not be identified for the TiN buffer layers (of roughly 15 and 80 nm), i.e., <100> and <111> oriented grains of TiN nucleated and grew. In contrast, the deposition of Ni-Ti on TiN with a topmost layer formed mainly by <111> oriented grains showed a preferential growth of <110> oriented grains of the Ni-Ti B2 phase from the beginning of the deposition, with a constant growth rate during the whole deposition.

A typical XRR result obtained in situ for a TiN layer after 3 min deposition is shown in Fig. 3. This technique allowed to calculate the growth rate for the TiN buffer layers (≈0.086 nm/s) since the thickness is obtained from the fitting procedure. The TiN layer exhibits a density of ≈97% (≈5.27 g/cm^3) of the bulk density (5.43 g/cm^3). From the fitting an average roughness of ≈0.9 nm was derived. However, although XRR provides valuable information it is in general useful only for lower film thicknesses. TiN buffer layers with thickness values larger than ≈100 nm could hardly be resolved by the

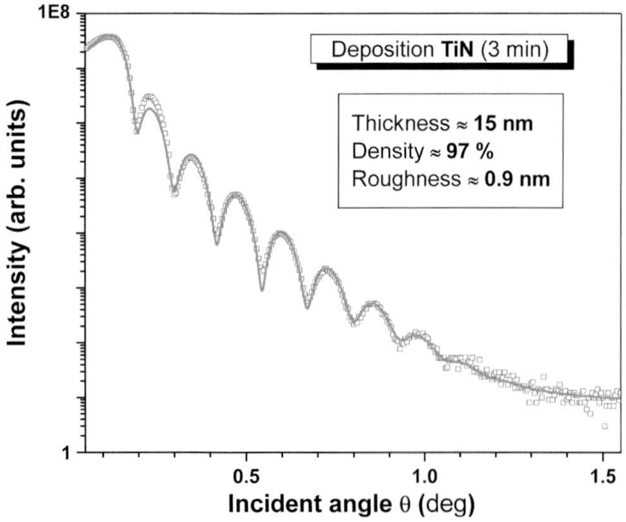

Fig. 3 Typical XRR spectra for a 3 min deposition of a TiN buffer layer together with the fitting curve (red line), obtained immediately after TiN growth and before the deposition of the Ni-Ti film on top

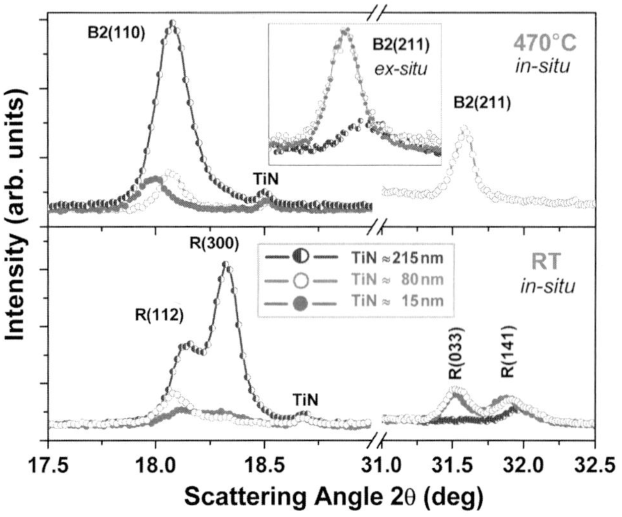

Fig. 4 XRD spectra with change of temperature obtained at ROBL for the Ni-Ti samples with TiN buffer layers; (top) at 470 °C, (bottom) at RT. The inset (top) shows the results recorded around the B2(211) peak at 100 °C (austenitic state) acquired at a laboratory source

reflectivity experiments because of interference with the SiO_2 buffer layer. Therefore, XRR technique did not allow a direct comparison of the roughness values of the thinner TiN buffers layers with the thicker layer (with ≈215 nm). Nevertheless, it is believed that the distinct surface morphology of the TiN layers with different thickness values has an important role on the adatoms surface mobility and shadowing effects. Most likely, the TiN film roughness is related with texture and film roughness is determined by crystal habit. The ROBL team has been working on TiN for several years (Ref 24) and these comments are also based in previous observations.

XRD obtained at RT and 470 °C (after deposition) by means of synchrotron radiation at ROBL for the Ni-Ti samples with TiN are shown in Fig. 4. The experimental set-up used at ROBL for the study of the Ni-Ti samples with TiN buffer layers of approximately 80 and 215 nm did not allow the study of the

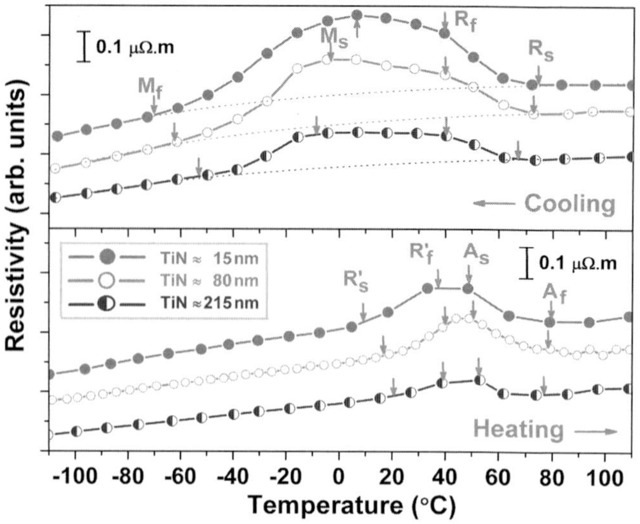

Fig. 5 Dependence of ER with temperature for the Ni-Ti samples with TiN buffers layers of thickness values of roughly 15, 80 and 215 nm during cooling (top) and heating (bottom). The dot line is used to emphasis the variation of the resistivity during the R-phase transformation (including distortion) on cooling for the different samples

B2(211) peak angular range at 470 °C. Thus, the samples have also been studied ex-situ (without angular range restrictions) at a laboratory source (diffractometer D-5005 from BRUKER AXS) at 100 °C (austenitic state) in order to record the B2(211) diffraction peak. The respective results are shown in the inset (top). These results show similar B2(211) peak shape and peak intensity for the Ni-Ti films deposited on TiN with total thickness values of roughly 15 and 80 nm. At RT the Ni-Ti film structure was mainly rhombohedral (R-phase) as shown by the bottom diffractograms of Fig. 4. With the increase of temperature to 470 °C it transformed to B2 phase (top diffractograms). During the transformation from the high temperature phase (B2) to the rhombohedral R-phase while cooling, the height of the B2(110) diffraction peak decreased, splitting into two resolvable R(112) and R(300) peaks. This was especially visible in the case of the Ni-Ti film deposited on top of a TiN layer with a thickness of ≈215 nm. Furthermore, it was perceptible a splitting of the B2(211) peak into the (033) and (141) peaks of the R-phase structure for the samples with TiN layers of about 15 and 80 nm.

The temperature dependence results of the ER of the Ni-Ti samples with TiN buffer layers of thickness approximately 15, 80 and 215 nm are shown in Fig. 5 for cooling (top curves) and heating (bottom curves). In the figure, R_s, R_f, M_s, M_f are the temperatures for the start and finish of the formation of the R-phase and martensite (B19′), on cooling, and R'_s, R'_f, A_s, A_f are the temperatures for the start and finish of the formation of the R-phase and the B2 phase, on heating. For the different samples, as the temperature decreases, an increase in the resistivity values was observed (B2 ⇒ R-phase transformation), followed by a decrease associated with the R-phase ⇒ B19′ transformation. During heating the transformation was also occurring in two steps, B19′ ⇒ R-phase ⇒ B2 as observed by the respective increase in the ER. However, during heating the increase in the ER was smaller than the one observed on cooling. Moreover, the increase of the ER during R-phase formation (on cooling and heating) is lower for the Ni-Ti sample with a TiN buffer layer of ≈215 nm.

According to the ER results obtained during temperature cycling, presented in Fig. 5, the distinct crystallographic orientations of the Ni-Ti films influence their phase transformation characteristics. The resistivity increase during R-phase transformation, on cooling and heating, was higher for Ni-Ti films with a higher fraction of grains of the B2 phase with (211) parallel to the film surface. Therefore, the use of a TiN buffer layer is an important finding for the control of the crystallographic orientations of the Ni-Ti SMA films and consequently to influence their phase transformation characteristics.

4. Conclusion

The in situ XRD measurements have shown that on naturally oxidized Si(100) substrates, without a substrate bias voltage, the Ni-Ti B2 phase starts by stacking onto (*h*00) planes (due to the native oxide layer), and as the thickness increases, evolves into a <110> preferential orientation. In the case of a Ni-Ti film deposited on TiN with a topmost layer formed mainly by <111> oriented grains, there is a preferential growth of <110> oriented grains of the Ni-Ti B2 phase from the beginning of the deposition, with a constant growth rate during the whole deposition. For the Ni-Ti films deposited on top of a TiN layer where a dominating orientation could not be identified (i.e., <100> and <111> oriented grains of TiN nucleated and grew), a different behavior is observed, that is, <110> oriented grains of the Ni-Ti B2 phase dominate at lower thickness values, while the B2(211) planes parallel to the substrate surface take over at higher thickness values (due to an inclination of (110) of the columnar grains relative to the substrate normal).

The ER results obtained during temperature cycling have shown that the distinct crystallographic orientations of the Ni-Ti films influence their phase transformation characteristics. The resistivity increase during R-phase transformation, noticeably on cooling, is higher for Ni-Ti films with a higher fraction of grains of the B2 phase with (211) parallel to the film surface.

Acknowledgments

The authors thank the FCT/MCTES for a Ph.D. scholarship (POCI 2010/FSE for R.M.S.M.) and U. Strauch for his technical assistance during the measurements at ROBL. Financial support from the ESRF for the experiments at ROBL and from the FCT/MCTES for the pluriannual financial support of CENIMAT/I3N is gratefully acknowledged by K.K.M., L.P., R.J.C.S. and F.M.B.F.

References

1. J.A. Thornton, Influence of Apparatus Geometry and Deposition Conditions on Structure and Topography of Thick Sputtered Coatings, *J. Vac. Sci. Technol.*, 1974, **11**, p 666–670
2. P.B. Barna and M. Adamik, Fundamental Structure Forming Phenomena of Polycrystalline Films and the Structure Zone Models, *Thin Solid Films*, 1998, **317**, p 27–33
3. L. Hultman and J.E. Sundgren, Structure/Property Relationships for Hard Coatings, *Handbook of Hard Coatings: Deposition Technologies, Properties and Applications*, R.F. Bunshah, Ed., Noyes Publications, 2001, p 108–180
4. C.V. Thompson, Structure Evolution During Processing of Polycrystalline Films, *Annu. Rev. Mater. Sci.*, 2000, **30**, p 159–190
5. J.D. Busch, A.D. Johnson, C.H. Lee, and D.A. Stevenson, Shape-Memory Properties in Ni-Ti Sputter-Deposited Film, *J. Appl. Phys.*, 1990, **68**, p 6224–6228
6. K.R.C. Gisser, J.D. Busch, A.D. Johnson, and A.B. Ellis, Oriented Nickel-Titanium Shape Memory Alloy Films Prepared by Annealing During Deposition, *Appl. Phys. Lett.*, 1992, **61**, p 1632–1634
7. A.P. Jardine, H. Zhang, and L.D. Wasielesky, Investigation into the Thin-Film Fabrication of Intermetallic NiTi, *Mater. Res. Soc. Symp. Proc.*, 1990, **187**, p 181–186
8. K.K. Ho, K.P. Mohanchandra, and G.P. Carman, Examination of the Sputtering Profile of NiTi Under Target Heating Conditions, *Thin Solid Films*, 2002, **413**, p 1–7
9. A. Ishida and V. Martynov, Sputter-Deposited Shape-Memory Alloy Thin Films: Properties and Applications, *MRS Bull.*, 2002, **27**, p 111–114
10. M.J. Vestel, D.S. Grummon, R. Gronsky, and A.P. Pisano, Effect of Temperature Devitrification Kinetics of NiTi Films, *Acta Mater.*, 2003, **51**, p 5309–5318
11. Y.C. Shu and K. Bhattacharya, The Influence of Texture on the Shape-Memory Effect in Polycrystals, *Acta Mater.*, 1998, **46**, p 5457–5473
12. J.A. Walker, K.J. Gabriel, and M. Mehregany, Thin-Film Processing of TiNi Shape Memory Alloy, *Sens. Actuators*, 1990, **A21–A23**, p 243–246
13. Y. Fu, H. Du, W. Huang, S. Zhang, and M. Hu, TiNi-Based Thin Films in MEMS Applications: A Review, *Sens. Actuators A*, 2004, **112**, p 395–408
14. W. Matz, N. Schell, G. Bernhard, F. Prokert, T. Reich, J. Claußner, W. Oehme, R. Schlenk, S. Dienel, H. Funke, F. Eichhorn, M. Betzl, D. Pröhl, U. Strauch, G. Hüttig, H. Krug, W. Neumann, V. Brendler, P. Reichel, M.A. Denecke, and H. Nitsche, ROBL—A CRG Beamline for Radiochemistry and Materials Research at the ESRF, *J. Synchroton. Radiat.*, 1999, **6**, p 1076–1085
15. F.E. Wang, S.J. Pickart, and H.A. Alperin, Mechanism of the TiNi Martensitic Transformation and the Crystal Structures of TiNi-II and TiNi-III Phases, *J. Appl. Phys.*, 1972, **43**, p 97–112
16. F.E. Wang, B.F. DeSavage, W.J. Buehler, and W.R. Hosler, The Irreversible Critical Range in the TiNi Transition, *J. Appl. Phys.*, 1968, **39**, p 2166–2175
17. W.B. Cross, A.H. Kariotis, and F.J. Stimler, Nitinol Characterization Study, NASA Report, CR-1433, 1969
18. W. Matz, N. Schell, W. Neumann, J. Bøttiger, and J. Chevallier, A Two Magnetron Sputter Deposition Chamber for In Situ Observation of Thin Film Growth by Synchrotron Radiation Scattering, *Rev. Sci. Instrum.*, 2001, **72**, p 3344–3348
19. http://www.bede.co.uk.
20. R.M.S. Martins, N. Schell, M. Beckers, K.K. Mahesh, R.J.C. Silva, and F.M.B. Fernandes, Growth of Sputter-Deposited Ni-Ti Thin Films: Effect of a SiO_2 Buffer Layer, *Appl. Phys. A: Mater. Sci. Proc.*, 2006, **84**, p 285–289
21. I.-J. Kim, H. Nanjo, T. Iijima, and T. Abe, Crystallization Process of TiNi Thin Films Sputtered at Elevated Temperatures on Pt/Si Oxide/Si and Si Oxide/Si Substrates, *Jpn. J. Appl. Phys.*, 2000, **39**, p 568–571
22. R.M.S. Martins, N. Schell, R.J.C. Silva, L. Pereira, K.K. Mahesh, and F.M.B. Fernandes, In Situ Study of Ni-Ti Thin Film Growth on a TiN Intermediate Layer by X-Ray Diffraction, *Sens. Actuators B*, 2007, **126**, p 332–337
23. J.-M. Zhang, F. Ma, and K.-W. Xu, Calculation of the Surface Energy of bcc Metals by Using the Modified Embedded-Atom Method, *Surf. Interface Anal.*, 2003, **35**, p 662–666
24. N. Schell, W. Matz, J. Bøttiger, J. Chevallier, and P. Kringhøj, Development of Texture in TiN Films by Use of In Situ Synchrotron x-ray Scattering, *J. Appl. Phys.*, 2002, **91**, p 2037–2044

JMEPEG (2009) 18:548–553
DOI: 10.1007/s11665-009-9425-7

Properties of Co-alloyed Ni-Fe-Ga Ferromagnetic Shape Memory Alloys

V.A. Chernenko, K. Oikawa, M. Chmielus, S. Besseghini, E. Villa, F. Albertini, L. Righi, A. Paoluzi, P. Müllner, R. Kainuma, and K. Ishida

(Submitted October 17, 2008; in revised form January 29, 2009)

It is experimentally found that in the $Ni_{54}Fe_{20-x}Co_xGa_{26}$ ferromagnetic shape memory alloys, Co variation from 0 to 9 at.% leads to: (i) almost linear change of martensitic transformation temperatures from −70 °C to 120 °C; (ii) a non-monotonous change of the Curie temperature, and (iii) a linear decrease of saturation magnetization from 60 to 43 Am²/kg. The selected alloys grown as single crystals have been (magneto-) mechanically tested. The superelastic effect has been measured to be about 4%. The magneto-strain shows a training effect which is an evidence of the effect of magnetic-field-induced twin-boundary motion.

Keywords intermetallics, material selection, mechanical testing

1. Introduction

In the ferromagnetic thermoelastic martensites, such as those exhibited by prototype off-stoichiometric Ni_2MnGa alloys, the ordinary magnetostriction can trigger the twin boundary motion whereby giant magnetic-field-induced strains are obtained (see, e.g., Ref 1 and references therein). The same level of strains can also be obtained in these alloys using thermal and/or mechanical activation, as in the usual non-ferromagnetic shape memory materials. Such multi-functionality, important for applications, motivates a worldwide activity related to the development of ferromagnetic shape memory alloys (FSMAs), particularly based on the off-stoichiometric Heusler alloys (Ref 1-3). The Ni-Fe-Ga alloys (Ref 4-6), like the other Mn-free Heusler FSMAs, are found to be much more ductile than the classical Ni-Mn-Ga. The improved mechanical properties of these alloys are mainly due to the presence of the γ-phase precipitates. The disadvantage of ternary Ni-Fe-Ga FSMAs is related to the impossibility to use them at room temperature owing to the reduced Curie temperature of the austenite. Preliminary studies show that Co affects significantly the transformation temperatures and magnetic properties of Ni-Fe-Ga alloys: both, the martensitic transformation temperature, T_M, and Curie temperature, T_C, increase and the saturation magnetization, I_s, decreases with increasing Co-content in the series of $Ni_{51}Fe_{22-x}Co_xGa_{27}$ alloys. On the other hand, an increase of Co concentration in the series of $Ni_{54-x}Fe_{19}Co_xGa_{27}$ alloys causes increase in T_C and I_s and decrease in T_M (Ref 7, 8). In both cases, Co-content varies between 0 and 6 at.%.

In this article, the investigation is extended to the probing Ni-Fe-Ga FSMAs with different base compositions and doped by Co up to 9 at.%. The influence of Co on the transformation behavior is determined and selected compositions are studied in a single crystalline form to demonstrate the functional capacity of these alloys.

2. Experimental

$Ni_{54}Fe_{20-x}Co_xGa_{26}$ alloys, with x = 0, 2, 3, 4, 6, 8, and 9 at.%, labeled as Co0…Co9, were prepared as 20-g buttons by arc-melting of high purity elements and subsequent five-times re-melting. The samples cut of each alloy were sealed in a quartz ampoule with argon atmosphere and solution treated at 1100 °C for 24 h followed by a water-quench. Then, the samples were sealed in the quartz capsule again and aged at 650 °C for 100 h followed by a water-quench. More details about the sample preparation can be found in Ref 6. Scanning electron microscopy observations of aged samples confirmed that the β-phase matrix contains γ-phase precipitates as observed in similar alloys in some studies (Ref 4-8). Using x-ray powder diffraction, it was found that the alloys in a bulk are highly textured. In order to perform structural analysis, the grinding procedure of powder preparation and the abovementioned heat treatment was applied. The forward (on cooling), T_M, and reverse (on heating), T_A, martensitic transformation (MT) temperatures were determined as the peak positions on the calorimetric curves measured with a rate of 5 °C/min by a

This article is an invited paper selected from presentations at Shape Memory and Superelastic Technologies 2008, held September 21-25, 2008, in Stresa, Italy, and has been expanded from the original presentation.

V.A. Chernenko, S. Besseghini, and **E. Villa**, IENI, CNR, Corso Promessi Sposi, 29, Lecco 23900, Italy; **K. Oikawa, R. Kainuma,** and **K. Ishida**, Department of Materials Science, Graduate School of Engineering, Tohoku University, Aoba-yama 6-6-02, Sendai 980-8579, Japan; **M. Chmielus**, Department of Materials Science and Engineering, Boise State University, Boise, ID 83725 and Helmholtz Centre Berlin for Materials and Energy, Glienicker str.100, 14109 Berlin, Germany; **P. Müllner**, Department of Materials Science and Engineering, Boise State University, Boise, ID 83725; **F. Albertini** and **A. Paoluzi**, IMEM, CNR, Parco Area delle Scienze 37/A, Parma 43010, Italy; **L. Righi**, Dipartimento di Chimica GIAF, Parco Area delle Scienze 17/A, Parma 43100, Italy. Contact e-mail: vladimir.chernenko@gmail.com.

Table 1 Compositions, heat treatment, crystal orientation and martensitic (T_M, T_A), and Curie (T_C) transformation temperatures of single crystals used for mechanical and magneto-mechanical tests

Single crystal	Composition, at.%				Heat treatment	Lattice orientation along tensile axis	Transformation temperatures, °C		
	Ni	Fe	Ga	Co			T_M	T_A	T_C
#1	52.4	23.2	24.4	···	As-grown	···	−45	−20	51
#2	53.0	12.0	26.9	8.1	1100 °C for 24 h +650 °C 100 h	···	20	65	42
#3	50.0	17.0	27.0	6.0	As-grown	⟨10 6 13⟩	45	55	70

standard DSC analyzer. The Curie temperature, T_C, and saturation magnetization, I_s, ($T = -196$ °C, $H = 1$T) were measured by an in-house made low-field magnetic susceptibility set-up and a vibrating sample magnetometer (VSM), respectively.

Two single crystalline alloys #1 and #2 of 80 g each were grown by Bridgman method in an alumina crucible using previously produced induction-melted ingots. The single crystalline wires of alloy #3 were grown by pulling-down of the melt from the orifice of the alumina crucible with a rate about 1 mm/min in the micro-PD furnace (Ref 9). The wires were 1.5 mm in diameter and up to 200 mm in length. The compositions of alloys #1 and #2 were determined with uncertainty of 0.2% by X-ray fluorescent analysis, while EDX was used for the analysis of alloy #3 made with uncertainty of 0.5%. Table 1 shows the compositions, heat treatments, orientations, and transformation temperatures of the single crystalline samples. The values of the transformation temperatures for alloys #2 and #3 indicate that they can be considered as possible room-temperature FSMAs.

A DMA Q800 analyzer was utilized to register from the same sample the temperature dependencies of elastic modulus E (T) and internal friction (IF), tan δ, in the dynamic tensile mode, as well as stress versus strain, σ-ε, at a constant temperature in the static tensile mode. The dynamic measurements were carried out at a frequency of 1 Hz, an oscillation strain amplitude of 10^{-4}, and temperature rate of 2 °C/min. In the static mechanical tests, the upper limit of load and the stress change rate were 18 N and 2.5 MPa/min, respectively.

The alloy #2 and wire #3 were tested at room temperature in the Dynamical Magneto-Mechanical Testing Device (DMMT). Prior to these tests, the samples were heated above MT temperature, loaded by constant stress of 20-50 MPa and cooled to the martensitic state. The samples were glued on one end to the sample holder specially designed to prevent a possible wire bending. The magneto-mechanical measurements are made with a constant magnetic field of 0.97 T which was rotated with 29 revolutions per minute during the strain measurements, and with up to 5,000 revolutions per minute during the period between the measurements. The test was continued up to one million magneto-mechanical cycles. Experimental details are given in (Ref 10, 11).

3. Structural Behavior and Transformation Characteristics

Previous structural studies have shown that the martensitic structure formed in Ni-Fe-Ga(Co) alloys depends on the degree

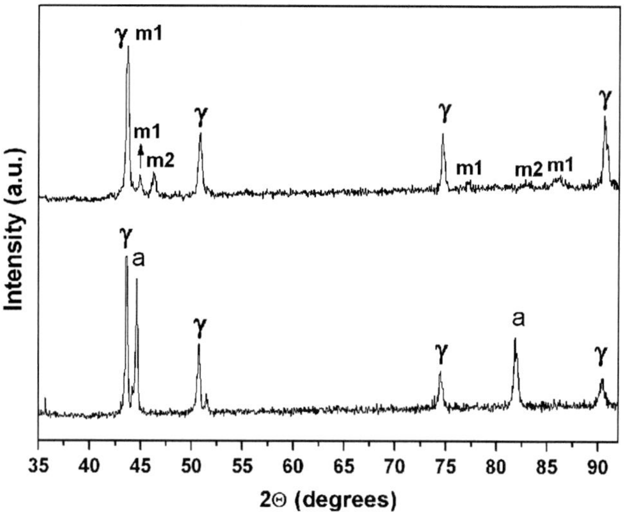

Fig. 1 XRD patterns of powdered and heat-treated sample of Co2 alloy at 20 °C (lower curve) and at −193 °C (upper curve) collected using CuK_α radiation. The peaks are attributed to two martensites, m1 and m2; austenite, a; and γ-phase

of B2/L2$_1$ atomic order (Ref 6) and/or the amount of Co (Ref 9). In particular, it was found that the lower degree of the atomic order and/or higher content of Co favor the formation of non-modulated 2M-martensite. It is also worth noting that a mixture of martensitic phases with modulated or non-modulated structure is frequently observed in FSMA Heusler alloys (Ref 12).

In this study, x-ray analysis was hindered by a strong texture of the alloys in bulk form. Moreover, the powder samples appeared to be enriched by γ-phase depressing peaks of the other phases as it is evident from Fig. 1. As an example, Fig. 1 shows x-ray diffraction patterns of the Co2 sample obtained at 20 °C (lower curve) and at −193 °C (upper curve). Obviously, the results such as shown in Fig. 1 do not allow a reliable structural analysis of the phases. Nevertheless, they contain some useful informations which can be deduced by taking into account the literature data (e.g., Ref 13). Particularly, the Co2 sample presents about 30 vol.% of austenitic phase at 20 °C with a lattice parameter of 0.5756 nm. The γ-phase has a unit cell parameter of 0.3601 nm. By lowering the temperature (to −193 °C), the austenitic reflections are replaced by few new peaks related to the two martensitic phases (m1 and m2 in Fig. 1). Presumably, crystal lattices of these martensites could be attributed to the basic monoclinic and tetragonal unit cells exhibiting different types of the lattice modulations (e.g., Ref 12, 13). With an increasing Co-content, alloys show a

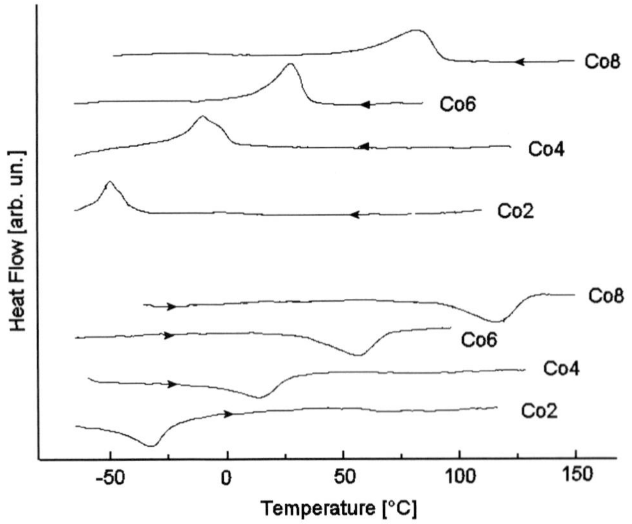

Fig. 2 Typical heating-cooling DSC curves for $Ni_{54}Fe_{20-x}Co_xGa_{26}$ alloys

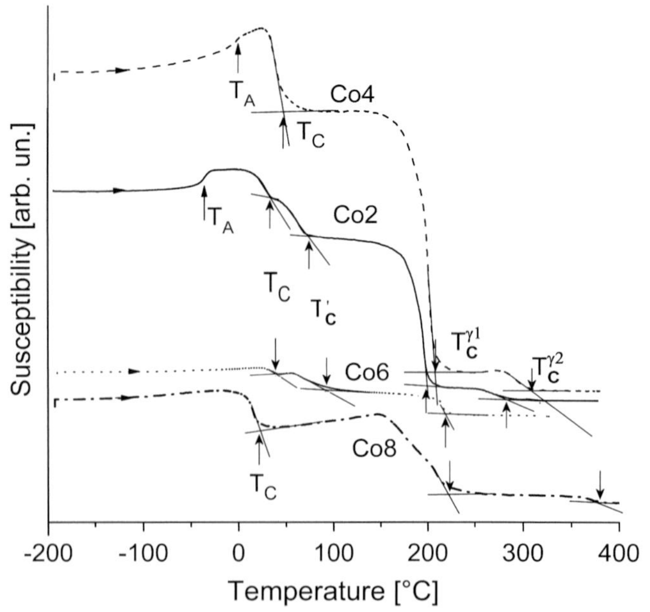

Fig. 3 Typical heating curves of low-field magnetic susceptibility for $Ni_{54}Fe_{20-x}Co_xGa_{26}$ alloys. The reverse MT temperature, T_A, and Curie temperatures T_C, T'_C, $T_C^{\gamma1}$, and $T_C^{\gamma2}$ are indicated by vertical arrows. The reverse MT temperature can only be detected when $T_A < T_C$. T_C is attributed to B2/L2$_1$-ordered matrix, $T_C^{\gamma1}$ and $T_C^{\gamma2}$ observed above 200 °C in all the alloys are tentatively attributed to the two γ-phases (γ1 and γ2), respectively. T'_C observed in Co2 and Co6 alloys can be related either to the magnetic inhomogeneities of matrix or different martensitic phases. One cannot exclude also a possible presence of the third γ-phase

larger amount of γ-phase. Alongside the predominant γ1-phase with a = 0.360 nm, an additional γ2-phase with a = 0.3544 nm appears.

Typical DSC heating and cooling curves and typical heating curves of low-field magnetic susceptibility, χ(T), for $Ni_{54}Fe_{20-x}Co_xGa_{26}$ alloys are shown in Fig. 2 and 3. Curves in these figures allow the determination of characteristic transformation temperatures, such as T_M and T_A corresponding to the maximum values on the cooling and heating DSC curves, respectively (Fig. 2) and tangentially defined Curie temperatures T_C corresponding to the B2/L2$_1$-ordered matrix phase and, at least, two to the γ-phases (Fig. 3). The DSC curves yield the average transformation heat of MT equal to 1.8 ± 0.6 J/g for all the alloy series. This average value is provided because the experimental composition dependence of the transformation heat is irregular.

The χ(T) curves show distinctive anomalies produced by the reverse MT only when condition $T_A < T_C$ holds, alloys Co2…Co4 in Fig. 3. The curves in Fig. 3 reflect a non-homogeneous character of the alloys exhibiting at least two additional anomalies above 200 °C for all the alloys studied which can naturally be attributed to the Curie temperatures of the two γ-phases found by x-ray diffraction. In the Co2 and Co6 alloys, an additional anomaly at T'_C appears which could be thought as produced by some magnetic inhomogeneities of matrix or a possible presence of the third γ-phase with its Curie temperature. Note that the multiple anomalies at χ(T) curves were observed for ternary Ni-Fe-Ga alloys (Ref 14), which were attributed to the Curie temperatures of γ-phases.

The characteristic temperatures, T_M and T_A, and Curie temperature, T_C, determined from Fig. 2 and 3 are plotted in Fig. 4 as a function of Co-content. As in the case of other Co-doped Ni-Fe-Ga alloys (Ref 7), T_M and T_A show a strong increase when Co substitutes Fe whereas, contrary to Ref 7, T_C exhibits a diffuse maximum at about 4 at.%. Note that a similar behavior of T_C is also observed for Ni-Mn-Ga alloys and their Co-doped analogs, which was explained by the influence of electron concentration and Mn occupation of different sites in the lattice (Ref 15). Furthermore, T_A and T_M dependencies in Fig. 4 demonstrate a considerable and monotonous increase of the temperature hysteresis of MT as a function of Co-doping. This evolution can be attributed to the increasing volume fraction of γ-phase precipitates and decreasing mobility of the interfaces. Incidentally, this increase does not seem to affect the Co-concentration dependence of total magnetic moment as the experimental results in the Inset to Fig. 4 evidence. Here,

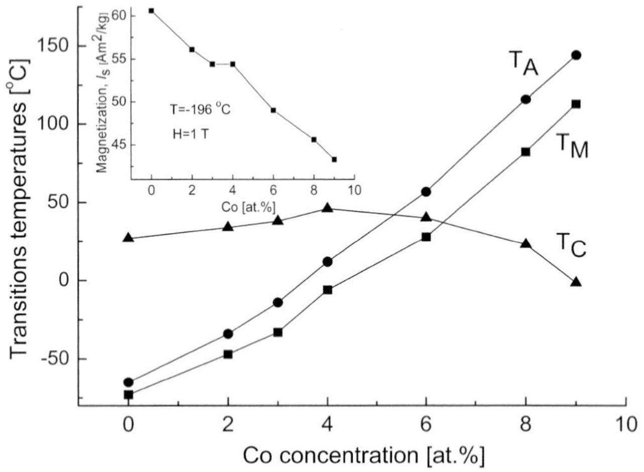

Fig. 4 Characteristic temperatures of martensitic transformation and Curie temperature as a function of Co-content for the $Ni_{54}Fe_{20-x}Co_xGa_{26}$ alloys. *Inset*: Co-concentration dependence of saturation magnetization value measured at −196 °C in 1 T magnetic field. Curves are guides for the eye

the linear dependence could mean that the contribution of γ-phase to total magnetic moment is relatively the same in all alloys studied.

The studies of Co-doping on the transformation characteristics have shown that in this particular Ni-Fe-Ga alloy one can design an alloy with room-temperature ferromagnetic martensite at a narrow Co-content window located in the vicinity of 4.5 at.% (Fig. 4).

4. Elastic and Superelastic Properties

Dynamic and quasi-static mechanical properties have been studied in tensile mode for the single crystalline samples shown in Table 1. The cooling/heating temperature dependencies of elastic modulus and tan δ are shown in Fig. 5 for two samples. The results for sample #3 are not shown due to the similar behavior as #1. In samples #1 and #3, the elastic modulus exhibits a low value in all the temperature ranges with a deep minimum at T_M (on the cooling curve) and at T_A (on the heating curve) corresponding to the highest rate of the temperature change of tan δ produced by MT. Besides, quite strong modulus softening of about 1.3 MPa/K is observed for alloys #1 and #3 in the cubic phase. In contrast, alloy #2 shows a high value of elastic modulus accompanying by a low level of internal friction. All the other features mentioned for samples #1 and #3 are not so pronounced in the alloy #2. Both, the almost three times larger modulus and multiple difference in tan δ between the samples and #2 and #3, suggest a strong stiffening of the matrix in alloy #2, presumably, due to heat treatment accompanied by a disperse hardening.

Quasi-static stress-strain, σ-ε, measurements at different constant temperatures have been performed previously for alloy #3 (Ref 9). In this study, Fig. 6 depicts the typical tensile stress-strain dependencies for alloy #1 obtained in the martensitic and austenitic states in the stress range up to 160 MPa which is our instrumental limitation. Incidentally, in this stress range, alloy #2 did not show a superelasticity effect. In the austenitic state, the σ-ε dependencies for alloy #1, indeed, exhibit a reversible superelastic strain of about 4.0% produced by a stress-induced martensitic transformation with a stress hysteresis of 50 MPa (Fig. 6). The superelastic strain value, although being influenced by the presence of non-transformed γ-phase (Ref 16, 17), is comparable with the crystallographically estimated value when tensile axis is along ⟨100⟩ (Ref 16). The plateau-like accumulation of the superelastic strain is steep in Fig. 6 because in the alloys containing precipitates, a considerable hardening occurs during MT (see also Ref 16, 17). The yield stress in the martensitic state is quite high due to a reduced mobility of interfaces which are possibly pinned by precipitates.

Figure 7 shows the temperature dependence of critical stress determined by a tangential method from the stress-strain curves (Ref 17). The data points in Fig. 7 can be approximated by the two straight lines with the slopes of 1.0 and 0.5 MPa/K. As far as the data points in Fig. 7 are reproducible, one can attribute the first line to the low-stress-induced martensitic phase being identical to thermally induced one, while the second line is assumed to belong to the high-stress-induced martensitic phase with different crystallographic structure. Previous structural studies revealed that the thermally induced martensitic phase in alloys similar to alloy #1 can be described as a monoclinic 7M-structure (Ref 18, 19). The crystallographic identification

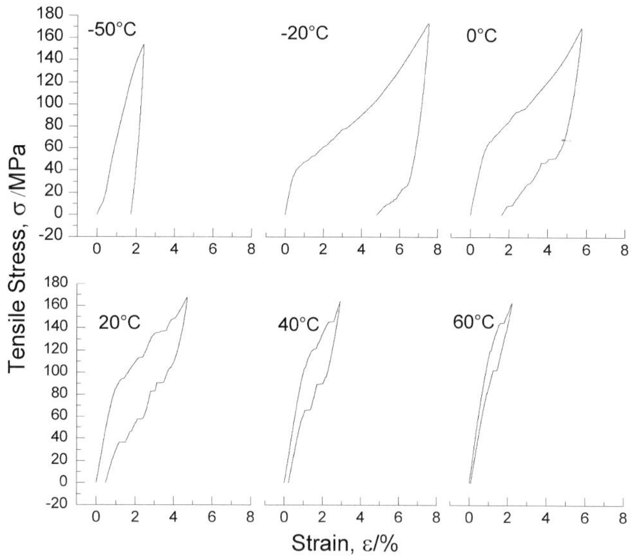

Fig. 6 Tensile stress-strain dependencies for alloy #1

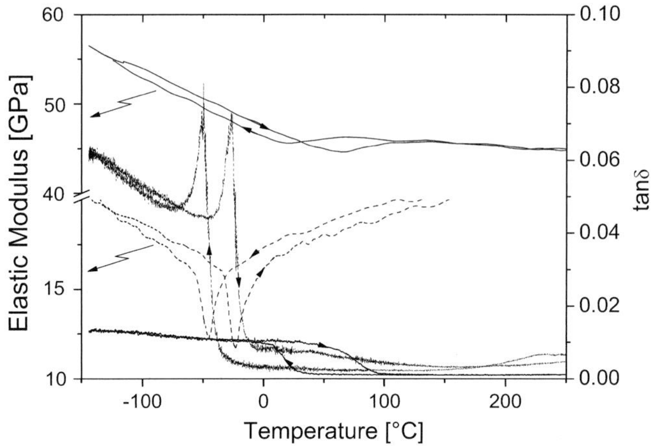

Fig. 5 Heating-cooling temperature dependencies of elastic modulus and internal friction of alloy #1 (dashed line) and alloy #2 (full line)

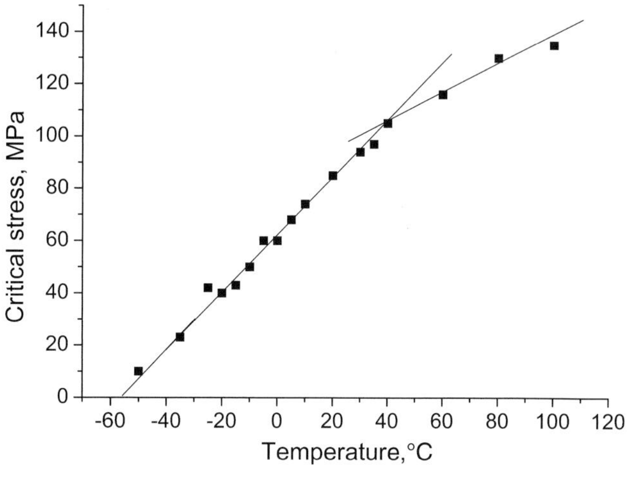

Fig. 7 Tensile critical stress vs. temperature phase diagram of alloy #1. Linear approximations of the data points are used

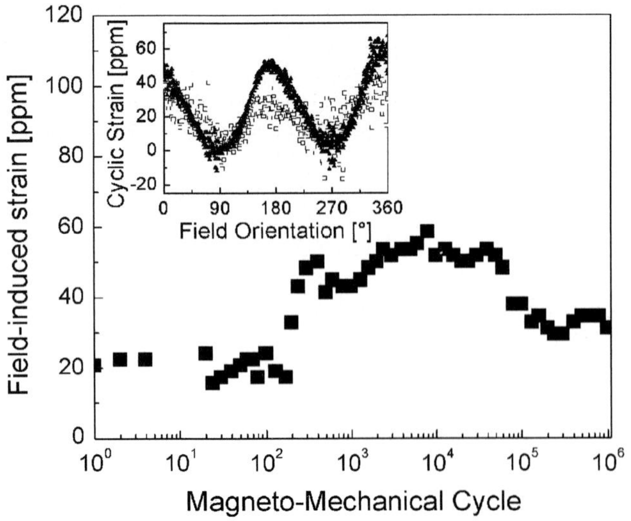

Fig. 8 Field-induced strain as a function of the number of magneto-mechanical cycles. The magnetic field-induced strain as a function of field direction at cycle 6 (open squares) and at cycle 16,000 (full triangles) are shown in the inset

of second stress-induced martensitic phase needs separate studies. With regard to the results for the single crystalline Ni-Fe-Ga alloy (Ref 18), the formation of a tetragonal $L1_0$ martensite is anticipated in this case.

Since the critical stress versus temperature dependence was shown experimentally to be a function of volume fraction of γ-phase (Ref 17), in this study, we refrain from the thermodynamic estimations using Clausius-Clapeyron relationship. The results of the mechanical testing confirm an assessment that the FSMAs can also be competitive in their functional properties typical for the ordinary shape memory alloys.

5. Magnetostrain Properties

Magnetostrain response was studied for the ferromagnetic martensitic phases of alloys #2 and #3 which were suitable for room-temperature-operating DMMT.

The magnetic field-induced strain (MFIS) of sample #3 was measured over 1 million magneto-mechanical cycles. The measured signal was with a maximum of 34 increments close to the detection limit. However, a cyclic strain change of a sample was detected and is shown in Fig. 8 for cycle 6 and 16,000. It can be seen that the shape of the detected MFIS periodic signal is not changing. The strain percent scattering recorded during the first 200 cycles was approximately ±10 ppm of the total cyclic strain and with increasing cycles numbers, it decreased to less than ±5 ppm. The highest strain corresponds to a magnetic field direction parallel to the axis of the wire #3, e.g., at 0° and 180°. As seen in Fig. 8, the MFIS during the initial 200 cycles was around 20 ppm and than tripled over the next 10,000 cycles to 60 ppm. After 40,000 cycles, a decrease of strain to about 40 ppm occurs and then remains constant at about 30 ppm over the next 950,000 cycles until the experiment was stopped at one million cycles.

In the magnetostrain measurements up to 6,000 field rotations, sample #2 did not show a strain evolution exhibiting cyclic strain of about 60 ppm.

The interesting feature of the results presented in Fig. 8 is a training effect, i.e., a strong dependence of MFIS on the number of magneto-mechanical cycles. Despite the low absolute value of magnetostrain, this behavior suggests a magnetic field-induced structural changes, among them the most probable in the case under this study is the phenomenon of magnetic field-induced twin-boundary motion. The training effect provides a preliminary evidence of the existence of magnetic shape memory effect in non-modulated $L1_0$-ordered 2M-martensitic phase which was considered to be formed in the alloy #3 (Ref 9). Although the martensitic structure was not determined directly in this alloy, its superelastic properties are in agreement with the lattice parameters of non-modulated martensitic phase (Ref 9).

6. Summary

In this study, we found that increasing Co-content in the alloy $Ni_{54}Fe_{20-x}Co_xGa_{26}$ results in a considerable and regular change of its properties such as MT temperature, transformation hysteresis, Curie temperature, and saturation magnetization.

Selected alloys grown as the single crystals demonstrate a 4% tensile superelastic strain in as-grown state which is of interest for the applications. The temperature behaviors of the elastic modulus and internal friction in DMA measurements are similar to Ni-Mn-Ga alloys. The cyclic magnetostrain shows a training effect which is an evidence of the effect of magnetic field-induced twin-boundary motion.

Acknowledgments

This work was supported by Grant-in-Aid for Scientific Research from the Ministry of Education, Science, Sports and Culture, Japan, the Core Research for Evolution Science and Technology (CREST), Japan Science and Technology Agency (JST), and Fondazione Cariplo (2004.1819-A109251). PM and MC acknowledge financial support through the Department of Energy (DOE), Office of Basic Energy Sciences (BES) under Grant DE-FG-02-07ER46396. The authors are grateful to A. Gambardella and G. Carcano for technical assistance.

References

1. V.A. Chernenko, Ed., Advances in Shape Memory Materials: Ferromagnetic Shape Memory Alloys, *Mat. Sci. Forum*, **583**, TTP, Switzerland, 2008, 302 p
2. E. Cesari, J. Pons, R. Santamarta, C. Segui, and V.A. Chernenko, Ferromagnetic Shape Memory Alloys: An Overview, *Archiv. Metall. Mater.*, 2004, **49**, p 779–789
3. E. Cesari, J. Pons, C. Segui, and V.A. Chernenko, New Ferromagnetic Shape Memory Alloy Systems, *Applied Crystallography*, H. Moraviec and D. Stróz, Eds., World Scientific, 2004, p 128–133, ISBN 981-283-761-7
4. T. Oikawa, T. Ota, Y. Ohmori, H. Tanaka, A. Morito, R. Fujita, K. Kainuma, K. Fukamichi, and K. Ishida, Magnetic and Martensitic Phase Transitions in Ferromagnetic Ni-Fe-Ga Shape Memory Alloys, *Appl. Phys. Lett.*, 2002, **81**, p 5201–5203
5. H. Morito, A. Fujita, K. Fukamichi, R. Kainuma, K. Ishida, and K. Oikawa, Magnetic-field-induced Strain of Fe-Ni-Ga in Single-variant State, *Appl. Phys. Lett.*, 2003, **83**, p 4993–4995
6. T. Omori, N. Kamiya, Y. Sutou, K. Oikawa, R. Kainuma, and K. Ishida, Phase Transformations in Ni-Ga-Fe Ferromagnetic Shape Memory Alloys, *Mater. Sci. Eng. A*, 2004, **378**, p 403–408

7. K. Oikawa, Y. Imano, V.A. Chernenko, F. Luo, T. Omori, Y. Sutou, R. Kainuma, T. Kanomata, and K. Ishida, Influence of Co Addition on Martensitic and Magnetic Transitions in Ni-Fe-Ga—Based Shape Memory Alloys, *Mater. Trans.*, 2005, **46**, p 734–737
8. H. Morito, K. Oikawa, A. Fujita, K. Fukamichi, R. Kainuma K. Ishida, and T. Takagi, Effects of Partial Substitution of Co on Magnetocrystalline Anisotropy and Magnetic-field-induced Strain in NiFeGa Alloys, *J. Magn. Magn. Mater.*, 2005, **290–291**, p 850–853
9. K Oikawa, R Saito, K Anzai, H Ishikawa, Y Sutou, T Omori, K Ishida, R Kainuma, A Yoshikawa, VA Chernenko, S Besseghini, and A Gambardella, Elastic and Superelastic Properties of NiFeCoGa Single Crystals Grown by Micro-pulling-down Method, *Mat. Trans.*, 2009, **50**, p 934–937
10. P. Müllner, V.A. Chernenko, and G. Kostorz, Large Cyclic Magnetic-field-induced Deformation in Orthorhombic (14 M) Ni-Mn-Ga Martensite, *J. Appl. Phys.*, 2004, **95**, p 1531–1536
11. M. Chmielus, V.A. Chernenko, W.B. Knowlton, G. Kostorz, and P. Mullner, Training, Constraints, and High-cycle Magneto-mechanical Properties of Ni-Mn-Ga Magnetic Shape-memory Alloys, *Eur. Phys. J. Spec. Top.*, 2008, **158**, p 79–85
12. J. Pons, V.A. Chernenko, R. Santamarta, and E. Cesari, Crystal Structure of Martensitic Phases in Ni-Mn-Ga Shape Memory Alloys, *Acta Mater.*, 2000, **48**, p 3027–3038
13. H. Morito, K. Oikawa, A. Fujita, K. Fukamichi, R. Kainuma, and K. Ishida, Enhancement of Magnetic-field-induced Strain in Ni-Fe-Ga-Co Heusler Alloy, *Scripta Mater.*, 2005, **53**, p 1237–1241
14. J.M. Barandiarán, J. Gutiérrez, P. Lázpita, V.A. Chernenko, C. Seguí, J. Pons, E. Cesari, K. Oikawa, and T. Kanomata, Martensitic Transformation in Ni-Fe-Ga Alloys, *Mater. Sci. Eng. A*, 2008, **478**, p 125–129
15. V. Sanchez-Alarcoz, J.I. Perez-Landazabal, V. Recarte, C. Gomez-Polo, and J.A. Rodriges-Velamazan, Correlation Between Composition and Phase Transformation Temperatures in Ni-Mn-Ga Ferromagnetic Shape Memory Alloys, *Acta Mater.*, 2008, **56**, p 5370–5378
16. R.F. Hamilton, H. Sehitoglu, C. Efstahiou, H.J. Maier, and Y. Chumlyakov, Pseudoelasticity in Co-Ni-Al Single and Polycrystals, *Acta Mater.*, 2006, **54**, p 587–599
17. Y. Tanaka, K. Oikawa, Y. Sutou, T. Omori, R. Kainuma, and K. Ishida, Martensitic Transition and Superelasticity of Co-Ni-Al Ferromagnetic Shape Memory Alloys with $\beta + \gamma$ Two-phase Structure, *Mat. Sci. Eng. A*, 2006, **438–440**, p 1054–1060
18. Y. Sutou, N. Kamiya, T. Omori, R. Kainuma, and K. Ishida, Stress-strain Characteristics in Ni-Ga-Fe Ferromagnetic Shape Memory Alloys, *Appl. Phys. Lett.*, 2004, **84**, p 1275–1277
19. J. Gutierrez, P. Lázpita, V. Siruguri, J.M. Barandiarán, P. Henry, V.A. Chernenko, and T. Kanomata, Neutron Study of the Martensitic Transformation in Ni-Fe-Ga Alloys, *Int. J. Appl. Electromagn. Mech.*, 2006, **23**, p 71–74

JMEPEG (2009) 18:554–557
DOI: 10.1007/s11665-009-9457-z

1059-9495/$19.00

Analysis of Variant Orientation Before and After Compression in Polycrystalline $Ni_{50}Mn_{29}Ga_{21}$ MSMA

C. Hürrich, H. Wendrock, M. Pötschke, U. Gaitzsch, S. Roth, B. Rellinghaus, and L. Schultz

(Submitted October 6, 2008; in revised form April 14, 2009)

This paper deals with the microstructure and plastic deformation of $Ni_{50}Mn_{29}Ga_{21}$ ferromagnetic shape memory alloys. In contrast to conventional shape memory alloys, plastic deformation in the martensitic phase, which is due to twin boundary motion, may be caused not only by mechanical stress but also by an external magnetic field. The polycrystalline sample was prepared by directional solidification with a texture parallel to the heat flow. Afterwards, a heat treatment follows for chemical homogenization and stress relaxation in the austenitic state. The configuration of the twin boundaries was analyzed before and after compressing the samples. The microstructure after compression was related to the magnetic properties.

Keywords casting, electron microscopy, mechanical testing

1. Introduction

Ni-Mn-Ga magnetic shape memory alloys with a chemical composition close to stoichiometric composition Ni_2MnGa arouse big interest in the last few years. In the martensitic state, plastic deformation is related to twin boundary motion. The motion of twin boundaries in the martensitic phase due to a magnetic field was demonstrated for the first time in 1996 by Ullakko et al. (Ref 1). Most of the previous investigations in magnetic field induced strain (MFIS) have been performed with single crystals (Ref 2, 3) because of well-defined orientation relationship and low defect density. In order to find MFIS by twin boundary motion, the magnetically induced shear stress must be larger than the twinning stress (Ref 4). The magnetically induced stress is determined by the magnetic crystal anisotropy, the lattice strain and the crystallographic relations between the twin variants. The magnetically induced stress is given by $\sigma_{max} = m \cdot K_u \cdot s^{-1}$, where m is the Schmid factor, K_u is the magnetocrystalline anisotropy constant and the twin shear $s = 0.5 \cdot \left(\frac{a}{c} - \frac{c}{a}\right)$, where a and c are the martensitic lattice parameters (Ref 2, 5, 6). To increase the strain and reduce the stress for twin boundary motion in polycrystals, mechanical training is necessary. Because of this training effect less stress is necessary for twin boundary motion. The influence of mechanical training on the stress-strain relationship in martensitic polycrystalline $Ni_{50}Mn_{30}Ga_{20}$ has been shown by Gaitzsch et al. (Ref 7). The influence of the magnetic field to the stress-strain behavior was determined by Gaitzsch et al. (Ref 6). In comparison to our work, the following are examples of other work on polycrystalline NiMnGa samples. Ullakko et al. (Ref 8, 9) reported on MFIS in polycrystals and by thermo-mechanical treatment. But there is no information about the change in microstructure, the data are similar to the single crystal. Up to now no further papers with more details about MFIS in polycrystals were found. There are only some papers that describe a technological route of texture development in polycrystals (directional solidification and plastic deformation) (Ref 10, 11). In the present paper, we investigate the microstructure and the surface topography of polycrystalline $Ni_{50}Mn_{29}Ga_{21}$ samples were analyzed before and after compression. Also, the magnetic domain structure was determined after compression. The influence of stress on the twin boundary arrangement is investigated in order to find possibilities to improve the MFIS in polycrystalline $Ni_{50}Mn_{29}Ga_{21}$.

This article is an invited paper selected from presentations at Shape Memory and Superelastic Technologies 2008, held September 21-25, 2008, in Stresa, Italy, and has been expanded from the original presentation.

C. Hürrich, H. Wendrock, M. Pötschke, U. Gaitzsch, S. Roth, B. Rellinghaus, and **L. Schultz**, IFW Dresden, Institute for Metallic Materials, P.O. Box 270116, 01069 Dresden, Germany. Contact e-mail: c.huerrich@ifw-dresden.de.

2. Experimental

The polycrystalline shape memory alloy $Ni_{50}Mn_{29}Ga_{21}$ was melted inductively and cast in a hot ceramic mold with a cold copper plate at the bottom of the cylindrical mold. With this method it is possible to achieve a coarse-grained and a textured microstructure by directional solidification (Ref 12). To achieve homogenization the samples were annealed for 48 h at 1000 °C under Ar/5% H_2 atmosphere. The texture was characterized by x-ray diffraction (XRD). Cubic samples with an edge length of 5 mm were cut erosively parallel to the direction of solidification which is preferably a [100] axis of the cubic high temperature phase (austenite). Afterwards, the samples were annealed for 14 h at 600 °C to relieve stress. This procedure had been shown to be effective by Gaitzsch et al. (Ref 13). The next step was grinding and polishing to achieve plane and parallel surfaces. To remove the working strain layer which forms due to grinding and polishing, an electrolytic etching with 25% (vol) HNO_3 in ethanol was used. The microstructure was analyzed by scanning electron microscopy (SEM), the orientation of the grains by electron backscatter diffraction (EBSD), and the magnetic domains by magnetic force microscopy (MFM).

3. Results and Discussion

The structure of $Ni_{50}Mn_{29}Ga_{21}$ alloys is tetragonal in the martensitic state (5M) at room temperature and cubic in the austenitic state (Ref 14). The Curie temperature is approximately 100 °C.

The surface topography of the cubic sample shows regions of parallel lamellae which overlap each other mainly at the interface between two neighboring regions. A typical area of the cube surface having different kinds of lamellae was chosen to investigate the orientation via EBSD and the topography with the SEM. Figure 1 shows the SEM and EBSD pictures of the same area. The fine lamellar structures have boundaries between the lamellae.

They are parallel arranged in certain regions. Each set of parallel lamellae arose from a single austenite grain. The boundaries between these lamellae are twin boundaries and their overlapping is observable in the SEM picture. The EBSD map (Fig. 1b) shows the lamellae which are colored according to their orientation. It can be seen that there are only three martensitic variants marked with black (34%), grey (18%), and light grey (8%) with an index rate of 60% and a step size of 0.3 μm. The [001] direction of the unit cell is parallel to the *x*-axis (marked with black), parallel to the *y*-axis (marked with grey), and parallel to *z*-axis marked with light grey. Here, the directions are indexed in the tetragonal structure of the martensite (*I4mmm*). Thus, the [110] direction based on the tetragonal structure is parallel to the [100] direction in the cubic high-temperature phase.

A shadowing effect occurs due to the surface roughness and a tilt angle of 70°. Therefore, there are regions which could not indexed by the EBSD system. These regions are marked white in Fig. 1(b). Figure 1(c) shows the change in the orientation of the *c*-axis between the lamellae. The period between two lamellae is approximately 1.5 μm. The line scan was taken along the black line in Fig. 1(b).

A cubic sample was compressed successively along the solidification direction. The maximum stress was 40 MPa and after unloading the remanent strain is about 2% (Fig. 2).

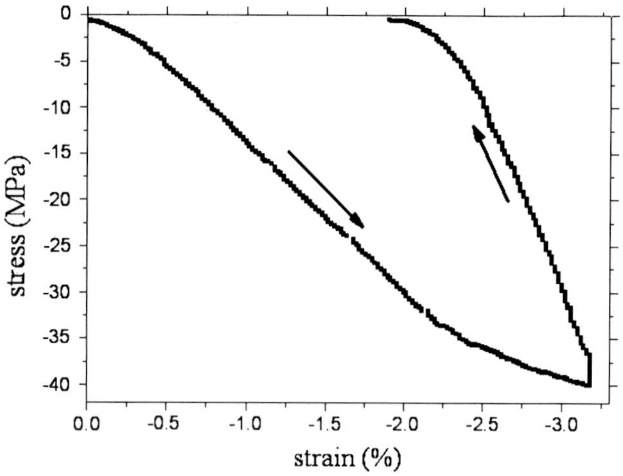

Fig. 2 Stress-strain curve of a 5M $Ni_{50}Mn_{29}Ga_{21}$ sample in solidification direction

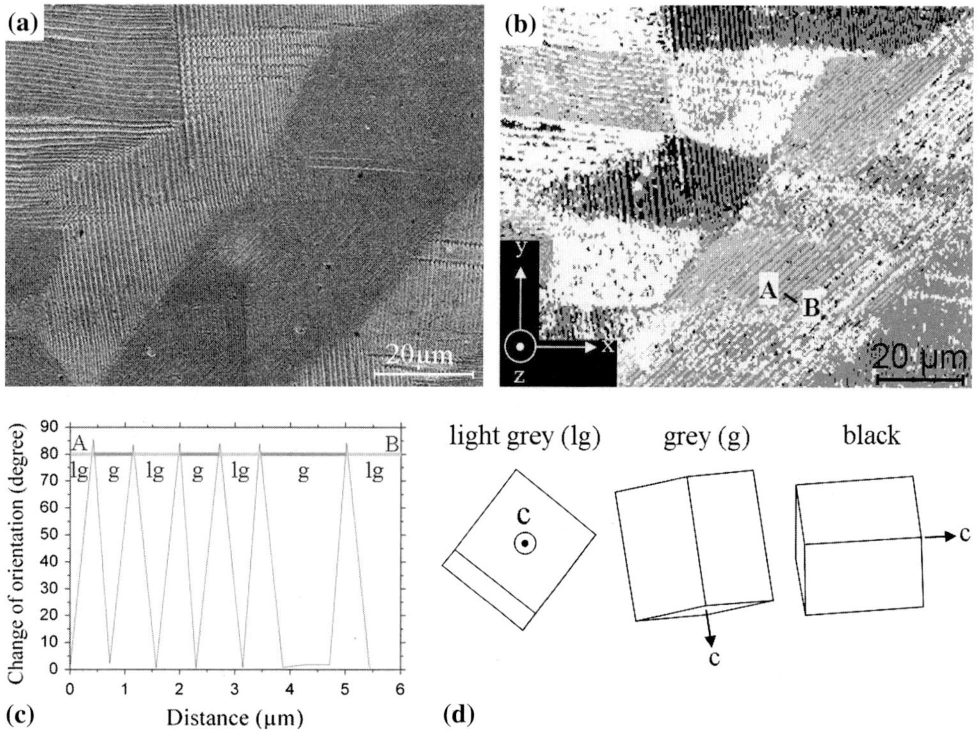

Fig. 1 Comparison of the topography by SEM (a) and orientation by EBSD (b, d) of the same area of NiMnGa with the change of orientation (c) along the black line in (b)

The twin boundaries are visible as height steps due to anisotropic etching. During the deformation process, twin boundary motion occurs, but the etched lamellae stay at the same place. Only a small tilting of the lamellae is observed because of the overall deformation of the sample. A very thin film develops on the surface which remains unchanged after etching the sample. After the deformation, a new preparation is not possible—to see the new microstructure—without changing the new twin structure. The reasons are compression force and shear force which affect the sample surface during polishing.

After compressing the sample, an EBSD picture taken from the same area is shown in Fig. 3 together with a line scan. The compressing direction is the same as the solidification direction. The twin distribution has changed dramatically. The grey colored areas where the *c*-axis is parallel to the compression direction increased and black colored regions, where the *c*-axis is perpendicular to it, are reduced. The index rate of Fig. 3 is 37%, with 31% grey, 5% black, and 1% light grey and the step size is 0.3 μm. There are only few black dyed regions. This means nearly the whole sample changed the orientation with the *c*-axis parallel to the compressing direction. A larger load would be necessary to remove the black dyed regions completely. The line scan shows that between all color changes the orientation of the *c*-axis changes by approximately 86°. This angle represents the orientation difference of the *c*-axis between the twins and can be calculated. From geometrical consideration it follows that the change in the orientation ϕ is given by $\phi = 2 \cdot \arctan\left(\frac{c}{a}\right)$ where the $\frac{c}{a}$ ratio is 0.94 for the 5M structure. A similar result was found by Cong et al. (Ref 15) for $Ni_{53}Mn_{25}Ga_{22}$ that each austenitic grain transforms to only two martensitic variants. After compression, the width of the martensitic variants with the *c*-axis parallel to the compression direction increased to about 7 to 11 μm.

Atomic force microscopy (AFM) and MFM images from the same area after compressing are shown in Fig. 4. The AFM (topography contrast) and MFM (phase contrast) are two

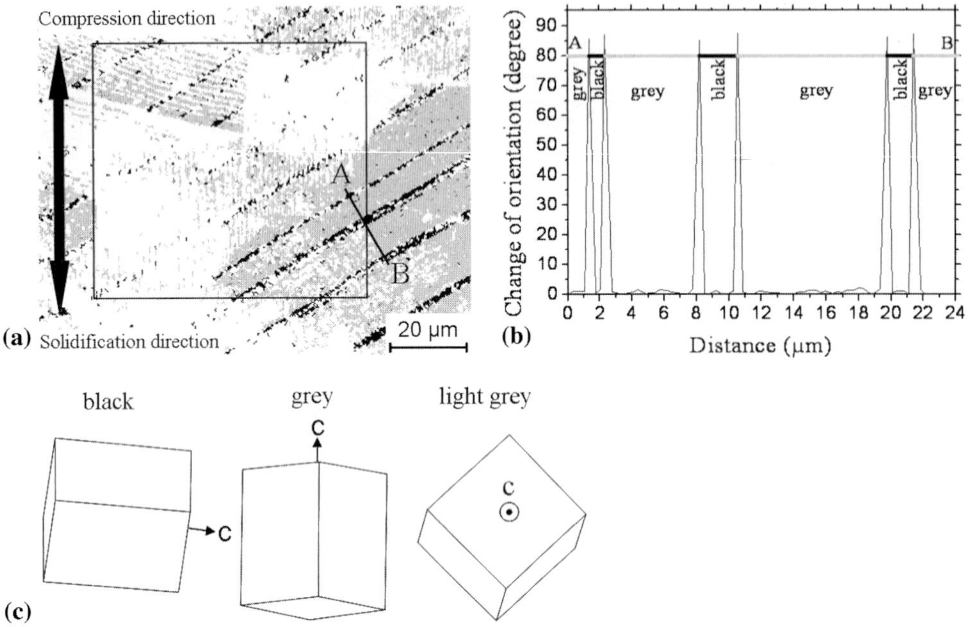

Fig. 3 (a) Orientation by EBSD after compression in solidification direction with (b) a line scan (black line in SEM picture) and (c) the direction of *c*-axes. The square marks the area which is shown in Fig. 4

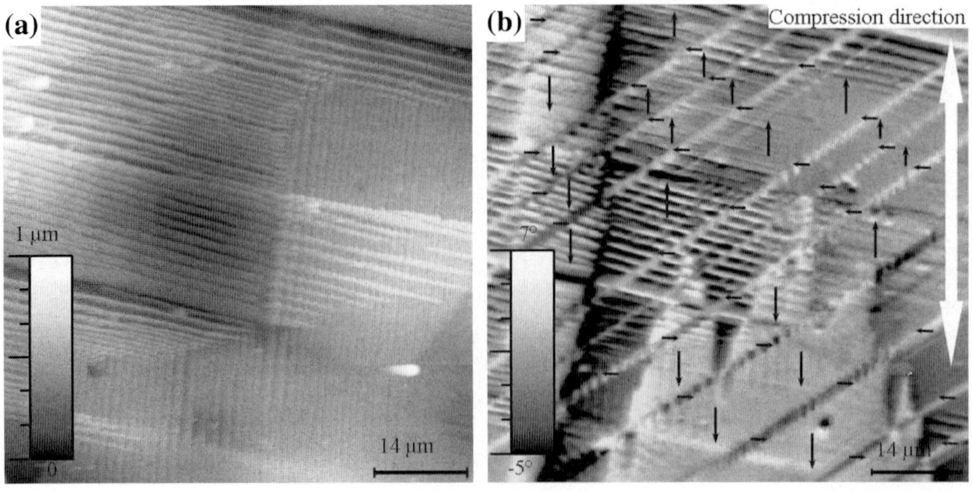

Fig. 4 (a) AFM picture (topography contrast) and (b) MFM picture (phase contrast) after compression of the same area like the EBSD image

different signals which were measured in the same scan. However, the AFM image taken after compression shows clearly those twins which were made visible by etching before compression and their overlapping while the MFM image reflects the twin distribution due to the magnetic anisotropy of the martensite after compression. The surface profile originated by etching shows the twin structure after compression and the height is 100 to 200 nm. This topography is also visible in the MFM picture but it is different from the domain contrast. The easy axis of magnetization is parallel to the *c*-axis. The EBSD image in Fig. 3(a) shows big areas with *c*-axis in *y* direction (grey), which is the compression direction. The square in Fig. 3(a) is the same area in Fig. 4(b). This area shows three big domains which are separated by 180° domain walls. The stripes corresponding to the black colored regions in Fig. 3 have another magnetic pattern. This contrast would be more intensive if there were 180° domains. The *c*-axis of the stripes is perpendicular to the compression direction. Two big domains are lighter than the background because the magnetization shows in the opposite direction which is also parallel to the compression direction. Also, the *c*-axis of the stripes can show in the opposite direction (arrows Fig. 4). Hence, no head-to-head configuration can be seen because of the stray field.

Comparing the MFM and the EBSD pictures, one can see that only with orientation sensitive methods the change of variant distribution by compression can be seen. SEM images are not suited for following the structure changes by, e.g., compression, because only the initial twin distribution is visible after compression due to topographic contrast produced by preparation. Both methods EBSD and domain observation by MFM show similar results.

4. Conclusion

The polycrystalline $Ni_{50}Mn_{29}Ga_{21}$ sample was prepared by directional solidification. The topography with its twin boundaries and the orientation of the twins fits very well before compression. After compression, the amount of those regions increases which have the *c*-axis parallel to the compression direction which is also the solidification direction in our case. It is possible to change the orientation of the *c*-axis in polycrystalline samples by compression. However, the surface topography did not change noticeable. Thus, observing the surface topography by optical microscopy or SEM is not suited for observing twin boundary motion during compression with this etch process. It is necessary to use orientation sensitive methods as EBSD or magnetic domain observation to follow the structural changes during compression.

Acknowledgments

The authors are grateful to A. Pöhl for technical assistance. The work was supported by DFG SPP 1239 code Ro 962/5.

References

1. K. Ullakko, J.K. Huang, C. Kantner, R.C. O'Handley, and V.V. Kokorin, Large Magnetic-Field-Induced Strains in Ni_2MnGa Single Crystals, *Appl. Phys. Lett.*, 1996, **69**(13), p 1966–1968
2. P. Müllner, V.A. Chernenko, M. Wollgarten, and G. Kostorz, Large Cyclic Deformation of a Ni-Mn-Ga Shape Memory Alloy Induced by Magnetic Fields, *J. Appl. Phys.*, 2002, **92**(11), p 6708–6713
3. R.C. O'Handley, S.J. Murray, M. Marioni, H. Nembach, and S.M. Allen, Phenomenology of Giant Magnetic-Field-Induced Strain in Ferromagnetic Shape-Memory Materials (Invited), *J. Appl. Phys.*, 2000, **87**(9), p 4712–4717
4. P. Müllner, V.A. Chernenko, and G. Kostorz, A Microscopic Approach to the Magnetic-Field-Induced Deformation of Martensite (Magnetoplasticity), *J. Magn. Magn. Mater.*, 2003, **267**, p 325–334
5. A. Sozinov, A.A. Likhachev, N. Lanska, and K. Ullakko, Giant Magnetic-Field-Induced Strain in NiMnGa Seven-Layered Martensitic Phase, *Appl. Phys. Lett.*, 2002, **80**(10), p 1746–1748
6. U. Gaitzsch, M. Pötschke, S. Roth, B. Rellinghaus, and L. Schultz, A 1% Magnetostrain in Polycrystalline 5M Ni-Mn-Ga, *Acta Mater.*, 2009, **57**, p 365–370
7. U. Gaitzsch, M. Pötschke, S. Roth, B. Rellinghaus, and L. Schultz, Mechanical Training of Polycrystalline 7 M $Ni_{50}Mn_{30}Ga_{20}$ Magnetic Shape Memory Alloy, *Scr. Mater.*, 2007, **57**, p 493–495
8. K. Ullakko, A. Sozinov, and P. Yakovenko, Large Magnetic-Field-Induced Strains in Ni-Mn-Ga Alloys due to Redistribution of Martensite Variants, Arxiv preprint cond-mat/0004211, 2000
9. K. Ullakko, Y. Ezer, A. Sozinov, G. Kimmel, P. Yakovenko, and V.K. Lindroos, Magnetic-Field-Induced Strains in Polycrystalline Ni-Mn-Ga at Room Temperature, *Scr. Mater.*, 2001, **44**, p 475–480
10. M. Pasquale, C.P. Sasso, S. Besseghini, F. Passaretti, E. Villa, and V.A. Chernenko, Effect of Texturing on the Magnetically Activated Properties of Polycrystalline NiMnGa Alloys, *J. Phys. IV France*, 2001, **11**, p 305–310
11. S. Besseghini, E. Villa, F. Passaretti, M. Pini, and F. Bonfanti, Plastic Deformation of NiMnGa Polycrystals, *Mater. Sci. Eng. A*, 2004, **378**, p 415–418
12. M. Pötschke, U. Gaitzsch, S. Roth, B. Rellinghaus, and L. Schultz, Preparation of Melt Textured Ni-Mn-Ga, *J. Magn. Magn. Mater.*, 2007, **316**, p 383–385
13. U. Gaitzsch, M. Pötschke, S. Roth, N. Mattern, B. Rellinghaus, and L. Schultz, Structure Formation in Martensitic $Ni_{50}Mn_{30}Ga_{20}$ MSM Alloy, *J. Alloys Compd.*, 2007, **443**, p 99–104
14. S. Roth, U. Gaitzsch, M. Pötschke, and L. Schultz, Magneto-Mechanical Behaviour of Textured Polycrystals of NiMnGa Ferromagnetic Shape Memory Alloys, *Adv. Mater. Res.*, 2008, **52**, p 29–34
15. D.Y. Cong, Y.D. Zhang, Y.D. Wang, M. Humbert, X. Zhao, T. Watanabe, L. Zuo, and C. Esling, Experiment and Theoretical Prediction of Martensitic Transformation Crystallography in a Ni-Mn-Ga Ferromagnetic Shape Memory Alloy, *Acta Mater.*, 2007, **55**, p 4731–4740

JMEPEG (2009) 18:558–562
DOI: 10.1007/s11665-009-9415-9

1059-9495/$19.00

A Study of the Effect of Diameter on the Fatigue Properties of NiTi Wire

Dennis W. Norwich and Audrey Fasching

(*Submitted September 11, 2008; in revised form February 19, 2009*)

It is a generally held belief that controlled strain fatigue life is independent of wire diameter. This has led to the widespread use of general S-N or ε-N curves for engineering metals. Independence of wire diameter on fatigue life may not hold true for Nitinol. If so, care must be used when applying generic Nitinol S-N or ε-N curves to product design. To provide a more comprehensive understanding of this issue, a comparison study was conducted on the fatigue properties of various diameters of NiTi (Ti-55.8/55.9wt.%Ni) wire drawn from the same starting material. ε-N curves were generated for various diameters. The wire diameter reduction from 0.030″ to 0.010″ increased the 0.8% strain 80k cycle survival rate from 35 to 96%. For the scope of this study, all factors that may have an effect on fatigue life except for wire diameter were held constant.

Keywords controlled strain, fatigue, Nitinol, rotating beam fatigue test, S-N curve, ε-N curve

1. Introduction

The use of Nitinol components in the medical device industry has increased as engineers and designers have become more familiar with the unique pseudoelastic (commonly called superelastic) and shape memory properties of the binary alloy (Ref 1). Other industries such as the automotive, military, and commercial sectors are also taking advantage of the benefits of Nitinol that outweigh the high relative cost of the material (Ref 2).

As more engineers and designers use Nitinol in their products, the need for published data on the properties and performance of the material becomes more important. For stainless steel alloys there are countless volumes of credible published data to assist in the design process. This is especially true of fatigue data, which has proven to be predictable and consistent for the many available steel alloys. For Nitinol, there is comparatively little fatigue or S-N or ε-N data that has been published and proven out through the design and test process (Ref 3). Engineers and designers are left with little credible data to use on which to base their designs.

For a specific stainless steel alloy, the published fatigue properties are generally independent of material diameter or condition. Thus, published data can be used to accurately predict the life cycle of a product. This paper will research the fatigue properties of different diameters of Nitinol wire drawn from the same starting material to determine if the same consistency of fatigue life holds true for Nitinol as it does for steel. Every effort was made to keep all factors other than wire diameter constant throughout this experiment. This reduces the chance of any property other than wire diameter affecting the fatigue life of the wire.

This article is an invited paper selected from presentations at Shape Memory and Superelastic Technologies 2008, held September 21-25, 2008, in Stresa, Italy, and has been expanded from the original presentation.

Dennis W. Norwich and **Audrey Fasching**, Memry Corporation, Bethel, CT 06801. Contact e-mail: dennis_norwich@memry.com.

2. Materials and Experimental Procedures

The fabrication process detailed below was designed to simulate that of a typical medical device that exhibits superelastic properties at room temperature. All of the samples for this study were produced from one lot of 0.052″ diameter wire (Ti-55.8/55.9wt.%Ni) which was cold worked and drawn to the following final diameters: 0.030″, 0.025″, 0.020″, 0.015″, and 0.010″. Each wire size was drawn with a similar drawing schedule to yield approximately 45% cold work on the final material. The test samples were shape set using the same heat treatment process: tooling, time, temperature, and heat source (Ref 4). After heat treatment, the parts were chemically etched, electropolished, and passivated. This surface treatment process reduced the diameter of each wire by 0.001″.

Fatigue testing was conducted using a rotating beam fatigue tester (RBT) with a zero mean 0.8% maximum alternating strain, as depicted in Fig. 1 and 2. Strain was determined by the diameter ratio methodology where ε = radius of wire/radius of curvature to neutral fiber. The samples were rotated at a constant speed guided in a test block machined with a radius to provide the level of strain desired. The fixture can test up to 10 samples at a time and counts revolutions on each sample individually using a laser counter. One hundred samples of each wire size were tested. The test was run to breakage or was suspended at 80,000 cycles. The results of this testing are shown in Fig. 3.

The RBT test is designed to provide comparative results for materials processed with different parameters at the same or varying strain levels. Tests were performed in a circulating 37 °C water bath at 1000 rpm to eliminate any heating effect

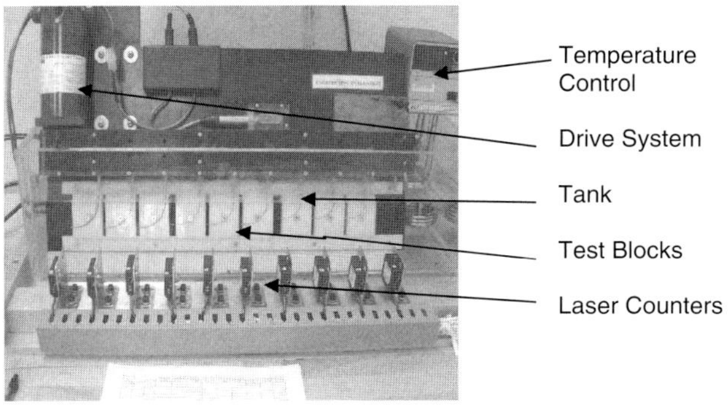

Fig. 1 Rotating beam fatigue tester

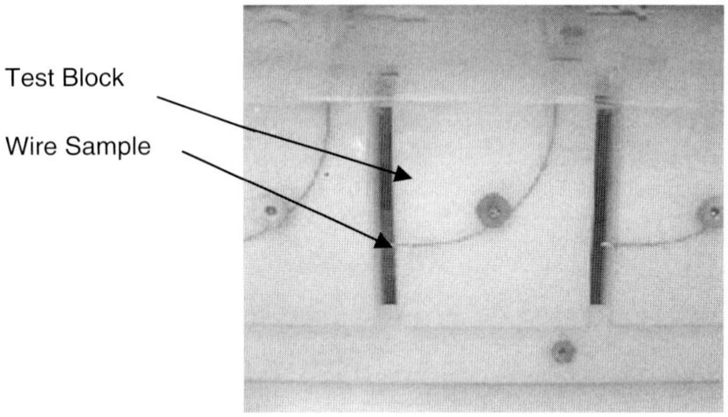

Fig. 2 Rotating beam test block

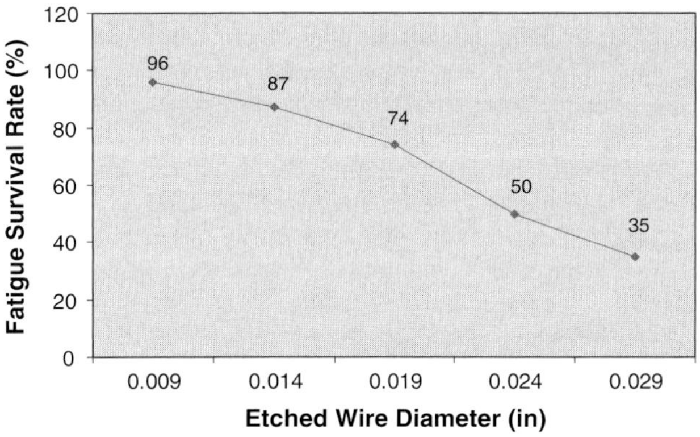

Fig. 3 Wire diameter vs. fatigue survival rate beyond 80k cycles at 0.8% strain

created in the specimen due to dissipative processes during cyclic loading (cyclic stress induced formation of martensite). The heat transfer from the specimen to the test environment helps to counteract this process (Ref 5).

Strain versus number of cycles to failure (ε-N) curves were generated for the 0.025″ and 0.010″ diameter wires. These wires were etched and electropolished to 0.024″ and 0.009″ final diameter and tested with the same rotating beam tester described above. Test blocks were made for each wire size to provide strains of 2.4, 1.8, 1.0, 0.8, and 0.65%. Three samples of each wire size were tested at each strain level. The test was

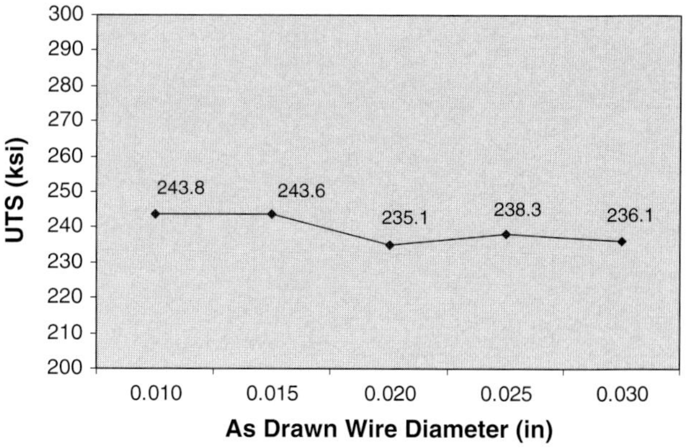

Fig. 4 Wire diameter vs. UTS as drawn

run to failure or suspended at 10 million cycles. The results of this testing are shown in Fig. 9.

Tensile tests were conducted on using an electromechanical tensile testing load frame with a 5 kN load cell as per ASTM F 2516-05. Tensile tests to failure (TTF) was performed on cold worked and shape set wire samples, while 6% strain tests were performed on only the shape set samples. Tensile testing was performed at 22 °C in air.

Differential scanning calorimeter (DSC) tests were conducted on cold worked and shape set parts per ASTM F 2004-3 to determine transformation temperature. Austenite finish temperature (A_f) was determined to be ~11 °C.

An outside laboratory was used to perform metallographic inclusion analysis on samples of each cold worked wire size. Metallographic mounts were created and examined using an SEM and an analytical computer program to measure and count inclusions. Five samples of each wire size were analyzed in both the longitudinal and the transverse directions for inclusion size, number of inclusions, and area fraction of inclusions.

3. Results and Discussion

The purpose of this study was to investigate the effect of diameter on the rotational bending fatigue life properties of Nitinol wire. Memry conducts an 80,000 cycle test at 0.8% strain as part of a wire fatigue qualification protocol. In this test, samples are fatigued at 0.8% strain until failure or 80,000 cycles. The results are reported as the percentage of parts exceeding the 80,000 cycle limit. Over years of data gathering, it has been noted that the smaller diameter wires survive this test at a much higher rate than the larger diameter wires. The results of this study corroborate and quantify this observation (see Fig. 3).

Although every measure was taken to ensure consistency in processing, basic tensile data were collected to confirm uniformity. Ultimate tensile strength (UTS) for the as-drawn wire for each wire size was determined and plotted in Fig. 4. The upper and lower plateaus (6% curves) for each heat-treated wire size are shown in Fig. 5 (Ref 6). A typical 6% stress-strain curve for a heat-treated test sample is shown in Fig. 6. Figure 7 depicts the stress-strain curves for each wire size to 2.4% strain, the maximum strain used for the fatigue study. Figure 8 is a

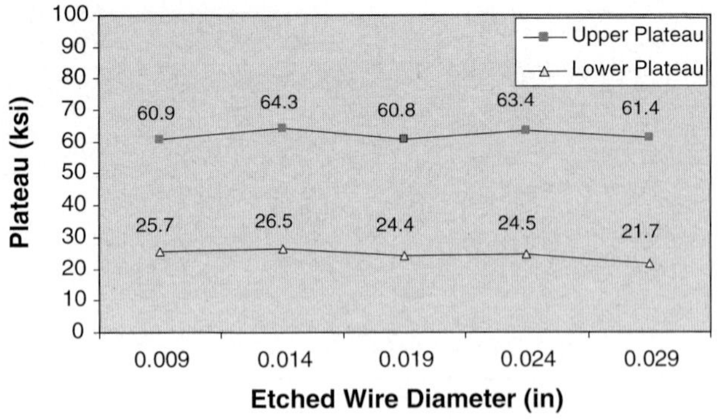

Fig. 5 Wire diameter vs. upper and lower plateau

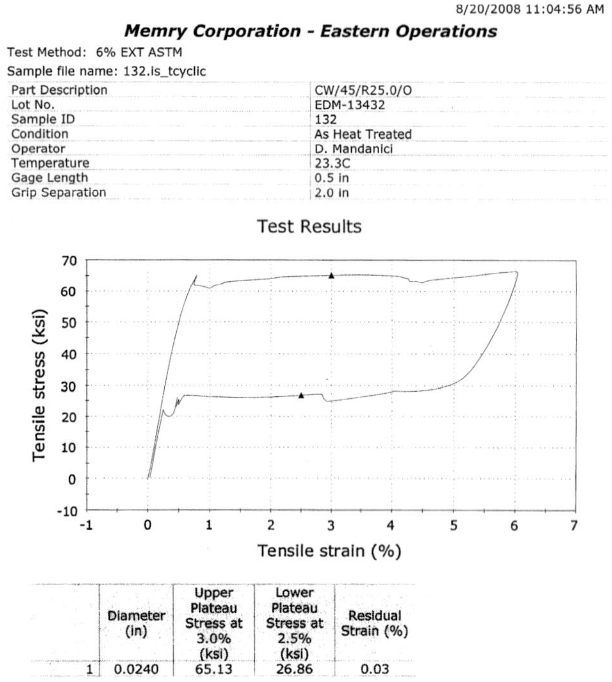

	Diameter (in)	Upper Plateau Stress at 3.0% (ksi)	Lower Plateau Stress at 2.5% (ksi)	Residual Strain (%)
1	0.0240	65.13	26.86	0.03

Fig. 6 The 6% curve of as heat-treated wire

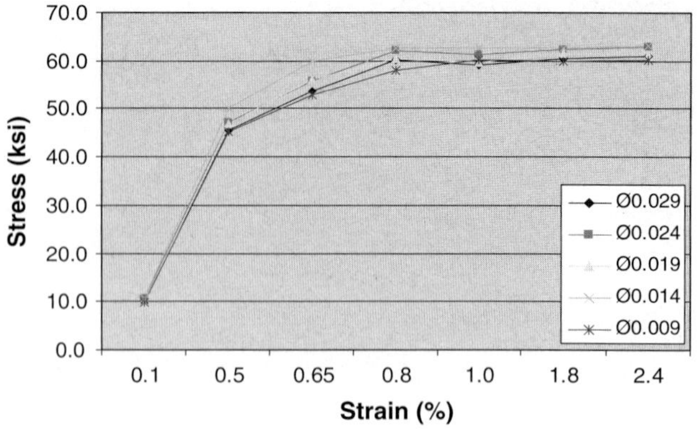

Fig. 7 Stress-strain curve for each wire size

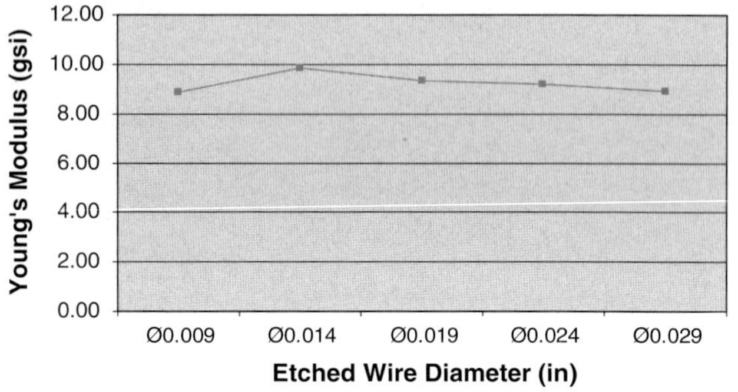

Fig. 8 Young's modulus vs. wire diameter

plot of Young's modulus for each wire size. The tensile data exhibit only small variations across the wire sizes. This validates the consistency of preparation of the test samples for each wire size.

ε-N curves were generated for the 0.024″ and the 0.009″ diameter wires used for this investigation. The results, shown in Fig. 9, were generated using three samples per strain level. The data indicate an increased fatigue life of the 0.009″ over the 0.024″ diameter wire. More data over a larger range of wire sizes will need to be collected to determine the extent of this effect.

An SEM image of a typical fatigue failure generated during this study is shown in Fig. 10. On the image, the fracture initiation site, likely an inclusion, is indicated by an arrow. This fracture surface is typical of fatigue failures, where the crack radiates out from the initiation point and is followed by a region of microvoid coalescence which indicates that final failure occurred by overload (Ref 7).

It is generally accepted that inclusions, especially those located at or near the surface of the wire, act as stress risers and initiation sites for fatigue cracks. Metallographic examination provides quantitative analysis of the inclusion size and content for the wire samples. The primary inclusions found in Nitinol alloys are titanium carbides (TiC) and intermetallic oxides ($Ti_4Ni_2O_x$). The titanium carbides tend to be small, uniform, and evenly distributed while the intermetallic oxides ($Ti_4Ni_2O_x$) are blocky and unevenly distributed throughout the matrix.

Inclusion analysis requires the examination of mounted and polished metallographic samples. Because inclusions can be unevenly distributed and analysis is conducted on a small area, the samples may not be representative of the entire matrix. The most useful data come from examination of multiple samples from the same material, but this is costly and time consuming.

Area fraction of inclusions and the largest inclusion size in both transverse and longitudinal directions are the most meaningful data gathered in the analysis when relating to fatigue properties of a material. A plot of the inclusion area fraction for each wire size in both the longitudinal and transverse directions is shown in Fig. 11. Since inclusions are fatigue initiation sites, a lower area fraction of inclusions should correlate to better fatigue performance. This plot did not correlate well with the cyclic fatigue data. The five largest inclusions for each wire size in both the longitudinal and transverse directions are plotted in Fig. 12. The trend lines show that transverse inclusions tend to get smaller as wire diameter decreases, while longitudinal inclusions tend to form stringers and elongate in the drawing direction. This phenomenon

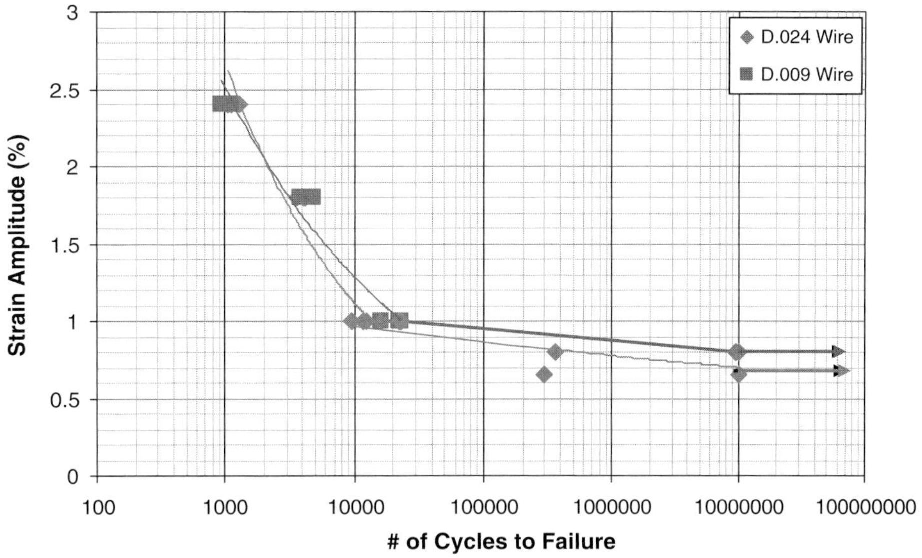

Fig. 9 ε-N curve 0.009″ vs. 0.024″ diameter heat-treated wire

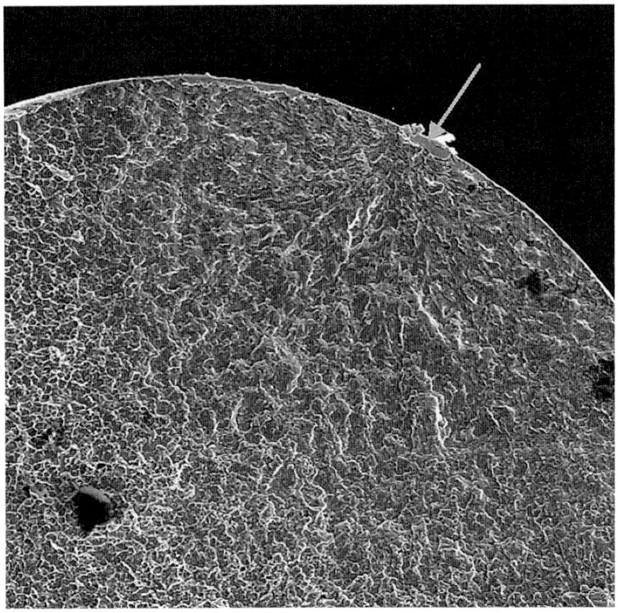

Fig. 10 SEM image of fatigue fracture surface 0.024″ diameter wire

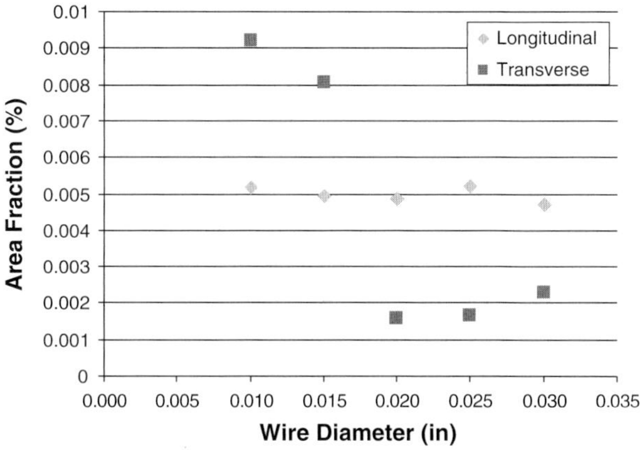

Fig. 11 Inclusion area fraction vs. wire diameter

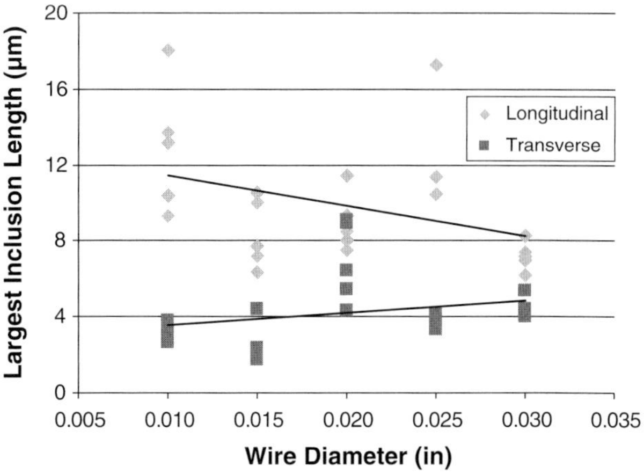

Fig. 12 Five largest inclusions vs. wire diameter

can be explained by the drawing process; each reduction breaks up and strings out individual inclusions. The smaller diameter wires have more reduction steps than the larger diameter wires and therefore the inclusions are more broken up and spread out (Ref 3). Larger, blocky inclusions are more likely to initiate fatigue cracks than small, symmetrical inclusions. This may explain the increase in fatigue life for the smaller diameter wires.

4. Summary

Final wire diameter has been shown to have an influence on the fatigue properties of Nitinol. The wire diameter reduction from 0.030″ to 0.010″ increased the 0.8% strain 80k cycle survival rate from 35 to 96% as shown in Fig. 3. For the scope of this study, all factors that may have an effect on fatigue life except for wire diameter were held constant. These factors include starting ingot material, wire drawing process, sample heat treatment and fixturing, post-processing surface treatment, and fatigue testing procedures. The ε-N data (Fig. 9) show slightly increased fatigue life for 0.009″ over the 0.024″ diameter wire.

Inclusion size and dispersion are generally accepted as factors that can negatively impact fatigue life. While the inclusion analysis performed for this experiment is not entirely conclusive, the trend lines in Fig. 12 do show the inclusions becoming smaller in the transverse direction and stringing out in the longitudinal direction. This makes logical sense and may contribute to greater fatigue life in the smaller diameter wires with all other factors held constant.

The results of this study along with other published fatigue data on Nitinol wire demonstrate that the fatigue performance of Nitinol is very product specific. Unlike stainless steel, there is no generic S-N or ε-N curve for Nitinol. There are many factors that can affect fatigue life that were held constant for the scope of this experiment. With a difference in wire diameter, there was a difference in fatigue performance. More data need to be collected to determine the significance of the findings in this paper. However, engineers and designers must recognize that no generic S-N or ε-N curve will provide an accurate prediction of the fatigue life of Nitinol. Each product, material, and set of process parameters must be individually tested and analyzed to provide accurate fatigue life data.

References

1. T.W. Duerig and R. Zadno, An Engineer's Perspective of Pseupoelasticity, *Engineering Aspects of Shape Memory Alloys*, T.W. Dureig, K.N. Melton, D. Stockel, and C.M. Wayman, Eds. (Butterworth-Heinemann), 1990, p 369–393
2. M.H. Wu and L.McD. Schetky, Industrial Applications for Shape Memory Alloys, *SMST-2000 Proceedings of the International Conference on Shape Memory and Superelastic Technologies*, S.M. Russel and A.R. Pelton, Eds. (Pacific Grove, CA), International Organization on SMST, 2001, p 171–182
3. M. Reinoehl, D. Bradley, R. Bouthot, and J. Proft, The Influence of Melt Practice on Final Fatigue Properties of Superelastic NiTi Wires, *SMST-2000 Proceedings of the International Conference on Shape Memory and Superelastic Technologies*, S.M. Russel and A.R. Pelton, Eds. (Pacific Grove, CA), International Organization on SMST, 2001, p 397–403
4. A. Pelton, J. DiCello, and S. Miyazaki, Optimization of Processing and Properties of Medical-Grade Nitinol Wire, *SMST-2000 Proceedings of the International Conference on Shape Memory and Superelastic Technologies*, S.M. Russel and A.R. Pelton, Eds. (Pacific Grove, CA), International Organization on SMST, 2001, p 361–374
5. M. Wagner, T. Sawaguchi, G. Kaustrater, D. Hoffken, and G. Eggeler, Structural Fatigue of Pseudoelastic NiTi Shape Memory Wires, *Mater. Sci. Eng. A*, 2004, **378**, p 105–109
6. T. Saburi, Ti-Ni Shape Memory Alloys, *Shape Memory Materials*, K. Otsuka and C.M. Wayman, Eds. (Cambridge University Press), 1998, p 49–96
7. B. James, S. Murray, and S. Saint, Fracture Characterization in Nitinol, *SMST-2003 Proceedings of the International Conference on Shape Memory and Superelastic Technologies*, A.R. Pelton and T. Duerig, Eds. (Pacific Grove, CA), International Organization on SMST, 2004, p 321–324

JMEPEG (2009) 18:563–568
DOI: 10.1007/s11665-009-9363-4

1059-9495/$19.00

Fatigue Performance of Nitinol Round Wire with Varying Cold Work Reductions

Jeremy E. Schaffer and David L. Plumley

(*Submitted October 8, 2008; in revised form November 13, 2008*)

The role of cold work reduction and its effect on fatigue properties are examined by means of rotary beam fatigue testing. In this study, processing parameters from the initial ingot to the final size before finish drawing were held constant. Fatigue specimens were prepared with between 20 and 55% cold work in five percent increments. These samples were subsequently heat treated, cut into test sections at room temperature, and subjected to strain-controlled fatigue testing in a temperature-controlled RO water bath at a constant 10 °C above the measured active austenitic finish temperature (A_f). An empirical analysis of the relationship between final reduction prior to heat treatment and fatigue lifetime is presented. Under alternating strain amplitudes greater than 1%, the 20 to 30% cold work level was found to deliver the greatest mean cycles to failure. Supplemental to the fatigue data, mechanocyclic and thermomechanical data are also presented as they evolve with prior retained cold work.

Keywords nitinol wire, nitinol processing, nitinol cold work, strain fatigue

1. Introduction

There are many processing parameters which can influence the mechanical, thermal, and bioreactive properties of nitinol wire and its suitability for particular applications (Ref 1-3). Variables may include: ingot composition; inclusion particle size and shape distribution; cold work history; annealing temperature, stress, atmosphere and resultant microstructural cell size; surface conditioning including oxide thickness, acidic oxide removal or mechanical polishing; shape-setting temperature and others. Use of nitinol as an implantable material requires an understanding of specific structure-property relationships. For long term implants, life expectancy before fracture under cyclic loading should be predictable given a set of processing and service conditions. Studies correlating material processing to fatigue behavior are therefore important in robust device design.

Significant research has been dedicated to understanding how Nitinol behaves in the body from the viewpoint of a biological host response (Ref 4-6); much less literature has been published which quantitatively correlates structure with mechanical properties. It could be because chemical studies are generally easier to carry out and interpret than the sometimes confounding mechanical response of Nitinol. Perhaps for this reason, there is an even greater dearth of information in the area of fine wire for medical devices such as guidewires, braided vascular stents, embolic filters, and other Nitinol wire-based devices. Biomedical engineers need to understand not only how such variables affect reversible inelastic behavior, but also be able to quantify their effect on fatigue life. Up to the time of this writing, with few exceptions (Ref 7-10), little work had been published on the fatigue behavior of fine Nitinol wires as a function of wire processing variables.

Plastic deformation through cold wire drawing of most metallic alloys results in increased mechanical strength that is attributable to an increase in lattice defects, many of which are dislocations. It is largely the strain fields associated with these dislocations that result in implant durability and strength or in some cases can lead to brittleness and failure. In the case of Nitinol, cold work and the associated dislocation network density can dramatically influence mechanical properties (Ref 1, 3). In this investigation, the goal was to elucidate the effect of cold work on mechanocyclic, mechanothermal, and thermocyclic properties. To accomplish this, nominally Ti-50.9 at.% Ni binary NiTi wires having an ingot A_s temperature of -17 °C were produced with variable retained cold work prior to heat treatment. In order to mimic field use, the nominally 50-nm thick process-intrinsic oxide barrier was left intact throughout testing.

2. Experimental

The initial material for the experiment was 6.35 mm diameter wire of Ti-50.8 at.% Ni which was repeatedly drawn and annealed to staged diameters in preparation for final full annealing. Interpass and final annealing processes were performed at a temperature of 650-850 °C at a sufficient dwell time to ensure complete recrystallization. Wires were conventionally wet-drawn using natural diamond drawing dies and

This article is an invited paper selected from presentations at Shape Memory and Superelastic Technologies 2008, held September 21-25, 2008, in Stressa, Italy, and has been expanded from the original presentation.

Jeremy E. Schaffer and **David L. Plumley**, Fort Wayne Metals Research Products Corporation, 9609 Ardmore Avenue, Fort Wayne, IN 46809. Contact e-mail: jeremy_schaffer@fwmetals.com, dave_plumley@fwmetals.com.

oil-based lubricants to a finished diameter of 0.32 mm. All of the wires were heat treated at 490 °C for a dwell time of 15 min in an air-stirred fluidized bed with alumina media and quenched in ambient air at nominally 25 °C. In order to reduce the effects of ingot compositional variability, the total quantity of material consumed was limited; total sample material comprised approximately 2 kg from a consecutive billet location.

After the 490 °C heat treatment, active sample transformation temperatures were analyzed using bend and free recovery (BFR) methodology using a proprietary digital video analysis apparatus per standard ASTM F2082-06 (Ref 11). Cyclic and monotonic uniaxial tensile properties were measured at 37 °C using an Instron Model 5565 Tensile Test Bench equipped with an environmental chamber. Near-surface microstructure and oxide thickness determination was performed using a combination of focused ion beam (FIB) milling and transmission electron microscopy (TEM). Fatigue behavior was characterized using rotary beam fatigue test equipment manufactured by Positool, Inc. at 3600 rpm in a temperature-controlled, continuously flowing water bath. Ten specimens from each cold work condition were tested at alternating strain amplitudes ranging from 0.9 to 2.5% to a maximum of 1×10^8 cycles.

Fatigue sample geometry was calculated based on the assumption of elastic loading with a central neutral axis using Eq 1. While this approach was used as a convention, the calculated strain amplitude and strain ratio are likely not a perfect representation of the true material state. Nitinol is known to exhibit a stress-strain response that is asymmetric in tensile versus compressive loading (Ref 1, 14). During initial cycling where compression requires greater strain to begin transformation, we expect a dynamic neutral axis shift away from the tensile surface. This behavior would alter the strain ratio and thus the effective crack opening strain range and may be an important and dynamic variable in bend-type testing of nitinol wire.

$$\varepsilon_a = d/2R \qquad \text{(Eq 1)}$$

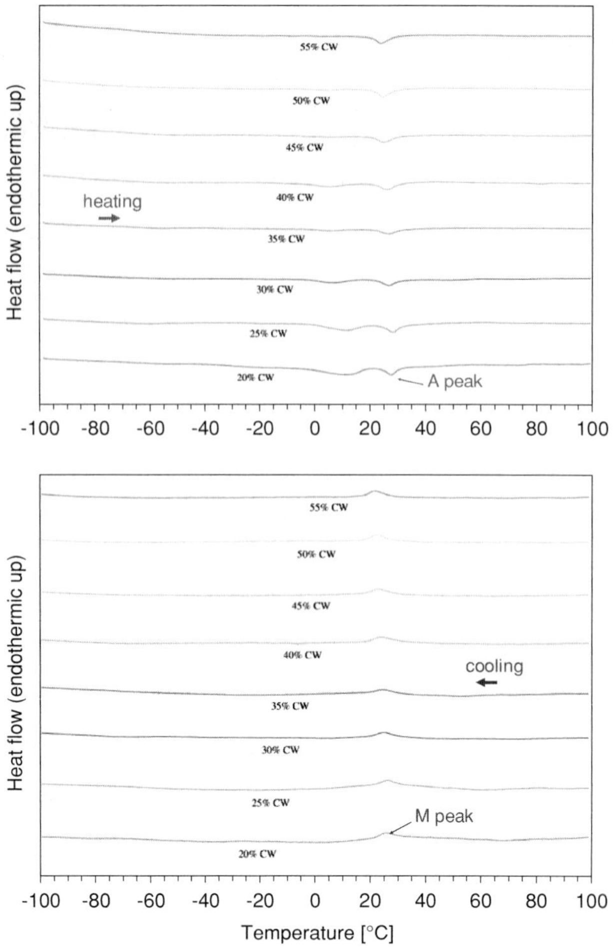

Fig. 1 DSC curves of heat-treated nitinol wires. Material was scanned in the active state over the range of −100 to 100 °C

3. Results and Discussion

3.1 DSC Analysis

Differential scanning calorimetry (DSC) was used to examine thermal phase transformation points in both the fully annealed as well as active heat-treated states. The martensitic and austenitic transformation ranges were found to be consistent to be within 10 °C for all the specimens tested, showing a small downward shift with increasing cold work similar to other work. Multistep transformations were evident in the heating curves for samples that were cold worked between 20 and 40%. The resultant transformations shown in Fig. 1 were described by the two-step transformation as: martensite (M) → R′ and R′ → B2 (Ref 12). Multiple peaks were not visually evident in the cooling direction or in the heating direction above ~40% cold work indicating likely the thermal superposition or single stage transformation (Ref 13).

3.2 BFR Analysis

All investigated samples were taken consecutively from a single nitinol parent ingot and were found to possess similar austenitic transformation temperatures. Bend and free recovery testing (BFR) was used in order to characterize the active or process-dependent, transformation temperature range. At 0.32 mm in the heat treated state, the austenitic start and finish temperatures (A_s and A_f) were derived from computer-generated plots that were collected using digital video analysis.

The BFR curve shown in Fig. 2 clearly exhibits at least three distinct regions: a martensitic flat line at the left, initial recovery to −10 °C, essentially rate constant recovery until 20.5 °C (A_s), and an increasing rate toward completion through the austenitic transformation. The actual three regions of recovery may comprise: complex transition from B19′ to an energetically preferred R-phase variant back to B19′ before transitioning to B2 austenite. Duerig described possible stress-induced *R* transformation at low temperatures in (Ref 14). Here, strain is initially imparted at 218 K, below M_f and within the range described by Duerig. Thus a mixed strain-stabilized *R*-phase, B19′ starting point is possible. The strain stabilization of *R*-phase in this temperature range may also explain the lack of a deconvoluted *R*-peak in the DSC results: subtract the strain imparted for BFR, and eliminate the *R*-phase. Verification of this phenomenon should be explored further via in situ TEM deformation studies.

3.3 Tensile Properties

The as-drawn material showed a large increase in both the engineering yield and ultimate strength as a function of cold work as exemplified in Fig. 3. Cold work was also accompanied by a large reduction in strain to rupture, similar to other published studies (Ref 1, 15). The modulus of elasticity was also found to increase in the as-worked material from a minimum at 30% cold work of 40 GPa to a maximum at 55% cold work of about 60 GPa. The fluctuation in elastic modulus has been reported in other studies and is ascribed to initial martensite reorientation and variant activity in the fully developing martensitic structure (Ref 16) (see Fig. 3 inset).

In the heat-treated condition, samples were cyclically and monotonically tensile tested at 37 °C. A positive correlation was found between ultimate strength, loading plateau strength, and stress hysteresis relative to prior retained cold work. No clear relationship was found between the permanent set after a 6% strain departure and the prior cold work (see Fig. 4 and 5).

3.4 Fatigue Properties

Ten specimens were cycled to failure from each condition using a rotary beam fatigue apparatus until failure or until a maximum of 10^8 cycles. The testing was carried out in reverse osmosis (RO)-purified water that was continuously flowing and controlled at a temperature of 37 ± 2 °C. A strain-life representation of the data is depicted in Fig. 6. At alternating strain levels above 1%, the samples in the 20 to 35% cold work range consistently lasted longer than the samples with greater than 40% cold work. The downward shift in life with cold work is readily evident in the normalized life plot shown in Fig. 7. A mixture of high cycle runouts and fractures was observed for all samples tested at the 0.90% alternating strain level.

3.5 Scanning Electron Microscopy

Scanning electron microscopy (SEM) was used to characterize the fatigue crack-initiating features by location, type and

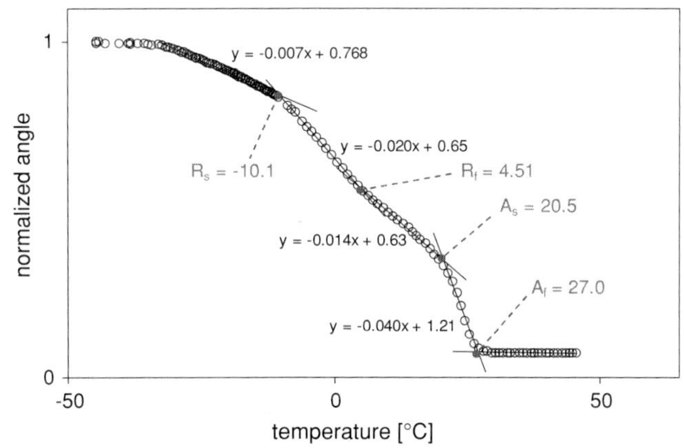

Fig. 2 Bend and free recovery curve as per ASTM F2082 for sample at 45% cold work. Note apparent multi-stage transformation as strained sample is recovered. Ordinate angle axis has been normalized to maximum bend angle (A/A_{max})

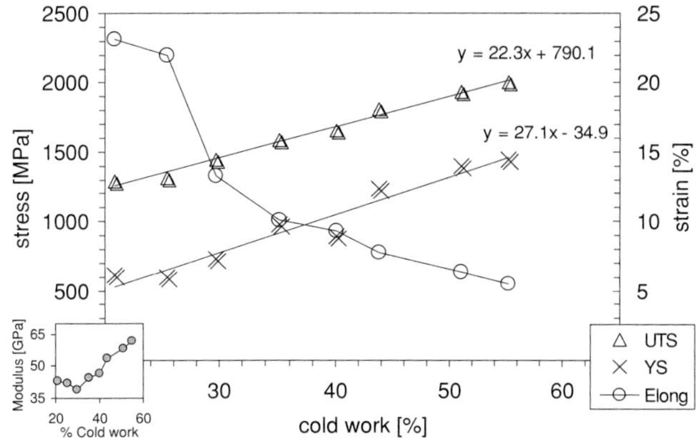

Fig. 3 Uniaxial tensile property data for eight cold work levels in the as-drawn condition where UTS is ultimate tensile strength, YS is 0.2% yield strength, and Elong is the strain to rupture

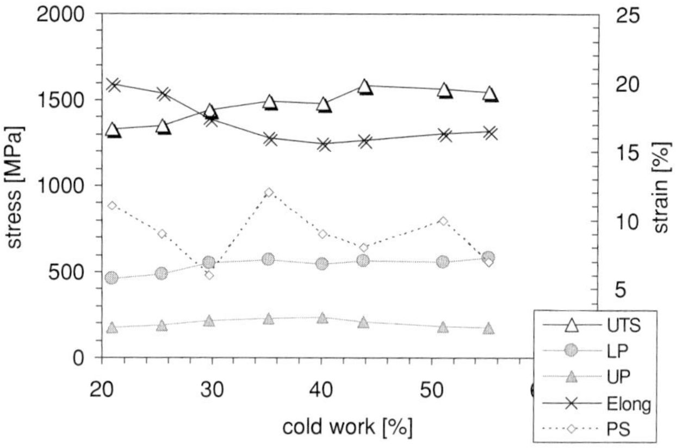

Fig. 4 Cyclic tensile property data for eight cold work levels after heat treatment, where LP is the loading, stress-induced B19′, plateau stress; UP is the unloading plateau stress; and PS is 100× permanent set after a single 6% superelastic cycle

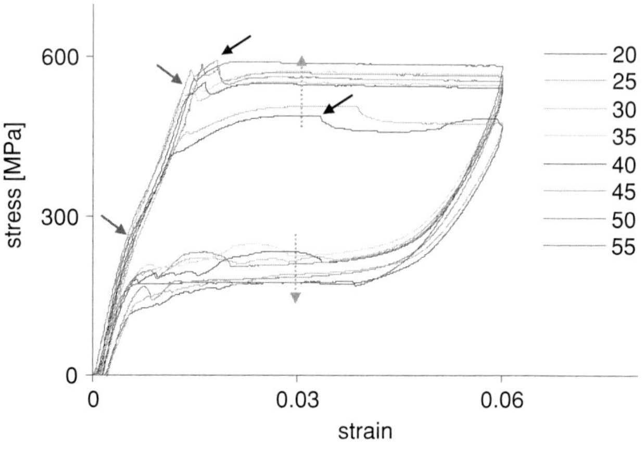

Fig. 5 Six percent strain cycle at 37 °C, 10^{-3} s^{-1} strain rate for all the samples. Gray vertical arrows show general plateau trends with increasing cold work. Black diagonal arrows indicate evidence of possible diffuse yield drop associated with the onset of martensite plate variants propagating into a mixed elastically strained austenite/R-phase matrix or the possible onset of detwinning of major B19′ variants (Ref 18). Red arrows show the onset of elastic deviation possibly due to microtwinning or the presence of a stress-induced R-phase plateau running from 0.6 to about 1.3%

geometry. Fracture analysis was only performed for fractures that occurred at the 0.9% strain level. Under this condition, two types of initiating features were found: melt-intrinsic inclusions and melt-extrinsic oxide defects. Representative initiating defects are shown in Fig. 8. A summary of defect counts versus cold work is shown in Fig. 9.

Prior studies completed for other medically relevant alloys have shown the importance of melt-intrinsic constituent particle distributions on total fatigue life, both in terms of defect size and proximity to the free wire surface (Ref 17). Crack-initiating defect states were measured here in a similar manner and compared with total life. The resultant data, plotted in Fig. 10,

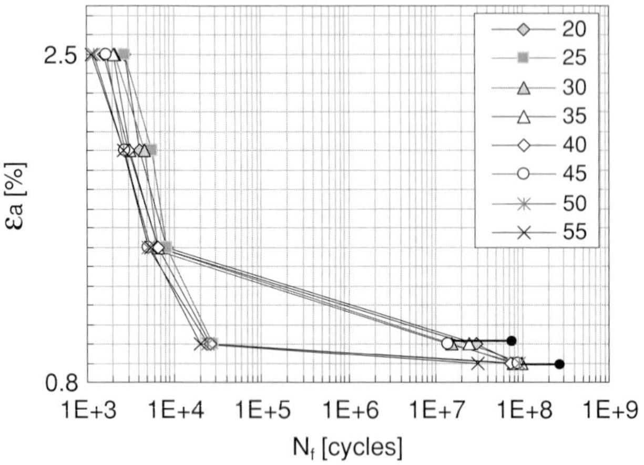

Fig. 6 Strain-controlled fatigue cycles to failure plotted on the abscissa as a function of ordinate alternating strain (ε_a). Lines terminating in a solid-filled circle represent an averaged mix of runouts and fractures

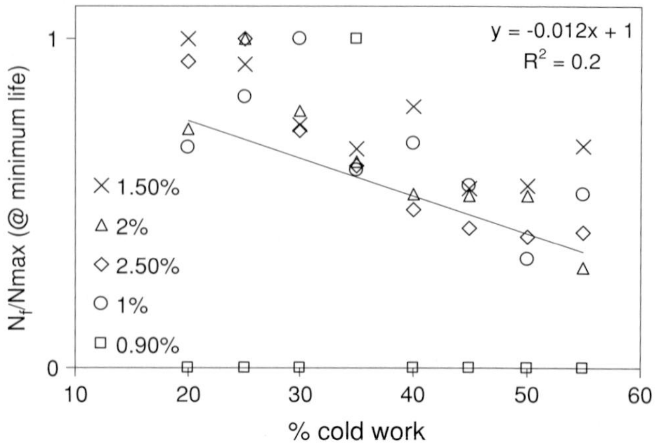

Fig. 7 This chart represents the following operations: (1) minimum cycles to fracture data were grouped for each cold work level, (2) the maximum cycle life value was taken at each strain level, and (3) data was normalized at each strain level by dividing into this maximum. The upper left "X" on the plot indicates that the 20% cold work level gave the greatest number of minimum cycles to failure (10 percentile) at the 1.5% strain level

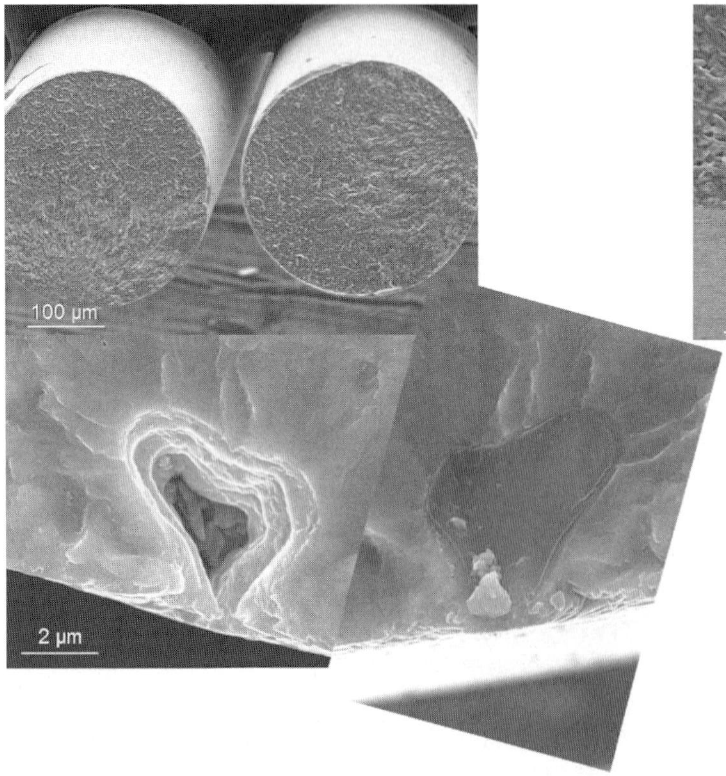

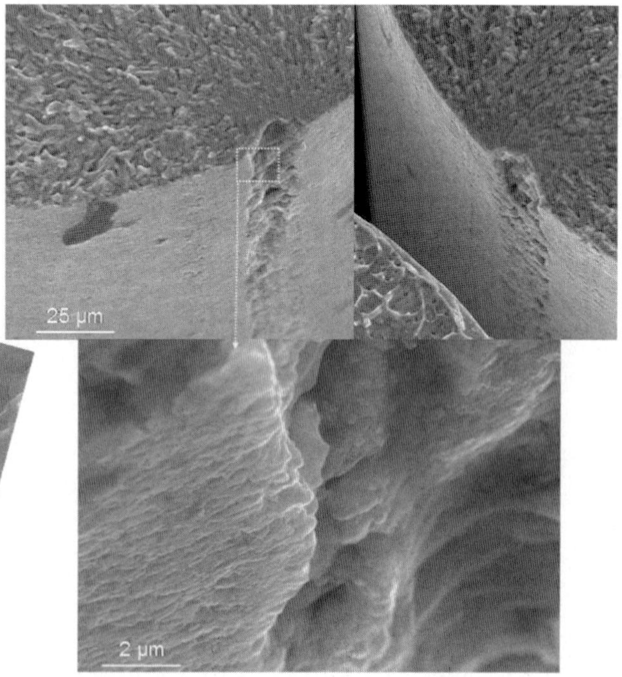

Fig. 8 Scanning electron microscopy (SEM) image collages showing (left) failure at a melt intrinsic 5.5-μm TiC inclusion in a 40% cold work sample ran at 0.9% alternating strain to 24,120 cycles and (right) failure an melt-extrinsic, 6-μm deep oxide defect in a 55% cold work sample tested to failure at 0.9% alternating strain in 32,976 cycles

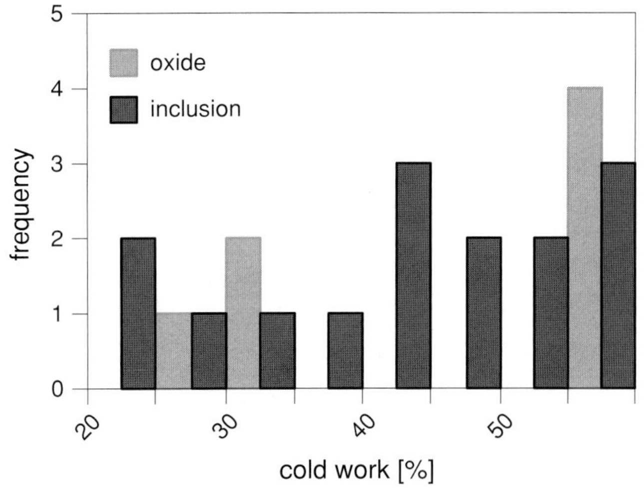

Fig. 9 Crack-initiating features collected from samples that fractured at 0.9% alternating strain. Types: oxide refers to extrinsic oxide surface defects similar to depiction in Fig. 8, right; inclusion refers to melt-intrinsic, primarily titanium carbide (TiC), constituent particles

does not show a definable trend. While contributing factors to this lack of correlation will not be made explicit within this manuscript, possible explanation for this phenomenon could be related to the relatively insignificant contribution of crack propagation to total fatigue life (Ref 17).

4. Conclusions

A multistage transformation was observed in BFR testing that was not seen as separate *R*-phase transformation peaks in the DSC data. It is postulated that this may have proceeded by low temperature strain-stabilization as: (1) matrix cooled to 218 K resulting in B19′ matrix, (2) strain accommodation in B19′ resulted in some energetically favorable *R*-phase stabilization, (3) free recovery during heating yielded separate recovery via distinct routes: *R*-phase $\rightarrow$ B2 and B19′ $\rightarrow$ B2 thus resulting in the observed multistage BFR character. Further in situ TEM analysis is needed to develop a full mechanistic explanation of this phenomenon.

Tensile testing of the heat-treated samples at +10 °C relative to the measured A_f and a strain rate of 10^{-3} s^{-1} showed a yield drop within the first 1 to 3% strain. Without further in situ TEM structural data, it is inconclusive whether this load drop was associated with inhomogeneous strain-stabilization and reorientation of B19′ variants, the onset of detwinning in some B19′ variants, or other mechanisms. A deviation from elastic B2 loading, possibly related to stress-induced *R*-phase transformation plateaus, was evident as departure from otherwise linear loading within the first 0.6 to 1.3% strain of all the heat-treated samples.

Fatigue life was maximized in the range of 20 to 35% retained cold work as measured at strain levels exceeding about 1%. Bimodal failure conditions were found for all the samples at the 0.9% alternating strain level where this conundrum was observed: for all cold work levels most samples ran to the 10^8 cycle limit (5/10 to 9/10) while all the levels also showed low level fractures (2.1×10^4 to 5.4×10^7 cycles) at a mixture of melt-intrinsic constituent particles and extrinsic surface oxide defects.

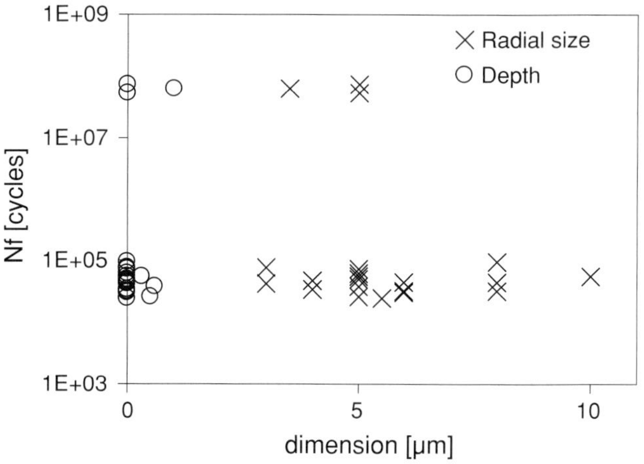

Fig. 10 Plot showing ordinate cycles to failure compared to abscissa radial inclusion size and depth from the surface normal stress-free interface (wire surface)

A bimodal lifetime distribution comprising fatigue lifetimes on the order of 10^4 cycles mixed with lifetimes greater than 10^8 cycles was observed at alternating strain amplitudes of 0.9%. One possible explanation for this phenomenon is related to defect seeding associated with the initial defect state (IDS). The IDS includes the statistical distribution of inclusions, surface defects, grain boundaries, etc., and may seed nitinol for bimodal failure lifetime distributions as follows. In essence, if a defect is seeded in the critical fatigue strain zone that initiates small crack regime growth, the short lifetime may not vary much by initial defect size because of rapid crack advancement of the seeded failure after a relatively constant incubation period. In critically loaded sections where such defect seeding is not present, effective loading may yield fatigue lifetimes that are orders of magnitude greater. Some level of protection may also be induced near larger defects due to proximal crack-tip B19′, which has been shown to possess lower crack growth susceptibility relative to the B2 phase (Ref 19). This protective crack tip phase transformation mechanism would not be present in most other alloys. These mechanisms are actively being investigated further in continuation of this study with unlimited runout cycle criteria.

References

1. K. Gall, J. Tyber, V. Brice, C.P. Frick, H.J. Maier, and N. Morgan, Tensile Deformations of NiTi Wires, *J. Biomed. Mater. Res.*, 2005, **75A**, p 810–823
2. S.W. Robertson, X.Y. Gong, and R.O. Ritchie, Effect of Product Form and Heat Treatment on the Crystallographic Texture of Austenitic Nitinol, *J. Mater. Sci.*, 2006, **41**, p 621–630
3. M.H. Wu, Effects of Phase Transformations on Fatigue Endurance of a Superelastic NiTi Alloy, *J. ASTM Int.*, 2007, **4**(3), p 1–6
4. E. McLucas, Y. Rochev, and W.M. Carroll, Analysis of the Effects of Surface Treatments on Nickel Release from Nitinol Wires and Their Impact on Candidate Gene Expression in Endothelial Cells, *J. Mater. Sci. Mater. Med.*, 2008, **19**, p 975–980
5. M. Assad, L.H. Yahia, C.H. Rivard, and N. Lemieux, In Vitro Biocompatibility Assessment of a Nickel-Titanium Alloy Using Electron Microscopy In Situ End-Labeling (EM-ISEL), *J. Biomed. Mater. Res.*, 1998, **41A**, p 154–161

6. J.C. Wataha, P.E. Lockwood, M. Marek, and M. Ghazi, Ability of Ni-Containing Biomedical Alloys to Activate Monocytes and Endothelial Cells In Vitro, *J. Biomed. Mater. Res.*, 1999, **45A**, p 251–257
7. M.M. Patel, Characterizing Fatigue Response of Nickel-Titanium Alloys by Rotary Beam Testing, *J. ASTM Int.*, 2007, **4**, p 1–11
8. M. Reinoehl, D. Bradley, R. Bouthot, and J. Proft, The Influence of Melt Practice on Final Fatigue Properties of Superelastic NiTi Wires, *SMST-2000: Proceedings of the International Conference on Shape Memory and Superelastic Technologies*, S.M. Russell and A.R. Pelton, Eds., International Organization on SMST, 2001, p 397–403
9. D. Tolomeo, S. Davidson, and M. Santinoranont, Cyclic Properties of Superelastic Nitinol: Design Implications, *SMST-2000: Proceedings of the International Conference on Shape Memory and Superelastic Technologies*, S.M. Russell and A.R. Pelton, Eds., International Organization on SMST, 2001, p 471–476
10. T. Sawaguchi, G. Kaustrater, A. Yawny, M. Wagner, and G. Eggeler, Crack Initiation and Propagation in 50.9 at pct Ni-Ti Pseudoelastic Shape-Memory Wires in Bending-Rotation Fatigue, *Metall. Mater. Trans. A*, 2003, **34A**, p 2847–2860
11. "Standard Test Method for Determination of Transformation Temperature of Nickel-Titanium Shape Memory Alloys by Bend and Free Recovery," F2082-06, ASTM International, p 1–7
12. "Standard Terminology for Nickel-Titanium Shape Memory Alloys," F2005-05, ASTM International, p 1–3
13. L. Tan and W.C. Crone, In Situ TEM Observation of Two-Step Martensitic Transformation in Aged NiTi Shape Memory Alloy, *Scripta Metall. Mater.*, 2004, **50**, p 819–823
14. T.W. Duerig, Some Unsolved Aspects of Nitinol, *Mater. Sci. Eng. A-Struct.*, 2006, **438**, p 69–74
15. P.H. Adler, W. Yu, A.R. Pelton, R. Zadno, T.W. Duerig, and R. Baressi, On the Tensile and Torsional Properties of Pseudoelastic NiTi, *Scripta Metall. Mater.*, 1990, **24**, p 943–947
16. W.M. Huang and H.K. Lim, Evolution of Energy Dissipation and the Young's Modulus in a Martensite NiTi Shape Memory Alloy Wire Upon Cyclic Loading, *J. Mater. Sci. Lett.*, 2003, **22**, p 1399–1400
17. J.E. Schaffer, A Probabilistic Approach to Modeling Microstructural Variability and Fatigue Behavior in ASTM F562 Medical Grade Wire, in *Proceedings of the 9th International Congress on Fatigue, Fatigue 2006*, Atlanta, Georgia, USA, Elsevier Inc., Saint Louis, MO, USA, 2006
18. T.W. Duerig and K.N. Melton, Diffuse Yield Drop and Snap Action in a NiTi Alloy, *Materials Research Society Proceedings, Materials Research Society*, Vol. 245, 1992, p 159–167
19. A.L. McKelvey and R.O. Ritchie, Fatigue Crack Growth Behavior in the Superelastic and Shape Memory Alloy Nitinol, *Metall. Mater. Trans. A*, 2001, **32A**, p 731–743

JMEPEG (2009) 18:569–574
DOI: 10.1007/s11665-009-9351-8

1059-9495/$19.00

Fracture Behaviour of Nickel-Titanium Laser Welded Joints

C. Maletta, A. Falvo, F. Furgiuele, G. Barbieri, and M. Brandizzi

(Submitted October 6, 2008; in revised form December 2, 2008)

In this study, the effects of Nd:YAG laser welding on the fracture behavior of Ni-rich nickel-titanium sheets are analyzed by experimental investigations. The welding was carried out in open air conditions by using a special shielding/clamping system to avoid the chemical contamination of the molten zone and the formation of hot cracks. Mechanical tests of standard dog bone-shaped and single edge crack specimens were carried out to measure the stress-strain response and the fracture resistance of both the base and the welded materials. Furthermore, scanning electron microscopy observations of the fracture surfaces were carried out in order to better understand the failure mechanisms. Finally, systematic comparative studies between base and laser-welded materials were carried out.

Keywords Fracture mechanics, Mechanical Testing, Shape Memory Alloys (SMAs), Welding

1. Introduction

Nickel-Titanium-based (NiTi) Shape Memory Alloys (SMAs) have seen growing applications in recent years in many branches of engineering, due to their special functional properties; namely Shape Memory Effect (SME) and Super-elastic Effect (SE) (Ref 1). These properties are due to a solid-state phase transformation between martensite and austenite, the so-called Thermoelastic Martensitic Transformation (TMT), which can be activated by thermal or mechanical loads. Furthermore, the demand for NiTi alloys is expected to rise considerably in the near future due to the reductions in manufacturing costs and improvements in product quality. However, due to the low formability of these alloys, suitable joining and cutting techniques must be used to obtain devices and components with complex geometries. In fact, the functional properties of NiTi alloys are strongly affected by the modifications induced by the working processes, in terms of thermo-mechanical effects and chemical contaminations (Ref 1), and a deeper understanding of these modifications could be useful for developing new industrial applications. In this field, obtaining welded joints with similar mechanical and functional properties of the base material, could create new possibilities for the realization of complex-shaped components.

This article is an invited paper selected from presentations at Shape Memory and Superelastic Technologies 2008, held September 21-25, 2008, in Stressa, Italy, and has been expanded from the original presentation.

C. Maletta, A. Falvo, and **F. Furgiuele**, Department of Mechanical Engineering, University of Calabria, Arcavacata Rende, CS, Italy; **G. Barbieri**, CR ENEA Casaccia, Rome, Italy; and **M. Brandizzi**, Consorzio CALEF, CR ENEA Trisaia, Matera, Italy. Contact e-mail: carmine.maletta@unical.it.

Laser welding is one of the most important joining techniques for this class of materials, and in particular the Nd:YAG source is suitable for welding low-thickness components due to an accurate control of the welding power, resulting in a reduced heat-affected zone (HAZ) (Ref 2-4). However, two major drawbacks arise during welding of NiTi alloys: (i) the chemical contamination of the joint, mainly due to the high oxidation rate of titanium, and (ii) the formation of hot cracks in the molten zone, due to the thermal shrinkage of the material during the cooling stage after welding. Therefore, an efficient protection of the molten zone, by inert gas, and an accurate positioning of the components to be welded, are of major concern in order to obtain welded joints with satisfactory mechanical and functional properties. For these reasons, some research activities have been carried out in the last few years to evaluate the effects of laser welding on the mechanical and functional properties of NiTi alloys (Ref 2-7).

Unfortunately, few experimental research programs have been devoted to the fracture mechanics of NiTi alloys (Ref 8-11), i.e., to study the mechanical behavior of cracked components. This topic is particularly interesting because the stress singularity around the crack tip activates the Stress-Induced Martensitic (SIM) transformation (Ref 8-13), which significantly changes the stress distribution in the crack tip region with respect to common engineering metals and, consequently, plays a critical role in the fracture toughness of the material. Furthermore, no research, to the knowledge of the authors, has been carried out to study the fracture mechanics of NiTi welded joints.

In this article, the effects of Nd:YAG laser welding on the fracture behavior of Ni-49.2 at.% Ti sheets were analyzed by experimental investigations. The welding was carried out in open air conditions, i.e., without using closed shielding chambers, by special dynamic shielding equipment and a positioning/clamping system to overcome the aforementioned drawbacks.

The stress-strain responses of both the welded and the base (reference) materials, were measured by using dog bone-shaped specimens, while the fracture resistance was analyzed by Single Edge Crack (SEC) specimens. The test specimens were cut from reference and previously welded sheets, using a water jet technique, and the pre-cracks in the SEC specimens were made by electro-discharge machining (EDM). Furthermore, Scanning Electron Microscopy (SEM) observations of the fracture surface were carried out in order to better understand the failure mechanisms of both the reference and the welded specimens. Finally, systematic comparisons of the results between the base and the laser-welded materials were carried out.

2. Materials and Experiments

Ni-49.2 at.% Ti sheets with oxide-free surfaces and a thickness of 0.75 mm (Type S, Memory Metalle, Germany) were used in this investigation. The material has a nominal austenite finish temperature $A_f = -7$ °C, and therefore, it exhibits a full austenitic structure at room temperature.

2.1 Laser-Welding Process

The welding process was carried out by a Nd:YAG laser source (Trumpf HL2006D) with a maximum power of 2 kW. Special systems for open-air laser welding, illustrated in Fig. 1, were used in order to overcome the main drawbacks which arise during welding NiTi alloys, such as the chemical contamination of the joint and the formation of hot cracks. In particular, a dynamic shielding system allows an efficient protection of the top side of the molten zone by an inert buffer state, which is generated by blowing argon gas through a perforated plate during welding. Furthermore, a special system was used, which allows an accurate positioning of the components to be welded and applies a compressive force to the joint in order to avoid the formation of hot cracks; in addition, an inert gas delivery system protects the back side of the molten zone from the oxidation.

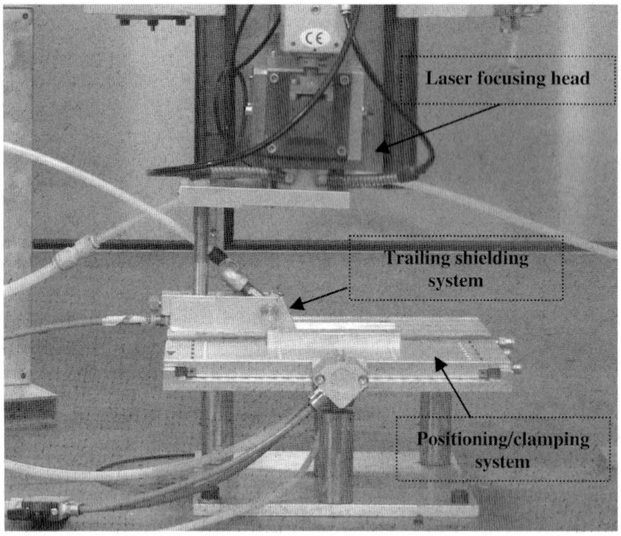

Fig. 1 Open-air laser welding equipments: (a) dynamic top shielding system; (b) positioning/clamping system

The welding was executed in continuous mode and a set of preliminary tests were carried out in order to identify the optimal values of the process parameters, as illustrated in the following sections.

2.2 Specimen Preparation and Mechanical Testing

Two types of specimens were used for mechanical tests, a dog bone-shaped for stress-strain measurements (see Fig. 2a) and a SEC for fracture tests (see Fig. 2b).

Due to the poor workability of NiTi alloys with conventional cutting techniques, the specimens for mechanical tests were made by a water jet cutting technique, which provides a high accuracy with no HAZ.

In Fig. 3, the two types of test specimens are shown, together with a comparison between reference and laser-welded ones. The welded specimens were cut from previously butt-welded sheets, with the weld bead perpendicular to the loading direction and in the middle section of both types of specimens. In particular, Fig. 3(a) illustrates the dog bone-shaped specimens, while Fig. 3(b) illustrates the SEC ones with an highlight of the pre-crack; this latter was made by EDM by using a copper wire with diameter $\phi = 150$ μm, which produces a notch radius of about 80 μm. As clearly illustrated in Fig. 3(b), the pre-cracks were made in the middle of the molten zone of the welded specimens.

The mechanical tests were carried out by using a universal testing machine (Instron 8500), equipped with a furnace (MTS 653) and a resistance extensometer with a gauge length of 10 mm. Both specimens were tested in isothermal conditions, by applying a displacement-controlled monotonic load, at a cross head rate of 0.2 mm/min. Furthermore, a special loading/clamping system was used to avoid spurious loading conditions and premature failure of the specimen in the pin-hole regions.

3. Results

3.1 Microscopic Observations and Vickers Hardness

A set of preliminary tests were carried out in order to identify the optimal values of the process parameters, such as

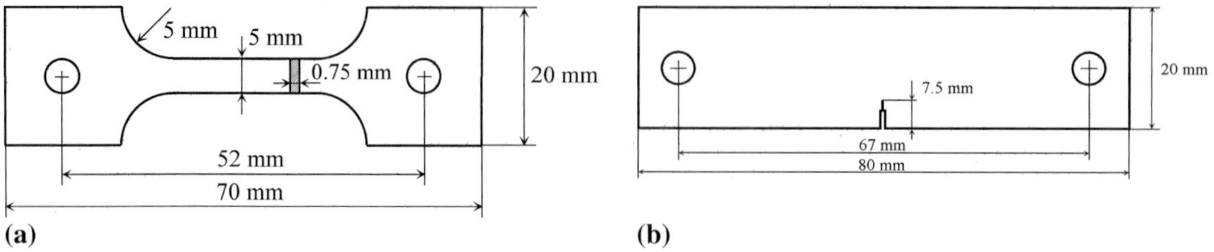

Fig. 2 Specimens for mechanical tests: (a) dog bone-shaped specimen for stress-strain tests; (b) single edge crack specimen for fracture tests

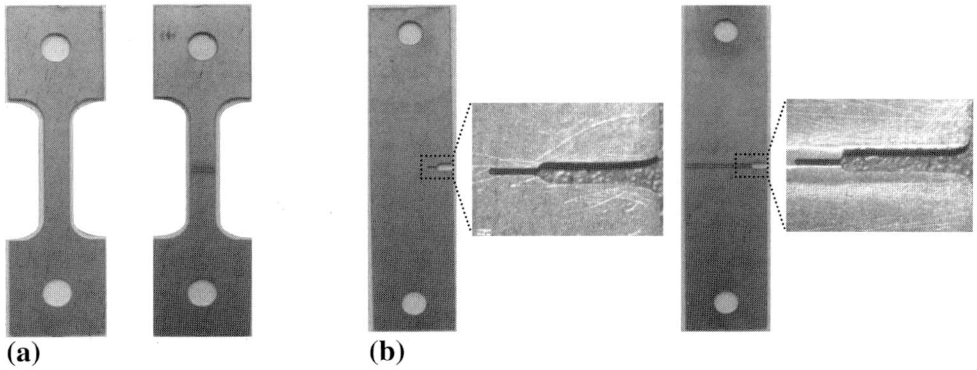

Fig. 3 Reference and welded specimens for mechanical tests: (a) dog bone-shaped specimen; (b) single edge crack with an highlight of the pre-crack

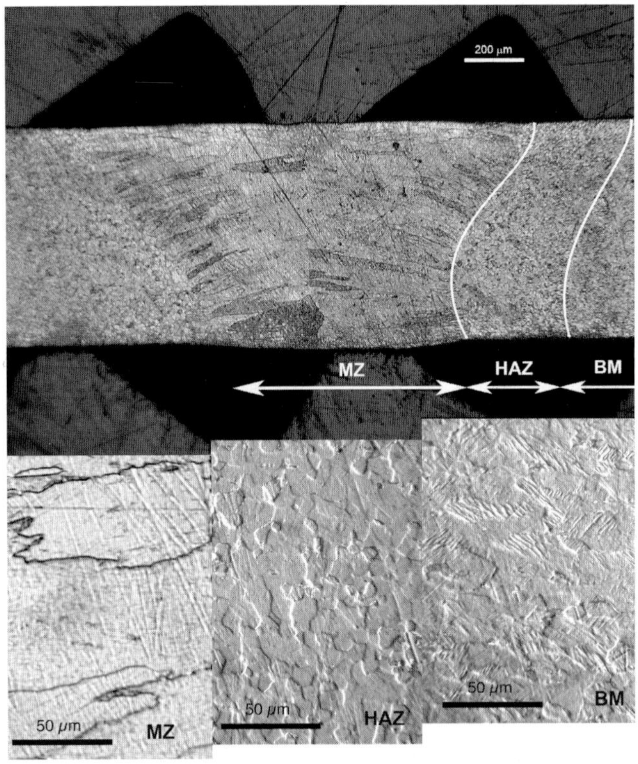

Molten zone	0.8~1.5 mm
Heat affected zone	~0.4 mm

Fig. 4 Light micrographs of the cross section of the weld bead

average power and welding rate. To this aim, microscopic observations and hardness tests were executed to evaluate the extension of the HAZ and molten zone (MZ), obtained by several values of the aforementioned welding parameters. Finally, the following values were chosen: average power of 850 W and welding rate of 2400 mm/min.

Figure 4 illustrates a light micrograph of the cross section of the weld bead together with micrographic observations of the MZ, HAZ, and base material (BM), carried out at higher magnifications. The figure shows a V-shaped MZ with root width of about 0.8 mm and maximum width of about 1.5 mm, and a HAZ of about 0.4 mm.

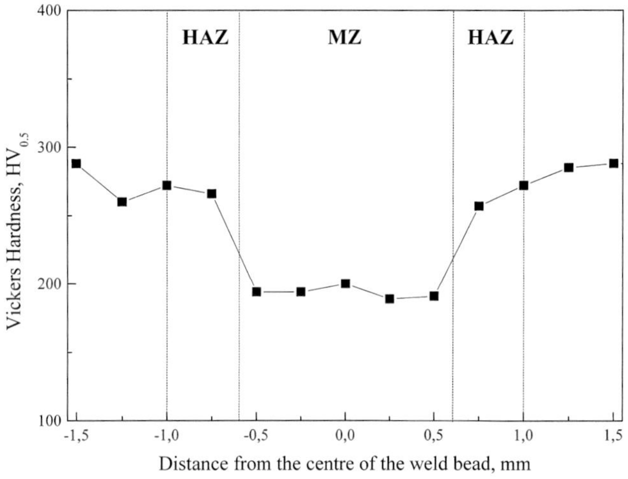

Fig. 5 Hardness profile of a cross section of the weld bead

Figure 5 illustrates the results of the hardness tests, carried out on a cross section of the weld bead by using a Vickers indenter and a weight of 0.5 kg ($HV_{0.5}$), which show an overall reduction of hardness in the molten zone with respect to the base material from about 280 to 190 $HV_{0.5}$.

3.2 Stress-Strain Measurements

Figure 6 illustrates a comparison between the stress-strain curves of the reference and laser welded specimens, obtained from isothermal uniaxial loading tests of dog bone shaped specimens, carried out at room temperature. Three different specimens for each type were analyzed and a high repeatability of the results was found as illustrated in Table 1; this latter reports the average values of the maximum stresses as well as the transformation stresses and strains, together with the standard deviations. Figure 6 and Table 1 clearly illustrate a strong reduction of the maximum stress in the welded specimen with respect to the reference, from about 1250 to 780 MPa, mainly due the modifications induced by the welding. Notwithstanding the reduction in the mechanical strength, the stress-strain transformation behavior of the material seems to be mainly unaffected by the welding, as a similar stress plateau is observed in both materials, i.e., similar values of the transformation stress and strain were measured. In fact, student-*t* test, at a 99% confidence level, indicates no significant differences

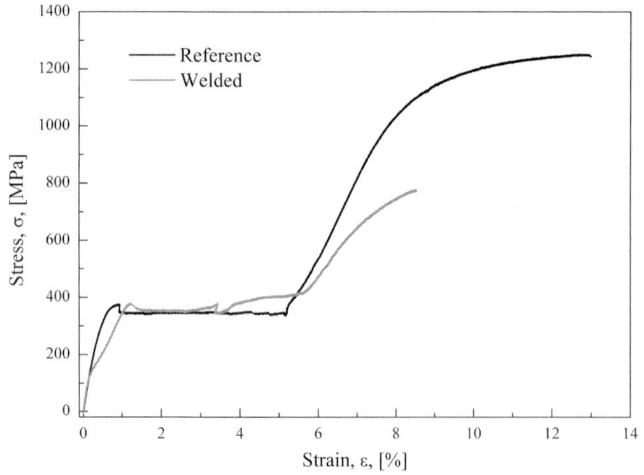

Fig. 6 Comparison between the stress-strain curves of the reference and welded specimens

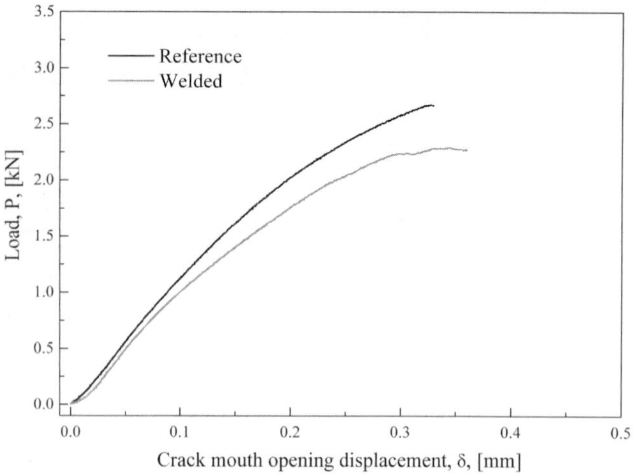

Fig. 7 Comparison between the load versus crack mouth opening displacement curves of the reference and welded specimens

Table 1 Main mechanical and functional parameters of the reference and welded specimens

	Transformation stress, MPa	Transformation strain, %	Maximum stress, MPa
Reference	345 ± 3	4.6 ± 0.1	1248 ± 7
Welded	355 ± 5	4.6 ± 0.3	776 ± 35

between transformation stress and strain in welded and reference specimens. However, it is worth noting that the curves in Fig. 6 illustrate the effective behavior of the joint over a gauge length of 10 mm, while the stress-strain response of the material in the MZ and in the HAZ is expected to be different from the measured one. This observation seems to be confirmed by more detailed comparisons between the stress-strain curves of the reference and laser-welded specimens, which show the following differences: (i) laser welded specimens exhibit a change of the slope in the austenitic region at a stress level lower than the nominal transformation stress and (ii) stress variations are observed in the stress-strain transformation curves of the welded specimens. Notwithstanding that full-field displacement measurements should be carried out in order to better understand the functional behavior of the material in the welded region, the observed differences should be attributed to a different transformation behavior of the material in the MZ and HAZ. In particular, the change of the slope in the early stage of loading and the stress variations in the transformation plateau should be attributed to non-homogeneous transformations, i.e., the stress-induced martensitic transformation in the welded region occurs in a different stress range with respect to the reference material.

3.3 Fracture Tests

Figure 7 illustrates a comparison between the load versus crack mouth opening displacement curves of the reference and laser-welded materials, obtained from isothermal tests of SEC specimens, carried out at room temperature. Three different specimens for each type were tested, and the results are illustrated in Table 2 in terms of maximum load and notch strength. The latter data were computed as the maximum nominal stress at failure according to the standard ASTM E338-91 (Ref 14). The table clearly illustrates that the welded specimens exhibit a reduction in the notch strength of about 15% with respect to the reference specimens. Furthermore, student-*t* test, at a 99% confidence level, indicates a significantly different strength between welded and reference specimens.

Table 2 Mechanical strength of the SEC specimens

	Maximum load, kN	Notch strength, MPa
Reference	2.65 ± 0.03	284 ± 3
Welded	2.26 ± 0.05	243 ± 5

The fracture path is almost straight along the symmetry plane of both specimens and it is always located in the HAZ of the welded specimens, although the non-symmetric shape of the weld bead causes non-pure mode I loading conditions.

Furthermore, the notch strength is calculated for a comparative analysis between the fracture resistance of the reference and the laser-welded materials, and no further considerations are made about the fracture resistance; in fact, the well-known equations to calculate the stress intensity factor cannot be applied to such class of materials, due to the stress-induced martensitic transformation occurring in the crack tip region, as a consequence of the high values of local stresses, which generate a more complex stress distribution with respect to common engineering metals.

3.4 SEM Observations

In order to better understand the fracture mechanisms of both reference and laser welded materials, the fracture surfaces of the SEC specimens were analyzed by SEM investigations. Figure 8 illustrates the fracture surface of the reference (Fig. 8a, c) and welded specimens (Fig. 8b, d). Figure 8(a) and (b) show the fracture surfaces, together with the pre-cracks (left side) which were made by EDM; the figure indicates a mostly ductile fracture surface in the reference material, while a more brittle fracture mechanism is observed in the welded specimen. This observation is qualitatively confirmed by SEM observation at higher magnification, as illustrated in the

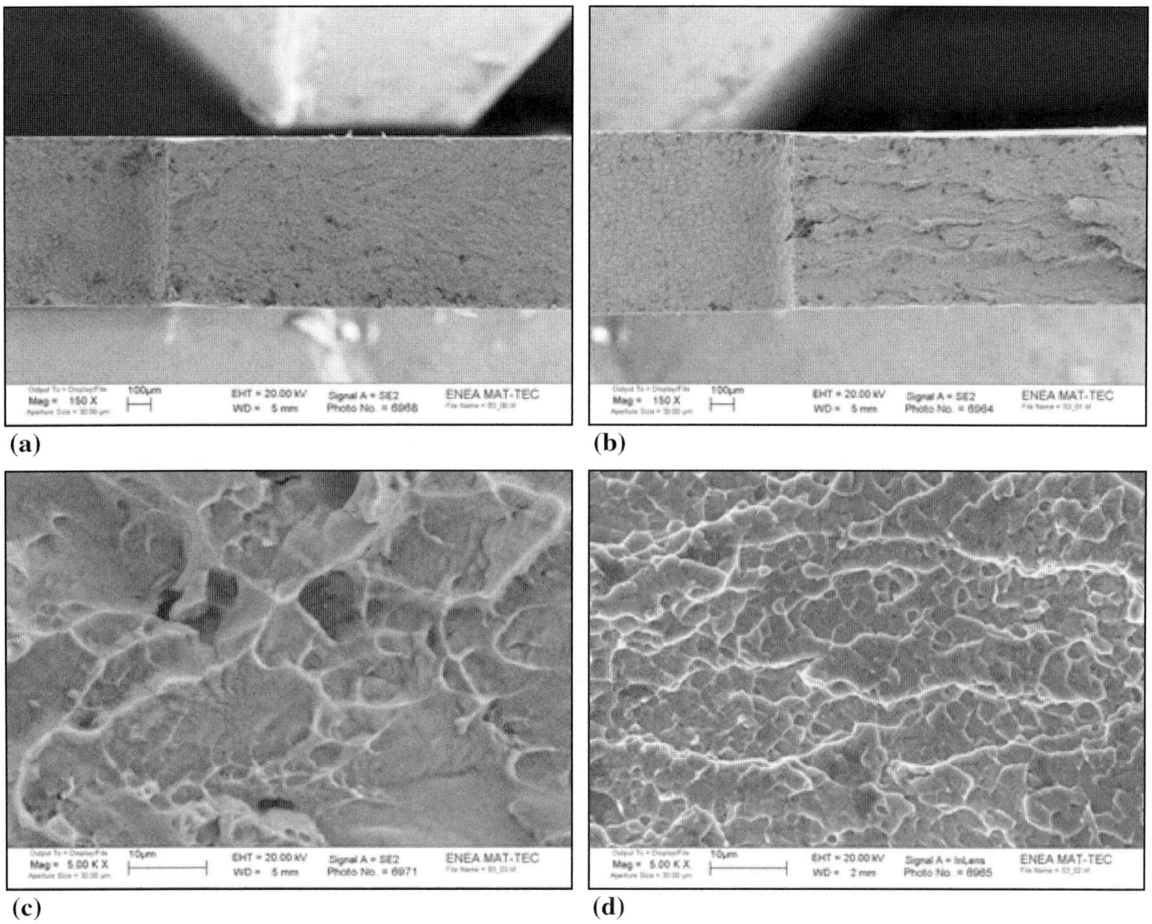

Fig. 8 SEM observation of the fracture surface: (a) reference and (b) welded (magnification: 150×); (c) reference and (d) welded (magnification: 5000×)

Fig. 8(c) and (d), which show dimple size of about 10-30 μm in the reference specimens and about 5 μm in laser-welded ones. Notwithstanding that the dimple sizes could be also affected by different grain size in the welded material, these results indicate a greater plasticization capability in the reference specimens than in the welded ones.

4. Conclusions

Ni-rich nickel-titanium sheets were welded in open-air conditions by a Nd:YAG laser source, and special shielding/positioning systems were used to avoid the chemical contamination of the joints and the formation of hot cracks. The effects of laser welding on the fracture behavior were analyzed by mechanical tests and systematic comparative studies between the base and the welded materials.

The mechanical tests of dog bone-shaped specimens indicate a reduction in the mechanical strength, while the overall functional behavior of the specimens seems to be mainly unaffected by the welding, as a similar stress-strain transformation plateau is observed in both materials.

The tests of single edge crack specimens show a reduction of about 15% in the notch strength of the welded specimens with respect to the reference ones. The fracture path is almost straight along the symmetry plane of both specimens, as a consequence of a pure mode I loading condition, and it is always located in the HAZ of the welded specimens. Furthermore, SEM observations show mostly ductile fracture surfaces in the base material, while a more brittle behavior is observed in the welded one.

Future studies should be carried out in order to better understand the functional behavior of the material in the welded region, by using full-field displacement measurements. Furthermore, for a complete fracture characterization of the material, the stress-induced martensitic transformation which occurs in the crack tip region should be better investigated, by numerical simulations and experimental measurements.

Acknowledgments

This research was supported by the CALEF consortium (Rotondella, Italy), within the framework of the LACER project. Furthermore, the authors wish to thank L. Pilloni (ENEA, Casaccia) and M. B. Alba (ENEA, Trisaia) for their valuable supports in microscopic observations.

References

1. K. Otsuka and X. Ren, Physical Metallurgy of Ti-Ni-based Shape Memory Alloys, *Prog. Mater. Sci.*, 2005, **50**(5), p 511–678

2. A. Tuissi, S. Besseghini, T. Ranucci, F. Squatrito, and M. Pozzi, Effect of Nd-YAG Laser Welding on the Functional Properties of the Ni-49.6at.%Ti, *Mater. Sci. Eng. A*, 1999, **273–275**, p 813–817
3. A. Falvo, F. Furgiuele, and C. Maletta, Laser Welding of a NiTi Alloy: Mechanical and Shape Memory Behaviour, *Mater. Sci. Eng. A*, 2005, **412**(1–2), p 235–240
4. A. Falvo, F. Furgiuele, and C. Maletta, Functional Behaviour of a NiTi Welded Joint: Two Way Shape Memory Effect, *Mater. Sci. Eng. A*, 2008, **481–482**(1–2), p 647–650
5. P. Schlossmacher, T. Haas, and A. Shussler, Laser-Welding of a Ni-Rich TiNi Shape Memory Alloy: Mechanical Behavior, *J. Phys. IV Coll*, 1997, **C 5**, p 251–256
6. P. Schlossmacher, T. Haas, and A. Shussler, Laser Welding of Ni-Ti Shape Memory Alloys, *SMST-94: Proceedings of the 1st International Conference on Shape Memory and Superelastic Technologies*, March 7–10, 1994 (Pacific Grove, CA), p 85–90
7. T. Haas and A. Schuessler, Welding and Joining of TiNi Shape Memory Alloys: Engineering Aspects and Medical Applications, *SMST-99: Proceedings of the 1st European Conference on Shape Memory and Superelastic Technologies*, Sept. 5–6, 1999 (Paris), p 103–114
8. J.H. Chen, W. Sun, and G.Z. Wang, Investigation on the Fracture Behavior of Shape Memory Alloy NiTi, *Metall. Mater. Trans. A*, 2005, **36**, p 941–955
9. S. Gollerthan, D. Herberg, A. Baruj, and G. Eggeler, Compact Tension Testing of Martensitic/Pseudoplastic NiTi Shape Memory Alloys, *Mater. Sci. Eng. A*, 2008, **481–482**, p 156–159
10. S.W. Robertson, A. Mehta, A.R. Pelton, and R.O. Ritchie, Evolution of Crack-tip Transformation Zones in Superelastic Nitinol Subjected to In Situ Fatigue: A Fracture Mechanics and Synchrotron X-ray Microdiffraction Analysis, *Acta Mater.*, 2007, **55**, p 6198–6207
11. M.R. Daymond, M.L. Young, J.D. Almer, and D.C. Dunand, Strain and Texture Evolution During Mechanical Loading of a Crack Tip in Martensitic Shape-memory NiTi, *Acta Mater.*, 2007, **55**, p 3929–3942
12. X.M. Wang, Y.F. Wang, A. Baruj, G. Eggeler, and Z.F. Yue, On the Formation of Martensite in Front of Cracks in Pseudoelastic Shape Memory Alloys, *Mater. Sci. Eng. A*, 2005, **394**, p 393–398
13. G.Z. Wang, A Finite Element Analysis of Evolution of Stress-Strain and Martensite Transformation in Front of a Notch in Shape Memory Alloy NiTi, *Mater. Sci. Eng. A*, 2007, **460–461**, p 383–391
14. "Standard Test Method of Sharp-Notch Tension Testing of High-Strength Sheet Materials," E338-91, *Annual Book of ASTM Standards 03.01*, 1996, p 360–365

JMEPEG (2009) 18:575–581
DOI: 10.1007/s11665-009-9474-y

1059-9495/$19.00

Influence of Carbon on the Microstructure of a Fe-Mn-Si-Cr-Ni Alloy

Khaled M. Mostafa, J. De Baerdemaeker, N. Van Caenegem, D. Segers, and Y. Houbaert

(Submitted August 21, 2008; in revised form April 28, 2009)

The influence of the addition of C to the Fe-Mn-Si-Cr-Ni base material is investigated at room temperature. Steel samples were deformed during a tensile experiment up to a strain of 17%. Light optical microscopy (OM) and x-ray diffraction (XRD) gave information about the different micro-structural phases that exist in the deformed and the undeformed alloys. The evolution of the defect structure is followed by positron annihilation techniques such as Doppler broadening of annihilation radiation spectroscopy (DBAR) and the positron annihilation lifetime spectroscopy (PALS). During deformation a martensitic ε-phase is induced. The size of the martensite plates increases with increasing deformation.

Keywords nondestructive testing, optical microscopy, shape memory alloys

1. Introduction

Fe-Mn alloys with 10-30% Mn are characterized by the reversible and self-accommodating face centred cubic (fcc) γ austenite to hexagonal close packed (hcp) ε martensite transformation (Ref 1). Ferrous shape memory alloys (SMA) based on Fe-Mn alloy system have received much attention since the eighties of the last century and there are definite signals from the shape memory materials research community that they may become a new class of one-way SMAs of great technical importance due to their cost-effectiveness (Ref 2). The shape memory effect (SME) in FeMnSi alloys was first reported by Sato et al. (Ref 3). They have found that a single crystal of a Fe-30%Mn-1%Si alloy exhibited a complete SME when deformed in the ⟨414⟩ direction. Sato et al. (Ref 4) and Murakami et al. (Ref 5) also succeeded in developing poly-crystalline FeMnSi SMAs. Moriya et al. (Ref 6) and Otsuka et al. (Ref 2) developed FeCrNiMnSi and FeCrNiMnSiCo SMAs, which are characterized by their corrosion resistance.

The large hysteresis and a poor shape recovery in the Fe-Mn binary system was addressed by making specific alloying additions of Si, C, Co, Ni, and Cr. The SME in FeMnSi alloys strongly depends on the alloy composition. The addition of C strengthens the austenite and increases the stacking fault energy (Ref 7, 8), while it decreases the Ms temperature (Ref 9, 10). Tian et al. (Ref 11) determined the effect of carbon on the Ms of FeMnAlCr alloy and suggested that C strongly depresses the $\gamma \rightarrow \varepsilon$ martensitic transformation. Tsuzaki et al. (Ref 9) examined the effect of carbon in the FeMnSi alloy. They reported that the SME is improved by the addition of carbon.

In order to reach a complete SME, three conditions must be fulfilled. First, the deformation must result in the stress-induced ε martensite formation only. Second, the yield stress of the γ matrix should be as high as possible to avoid an initial deformation by slip. Finally, the shape strain of stress-induced martensite should be completely reversible, which means that the martensite interface must remain mobile at all times. The forward and reverse transformations occur by the movement of planar martensite interfaces, which contain the transformation dislocations. The required martensite morphology for ferrous SMA is of the thin plate type.

Because the SMAs posses low stacking faults energy, the nucleation of the martensite may occur by means of the stacking fault mechanism (Ref 12). The transformation results in thin ε martensite plates, which have specific crystallographic orientations with respect to each other. This is due to the fact that the martensitic $\gamma \rightarrow \varepsilon$ transformation is obtained by the glide of isolated Shockley partial dislocations on every second $\{111\}_\gamma$ plane. There is no macroscopic deformation of the single crystal when the $\gamma \rightarrow \varepsilon$ transformation is induced thermally. The deformation results in the dominance of ε martensite variants, which are suitably oriented with respect to the applied stresses, and a decrease of the volume fraction of the other ε variants in each original γ grain. It must be mentioned that FeMnSi-type ferrous SMAs always contain retained austenite at room temperature. The ε martensite forms dense regions of transformed regions, which still contain the parent γ phase. At the intersection of two variants of ε martensite, α′ martensite was often observed (Ref 13). It is very likely that the formation of α′ martensite is the main

This article is an invited paper selected from presentations at Shape Memory and Superelastic Technologies 2008, held September 21-25, 2008, in Stresa, Italy, and has been expanded from the original presentation.

Khaled M. Mostafa, Department of Subatomic and Radiation Physics, Ghent University, Proeftuinstraat 86, B-9000 Ghent, Belgium and Department of Metallurgy and Materials Science, Ghent University, Technologiepark 903, B-9052 Ghent, Belgium, and **J. De Baerdemaeker** and **D. Segers**, Department of Subatomic and Radiation Physics, Ghent University, Proeftuinstraat 86, B-9000 Ghent, Belgium; and **N. Van Caenegem** and **Y. Houbaert**, Department of Metallurgy and Materials Science, Ghent University, Technologiepark 903, B-9052 Ghent, Belgium. Contact e-mail: khaled.mostafa@ugent.be.

cause for relatively low recovery of the ferrous SMAs. The α' regions impede the reverse motion of the Shockley partial dislocations. One of the challenging problems of the Fe-Mn-based alloys is the relatively low, fully reversible pre-strain, usually less than 2% in contrast to over 6% in the Ni-Ti and Cu-based alloys.

During the last few years, the positron annihilation technique has been widely used to study defects and phase transition in SMAs. Most of the work had been done on the Ni-Ti and Cu-based alloys (Ref 14-16). Very few publications have been published about the Fe-Mn-based alloys (Ref 17, 18).

In the present study, the influence of deformation and carbon addition on Fe-Mn-Si-Cr-Ni alloys is studied by a combination of optical microscopy (OM), x-ray diffraction (XRD), and positron annihilation spectroscopy (PAS) techniques. OM and XRD give accurate information on the different phases present in the material. PAS techniques were used to study the defect behavior in the samples and to see if any effect of the phase transformation during deformation could be detected.

The positron annihilation technique is known to be a highly sensitive method for open volume defects (Ref 19). The defect structure after plastic deformation of metallic samples can also be investigated with the positron annihilation technique (Ref 20). It is expected that PAS can help in understanding more about the stress-induced martensitic transformation in FeMn-based alloys. Two basic techniques are used: positron annihilation lifetime spectroscopy (PALS) and Doppler broadening of annihilation radiation spectroscopy (DBAR).

Positrons from a radioactive source are injected into the sample. After thermalization, the positron diffuses through the solid and gets trapped by defects present in the sample. Defects capable of localizing the positron are vacancies, vacancy agglomerates, voids and dislocations. PAS has proven to be a uniquely sensitive probe for nanometre defects. The high sensitivity of positrons to defects stems from two facts: their attraction to atomic-size defects and their long diffusion length, which is about a few hundred nanometres in most materials (Ref 21). PALS can quantify the size of open volume defects as well as the defect concentration and is based on the precise measurement of the lifetime of a positron in a solid. The concentration can be deduced from the fraction of positrons that annihilate from a trapped state. The defect size is directly related to the position lifetime; the larger the defect, the lower the local electron density is and consequently the longer the positron lifetime will be (Ref 22).

DBAR is based on the measurement of the linear momentum of the annihilating electron-positron pair. The photons created during the electron-positron annihilation are detected by a germanium detector. The shape of the resulting photo-peak reflects the momentum distribution of the original electron-positron pair. This distribution is determined by the size and the nature of the defects. Measurements of the Doppler broadening of positron-electron annihilation radiation are generally characterized by the S (shape) parameter, defined as the ratio between the central part of the annihilation spectra and the total spectra. This parameter reflects the positron annihilation with valence electrons (low momentum). In general, a high value of S indicates positron annihilation in open volume defects. A second useful parameter for the analysis of DBAR is the W (wing) parameter, which reflects the positron annihilation with high momentum electrons (core electrons). It is defined as the ratio of counts in two side windows and the total number of counts.

The plastic deformation of metals and alloys produces changes in the density and distribution of defects. In plastic deformed metals and alloys, the positrons are captured by dislocations and vacancies. In polycrystalline samples, the deformation becomes complex due to the various interactions between dislocations and the grain boundaries (Ref 23). The interaction of dislocations during the deformation can lead to a raise of the formation of jogs and point defects. Some measurements of positron annihilation have been made for martensitic transformations in metals and alloys (Ref 24-27). Dislocations produced during the transformation must be considered when martensitic transformation is studied by positron annihilation. Segers et al. studied together different types of SMAs (Ref 17, 28, 29).

2. Experimental

Two different Fe-Mn-Si-Cr-Ni alloys were prepared for this study. Their chemical composition is given in Table 1. They were cast in an air furnace, air cooled, reheated to 1200 °C and hot rolled on a laboratory mill from 20 to 2 mm and air cooled. In order to austenitize the samples; they were heated to 1100 °C for 15 min and water quenched to room temperature. The specimens examined by optical microscopy were first mechanically polished and then electrolytically polished in a solution of 20% perchloric acid and 80% butylcellosolve. Afterwards, the specimens were color etched in an aqueous solution of 1.2% $K_2S_2O_5$ and 0.5% NH_4HF_2. Each phase could clearly be distinguished on the basis of its color and morphology.

Positron lifetime measurements were performed at room temperature using a fast-fast lifetime spectrometer. Each spectrum contained at least 10^6 counts and was analyzed with the program LT by Kansy (Ref 30). As a positron source $^{22}NaCl$ of about 10 μCi was sealed between two kapton foils with a thickness of 7.5 μm, which had to be surrounded by two identical samples in the so-called sandwich configuration (Ref 31). The resolution of the setup is 200 ps, while the source contribution was 14% with a single lifetime of 384 ps.

The tensile tests were carried out on LRX–PLUS (LLOYD Instruments) tensile test machine with maximum force of 5 kN. The shape and dimensions of the specimens are shown in Fig. 1. Tensile test was performed to make elongation of the sample. The elongation rate was 0.1 mm/min. During the deformation of the sample, the Doppler broadening (DB) of the 511 keV annihilation line is measured using high-purity germanium detector with a resolution of 1.2 keV at the 514 keV line of ^{85}Sr. The results are analyzed in terms of the so-called *S*- and *W*-parameter. The *S* parameter is obtained for each sample from the measured 511-keV peak by dividing the counts in the central region by the total counts in the peak after subtracting background. The energy window of the central

Table 1 Chemical composition of Fe-Mn-Si-Cr-Ni-C in wt.%

Element	Fe	Mn	Si	Cr	Ni	C
FeMnSiCrNi	67.34	12.6	6	9.27	4.74	0.05
FeMnSiCrNiC	66.84	12.56	6.09	9.44	4.89	0.18

region was chosen so that S parameter equals to 0.5 in the well-annealed Fe sample.

3. Results and Discussion

3.1 OM and XRD

Figure 2 shows the microstructure of the Fe-Mn-Si-Cr-Ni-C alloys before and after deformation. The micro-structural analysis for the undeformed samples revealed the presence of large austenite grains (γ phase), residual ferrite grains and ε phase. The amount of residual ferrite grains between the austenite is less in the alloy without carbon.

This is also measured with the XRD (Fig. 3). The ε martensite is clearly visible in the alloy without carbon and is less in the carbon alloyed one. It appears as thin parallel lines with dark contrast inside the austenite grains. This small quantity of ε martensite before the deformation process is probably induced during grinding or polishing or can be induced thermally. By deformation, more ε martensite is

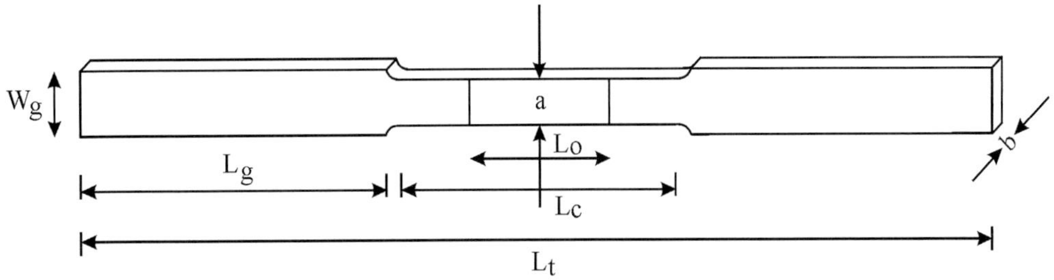

Specimen type	a	Lt	Wg	Lg	Lc	Lo	b
FeMn	7	150	10	50	45	23	1

Dimensions in mm

Fig. 1 Shape and dimensions of the tensile test sample

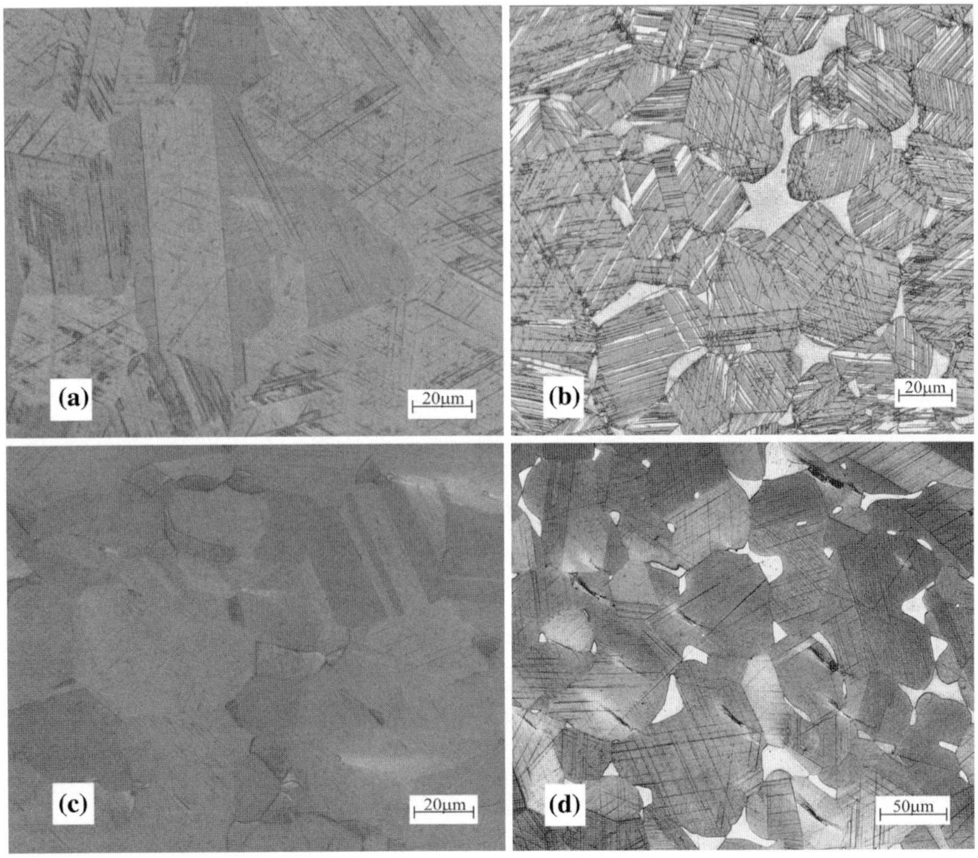

Fig. 2 Light optical microstructures (OM) for the alloys: (a) Fe-Mn-Si-Cr-Ni undeformed, (b) deformed, (c) Fe-Mn-Si-Cr-Ni-C undeformed, and (d) deformed

transformed from the austenite. There is an increase in the intensity of the $(002)_\varepsilon$ peak. In the OM, the ε phase appears as broader white lines in the alloy without carbon and clearly thinner black lines in the alloy with carbon. From the XRD result we can also conclude that the alloy without carbon contains more ε martensite phase.

Comparing the deformed and the non-deformed samples through the OM pictures, one can see that the stress induced $\gamma \rightarrow \varepsilon$ transformation proceeds by the growth and broadening of the existing ε martensite plates rather than the formation of new plates (Ref 32, 33).

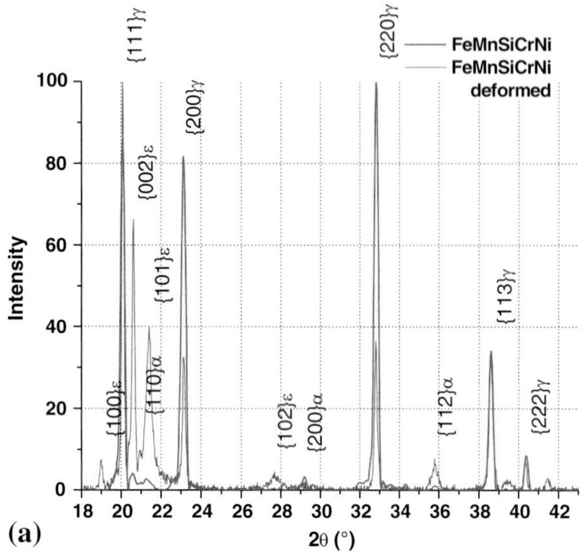

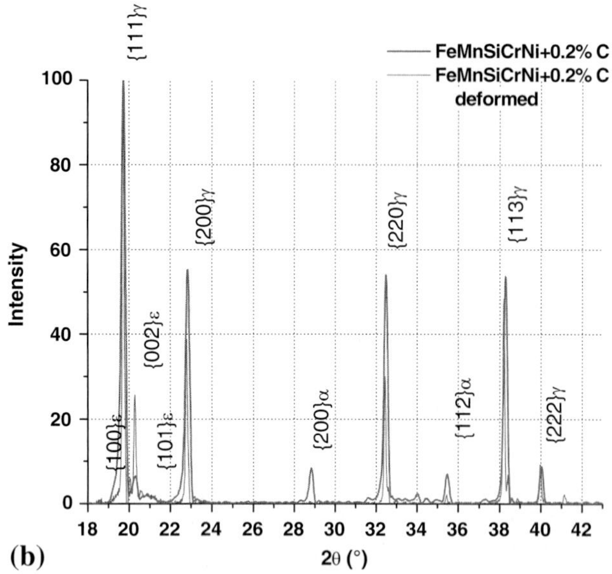

Fig. 3 XRD for the alloys: (a) undeformed and deformed Fe-Mn-Si-Cr-Ni; (b) undeformed and deformed Fe-Mn-Si-Cr-Ni-C

3.2 Positron Annihilation

Positron annihilation lifetime data are presented in Table 2. For the undeformed samples of Fe-Mn-Si-Cr-Ni-C, the lifetime for positrons was 107 ps which is the same in case of defect-free Fe (Ref 34). In the widely used two-state trapping model (Ref 35), it is assumed that positrons annihilate either in the bulk portion or get trapped in defects before annihilation. In the latter case, they survive for more time in the material, which is given by the longer lifetime. This variation of the lifetime values is relevant to defect type (dislocation, mono-vacancy, cluster of vacancy, micro-void). In pure Fe, the lifetime value for positrons trapped in dislocations is around 160 ps and for positrons trapped in mono-vacancies, it is around 180 ps (Ref 36-39).

For the deformed samples, the positron annihilation lifetime spectra have two components for the lifetime. The shortest lifetime components τ_1 for the two samples are 61 and 67 ps, which are due to the annihilation of non-trapped positrons. The second lifetime component for the two samples is 151 and 150 ps, which are related to the positron annihilation in dislocations. The concentrations of dislocations in both deformed samples are calculated using the same equations and parameters used by Park et al. (Ref 40). One can conclude that the alloy without carbon has more dislocations than the one of carbon.

3.3 Doppler Broadening (DB) and Stress-Strain Relation

There are different stages of deformation of the SMAs, which have different deformation mechanisms. The first stage happens in the range of smaller strains, the self-accommodation mechanism of martensite plates, not the dislocation slip mechanism, is responsible for the deformation of the SMAs. The orientations of martensite plates are adjusted or reoriented in accordance with the direction of the applied stress, which results in macroscopic deformation, but does not result in an increase of micro defects. Even if defects as twin boundaries and stacking faults may be created during deformation, they are thought to be unable to attract positrons. Once the strain increases beyond several percent, another deformation mechanism will take place. Dislocation slip produces many defects, capable to trap the positrons. The S parameter can be written as

$$S = \sum_i P_i S_i + \sum_i P_{d_i} S_{d_i}$$

where P_i is the probability of the annihilation of positrons in phase i and S_i is the respective line shape parameter. S_{d_i} counts for the part of positrons annihilated due to the defects

Table 2 Positron annihilation lifetime in Fe-Mn-Si-Cr-Ni-C samples

Sample	τ_1, ps	I_1, %	τ_2, ps	I_2, %	Concentration of dislocations, 10^{14} m^{-2}
FeMnSiCrNi	107	100	...	...	
FeMnSiCrNi 15% deformed	61	21	151	79	1.287
FeMnSiCrNiC	107	100	...	...	
FeMnSiCrNiC 15% deformed	67	24	150	76	1.046

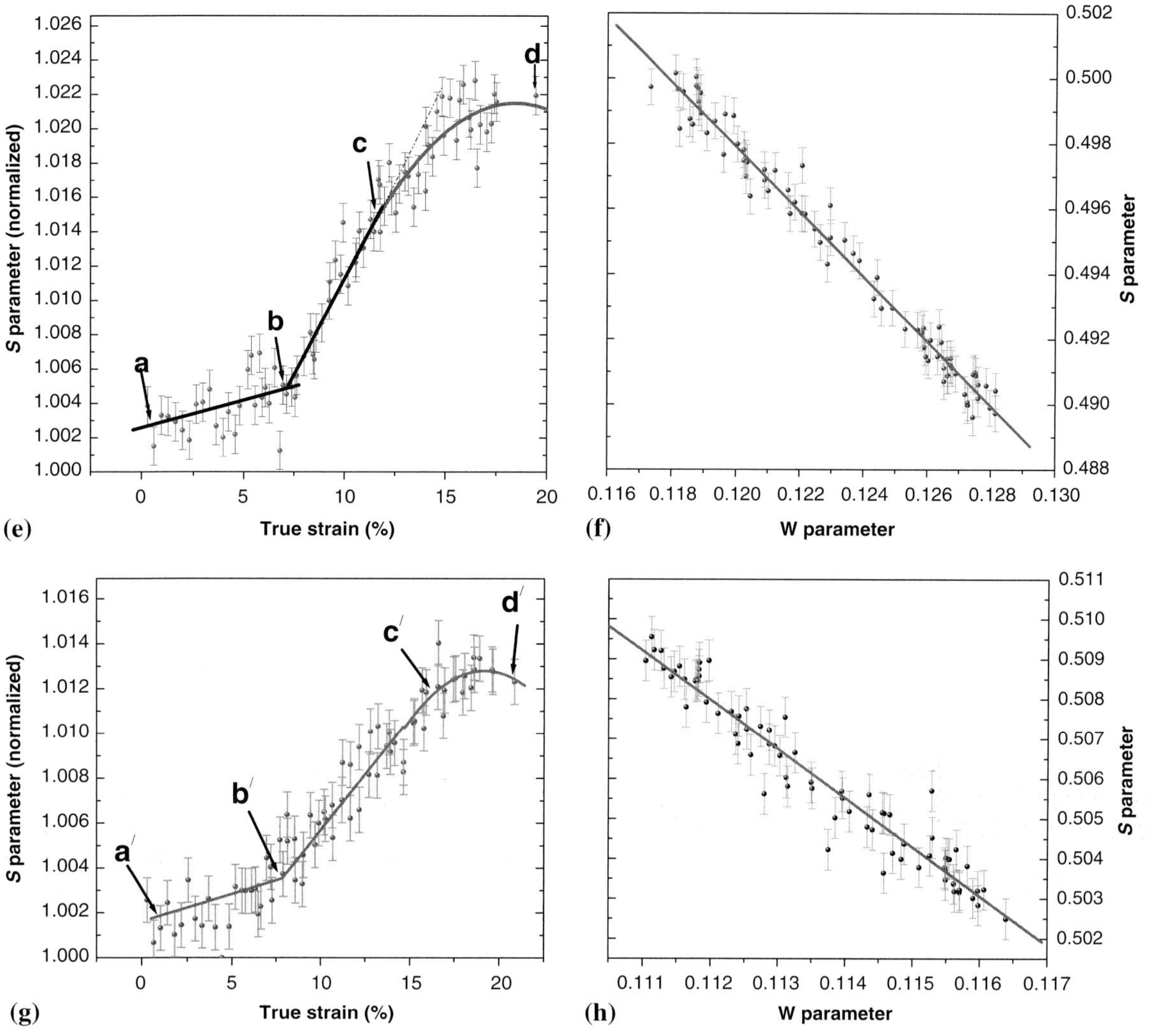

Fig. 4 Plot of S parameter (normalized to the S values of the corresponding un-deformed samples) as function of strain for (e) FeMnSiCrNi and (g) FeMnSiCrNiC alloys and the S-W relation for (f) FeMnSiCrNi and (h) for FeMnSiCrNiC alloys

of phase i with its respective probability P_{d_i} (Ref 17). In Fig. 4 as the degree of deformation is smaller than 5%, the change in S parameter (normalized to the S values of the corresponding un-deformed samples) is very small to be recognized and for deformations higher than 5% the S parameters increase remarkably. An increase of defect density leads to an increase of the S-parameter in both samples, which can be recognized through the comparison of the data of the deformed and the undeformed samples. For the common metals and alloys, the positron parameters increase markedly at the early stage of increasing strain, and rapidly became saturate when the samples are subjected to plastic deformation (Ref 41-43).

3.3.1 Effect of Carbon on the *S* Parameter. In both samples (low carbon, high C sample) the pattern of the S parameter (normalized to the S values of the corresponding un-deformed samples) versus the deformation gave the S-line shape curve (Fig. 4e, g). At the first stage increasing the true strain up to 7% (a $\rightarrow$ b) and (a′ $\rightarrow$ b′) respectively leads to a slight increase in the value of the S parameter. The next stage is starting from 7% up to 11% (b $\rightarrow$ c) in case of the low C sample and between 7% and 16% (b′ $\rightarrow$ c′) of the high C sample. The value of S parameter increases greatly with increasing strain. This is also observed in CuZnAl SMAs (Ref 16). The behavior of the S parameter in case of the high carbon sample differs from that of the low carbon sample. The S parameter increases with increasing strain but the change in the S parameter is not that high as the one of the low carbon sample. This means that the carbon content added to the alloy has an effect in the change of the microstructure, the concentration, and the type of defects created in the alloys.

During deformation the transformation $\gamma \rightarrow \varepsilon$ proceeds and α' martensite is formed at the intersections of the ε plates, which causes strain hardening. As shown by the XRD measurement, the low carbon alloy forms much more ε martensite during deformation, which results in a higher strain hardening compared to the high C alloy. The uniform elongation is larger for low C alloy. This might be correlated with the fact that due to the formation of the ε phase or twinning the microstructure is progressively subdivided in much smaller sized units in the C-added alloy. Consequently, the effective grain size is reduced. In such steels with a smaller grain size, more stress is needed for the movement and multiplication of dislocations. They are impeded by the grain boundaries, which lead to a lower strain hardening and a smaller uniform elongation. This can be recognized in the behavior of the *S* parameter in both samples. The *S* parameter increases with increasing deformation and in case of low carbon steel, it is clear that the change in the *S* parameter is more than the one of high C sample.

From the *S*-*W* plots (Fig. 4f, h), the relation is giving one straight line (lines are drawn as a guide for the eyes). This can be ascribed to the existence of one kind of defect. This defect could be dislocations, which is correlated with lifetime results.

4. Conclusion

From the results of the positron annihilation, the Doppler broadening parameter *S* increases as a result of the trapping of positrons in defects. In the early stages of deformation, no significant increases of the *S* parameters for both alloys are observed. The increase of *S* is caused by an increasing number of lattice defects (dislocations). The carbon content in the alloys has a significant effect in the value of the *S* parameter, which means the concentration of defects.

References

1. T. Maki, *Shape Memory Materials*, K. Otsuka and C.M. Wayman, Ed., Cambridge University Press, UK, 1998, p 117–132
2. H. Otsuka, H. Yamada, T. Maruyama, H. Tanahashi, S. Matsuda, and M. Murakami, Effects of Alloying Additions on Fe-Mn-Si Shape Memory Alloys, *Trans. ISIJ*, 1990, **30**, p 674
3. A. Sato, E. Chishima, K. Soma, and T. Mori, Shape Memory Effect in $\gamma \rightarrow \varepsilon$ Transformation in Fe-30Mn-1Si Alloy Single Crystals, *Acta Metall.*, 1982, **30**, p 1177
4. A. Sato, Y. Yamaji, and T. Mori, Physical Properties Controlling Shape Memory Effect in FeMnSi Alloys, *Acta Metall.*, 1986, **34**, p 287
5. M. Murakami, H. Otsuka, H.G. Suzuki, and S. Matsuda, Complete Shape Memory Effect in Polycrystalline Fe-Mn-Si Alloys, *Proceeding of the International Conference on Martensitic transformations (ICOMAT-86)*, Japan Institute of Metals, 1986, p 985
6. Y. Moriya, T. Sampei, and I. Kozasu, Annual Meeting at Yokohama, *Conference Abstract of Spring Meeting of JIM*, 1989, p 222
7. L.M. Kaputkina and V.G. Prokoshkina, Martensitic Transformations Structure and Strengthness of Processed High-Nitrogen and High-Carbon Ferrous Alloys, *J. Phys. IV Fr.*, 2003, **112**, p 263
8. D. Dulieu and J. Nutting, Metallurgical Development in High Alloy Steels, *Iron Steel Inst.*, 1964, **86**, p 82
9. K. Tsuzaki, Y. Natsume, Y. Tomota, and T. Maki, Effect of Solution Hardening on the Shape Memory Effect of FeMn Based Alloys, *Scr. Metall. Mater.*, 1995, **33**, p 1087
10. P. Dai, Effect of Solution Hardening on the Shape Memory Properties of Fe-Mn-Si Based Alloys, *J. Mater. Sci. Lett.*, 2000, **19**, p 111
11. X. Tian, Y.S. Zhang, and C.X. Shi, Relationship Between $\gamma \rightarrow \varepsilon$ Transformation Temperature $M_{\varepsilon S}$ and Composition of Metastable Austenitic Region in Fe-Mn-Al-Cr System, *Acta Metall. Sinica*, 1986, **22**, p A101 (in Chinese)
12. H. Inagaki, Shape Memory Effect of Fe-14% Mn-6% Si-9% Cr-6% Ni Alloy Polycrystals, *Z. Metallkd.*, 1992, **83**, p 90
13. G.J. Arruda, V.T.L. Buono, and M.S. Andrade, The Influence of Deformation on the Microstructure and Transformation Temperatures of Fe-Mn-Si-Cr-Ni Shape Memory Alloys, *J. Mater. Sci. Eng.*, 1999, **272–275**, p 528
14. I. Hurtado, D. Segers, L. Dorikens-Vanpraet, C. Dauwe, and J. Van Humbeeck, Positron Annihilation Measurements in Cu-Al-Ni Based Shape Memory Alloys, *J. Phys. IV Colloq.*, 1995, **C8**, p 949
15. I. Hurtado, D. Segers, J. Van Humbeeck, L. Dorikens-Vanpraet, and C. Dauwe, Evolution of the Positron Annihilation Lifetime for Ageing in Beta-Phase Cu-Al-Ni-(Ti)-(Mn) Shape Memory Alloys, *Scr. Metall.*, 1995, **33–35**, p 741
16. W. Jing-cheng, S. Zi-chang, and Z. Hao, An Anomaly of the Parameters of Positron Annihilation for Plastically Deforming Shape Memory Alloys, *J. Scr. Metall. Mater.*, 1990, **24**, p 1511
17. J. Van Humbeeck, G. Ghosh, L. Delaey, D. Segers, M. Dorikens, and L. Dorikens Vanpraet, Defect Accumulation During the Martensitic Transformation in Fe-Mn-Ni Shape Memory Alloys, Positron Annihilation, *Proceeding of the ICPA8 Conference*, 1989, p 395
18. T. Liu, R.D. Xia, G.W. Liu, Z.T. Zhao, R.Z. Ma, W.H. Wang, and Y.H. Guo, A Positron Annihilation Study of Stress-Induced Martensite Transformation in Fe-Mn Based Alloys, *J. Mater. Sci. Lett.*, 1998, **17**, p 887
19. P. Hautojarvi, A. Vehanen, A. Dupasquier, M.J. Manninen, P.E. Mijnarends, R.M. Nieminen, and R.N. West, *Positrons in Solid*, Springer-Verlag, Berlin, 1979, p 1–23
20. K. Petersen, *Proceedings of the "Enrico Fermi"*, W. Brandt, Ed., International School of Physics, 1983, p 298
21. R. Krause-Rehberg and H.S. Leipner, *Positron Annihilation in Semiconductors*, Springer, Berlin, 1999
22. P. Hautojarvi, J. Heinio, M. Mannien, and R. Nieminen, The Effect of Microvoid Size on Positron Annihilation Characteristics and Residual Resistivity in Metals, *Philos. Mag.*, 1977, **35**, p 973
23. C. Hidalgo, S. Linderoth, and N. de Diego, The Recovery of Deformed Zinc Studied by Positron Annihilation Spectroscopy, *Philos. Mag.*, 1986, **A64**, p L61
24. T.D. Troev, K. Hinode, S. Tanigawa, and M. Doyama, Positron Study of the Martensitic Transformation in Fe-29.5at.%Ni, *Appl. Phys.*, 1977, **13**, p 105
25. S. Tanigawa, K. Hinode, T. Yamauchi, and M. Doyama, *New Aspects of Maternsitic Transformation*, Japan Institute of Metals, Tokyo, 1976, p 123–128
26. P. Hautojärvi, T. Judin, A. Vehanen, J. Yli-Kauppila, J. Johansson, J. Verdone, and P. Moser, Annealing of Vacancies in Electron-Irradiated α-Iron, *Solid State Commun.*, 1979, **29**, p 855
27. P. Hautojärvi, J. Johansson, T. Judin, P. Moser, M. Puska, A. Vehanen, and J. Yli-Kauppila, *Proceedings of the Fifth International Conference on Positron Annihilation*, Lake Yamanaka, Japan, 1979, R.R. Hasiguti and K. Fujiwara, Ed. (Sendai, Japan), Japan Institute of Metals, 1979, p 737
28. D. Segers, I. Hurtado, L. Dorikens-Vanpraet, and J. van Humbeeck, Positron Doppler Broadening Measurements in Cu-Al-Ni Based Shape Memory Alloys, *J. Phys. IV*, 1995, **5**, p C1–C163
29. D. Segers, J. Van Humbeeck, L. Delaey, M. Dorikens, and L. Dorikens, The Study of Defects in Some CuZnAl Alloys, Positron Annihilation, *Proceedings of the Seventh International Conference* (Singapore), World Scientific, 1985, p 880
30. J. Kansy, Microcomputer Program for Analysis of Positron Annihilation Lifetime Spectra, *Nucl. Instrum. Methods*, 1996, **374**, p 235
31. D. Segers, S. Van Petegem, J.F. Loffler, H. Van Swygenhven, W. Wagner, and C. Dauwe, Positron Annihilation Study of Nanocrystalline Iron, *J. Nano Struct. Mater.*, 1999, **12**, p 1059
32. H. Li, D. Dunne, and N. Kennon, Factors Influencing Shape Memory Effect and Phase Transformation Behaviour of Fe-Mn-Si Based Shape Memory Alloys, *Mater. Sci. Eng. A*, 1999, **273–275**, p 517–523
33. B. Bergersen and M.J. Stott, The Effect of Vacancy Formation on the Temperature Dependence of the Positron Lifetime, *Solid State Commun.*, 1969, **7**, p 1203
34. J.M. Campillo Robles and F. Plazaola, Collection of Data on Positron Lifetimes and Vacancy Formation Energies of the Elements of the Periodic Table, *Defect Diffus. Forum*, 2003, **213–215**, p 141
35. R.N. West, Positrons in Solids, in *Topics in Current Physics*, Vol. 12, P. Hautojarvi, Ed., Springer, Berlin, 1979, p 89
36. T.E.M. Staab, R. Krause-Rehberg, and B. Kieback, Review Positron Annihilation in Fine-Grained Materials and Fine Powders—An Application to the Sintering of Metal Powders, *J. Mater. Sci.*, 1999, **34**, p 3833
37. M.J. Puska, P. Lanki, and R.M. Nieminen, Positron Affinities for Elemental Metals, *Phys. Condens. Matter*, 1989, **1**, p 6081
38. Y. Kamimura, T. Tsutsumi, and E. Kuramoto, Calculations of Positron Lifetimes in a Jog and Vacancies on an Edge-Dislocation Line in Fe, *Phys. Rev. B*, 1995, **52**, p 879
39. K.M. Mostafa, J. De Baerdemaeker, P.R. Calvillo, N. Van Caenegem, Y. Houbaert, and D. Segers, A Study of Defects in Iron Based Alloys by Positron Annihilation Techniques, *J. Acta Phys. Polon. A*, 2008, **113–115**, p 1471

40. Y.-K. Park, J.T. Waber, M. Meshii, C.L. Snead Jr., and C.G. Park, Dislocation Studies on Deformed Single Crystals of High-Purity Iron Using Positron Annihilation: Determination of Dislocation Densities, *Phys. Rev. B*, 1986, **34**, p 823
41. I.K. Mackenzie, A. Eady, and R.R. Gingerich, The Interaction Between Positrons and Dislocations in Copper and in an Aluminum Alloy, *Phys. Lett. A*, 1970, **33**, p 279
42. J.H. Yang and M. Wayman, Development of Fe-Based Shape Memory Alloys Associated with Face-Centered Cubic-Hexagonal Close-Packed Martensitic Transformations, *Metall. Trans.*, 1992, **23A**, p 1445–1454
43. K.M. Mostafa, N. Van Caenegem, J. De Baerdemaeker, D. Segers, and Y. Houbaert, Study of Defects and Phase Transformation in a Fe-Mn-Si-Cr-Ni Alloy by Positron Annihilation Method, *Phys. Status Solidi (c)*, 2007, **4–10**, p 3554

JMEPEG (2009) 18:582–587
DOI: 10.1007/s11665-009-9369-y

1059-9495/$19.00

Structure-Property Relationships in Conventional and Nanocrystalline NiTi Intermetallic Alloy Wire

Jeremy E. Schaffer

(Submitted October 8, 2008; in revised form December 3, 2008)

Homogeneous 177-μm diameter nanocrystalline Nitinol wires were produced and compared to microcrystalline Nitinol from equivalent ingot stock. Property measurement was carried out using cyclic tension testing, strain-controlled fatigue testing and bend and stress-free recovery testing (BFR). A B2 cubic structure of 5- to 60-nm grain size was confirmed by transmission electron microscopy (TEM). The fatigue strain capability at 10^7 cycles is 30% greater in the nanocrystalline versus microcrystalline annealed wire. Grain size and residual strain are positively correlated in cyclic uniaxial tension testing within the superelastic temperature and strain regime. A transformation-induced loading plateau that is 40% longer than in standard superelastic polycrystalline wire is also reported for the first time.

Keywords long-plateau Nitinol, medical wire, nanocrystalline Nitinol, Nitinol processing, Nitinol wire

1. Introduction

Nitinol medical devices are designed with functional shapes which are driven into living anatomy through tiny incisions, forced through millimeter scale conduit and finally thermally or mechanically placed to function and ideally enhance the life of the patient in a mechanically and chemically dynamic biological flora. It is critical that medical device engineers have the proper tools to discern optimal design with Nitinol including knowledge of relevant microstructural-property relationships.

Grain size has been known as an important variable in defining material strength, fracture toughness, and fatigue strength since the pioneering work of E.O Hall and N.J. Petch in the late 1940s (Ref 1). In fine wire structural applications, such as medical lead wire assemblies, coarse microstructure in Co-Ni-Cr alloys has been shown to negatively impact mechanical fatigue performance (Ref 2).

Researchers have recently begun to recognize the unique properties of ultra fine-grained (UFG) metals (Ref 3, 4). Pelton and colleagues first reported UFG Nitinol in 1986 during in situ crystallization of amorphized Nitinol thin films (Ref 5). UFG Nitinol studies have become more common in the primary literature since this pioneering work. Sun and Li reported on the unique phase transformation and localization in nano-grained 1-1.5 mm diameter Nitinol microtubes (Ref 6). Sergueeva et al. presented an in depth study of the nanocrystallization process in 12 mm Nitinol disks, which underwent amorphization by severe plastic deformation (SPD) (Ref 7). Others researchers (Ref 8-12) have examined certain aspects of nanostructure generation and properties, while few have examined detailed structure-function relationships in fine (less than 250 μm diameter) Nitinol wire over a range of grain sizes from nanometers to micrometers. The goal of this study was to begin to quantify the grain size effect at fine wire diameters, which are of increasing importance in medical device design for example in self-expanding stents, blood filters, and tissue scaffold platforms.

This article is an invited paper selected from presentations at Shape Memory and Superelastic Technologies 2008, held September 21-25, 2008, in Stressa, Italy, and has been expanded from the original presentation.

Jeremy E. Schaffer, Fort Wayne Metals Research Products Corporation, 9609 Ardmore Avenue, 46809 Fort Wayne, IN; and **Jeremy E. Schaffer**, Weldon School of Biomedical Engineering, Purdue University, 47906 West Lafayette, IN. Contact e-mail: jeremy_schaffer@fwmetals.com.

2. Experimental

Approximately 2 mm Nitinol wire with an ingot A_S of 243 K, comprising Ti-50.9 at.% Ni was equivalently drawn and annealed to a diameter of 230 μm. At this stage, wires were continuously strand annealed at 923-1123 K at a sufficient dwell time to ensure complete recrystallization with a nominally 2-μm grain size. Samples were subsequently wet-drawn using natural diamond drawing dies and oil-based lubricants to a finish diameter of 177 μm with nominally 40% retained cold work. Wires were then continuously annealed at constant stress for less than 60s to effect recrystallization while minimizing potential Ni-rich precipitate formation.

Near-surface microstructure and oxide thickness determination was performed using thin foil preparation and extraction by focused ion beam (FIB, FEI Nova 200 NanoLab) and transmission electron microscopy (TEM, JEOL 2000EX, 200 keV). The preparation method is represented in Fig. 1. The targeted sample surface is initially masked with a thin layer of vapor platinum compound deposit to protect the surface from ion implantation damage (Ref 13). Grain size distributions were calculated using digital image analysis (ImageJ 1.40f, National Institutes of Health, USA) as shown schematically in Fig. 2.

Active sample transformation temperatures were analyzed using bend and free recovery (BFR) methodology using a

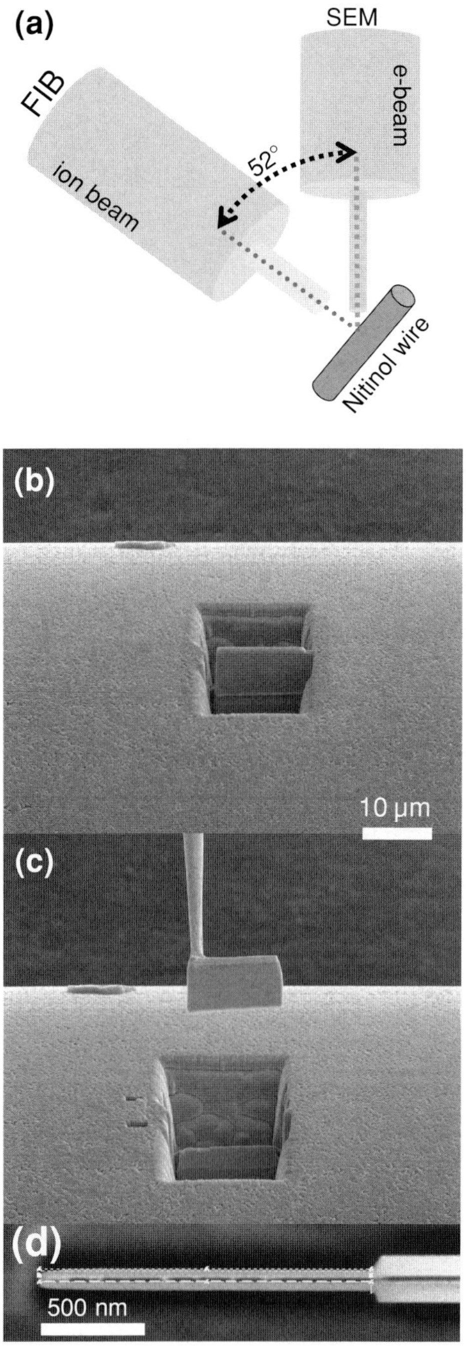

Fig. 1 (a) Schematic of dual beam SEM/FIB equipment used for thin foil TEM sample preparation, gallium ion source for milling is targeted at a common eucentric height shared by the electron beam for imaging. Gallium ions are accelerated using a 30 kV potential and electromagnetically focused to an area that can be controlled by feedback image data gathered from the far less damaging scanning electron beam. (b) Cross section "rough" cuts to expose lift out foil at a nominal thickness of 1000 nm. (c) Foil lift out using a 2-μm tungsten probe. (d) After attachment to a SEM grid post, sample is thinned at reduced ion current to a thickness of about 100 nm or less

proprietary digital video analysis apparatus per standard ASTM F2082-06 (Ref 14). Cyclic and monotonic uniaxial tensile properties were measured at an ambient temperature of 298 K at a strain rate of 10^{-3} s^{-1} using an Instron Model 5565 Tensile Test Bench equipped with pneumatic grips. Fatigue behavior was characterized using rotary beam fatigue test equipment manufactured by Positool, Inc. at 3600 r/min in ambient 298 K air. Specimens from each grain size condition were tested at alternating strain levels ranging from 0.5 to 2.5% to a maximum of about 10^8 cycles.

3. Results and Discussion

3.1 Microstructural Data

Nitinol wire was successfully produced at 177 μm diameter with many levels of grain nucleation and growth. Five distinct microstructures were chosen for subsequent thermomechanical and structural fatigue testing comprising median grain sizes of: 50 nm, 100 nm, 2 μm, 5 μm, and 10 μm. Selected area electron diffraction patterns were acquired in order to observe texture states and to look for evidence of precipitate reflections, an example of which is shown in Fig. 2(d-f). No evidence of lenticular precipitates was found in the diffraction data.

The nanocrystalline samples were confirmed to be homogeneous within the first 10 μm of material below the drawn oxide surface using TEM. Observed grain structures are presented in Fig. 3. Empirical cumulative distribution functions were plotted for each grain size as given in Fig. 4. Average oxide thickness for all samples was found to be about 120 nm, an example of which is given in Fig. 5. Further experimentation is underway to understand microstructure along the central portion of the wires.

3.2 Transformation Temperatures

Active transformation temperatures were measured using BFR methods in accordance with ASTM standard F2082. All samples analyzed exhibited an A_s temperature of 240 ± 5 K, in good agreement with ingot DSC data. Plots of recovery data for each sample are given in Fig. 6. The samples that possessed equiaxed grain sizes greater than 100 nm showed a clear single stage transformation associated with recovery from the martensitic to austenitic structure, and an active A_f of 250 ± 5 K. At the 50-nm grain size, samples exhibited a right-shifted recovery. The assumption of an entirely austenitic transformation placed calculations for the A_f of the 50 nm samples at 277-279 K. Further data is needed, particularly in situ TEM analysis, to provide a mechanistic description of this behavior. The extended thermal recovery may be related to high levels of strain energy stored in the nanocrystalline material due to a relatively large grain-boundary density. Martensite variant activity may also differ in the nanostructures during recovery, which could lead to distinct behavior in BFR testing.

3.3 Tensile Testing

The evolution of axial properties as a function of grain size followed a path expected from the classic Hall-Petch relationship: fine-grained material exhibited increased strength and reduced plasticity during cyclic testing. Single cycle uniaxial tension testing to a reversal point of 4% strain showed a positive correlation between grain size and isothermal residual strain, defined by the residual strain after load release without raising sample temperature. This relationship is shown in Fig. 7(c). The ultimate tensile strength measured in axial monotonic tension testing increased with reduced grain size as shown in Fig. 7(d).

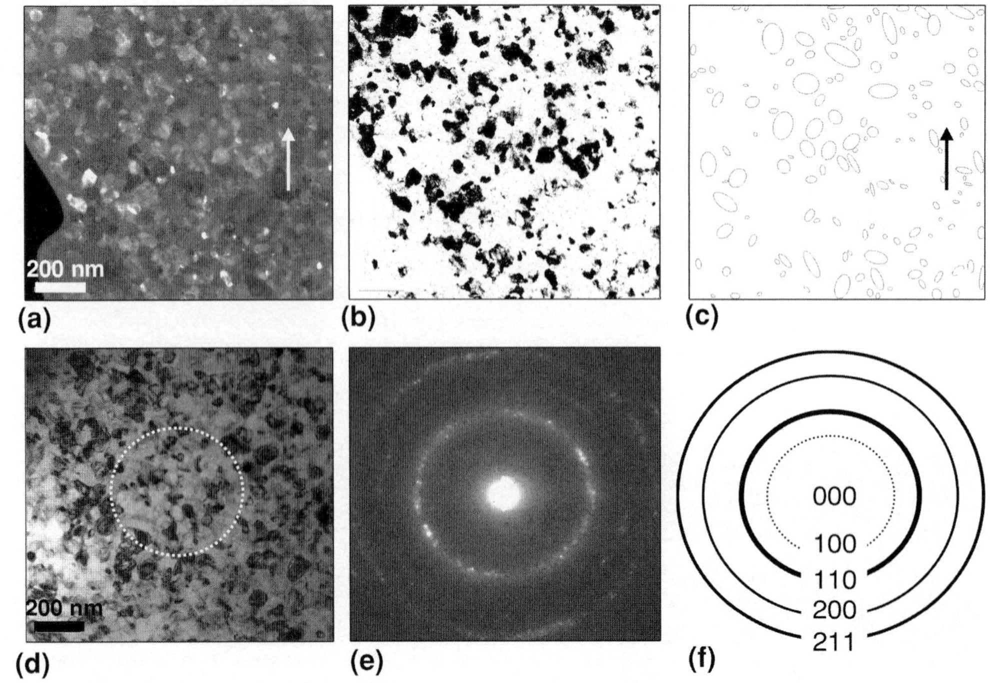

Fig. 2 (a) Dark field (DF) TEM image of sample with median 50-nm grain size. (b) Binary conversion of preceding DF-TEM image for digital grain size analysis. (c) Ellipsoidal representation of analyzed grain field showing lack of residual grain elongation in the drawing (arrow) direction. (d) Bright field (BF) TEM image of equivalent material (different area). (e) Selected area diffraction pattern (SADP) of area shown in (d) showing bright B2 cubic (110) polycrystalline ring. (f) Calculated powder DP for B2 Nitinol

Fig. 3 BF-TEM images showing three of five examined grain sizes, left to right median sizes: 50 nm, 100 nm, 2 μm. Red arrow indicates wire drawing direction. All samples were found to possess equiaxed microstructure with little apparent signature from remnant cold work. Of note, the dark fringes in the far right image are associated with Moire interference and tilt grain-boundary thickness fringes (Ref 15)

The initial onset of isothermally recoverable transformation strain and significant reversion stress at the 2-μm grain size was attributed to a reduction in the irreversible plastic behavior during stress-induced transformation. The finer, more plasticity resistant microstructures were not as susceptible to the stabilization of martensite by dislocation network formation. This behavior was consistent with the discussion by Duerig in Ref 16.

A large increase in both ultimate and unloading plateau strength was found in the samples comprising 100 and 50-nm grain sizes, respectively. Waitz et al. found competing accommodation mechanisms in martensitic NiTi, with single variant B19′ plates forming below about 80 nm and compound twinned variants forming in grains larger than about 100 nm (Ref 17). A similar grain size threshold mechanism may contribute to constitutive behavior differences found in the present samples.

The length of the stress-induced martensite load plateau was found to vary with the microstructural state. Figure 7(b) shows a typical superelastic plateau; the onset of elastic loading of the B19′ matrix occurred at 7% strain. In the precipitate-free structure with a 100-nm equiaxed grain size, the transformation plateau terminated at 10.1% strain, with 9.6% strain recovery upon isothermal unloading. Theoretical recoverable strains as high as 10.5% have been calculated in preferred transformation orientations such as [123] by Sehitoglu et al. in Ref 18. Until now, near-theoretical strain recovery has not been observed in isothermal tension testing of fine wire samples. The long-plateau behavior observed here is likely related to favorable texture that was not apparent in the present electron diffraction data sets. One limiting factor that should be resolved is the small amount of analyzed surface material; understanding of full section texture will be critical going forward. Further texture measurement is underway aimed at providing a mechanistic description of the long-plateau phenomenon.

Residual stress distributions on the microstructural level have been shown to provide B19′ phase stabilization (Ref 19).

It is possible that grain constraint effects in the nanocrystalline regime are another important factor in determining initial phase distributions, transformation paths, and overall constitutive behavior. It is unknown whether this could impact effective transformation plateau length.

As shown in Fig 7(a), the unloading stiffness was found to decrease significantly from a nearly grain-size-invariant 42 GPa to 29 GPa at the 50-nm grain size. In 2006, Duerig demonstrated a similar unloading modulus range for various strain values and gave some explanation related to anelastic reversion of detwinned martensite (Ref 16). Here we see variability as a function of grain size with step behavior between 50 and 100 nm. This behavior may be the result of energetic competition between single and compound martensite variants and will require in situ TEM deformation studies for further illumination. If properly controlled, this discovery could be meaningful in the context of a wire stent in a pulsating blood vessel where relative implant stiffness is known to affect overall device performance (Ref 20, 21).

3.4 Fatigue Testing

Rotary beam fatigue test results show increased damage accumulation rates with increasing grain size. This was particularly evident under conditions where lives were on the order of 10^5-10^7 cycles. Figure 8(a) shows strain-life curves for each material. Wire with a 50-nm grain size resisted fracture at the 0.9% alternating strain to beyond 10^7 cycles. At a 10-μm grain size, the 10^7 cycle strain level was 0.6%; other grain sizes fell in order between these values with coarse structures yielding reduced resistance to damage (Fig. 8b).

4. Conclusions

1. Isothermal recoverable strains of 9.6%, greater than 90% of the theoretical limit, were measured in nanocrystalline Nitinol wire.
2. Zero residual strain in a 4% strain cycle was observed in an equiaxed, precipitate-free, and untrained material with a median 50-nm grain size.
3. A 30% increase in fatigue damage resistance was observed in nanocrystalline versus microcrystalline wire.
4. Thermally extended recovery during BFR testing was found in samples with a median grain size less than about 100 nm.
5. The unloading stiffness was found to decrease stepwise from 42 GPa to 29 GPa moving from 100 nm to the 50-nm grain size.

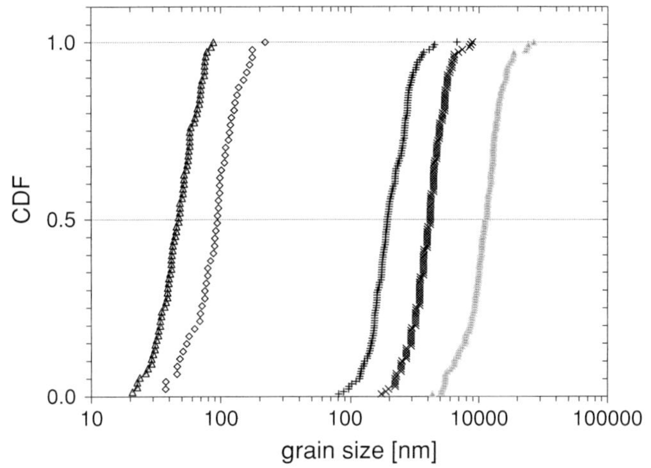

Fig. 4 Cumulative distribution function representation of grain size distributions for, median grain sizes, left to right: 50, 100, 2000, 5000, and 10000 nm

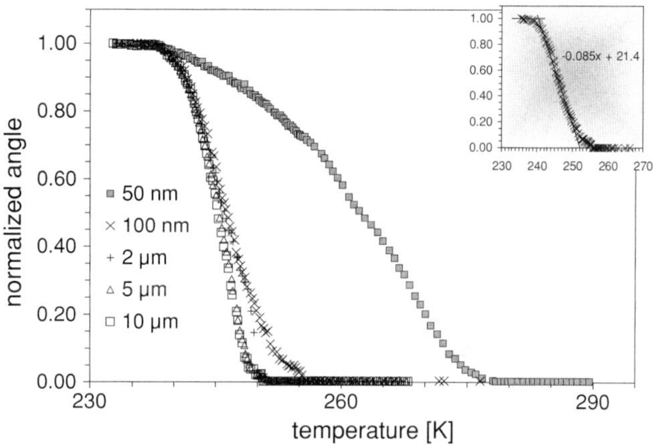

Fig. 6 Digital vision BFR test data for five distinct grain sizes. It is postulated that active austenitic finish temperature shift is due to the relatively high grain-boundary content in the nanocrystalline material

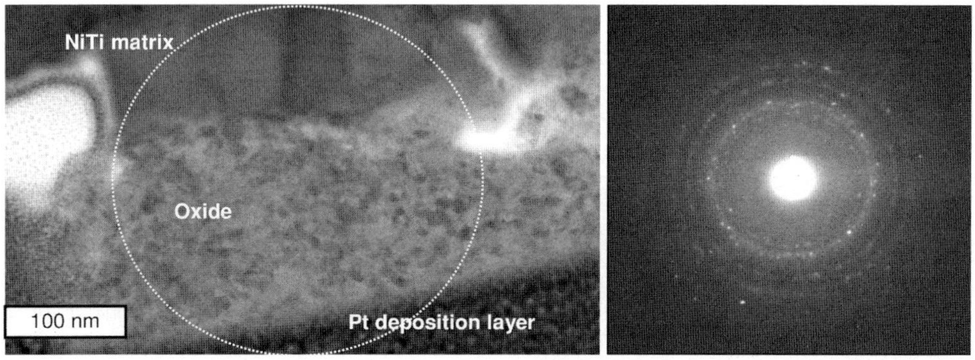

Fig. 5 (left) NiTi oxide layer. (right) SADP taken from the indicated area showing powder diffraction rings associated with nanocrystalline oxide

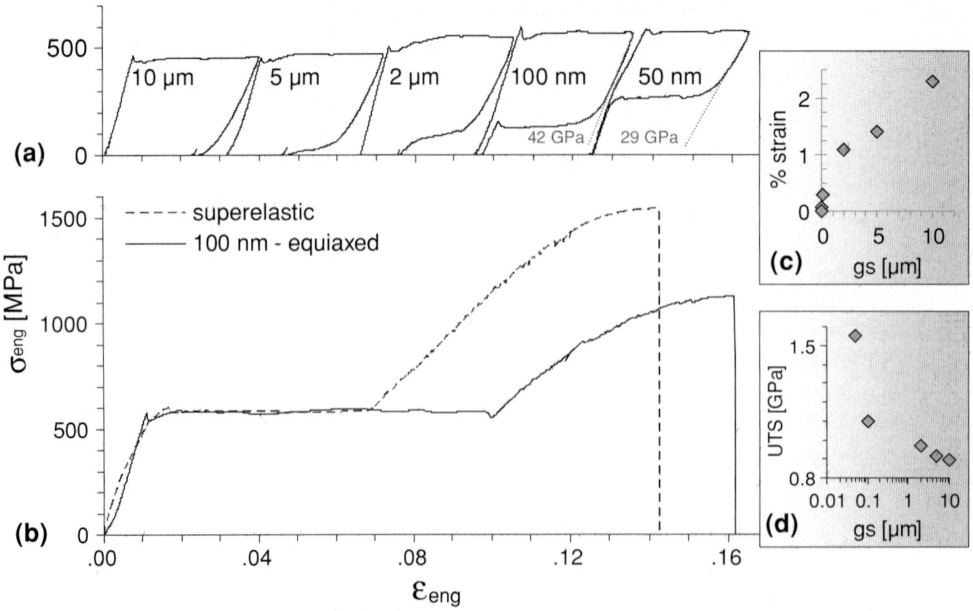

Fig. 7 Engineering stress-strain data ran at a temperature of 298 K in quiescent air at a strain rate of 10^{-3} s^{-1}. (a) 4% hysterisis loops as a function of equiaxed grain size. Unloading moduli and values shown in red; (b) monotonic loading curves for standard superelastic material and wire with a homogeneous 100 nm, precipitate-free matrix; (c) residual strain as a function of grain size; (d) ultimate tensile strength as a function of grain size

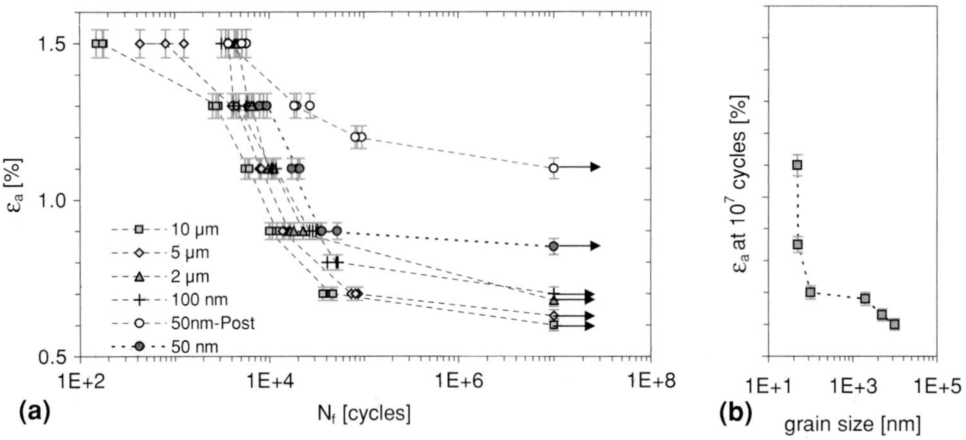

Fig. 8 (a) Strain-life representation of rotary beam fatigue data. Test conditions: $R = -1$, $T = 298$ K, $f = 60$ s^{-1}, environment: quiescent air, $N = 5$ at each strain level. (b) Strain limit at 10^7 cycles. Note: pictured with maximum strain level error of ±3%, cycle count error <0.1%

Acknowledgments

The author gratefully acknowledges the staff of Fort Wayne Metals for providing financial support to this project. Special thanks also go to David Snider, Mike McGuire, Dave Plumley, and Chris Hoffman for help with material and data acquisition.

References

1. E.O. Hall, The Deformation and Ageing of Mild Steel: III Discussion of Results, *Proc. Phys. Soc. B*, 1951, **64**, p 747–753
2. J.E. Schaffer, "A Hierarchical Initiation Mechanism Approach to Modeling Fatigue Life Variability in 35Co–35Ni–20Cr–10Mo Medical Grade Fine Wire", Master of Science in Mechanical Engineering Thesis Manuscript, Purdue University, West Lafayette, IN, 2007
3. T. Hanlon, E.D. Tabachnikova, and S. Suresh, Fatigue Behavior of Nanocrystalline Metals and Alloys, *Int. J. Fatigue*, 2005, **27**, p 1147–1158
4. H.W. Hoeppel, M. Kautz, C. Xu, M. Murashkin, T.G. Langdon, R.Z. Valiev, and H. Mughrabi, An Overview: Fatigue Behaviour of Ultrafine-Grained Metals and Alloys, *Int. J. Fatigue*, 2006, **28**, p 1001–1010
5. A. Pelton, P. Moine, M.A. Noack, and R. Sinclair, Crystalline-to-Amorphous Transitions in Ti-Ni Alloys, *Mater. Res. Soc. Symp. Proc.*, 1986, **62**, p 291–302
6. Q.P. Sun and Z.Q. Li, Phase Transformation in Superelastic NiTi Polycrystalline Micro-tubes Under Tension and Torsion from Localization to Homogeneous Deformation, *Int. J. Solids Struct.*, 2002, **39**, p 3797–3809
7. A.V. Sergueeva, C. Song, R.Z. Valiev, and A.K. Mukherjee, Structure and Properties of Amorphous and Nanocrystalline NiTi Prepared by Severe Plastic Deformation and Annealing, *Mater. Sci. Eng. A*, 2003, **339**, p 159–165

8. X. Xu and N. Thadhani, Shock Synthesis and Characterization of Nanostructured Nitinol Alloy, *Mater. Sci. Eng. A*, 2004, **384**, p 194–201
9. P.L. Potapov, W. Tirry, and D. Schryvers, Cross-section Transmission Electron Microscopy Characterization of the Near-surface Structure of Medical Nitinol Superelastic Tubing, *J. Mater. Sci.: Mater. Med.*, 2007, **18**, p 483–492
10. T. Mousavi, F. Karimzadeh, and M.H. Abbasi, Synthesis and Characterization of Nanocrystalline NiTi Intermetallic by Mechanical Alloying, *Mater. Sci. Eng. A*, 2008, **487**, p 46–51
11. T. Sawaguchi, G. Kaustraeter, A. Yawny, M. Wagner, and G. Eggeler, Crack Initiation and Propagation in 50.9 at. pct Ni-Ti Pseudoelastic Shape-memory Wires in Bending-rotation Fatigue, *Metal. Mater. Trans. A*, 2003, **34**, p 2847–2860
12. G. Eggeler, E. Hornbogen, A. Yawny, A. Heckmann, and M. Wagner, Structural and Functional Fatigue of NiTi Shape Memory Alloys, *Mater. Sci. Eng. A*, 2004, **378**, p 24–33
13. W.Y. Kwong and W.Y. Zhang, Electron-beam Assisted Platinum Deposition as a Protective Layer for FIB and TEM Applications, *Proceedings of IEEE International Symposium on Semiconductor Manufacturing*, ISSM September, IEEE, New York, NY, USA, p 469–471
14. ASTM F2082-06, Standard Test Method for Determination of Transformation Temperature of Nickel-Titanium Shape Memory Alloys by Bend and Free Recovery, in *Annual Book of Standards*, ASTM International, Philadelphia, PA, 2006, p 1–7
15. D.B. Williams and C.B. Carter, *Transmission Electron Microscopy III Imaging*, Springer Science + Business Media, LLC, New York, NY, USA, 1996, p 444–450
16. T.W. Duerig, Some Unsolved Aspects of Nitinol, *Mater. Sci. Eng. A*, 2006, **438–440**, p 69–74
17. T. Waitz, W. Pranger, T. Antretter, F.D. Fischer, and H.P. Karnthaler, Competing Accommodation Mechanisms of the Martensite in Nanocrystalline NiTi Shape Memory Alloys, *Mater. Sci. Eng. A*, 2008, **481–482**, p 479–483
18. H. Sehitoglu, R. Hamilton, D. Canadinc, X.Y. Zhang, K. Gall, I. Karaman, Y. Chumlyakov, and H.J. Maier, Detwinning in NiTi Alloys, *Metal. Mater. Trans. A*, 2003, **34**, p 5–13
19. X. Huang, G.J. Ackland, and K.M. Rabe, Crystal Structures and Shape-Memory Behaviour of NiTi, *Nat. Mater.*, 2003, **2**, p 307–311
20. J. Black, *Biological Performance of Materials*, 4th ed., CRC Press Taylor & Francis Group, Boca Raton, FL, USA, 2006, p 195–205
21. D. Stoeckel, A. Pelton, and T. Duerig, Self-expanding Nitinol Stents: Material and Design Considerations, *Eur. Radiol.*, 2004, **14**(2), p 292–301

JMEPEG (2009) 18:588–593
DOI: 10.1007/s11665-009-9428-4

Development of Texture and Microstructure During Cold Rolling and Annealing of a Fe-Based Shape Memory Alloy

Bikas C. Maji, Madangopal Krishnan, Vijay Hiwarkar, Indradev Samajdar, and R.K. Ray

(Submitted October 13, 2008; in revised form February 21, 2009)

Texture evolution and microstructural changes during cold rolling and annealing of Fe-14Mn-6Si-9Cr-5Ni shape memory alloy have been investigated. The starting solution-annealed material has a nearly random texture with microstructure composed of equiaxed austenite grains with ε martensite plates inside. Cold rolling induces a strong alloy type texture with Brass {011}⟨211⟩ and Goss {011}⟨100⟩ as major components. Annealing of the cold-deformed material produces a nearly random texture. The microstructural investigation reveals that with increase in cold deformation, the amount of stress-induced ε and α′ martensite volumes increase. The electron back-scattered diffraction (EBSD) phase mapping shows that reversion of the ε martensite begins only after recrystallization sets in at a temperature of 1073 K.

Keywords EBSD, shape memory alloy, texture

1. Introduction

The Fe-Mn-Si-based shape memory alloys have been extensively investigated for the past few decades due to their potential application as low-cost shape memory alloys. The shape memory effect (SME) in these alloys owes its origin to the γ to ε martensitic transformation. The SME in these alloys was first discovered by Sato et al. (Ref 1) who investigated single-crystal Fe-30Mn-1Si alloy in early 1980s. They showed that the SME in these material is strongly dependent on the crystallographic orientation in such a way that up to 8% tensile strain, a nearly perfect recovery (100%) could be obtained along the ⟨441⟩ direction. By contrast, a remarkably suppressed recovery (~15%) was noticed along the ⟨100⟩ direction. The SME observed in polycrystalline Fe-Mn-Si-based shape memory alloys, developed so far, is also limited to a maximum 3% recoverable strain only, which is much smaller than what has been reported for the single crystal. Therefore, introducing proper texture in the parent phase during processing could be an effective approach for improving the SME of these alloys. However, so far no systematic study on development of texture and microstructure during cold deformation and annealing of Fe-based shape memory alloys has been reported in published literature. In the present work, an attempt has been made to study the development of texture and microstructure during cold rolling and annealing in Fe-14Mn-6Si-9Cr-5Ni shape memory alloy.

This article is an invited paper selected from presentations at Shape Memory and Superelastic Technologies 2008, held September 21-25, 2008, in Stresa, Italy, and has been expanded from the original presentation.

Bikas C. Maji and **Madangopal Krishnan**, Materials Science Division, Bhabha Atomic Research Centre, Mumbai 400 085, India; **Vijay Hiwarkar** and **Indradev Samajdar**, Department of Metallurgical Engineering and Materials Science, Indian Institute of Technology, Powai, Mumbai 400 076, India; and **R.K. Ray**, R&D Division, Tata Steel, Jamshedpur 831 001, India. Contact e-mail: madangk@barc.gov.in.

2. Experimental

The Fe-14Mn-6Si-9Cr-5Ni (in mass%) alloy used in this study was prepared by nonconsumable vacuum arc melting using high purity Fe, Mn, Si, Cr, and Ni. The arc-melted button was hot rolled at 1173 K into a strip of 2 mm thickness. This strip was solution annealed at 1323 K for 1 h and further cold rolled at room temperature to 6, 20, 35, and 45% reduction in thickness. Annealing study was carried out in the temperature range of 873-1373 K using the maximum deformed (45% reduction in thickness) specimen. The forward and reverse martensitic transformation temperatures for this alloy as measured by differential scanning calorimeter (DSC) after solution annealing treatment are found to be $M_s = 291$ K, $M_f = 257$ K, $A_s = 393$ K, and $A_f = 413$ K. The microstructural evolution was characterized using an optical microscope, x-ray diffractometer, and a FEI Quanta 200 HV tungsten filament scanning electron microscope (SEM) equipped with TSL (TexSEM Laboratories, Incorporated) orientation imaging microscopy (OIM) system. The x-ray diffractometry (XRD) and bulk texture measurement were carried out using MoK_α ($\lambda = 0.7093$ Å) x-radiation in rectangular specimens of 20 mm × 15 mm size which were mechanically ground and polished up to the mid thicknesses level. The XRD measurements were carried out in 2θ range of 15°-55° and at a scanning speed 0.6°/min. Both pole figure and orientation distribution function (ODF) methods were employed to study the texture development in the parent austenite phase. The optical microstructures of these specimens were revealed by electroetching, using a solution of $CH_3OH + HCl$ (90:10 volume ratio) and 6 V dc potential supply. The electron back scatter diffraction (EBSD) measurements were conducted at 20 kV with a step size between 0.5 and 0.2 μm in an area about 400 μm × 400 μm.

3. Results

3.1 Microstructure and Texture of the Starting Material

Figure 1(a) shows the optical micrograph of the starting material which was prepared by solution annealing of the hot-rolled material at 1323 K for 1 h followed by water quenching. The microstructure reveals equiaxed austenite grains (~70 μm in size) with some parallel lines inside. These lines are mostly ε martensite plates which must have been formed during water quenching of the specimen. Apart from these, some annealing twins with straight boundaries are also observed within the microstructure. The {111} pole figure of the austenite phase is shown in Fig. 1(b). The pole figure indicates that the material has a nearly random texture, with some isolated weak peaks.

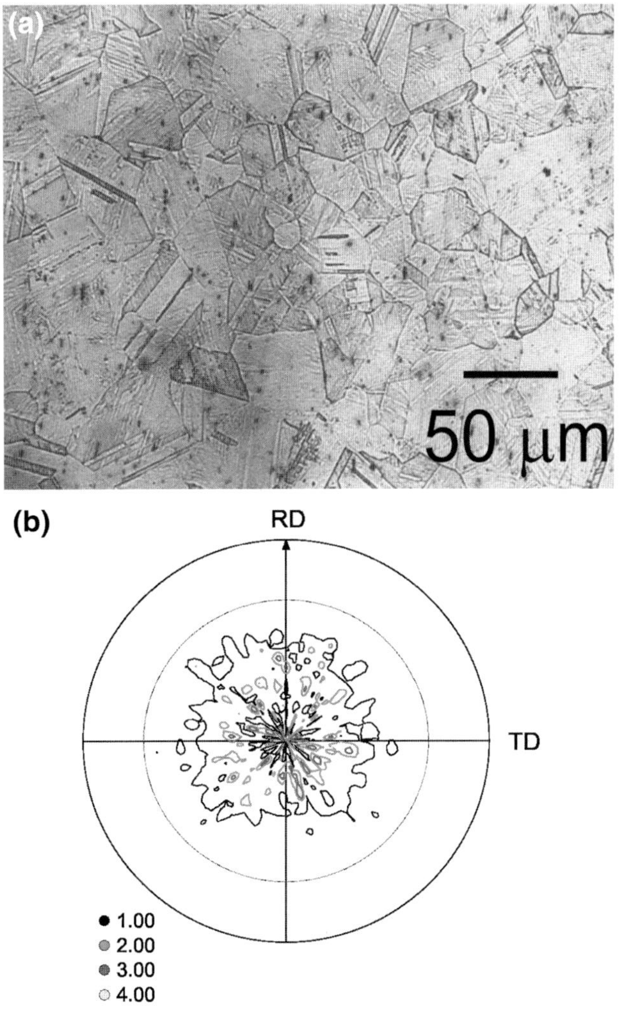

Fig. 1 (a) Optical micrograph and (b) the {111} austenite pole figure of the starting material solution annealed at 1323 K for 1 h

3.2 Microstructural Changes During Cold Rolling

Figure 2 displays the XRD patterns obtained from the mid-thickness levels of the specimens at different stages of cold-rolling reduction. In this figure, the gradual changes in phase constitutions during the progress of cold rolling are compared with that of the starting solution-annealed specimen. After 6% cold rolling (Fig. 2b), mainly, the austenite and ε martensite phase which form due to stress-induced transformation of the parent austenite are present. However, after 20% cold rolling, formation of α′ martensite can be clearly identified by the presence of distinct XRD peaks. The α′ martensite possibly forms due to the interaction of different ε martensite variants during cold deformation (Ref 2, 3). Another interesting observation is that the (002) peak of the parent austenite phase disappears after 20% cold rolling, indicating that the material is getting textured. The intensities of the α′ martensite peaks increase with further increase in cold reduction. At higher amount of strain (more than 20%), peak broadening has been noticed which is possibly associated with accommodation of heavy strain that leads to fragmentation of grains (Ref 4, 5). Some of the peaks could be due to the merging of peaks arising from the two different types of martensite phases.

Optical microscopy of longitudinal sections of the cold-rolled specimens was done. Figure 3(a) shows the microstructure of a specimen cold rolled to 6%. It can be seen that a large amount of ε martensite plates have formed within the austenite grains, and in some places, the fragmentation of grains due to localized deformation has begun. The microstructure after 20% cold rolling becomes highly inhomogeneous, and austenite grains get further fragmented. Beyond 35% reduction in thickness, shear bands and deformation bands are observed to occur. Figure 3(b), taken from a specimen after 45% cold rolling, clearly shows that austenite grains become highly elongated along the rolling direction and extremely thin in the thickness direction. A large number of deformation bands can also be noticed.

3.3 Texture Development During Cold Rolling

Texture development has been investigated by measuring three incomplete pole figures, {111}, {110}, and {311}, using

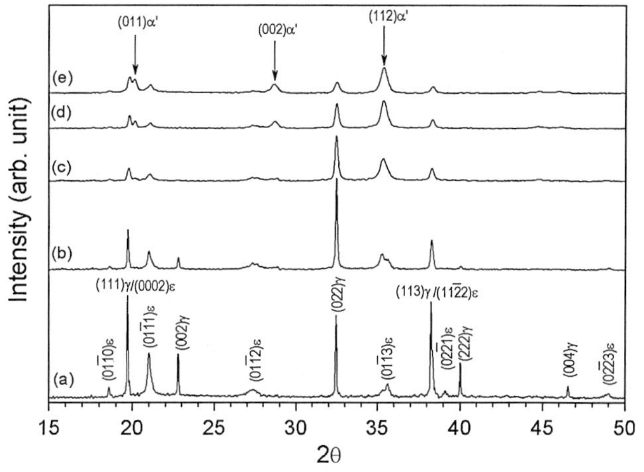

Fig. 2 XRD patterns obtained from specimens after (a) solution annealing, (b) 6%, (c) 20%, (d) 35%, and (e) 45% cold-rolling reductions shows the gradual change in phase constituents

Fig. 3 Optical micrographs of the longitudinal sections after (a) 6% and (b) 45% reduction in thickness. Formation of shear band and deformation bands are noticed after 45% cold rolling

the Schultz (Ref 6) reflection technique. Three-dimensional ODFs were calculated on the basis of Bunge's series expansion method (Ref 7). Figure 4(a) shows the measured {111} pole figure obtained from specimens cold rolled by 20%. It can be seen that the spread of intensities is close to the {110}⟨112⟩ Brass orientation. Further reduction in thickness is seen to lead to sharpening of the texture intensity. However, no other orientations appear in the pole figure till 45% reduction. An orientation is most adequately represented as a point in three-dimensional orientation space, for instance, by the three Eulerian angles, ϕ_1, ϕ, and ϕ_2, in the reduced Eulerian space. Traditionally, the ODF is represented in sections of ϕ_2 = const. for fcc materials. The $\phi_2 = 0°$, 45°, and 65° sections of the ODFs of the fcc materials are very useful for showing the typical texture components that are present. Figure 4(b) shows these ODF sections of the 45% cold-rolled specimens. Presence of a strong Brass component {110}⟨112⟩ at $\phi_1 = 35°$, $\phi = 45°$ in the $\phi_2 = 0°$ section and $\phi_1 = 55°$, $\phi = 90°$ in the $\phi_2 = 45°$ section can be seen in this figure. Apart from the Brass orientation, the presence of Copper orientation {112}⟨111⟩ is also observed at $\phi_1 = 90°$, $\phi = 35°$ in the $\phi_2 = 45°$ section. At $\phi_1 = 60°$, $\phi = 35°$ in the $\phi_2 = 65°$ section, another texture component with a broad intensity spread is noticed. This could be the {123}⟨634⟩ S orientation which normally appears in between Brass and Copper orientations. In addition to the ODF sections, the orientation density along certain fibers in Euler space contain the most important texture components and give a good overview of the rolling texture development. For fcc phase a clear view of the texture development can be obtained by plotting the ODF intensity along the α and β-fibers. Figure 5 shows the intensity change in (a) α-fiber and (b) β-fiber with deformation obtained from specimens cold rolled to 20% and 45% reduction, respectively. The major texture components in

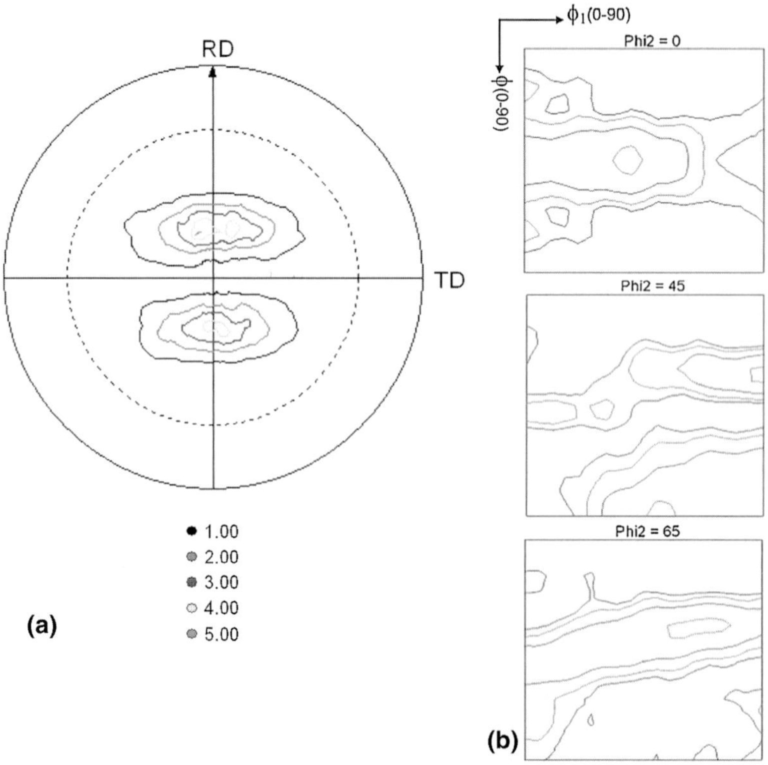

Fig. 4 (a) The {111} pole figure of the 20% cold-rolled specimen and (b) the ODFs sections of the 45% cold-rolled material

the α-fiber is Brass, with which a minor Goss {110}⟨001⟩ is also observed. It is seen that while Brass orientation markedly intensifies with increasing reduction, Goss orientation remains at a relatively low intensity. The β-fiber plots, Fig. 5(b), for 20% and 45% reductions are remarkably uniform, though the overall texture intensity of the fiber tends to increase with increasing rolling reduction.

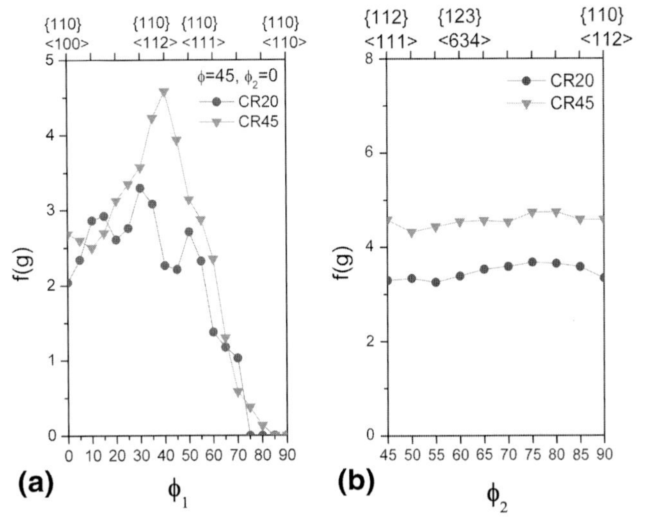

Fig. 5 The variation of texture intensity along (a) the α-fiber and (b) the β-fiber with increasing cold deformation

3.4 Textural and Microstructural Changes During Annealing

The development of texture during annealing was measured after isochronal annealing treatments of 1 h at 973, 1123, 1173, and 1273 K. Figure 6 shows a gradual change in the {111} pole figures with increasing annealing temperature. The figure also indicates that noticeable change occurs only above annealing temperature of 1123 K (Fig. 6b). Further increase in annealing temperature only leads to spreading of texture intensity (Fig. 6c). The overall texture intensity shows a gradual decrease with increase in annealing temperature and annealing at 1273 K (Fig. 6d) results in a nearly random texture.

Figure 7 displays a few optical micrographs of the annealed specimens. Figure 7(a) shows that cold-worked structure persists till 973 K. The recrystallization process starts at 1123 K, as evidenced in Fig. 7(b). However, in this microstructure some cold-worked areas are still observed. Annealing at 1173 K (Fig. 7c) is observed to give rise to a completely recrystallized microstructure.

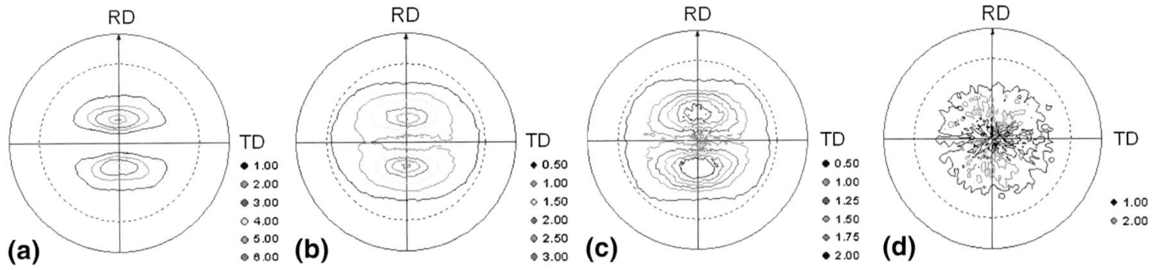

Fig. 6 The {111} pole figures obtained from specimens after 1 h isochronal annealing at (a) 973 K, (b) 1123 K, (c) 1173 K, and (d) 1273 K

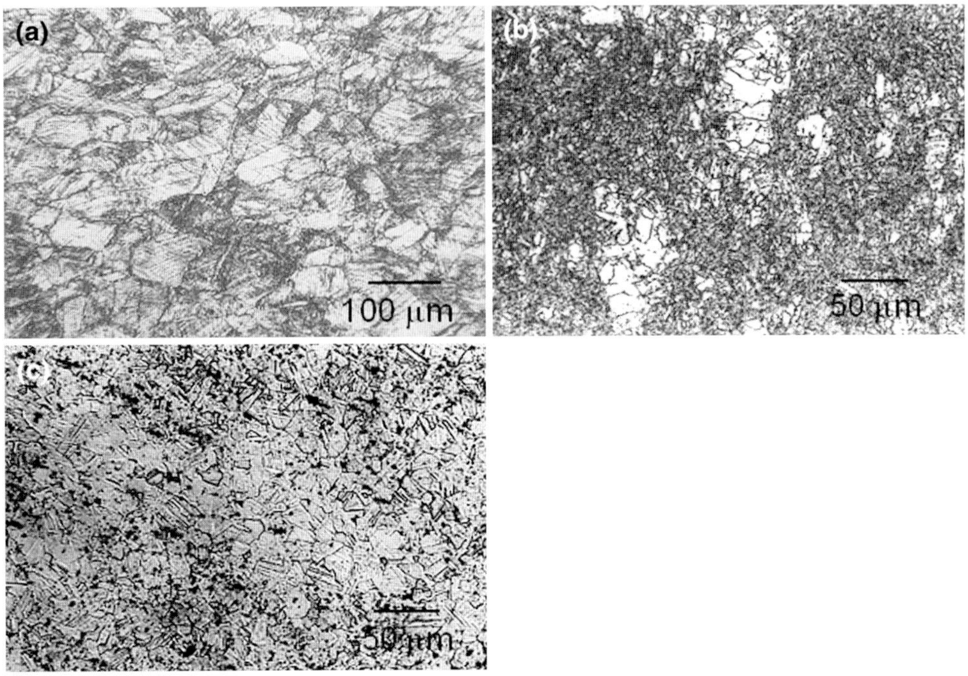

Fig. 7 The optical micrographs of specimens annealed at (a) 973 K, (b) 1123 K, and (c) 1173 K for 1 h

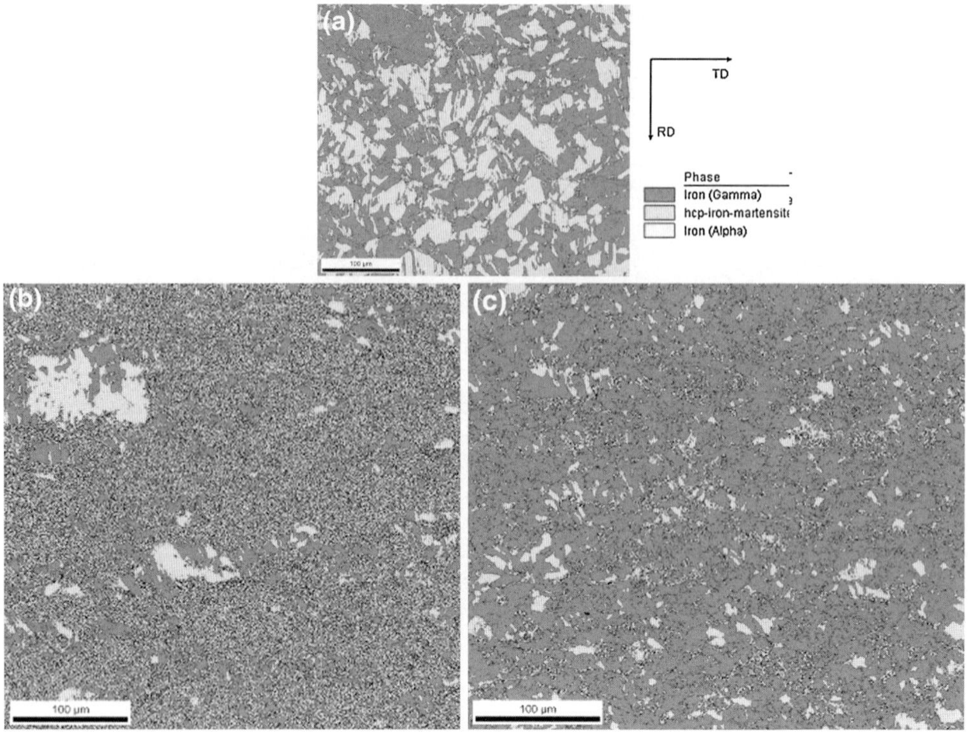

Fig. 8 EBSD phase maps of specimens after (a) 6% cold rolling, (b) annealing at 973 K for 1 h, and (c) annealing at 1123 K for 1 h

3.5 EBSD Phase Mapping

The EBSD phase maps were obtained at each stage of cold reduction. Figure 8(a) shows a typical map obtained from a specimen cold deformed by 6%. The map essentially reveals that at lower deformation levels, austenite phase transforms to ε martensite. However, fragmentation of grains starts beyond 20% cold deformation. The relative changes in volume fractions of different constituent phases with increased cold reduction, as measured by the EBSD phase mapping, are given in Table 1.

EBSD phase maps obtained after isochronal annealing show interesting results. For example, after 1 h isochronal annealing of the 45% deformed sample at 973 K (Fig. 8b) does not show much of austenite phase. On the other hand, for the specimen annealed at 1123 K that was initially deformed to similar extent (Fig. 8c) clearly shows a large volume fraction of austenite. It is also observed that the 1123 K annealed specimen is mostly recrystallized, whereas recrystallization has not initiated in the 973 K one. This suggests that austenite reversion takes place only after recrystallization starts.

4. Discussion

The cold-rolling texture of this material is quite typical of low stacking fault energy fcc alloys that are cold deformed by moderate (<50%) extents. It was observed that this material could be cold rolled only up to 45%, before cracks could begin appearing. It is possible that the large amount of Mn and Si present in this alloy cause a high rate of work hardening. The room temperature stacking fault energy was calculated using Eq 1 of Ref 8 and found to be ~3.41 mJ/m^2. The cold-rolling texture of heavily deformed (~95%) fcc material of comparable stacking fault energy usually shows a large volume fraction of

Table 1 Relative change in volume fraction of constituent phases with cold rolling

Cold rolling amount (% reduction in thickness)	Austenite (vol. fraction)	ε martensite (vol. fraction)	α′ martensite (vol. fraction)
6	0.55	0.44	0.01
20	0.22	0.64	0.14
35	0.11	0.72	0.17
45	0.07	0.75	0.18

the Brass {110}⟨112⟩ component. This has been seen in a 95% cold-rolled Ni-60Co alloy with stacking fault energy of ~13 mJ/m^2 (Ref 9). However, after 40% cold rolling, the same alloy shows a rather uniform β-fiber (Ref 10)—the type that has been observed in this study after 45% cold reduction. Thus, it can be said that the texture evolution in this alloy during cold rolling is similar to that of fcc materials with low SFE.

Annealing of the 45% cold-rolled material shows a gradual change in the texture evolution. The overall texture intensity decreases with increase in annealing temperature, although, the basic nature of the texture does not change. After the material is fully recrystallized, above annealing temperatures of 1173 K, texture intensity decreases drastically, although, the main Brass component of texture persists. This clearly indicates a nucleation-dominated recrystallization process. The inverse pole figure (Fig. 9) taken from the recrystallized sample substantiates this point. Recrystallization at higher temperatures leads to grain growth. In fact, the average grain size after 1273 K annealing is 6 to 7 times that after 1173 K annealing (fully recrystallized). The pole figure correspondingly shows a random distribution of discrete spots. It is interesting to note

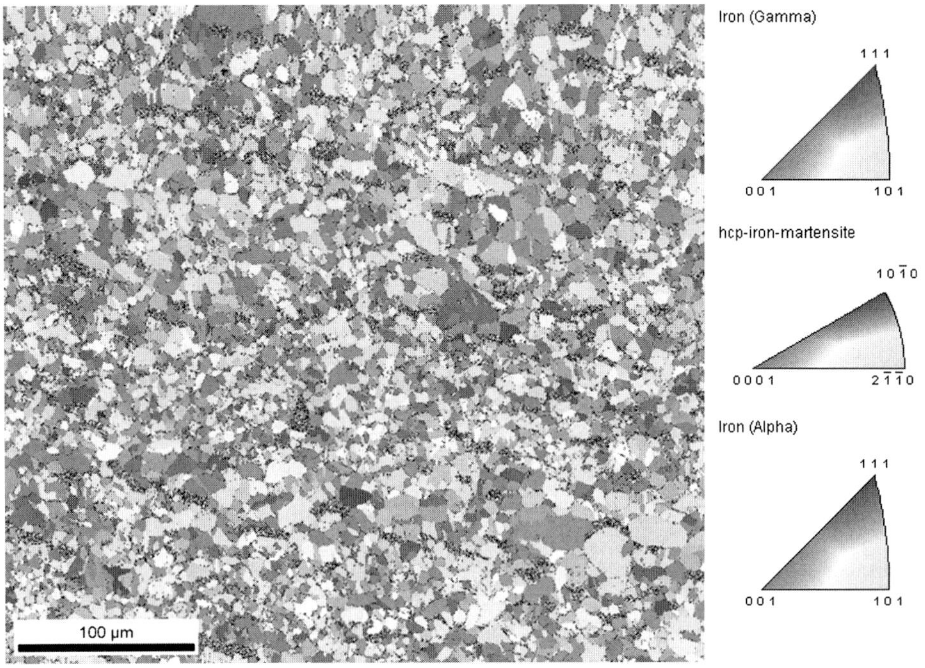

Fig. 9 Inverse pole figure map of the fully recrystallized specimen (annealed at 1173 K for 1 h) shows that the orientations of freshly nucleated recrystallized grains are random

that the overall texture intensity does not show any change during grain coarsening.

The textural changes in this alloy are accompanied by the γ-ε-α' martensitic phase transformation. The parent austenite transforms to hcp ε martensite, which further transforms to bcc α' martensite through the interaction of different ε variants (Ref 2, 3). The observation, that the reverse transformation occurs only after recrystallization begins, suggests that the driving force for the reverse transformation is high. In fact, the extent of reversion increases with progress of recrystallization. Further studies will be required to determine the precise reason for this behavior.

5. Summary

1. Cold deformation induces a strong alloy-type texture with Brass {110}⟨112⟩ as the major component.
2. Cold deformation above 6% results in formation of strain-induced α' martensite.
3. Complete recrystallization of cold-deformed material results in random texture.
4. Reverse transformation of stress-induced ε martensite starts only after initiation of recrystallization.

References

1. A. Sato, E. Chishima, K. Soma, and T. Mori, Shape Memory Effect in γ-ε Transformation in Fe-30Mn-1Si Alloy Single Crystals, *Acta Metall.*, 1982, **30**(6), p 1177–1183
2. J.H. Yang and C.M. Wayman, Development of Fe-Based Shape Memory Alloys Associated with Face-Centered Cubic-Hexagonal Closed-Packed Martensitic Transformations: Part III. Microstructures, *Metall. Trans. A*, 1992, **23**(5), p 1445–1454
3. B.C. Maji and M. Krishnan, The Effect of Microstructure on the Shape Recovery of Fe-Mn-Si-Cr-Ni Stainless Steel Shape Memory Alloy, *Scr. Mater.*, 2003, **48**, p 71–77
4. B.D. Cullity, *Elements of X-ray Diffraction*, 2nd ed., Addison-Wesley, Reading, MA, 1978, p 285–292
5. B.L. Averbach and B.E. Warren, The Effect of Cold Work in Metals on Powder Pattern Intensities, *J. Appl. Phys.*, 1949, **20**, p 1066–1069
6. L.G. Schulz, A Direct Method of Determining Preferred Orientation of a Flat Reflection Sample Using a Geiger Counter X-ray Spectrometer, *J. Appl. Phys.*, 1949, **20**, p 1030–1033
7. H.J. Bunge, *Texture Analysis in Materials Science*, Butterworth, London, 1982
8. H. Li, F. Yin, T. Sawaguchi, K. Ogawa, X. Zhao, and K. Tsuzaki, Texture Evolution Analysis of Warm-Rolled Fe-28Mn-6Si-5Cr Shape Memory Alloy, *Mater. Sci. Eng. A*, 2008, **494**, p 217–226
9. R.K. Ray, Rolling Texture of Pure Nickel, Nickel-Iron and Nickel-Cobalt Alloys, *Acta Metall. Mater.*, 1995, **43**(10), p 3861–3872
10. R.K. Ray, R&D Division, Tata Steel, Personal communication, 2008

JMEPEG (2009) 18:594–599
DOI: 10.1007/s11665-009-9493-8

1059-9495/$19.00

Thermomechanical Properties of Porous NiTi Alloy Produced by SHS

P. Bassani, P. Giuliani, A. Tuissi, and C. Zanotti

(Submitted October 24, 2008; in revised form April 21, 2009)

Samples obtained from relatively large powders (<150 μm), with total porosity in the range 30-68%, were characterized mainly from a morphological point of view. Total porosity, as well as pore size, shape and distribution, was analyzed. Sample microstructure was also investigated, indicating that the main phase produced during the self-propagating high-temperature synthesis (SHS) reaction is Ti reach NiTi phase, as confirmed by calorimetric analyses. Moreover, the presence of secondary phases, suggested by the low transformation enthalpy, was confirmed by SEM observations. In fact, EDS microanalyses and EBSD mapping helped in the identification of such secondary phases, such as Ni_3Ti, Ti_2Ni and $Ti_4Ni_2O_x$. Other samples were successively produced starting from the same powders but introducing a different powder compression methodology and operating conditions. In this way, the obtained samples showed higher porosity featured by more uniform size, shape and distribution while, from a micro-structural point of view, no significant differences were observed. Mechanical compression tests were carried out at room temperature and, on selected samples, also above A_f in order to highlight the influence of pore shape and distribution. Results obtained at room temperature show that the mechanical properties decrease with the porosity augmentation. For higher temperatures, the samples presented a pseudoelastic behavior. Dilatometric tests were also performed and the results well indicated the martensite to austenite transformation at the same temperature showed by the DSC analyses. Thermal analysis was completed by evaluating the thermal diffusivity temperature and porosity dependence using an experimental-numerical approach especially developed.

Keywords advanced characterization, intermetallics, mechanical testing, metallography

1. Introduction

Porous NiTi has been intensively studied in the last years due to the high interest as a possible material for biomedical applications; it offers good biocompatibility, mechanical properties closer than other metallic material to those of bone, shape recovery behavior that can help in insertion operation, excellent compatibility with magnetic resonance and computer tomography scanning. Several studies have been performed on mechanical properties and corrosion characteristics of bulk and porous NiTi (Ref 1, 2) in view of these field of applications. Interest in porous NiTi is still acute, due to continuing research on biomedical field application/optimization but also due to possible application as vibration dampers or shock absorber.

For the latter application not only mechanical behavior is important but also a comprehensive view of thermomechanical properties is of interest. Self-propagating high-temperature synthesis (SHS) was proposed as a suitable route for the production of porous NiTi alloys with different degree of porosity. In fact, by controlling the compaction procedure of the starting powders and some parameters such as mixing ratio of Ni and Ti powders, powder size and shape, sample size and shape, it is possible to adjust the product porosity. Moreover, the operating conditions driving the SHS process (ignition technique, preheating temperature, atmosphere and gas (argon) pressure) can influence the characteristics of the final porosity. In order to obtain samples featured with greater porosity, a variation in powder processing was introduced: the starting material was loaded inside a ceramic tube and then reacted. Microstructural and thermomechanical tests were performed in order to characterize the properties of the samples produced and also to verify if any abrupt variation occurred during the SHS procedure.

2. Experimental

It is necessary to define the sample porosity, ρ, used for presenting our results.

$$\rho\% = \left(1 - (\rho_p/\rho_f)\right) * 100$$

where ρ_f is the density of the full dense while ρ_p is the density of the porous NiTi.

Samples of porous NiTi were prepared by the SHS technique starting from relatively large size powders (~150 μm). Powders were dry mixed, with equiatomic

This article is an invited paper selected from presentations at Shape Memory and Superelastic Technologies 2008, held September 21-25, 2008, in Stresa, Italy, and has been expanded from the original presentation.

P. Bassani and **A. Tuissi**, National Research Council, Institute for Energetics and Interphases, Lecco, Italy; and **P. Giuliani** and **C. Zanotti**, National Research Council, Institute for Energetics and Interphases, Milan, Italy. Contact e-mail: p.bassani@ieni.cnr.it.

composition of Ni and Ti, compacted in cylindrical shape with different compacting pressure and then reacted, obtaining porosity up to 45%. Details regarding the sample preparation by SHS technique can be found in Ref 3.

In order to increase the final porosity of the reacted sample, another preparation route was optimized: after mixing, the powders were placed inside a ceramic tube and only slightly shacked or pressed. Porosities in the range 48-68% were thereby obtained.

Tests were replicated in order to verify the repeatability of our results and to obtain various samples with the same characteristics for subsequent tests.

Selected samples were prepared for porosity measurements and microstructural investigations: the cylinders obtained were sectioned longitudinally in two parts: one was further sectioned longitudinally at ½ R and the other was sectioned perpendicularly to the cylinder axis, obtaining three transverse sections. Surfaces were first roughly polished for image analysis, and after that, selected sections were cold mounted in epoxy resin, polished with diamond paste and finally with colloidal silica, in order to obtain surfaces suitable for EDS and EBSD (Electron Back Scattered Diffraction) analyses, performed with a LEO 1430 SEM equipped with INCA Energy 300 and INCA Crystal detectors. For details on EBSD sample preparation and observation see Ref 4. DSC analyses were carried out on specimen cut with a metallographic diamond saw from different position in the samples. Analyses were performed with a Seiko DSC, with heating/cooling rate of 10 °C/min, on specimen of about 30 mg in mass.

Samples for thermal diffusivity evaluation, dilatometric and mechanical tests were obtained simply by machining top and bottom surfaces of the products in order to obtain cylinders approximately 12 mm height and flat, parallel bases. Thermal diffusivity was estimated using a simple, 1D, numerical code that simulates the heat transfer phenomena in a multi-layers cylindrical sample. Thermal diffusivity dependence on temperature and porosity has been changed in order to reproduce experimental data concerning heating and cooling transients. The experimental procedure and the numerical approach are described in Ref 5. Mechanical tests were performed according to ASTM E9-89a (Ref 6), on an MTS 810 universal testing machine, equipped with a 250 kN force transducer. Test to failure was performed at room temperature, while cyclic loading/unloading with increasing deformation up to 6% was performed both at room temperature and at 110-120-130 °C, according to ASTM E209-00 (Ref 7). In all cases test parameters were selected in order to have a constant strain rate of about 10^{-6} [s^{-1}].

3. Results and Discussion

3.1 Microstructural Characterization

By employing the SHS route we were able to produce samples with porosity ranging from 10 to 68%. For porosity less than 45%, pores are often not interconnected, and especially the larger ones lie on planes parallel to the specimen base. By increasing porosity a significant interconnection is obtained, without preferential orientation. Image analyses of the percent of porosity on the various section of specimen confirmed substantial homogeneous distribution of the resultant pores. The pore morphology of samples produced by SHS has been already reported by the authors (see Ref 4).

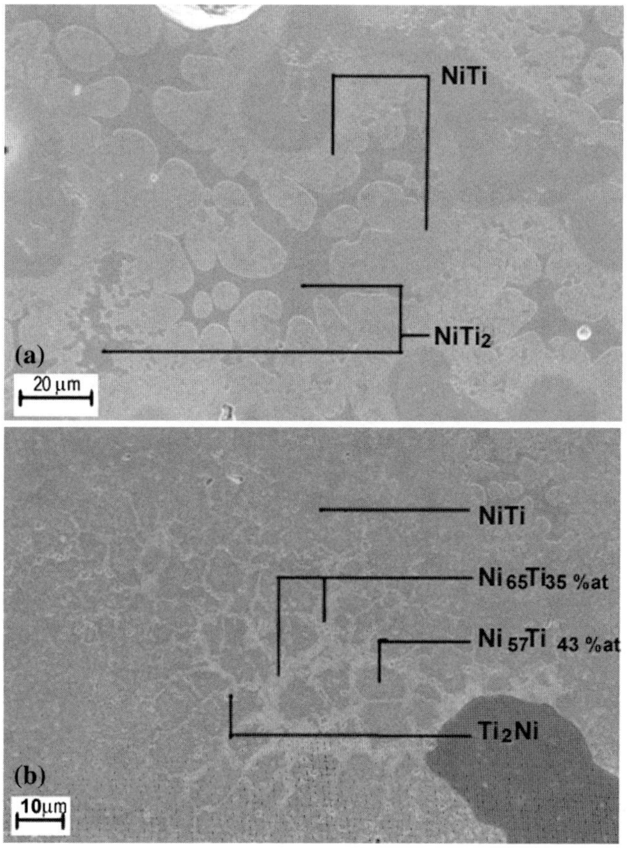

Fig. 1 SEM images with indication of EDXS results on typical morphology: (a) rounded concave agglomerates (mainly Ti_2Ni or $Ti_4Ni_2O_x$); (b) branched agglomerates (mainly Ni_3Ti and NiTi)

Optical microscope investigations on microstructure did not show any differences inside the specimens and between samples of differing total porosity. In all cases, in a matrix of NiTi, the presence of secondary phases was observed, mainly grouped in small region with two main morphologies (Fig. 1a, b). EDS analyses allowed us to identify possible candidates for the observed structure. EBSD analyses provided further information on structure and eventual grain size of the observed phases. In one case (Fig. 1a and 2a), rounded, concave particles are constituted mainly by Ti_2Ni or $Ti_4Ni_2O_x$. In the other case branched structures resulted in a mix of eutectic constituent formed by Ni_3Ti and NiTi, mixed with Ti_2Ni or $Ti_4Ni_2O_x$, as polygonal particles (Fig. 1b and 2b). It was not possible with the investigations brought about to define the oxygen content of the $Ti_4Ni_2O_x$ particles: so it was preferred not to make any assumption about the exact composition of those phases.

3.2 DSC Investigations

Calorimetric analyses results are summarized in Fig. 3 and 4. As indicated in Fig. 3, transformation temperatures are substantially independent of total porosity and can be associated with a Ti-rich NiTi matrix. Enthalpy values shown in Fig. 4 and estimated using small samples (20-30 mg in mass) are less than a corresponding annealed bulk material (26.1 J/g). A correlation between transformation enthalpies and porosity, referred to the sample having greater size, can only be argued.

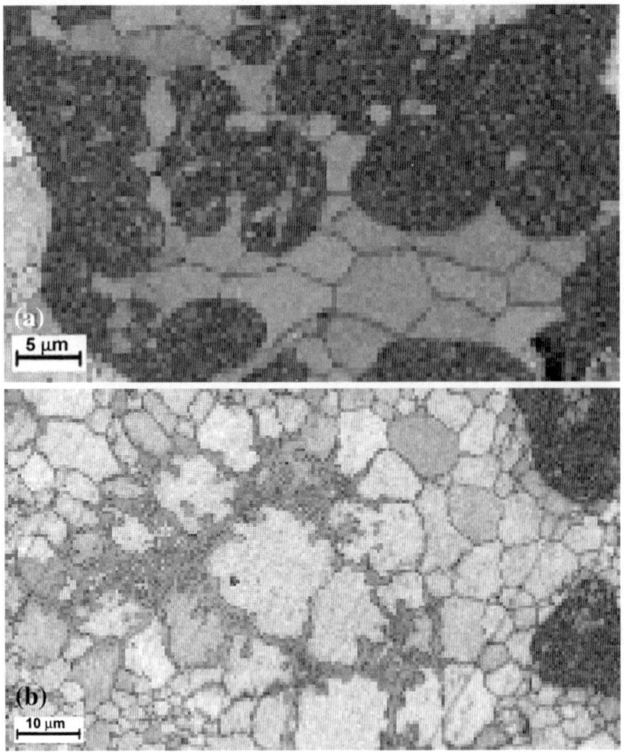

Fig. 2 Pattern quality map obtained from EBSD analysis on two regions with morphology similar to those of Fig. 1. Strict correspondence between quality pattern and phase identification maps (not reported) was observed. Ti_2Ni (medium dark grey area), Ni rich-NiTi (light grey), martensite matrix (dark grey). Mapped area: (a) 50 × 28 µm; (b) 100 × 60 µm

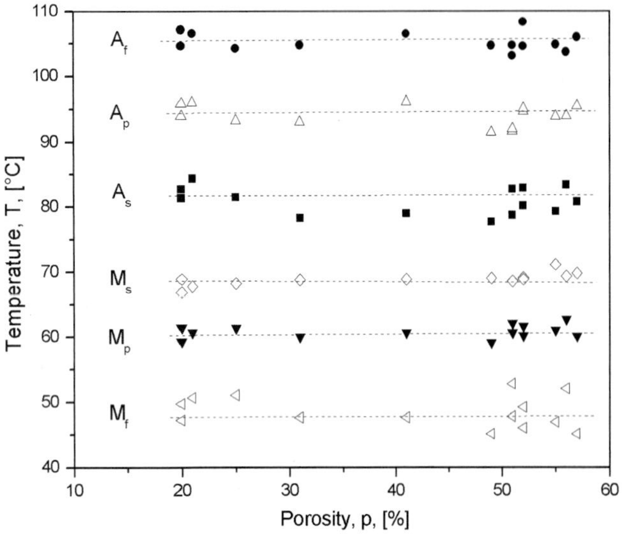

Fig. 3 DSC transformation temperatures for samples with different porosity: no porosity dependence can be argued

In fact, the general trend of the data indicates that transformation enthalpy increases slightly with the sample porosity augmentation. Heating/cooling enthalpies fall in a range from 4 to 8 J/g and show minimum values for porosity of about 30%. This behavior is in agreement with results reported in Ref 8, in which the porosity ranged from 25 to 45%.

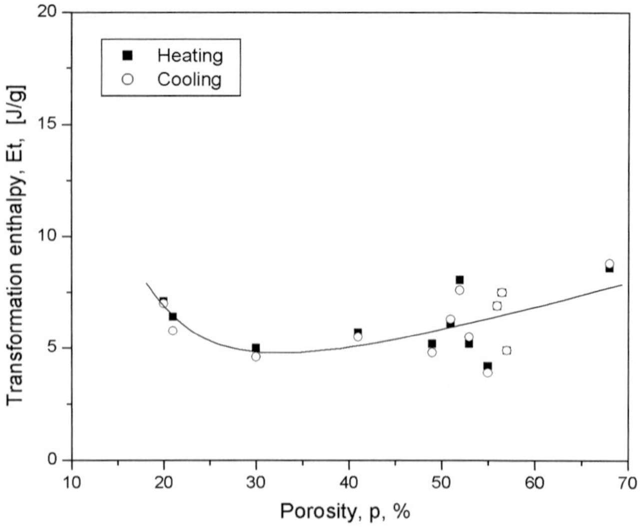

Fig. 4 DSC transformation enthalpies as a function of porosity

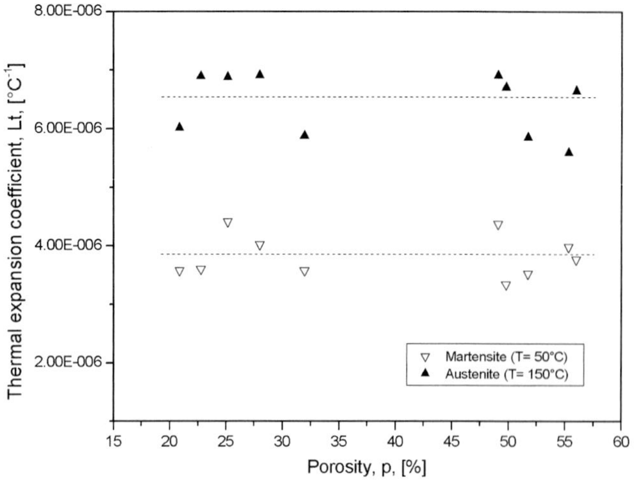

Fig. 5 Linear expansion coefficient for martensitic and austenitic conditions vs. porosity

The observed scatter can be attributed in part to local inhomogeneities, with a greater/lesser presence of secondary phases. The low mean value can be related in part only to the presence of secondary phases that subtract mass to transformation. In particular, as observed with the help of EDS-EBSD analyses, also Ni-rich NiTi regions were observed that did not result in any calorimetric event, evidencing a stabilized austenitic structure.

3.3 Dilatometric Tests

Transformation temperatures from dilatometric tests mimic very well those results from the more commonly used DSC analysis. An attempt to derive linear expansion coefficients above and below the transformation region has been performed. Unfortunately only a qualitative result could be obtained. Scattered values were obtained not only from one sample to another with the same porosity, but also during repeated tests on the same sample. An overview of the results is reported in Fig. 5: no correlation with the porosity was observed, and an

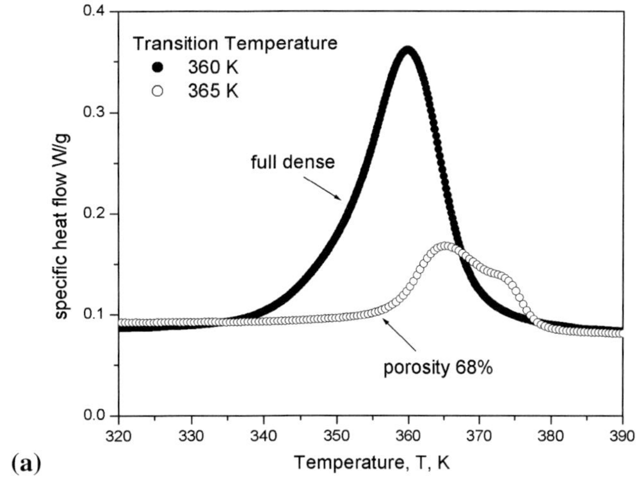

Fig. 6 Comparison between heating ramp of DSC analyses on bulk and porous NiTi (a) and corresponding thermal diffusivity (b)

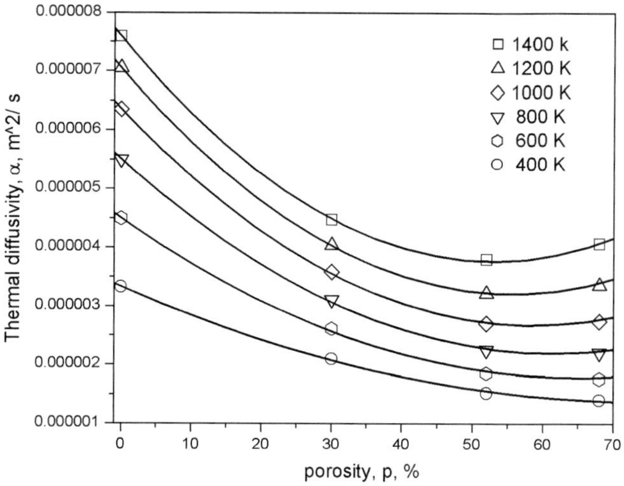

Fig. 7 Thermal diffusivity vs. porosity for different temperatures

average value around 6.0×10^{-6} and 3.7×10^{-6} for austenite and martensite respectively so derived is in good agreement with literature value for bulk material.

3.4 Thermal Diffusivity Evaluation

In the following, a general definition of thermal diffusivity is reported and the effective thermal diffusivity in the case of porous matter is explained.

$$\text{Thermal diffusivity: } \alpha(T) = K_c(T)/(C_p(T) * \rho)$$

where $K_c(T)$ is the thermal conductivity, $C_p(T)$ the heat capacity and ρ the density.

$$\text{Effective thermal diffusivity: } \alpha_e = \alpha_s + \alpha_f + \alpha_r$$

where α_s concerns the solid (NiTi), α_f the fluid (argon) and α_r the radiative transport phenomena inside the porous sample.

Generally, heat transfer phenomena, associated to the presence of gas (argon) in the pores, is negligible at low operating pressure (0.1 MPa). Also, no effect for the temperature increasing has to be taken into account when porosity was less than ~90%.

On the other hand, of more importance is the radiative transport phenomenon inside the porous samples. In fact, the heat transfer by radiation depends on the random, open-cell porosity distribution (Ref 9-11) and this influence become greater and greater with increased porosity and temperature. Thermal diffusivities were therefore estimated based on these considerations and in accord with the DSC data obtained for porous samples.

For what concerns DCS data a comparison between full dense and porous samples is reported in Fig. 6(a) (heating case), indicating that the peak value decreases as porosity increases. This occurrence has been taken into account carrying out simulations and results, presented in Fig. 6(b), were obtained. This figure illustrates that the effective thermal diffusivity of porous sample is lower with respect to the value of the full dense material but the minimum, reached close to the transition temperature, is essentially the same.

Finally, data obtained by the numerical computations, relevant samples characterized by different porosity values (30, 48 and 68%) are reported and compared with the full dense data in Fig. 7.

The general trend of the reported curves indicates that the α decreases with the porosity increase while it increases with the temperature augmentation.

Thermal diffusivity dependence on porosity indicates that, at low temperature (400 K), it decreases with the porosity augmentation, while in the case of high temperature (1400 K) this behavior change approaching the porosity value of 50%. As mentioned above, this occurrence can be explained considering that radiative transport phenomena depend on the random open-cell porous samples and their effect becomes larger and larger with porosity and temperature increase.

3.5 Mechanical Tests

The results of compression until failure, as expected, confirmed that increasing the porosity reduces the mechanical properties of the material: maximum stress values of about 350 MPa were reached for samples with approximately 30% porosity, while samples with over 50% porosity hardly achieved 100 MPa. This effect was in part expected, as increasing porosity results in a reduction of resistant cross-sectional area. Not only the degree of porosity but, of course, the morphology of the void influences the mechanical results: a more homogeneous distribution of voids could contribute to a

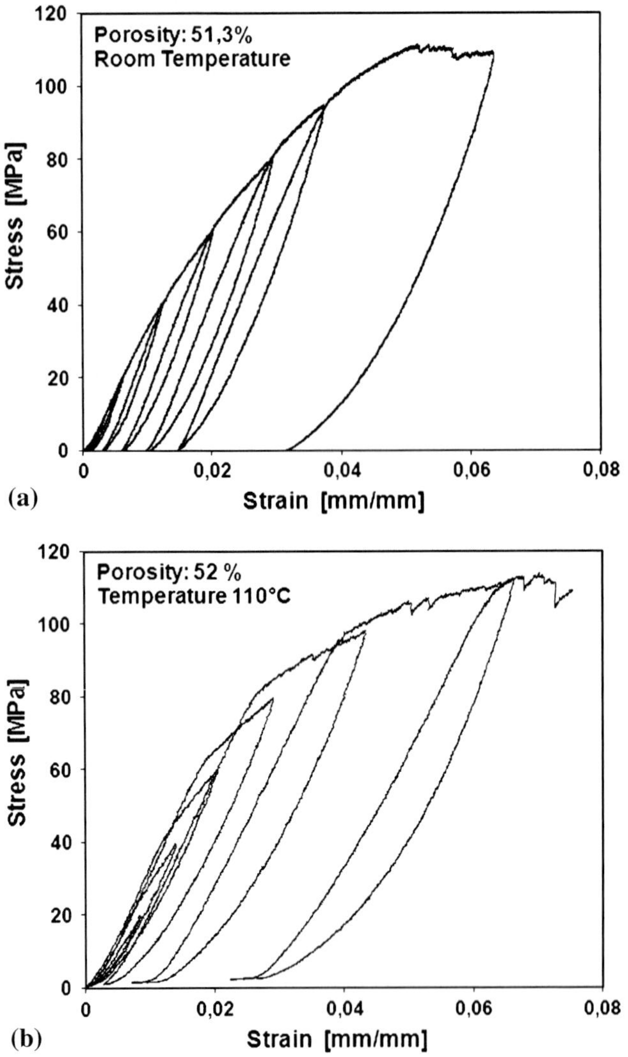

Fig. 8 Cyclic compression curves on two samples with same porosity at (a) room temperature and (b) 110 °C. In the latter condition, the sample exhibits pseudoelastic effect

better distribution of force on the section of the sample, with respect to samples with a localized concentration of voids, resulting in improved mechanical properties. Moreover, failure was observed to occur always as a localized collapsing on a plane roughly at 30° from horizontal. The previous assumption, together with the observation of localized deformation region in failed specimens, can justify some irregularities in stress-porosity dependence of mechanical behavior.

Samples with approximately 53% porosity were also tested with cyclic loading/unloading, at room temperature and at 110, 120 and 130 °C in order to check the presence of pseudo-elastic effects. Typical results are reported in Fig. 8(a) and (b). At room temperature, the specimens suffered progressive accumulation of deformation, with reloading curves copying the previous unloading curve, while at higher temperature a typical recovery during unloading was observed. The maximum recovery for each unloading cycle of about 0.5% is observed at 110 °C. At higher temperatures the strain recovery was reduced. This is reasonable because with increasing the temperature the permanent deformation of austenite is favored with respect to martensite-induced formation. Moreover, it should be noted that the NiTi matrix has not been subjected to any treatment with the aim of increasing the superelastic window.

4. Conclusions

SHS technique can be employed satisfactorily for producing NiTi porous sample with tailored porosity.

Calorimetric and dilatometric measurements show no dependence from degree of porosity, in agreement with microstructural observation, that reported common features in all samples. Mechanical properties were clearly affected by the degree of porosity; nonetheless, pseudo-elastic behavior was observed also for higher porosity.

Thermal diffusivity measurements on the contrary show a clear dependence not only on temperature, but also on porosity: at lower temperature thermal diffusivity decreases, due to reduction of material area available for conductive thermal transfer. At higher temperatures, up to 53% of porosity, reduction was observed, while at greater porosity thermal diffusivity increases due to the increasing contribute of radiative transmission through interconnected porosity inside the sample.

Further optimization of the SHS technique could be obtained with increased transforming fraction and further control over porosity size and distribution. As an SHS sample's porosity is related, among other parameters, to the initial porosity of the compacted powders, in order to obtain greater values of porosity looser powders need to be employed, with clear limitation in handling procedures. Optimization of SHS techniques applied to confined loose powders resulted in up to 68% porous NiTi samples. This technique could lead to the production of near net shape components for small to medium component application.

Acknowledgment

The authors thank Mr. G. Negro for technical support during experimental tests and numerical analyses.

References

1. S. Shabalovskaya, Surface, Corrosion and Biocompatibility Aspects of Nitinol as an Implant Material, *Biomed. Mater. Eng.*, 2002, **12**, p 69–109
2. A. Bansiddhi, T.D. Sargeant, S.I. Stupp, and D.C. Dunand, Porous NiTi for Bone Implants: A Review, *Acta Biomater.*, 2008, **4**, p 773–782
3. C. Zanotti, P. Giuliani, P. Bassani, F. Passaretti, and A. Tuissi, Characterization of Porous NiTi Alloys Produced by SHS, *Proceedings of SMST 2006 International Conference on Shape Memory and Superelastic Technologies*, B. Berg, M.R. Mitchell, and J. Proft, Ed., May 7-11, 2006 (Asilomar, Pacific Grove, CA), p 373–380
4. A. Tuissi, P. Bassani, and F. Passaretti, EBSD Microstructural Characterization of Ni50.8Ti49.2 Shape Memory Alloy, *Proceedings of SMST 2006 International Conference on Shape Memory and Superelastic Technologies*, B. Berg, M.R. Mitchell, and J. Proft, Ed., May 7-11, 2006 (Asilomar, Pacific Grove, CA), p 741–748
5. C. Zanotti, P. Giuliani, G. Riva, A. Tuissi, and A. Chrysanthou, Thermal Diffusivity of Ni-Ti SMAs, *J. Alloys Compd.*, 2009, **473**, p 231–237. doi:10.1016/j.jallcom.2008.05.040
6. Standard Test Method of Compression Testing of Metallic Materials at Room Temperature. ASTM (American Society for Testing and Materials) E9-89a (Reapproved 2000). ASTM International, 2000

7. Standard Practice for Compression Test of Metallic Materials at Elevated Temperatures with Conventional or Rapid Heating Rates and Strain Rates. ASTM (American Society for Testing and Materials) E209-00. ASTM International, 2000
8. B. Yuan, X.P. Zhang, C.Y. Chung, and M. Zhua, The Effect of Porosity on Phase Transformation Behavior of Porous Ti-50.8 at.% Ni Shape Memory Alloys Prepared by Capsule-Free Hot Isostatic Pressing, *Mater. Sci. Eng. A*, 2006, **438–440**, p 585–588
9. C.Y. Zhao, S.A. Tassou, and T.J. Lu, Analytical Considerations of Thermal Radiation in Cellular Metal Foams with Open Cells, *Int. J. Heat Mass Transf.*, 2008, **51**, p 929–940
10. M. Wang and N. Pan, Modeling and Prediction of the Effective Thermal Conductivity of Random Open-Cell Porous Foams, *Int. J. Heat Mass Transf.*, 2008, **51**, p 1325–1331
11. X. Huai, W. Wang, and Z. Li, Analysis of Effective Thermal Conductivity of Fractal Porous Media, *Appl. Therm. Eng.*, 2007, **27**, p 2815–2821

JMEPEG (2009) 18:600–602
DOI: 10.1007/s11665-009-9401-2

1059-9495/$19.00

Effect of Impact-Induced Strain on the Stress-Induced Martensitic Transformation of Superelastic NiTi Shape-Memory Alloy Wires

J. Zurbitu, G. Castillo, I. Urrutibeascoa, and J. Aurrekoetxea

(Submitted September 15, 2008; in revised form February 4, 2009)

The stress-induced martensitic (SIM) transformation of NiTi shape-memory alloy wires has been studied as a function of the maximum strain induced during tensile deformation at impact and quasi-static strain rates. The SIM transformation stresses are higher at impact than at quasi-static strain rates. Only the lower plateau strength is sensitive to the maximum strain achieved during transformation when this is higher than necessary to complete the SIM transformation. For the same maximum strain achieved, the deformation energy (E_d) and recoverable strain energy (E_r) are greater at impact than at quasi-static strain rates, and the dissipated energy (W_d) is slightly lower at impact, reaching values close to those obtained at quasi-static strain rates.

Keywords impact, mechanical testing, NiTi, polymer matrix composites, shape-memory alloys

1. Introduction

Due to its high damping capacity, the superelastic (SE) property of shape-memory alloys (SMAs) in the form of wires has been required to be used as reinforcement in composite materials for improving damage resistance and damage tolerance (Ref 1) or as components for seismic protection devices (Ref 2). The effects of the maximum strain achieved during the stress-induced martensitic (SIM) transformation at quasi-static loading strain rates on the SE property is known in terms of transformation stresses (Ref 3, 4) or dissipated and recoverable strain energy (Ref 4). Nevertheless, the fate of these features when the loading occurs at higher strain rates, that is, at impact strain rates, is unknown. The aim of this work is to study the effect of the maximum strain achieved during an impact event on the SIM transformation of SE-NiTi-SMA wires.

2. Experimental Procedure

For the experimental tests carried out in this work, a NiTi-SMA (56 wt.% nickel) has been selected. This is a commercially available alloy in the form of wire, 0.5 mm in diameter (ref. NT09), purchased from AMT n.v. Herk-de-Stad, Belgium.

This article is an invited paper selected from presentations at Shape Memory and Superelastic Technologies 2008, held September 21-25, 2008, in Stresa, Italy, and has been expanded from the original presentation.

J. Zurbitu, G. Castillo, I. Urrutibeascoa, and **J. Aurrekoetxea,** Mechanical and Industrial Production Department, Mondragon Unibertsitatea, Loramendi 4, 20500 Arrasate-Mondragón, Spain. Contact e-mail: jzurbitu@eps.mondragon.edu.

The austenitic finish temperature (A_f), determined by tensile tests carried out at different temperatures, is about 240 K; hence, this alloy is in the austenitic phase at room temperature and shows SE behavior under stress. These specimens have been subjected to a loading-unloading cycle at impact strain rates, achieving different maximum strains.

Instrumented tensile impact tests have been carried out to obtain the stress-strain curves at impact strain rates (of the order of 10 s^{-1}), with different energies to obtain the different maximum strains not only for incomplete, but also for complete SIM transformations (from 1% until failure occurs 11%). The experimental technique is described in detail in Ref 5. The sample is fixed between a mobile and a fixed grip. An impactor hits the mobile grip and the tensile force is transmitted to the sample. The force during the deformation is measured at the fixed grip by a piezoelectric sensor and the stress is calculated by dividing the impact force curve by the initial section of the wire. The displacement is calculated as the integration of the deformation velocity measured during the impact test with laser-based noncontacting measurement equipment and the strain is obtained by dividing this displacement by the initial length of the sample. Moreover, quasi-static tensile tests (at strain rates of 10^{-4} s^{-1}) for the same maximum strains have also been carried out to facilitate comparison of the results at low and high strain rates. These tests have been carried out in a uniaxial screw-driven testing machine Instron 4206. Displacement-controlled tension tests have been conducted by monitoring the force with a 50-kN load cell and measuring the strain by the crosshead displacement. Both quasi-static and impact experiments have been carried out at room temperature.

3. Results and Discussion

3.1 SE Stress-Strain Response Based on the Maximum Induced Strain

The SE behavior of NiTi is shown in Fig. 1 for different loading-unloading cycles in which different maximum strains

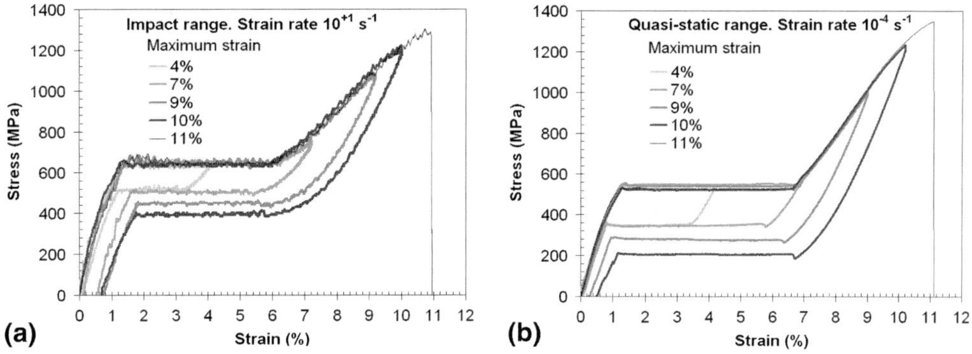

Fig. 1 Stress-strain response of NiTi obtained for different maximum strains. (a) At impact strain rates and (b) At quasi-static strain rates

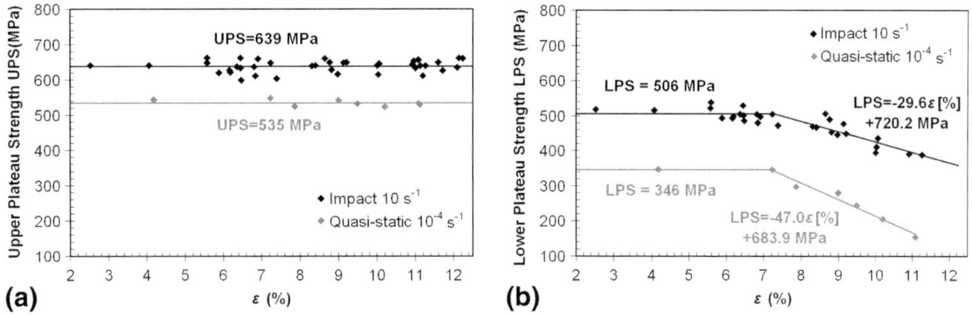

Fig. 2 Plateau strength as function of the maximum strain achieved during deformation of NiTi. (a) UPS and (b) LPS

are achieved. With the aim of studying and comparing the SIM transformation stresses as a function of the maximum strain achieved during deformation, it has been used for each test a specimen without any previous deformation. This avoids any accumulation of defects which are favorable for the following transformations, leading to a decreasing of the transformation stresses. The waved response shown in the impact curves is due to local oscillations in the stress-strain curve associated with the dynamic response of the impact test, which are superimposed on the mean response of the material.

Experiments carried out at impact strain rates are shown in Fig. 1(a), and in Fig. 1(b) tests at quasi-static strain rates are shown. In both cases, the appearance of plateau zones indicates that SIM transformation is observed during deformation, yielding incomplete transformations (4% of maximum strain), complete SIM transformations (7%), complete transformations and elastic deformation of the martensitic phase (9%), and failure of specimens (11%). The residual plastic strain in impact loaded specimens is slightly higher at impact than at quasi-static strain rates (Fig. 1). This may be attributed in part due to higher stresses developed during the impact tests, that generates more defects and causes more plastic deformation, but also due to a very small slippage of the specimen at the grips, difficult to avoid due to the dynamic nature of the impact experiment.

3.2 The Effect of Maximum Strain Achieved on the Transformation Stresses

The upper plateau strength (UPS), measured as the forward transformation stress at 3% strain during the loading process, and the lower plateau strength (LPS), measured as the reverse transformation stress at 2.5% strain during unloading are plotted as functions of the maximum strain achieved during deformation (Fig. 2). The local oscillations observed in the impact stress-strain curves, Fig. 1, contribute to a higher scatter data points at impact than at quasi-static strain rates in Fig. 2.

The UPS is around 20% higher at impact than at quasi-static strain rates (Fig. 2a) as a result of the self-heating process, which is due to the exothermic character of the forward SIM transformation. The LPS is also higher at impact but depends mainly on the maximum deformation achieved until the unloading process starts (Fig. 2b). At strains lower than 7%, when the SIM transformation is not complete, the LPS remains constant and is strain independent. However, when the impact energy is sufficient for completing the SIM transformation, at strains higher than 7%, this trend changes and the reverse SIM transformation stress diminishes linearly as the strain increases. This may be attributed to the high stresses generated during the loading process, which generates internal stresses that are favorable for reverse SIM transformation, similar to the stress reduction that occurs while cycling SE-NiTi (Ref 6, 7). In the present work, a similar trend has been found at quasi-static strain rates, with the difference that the stresses at impact are higher, and the rate at which the LPS decreases with strain is lower at impact strain rates (Fig. 2b). The quasi-static results shown here match well with those observed in reviewed literature (Ref 3, 4).

3.3 The Effect of Maximum Strain Attained on the Deformation, Dissipated and Recoverable Energies

The deformation energy (E_d), the energy required to deform the material until the maximum strain, obtained as the integration of the force during loading along the deformation

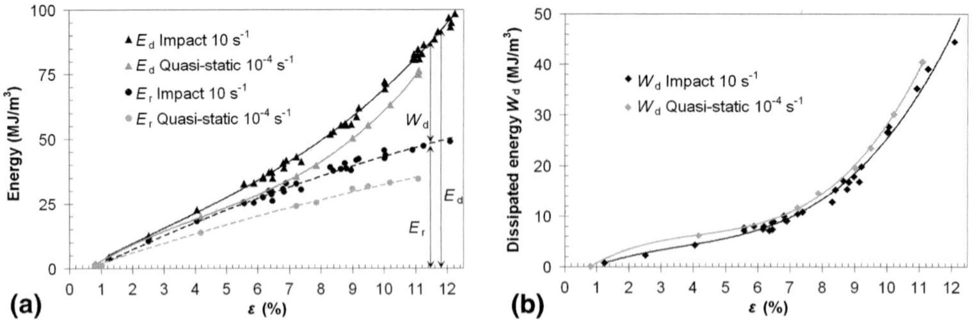

Fig. 3 Significant energies as function of the maximum strain achieved during deformation of NiTi. (a) Deformation energy (E_d) and recoverable strain energy (E_r). (b) Dissipated energy (W_d)

of the sample from zero to the maximum strain; and the recoverable strain energy (E_r), the energy recovered during unloading from the maximum strain until the zero stress state, as the integration of the force during unloading along the deformation of the sample from the maximum strain to zero, are plotted as function of the maximum strain achieved during deformation (Fig. 3a).

Each test is represented in the graph with its respective mark, and the superimposed smooth line shows the average tendency. Results from the impact tests show that E_d and E_r increase with the maximum strain reached and are higher at impact than at quasi-static strain rates, with average higher values at impact of 7.8 and 11 MJ/m^3 of SIM transformed material, respectively. Similarly, the dissipated energy (W_d), the difference between E_d and E_r, is plotted as a function of the maximum strain (Fig. 3b). W_d also increases throughout the strain range studied, but the value is slightly lower at impact strain rates. For the same maximum strain attained, the average difference between the impact and quasi-static values of W_d is 3.2 MJ/m^3 of SIM transformed material. This implies similar values of W_d at impact and at quasi-static strain rates for complete SIM transformations. Nevertheless, the difference may be higher in relative terms for incomplete SIM transformations.

4. Conclusions

This study addresses the effect of the maximum strain achieved at impact strain rates, compared with quasi-static strain rates, on the SE behavior of NiTi in terms of stress-induced martensitic transformation stresses; and deformation, recoverable and dissipated energies. The results are summarized as follows:

1. The UPS is higher at impact than at quasi-static strain rates because of self-heating due to the exothermic character of the forward SIM transformation.
2. The LPS is also higher at impact and depends on the maximum strain achieved during deformation. For strains lower than that necessary for completing the SIM transformation, LPS remains constant. Nevertheless, for higher values of strain, LPS diminishes linearly as the strain increases.
3. Deformation energy (E_d), recoverable strain energy (E_r), and dissipated energy (W_d) increase as the maximum strain achieved during deformation becomes greater.
4. E_d and E_r are higher at impact than at quasi-static strain rates, but W_d is lower. For the latter parameter, the results of the quasi-static and impact states are similar.

Acknowledgment

The authors thank the Basque Government for its financial support for the study (PI2008-07, IE05-150).

References

1. K.A. Tsoi, R. Stalmans, J. Schrooten, M. Wevers, and Y. Mai, Impact Damage Behaviour of Shape Memory Alloy Composites, *Mat. Sci. Eng. A Struct.*, 2003, **A342**(1–2), p 207–215 (in English)
2. M. Dolce and D. Cardone, Mechanical Behaviour of Shape Memory Alloys for Seismic Applications 2. Austenite NiTi Wires Subjected to Tension, *Int. J. Mech. Sci.*, 2001, **43**(11), p 2657–2677 (in English)
3. P.H. Lin, H. Tobushi, K. Tanaka, T. Hattori, and M. Makita, Pseudoelastic Behaviour of TiNi Shape Memory Alloy Subjected to Strain Variations, *J. Intel. Mat. Syst. Str.*, 1994, **5**(5), p 694–701 (in English)
4. H. Tobushi, K. Tanaka, T. Hori, T. Sawada, and T. Hattori, Pseudoelasticity of TiNi Shape Memory Alloy (Dependence on Maximum Strain and Temperature), *JSME Int. J. A Mech. M.*, 1993, **36**(3), p 314–318 (in English)
5. J. Zurbitu, L. Aretxabaleta, G. Castillo, I. Urrutibeaskoa, and J. Aurrekoetxea, Técnicas de Impacto Instrumentado Sobre Hilos de Aleaciones con Memoria de Forma (Instrumented Tensile Impact Techniques Applied to Shape Memory Alloy Wires), *X Congreso Nacional de Materiales*, 2008, **2**, p 1073–1076 (in Spanish)
6. S. Miyazaki, T. Imai, Y. Igo, and K. Otsuka, Effect of Cyclic Deformation on the Pseudoelasticity Characteristics of Ni-Ti Alloys, *Acta. Metall. Mater.*, 1986, **17**, p 115–120 (in English)
7. S. Miyazaki, Thermal and Stress Cycling Effects and Fatigue Properties of Ni-Ti Alloys, *Engineering Aspects of Shape Memory Alloys*, T.W. Duerig, K.N. Melton, D. Stoeckel and C.M. Wayman, Eds. (London), Butterworth-Heinemann Ltd, 1990, p 394–413

JMEPEG (2009) 18:603–611
DOI: 10.1007/s11665-009-9358-1

Factors Controlling Superelastic Damping Capacity of SMAs

L. Heller, P. Šittner, J. Pilch, and M. Landa

(Submitted October 26, 2008; in revised form December 2, 2008)

In this paper, questions linked to the practical use of superelastic damping exploiting stress-induced martensitic transformation for vibration damping are addressed. Four parameters, particularly vibration amplitude, prestrain, temperature of surroundings, and frequency, are identified as having the most pronounced influence on the superelastic damping. Their influence on superelastic damping of a commercially available superelastic NiTi wire was experimentally investigated using a self-developed dedicated vibrational equipment. Experimental results show how the vibration amplitude, frequency, prestrain, and temperature affect the capacity of a superelastic NiTi wire to dissipate energy of vibrations through the superelastic damping. A special attention is paid to the frequency dependence (i.e., rate dependence) of the superelastic damping. It is shown that this is nearly negligible in case the wire is in the thermal chamber controlling actively the environmental temperature. In case of wire exposed to free environmental temperature in actual damping applications, however, the superelastic damping capacity significantly decreases with increasing frequency. This was explained to be a combined effect of the heat effects affecting the mean wire temperature and material properties with the help of simulations using the heat equation coupled phenomenological SMA model.

Keywords shape memory alloys, superelastic damping, thermomechanical testing

1. Introduction

Taking advantage of the relatively large hysteresis, strain and stress amplitude as well as recentering capability, the superelastic deformation of shape memory alloys (SMAs) has been widely promoted as a phenomenon to be used for passive damping purposes (Ref 1-3). This superelastic damping capacity (SDC) of SMAs is commonly quantified based on the enclosed area of the superelastic stress-strain loop measured in quasistatic tensile test at room temperature. However, the relevance of such a quantification is questionable with regard to dynamic conditions the damping applications are subjected to. Many publications in the literature dealing with the high damping capacity of SMA are based on DMA studies of the internal friction damping performed at extremely low vibration amplitudes (Ref 4, 5). In reality, as everyone knows, the stresses and strains involved in SMA damping can be rather large and the dissipation energy through the stress-induced martensitic transformation (MT) may in fact reach even much higher values compared to the internal friction damping. In spite of the great potential of the superelastic damping, there seems to be only a limited knowledge of dynamic superelastic deformations of SMA wires subjected to dynamical excitations with higher strain amplitudes and frequencies. Most of existing literature deals with the superelastic hysteresis at rather low range of strain rates (Ref 6, 7). Although the literature focused on the engineering and applied mechanics provides studies of SMAs damping capacity in a wider range of amplitudes and frequencies (Ref 8), the quasistatic superelastic responses are still typically considered in modelling as well as in the design of passive damping applications based on SMA elements.

The present paper is aimed at analyzing the SDC of SMAs with particular respect to working conditions the SMA damper is subjected to in real applications. We consider a superelastic wire as the simplest representation of the SMA damper. For such a damper, we define four working conditions (see Fig. 1) that are supposed to have an important impact on the SDC. The first parameter represents the vibration amplitude that is basically given by the external load and geometry of the SMA damper. SMA dampers are installed with a predefined prestrain which is the second working condition. As the SMA behavior is strongly temperature dependent, the temperature of the wire surroundings is the third important parameter to be considered. The frequency of the external load to which the SMA damper is exposed represents the fourth condition one has to consider when designing a SMA damper. The influence of those four parameters on the SDC was analyzed by an experimental study carried out on a commercially available superelastic thin NiTi wire. The role of frequency (i.e. strain rate) is analyzed more deeply in the last part of this paper using a phenomenological SMA model coupled with a lumped-capacity heat equation. The rate dependence of the

This article is an invited paper selected from presentations at Shape Memory and Superelastic Technologies 2008, held September 21-25, 2008, in Stressa, Italy, and has been expanded from the original presentation.

L. Heller, P. Šittner, and **J. Pilch**, Institute of Physics, Academy of Sciences of the Czech Republic, Prague 18221, Czech Republic; and **M. Landa**, Institute of Thermomechanics, Academy of Sciences of the Czech Republic, Prague 18000, Czech Republic. Contact e-mail: heller@fzu.cz.

superelastic damping is finally rationalized and explained through simulations made with this model. It comes out of this analysis that the heat transfer through the wire surface (i.e. the ability of the environment to take the heat out of the wire) represents a fifth important condition which, however, was not particularly considered in present experimental investigations.

The working conditions introduced above represent a set of parameters within which one shall map the SDC of a given SMA damper in order to find optimal working conditions maximizing its damping efficiency as well as boundaries in the parameter space where the SDC becomes negligible. Results of such a mapping will help the potential designer of SMA damper who needs to take into account further requirements on the parameters additional to the maximal SDC of the SMA damper.

2. Experimental Identification of the Superelastic Damping

In order to realize experimentally the mapping of the SDC in the space of the four working conditions (vibration amplitude, frequency, prestrain, and temperature), a dedicated experimental equipment called AVUT (Fig. 2) was designed and built in the frame of AVALON project (Ref 8) focused on development of hybrid NiTi textiles with integrated thin NiTi wires. The equipment consists of an electromagnetic shaker mounted on a movable base that allows to prestrain the wire fixed between the shaker and the frame. The excitation provided by the shaker is measured by a piezoelectric accelerometer at the moving end of the studied wire where a load cell is also mounted ensuring the measurement of the mean force in the wire. The alternating

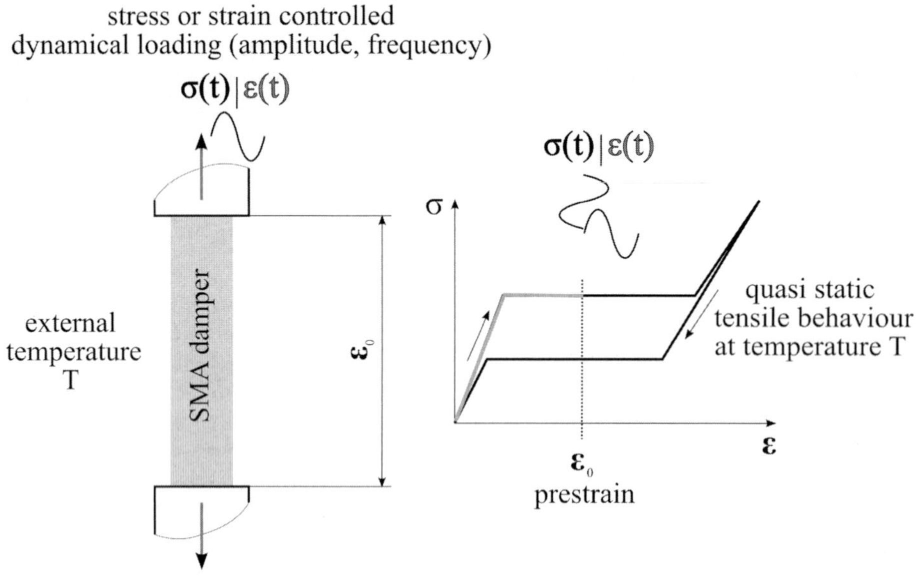

Fig. 1 Schematic representation of a simple 1D SMA damper and four working conditions to be considered in its design represented by the vibration amplitude, frequency, prestrain, and temperature

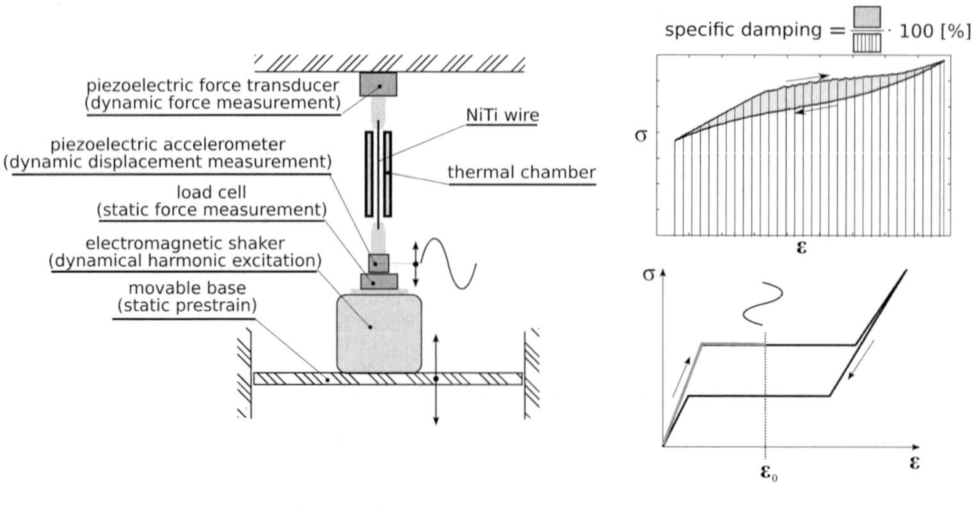

Fig. 2 Scheme with main components of the testing device for identification of the SDC and its dependence on working conditions. The upper right corner shows the definition of specific damping used for evaluating the damping capacity of the SMA wire

component of the force in the wire is measured through a piezoelectric force transducer attached to the fixed end of the wire. The temperature of the wire surroundings is controlled by a thermal chamber based on the Peltier elements. The system allows to measure the dynamic response of the SMA wire to a harmonic excitation with controlled strain amplitude, prestrain, frequency, and temperature. Three signals are simultaneously recorded in the experiments—the mean and alternating components of the force in the wire and the acceleration of the excited end of the wire. The experimental results in terms of stress-strain loop, its area, and corresponding specific damping (see Fig. 2 for its definition) are extracted from measured data.

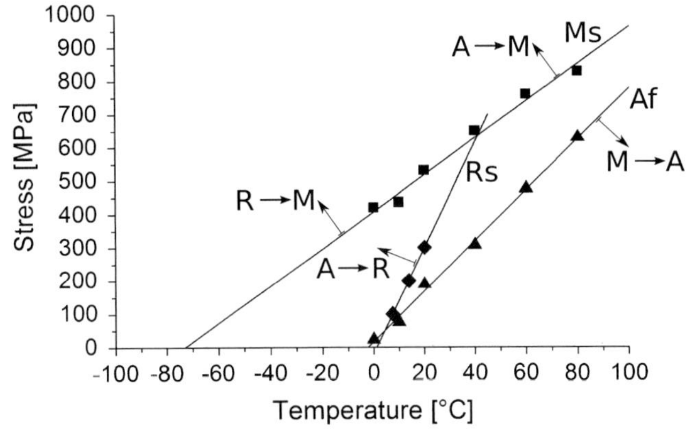

Fig. 3 Non-equilibrium stress-temperature diagram of the tested superelastic NiTi wire where the transformation lines represent the temperature dependence of transformation stresses for austenite to martensite, martensite to austenite, and austenite to R-phase transformation

3. Effect of Working Conditions on the Superelastic Damping

The AVUT equipment was used to investigate the effect of the working conditions on the SDC of a commercially available superelastic NiTi wire (55 wt.% Ni) of diameter 100 μm. The wire supplied in already heat-treated conditions was first analyzed and characterized under quasistatic loadings using the approach described in detail in Ref 9. This approach led to a quantification of thermomechanical properties of the wire via evaluation of 23 material parameters (selected ones are listed in Table 1) and construction of the non-equilibrium stress-temperature diagram (Fig. 3). As a specific property of this particular wire, one can notice that the superelastic hysteresis decreases with increasing temperature (as evidenced by the converging M_s and A_f transformation lines in the stress-temperature diagram).

The dynamical testing of the superelastic NiTi wire was performed in such a way that a virgin sample of gauge length 50 mm was excited at given working conditions until its failure. A characteristic wire response in terms of evolution of displacement and specific damping in case of test in which the strain amplitude, frequency, and environment temperature was prescribed is shown in Fig. 4. One can see that the wire response is dramatically changing in the first 2 s (~100 cycles) in a region denoted "unsteady vibrations". Later on in the stable vibration region, however, the SDC evolves as well although in much lesser extent. There are basically two factors contributing to the evolution of damping response with the number of cycles. These are the heat generated/absorbed by the vibrating wire and microstructural changes in the wire leading to variation of its stress-strain response.

The specific damping capacity (see Fig. 2 for its definition) decreases in the unsteady vibration stage down to a characteristic value (~12% in Fig. 4) as the wire vibrations settle into

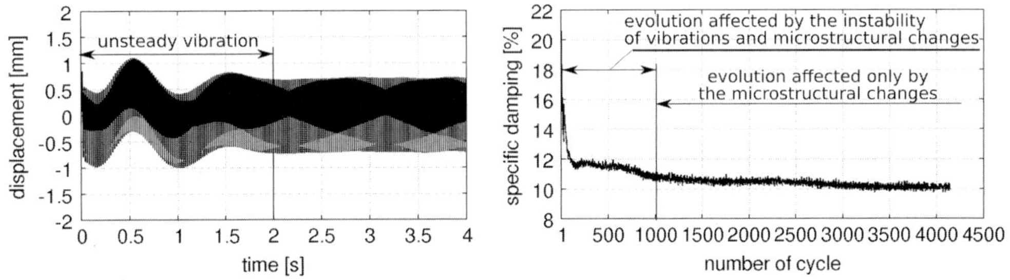

Fig. 4 Vibration test in which the strain amplitude (2%), frequency (50 Hz), and environment temperature (295 K) was prescribed. On the left-hand side, a typical time response in terms of displacement of the excited wire tip is shown with depicted unsteady stage of vibrations; on the right-hand side, a characteristic evolution of specific damping with number of cycles is given

Table 1 Selected material parameters of the tested wire obtained from quasistatic thermomechanical testing

NiTi 100 μm													
M_s', °C	A_f', °C	R_s, °C	$S_{A\text{-}M}$, MPa/°C	$S_{A\text{-}R}$, MPa/°C	$S_{M\text{-}A}$, MPa/°C	σ_{UTS}, MPa	σ_Y, MPa	σ_{tr}, MPa	Δh_σ, MPa	ε_{tr}, %	ε_f, %	E_m, GPa	E_a, GPa
−72	0	−11	5.7	16.5	7.5	1600	1330	550	340	5.2	13	21	54

M_s'/A_f' effective martensite/austenite start/finish temperature; R_s R-phase start temperature; $S_{A\text{-}M}/S_{A\text{-}M}/S_{A\text{-}R}$ temperature dependences of transformation stress for A-M/M-A/A-R; σ_{UTS} ultimate tensile strength; σ_Y yield stress; σ_{tr} transformation yield stress of austenite at RT; Δh_σ superelastic stress hysteresis at RT; ε_{tr} maximum recoverable transformation strain; ε_f strain at failure; E_a/E_m Young modulus of austenite/martensite the steady state meaningful for the damping applications. Since this is observed on the virgin wire as well as on the already cycled wires, we believe this unsteady region is mainly due to the heat effects as will be discussed below. The cyclic tensile loading is, however, also accompanied by microstructural changes taking place in the SMA wire which are known to be most pronounced at the early stage of cycling. Both effects thus affect the SDC particularly in the unsteady vibration range and can be hardly distinguished. The effect of microstructural changes on the damping capacity of the SMA wire can, however, be very well identified in the steady vibrations range (see Fig. 4), where the heat effects are completely stable. In order to evaluate the effect of the four parameters on the SDC of the studied NiTi wire, the specific damping and energy dissipation within one period of motion in the steady vibration range at the 1000th cycle were taken as characteristic values.

In this paper, we basically attempt to demonstrate the importance of the effect of the used working conditions when applying the superelastic damping as a damping mechanism in passive damping elements. Hence, the SDC of the studied wire was mapped in the space of the four parameters using a simplified approach supposing that a change in one working parameter does not affect the influence of other parameters on the SDC. This is not correct for many combinations of parameters. However, we have tried to select parameters in the ranges where this is approximately true. Therefore, the influence of one working parameter was studied by varying its value around a basic level and measuring the corresponding SDC while keeping the remaining working parameters constant. The basic level of all four parameters was chosen as follows: prestrain 4% in the middle of the plateau, amplitude 1.2%, temperature 20 °C, frequency 50 Hz.

3.1 The Influence of Prestrain

SMA elements used in dampers are usually installed with a defined prestrain for two main reasons. In case of tension, the prestrains avoid the compressional load that would lead to buckling and damage. In addition, when a superelastic SMA element is prestrained to a partially transformed state, the strain amplitude necessary to initiate the superelastic damping mechanism and hence significant damping is much lower than in the case of an unprestrained SMA element. However, the level of prestraining affects presumably the SDC. This expected effect was studied by measuring the SDC at four different levels of prestrain distributed along the plateau as schematically shown in Fig. 5. Corresponding stress-strain loops measured at 1000th cycle are displayed in Fig. 6.

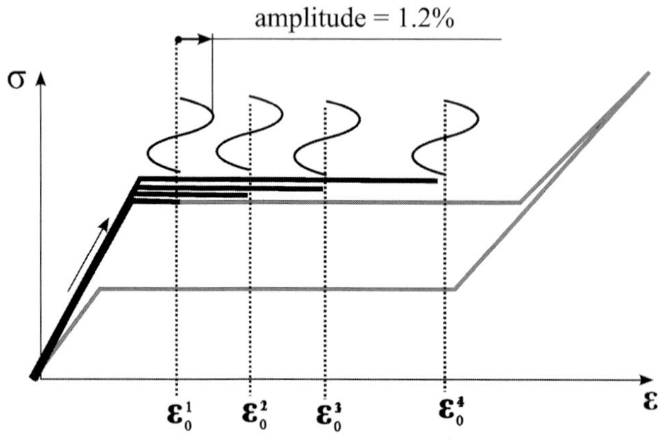

Fig. 5 Schematic representation of four prestrains at which the SDC was evaluated

As one can notice in Fig. 6, the stress-strain loop area decreases with increasing prestrain that is more evident when plotting directly the loop areas (Fig. 7) and specific damping versus prestrain (Fig. 8). The dissipated energy per cycle is very low in the elastic range, increases with increasing prestrain reaching a local maximum at the first third of the plateau range, and decreases as much as 60% toward the end of the plateau. The specific damping capacity shows basically a similar trend.

3.2 The Influence of Vibration Amplitude

A vibrating prestrained SMA element always deforms partially elastically and partially due to stress-induced MT

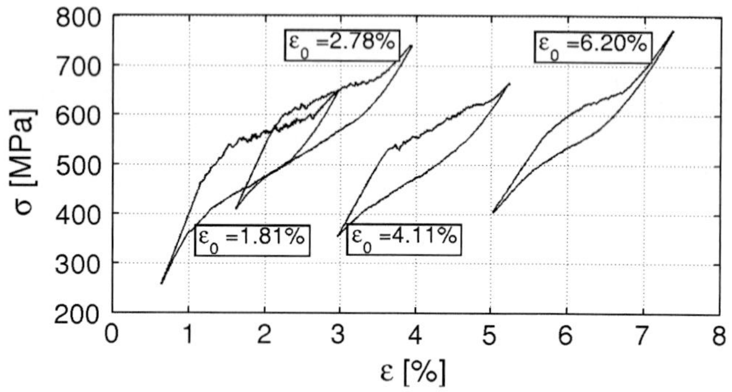

Fig. 6 Stress-strain loops measured at four prestrains while keeping the other working conditions at the basic level, i.e. amplitude 1.2%, temperature 20 °C, and frequency 50 Hz

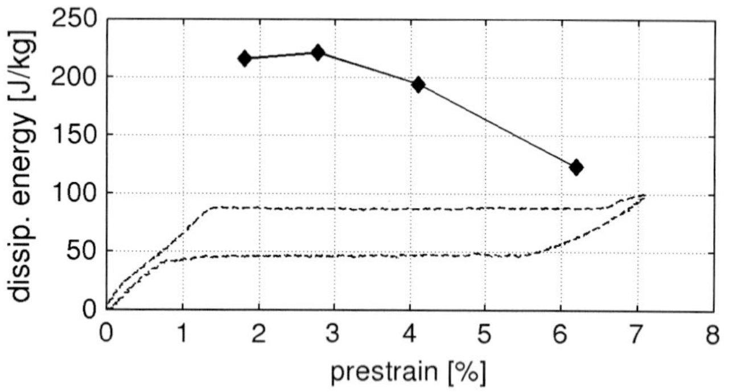

Fig. 7 Plot of dissipated energy vs. prestrain

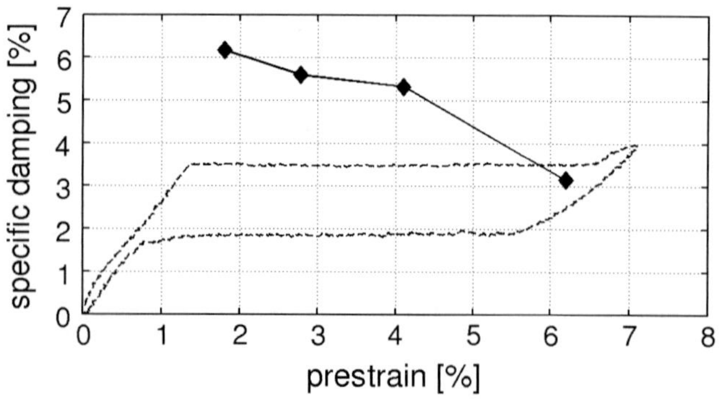

Fig. 8 Plot of specific damping vs. prestrain

responsible for damping. With increasing vibration amplitude, the fraction of the latter and consequently the damping capacity increases. However, in the lower end, the mechanical excitation has to be sufficiently large in order to give rise to deformation beyond this elastic range. In other words, if the superelasticity is being used for damping purposes, the amplitude of vibrations has to be at least slightly higher than the elastic deformation the prestrained SMA element is capable of at the given prestrain, frequency, and temperature. Although this is generally known, it is violated in many SMA damping applications, since the designers tend to decide for lower amplitudes due to the fatigue constraints.

At any circumstances, knowledge of the evolution of the SDC with increasing vibration amplitude is important for the efficient use of SMAs in damping applications. We evaluated the effect of increasing strain amplitude on the SDC for five different strain amplitudes as illustrated in Fig. 9 and 10. The transition from almost purely elastic deformation to the superelastic one when increasing vibration amplitude is clearly seen on the evolution of stress-strain loops shown in Fig. 10. Such a transition is reflected by an exponential increase of the dissipated energy with increasing strain amplitude (see Fig. 11) which saturates beyond the 3% level. The corresponding specific damping evolution increases as seen in Fig. 12 and reaches maximum at 3%. Minimum strain amplitude of 0.5% is needed for obtaining a reasonably high superelastic damping from the present wire at the given prestrain, frequency, and temperature.

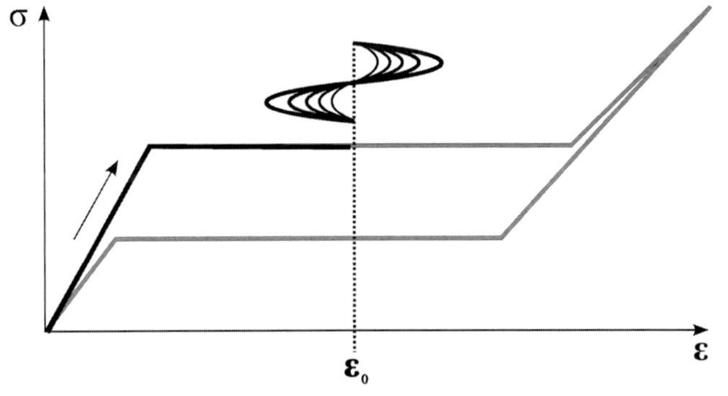

Fig. 9 Schematic representation of five amplitudes at which the SDC was evaluated

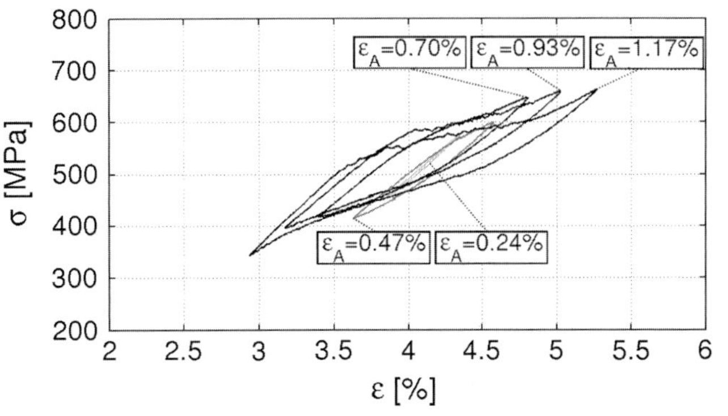

Fig. 10 Stress-strain loops measured at five amplitudes while keeping the other working conditions at the basic level, i.e. prestrain 4%, temperature 20 °C, and frequency 50 Hz

4. The Influence of Temperature

The temperature dependence of SMA behavior is generally assumed in term of Clausius-Clapeyron relation that is, however, directly applicable only to a single crystal specimen. According to this relation, no temperature dependence of the superelastic hysteresis shall be expected. However, when dealing with polycrystalline material, the Clausius-Clapeyron relation represents a certain approximation. As it was mentioned above, the stress-temperature diagram of the studied wire shows the M_s and A_f transformation lines that are not parallel as predicted by Clausius-Clapeyron relation but converge one to the other, i.e. the stress hysteresis decreases with increasing temperature. Hence, the temperature can be considered in some cases as a relevant working condition influencing the SDC. In the present work, the sensitivity of the SDC to temperature changes was analyzed in a range between 10 and 65 °C. At each temperature, the wire was prestrained to the middle of the plateau, i.e. the prestrain was not constant but linearly increasing with increasing temperature to compensate the increase of the plateau stress as illustrated in Fig. 13. The stress-strain loops for different temperatures obtained under dynamic loading (see Fig. 14) reveal clearly the tendency to decrease the hysteresis with increasing temperature as in the case of quasistatic tensile loading (see stress-temperature diagram in Fig. 3).

The dramatic effect of the temperature on the SDC is evidenced by temperature evolutions of the dissipated energy and specific damping shown in Fig. 15 and 16. One can notice that the temperature change from 10 to 65 °C is accompanied by a decrease of the dissipated energy per cycle and specific

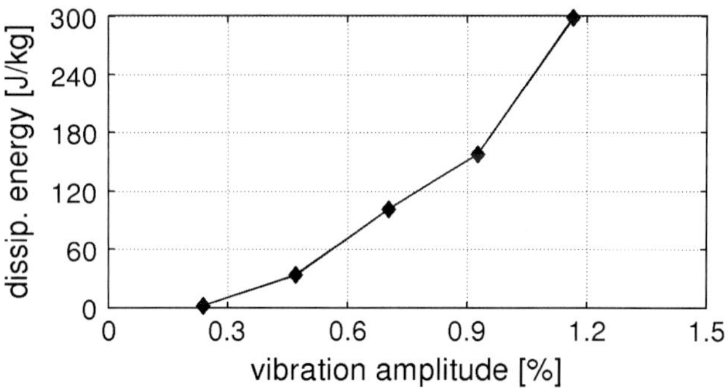

Fig. 11 Plot of dissipated energy vs.vibration amplitude

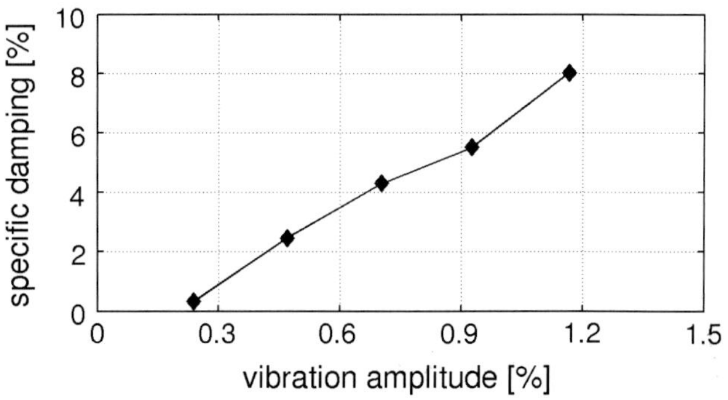

Fig. 12 Plot of specific damping as a function of the vibration amplitude

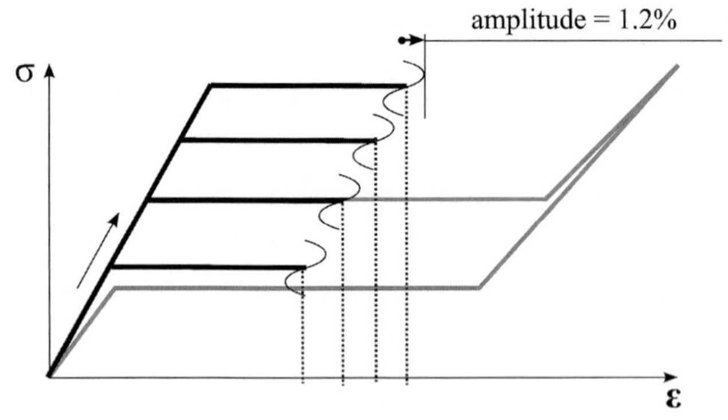

Fig. 13 Schematic representation of stress-strain curves corresponding to prestrains at four temperatures at which the SDC was evaluated

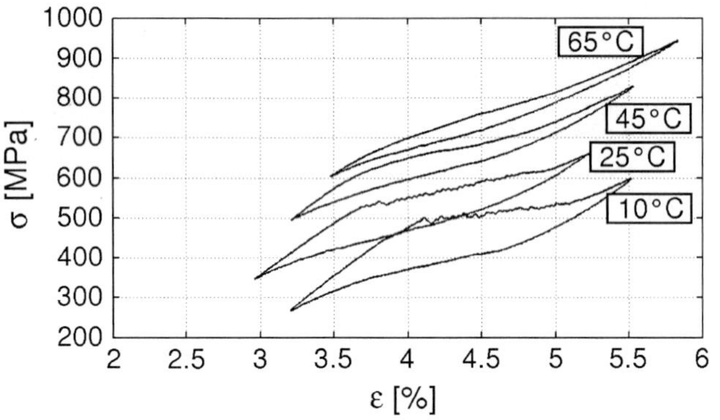

Fig. 14 Stress-strain loops measured at four temperatures while keeping the other working conditions at the basic level, i.e. prestrain 4%, amplitude 1.2%, and frequency 50 Hz

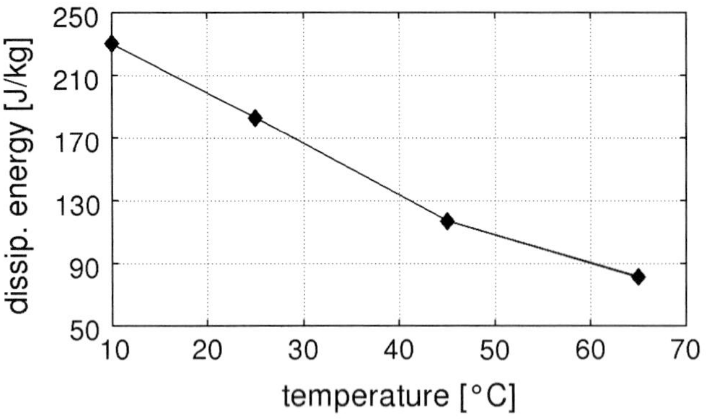

Fig. 15 Plot of dissipated energy vs. temperature of surroundings

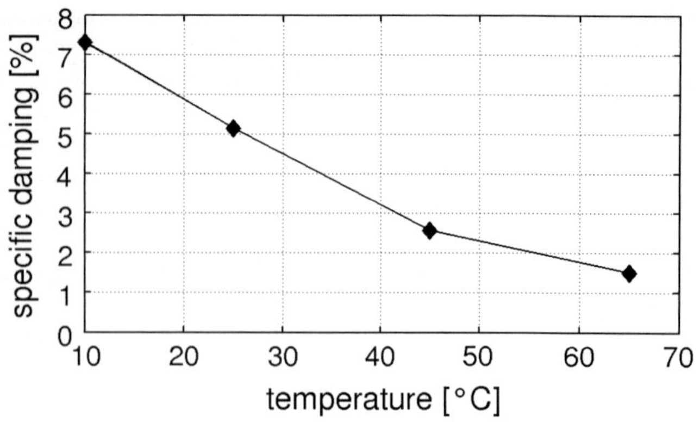

Fig. 16 Plot of specific damping as a function of temperature of surroundings

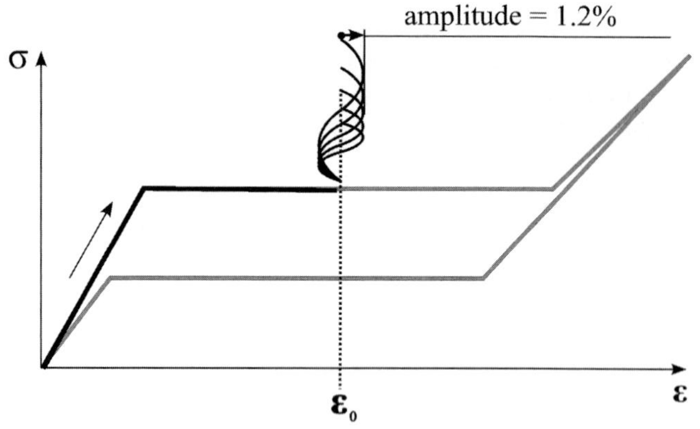

Fig. 17 Schematic representation of five frequencies at which the SDC was evaluated

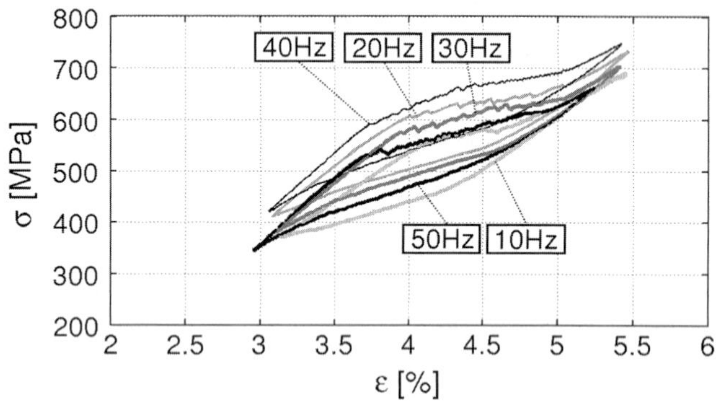

Fig. 18 Stress-strain loops (1000th) measured at five frequencies while keeping the other working conditions at the basic level, i.e. prestrain 4%, amplitude 1.2%, and temperature 20 °C

damping as large as 65 and 85%, respectively. However, since the heat and microstructural effects may also be involved in this decrease, we do not expect it to be solely due to the above discussed decrease of the stress hysteresis with increasing temperature.

5. The Influence of Frequency

SMA dampers are commonly considered for damping low frequency vibrations. Can we use it for higher frequencies as well? How the frequency affects the SDC and where is the high frequency limit? The frequency is yet another working condition with respect to which the SDC of SMA elements should be evaluated. We investigated the SDC of the present wire in the frequency range 10-50 Hz (Fig. 17). The results are shown in Fig. 18-20. The stress-strain loops do not display substantial differences in this frequency range as can be seen in Fig. 18. Neither the stress-strain loop areas nor the damping capacity seems to be significantly affected by the frequency change in this frequency range.

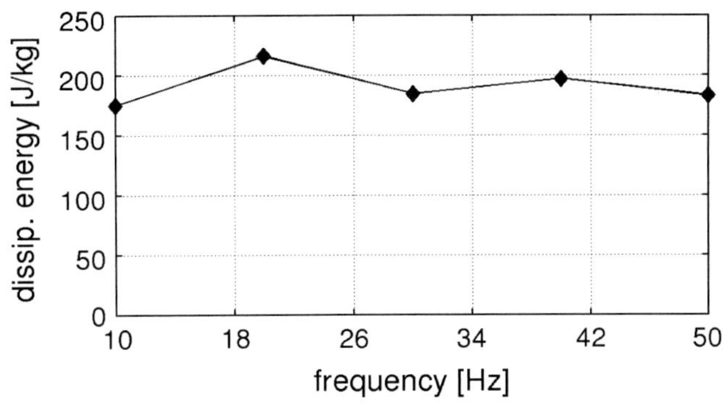

Fig. 19 Plot of dissipated energy vs. frequency

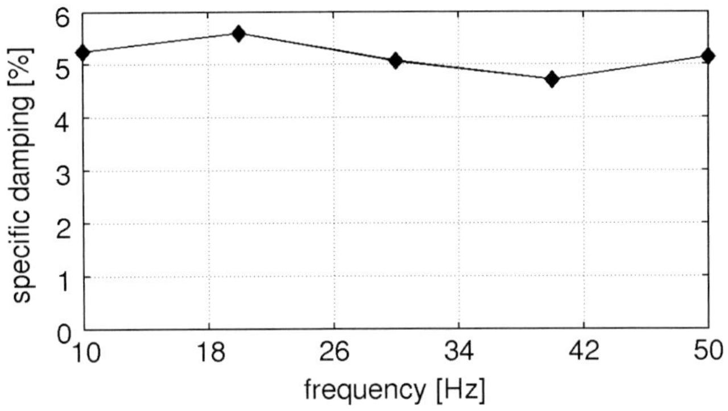

Fig. 20 Plot of specific damping as a function of frequency

6. The Role of Heat Effects in the Damping Capacity of SMAs

As reported above, the increasing frequency does not seem to affect the SDC which suggests that the superelastic damping is rate independent. However, such a conclusion is in clear contradiction with the results obtained on the same wire (Fig. 21) we recently published in Ref 9. In this paper, the superelastic damping capacity was investigated using the same experimental equipment except for the thermal chamber that was not used. The wire was simply exposed to the ambient temperature environment. In that case, a dramatic decrease of dissipated energy with increasing frequency was found. The fact that the thermal conditions of the wire surroundings influence the sensitivity of superelastic damping to the frequency (deformation rate) suggests that the thermal effects accompanying the MT (heat exchange with the environment) may be responsible for such a sensitivity. The mechanism how this happens, however, was not clear.

7. Modelling of the Heat Effects

In order to find out in which way the heat effects influence the superelastic damping of NiTi wire in tension, an earlier proposed phenomenological SMA model RLOOP (Ref 10) has been coupled with a lumped-capacity heat equation and used to simulate the fast superelastic deformation in Ref 11. The RLOOP model was originally developed (see Ref 10 for a detailed description) to simulate the key SMA responses to

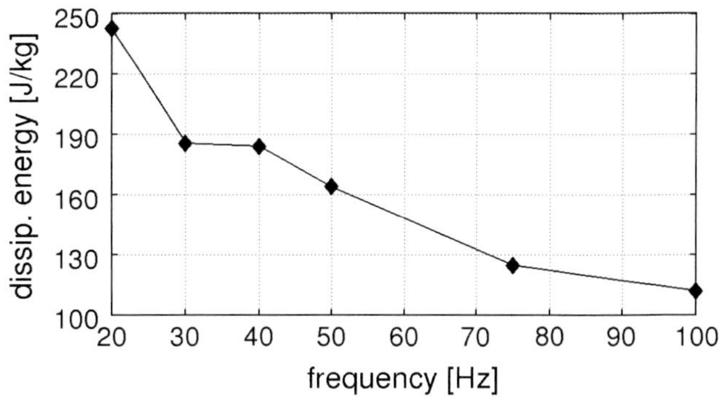

Fig. 21 Graph of dissipated energy vs. frequency published in Ref 9. The curve was measured on the same NiTi wire as the one analyzed in this paper. The measurement was performed at the room temperature without the use of the thermal chamber. The other measurement conditions were set up as follows: prestrain 4%, amplitude 1.2%, and frequency 50 Hz

general thermomechanical cyclic loads in tension, as e.g. actuator or recovery stresses generated in SMA composites. It assumes the existence of a general thermodynamic driving force for the MT when constructing a governing kinetics equation for the evolution of the martensite fraction ζ. The martensite phase fraction rate, $\dot{\zeta}$, is assumed to be a function of the driving force and its rate. The important features of the RLOOP algorithm are the rate independence of the simulated stress-strain-temperature responses and temperature independent hysteresis. The coupling of the RLOOP with the heat equation was done in such a way that the temperature T originally assumed as an independent parameter becomes an unknown variable coupled with the algorithm through the heat equation (see Ref 11 for details).

The applied lumped-capacity heat equation is based on the assumption of a homogeneous temperature field. Such an assumption is in contradiction to the commonly observed shear band propagation where the latent heat exchange occurs locally (Ref 12). Nevertheless, as shown in Ref 11, even such a simplified approach seems to capture the phenomena observed experimentally during superelastic cycling. Assuming a homogeneous temperature field, the wire temperature T in the RLOOP model becomes a variable governed by the heat equation. For a wire undergoing superelastic cycling at ambient environment with a constant temperature of the surrounding air $T_{\rm ext}$, the heat equation has to account for three most important heat sources—(i) the latent heat absorption/generation due to the stress induced MT, (ii) the heat generation due to the friction, and (iii) the convective heat exchange between the wire and its surrounding (see Fig. 22, Eq 1).

$$c_v\dot{T} = q_{\rm lat} + q_{\rm conv} + q_{\rm fric}$$

$$q_{\rm lat} = T\Delta s\dot{\zeta},\quad q_{\rm conv} = -\frac{h}{\rho L}(T T_{\rm ext}), \qquad \text{(Eq 1)}$$

$$q_{\rm fric} = A\dot{\zeta}$$

The latent heat flow $q_{\rm lat}$ is defined as a product of the martensite fraction rate $\dot{\zeta}$, wire temperature T and the difference between the specific entropies of austenite and martensite Δs at a referential temperature. The heat flow $q_{\rm fric}$ expresses the heat assumed to be generated during the transformation process additionally to the latent heat. It is called "friction" and considered to take place during both the forward and reverse

MT. Due to lack of information about the friction heat flow, it is expressed as a linear function of the rate martensite fraction change. Although the heat exchange by convection q_{conv} represents a complex heat transfer mode, it is expressed in a commonly used way using the heat transfer coefficient h, the wire length L, and the temperature difference with respect to the temperature of surrounding T_{ext}. The wire is supposed to have a volume V, the specific mass ρ and the specific heat capacity c_v that is assumed to be the same for both phases. The model captures nearly all the apparently curious phenomena observed experimentally—i.e. (i) the variation and stabilization of the superelastic loop within the unsteady vibration range (Ref 11), (ii) the rate dependence of the hysteresis at strain rates for which the thermal conditions of cycling pass from isothermal to adiabatic (Ref 13).

The heat equation coupled RLOOP model was used to simulate the effect of the excitation frequency on the superelastic damping at ambient environment (Fig. 21). The heat convection considered in the heat equation allows only for limited heat transfer across the wire interface. The parameters of the heat equation that were used in the simulation of heat effects are listed in Table 2. The geometrical parameters of the wire were chosen according to the experimentally tested

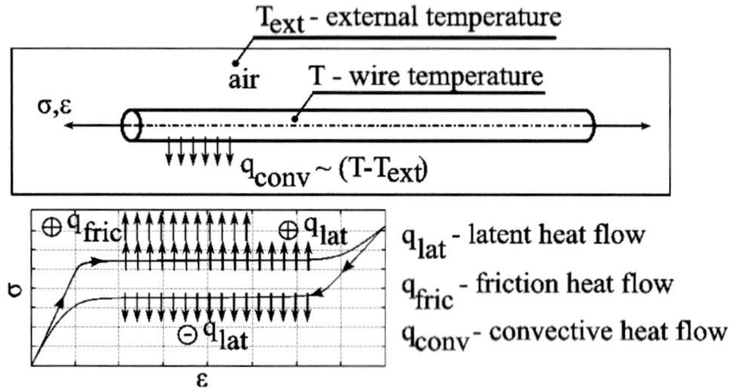

Fig. 22 All assumed heat processes taking place during the superelastic cycling

specimen, i.e. diameter 0.1 mm, gauge length 50 mm. The coefficient A in Eq 1 is computed according to Eq 2 based on the assumption that the hysteresis area of the stabilized superelastic loop at isothermal conditions ($h \rightarrow \infty$) represents the integral friction heat within one cycle (no energy dissipated to defects is assumed).

$$A = \frac{1}{\rho} \frac{\oint_{1\,\mathrm{cycle\ at}\ T=T_{\mathrm{ext}}} \sigma \dot{\varepsilon} dt}{\oint_{1\,\mathrm{cycle\ at}\ T=T_{\mathrm{ext}}} \dot{\zeta} dt} \qquad \text{(Eq 2)}$$

The simulation results are shown in Fig. 23. The model predicts that although the environmental temperature is constant, the temperature of the wire T evolves periodically during a single cycle (the left-hand side of Fig. 23) and the mean temperature increases toward stabilization with increasing number of cycles (the right-hand side of Fig. 23).

It should be noticed that the difference between the environment temperature and the stabilized value of the mean temperature of the wire dramatically increases with increasing frequency. The stabilization itself is due to the tendency of the system toward a heat balance over one cycle. It spreads into many cycles (~50-100) due to the limited constant rate of the heat exchange by convection requiring more cycles to reach the heat balance over one cycle (Ref 11). The simulated frequency dependence of the dissipated energy and SDC is shown in Fig. 24. Surprisingly, the model does not predict the experimentally observed decrease of the SDC with increasing frequency. One possible explanation is that the RLOOP simulation does not reflect the converging M_s A_f lines (Fig. 3) and thus cannot simulate the effect of temperature on the energy dissipation. However, when the simulation results are recalculated taking into account both the experimentally identified temperature dependency of the dissipated energy (Fig. 15) and the mean wire temperature evolution with respect to frequency, one obtains an exponentially decreasing dissipation energy with increasing frequency (Fig. 24) as it was observed experimentally in Ref 9 (Fig. 21). Apparently the decrease of the SDC with increasing frequency may be partially

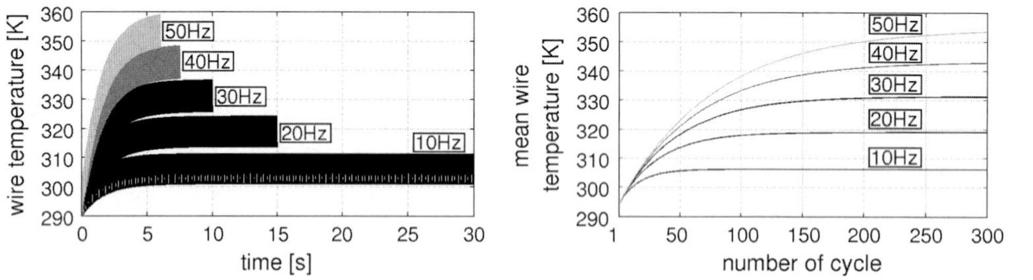

Fig. 23 Plot of the instantaneous wire temperature variation during cycling (left-hand side) and the mean wire temperature over one cycle vs. number of cycle (right-hand side)

Table 2 Parameters of RLOOP model and heat equation

M_s, K	E_a, GPa	σ_{re}, MPa	s, MPa/°C	G, 1/K	m	$k1$	c_v, J/kg/K	ρ, kg/m³	T_{ext}, K	A_s, K	E_m, GPa	ε_{tr}, %	C, MPa/K	l	n	$k2$	Δs, J/kg/K	h, W/m²/K
210	50	100	5.5	0.25	1	1	326	6450	293	250	21	5.2	50	1.1	5	1	54	35

M_s/A_s martensite/austenite start equation; E_a/E_m Young modulus of austenite/martensite; σ_{re} reorientation stress below M_f; s slope of the temperature dependence of the transformation start stresses; ε_{tr} transformation strain; G, c, l, m, n, $k1$, $k2$ fitting parameters; c_v specific heat capacity; Δs difference between specific entropies of austenite and martensite at referential temperature; h heat transfer coefficient; T_{ext} temperature of surroundings

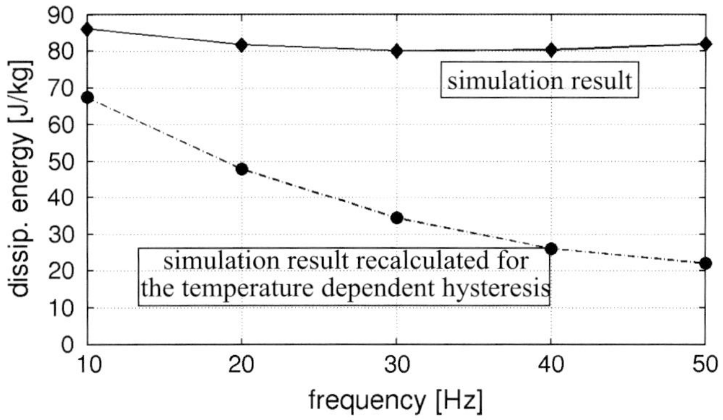

Fig. 24 Frequency evolution of dissipated energy predicted by simulation and recalculated using the experimentally identified temperature dependence of dissipated energy shown in Fig. 15

due to the combination of the temperature and frequency effects. For the discussion of the low frequency maximum see Ref 11, 13.

On the other hand, such a dramatic increase of the mean wire temperature with increasing frequency cannot be expected in case of the present experiment in which the temperature of the wire surroundings is actively controlled by the thermal chamber. The thermal chamber represents a heat sink absorbing the heat much more efficiently though the conditions are still very far from isothermal in the sense that the temperature variations within onc cycle are still present.

The simulation and experimental results presented above thus lead us to a conclusion that the exponential decrease of the dissipated energy with frequency (Fig. 21) obtained without an active thermal control of surroundings is most likely due to the combined effect of the heat effects and material properties showing converging M_s and A_f lines (Fig. 3). In other words, as the stabilized mean wire temperature increases with frequency due to the heat effects, the dissipated energy in one cycle decreases due to the temperature dependence of the superelastic hysteresis being a specific property of the studied wire. Such a property was experimentally evidenced in the quasistatic regime by the constructed stress-temperature diagram (Fig. 3) and in the dynamic regime by the identified temperature dependence of the dissipated energy (Fig. 15). If the temperature of the wire is actively controlled by the environmental chamber, the dramatic increase of the mean temperature of the wire in the stabilized regime does not occur and the effect of increasing frequency on the dissipated energy and SDC is negligible. Nevertheless, since the SMA dampers are not commonly put in thermal chambers and wires thicker than 0.1 mm are commonly used, it is essential to realize the heat transfer across the wire is in fact a very important fifth condition affecting the superelastic damping of SMA elements.

8. Conclusions

The superelastic damping of NiTi wires was investigated with regard to their use in passive damping applications. Four parameters—vibration amplitude, prestrain, temperature of surroundings and frequency—were identified as having the most pronounced influence on the superelastic damping. The influence of those parameters on the superelastic damping of a commercially available superelastic NiTi wire was experimentally investigated using a self-developed dynamical tester. The experimental investigation was carried out assuming that the influence of each of the four considered parameters is independent of the variation of the others. Since this is not true generally, a limited set of parameters around basal conditions corresponding to the prestrains within the plateau range, strain amplitudes $<2\%$, $T = 20\text{-}65$ °C, and frequencies 10-50 Hz.

The superelastic damping capacity of a commercial NiTi wire (d-0.1 mm)

- decreases by as much as 60% with increasing prestrain within the plateau range
- increases dramatically with increasing strain amplitude, a minimum vibration amplitude of 0.5% is required to initialize the superelastic damping
- decreases by as much as 60% with increasing temperature in the interval 10-65 °C. This was attributed to the temperature dependent stress hysteresis.
- remains approximately constant with increasing frequency in the interval 10-50 Hz in case the wire is in the thermal chamber controlling actively the environmental temperature. In case of free wire (no thermal chamber) used in actual damping applications, the superelastic damping capacity, however, significantly decreases with increasing frequency. This was explained as a combined effect of the heat effects affecting the mean wire temperature and material properties with the help of simulations using the heat equation coupled phenomenological SMA model.

References

1. Y. Liu, Z. Zeliang Xie, and J. Humbeeck, Cyclic Deformation of NiTi Shape Memory Alloys, *Mater. Sci. Eng. A*, 1999, **273–275**, p 673–678
2. D. Wolons, F. Gandhi, and B. Malovrh, Experimental Investigation of the Pseudoelastic Hysteresis Damping Characteristics of Shape Memory Alloy Wires, *J. Intell. Mater. Syst. Struct.*, 1998, **9**, p 116–126
3. F. Gandhi and G. Chapuis, Passive Damping Augmentation of a Vibrating Beam Using Pseudoelastic Shape Memory Alloy, *J. Sound Vib.*, 2002, **250**(3), p 519–539
4. F. Mazzolai, A. Biscarini, B. Coluzzi, G. Mazzolai, E. Villa, and A. Tuissi, Low-Frequency Internal Friction of Hydrogen-Free and Hydrogen-Doped NiTi Alloys, *Acta Mater.*, 2007, **55**, p 4243–4252
5. S.K. Wua and H.C. Linb, Damping Characteristics of TiNi Binary and Ternary Shape Memory Alloys, *J. Alloys Compd.*, 2003, **355**, p 72–78
6. S. Nemat-Nasser and W. Guo, Superelastic and Cyclic Response of NiTi SMA at Various Strain Rates and Temperatures, *Mech. Mater.*, 2006, **38**, p 463–474
7. M.C. Piedboeuf and R. Gauvin, Damping Behaviour of Shape Memory Alloys: Strain Amplitude, Frequency and Temperature Effects, *J. Sound Vib.*, 1998, **215**(5), p 885–901
8. http://www.avalon-eu.org
9. L. Heller, et al., Quasistatic and Dynamic Functional Properties of Thin Superelastic Wires, *Eur. Phys. J. Spec. Top.*, 2008, **128**, p 7–15
10. P. Sittner, R. Stalmans, and M. Tokuda, An Algorithm for Prediction of the Hysteretic Responses of Shape Memory Alloys, *Smart Mater. Struct.*, 2000, **9**, p 452–465
11. L. Heller, P. Sittner, and J. Pilch, Impact of Heat Effects on Superelasticity, *Proceedings of ICOMAT 2008*, 29/6 to 5/7/2008, Santa Fe, New Mexico, submitted
12. P. Sittner, Y. Liu, and V. Novak, On the Origin of Luders-Like Deformation of NiTi Shape Memory Alloys, *J. Mech. Phys. Solids*, 2005, **53**, p 1719–1746
13. J. Van Humbeeck and L. Delaey, The Influence of Strain-Rate, Amplitude and Temperature on the Hysteresis of a Pseudoelastic Cu-Zn-Al Single Crystal, *J. Phys.*, 1981, Col C5, Tome 42, p C5-1007-1011

JMEPEG (2009) 18:612–619
DOI: 10.1007/s11665-009-9453-3

Application of SMA Composites in the Collectors of the Railway Pantograph for the Italian High-Speed Train

A. Tuissi, P. Bassani, R. Casati, M. Bocciolone, A. Collina, M. Carnevale, A. Lo Conte, and B. Previtali

(Submitted October 14, 2008; in revised form February 20, 2009)

In high-speed train operation the flexural modes of the collector play a significant role in the vibrations for the frequencies over 40-50 Hz. In a previous work it has been established that a possible way to increase the modal damping of these flexural modes, without deep modification of the collector structure, is to increase the specific damping of the lateral horns, usually made of glass fiber polymer. Ni-Ti alloy yarns can be used as "smart fibers" embedded in this conventional material in order to make new horns with increased damping capacity, with a configuration of laminated composite material. The first step of the work herein presented consists in setting, through a proper thermal treatment, martensitic structure within the pantograph working temperature range, in order to obtain damping capabilities at low amplitude strain in the range 10^{-4}-10^{-3}. Afterwards a series of dynamic tests aimed at identifying the damping capacity of the NiTi wires has been undertaken. A finite element (FE) model of the SMA composites horn has been finally validated, comparing the results of dynamic numerical analysis with the results of measurements.

Keywords composite material, damping, shape memory alloy

1. Introduction

In high-speed train operation, dynamic behavior of the pantograph heavily influences the quality of current collection. In particular, the flexural modes of the collector play a significant role in the vibrations for the frequencies over 40-50 Hz. In this range, the vibration level is high and consequently high are the levels of sparking and the wear rate of the contact wire and of the contact strip due to electric arcs.

The sources of excitation of these vibrations are mainly two: the direct dynamic interaction of the collector with the contact wire of the overhead line, and the air flow acting on the pantograph. In the latter, both turbulence effects and non-stationary aerodynamic phenomena, in particular, vortex shedding are involved.

The influence of collector deformability has been confirmed through wind tunnel tests, as shown in Fig. 1 where the spectral analysis of the contact force measured during a run-up and run-down test on a complete pantograph is reported. The contribution of the flexural modes of the collector can be clearly identified at 60 and 137 Hz, corresponding to the first and third flexural modes of vibration of the collector.

In a previous work (Ref 1) it has been established that a possible way to increase the modal damping of the flexural mode of the collectors, without deep modification of collector structure, is to increase the specific damping of the lateral horns, whose primary function is to prevent an irregular positioning of the contact wire with respect to the collector.

Since the material of the lateral horns is usually glass fiber polymer, martensitic Ti-Ni alloy yarns, thanks to the high damping capacity of the NiTi in its martensitic microstructure, can be used as "smart fibers" embedded in this conventional material in order to make new horns with increased damping capacity, with a configuration of laminated composite material like the ones reported in (Ref 2-4).

The first step of the work herein presented consists in the setting of the material structure in order to obtain martensitic phase in the working temperature range through a proper thermal treatment. Afterwards, a series of dynamic tests aimed at identifying the damping properties of the NiTi wires, under uniaxial vibration tests, has been undertaken. The same wires have been used to produce composite SMA prototype.

Moreover, a finite element (FE) model of the SMA composites horn has been subsequently validated, comparing the results of dynamic numerical analysis with the results of measurements (Fig. 1).

The paper is organized as follows:

- modal analysis of a current collector by means of testing and FE analysis;
- treatment of NiTi wires, experimental damping characterization, and modeling of its behavior;
- fabrication process of the horn as laminated composite;
- simulation of the dynamic response of the horn;

This article is an invited paper selected from presentations at Shape Memory and Superelastic Technologies 2008, held September 21-25, 2008, in Stresa, Italy, and has been expanded from the original presentation.

A. Tuissi, P. Bassani, and **R. Casati**, CNR, IENI, Unità di Lecco, Corso Promessi Sposi 29, 23900 Lecco, Italy; and **M. Bocciolone, A. Collina, M. Carnevale, A. Lo Conte**, and **B. Previtali**, Dipartimento di Meccanica, Politecnico di Milano, Via La Masa 1, 20156 Milano, Italy. Contact e-mail: andrea.collina@polimi.it.

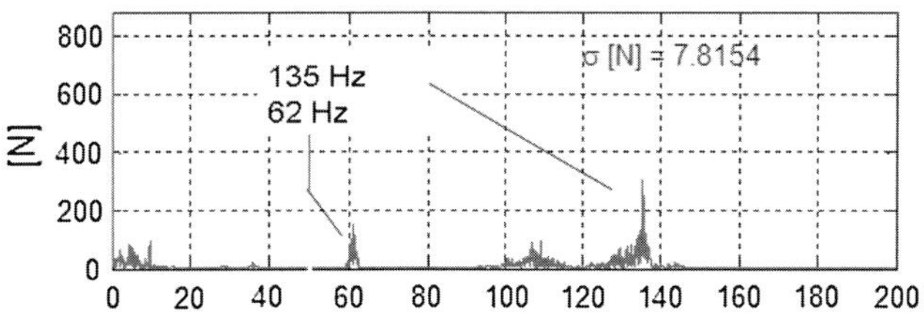

Fig. 1 Contribution of deformable collector modes to the spectrum of the contact force related to the effect of incident flow: wind tunnel measurement results

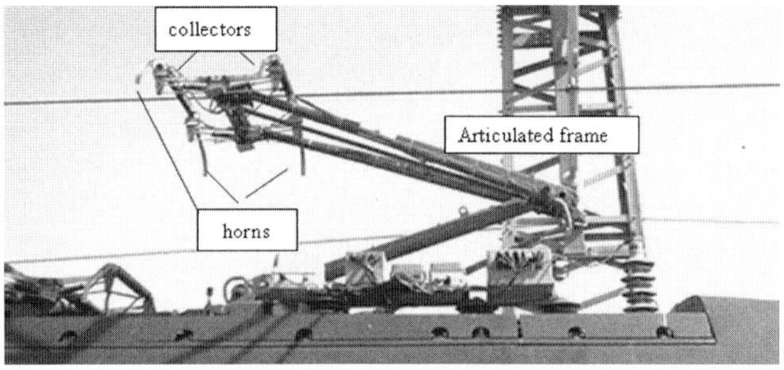

Fig. 2 High-speed pantograph in contact with the catenary

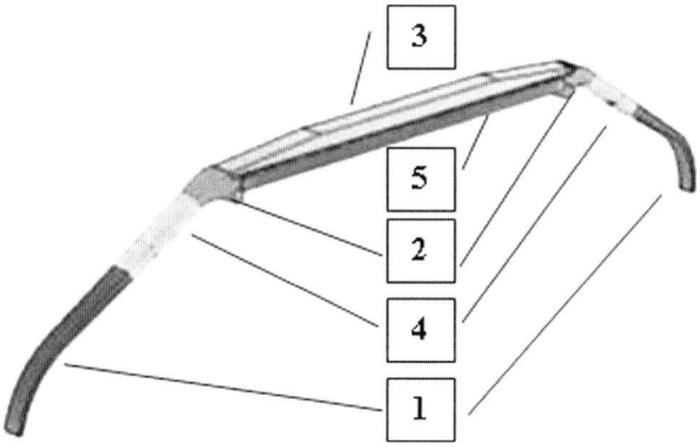

Fig. 3 Finite element model of the collector A: different parts are shown in colors (refer to Table 2)

- dynamic testing of the horn;
- remarks and future activities.

2. Modal Analysis of the Current Collector

When considering current collectors for high-speed pantographs (see Fig. 2), common feature can be found in their structure. A current collector of a pantograph is composed at least of the following parts:

- one or two contact strips (copper or graphite or graphite with copper powder) that have the function of a sliding contact for collecting the current from the overhead line. This is the part that undergoes wear during the normal operation and excessive wear in case of arcing and continuous sparking;
- a steel or aluminum supporting structure for each contact strip (or couple of contact strip) that has a structural function and is connected to the articulated frame of the pantograph through the suspensions of the collector;
- lateral horns, whose function is to prevent the contact wire sliding under the collector, that are realized with polymeric material or fiber glass.

Analyzing the dynamic behavior of a collector, it can be found that low damped flexural modes of the collector are present starting from 50 Hz. In order to evaluate which components are more suitable for introducing damping, a FE model of the current collector has been set-up, able to reproduce the experimentally observed frequencies and mode shapes (Fig. 3 and Table 1). Hexahedral elements (C3D8R) with reduced integration have been used to the solid parts (contact strip, connecting blocks, and lateral horns), whereas quadrangular elements (S4R) have been adopted for the tubular parts (central supporting beams and lateral tubular beams). The contact between the different parts has been reproduced forcing the congruence between the surfaces of adjacent parts.

The model has been validated by means of the comparison with the experimental results from forced frequency response and impact tests. It can be observed that the maximum

Table 1 Results of the FE analysis and comparison with experimental data for the collector

Mode no.	FE calculated mode shape	Exp. freq., Hz	FE freq., Hz
1		60.1	60
2		76.9	76.2
3	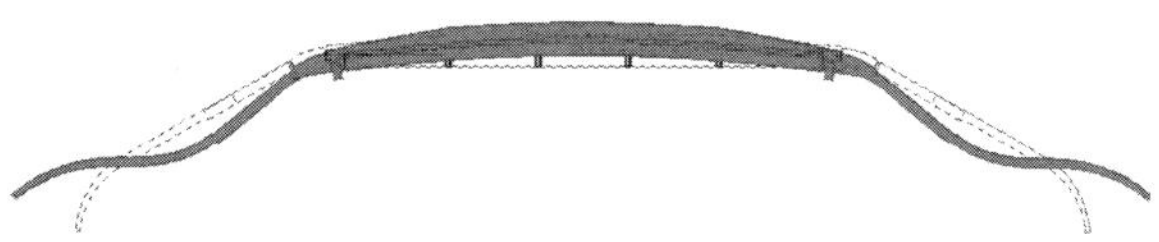	136	136

amplitudes occur in correspondence of the horns that play an important role in the location of the first flexural frequencies of the collector.

Analyzing the modal shape of the flexural modes, it can be shown that the horns contribute with a significant portion, from 42% up to 70% of the total strain energy associated to the first modes, as reported in Table 2.

For this reason it is proposed to use the horns as elements to increase the damping of the entire collector. The implementation of horns with laminated composite does not modify the structure of the collector; therefore, it can be easily implemented in the current design without need of architectural modification, and eventually proposed for a retrofit of the existing collectors.

3. NiTi Wires Realization, Characterization, and Modeling

As is known, NiTi alloys are characterized by a transformation between an austenitic phase B2 stable at high temperature and a martensitic B19′ at low temperature. The corresponding characteristic temperatures can be properly set both through modifications of their composition and thermomechanical treatments.

In the literature it is well documented that the martensite of NiTi shows a damping capacity of at least an order of magnitude higher than classic structural metallic materials properties (Ref 5, 6).

Considering the expected low strain level of the application and the service temperature, it has been decided to use $Ni_{49}Ti_{51}$ (at.%) alloy in order to have a martensitic temperature range below 90 °C. The material was melted by VIM furnace in a controlled argon atmosphere. The ingot was hot forged, hot

Table 2 Fraction of elastic energy distribution for the ATR95 FE calculated modes

		Mode no.		
Parts of the collector	**Color legend**	**1**	**2**	**3**
Horns	1	0.46	0.7	0.42
Connecting bloks	2	0.13	0.09	0.05
Graphite strip	3	0.12	0.02	0.26
Lateral tubular beams	4	0.16	0.16	0.01
Central beam	5	0.13	0.03	0.26
Total		1	1	1

The color legend of Fig. 3 is recalled to distinguish the different parts

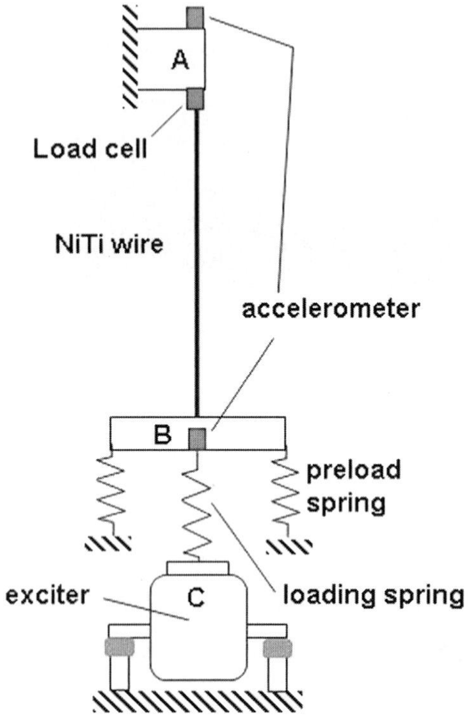

Fig. 5 Sketch of the experimental set-up

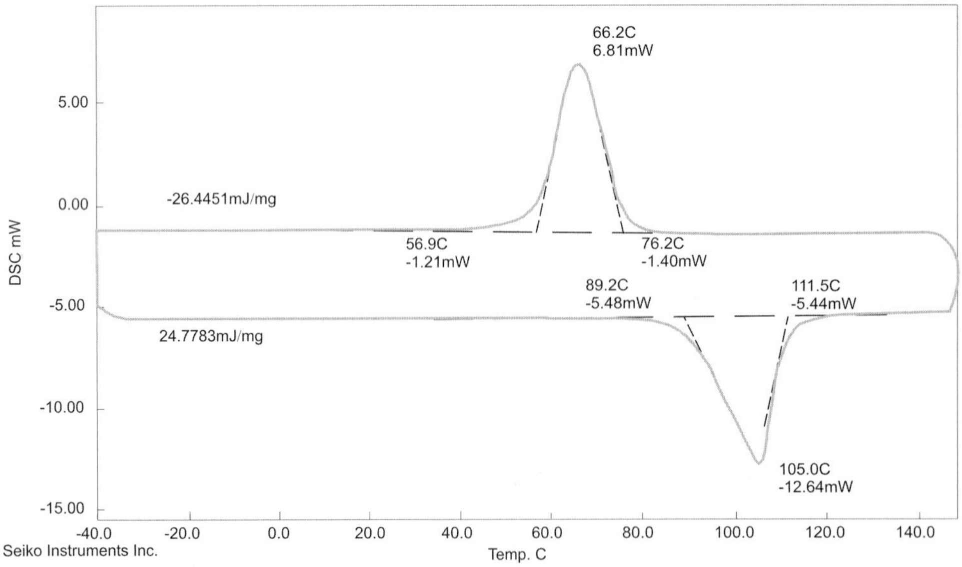

Fig. 4 DSC analysis of the NiTi wire material

rolled, and cold drawn down to 0.5 mm in diameter and then heat treated at 700 °C for 20 min in order to obtain an A_s of 90 °C as shown in the calorimetric curve (Fig. 4), from which the characteristic temperatures can be inferred: M_f 56.9 °C, M_s 76.2 °C, A_s 90 °C, and A_f 111.5 °C.

In order to determine the damping properties of the NiTi wires, ordinary vibration tests at ambient temperature in the frequency range 50-150 Hz have been carried out. The sketch of Fig. 5 shows the principle of the test set-up: the upper end of the wire is fixed and connected to a piezoelectric load cell, while the lower end is connected to an electro-dynamic shaker through a soft spring, in order to decouple the wire from the actuator. The wire is preloaded by a couple of springs. Accelerometers are placed in both extremities of the wire connections in order to measure the relative elongation of the wire, eliminating the effect of the residual vibration of the upper fixation point, which is connected to a supporting structure placed above the shaker. Since the forces that are involved are very low, a particular attention was devoted to the implementation of the set-up. In particular, the use of piezo-load cells has been preferred with respect to estensimetric ones in order to enhance the stiffness of the connecting point.

The tests have been carried out applying a sinusoidal waveform at 60 Hz, corresponding to the frequency of the first flexural mode of the collector of the pantograph for which the application is conceived. The range of deformations that have been investigated are between 0.5×10^{-4} and 10^{-3}. The parameter tan δ is defined as:

$$\tan\delta = \frac{Im(\overline{F}/\overline{X})}{Re(\overline{F}/\overline{X})} \qquad \text{(Eq 1)}$$

In Eq 1 $\overline{F}$ is the complex number that represents the force at the test frequency, whereas $\overline{X}$ is the complex number representing the amplitude of the relative motion of wire extremities. Figure 6 reports the parameter tan δ at 60 Hz as a function of the axial deformation of the wire. The latter is calculated through the difference of the acceleration of wire extremities, obtaining the relative displacements. Every point represents the results of each single test. Since the damping parameter is based on the phase difference between force and mobility signal, it is fundamental to detect the presence of eventual spurious effects due to the measurement chain, associated to the dynamic response of the transducers and the acquisition data system. An inertial test with the accelerometer placed on a calibrated mass, compared with the measurement of the load cell mounted under the mass, has been performed. The frequency response function between the load cell signal and the accelerometer signal is obviously expected to be the value of the mass, with a zero phase lag, but as a matter of fact a phase lag introduced by instrumentation has been found. The estimated instrumental phase lag at 60 Hz, equal to 2° 25′, is then subtracted from the phase lag value measured in the test on the wires.

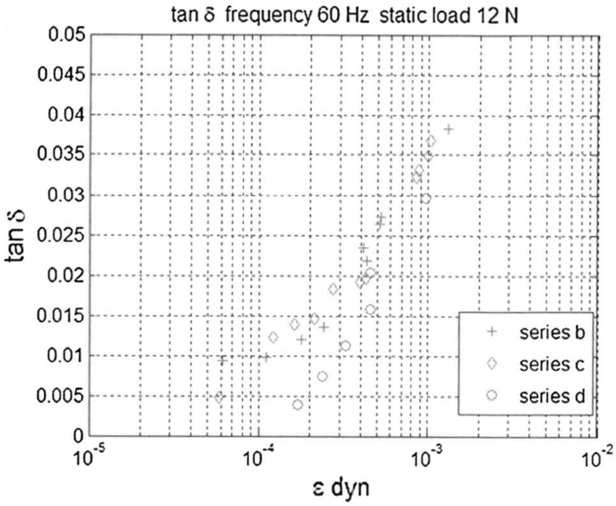

Fig. 6 Values of tan δ for the tested NiTi wires at preload of 12 N

The results obtained from the tests on the NiTi wires showed an amplitude dependence of the damping capacity, expressed as tan δ (see Fig. 6). Using the data obtained from the above-mentioned tests, a simple model has been developed to reproduce the dependence of the tan δ parameter on the strain amplitude.

State-of-the-art modeling of SMA focuses on the evolution of austenite-martensite transformation, through thermo-mechanical model (see, for instance, Ref 7, 8), for strain values of the order of 1%÷5%, including strain-rate dependence for low frequency. As mentioned earlier, in the present application lower strain amplitudes are induced, and the intrinsic damping of the martensite due to twin boundary movement is exploited.

For the purpose of phenomenological modeling of the damping of the examined wires, a friction-type model is adopted, following the definitions given in Ref 9. Different elementary hysteresis elements are combined in parallel, giving the non-linear strain amplitude dependence of the damping (Fig. 7). In addition, a fraction of the linear-dependent portion of damping is added with a viscous element. The identification is carried out following a procedure similar to the one described in Ref 10 for a related problem. (σ_i, ε_i) are the n points of the experimental stress-strain cycle, with strain amplitude A, used for the identification procedure. Equating the stress variation ($\sigma_{i-1} - \sigma_i$) across the ($\varepsilon_{i-1} - \varepsilon_i$) intervals to the stress variation obtained by the model used to reconstruct the stress-strain cycle, the following equation holds:

$$\begin{bmatrix} (\varepsilon_1-\varepsilon_0)E_1 & 0 & \cdots & 0 \\ (\varepsilon_1-\varepsilon_0)E_1 & (\varepsilon_2-\varepsilon_1)(E_1+E_2) & \cdots & 0 \\ \cdots & \cdots & \cdots & 0 \\ (\varepsilon_1-\varepsilon_0)E_1 & (\varepsilon_2-\varepsilon_1)(E_1+E_2) & & (\varepsilon_n-\varepsilon_{n-1})(E_1+E_2+\cdots+E_n) \end{bmatrix} + \begin{Bmatrix} \sigma_{V,1}-\sigma_{V,0} \\ \sigma_{V,2}-\sigma_{V,1} \\ \\ \sigma_{V,n}-\sigma_{V,n-1} \end{Bmatrix} = \begin{Bmatrix} \sigma_1-\sigma_0 \\ \sigma_2-\sigma_1 \\ \\ \sigma_n-\sigma_{n-1} \end{Bmatrix} \qquad \text{(Eq 2)}$$

where $\sigma_{V\hat{i}}$ are the viscous contributions, which are set a priori considering the tan δ value of the lowest strain amplitude. From the previous expression the E_i unknowns can be calculated, while the limiting equivalent stress for each element σ_{ci} can be simply calculated as

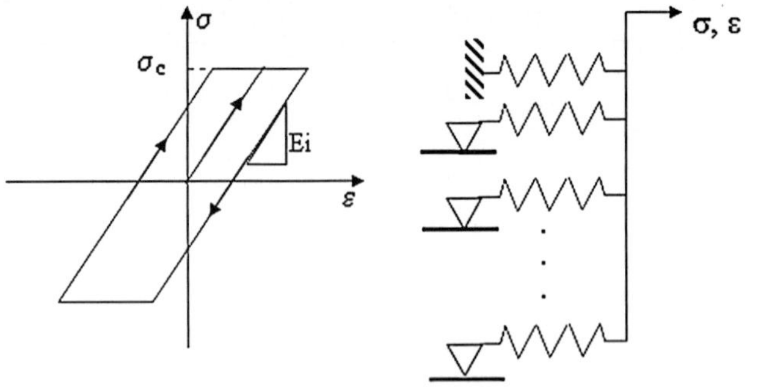

Fig. 7 Basic hysteresis model and combined in parallel used for the simulation of material behavior

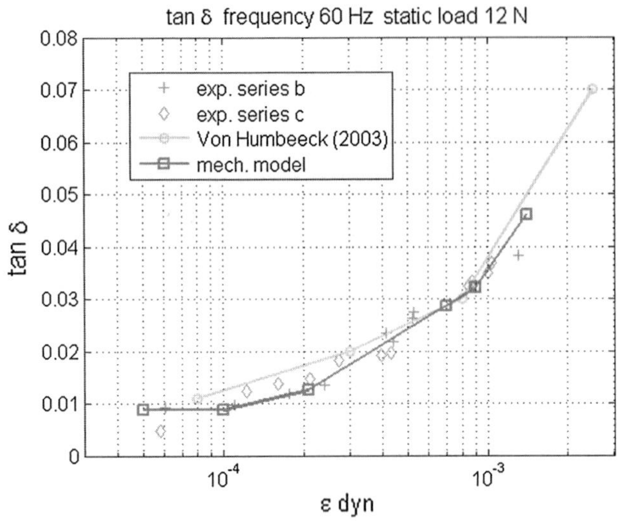

Fig. 8 Comparison of tan δ values obtained from the model with the measured ones as a function of strain amplitude

$$\sigma_{ci} = \frac{1}{2}E_i(A - \varepsilon_i) \quad i = 2, n \tag{Eq 3}$$

In the presented application, $n = 6$.

The identification of the parameters E_i and σ_{ci} is carried out considering the stress-strain amplitude at the highest strain amplitude (1.4×10^{-3}). The model is then applied for the simulation of the behavior at lower strain amplitudes. From the stress-strain loop simulated at each amplitude, the corresponding tan δ values are calculated. Figure 8 shows the dependence of tan δ on the strain amplitude, compared with the measured results and with the values from Ref 5 in the same strain amplitude, although for a lower annealing temperature. The adopted modeling refers obviously to a constant ambient temperature since it does no include any temperature effect.

4. Fabrication Process of the Laminated Composite and the Horn

As mentioned earlier, the proposed laminated composite consists of two external and one internal mat. The two external mats are two composite mats made by NiTi wires and epoxy fiberglass fabric, while the internal mat is made of epoxy fibreglass fabric. The outer surfaces of the composite mats are then covered by two epoxy fibreglass layers (see Fig. 9 and Table 3).

The fabrication process is divided into two steps. First, the fabrication of the composite mat having the horn shape is carried out using a metallic sheet in order to give the horn curvature. In particular, the first epoxy fiberglass fabric is positioned on the tool, the series of 13 NiTi wires are lined above the fabric and the second fiberglass fabric covers the NiTi wires (Fig. 10). Then, the eventual air bubbles are eliminated making use of a vacuum bag and the composite mat is cured in autoclave.

In the second step, the laminated composite is assembled, joining the two composite mats with NiTi wires with the inner part in epoxy fibreglass and adding the two external clad layers. In particular, the horn mold is filled with one epoxy fiberglass layer, the lower composite mat, about 30 layers of epoxy fiberglass fabric, the upper composite mat, and one final epoxy fiberglass layer.

After the cure cycle in autoclave the "horn" is finished as shown in Fig. 11.

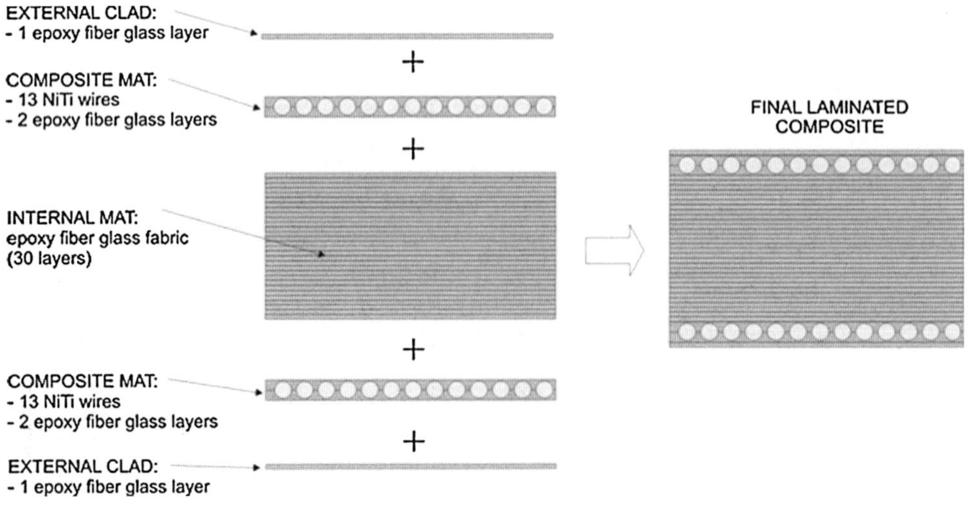

Fig. 9 Cross section of the laminated composite

Table 3 Characteristic of the matrix

Matrix	Epoxy resin
Fiber	Glass EC9 5x136 tex (weft)
	Glass EC9 68 tex (warp)
Resin content	32%
Cure temperature	120 °C
Autoclave pressure	6 bar
Autoclave time	6 h

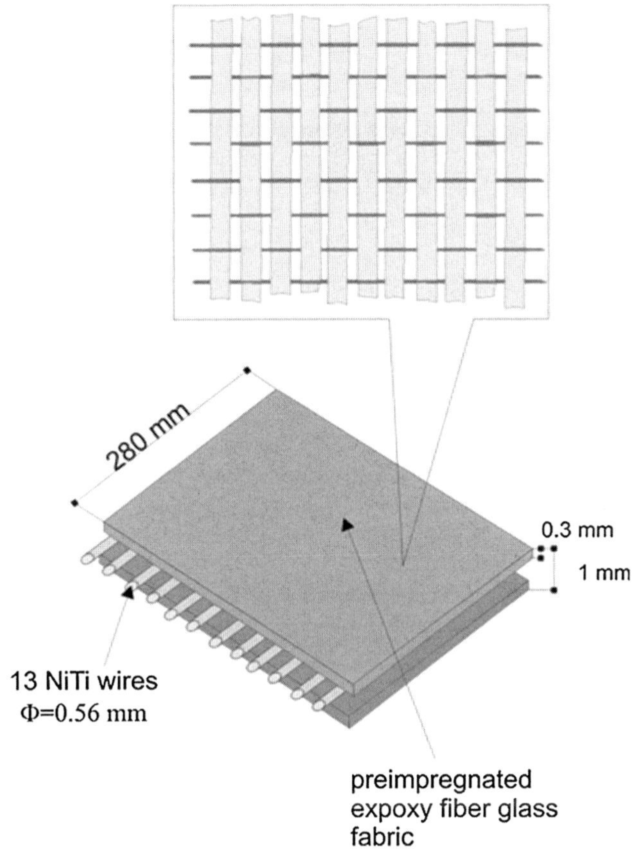

Fig. 10 Structure of the external mat

With the same procedure a laminated horn of only epoxy fiberglass has been fabricated, substituting the NiTi wires in the two external mats with two additional layers of epoxy fiberglass fabric, in order for the two horns to have the same height. The improvement due to NiTi wires in the composite horn has been established by comparison as shown in the following.

5. Finite Element Model of the Horn

A FE model of the horn fitted with the wires has been implemented in order to perform the dynamic analysis including the behavior of the material by means of the material behavior model previously presented. In Fig. 12, the FE model representing the horn and including the NiTi wires is shown corresponding to the prototype horn made as described previously. In the dynamic simulation, the horn is clamped at one extremity, and the initial excitation is a velocity field corresponding to the first flexural mode, so that the dynamic response reproduces a decay after an impulse application.

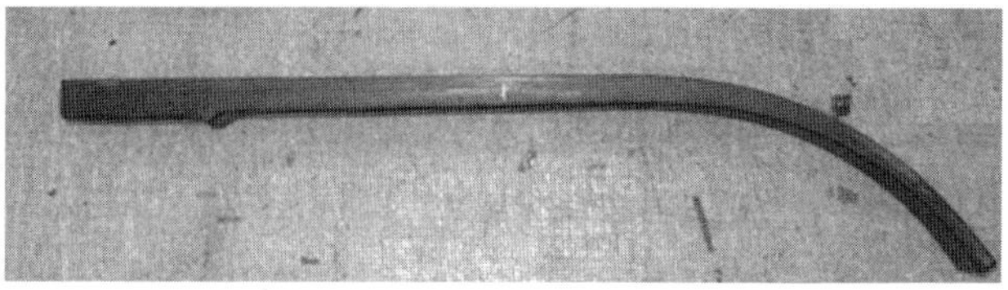

Fig. 11 Final appearance of the horn in laminated composite

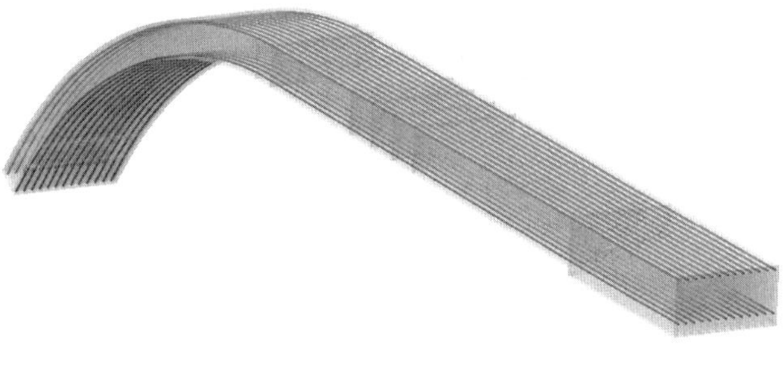

Fig. 12 Finite element modeling of the horn, including the NiTi wires

Table 4 Results of the simulation of decay test

Configuration	Strain amplitude max	Frequency, Hz	Non-dimensional damping, %
Fiberglass only	...	101.4	0.27
NiTi 13 wires	0.5×10^{-3}	105.5	0.298
	0.75×10^{-3}	105.5	0.304
	1×10^{-3}	105.5	0.336
	1.25×10^{-3}	105.5	0.372
NiTi 20 wires	0.5×10^{-3}	109.9	0.319
	0.75×10^{-3}	109.9	0.329
	1×10^{-3}	109.9	0.37
	1.25×10^{-3}	109.9	0.419

Based on a DMTA analysis, the fiberglass material characteristics are assumed as follows: Young modulus 1.82×10^{10} Pa, with a non-dimensional damping $h = 0.0027$ associated to the first flexural mode. The behavior of the NiTi wires is included as a user-defined subroutine for the related material, with the actual strain condition as input and the corresponding stress in output. In this way, the strain dependence of the model accounts for the different values of strain along the horn for flexural vibrations. Moreover, different values of the vibration and consequently different strain amplitudes are considered in order to verify the dependence of the NiTi modeling included in the FE calculation. Table 4 reports the results in terms of non-dimensional damping obtained respectively for the matrix material only for the configuration with 13 wires for each of the two external mat (corresponding to the real prototype) and with a configuration with 20 wires for each external mat. The non-dimensional damping h%, calculated from the logarithmic decrement (see Fig. 13), increases with amplitude of the strain, as expected from the model adopted for the SMA. Simulated values of h% damping are in the range from 0.27%, related to the fiberglass matrix alone, up to 0.37%, at the maximum amplitude in the configuration with 13 NiTi alloy wires.

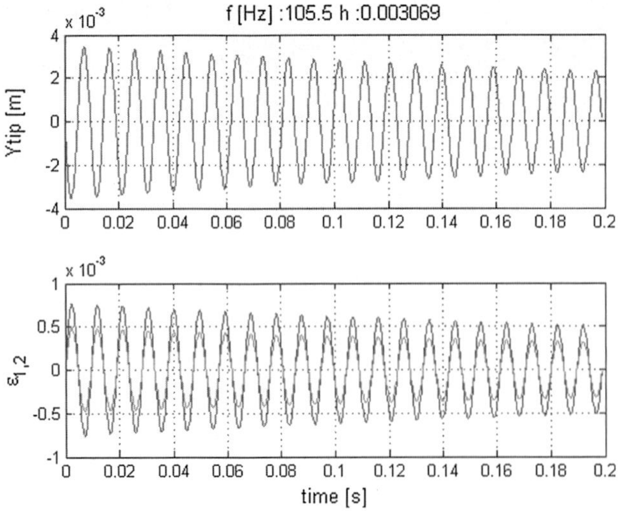

Fig. 13 Simulated decay. Upper graph: motion at the tip of the horn (with NiTi wires) in the vertical direction. Lower graph: strain in the section at 30 and 110 mm from the clamped section

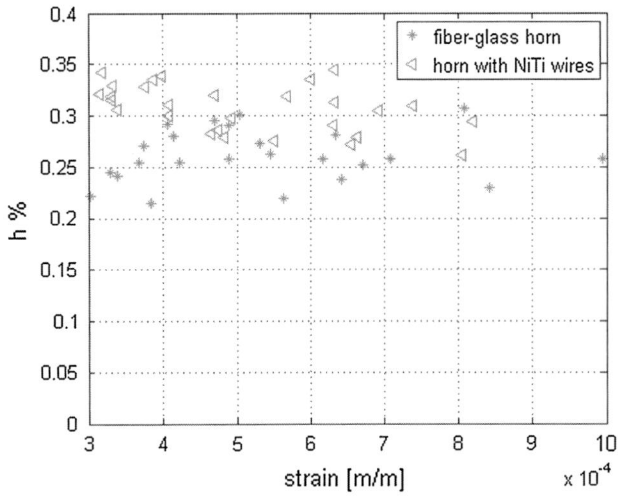

Fig. 14 Comparison of experimental non-dimensional damping (*r*/*rc*) related to the first flexural frequency obtained from the decay response test on fiberglass only and NiTi composite horns

A secondary effect of the NiTi wires is that the flexural frequency increases with the number of wires, indicating the influence of the higher stiffness of the wires.

6. Dynamic Testing of the Horn

Experimental decay tests have been carried out in a single cantilever configuration on both the fabricated horns (epoxy fiberglass with NiTi wires and only epoxy fiberglass), in order to identify the non-dimensional damping values related to their first flexural mode. The horns have been given an initial displacement condition and the decay response has been measured by means of a triangulation laser MEL model M5L/10 pointing at a distance of 155 mm from the clamp. The laser sensor was used in order to provide a non-contact measurement. Strains at the distance of 30 and 170 mm from the clamp have been measured by means of strain gages in a half bridge configuration.

The results of experimental non-dimensional damping $h = r/rc$ associated to the first flexural mode, calculated from the logarithmic decrement over five periods, are reported in Fig. 14 for both the horns, as a function of the mean strain amplitude of the analyzed periods, measured at the distance of 30 mm from the clamp. Due to epoxy fiberglass initial values of tan δ 0.54% and to tan δ dependence on strain for the NiTi wires (Fig. 6), damping improvement can be expected when strains along the horn are higher than 1E-4. The mean value of the non-dimensional damping h for experimental data has increased from 0.0027 in the case of fiberglass laminated horn to the value of 0.003 for the horn containing NiTi wires, which means an increase of 11% in the non-dimensional damping of the first flexural mode.

7. Conclusions

A composite material made with fiberglass mats and NiTi wires has been proposed for an application on the current collector for pantographs, with the aim of increasing its damping capacity related to the first flexural mode, keeping the original configuration, and therefore without any need of structural modification in the architecture of the collector.

The developed NiTi wires have been produced and dynamically tested at 60 Hz for the determination of damping parameter tan δ. According to the obtained results, a simple phenomenological model with strain amplitude dependence has been developed and applied to the simulation of a FE model of the horn.

A first prototype of the horn has been made too in order to assess its fabrication process, and it has been tested dynamically in laboratory. NiTi wires collaborate in the composites since the experimental data obtained for non-dimensional damping h of the first flexional mode increase of 11%.

Acknowledgment

The authors thank Mako Shark S.r.l–Viale Montecuccoli 18, 23843 Dolzago (LC), Italy, for the realization of the laminated horns.

References

1. M. Bocciolone, A. Collina, A. Lo Conte, B. Previtali, A. Tuissi, and P. Bassani, Characterisation and Modeling of a Shape Memory Material for the Increase of the Strctural Damping of Collectors for High Speed Pantographs, XXXV Convegno Nazionale AIAS, 13–16 Settembre 2006, Università Politecnica delle Marche (in italian)
2. R. Zhang, Q.-Q. Ni, A. Masuda, T. Yamamura, and M. Iwamoto, Vibration Characteristics of Laminated Composite Plates with Embedded Shape Memory Alloys, *Compos. Struct.*, 2006, **74**, p 389–398
3. K.-t. Lau, L.-m. Zhou, and X.-M. Tao, Control of Natural Frequencies of a Clamped–Clamped Composite Beam with Embedded Shape Memory Alloy Wires, *Compos. Struct.*, 2002, **58**, p 39–47
4. Y. Matsuzaki, T. Ikeda, and C. Boller, Technological Development of Passive and Active Vibration Control: Analysis and Test, *Smart Mater. Struct.*, 2005, **14**, p 343–348

5. J. Van Humbeeck, Damping Capacity of Thermoelastic Martensite in Shape Memory Alloys, *J. Alloys Compd.*, 2003, **355**, p 58–64
6. S.K. Wua and H.C. Lin, Damping Characteristics of TiNi Binary and Ternary Shape Memory Alloys, *J. Alloys Compd.*, 2003, **355**, p 72–78
7. F. Auricchio, D. Fugazza, and R. Des Roches, Numerical and Experimental Evaluation of the Damping Properties of Shape-Memory Alloys, *ASME J. Eng. Mater. Technol.*, 2006, **128**, p 312–319
8. T. Ikeda, F.A. Nae, H. Naito, and Y. Matsuzaki, Constitutive Model of Shape Memory Alloys for Unidirectional Loading Considering Inner Hysteresis Loops, *Smart Mater. Struct.*, 2004, **13**, p S916–S925
9. J. Van Humbeek and S. Kustov, Active and Passive Damping of Noise and Vibrations Through Shape Memory Alloys: Applications and Mechanisms, *Smart Mater. Struct.*, 2005, **14**, p S171–S185
10. A. Falvo, F. Furgiuele, and C. Maletta, Hysteresis Modeling of Two-Way Shape Memory Effect in NiTi Alloys, *Meccanica*, 2008, **43**, p 165–172

JMEPEG (2009) 18:620–625
DOI: 10.1007/s11665-009-9434-6

1059-9495/$19.00

Characterization of Transformation Temperatures with the Bend and Free Recovery Technique: Parameters and Effects

Masao Drexel, Jim Proft, and Scott Russell

(*Submitted September 16, 2008; in revised form February 13, 2009*)

This study was conducted to investigate factors such as the deformation strain, straining temperature, and the number of testing cycles on the measurement of transformation temperatures using the bend and free recovery (BFR) technique. Ti-56.0%Ni wire with approximately 40% cold work and a 2 mm diameter was heat treated in an air furnace for 10 min at 490 °C to obtain an A_f of approximately 21 °C. Wire specimens were deformed with one of two mandrels to apply an outer fiber strain of 2.4 or 5.8%. Deformation was performed at one of four straining temperatures: 0, −30, −50, or −65 °C. Specimens were tested ten times to investigate the effect of repeated testing. The resulting BFR curves were analyzed to determine the trends in the R-phase start (R'_s) and austenite finish (A_f) temperatures. For specimens strained at both 2.4 and 5.8% there was no detectable change in A_f resulting from changes in the deformation temperature. Increasing the deformation strain from 2.4 to 5.8% tended to increase the measured A_f by approximately 1 °C independent of deformation temperature. Repeat testing seemed to result in a slight increase in A_f but the intrinsic scatter of the BFR data made it impossible to conclusively identify a trend.

Keywords intermetallics, mechanical testing, nondestructive testing

1. Introduction

When working with Nitinol, the ability to control and monitor transformation temperatures is necessary. Heat treatments can be used to alter transformation temperatures through the activation of Ni-rich precipitation and dissolution processes allowing the tailoring of the thermo-mechanical properties (Ref 1). These same heat treatments can provide enough energy to activate recovery and recrystallization processes, which will further alter the material properties (Ref 2).

The bend and free recovery (BFR) technique, as described in ASTM F2082-06 *Standard Test Method for Determination of Transformation Temperature of Nickel-Titanium Shape Memory Alloys by Bend and Free Recovery*, is widely used in the medical device industry to determine the transformation temperatures of incoming material, partially processed devices, and finished devices (Ref 3). The standard outlines a procedure for determining the austenite start (A_s) and austenite finish (A_f) temperatures for one-stage transformations and the rhombohedral, or R-phase, start (R'_s) and R-phase finish (R'_f) temperatures in the case of two-stage transformations. The current version of F2082-06 leaves some questions regarding testing parameters unanswered. The deformation temperature suggested for all room-temperature superelastic materials is −55 °C or lower. At these temperatures the standard refers to the material as "nominally fully martensitic". Studies have found that cold-worked and heat-treated material can have martensite start (M_s) and martensite finish (M_f) temperatures below −60 and −100 °C, respectively, suggesting that some samples cooled to −55 °C will not be in the fully martensitic state and most, if not all, of the martensite that is present after deformation will have been stress-induced (Ref 4). The F2082-06 standard also does not address the possibility of repeatedly testing the same part and any changes in the measured transformation temperatures that may or may not result. And finally, the standard asks for a deformation strain of between 2 and 2.5%. A previous study by Chen et al. has suggested that higher deformation strains will result in an A_f increase (Ref 5). Device manufacturers are often limited in their ability to control strain levels when testing devices with complicated geometries. This study was conducted to help understand the effects of deformation temperature, repeated testing, and deformation strain on the resulting transformation temperatures measured.

2. Experimental Methodology

As-drawn Ti-56.0at.%Ni wire of 2 mm diameter with approximately 40% cold work was heat-treated at 490 °C for 10 min. Heat treatment was performed in an air furnace followed by a water quench. Transformational properties were characterized by the BFR technique based on ASTM F2082-06. Figure 1 shows a schematic illustration of a typical BFR tester.

This article is an invited paper selected from presentations at Shape Memory and Superelastic Technologies 2008, held September 21-25, 2008, in Stresa, Italy, and has been expanded from the original presentation.

Masao Drexel, Jim Proft, and **Scott Russell**, Confirmd LLC, San Carlos, CA. Contact e-mail: masao@confirmd.com.

The F2129-06 test protocol outlines a procedure for cooling a room-temperature superelastic specimen in a liquid bath (a 90% isopropyl alcohol bath was used in this case) to a temperature of less than −55 °C, deforming the specimen to an outer fiber strain between 2 and 2.5%, then heating the bath at a rate of no more than 4 °C/min. Temperature and displacement are monitored as the specimen recovers its original, undeformed, shape due to the shape memory effect (SME). A typical two-stage BFR curve from this study fit for determination of R'_s and A_f is shown in Fig. 2.

To investigate the effects of various parameters on the resulting transformation temperature measurements, deviations from F2082-06 were taken:

- Instead of cooling to below −55 °C, deformations were performed at −65, −50, −30, and 0 °C.
- Instead of the prescribed 2 to 2.5% outer fiber strain, strains of either 2.4 or 5.8% were applied to the wire specimens before testing.

Additionally, there were three repeat specimens for each test condition and each specimen was tested 10 times to investigate the effects of repeat testing on the measured transformation temperatures. The six specimens deformed at −65 °C were only tested twice each because the alcohol bath was viscous and opaque at this temperature, making testing difficult and time consuming. Table 1 outlines the experimental matrix.

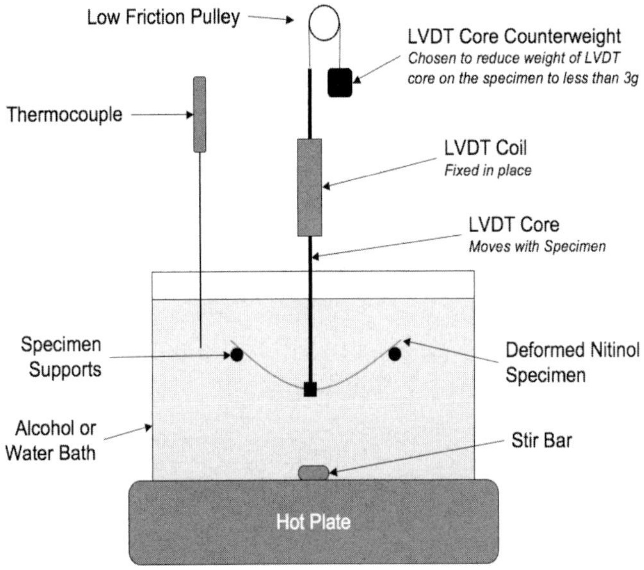

Fig. 1 BFR test apparatus

3. Results

The thermal energy provided by a 10-min heat treatment at 490 °C has been shown to enable the SME through the activation of annealing processes, maintaining a significant amount of residual cold work, and increasing the A_f to around

Table 1 Experimental matrix

Deformation temperature, °C	Deformation strain	Specimens	Repeat testing cycles	Number of tests
0	2.4	3	10	30
	5.8	3	10	30
−30	2.4	3	10	30
	5.8	3	10	30
−50	2.4	3	10	30
	5.8	3	10	30
−65	2.4	3	2	6
	5.8	3	2	6
Total		24		192

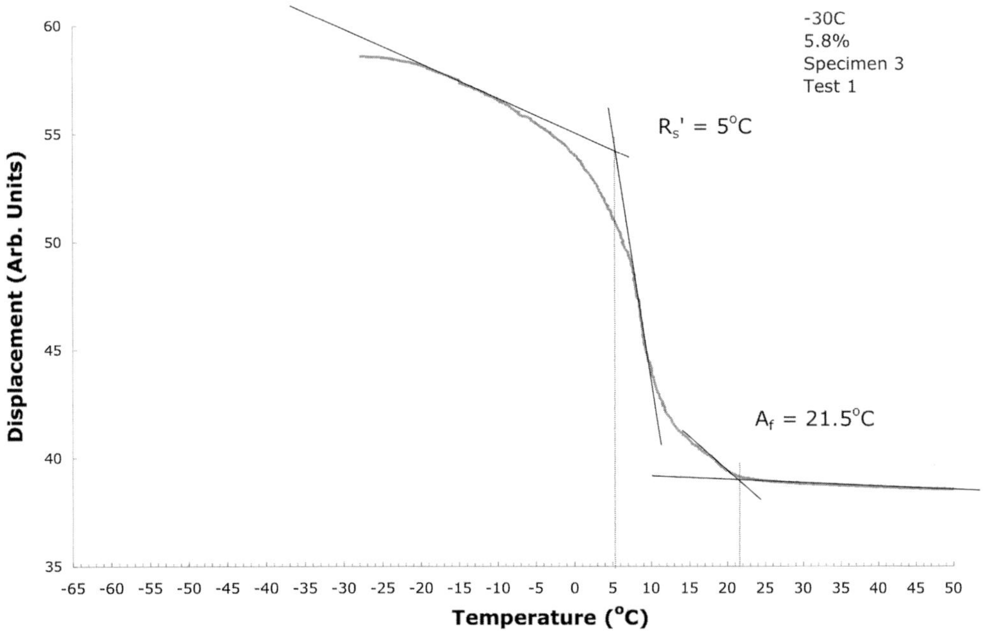

Fig. 2 Typical two-stage transformation with tangent lines fit for determination of R'_s and A_f

Table 2 A_f data summary

Strain, %	Def. Temp., °C	Average A_f of three specimens for each of 10 repeat tests (standard deviation of three specimens)										Average of all tests	Standard deviation of all tests
		1	2	3	4	5	6	7	8	9	10		
2.4	0	20.5 (0.5)	20.5 (0.5)	20.8 (1.0)	20.5 (0.9)	21.0 (1.0)	20.2 (0.3)	20.7 (0.3)	20.7 (0.8)	20.5 (1.3)	20.7 (0.6)	20.6	0.7
5.8	0	21.5 (0.5)	22.2 (0.3)	22.0 (0.0)	21.8 (0.3)	22.0 (0.0)	22.0 (0.0)	21.5 (0.5)	22.0 (0.5)	21.7 (0.6)	22.0 (0.5)	21.9	0.4
2.4	−30	20.2 (0.3)	21.2 (0.3)	21.5 (0.5)	21.2 (0.8)	21.5 (0.5)	20.7 (0.6)	20.8 (0.3)	21.3 (0.6)	20.8 (0.8)	21.0 (0.0)	21.0	0.6
5.8	−30	21.8 (1.0)	22.3 (0.8)	22.3 (1.0)	22.0 (0.5)	22.0 (1.0)	22.2 (0.3)	21.3 (0.6)	21.7 (0.8)	21.7 (0.8)	22.0 (0.5)	22.2	0.7
2.4	−50	20.2 (0.3)	20.7 (0.6)	20.5 (0.5)	20.0 (0.5)	20.2 (0.3)	20.2 (0.6)	20.3 (0.3)	20.0 (0.5)	20.0 (0.0)	20.0 (1.0)	20.4	0.5
5.8	−50	21.0 (1.0)	21.7 (0.6)	22.0 (1.0)	22.0 (1.0)	22.0 (1.0)	22.5 (0.5)	21.3 (0.3)	22.0 (0.5)	21.7 (0.6)	22.0 (1.0)	21.7	0.8
2.4	−65	20.3 (0.6)	20.5 (0.9)	…	…	…	…	…	…	…	…	20.4	0.7
5.8	−65	21.0 (1.0)	21.0 (1.3)	…	…	…	…	…	…	…	…	21.0	1.0

Table 3 R'_s data summary

Strain, %	Def. Temp., °C	Average R'_s of three specimens for each of 10 repeat tests (standard deviation of three specimens)										Average of all tests	Standard deviation of all tests
		1	2	3	4	5	6	7	8	9	10		
2.4	−30	1.7 (2.3)	2.3 (3.1)	3.0 (2.6)	3.0 (3.0)	3.0 (2.6)	2.3 (4.7)	2.0 (3.5)	2.0 (3.6)	1.7 (3.2)	2.3 (3.8)	1.7	2.8
5.8	−30	7.3 (3.2)	7.7 (3.1)	7.7 (3.5)	8.0 (3.0)	7.7 (3.1)	6.7 (2.9)	4.3 (4.5)	5.0 (3.6)	5.3 (3.2)	5.7 (3.1)	7.5	3.1
2.4	−50	−0.3 (1.2)	−0.7 (0.6)	−0.3 (1.2)	−0.7 (0.6)	0.0 (1.0)	0.3 (1.5)	0.3 (1.2)	0.3 (0.6)	−0.3 (0.6)	0.0 (1.7)	0.2	1.0
5.8	−50	5.7 (4.0)	6.7 (4.0)	6.7 (4.2)	6.7 (3.2)	6.3 (3.8)	7.0 (2.6)	6.7 (2.3)	7.0 (3.5)	6.3 (4.6)	6.3 (3.8)	5.7	3.1
2.4	−65	−3.0 (3.0)	−2.7 (3.2)	…	…	…	…	…	…	…	…	−2.8	2.8
5.8	−65	1.7 (3.8)	2.7 (4.6)	…	…	…	…	…	…	…	…	2.2	3.8

21 °C through the precipitation of Ni-rich phases (Ref 1, 2). These heat treatment parameters were chosen because they are similar to those often used by medical device manufactures who want a balance of strength, mechanical stability, and superelastic recovery. Heat treatments of as-drawn wire providing approximately this amount of thermal input often generate a microstructure that stabilizes the R-phase (Ref 2, 6, 7). R-phase stabilization can result in a two-stage transformation during heating. The F2082-06 standard identifies the starting temperature of a two-stage transformation, i.e. the start of the martensite-to-R-phase transformation, as R'_s, and the temperature corresponding to the end the transformation, i.e. the end of the R-phase-to-austenite transformation, as A_f.

The average R'_s and A_f values for each test condition and the corresponding standard deviations are presented in Table 2 and 3. Also provided in Table 2 and 3 are the average and standard deviations for each deformation strain and deformation temperature, neglecting the effects of repeat testing. R'_s values could not be determined for specimens deformed at 0 °C because there was no onset to these transformation curves to allow fitting of an upper tangent line. The M_s temperature of this material was determined to be −63 °C with differential scanning calorimetry (DSC), indicating that nearly all the martensite phase in these specimens was stress-induced and not thermally induced.

Determining trends with the BFR technique is limited by the intrinsic resolution of the test. For this data set the standard deviation values for the samples ranged from 0 to 1.3 °C with an average of 0.6 °C for A_f data and from 0.6 to 4.7 °C with an average of 2.8 °C for R'_s data (with a sample size of three).

Although there was a slight increase in the average A_f and R'_s values after ten repeat tests (0.6 and 0.1 °C, respectively), these increases were too small to conclusively identify as a trend. A larger sample size and increased number of tests may confirm the trend.

Comparing the average A_f from the first test of each deformation temperature and strain condition illustrates an increase in A_f with increased deformation strain, shown in Fig. 3. Although the error bars overlap for several of the strain and deformation temperature conditions, there is a consistent increase in A_f with increased deformation strain at all deformation temperatures. These A_f increases range from 0.7 °C at a deformation temperature of −65 to 1.6 °C at a deformation temperature of −30 °C. The average increase in A_f with increased deformation strain, from 2.4 to 5.8%, for the first test at all deformation temperatures was 1.0 °C.

Also shown in Fig. 3, there was no statistically significant change in A_f with increasing deformation temperature for either level of deformation strain. The A_f differences that can be seen in Fig. 3 are less than the noise level of the BFR technique. In an effort to increase the sample size, a similar plot was constructed using all the data from each specimen, i.e. ignoring any effects from repeat testing, and the same results were observed.

The measured R'_s values increased with both increasing strain and increasing deformation temperature, shown in Fig. 4. Again these increases have overlapping error bars but the consistency and underlying reasoning, as will be described in the next section, are considered sufficient to verify them. The average increase in R'_s as a result of increased strain, from 2.4 to 5.8%, was 5.6 °C. The increase in R'_s as a result of increasing the deformation temperature from −65 to −30 °C was about 5 °C for both deformation strains.

4. Discussion

The small standard deviation of the A_f values measured in this study is evidence that BFR testing is a reliable and repeatable means of characterizing the transformation temperatures of Nitinol. The higher amount of deviation in the R'_s data is largely a result of the difficulty in fitting a straight tangent line in a consistent manner to the curved onset of the transformation.

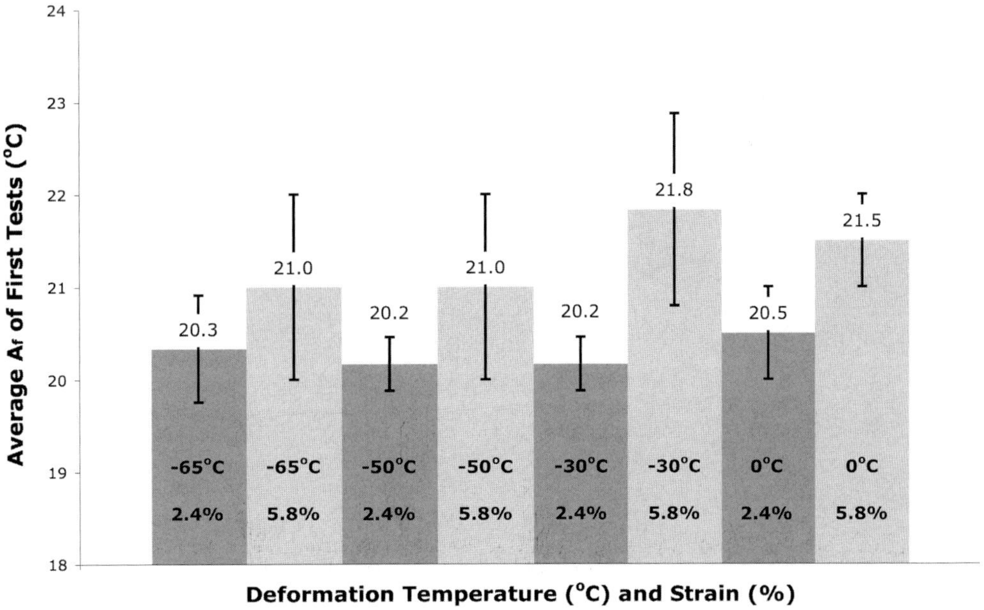

Fig. 3 Average A_f of the first test for each strain and deformation temperature condition. Two standard deviation error bars are shown

The increases in R_s' and A_f can be explained by considering a change in the character of the BFR curves. This change in character resulted from the larger amount of stress-induced martensite in specimens with 5.8% deformation strain. Figure 5 shows the BFR curves from specimens deformed 2.4 and 5.8% at −65 °C. As expected, the more heavily strained specimen has a larger amount of displacement, about twice as large in this case. This increase amount of stress-induced martensite in the 5.8% deformed specimens resulted in a steeper recovery curve in the martensite-to-R-phase region, which resulted in consistently fitting a steeper tangent line to the curve—consistently increasing the measured R_s' values. Although the A_f temperatures of the two BFR curves shown in Fig. 5 seem to be the same this is because the shift in A_f is subtle. When these BFR curves are fit in practice, they are usually scaled to fill the entire chart area, as shown in Fig. 6. When the 2.4% displacement curve is scaled to fill the entire displacement axis, the R-phase-to-austenite transformation is distorted, increasing its slope, as a result of the relatively small displacement from the martensite-to-R-phase transformation when compared with the 5.8% displacement curve. This change in scaling also resulted in an increase in the slope of onset and end of the transformations.

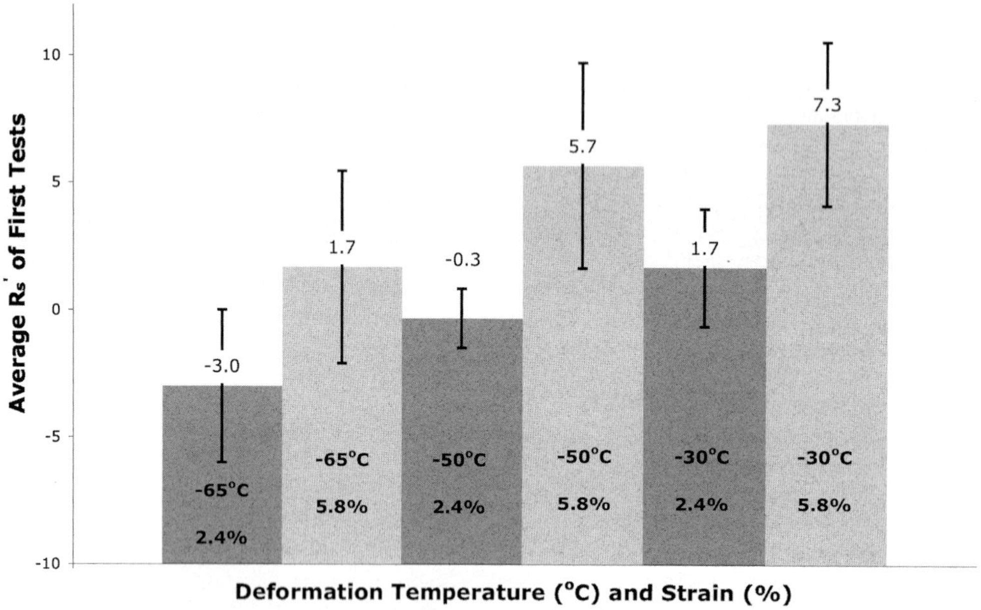

Fig. 4 Average R_s' of the first test for each strain and deformation temperature condition. Two standard deviation error bars are shown

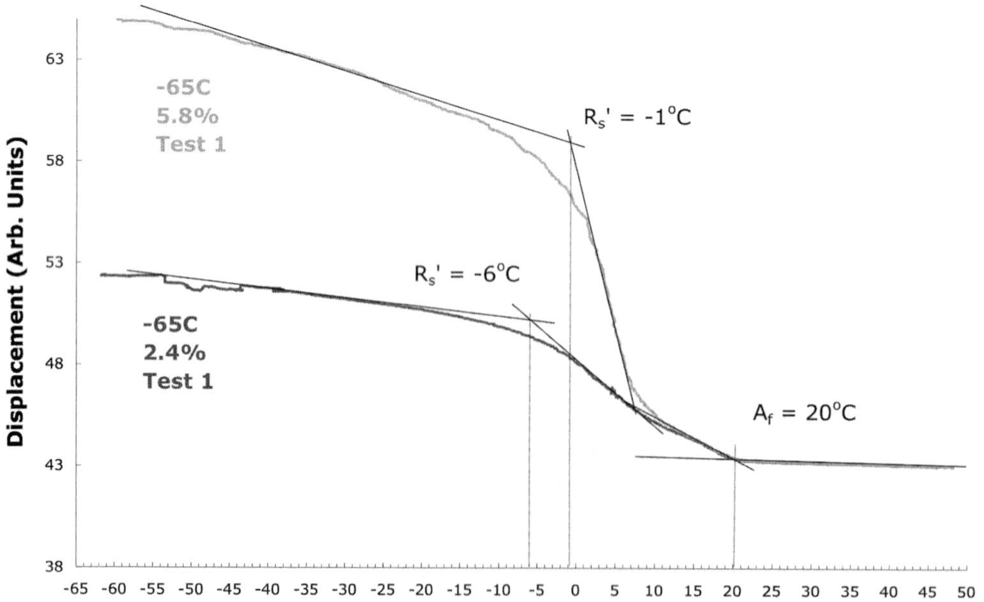

Fig. 5 BFR curves from specimens with 2.4 and 5.8% deformation strain

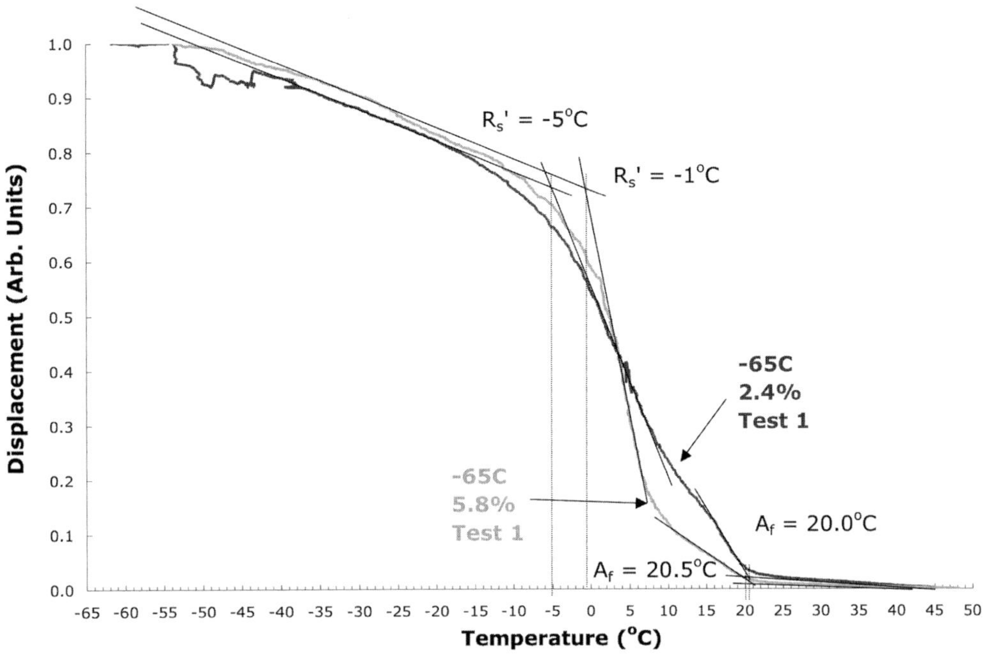

Fig. 6 Rescaled BFR curves from specimens with 2.4 and 5.8% deformation strain

These differences, resulting from scaling of the curves and a change in the character of the curves, seem to be more a result of the test than a material difference. To minimize this test-induced variability, it is recommended that consistent scaling and straining conditions be used.

The increase in R_s' with increasing deformation temperature originated from the length of the BFR curve onset. When more of the curved transformation onset is outlined, it is natural to fit the upper trend line at a lower temperature point along this curve, a point approximately halfway between the test starting temperature and the R_s' temperature is common. Therefore, one should again take care to be consistent in their test parameters and fitting technique when measuring and comparing R_s' values.

5. Conclusions

The conclusions from this study can be summarizes as follows:

- Increasing the outer fiber strain from 2.4 to 5.8% changed the character of the BFR curves causing a shift in the measured A_f and R_s' values by about 1 and 5 °C, respectively.
- Deformation temperature had no significant effect on the measurement of A_f and the effects on R_s' were mainly due to differences in trend line fitting.
- A slight increase in the transformation temperatures occurred after repeated testing but the intrinsic variability of the test did not allow a confident verification of any trend.
- The BFR technique is indeed a useful technique for measuring the transformation temperatures of Nitinol. If care is taken to minimize the effects of factors such at those outlined in this paper, i.e. deformation temperature, strain, repeated testing, and plot scaling, standard deviations of below 1 °C can be obtained.

Acknowledgment

The authors thank Mike Connally and NDC for running a DSC test to help complete this study.

References

1. A. Pelton, J. DiCello, and S. Miyazaki, Optimization of Processing and Properties of Medical-Grade Nitinol Wire, *SMST-2000 Conference Proceedings*, S. Russell and A. Pelton, Eds., 30 April to 4 May, 2000 (Asilomar), SMST, 2001, p 361–374
2. M. Drexel, G. Selvaduray, and A. Pelton, The Effects of Cold Work and Heat Treatment on the Properties of Nitinol Wire, *SMST-2006 Conference Proceedings*, B. Berg, M. Mitchell, and J. Proft, Eds., May 7–11, 2006 (Asilomar), ASM International, 2008, p 447–454
3. "Determination of Transformation Temperature of Nickel-Titanium Shape Memory Alloys by Bend and Free Recovery," F 2082-06, *Annual Book of ASTM Standards*
4. S. Gupta and F. Sczerzeni, Effect of Cold Work on Mechanical Properties and on TTR's of a Nickel Titanium Shape Memory Alloy, *SMST-97 Conference Proceedings*, A. Pelton, D. Hodgson, S. Russell, and T. Duerig, Eds., March 2–6, 1997 (Asilomar), SMST, 1997, p 41–46
5. J. Chen, T. Duerig, A. Pelton, and D. Stoeckel, An Apparatus to Measure the Shape Properties of Nitinol Tubes for Medical Applications, *J. Phys. IV*, 1995, **5**(2), p 1247–1252
6. K. Otsuka, Introduction to the R-phase Transition, *Engineering Aspects of Shape Memory Alloys*, T. Duerig, K. Melton, D. Stockel, and C. Wayman, Eds., Buttersworth-Heinemann, 1990, p 36–45
7. S. Miyazaki, Thermal and Stress Cycling Effects and Fatigue Properties of Ni-Ti Alloys, *Engineering Aspects of Shape Memory Alloys*, T. Duerig, K. Melton, D. Stockel, and C. Wayman, Eds., Buttersworth-Heinemann, 1990, p 369–413

JMEPEG (2009) 18:626–630
DOI: 10.1007/s11665-009-9483-x

1059-9495/$19.00

Implementation of the Müller-Achenbach-Seelecke Model for Shape Memory Alloys in ABAQUS

Frank Richter, Oliver Kastner, and Gunther Eggeler

(Submitted September 15, 2008; in revised form April 15, 2009)

Temperature changes caused by latent phase transformation heats are an integral part of the behavior of shape memory alloys. The ensuing thermomechanical coupling between the mechanical and thermal fields is covered by the Müller-Achenbach-Seelecke (MAS) model. Its versatility when implemented as a standalone program has been documented extensively in the literature (S. Seelecke and I. Müller, *Appl. Mech. Rev.*, 57(1), 2004, p 23-46; M. Achenbach, *Int. J. Plast.*, 5(4), 1989, p 371-395; I. Müller and S. Seelecke, *Mathem. Comp. Mod.*, 34(12-13), 2001, p 1307-1355). This model has been evaluated within various simulation environments, standalone programs as well as in commercial tools like FEMLAB and ANSYS. Here we present an application of the MAS model within the finite-element (FEM) simulation software ABAQUS. The MAS constitutive equation of state for SMA is ported into ABAQUS via a user material interface. We present the results of simulation examples using this computer model and validate them by comparison with reference solutions. Using ABAQUS finite elements allowing for temperature as a degree of freedom, the effects of the release and the absorption of latent heats in a fully coupled simulation are demonstrated. Further, a FEM implementation of the model extension to polycrystalline materials is presented as these are of greater relevance with regard to engineering applications. The results show that the incorporation of the MAS model into the ABAQUS environment provides a powerful tool useful in the framework of engineering design studies, especially in situations which require nonisothermal conditions and phase transitions.

Keywords ABAQUS, beam bending, FEM-simulation, Müller-Achenbach-Seelecke model, shape memory alloys, thermomechanical coupling

1. Introduction

Shape memory alloys (SMA) are increasingly being employed as sensors and actuators. To assign them a function their behavior should be accessible to physical interpretation to utmost precision, hence reliable modeling techniques are mandatory. Depending on temperature, the constitutive behavior of SMAs is characterized by either pseudoplasticity (at low temperature) or by pseudoelasticity (at high temperature). Ideally, the constitutive model for a SMA should be so versatile as to be able to capture both pseudoplasticity and pseudoelasticity; hence, accessing the entire range of SMA behavior. The model should reflect the complex, nonlinear hysteretic, and thermomechanically coupled material behavior of SMAs on a physically sound basis.

Thermomechanical coupling is inherent to SMA applications as phase changes are inevitably accompanied by temperature effects caused by latent heat. Consequently, mechanical and thermal equations need to be solved simultaneously. A model capable of this truly thermomechanically coupled approach is the one named after Müller, Achenbach, and Seelecke (Ref 1, henceforth 'MAS model') which reflects all the salient characteristics of SMA. In the original formulation, single-crystalline materials are considered. An extension to polycrystalline materials was recently reported (Ref 13).

The present work summarizes some results obtained by the implementation of this model in the Finite-Element-method (FEM) environment ABAQUS® (ABAQUS, Inc., Dassault Systèmes Simulia Corp., Providence, RI, USA; release 6.7), indicating harmony to reference solutions taken either from alternative implementations of the model or from basic mechanics. Both the single crystal and the polycrystal formulation of the model are considered.

2. MAS Model and FEM Implementation

Crystallographic observations have revealed that a SMA under uniaxial loading exhibits a layered crystalline structure. Thus, the MAS model considers ideal mesoscopic crystal layers in one of the three model phases austenite (A) and two martensitic twin phases (M+, M−) (Ref 1). The model derivation adheres strictly to thermodynamics and roots in the idea of a three-well potential energy with minima indicating the

This article is an invited paper selected from presentations at Shape Memory and Superelastic Technologies 2008, held September 21-25, 2008, in Stresa, Italy, and has been expanded from the original presentation.

Frank Richter, Oliver Kastner, and **Gunther Eggeler**, Department of Materials Science, Institute for Materials, Ruhr-University Bochum, 44780 Bochum, Germany. Contact e-mails: Frank.Richter@ruhr-universitaet-bochum.de, Oliver.Kastner@ruhr-universitaet-bochum.de, and Gunther.Eggeler@ruhr-universitaet-bochum.de.

stable locations of these three phases. The model yields a constitutive equation for the stress σ as a function of external strain ε and the internal variables x_A, x_{M+}, and x_{M-} which express the phase fractions of the model phases,

$$\sigma = E_M \cdot \frac{\varepsilon - (x_{M+} - x_{M-}) \cdot \varepsilon_T}{x_{M+} + x_{M-} + \frac{E_M}{E_A} \cdot x_A}. \quad \text{(Eq 1)}$$

The elasticity of the pure phases is expressed by the Young's moduli E_M and E_A and a martensitic transformation causes a residual strain ε_T. According to the MAS model, the phase fractions are implicitly functions of the temporal evolution of the stress-strain field and the temperature. These are determined by a set of rate equations representing an initial value problem (coupled ODE system) solved numerically for given strain and temperature. The numerically stiff feature of this ODE requires a stable integration scheme for robust solution. Here we use an implicit Runge-Kutta scheme of order five with an adaptive time step size called RADAU IIa (Ref 2). The temperature evolution within the specimen is due to latent heat emission and absorption and the according heat exchange with the ambiance. Assuming temperature homogeneity in simple geometries, the temperature can be calculated from the integral energy balance which represents an additional differential equation to be solved along with the rate equations for the phase fractions. Previously, this was accomplished by a standalone FORTRAN program (Ref 3, 4). An online version of this model considering a SMA wire is illustrated on the Internet (Ref 5).

ABAQUS/Standard allows user implementations of constitutive equations of state through a FORTRAN subroutine called UMAT (User Material), where the state of the nodal field quantities (here, displacements and temperature) can be accessed. UMAT is called by ABAQUS within its own numerical procedure while striving for a converged solution. Here, we have used the UMAT interface to adopt the standalone implementation of the MAS model into the ABAQUS environment. Further, the user has to provide the Jacobian stiffness matrix which for the MAS model reads

$$\frac{d\sigma}{d\varepsilon} = \frac{\partial\sigma}{\partial\varepsilon} + \frac{\partial\sigma}{\partial x_{M+}} \cdot \frac{dx_{M+}}{dt} \cdot \frac{dt}{d\varepsilon} + \frac{\partial\sigma}{\partial x_{M-}} \cdot \frac{dx_{M-}}{dt} \cdot \frac{dt}{d\varepsilon}, \quad \text{(Eq 2)}$$

owing to the dependence of the phase fractions on time t and the identity $x_A = 1 - (x_{M+} + x_{M-})$ (Ref 6, 7) while ε_T is a constant.

Whenever ABAQUS increments the nodal displacements and temperature, the actual evolution of the phase fractions x_A, x_{M+}, and x_{M-} are calculated from the rate equations by the use of the independent ODE solver RADAU. The phase fractions are calculated w.r.t. the nodal displacements fed in by ABAQUS in every iteration during a load step, when the displacements are changed from $u_i(t)$ to $u_i(t + \Delta t) = u_i(t) + \Delta u_i$. For instants τ with $\tau \in [t, t + \Delta t]$ the independent ODE solver then solves an initial value problem for the phase fractions where $u_i(\tau)$ is linearly interpolated between $u_i(t)$ and $u_i(t + \Delta t)$. Upon convergence, the stresses and the latent heat production due to phase transformations are returned to ABAQUS. Thermomechanical coupling is achieved by ascribing the actual specific latent heat production to the ABAQUS variable RPL within UMAT, hence serving as production term to the global energy balance. Within ABAQUS, the internal heat flux is expressed by Fourier's law and the surface heat flux by Dirichlet or Neumann boundary conditions, respectively, as usual. Note that not all of the ABAQUS elements are capable of the thermomechanical coupling. Rather, some elements employed in this study are strictly mechanical beam elements. In such cases, isothermal conditions are imposed.

Table 1 Details on the numerical implementation and performance of the simulation of the polycrystalline wire (Fig. 4 to 6)

ABAQUS	
Spatial discretization	$8 \times 8 \times 4$ elements (width $\times$ thickness $\times$ length)
Element type	C3D8T (8 nodes at the corners of the element, temperature degree of freedom)
Straining	6% strain in 6 s, 94 s hold time, release of strain in 6 s, 94 s hold time (defines the four load steps)
Initial conditions	$x_A = 1$, $x_{M+} = x_{M-} = 0$, temperature $T = 293$ K
Initial time increment	10^{-5} s in all four steps, set by user
Maximum permitted time increment	10^{-2} s (steps 1 and 3), 10^{-1} s (steps 2 and 4); set by user
Convergence criteria	Default ABAQUS values
RADAU	
Dimension of the system	3 (referring to the rate equations for the state variables x_A, x_{M+}, x_{M-})
Relative error tolerance	10^{-6}
Absolute error tolerance	10^{-6}
Initial step size	10^{-3} s

Convergence performance as reported by the ABAQUS status file for this particular simulation: 3136 increments, out of which 3130 converge in one iteration. A time increment of 10^{-1} s was accepted 1872 times during hold times and a time increment of 10^{-2} s was accepted 1196 times during straining and unstraining. The smoothly converged simulation demonstrates the numerical stability of the interplay between ABAQUS and RADAU

The numerical stability of the computer model is achieved by proper choices of the convergence criteria offered by both ABAQUS and the embedded ODE solver. Table 1 provides an overview over the most important quantities as set for the particular simulation explained in Section 3.3.

The FEM-embedded version of the MAS model features advantages of the FEM method as opposed to the standalone version: the deformation is no longer limited to tension, the geometry and spatial discretization are arbitrary, internal heat conduction affects thermal gradients, mechanical and thermal interaction with other parts of the model and the ambiance can be incorporated. At present, verification of the FEM results is restricted to close reproduction of the outcome of the standalone program as well as by an implementation in FEMLAB® (COMSOL AB, Stockholm, Sweden, Ref 7).

3. Results

3.1 Beam Bending

Beam bending is more interesting than simple axial tension as in this case the magnitude and the sign of the uniaxial normal stress vary through the beam thickness. Planar ABAQUS Euler-Bernoulli beam elements are attractive as they feature a single stress and strain component along and parallel to the beam axis. However, the ABAQUS element library does not offer a beam element for coupled temperature-displacement procedures.

In this subsection, we use the ABAQUS Euler-Bernoulli-beam element B23 and restrict all simulations to strictly isothermal conditions. Using this model, the bending of a straight horizontal SMA cantilever is simulated and the results are compared to a reference solution (a) and to the stress contour produced by a hypothetical ideal elastic-plastic material (b). The simulations allow for geometrically nonlinear behavior.

(a) The bending of a straight SMA cantilever which is fully restrained at one end, but loaded by a vertical point force at its free end is presented in Ref 7. Here, the force is defined as a follower load and the constitutive behavior is pseudoplastic according to the temperature set. The initial phase fractions of M+ and M− are equal. Figure 1 depicts the profile of the M+ fraction over the cross section of the beam in three-dimensional space at the instant of maximum deflection. Upon bending, the solution exhibits the typical compression/tension profile of stress for Euler-Bernoulli beams in the elastic regime modulated by pseudoplastic behavior in the transformation regime (Ref 8). Accordingly, the phase fractions of the martensitic variants evolve along the cantilever axis and through the thickness. The M+ fraction increases where the material is subjected to tensile stress (i.e., $y <$ y-position of the neutral fiber in Fig. 1) whereas in compressed material parts, the M− fraction is growing at the expense of the M+ fraction. The neutral fiber is indicated by the feet of the respective data points on the (x, y)-plane, showing the actual deflection in this plane. Analogous graphs can be constructed for the M− fraction, the stress and the strain. Such load-induced phase changes and the according formation of martensite plates have indeed been verified experimentally (Ref 9).

(b) The FEM solution for the transformation zone in a SMA cantilever can be compared to an analytical solution of the yield zone in an ideal elastic-plastic material (Ref 10). Figure 2 shows the solutions for the SMA model and the ideal elastic-plastic cantilever. Both theories predict the same boundary line of the transformed zone (solid black line in Fig. 2) thus verifying the numerical validity of the computer model. Note the analytical solution is capable of this boundary line only while the MAS model additionally provides the spatial distribution of the M+ phase. Both solutions were calculated on the basis of the geometrically nonlinear theories.

3.2 Torsion of a Pseudoelastic SMA Tube

Conveniently, torsion is characterized by a uniaxial state of torque stress. The free oscillation of a pseudoelastic SMA pendulum at constant temperature ($T = 353$ K) has been studied in Ref 11 and 12 using the standalone implementation of the MAS model. We adopt the setting from Ref 11 to study the behavior of the FEM-based computer model where the earlier results serve as reference solution. The torsion pendulum is represented by a thin-walled pseudoelastic NiTi tube (2 mm diameter, 5 cm length, 50 μm wall thickness). Supported at one end, this tube is attached to a weight at the opposite end which produces the required inertia upon free oscillation. The oscillation started from a pre-twisted state where the according torque induces a purely martensitic state. In the FEM-based model, the tube is discretized by 10 ABAQUS beam B33 elements along the axis. The FEM-based solution and the

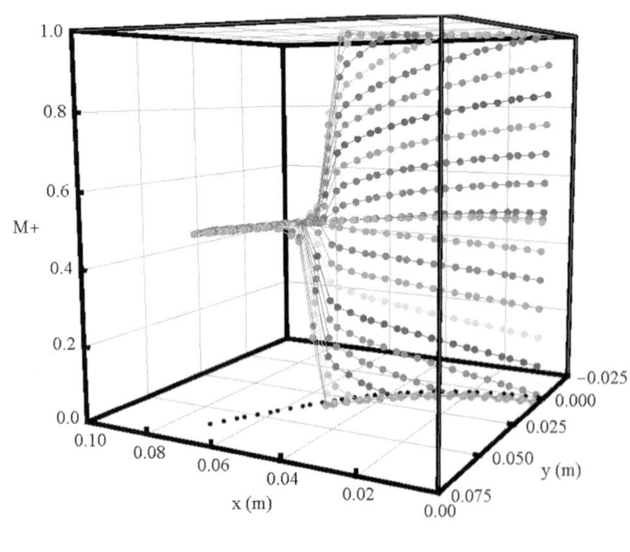

Fig. 1 Spatial distribution of the martensite M+ phase in a cantilever under vertical loading. The straight cantilever is clamped at $x = 0$ in the pseudoplastic state and bent by an external force acting in the positive y-direction at $x = 0.1$ m. Plotted is the spatial distribution of the M+ phase content on and parallel to the beam axis as a function of spatial coordinates at the instant of maximum bending. The spatial resolution through the beam thickness is provided by 27 section points equidistant through the thickness. The *black dots* on the base plane show the beam curvature. Original beam dimensions: Length = 0.1 m, thickness = 0.01 m (Ref 7)

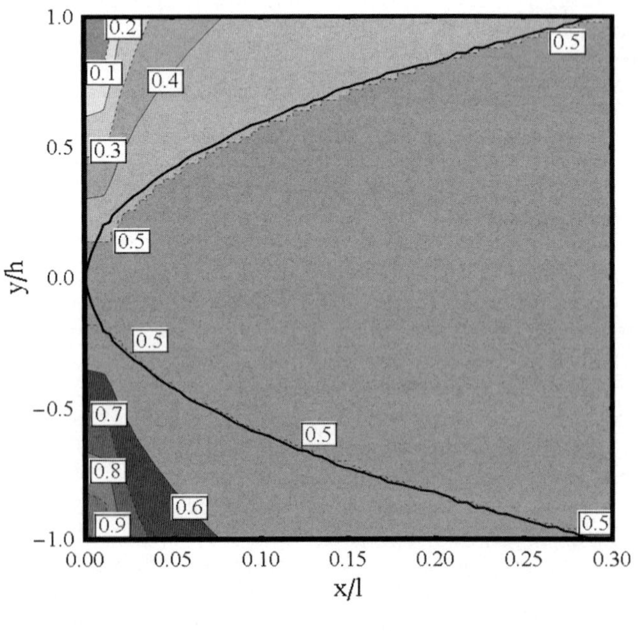

Fig. 2 Contour plot of the M+ fraction (plotted on the undeformed configuration; only 30% of the beam from the clamped end are depicted), defining the transformation zone in a pseudoplastic cantilever at vertical tip loading in the positive y-direction. The load equals 1.825 times the yield load as detected for the geometrically linear case. The *black solid line* indicates the analytical solution for the plastic deformation boundary obtained from an ideally elastic-plastic material (digitized from Ref 10)

reference solution are superposed in Fig. 3 where the black dots indicate the reference solution and lines reflect the FEM-based torsion at three axial nodes (located at the tube ends and in the center, respectively) as function of time. The reference solution coincides with the torsion/time function of the node at the oscillating tube end. At the center node, the torsion is proportionally less while the opposite end remains at rest. During the first 50 s, the inertia of the weight produces sufficient torque to loop the material through the pseudoelastic hysteresis. Accordingly, hysteresis-related energy dissipation causes the significant damping visible in Fig. 3. Eventually, the material behaves purely elastic, effecting undamped harmonic oscillations.

3.3 Tensile Test of a Polycrystalline SMA Wire

Seelecke and Heintze (Ref 13, 14) have contributed a polycrystalline formulation of the MAS model which overcomes some of the physical limitations of the original single-crystalline model. Their method of 'parameterization' renders the model more flexible but preserves the computational simplicity of the single crystal implementation with minor adaptations. The polycrystal model takes into account the averaged excess stresses occurring in a domain-structured material (originating from grain orientations, local transformation stresses, and other local effects). It has been ported into the ABAQUS environment and was used to simulate uniaxial straining of a pseudoelastic SMA wire. The results are compared to the original work (Ref 13). The FEM model employs the ABAQUS element type C3D8T to discretize a wire with dimensions 1.07 mm × 0.8 mm × 6.35 cm (width × thickness × length). This element type is capable of 3D problems, but for the special case of uniaxial loading the application of the one-dimensional polycrystalline MAS model is possible. The significant advantage of the C3D8T element over the formerly used beam-type elements is that this element type permits true thermomechanical coupling. Figures 4 to 6 show the result of such tensile testing with a tetragonal wire geometry. The wire is loaded and unloaded in displacement control mode so as to simulate a complete pseudoelastic loading cycle, see inset in Fig. 4. Temperature effects occur during yielding (0-6 s) and recovery (100-106 s) of the transformation strain (see Fig. 4, inset), where the release and the absorption of latent transformation heat cause self-heating and self-cooling, respectively. These effects are compensated by heat exchange with an ambient medium at a temperature of 293 K. The temperature profile resulting from these heat transfers is resolved in the simulation: Fig. 5 shows a contour plot of the temperature field at the tip face of the wire at maximum load (after 6 s). Note the minute thermal conductivity was chosen deliberately to create a notable thermal gradient throughout the tiny volume of the wire. Figure 4 shows the temperature evolution at three node sets which are localized along the volume center, one face center, and one edge of the wire (incorporating all nodes along the length). Even though

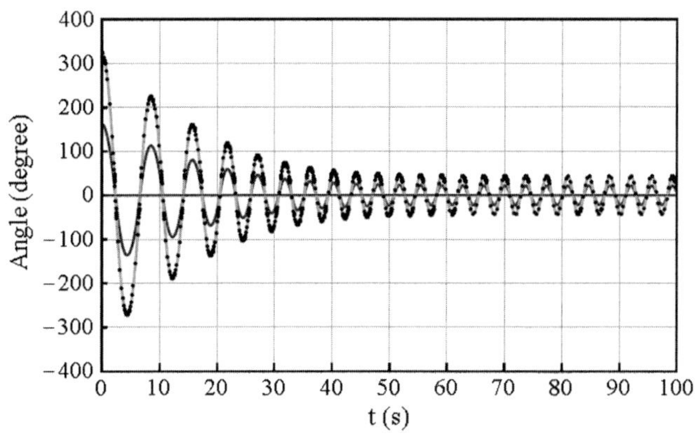

Fig. 3 Free oscillation of a pseudoelastic NiTi torsion pendulum. Comparison of a reference solution from the literature (*black dots*, Fig. 3 in Ref 11) with the FEM-embedded version of the MAS model (*colored lines*). Initially the pendulum is twisted by 324°. *Red line*: Fixed tube end (no angular displacement), *blue line*: node at mid-length, *green line*: torqued end. The reference solution refers to the latter; the red and blue solutions illustrate the axial evolution of the torsion

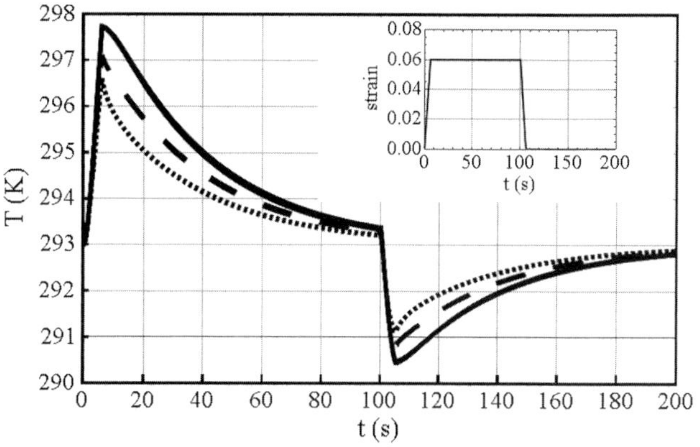

Fig. 4 Temperature as a function of time for three node sets located at the volume center (*solid line*), face center (*dashed line*), and edge (*dotted line*). Inset: imposed strain as function of time

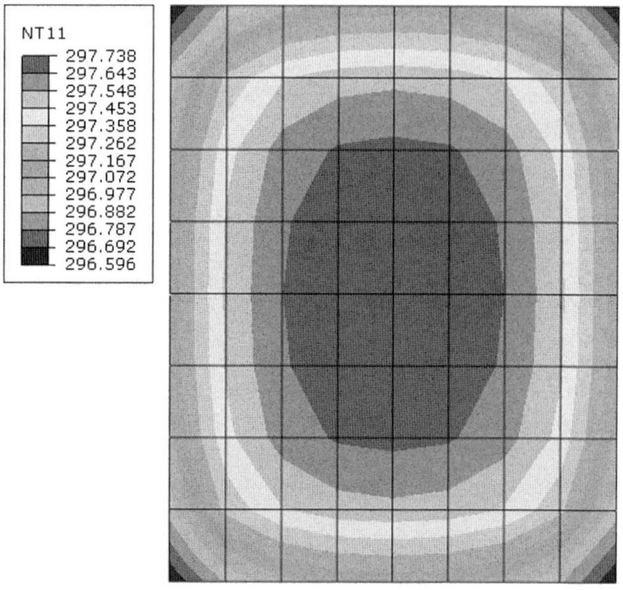

Fig. 5 Contour plot of the temperature field ('NT11,' in K) at the tip of the fully transformed wire at maximum load (after 6 s in Fig. 4), showing a pronounced temperature gradient which indicates the internal heat flux

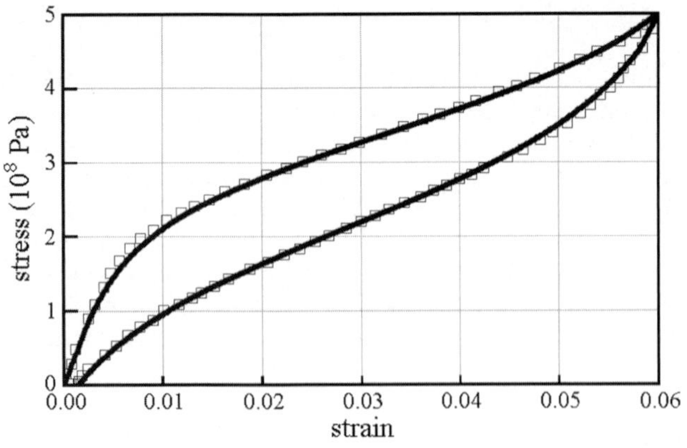

Fig. 6 Simulated stress/strain curve of a pseudoelastic, tetragonal SMA wire under uniaxial straining. *Black dots*: FEM-based simulation, *squares*: reference solution taken from the literature (Fig. 4.11 in Ref 13)

the detected temperature variations are small, the heat flux driven by latent heat produces a distinct temperature field with notable gradients. Such effects are augmented in special geometric situations, under fast straining or, e.g., in composite materials where the heat flow could be restricted. The according stress-strain curve for this example of a polycrystalline model is shown in Fig. 6. Obviously, the FEM-embedded version of the model exhibits convincing coincidence with results obtained by the standalone version of the polycrystalline model in the literature.

4. Conclusions

The ABAQUS-version of the MAS model is capable of achieving a good match to digitized data from reference solutions or, where applicable, to analytical computations. In summary, it can be stated that the MAS model is well suited to describe the constitutive behavior of SMA for both model formulations, single and polycrystalline materials. A prime advantage is its versatility, including the inherent thermomechanical coupling and the extension to polycrystalline materials.

Acknowledgments

Financial support by the German Research Foundation (DFG) Grant no. KA 2304/2-1 is much appreciated. We thank Prof. S. Seelecke (North Carolina State University, USA) and O. Heintze (PhD, German Aerospace Center (DLR), Braunschweig, Germany) for valuable discussions about the model.

References

1. S. Seelecke and I. Müller, Shape Memory Alloy Actuators in Smart Structures: Modeling and Simulation, *Appl. Mech. Rev.*, 2004, **57**(1), p 23–46
2. E. Hairer and G. Wanner, *Solving Ordinary Differential Equations II: Stiff and Differential-Algebraic Problems*, Springer Series in Computational Mathematics 14, 2nd ed., Springer, Berlin, 1996
3. M. Achenbach, A Model for an Alloy with Shape Memory, *Int. J. Plast.*, 1989, **5**(4), p 371–395
4. I. Müller and S. Seelecke, Thermodynamic Aspects of Shape Memory Alloys, *Math. Comp. Mod.*, 2001, **34**(12–13), p 1307–1355
5. Interactive Webpage for the Simulation of Shape Memory Wires, www.smaterial.com
6. J.P. Frautschi, "Finite Element Simulations of Shape Memory Alloy Actuators in Adaptive Structures," M.Sc. Thesis, North Carolina State University, USA, 2003
7. Q. Li, "Modeling and Finite Element Analysis of Smart Materials," Ph.D. Thesis, North Carolina State University, USA, 2006
8. T. Atanackovic and M. Achenbach, Moment-Curvature Relations for a Pseudoelastic Beam, *Cont. Mech. Thermodyn.*, 1989, **1**(1), p 73–80
9. J. Rejzner, C. Lexcellent, and B. Raniecki, Pseudoelastic Behaviour of Shape Memory Alloy Beams Under Pure Bending: Experiments and Modelling, *Int. J. Mech. Sci.*, 2002, **44**(4), p 665–686
10. H. Parisch, Large Displacement of Shells Including Material Nonlinearities, *Comput. Meth. Appl. Mech. Eng.*, 1981, **27**(2), p 183–214
11. O. Heintze, O. Kastner, H. Sahota, and S. Seelecke, The Role of Thermomechanical Coupling in the Dynamic Behavior of Shape Memory Alloys, *IUTAM Symposium on Smart Structures and Structronic Systems: Proceedings of the IUTAM Symposium*, Magdeburg, Germany, U. Gabbert and H.S. Tzou, Ed., Sept 26–29, 2000, Springer, 2001, p 145–152
12. S. Seelecke, Modeling the Dynamic Behavior of Shape Memory Alloys, *Int. J. Non-Lin. Mech.*, 2002, **37**, p 1363–1374
13. O. Heintze, "A Computationally Efficient Free Energy Model for Shape Memory Alloys—Experiments and Theory," Ph.D. Thesis, North Carolina State University, USA, 2004
14. O. Heintze and S. Seelecke, A Coupled Thermomechanical Model for Shape Memory Alloys—From Single Crystal to Polycrystal, *Mater. Sci. Eng. A*, 2008, **481–482**, p 389–394

JMEPEG (2009) 18:631–637
DOI: 10.1007/s11665-009-9395-9

Numerical Study on the Influence of Material Characteristics on Ni-Ti Endodontic Instrument Performance

Lorenza Petrini, Silvia Necchi, Silvio Taschieri, and Francesco Migliavacca

(*Submitted September 24, 2008; in revised form January 20, 2009*)

Ni-Ti rotary endodontic instruments (*files*) are used in dentistry during the endodontic treatment to shape the root canal of the tooth while removing the pulp when infected. Up to now, the studies for evaluating their performances and drawbacks were mainly limited to experimental tests on product flexural and torsional resistance. This work exploits computational analyses for investigating the effects of materials with different mechanical properties on the behavior of rotary endodontic instruments. The aim is to understand the appropriate material choice to reduce the criticality of the treatment in particular clinical conditions. In particular, the interaction between an accurately modeled rotating file and differently shaped root canals during the clinical procedure was studied performing finite element analyses. Strains induced by the treatment on a file made of a "standard" Ni-Ti alloy (characterized by average properties of the pseudoelastic behavior), a "long" Ni-Ti alloy (characterized by wide transformation region), a "super" Ni-Ti alloy (characterized by an extended Hookian behavior without transformation region), and stainless steel were compared. The results accurately show the advantages of the use of Ni-Ti alloy with respect to stainless steel and the better performance of the "long" alloy in all the tested case.

Keywords endodontic files, finite element method, ProTaper, pseudoelasticity, shape memory alloys

1. Introduction

Endodontics is the field of dentistry dealing with problems of the pulp, the living tissue contained in the root canal of the tooth (Fig. 1a). Rotary endodontic instruments (also called *endodontic files*) are used during the endodontic treatment (Fig. 1b-d) to shape the root canal while removing the pulp when infected. A good shaping of the canal is mandatory to allow its efficient cleaning and obturation and to ensure the success of the treatment. Endodontic instruments must show a good flexibility and a cutting profile to be able to penetrate in the root canal and to remove dentin; therefore, they result in a sort of thin and flexible drill bit (Fig. 1c).

The first endodontic files were manufactured in stainless steel and used manually. Since the end of 1980s, nickel-titanium engine-driven rotary instruments are commercially distributed and used increasingly in the endodontic practice. Indeed, Ni-Ti alloys (a family of shape memory alloys—SMA) confer the files a great flexural ability thanks to their pseudoelastic behavior, i.e., a nonlinear elastic behavior that allows the material to undergo mechanically induced high deformations without plastic residual strain after stress removal. This characteristic behavior of SMA is due to a stress-induced reversible solid-solid phase transformation between two different crystallographic structures, austenite and martensite, which takes place when the working temperature is greater than a specific value for each alloy (called *austenite finish temperature*, A_f). The use of Ni-Ti on behalf of stainless steel is a great improvement for the treatment, as shown in this work. The unique drawback is the proneness of the Ni-Ti device to break without early warning sign. Accordingly, it could be very useful for the clinician to know if and when a repeated use of the files is allowed. Quite a lot of works devoted to give some indications to dentists are present in the literature on Ni-Ti instruments. In particular, many experimental studies were dedicated to evaluate their performance in terms of flexibility and torsional resistance (Ref 1-3). On the other hand, from the computational point of view, only three studies (Ref 4-6) based on finite element analyses (FEA) were conducted, aiming to evaluate some aspects of the mechanical behavior of the instruments (e.g., the stress distribution) related to their critical condition during root canal instrumentation and not assessable through laboratory or in vivo tests. In particular, Berutti et al. (Ref 4) and Xu and Zheng (Ref 5) proposed simple models of different Ni-Ti instruments not accurate in terms of files geometries, loading conditions, and materials constitutive models. To simulate realistic clinical conditions, recently Necchi et al. (Ref 6) created a more truthful model of a

This article is an invited paper selected from presentations at Shape Memory and Superelastic Technologies 2008, held September 21-25, 2008, in Stresa, Italy, and has been expanded from the original presentation.

Lorenza Petrini, Structural Engineering Department, Politecnico di Milano, Milan, Italy; **Silvia Necchi** and **Francesco Migliavacca**, Laboratory of Biological Structure Mechanics, Structural Engineering Department, Politecnico di Milano, Milan, Italy; and **Silvio Taschieri**, IRCCS Istituto Ortopedico Galeazzi, Health Technologies Department, Università degli Studi di Milano, Milan, Italy. Contact e-mail: lorenza.petrini@polimi.it.

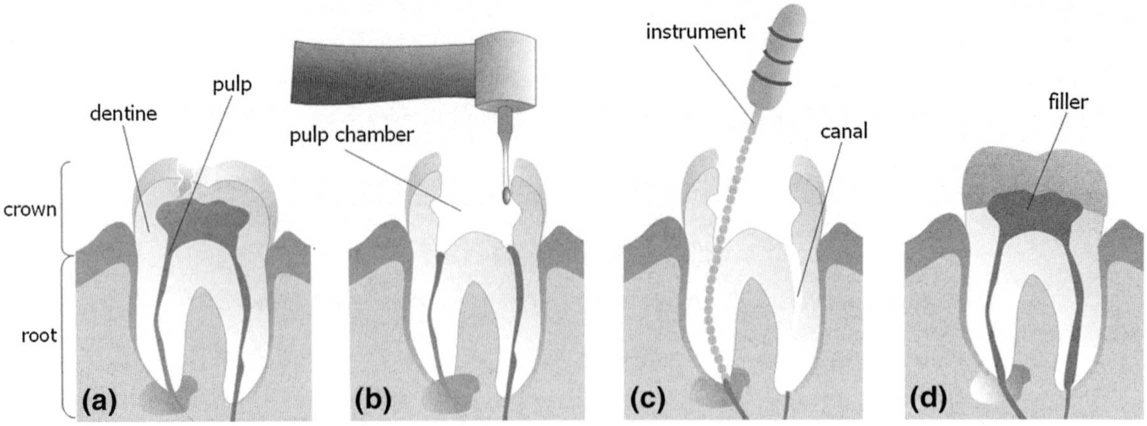

Fig. 1 Sketch of a tooth section with description of the endodontic treatment: unhealthy tooth (a), drilling (b), shaping (c), and obturation (d)

specific Ni-Ti instrument operating in a root canal. To reach this goal, the authors accurately reproduced the geometry of the instrument and different root canals. Furthermore, they used an advanced constitutive model for the Ni-Ti alloy. The stress that the instrument undergoes during clinical procedures (composed of continuous insertion and removal of the instrument in and out of the canal) was studied to evaluate the effects of the canal shape generated in the instrument.

In this work, the same authors intended to deepen the research previously presented on the subject by comparing the mechanical performance of three different Ni-Ti alloys when used to produce rotary endodontic instruments. Attention was paid to understand if an appropriate material choice may reduce the criticality of the treatment in particular clinical conditions. In fact, Ni-Ti alloys are characterized by strong dependence of the mechanical behavior on composition and working process. As a consequence, they give the possibility of "designing" the material according to the required device properties. Nevertheless, material and products are prone to anomalous mechanical behavior deriving from uncontrolled device production phases. The improvement in the technique due to SMA versus stainless steel employment was also underlined.

2. Numerical Modeling

The software Rhinoceros 2.0 Evaluation (Robert McNeel & Associates, Seattle, WA) was used for creating the geometrical model, while the commercial code ABAQUS 6.5-1/Standard (SIMULIA, Providence, RI) was employed for the computational analyses. In the following, the FE (finite element) model characteristics, in terms of geometry, material, boundary, and interaction conditions are described.

2.1 Geometrical Models and Meshes

The same 3D model of a ProTaper finisher file F1 (manufactured by Dentsply Maillefer Instruments—Ballaigues—Switzerland and used in final step of the shaping sequence) proposed in Necchi et al. (Ref 6) was used in this work. Figure 2 shows the main geometrical features of the instrument reproduced in the model. The handle part of the file does not serve any specific function in the shaping procedure, and therefore it was neglected in the numerical analyses. The created 3D geometrical model of the file was meshed using 10-nodes tetrahedral elements. A grid sensitivity study was performed to choose the most convenient number of elements (in terms of computational time and results accuracy) that was found in 5241, which assured an accurate description of sharp angles and curves.

The canals were modeled as simple rigid surfaces shaped as pipes (as described in the previous study), following a canal classification proposed in the literature (Ref 7), and based on radius, angle, and position of curvature. In this work, four types of canal geometries were considered (Fig. 3), characterized by a radius r of 2 or 5 mm, an angle α of 45° or 30°, and an apical or middle position (p) of the curvature. The 3D models of the canals were meshed using 4-node bilinear elements. A smooth discretization was used to prevent computational problems in the simulation during the contact between the canal wall and the file.

2.2 Materials Definition

Three Ni-Ti alloys, a *standard*, a *long*, and a *super* alloy, were considered, presenting a pseudoelastic behavior characterized by different values of hardening coefficient, limit transformation strain, and single-variant martensitic yielding stress and strain. In particular, the *standard alloy* resembles an alloy with mean properties and a transformation region amplitude of 7%. The *long alloy* resembles an ad hoc-designed material characterized by a wider transformation region (amplitude 9%) with respect to the *standard* one. The *super alloy* resembles a martensitic Ni-Ti alloy which underwent work-hardening during device production and hence shows an extended Hookian behavior without transformation region at any temperature (Ref 8). The stress-strain diagram describing the pseudoelastic behavior of the studied alloys is shown in Fig. 4. The mechanical parameters used for defining this behavior are listed in Table 1 and shown in Fig. 5. They are: the *transformation starting* and *finishing stress* (σ_{ts} and σ_{tf}, i.e., the stress values at which, at the working temperature, the transformation between austenite and single-variant martensite starts and finishes, respectively), the *limit transformation strain* (ε_L, i.e., the amplitude of the transformation strain interval), and the *martensitic yielding stress* and *strain* (σ_y^M and ε_y^M, the so-called *pseudoelastic limits*, i.e., the limit values over which the pseudoelastic behavior finishes). The respective values were arbitrarily imposed due to the impossibility of retrieving any information about the real materials used by the

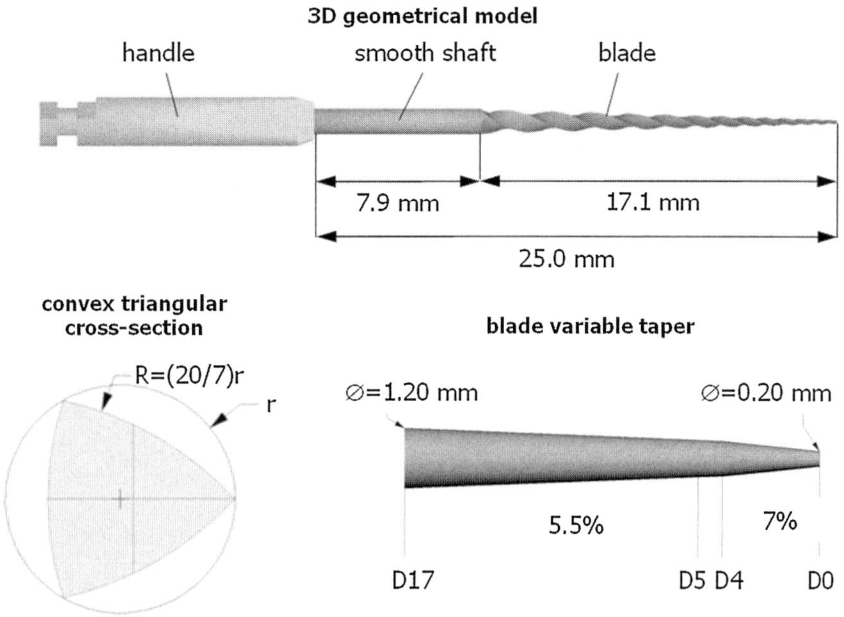

Fig. 2 Main geometrical features of the ProTaper F1 reproduced in the model

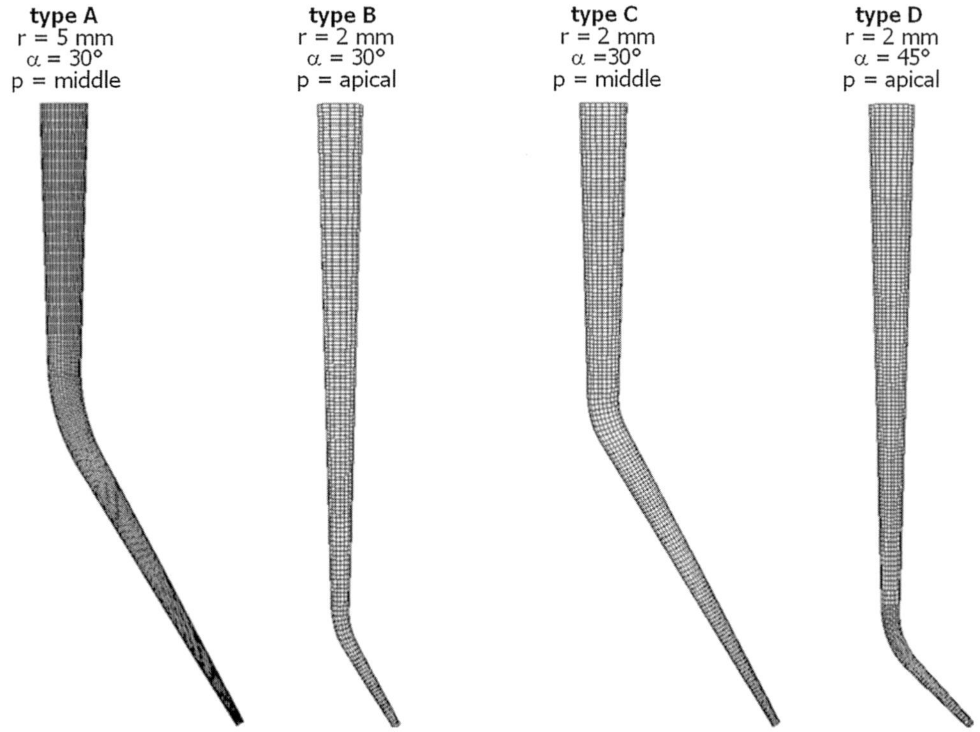

Fig. 3 Root canal geometries (meshed models)

instrument manufacturers. An ad hoc computational subroutine (Ref 9) was used to reproduce the complex thermomechanical behavior of the Ni-Ti alloys. Values of 25 and 10 °C were chosen for the working temperature and the temperature A_f, respectively. Since the used subroutine is not able to describe the plastic behavior following yielding of martensite, in this study, the simulation results were considered reliable only when stress and strain values were lower than the pseudoelastic limits. Moreover, the accumulation of plastic deformation due to cyclic loading in the pseudoelastic ranges was not accounted for.

AISI 316L stainless steel, widely used in biomedical applications, was employed as comparison material. The mechanical parameter values (Ref 10) are listed in Table 1 and the corresponding stress-strain diagram is reported in Fig. 4. For describing this behavior, an elastoplastic constitutive

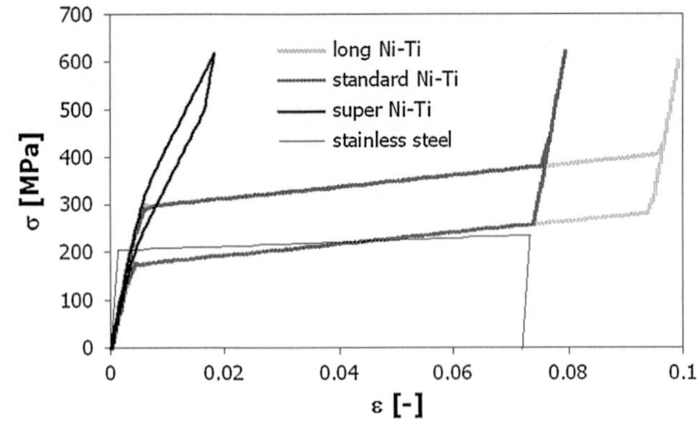

Fig. 4 Stress-strain diagram of the considered "ideal" Ni-Ti alloys and of a typical stainless steel

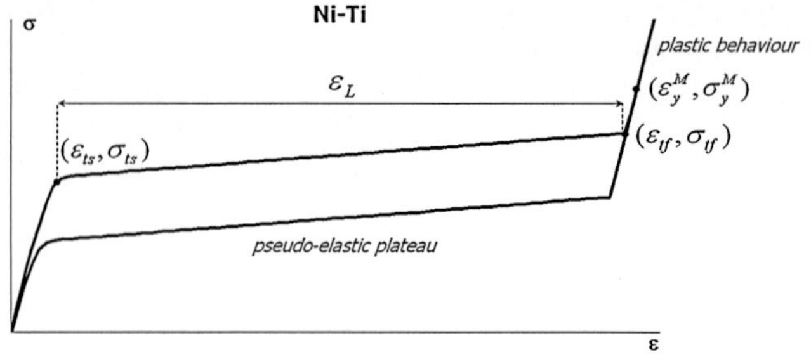

Fig. 5 Stress-strain diagram of the characteristic pseudoelastic behavior of SMA (ts = transformation starting, tf = transformation finishing, y = yield, L = limit)

Table 1 Values of stainless steel and Ni-Ti alloy characteristic mechanical parameters used in the numerical analyses

		Ni-Ti			
		Standard	Long	Super	Stainless steel AISI 316L
E, MPa	Young modulus	70,000	70,000	70,000	193,000
ν	Poisson ratio	0.3	0.3	0.3	0.3
σ_y^S, MPa	Yield stress	...	...	...	205
ε_y^S, %	Yield strain	...	...	...	0.1
σ_f, MPa	Fracture stress	...	...	...	515
ε_f, %	Fracture strain	...	...	...	60
H, MPa	Hardening parameter	1,000	1,000	30,000	520
σ_{ts}, MPa	Transformation starting stress	300	300	...	...
σ_{tf}, MPa	Transformation finishing stress	380	400	...	...
ε_L, %	Limit transformation strain	7	9	...	...
σ_y^M, MPa	Martensitic yielding stress	480	550	1,975	...
ε_y^M, %	Martensitic yielding strain	7.7	9.7	7.0	...

model with kinematic hardening, provided by the computational code library, was adopted.

2.3 Boundary and Interaction Conditions

The boundary and interaction conditions were imposed to simulate the clinical operating conditions (see Ref 6 for details). The modeled instruments were forced into the root canals until the apex was reached (*insertion* step) and immediately retrieved (*removal* step), while maintaining a continuous rotation around their longitudinal axis.

To avoid excessively long computational time, a rotational speed having an order of magnitude lower than the real one was imposed to the instruments. Moreover, as a first approach to the problem, neither friction nor machining action was considered between the instrument blade and the canal wall. A "soft" contact with an exponential pressure-overclosure model was imposed to simulate their interaction. Hence, the torsional stresses induced in the file during the procedure were neglected.

3. Results

The instruments performance was studied by analyzing the strain values reached during the phases of file insertion and removal into and out of the canals. The strain is a more accurate quantity to check. Indeed, the nonlinear behavior of both materials is mainly characterized by high increments of strain but small variations of stress for the presence of a wide region with low hardening modulus (corresponding to the transformation form austenite to single-variant martensite for Ni-Ti and to material plasticization for steel) in their stress-strain curves. In the following, this region will be indicated as "plateau."

For results evaluation, *equivalent strains*, given by the combination of the strain tensor components in all the directions ($\varepsilon^{eq} = \sqrt{(2/3)\boldsymbol{\varepsilon} : \boldsymbol{\varepsilon}}$), along with the *logarithmic maximum principal strain* ($\ln \lambda_{max}$) were calculated allowing to obtain scalar values to be compared with the limits derived from uniaxial tests.

In particular, the reference variables considered for the Ni-Ti alloys were ε_{tr}^{eq} (i.e., the instantaneous equivalent strain hoarded in the transformation plateau) and $\ln \lambda_{max}$ (i.e., the instantaneous maximum principal total strain the instruments undergoes, sum of the elastic, and the transformation component). The undeformed shape recovery after the removal step was considered assured if $\varepsilon_{tr}^{eq} \leq \varepsilon_L$. In the limit case of $\varepsilon_{tr}^{eq} = \varepsilon_L$, that means in the case of completed transformation, the additional condition $\ln \lambda_{max} \leq \ln(1 + \varepsilon_y^M)$ was required to be satisfied.

In the stainless steel case, ε_{pl}^{eq} (i.e., the instantaneous equivalent permanent deformation that appears whenever the yielding strain ε_y is reached) and $\ln \lambda_{max}$ (i.e., the instantaneous maximum principal total strain the instruments undergoes, sum of the elastic, and the plastic component) were considered.

The strain levels were analyzed mainly at the end of the insertion step, because it turned out as the more demanding condition for the materials. Table 2 provides the values obtained for the *logarithmic maximum principal strain* ($\ln \lambda_{max}$) and transformation (ε_{tr}^{eq}) or plastic (ε_{pl}^{eq}) strain in this configuration as a function of the considered material and canal type.

In the case of the Ni-Ti alloys, the $\ln \lambda_{max}$value ranges are 6.6 to 8.4%, 6.6 to 9.0%, and 5.2 to 7.9% for the *standard*, the *long*, and the *super* alloy, respectively. Correspondingly, ε_{tr}^{eq} values range between 5.8 and 7.0% and between 5.8 and 8.6% for the *standard* and the *long* alloy (no equivalent transformation strain is produced in the case of *super* alloy).

Critical values of $\ln \lambda_{max}$ and ε_{tr}^{eq} were observed for both the *standard* and *super* alloys during insertion in canal type C (very sharp curvature in the medial part) which turned out to be the worst working condition. In this case, the recovery of the undeformed shape was not possible. Contrarily, the instrument with *long* alloy resulted as working in the material pseudo-elastic domain. Moreover, in every considered case, the *long* alloy furnished the better performance in terms of difference between the reached strain levels and the material critical thresholds ($\ln(1+\varepsilon_y^M)$ and ε_L for $\ln \lambda_{max}$ and ε_{tr}^{eq}, respectively). Figure 6 shows the distribution of the total strain at the end of the insertion step in canal type C. Analyzing the answer of the different files to the variation of the curvature parameters, the *super* alloy results to be the most sensitive material in terms of reached $\ln \lambda_{max}$ values. Indeed, the percentage increases for the *standard*, *long*, and *super* alloys, respectively, were: (a) 27, 36, and 52% when shifting the radius from 5 to 2 mm (canal type A to type C); (b) 3, 18, and 21% changing the angle from 30° to 45° (canal type B to type D); and (c) 17, 23, and 36% when moving the curvature position from apical to medial (canal type B to type C). In terms of ε_{tr}^{eq}, the comparison between the *standard* and the *long* alloy indicates the second as the more sensitive to the variation of the canal parameters. In fact, the following respective increments were recorded for these alloys: (a) 21% and +48% when shifting the radius from 5 to 2 mm (canal type A to type C); (b) 3% and +5% changing the angle from 30° to 45° (canal type B to type D); and (c) 9 and 16% when moving the curvature position from apical to medial (canal type B to type C).

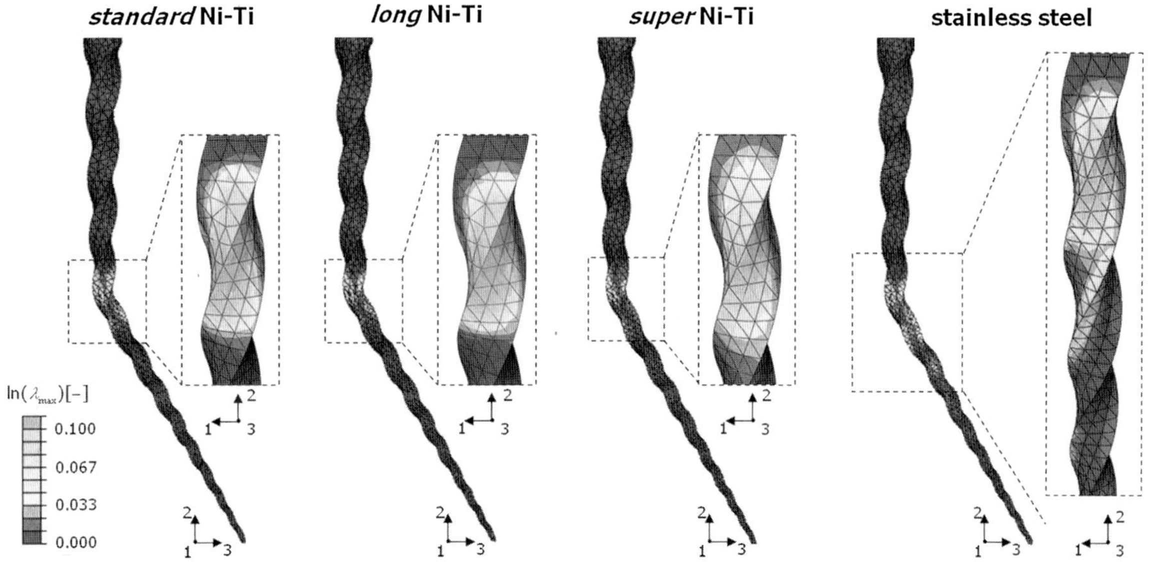

Fig. 6 End of insertion step, canal type C: logarithmic maximum principal strain levels reached in Ni-Ti and stainless steel instruments

Table 2 Percentage values obtained for the specific variables concerning Ni-Ti (logarithmic maximum principal strain $\ln \lambda_{max}$ and equivalent transformation strain ε_{tr}^{eq}) and stainless steel (logarithmic maximum principal strain $\ln \lambda_{max}$ and equivalent plastic strain ε_{pl}^{eq}) at the end of the insertion step

				Ni-Ti						
	Parameters			Standard		Long		Super	Stainless steel	
Canals	r	α, °	p	$\ln \lambda_{max}$	ε_{tr}^{eq}	$\ln \lambda_{max}$	ε_{tr}^{eq}	$\ln \lambda_{max}$	$\ln \lambda_{max}$	ε_{pl}^{eq}
A	5	30	m	6.6	5.8	6.6	5.8	5.2	7.1	7.1
B	2	30	a	7.2	6.4	7.3	7.4	5.8	10.2	10.0
C	2	30	m	8.4	7.0	9.0	8.6	7.9	10.9	11.5
D	2	45	a	7.4	6.6	8.6	7.8	7.0	12.1	11.9

The canal parameters refers to the radius (2 or 5 mm), the angle (30° or 45°), and the position (apical or medial) of the curvature

Table 3 **Percentage of variation in the strain variables as function of the curvature parameters for the Ni-Ti and the stainless steel instruments**

		Ni-Ti						
		Standard		Long		Super	Stainless steel	
Canals	Parameters	ln λ_{max}, %	ε_{tr}^{eq}, %	ln λ_{max}, %	ε_{tr}^{eq}, %	ln λ_{max}, %	ln λ_{max}, %	ε_{pl}^{eq}, %
A vs. C	r: 2 vs. 5 mm	+27	+21	+36	+48	+52	+54	+62
D vs. B	α: 45° vs. 30°	+3	+3	+18	+5	+21	+19	+19
C vs. B	p: m vs. a	+17	+9	+23	+16	+36	+7	+15

In the case of stainless steel, the instrument underwent critical total strain levels, reaching values of the *logarithmic principal strain* (ln λ_{max}) well beyond the yielding strain (ε_y^S), varying between 7.1 and 12.1%. ε_{pl}^{eq} ranged between 7.1 and 11.9%. From the studied cases, the most critical condition resulted in the canal type D. As a function of the curvature parameters, ln λ_{max} and ε_{pl}^{eq} increased of: 54 and 62% shifting the radius from 5 to 2 mm, 19% changing the angle from 30° to 45°, 7 and 15% moving the curvature position from apical to medial.

The depicted results (Table 3) clearly indicate that, for all the considered material, in the canal comparison, the primary curvature parameter influencing the mechanical behavior of the instrument (higher variations of strain) is the radius of curvature. Differently, the second dominant parameter is judged the position of curvature for the Ni-Ti alloys and the angle of curvature for the stainless steel.

The values of ln λ_{max} obtained for the stainless steel file were higher in every canal compared to the Ni-Ti instrument (see Fig. 6). Moreover, the stainless steel instruments showed a lower ability to conform to the canal shape during an entire insertion-removal cycle. In particular, the tip of the stainless steel file plasticized just after the first contact with the canal wall, determining successive irregular interactions between the file and the wall, and inducing the instrument to work in the plastic range of the material for the larger portion of the treatment duration. The Ni-Ti file, instead, bends uniformly along the blade, following the original curvature of the canal. This behavior was shown independently by the canal geometry.

4. Discussion and Conclusions

FEA are recognized to have an important role in optimizing the behavior of biomedical devices. In this work, focused on the study of the mechanical performance of a Ni-Ti rotary endodontic instrument, attention was mainly paid to understand the effects of the materials choice in the endodontic treatment and those of a not correct production on the performance of the file. Three types of alloys were considered. The clinical operating conditions were simulated by imposing a combined movement of translation and rotation of the instrument into and out of canal root. Four geometries of canals, differing in their curvature features, were taken into account. Flexural strains varying cyclically during rotation of the endodontic instrument were analyzed.

The results from the simulations suggest that Ni-Ti alloys are good materials for the production of engine-driven rotary instruments, especially if compared to stainless steel. As expected, the Ni-Ti files showed high flexibility during the transformation of the alloy from austenite to single-variant martensite. Considering the *standard* alloy, only in the case of a very sharp curvature in the medial part of the canal, the plastic domain was reached. In the case of *super* alloy, the martensitic yielding strain was reached when the instrument operated in canals with a small radius curvature having a sharp angle in the medial position or a more obtuse angle in the apical position. On the contrary, the *long* alloy always worked in its pseudo-elastic domain. The stainless steel instrument plasticized after the first contact with the canal wall and continued building up plasticity. These results suggest that the use of a Ni-Ti alloy with a wide pseudoelastic plateau should be advisable to prevent damages to the instruments causing clinical failures. Moreover, they confirmed the literature outcomes (Ref 5, 6, 11-15) indicating that sharp medial curvatures of the canals represent the most demanding conditions for the instruments.

Three main limitations can be recognized in the proposed model. First, in the implemented Ni-Ti constitutive model, the material was assumed to be indefinitely elastic (once completed the transformation from austenite and single-variant martensite) and the accumulation of plastic deformation due to cyclic loading in the pseudoelastic ranges was not accounted for. This limitation can cause underestimation of deformation levels. Moreover, hypothetical values were imposed to the mechanical parameters of the alloys: the knowledge of the exact physical and mechanical characteristics of the Ni-Ti alloys used by manufacturers could improve the results.

Second, a reduced rotational speed was imposed to the instrument: in real conditions, it implies a smaller number of cycles and a lower strain rate, affecting the fatigue behavior of the instruments. Nevertheless, in our simulations, this restriction affected the results of stainless steel, providing low values for the accumulated plastic deformations, while it was not significant for SMA, because of the omission of the plastic deformation accumulation phenomenon and strain rate influence in the material model.

Last, the study of torsional stresses induced in the instruments during instrumentation or generated during critical clinical events such as locking of the file tip into the dentine was not taken into account. Overcoming this limitation could allow to describe more precisely the file behavior and to give indications about the failure mode which seems to be given by the interaction (Ref 3, 16-18) between flexural and torsional fatigue (Ref 7, 19-24) and overloading (Ref 25, 26).

However, even under the highlighted limitations, the developed model demonstrated to be a useful tool to compare the behavior of Ni-Ti instrument differing in their material characteristics and can be considered promising in helping the instrument design.

Acknowledgments

The authors would like to thank Daniele Aspesi, MEng, for his contribution to the numerical analyses. Part of this work has been carried out under the financial auspices of the Italian Ministry for Research and Higher Education (MiUR—Ministero dell'Universita e della Ricerca) through the PRIN 2006 Project: "Shape memory alloy active microactuators and devices for biomedical applications: constitutive modeling, structural analysis, design, use of laser techniques for prototype implementation and experimental validation". Such support is gratefully acknowledged by the authors.

References

1. M. Hülsmann, O.A. Peters, and P.M.H. Dummer, Mechanical Preparation of Root Canals: Shaping Goals, Techniques and Means, *Endod. Top.*, 2005, **10**, p 30–76
2. S. Taschieri, S. Necchi, G. Rosano, M. Del Fabbro, R. Weinstein, and P. Machtou, Avantages et limites des instruments en nickel-titane pour la préparation canalaire endodontique. Revue de la littérature récente (Advantages and limits of nickel-titanium instruments for root canal preparation. A review of the current literature), *Schweiz. Monatsschr. Zahnmed.*, 2005, **115**, p 1000–1005 (in French)
3. M.G. Bahia, M.C. Melo, and V.T. Buono, Influence of Cyclic Torsional Loading on the Fatigue Resistance of K3 Instruments, *J. Endod.*, 2008, **41**, p 883–891
4. E. Berutti, G. Chiandussi, I. Gaviglio, and A. Ibba, Comparative Analysis of Torsional and Bending Stresses in Two Mathematical Models of Nickel-Titanium Rotary Instruments: ProTaper Versus ProFile, *J. Endod.*, 2003, **29**, p 15–19
5. X. Xu and Y. Zheng, Comparative Study of Torsional and Bending Properties for Six Models of Nickel-Titanium Root Canal Instruments with Different Cross-Sections, *J. Endod.*, 2006, **32**, p 372–375
6. S. Necchi, S. Taschieri, L. Petrini, and F. Migliavacca, Mechanical Behaviour of Ni-Ti Rotary Endodontic Files in Simulated Clinical Conditions: A Computational Study, *Int. Endod. J.*, 2008, **41**, p 939–949
7. J.P. Pruett, D.J. Clement, and D.L. Carnes, Cyclic Fatigue Testing of Nickel Titanium Endodontic Instruments, *J. Endod.*, 1997, **23**, p 77–85
8. G.R. Zadno and T.W. Duerig, Linear and Non-Linear Superelasticity in NiTi, *MRS Shape Memory Mater.*, 1989, **9**, p 201–206. Also available on www.nitinol.com
9. F. Auricchio and L. Petrini, A Three-Dimensional Model Describing Stress-Temperature Induced Solid Phase Transformations: Solution Algorithm and Boundary Value Problems, *Int. J. Numer. Methods Eng.*, 2004, **61**, p 716–737
10. F. Auricchio, M. Di Loreto, and E. Sacco, Finite-Element Analysis of a Stenotic Artery Revascularization Through a Stent Insertion, *Comput. Methods Biomech. Biomed. Eng.*, 2001, **4**, p 249–263
11. Y. Haikel, R. Serfaty, G. Bateman, B. Senger, and C. Allemann, Dynamic and Cyclic Fatigue of Engine-Driven Rotary Nickel-Titanium Endodontic Instruments, *J. Endod.*, 1999, **25**, p 434–440
12. G. Zelada, P. Varela, B. Martín, J.G. Bahíllo, F. Magán, and S. Ahn, The Effect of Rotational Speed and the Curvature of Root Canals on the Breakage of Rotary Endodontic Instruments, *J. Endod.*, 2002, **28**, p 540–542
13. U.M. Li, B.S. Lee, C.T. Shih, W.H. Lan, and C.P. Lin, Cyclic Fatigue of Endodontic Nickel Titanium Rotary Instruments: Static and Dynamic Tests, *J. Endod.*, 2002, **28**, p 448–451
14. B. Martín, G. Zelada, P. Varela, J.G. Bahíllo, F. Magán, S. Ahn, and C. Rodríguez, Factors Influencing the Fracture of Nickel-Titanium Rotary Instruments, *J. Endod.*, 2003, **36**, p 262–266
15. H.P. Lopes, E.J. Moreira, C.N. Elias, R.A. de Almeida, and M.S. Neves, Cyclic Fatigue of ProTaper Instruments, *J. Endod.*, 2007, **33**, p 55–57
16. C.J. Ullmann and O.A. Peters, Effect of Cyclic Fatigue on Static Fracture Loads in ProTaper Nickel-Titanium Rotary Instruments, *J. Endod.*, 2005, **31**, p 183–186
17. M.G. Bahia, M.C. Melo, and V.T. Buono, Influence of Simulated Clinical Use on the Torsional Behavior of Nickel-Titanium Rotary Endodontic Instruments, *Oral Surg. Oral Med. Oral Pathol. Oral Radiol. Endod.*, 2006, **101**, p 675–680
18. F.O.G. Barbosa, J.A.C. Ponciano, and M.C.P. Araújo, Influence of Previous Angular Deformation on Flexural Fatigue Resistance of K3 Nickel–Titanium Rotary Instruments, *J. Endod.*, 2007, **33**, p 1477–1480
19. B. Sattapan, J.E. Palamara, and H.H. Messer, Torque During Canal Instrumentation Using Rotary Nickel-Titanium Files, *J. Endod.*, 2000, **26**, p 156–160
20. D. Fife, G. Gambarini, and L.R. Britto, Cyclic Fatigue Testing of ProTaper NiTi Rotary Instruments After Clinical Use, *Oral Surg. Oral Med. Oral Pathol. Oral Radiol. Endod.*, 2004, **97**, p 251–256
21. G. Yared, In Vitro Study of the Torsional Properties of New and Used ProFile Nickel Titanium Rotary Files, *J. Endod.*, 2004, **30**, p 410–412
22. M.G. Bahia and V.T. Buono, Decrease in the Fatigue Resistance of Nickel-Titanium Rotary Instruments After Clinical Use in Curved Root Canals, *Oral Surg. Oral Med. Oral Pathol. Oral Radiol. Endod.*, 2005, **100**, p 249–255
23. N.M. Grande, G. Plotino, R. Pecci, R. Bedini, V.A. Malagnino, and F. Somma, Cyclic Fatigue Resistance and Three-Dimensional Analysis of Instruments from Two Nickel–Titanium Rotary Systems, *Int. Endod. J.*, 2006, **39**, p 755–763
24. G.S.P. Cheung and B.W. Darvell, Fatigue Testing of a NiTi Rotary Instrument. Part 1: Strain-Life Relationship, *Int. Endod. J.*, 2007, **40**, p 626–632
25. S.B. Alapati, W.A. Brantley, T.A. Svec, J.M. Powers, J.M. Nusstein, and G.S. Daehn, SEM Observations of Nickel-Titanium Rotary Endodontic Instruments that Fractured During Clinical Use, *J. Endod.*, 2005, **31**, p 40–43
26. A.P. Spanaki-Voreadi, N.P. Kerezoudis, and S. Zinelis, Failure Mechanism of ProTaper Ni-Ti Rotary Instruments During Clinical Use: Fractographic Analysis, *Int. Endod. J.*, 2006, **39**, p 171–178

JMEPEG (2009) 18:638–648
DOI: 10.1007/s11665-009-9421-y

1059-9495/$19.00

Conceptual Design and Simulation of a Compact Shape Memory Actuator for Rotary Motion

I. Spinella, G. Scirè Mammano, and E. Dragoni

(Submitted September 15, 2008; in revised form February 23, 2009)

This work describes the conceptual design, the modelling, the optimization, the detail design and the virtual testing of a shape memory actuator purposely conceived to maximize torque and angular stroke while limiting overall size and electric consumption. The chosen design, achieved by means of a Quality Function Deployment approach, features a fully modular concept in which an arbitrary number of identical modules are assembled to produce the desired angular stroke and output torque. The basic module contains shape memory springs that actuate the device and also a conventional spring that reduces the torque ripple. Following the concept generation stage, a thermo-electromechanical model is developed and a numerical optimization performed, aimed at minimizing the electrical consumption of the actuator. Finally, the device is designed in detail and the actuator is tested virtually. Thanks to the proposed modular construction and the use of a conventional balancing spring, the device shows better performances than known rotary shape memory actuators in terms of rotation, torque and customization.

Keywords conceptual design, modeling, rotary actuator, shape memory alloys, simulation

1. Introduction

Shape memory alloys (SMAs) are increasingly exploited to build actuators and micro-actuators for many applications. The chief advantages of these smart materials are the very high power versus mass and force versus mass ratios and the small amount of active material needed for the actuation. Moreover, SMAs allow to simplify the kinematic chain, to obtain a silent and clean actuation and to work in corrosive environment.

Although the use of SMAs as actuators is rising, these materials are basically confined to the field of linear devices. By contrast, SMA technology is virtually unexplored for the construction of compact rotary actuators able to provide high output torque and significant rotations. Furthermore, both in linear and in rotary SMA actuators, the electrical consumption and the bandwidth of the SMA actuator are seldom considered as constraints of the design, although the correct choice of these parameters is fundamental to obtain a useful device.

With regard to linear SMA actuators, many applications have been developed in the recent years and some works adopt rational approaches to design the devices. In Ref 1, Colli et al. propose the system design of a linear SMA actuator for automotive tumble flap: the rational Quality Function Deployment (QFD) approach is adopted for the design of the actuator, but the electrical power required for the actuation is excessive and the bandwidth is poor. Huang (Ref 2) shows possible selection criteria for SMA actuators, but no criteria are given about the required electrical power and about the bandwidth. In Ref 3, Reynaerts and Van Brussel propose a design methodology that considers even the cooling time of the actuator, but the design is focused just on SMA wires. The modelling and the simulation of some linear SMA actuators is proposed by Lu et al. (Ref 4), but no rational approaches are adopted to design these actuators. A complete and rational design of a linear actuator based on SMA wires is proposed by Khidir et al. (Ref 5). Even cooling time aspects are considered in this efficient actuator, which is a good example of linear SMA device.

With regard to rotary SMA actuators, just a few applications have been developed. Abadie et al. (Ref 6) develop a rotary SMA actuator cooled by Peltier elements. This actuator does not seem to have many practical applications, because its output torque and angular stroke are poor. Moreover, no electrical efficiency aspects are considered. The same limitations affect the device developed by Gabriel et al. (Ref 7). Kuribayashi (Ref 8) propose another small rotary actuator. The design approach is a rational one, but the same problems highlighted in Ref 6, 7 affect this device. A rotary actuator that supplies a considerable amount of output torque is proposed by Grant and Hayward (Ref 9), but this SMA device is not compact enough to compete with conventional actuators successfully.

The analysis of the state of the art shows that there is a lack of work focused on the rational design of compact SMA rotary actuators, able to supply a considerable amount of output torque and to exploit a wide angular stroke, minimizing at the same time the electrical consumption. Aimed at filling this gap, this work deals with the design, the optimization and the dynamic simulation of a rotary SMA actuator specifically studied to minimize the electrical consumption. Simultaneously, the output

This article is an invited paper selected from presentations at Shape Memory and Superelastic Technologies 2008, held September 21-25, 2008, in Stresa, Italy, and has been expanded from the original presentation.

I. Spinella, G. Scirè Mammano, and **E. Dragoni**, Dipartimento di Scienze e Metodi dell'Ingegneria, Università di Modena e Reggio Emilia, Reggio Emilia, Italy. Contact e-mail: igor.spinella@unimore.it.

Nomenclature			
α_D	extreme angular position of the cursor	K_{SC}	cold stiffness of active springs
α_P	angular pre-stretch of each active spring	K_{SH}	hot stiffness of active springs
α_{tot}	total angular stroke of the actuator	L_{0S}	free length of the SMA springs
$\gamma_{adm\,s}$	admissible shear strain of the shape memory material	L_{0T}	free length of the conventional spring
$\gamma_{max\,s}$	maximum shear strain of the shape memory material	m	mass of the SMA spring
ν	Poisson ratio of the conventional spring	M_F	martensite finish temperature
ρ	electrical resistivity of the SMA springs	M_S	martensite start temperature
ξ	SMA density	n	total number of elastic units
τ_{adm}	admissible shear stress of the conventional spring	N_S	number of SMA springs active coils
X_{AM}	enthalpy of the transformation from austenite to martensite	N_T	number of conventional spring active coils
		O	centre of the basic module of the actuator
X_{MA}	enthalpy of the transformation from martensite to austenite	P	electric power
		P_{adm}	admissible value of the electric power
A_F	austenite finish temperature	P_M	maintenance electric power
A_S	austenite start temperature	P_P	peak electric power
b	arm of the force of the elastic unit	R	electrical resistance of the SMA spring
B	mobile cursor of the basic module of the actuator	S	surface of the SMA spring
c	generic specific heat of the SMA	s_1	dimensionless parameter to model the SMA properties
c_A	austenitic specific heat of the SMA	S_1	shape memory spring 1 of the elastic group 1
c_{AM}	austenite to martensite specific heat of the SMA	S_2	shape memory spring 2 of the elastic group 1
c_M	martensitic specific heat of the SMA	S_3	shape memory spring 1 of the elastic group 2
c_{MA}	martensite to austenite specific heat of the SMA	S_4	shape memory spring 2 of the elastic group 2
C_S	spring index of the SMA springs	t_{adm}	admissible cooling time of the SMA springs
C_T	spring index of the conventional spring	t_{cool}	cooling time of the SMA springs
d_S	wire diameter of the SMA springs	t_{ON}	activation time
d_T	wire diameter of the conventional spring	t_R	rising time of the SMA springs
D_{adm}	admissible diameter of the rotary actuator	T_a	ambient temperature
D_S	pitch diameter of the SMA spring	T_D	external dissipative torque
D_T	pitch diameter of the conventional spring	T_M	maintenance temperature of the SMA springs
E_C	elastic modulus of the conventional spring	T_P	peak temperature of the SMA springs
E_D	energy dissipated in an actuation cycle	T_{max}	maximum output torque
f_{max}	maximum deflection of the SMA springs	T_{min}	minimum output torque
f_{min}	minimum deflection of the SMA springs	V_{adm}	admissible voltage
G_A	austenite shear modulus	V_M	maintenance voltage
G_M	martensite shear modulus	V_P	peak voltage
h	total convective heat transfer coefficient	$\mathbf{X}$	vector of design variables

torque, the angular stroke and the total size are chosen to obtain an SMA device that can compete with conventional rotary actuators. A QFD approach has been adopted to ensure a quality device, which is able to respect the customer needs for its lifetime. The disclosed rotary SMA actuator shows good performance in terms of output torque, angular stroke, total size and electrical consumption. Moreover, capitalizing on the modular architecture, the angular stroke and the output torque are easily adjusted to meet the requirements of many specific applications.

The development of the actuator is carried out in three steps. In the first step, a structured conceptual design procedure is performed, following the QFD approach. Output of the first step is a concept able to satisfy all the identified customer needs. The second step is focused on the modelling and on the numerical optimization of the concept, which is the output of the first step. Finally, in the third step, the detail design and the virtual testing of the device are accomplished to ensure that the performances of the actuator are able to satisfy the customer needs. The following presentation of the design is organized according to the above division of the work.

2. Conceptual Design

The conceptual design of the actuator was accomplished following the structured QFD approach, as described by Ulrich and Eppinger (Ref 10). Main steps of the conceptual design process are the following:

- definition of the House of Quality;
- functional decomposition;
- generation, evaluation and selection of concepts.

2.1 House of Quality

The House of Quality (HOQ) is a graphical way to arrange the customer needs, the metrics, the competitor performances and the target specifications. Figure 1 shows the complete HOQ of the SMA rotary actuator.

The customer needs (left column of the HOQ) were identified for the commodity field of automatic roll-up shutters. The conventional actuators of this commodity field show

N	Customer Needs	Importance	1 Maximum Dimension	2 Thickness	3 Torque density	4 Angular stroke	5 Price	6 Bidirectional	7 SMA technology	8 Nominal torque	9 Power consumption	10 Ripple torque	11 Adjustable nominal torque	12 Adjustable angular stroke	13 Lifetime	14 Maintenance	15 Degree of protection	16 Sound power level	17 Nominal voltage	18 Angular Speed	Competitors: BELIMO CM 24	BELIMO LF24	ARIS N1	SIEMENS 121.1
	House of Quality	Metrics / Importance	5	5	5	5	5	5	5	3	3	3	3	3	3	3	1	1	1	1				
2	It is compact	5	•	•	•																+	~	-	+
3	It supplies a high angular stroke	5				•								o							+	-	+	-
4	It is inexpensive	5	o				•														?	?	?	?
5	It is bidirectional	5						•													+	+	+	+
6	It exploits SMA technology	5							•												-	-	-	-
7	It supplies a high output torque	3								•											~	+	+	+
8	It is with low power consumption	3									•										+	~	-	+
9	It supplies a smooth motion	3										•									+	+	+	+
10	It is modular	3											•	•							-	-	-	~
11	It is reliable	3													•	•	•				+	~	~	~
12	It is safe	1															•				+	+	+	+
13	It is silent	1																•			+	-	?	?
14	It exploits electric energy	1																	•		+	+	+	+
15	It is fast	1																		•	-	~	+	-
	Measurement Unit		mm	mm	Nm/dm^3	deg	€	Bin	Bin	Nm	W	Nm	Bin	Bin	Cycles	List	IP	dB	V	deg/s				
	BELIMO CM 24		158	16	14.1	315	nd	YES	NO	2	0.5	nd	NO	NO	100k	Free	54	35	AC-DC 24	1.2				
	BELIMO LF24		155	64	4.1	95	nd	YES	NO	4	5	nd	NO	NO	60k	Free	54	62	AC-DC 24	2.4				
	ARIS N1		146	120	2.2	330	nd	YES	NO	5	50	nd	NO	NO	nd	Free	54	nd	AC 230	6				
	SIEMENS 121.1		192	60	7.5	90	nd	YES	NO	7	3.5	nd	YES	NO	nd	nd	54	nd	AC-DC 24	1				
	Target Value		≤150	≤60	≥ 4	≥60	100	YES	YES	≥2	≤10	≤30%	YES	YES	50k	Free	54	35	DC 24	≥5				

Competitors performances	
Symbol	Performance
+	Good
~	Neutral
-	Bad

Interrelationship among customer needs and metrics	
Symbol	Interrelationship
•	Strong
o	Weak

Fig. 1 House of Quality of the SMA rotary actuator

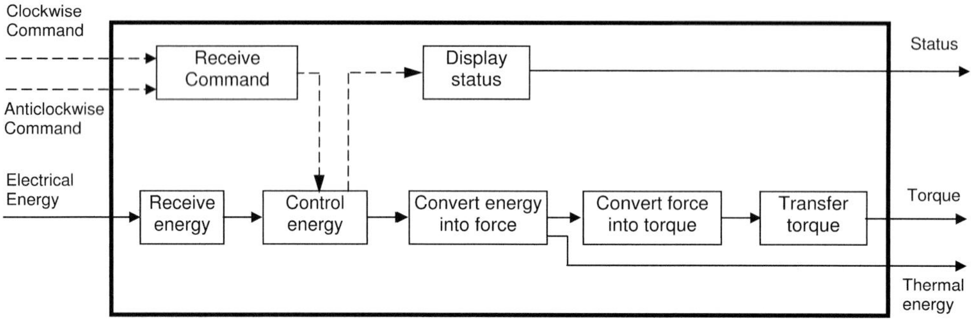

Fig. 2 Functional diagram of the SMA rotary actuator

excellent performances. For this reason, the design of an SMA device able to compete with these products is a good way to show the potential of the SMAs in the field of rotary actuators. Associated with the customer needs, a set of metrics was developed (top row of the HOQ), aimed at defining quantitative characteristics that help to lead the design toward the achievement of the customer needs. Then, the performances of the competitors found in the field of automatic roll-up shutters were studied (relative to both customer needs and metrics) to define the target specifications of the SMA rotary actuator (right column and lower row of the HOQ, respectively).

Relevant target specifications reported in the HOQ are total angular stroke = 60°, nominal output torque = 2 Nm, maximum dimension = 150 mm.

2.2 Functional Decomposition

The functional diagram is a graphical way to show the critical functions of a system and the relationships among these functions. Figure 2 shows the high-level functional diagram of the SMA rotary actuator. The diagram is focused on the critical functions that were identified: "Convert energy into force" and "Convert force into torque".

2.3 Generation, Evaluation and Selection of Concepts

A concept generation stage process was exploited to find solutions to each critical function. This process was undertaken by means of external research (to take inspiration from former solutions) and by means of internal research (to generate original ideas). Group-oriented techniques as brainstorming and 6-3-5 method were used (Ref 11). Finally, a combination of the concept fragments led to a portfolio of global concepts.

The evaluation and the selection of the global concepts were performed in three steps. Firstly, the global concepts were evaluated by means of a qualitative screening matrix. Then, a critical revision and an improvement of some global concepts

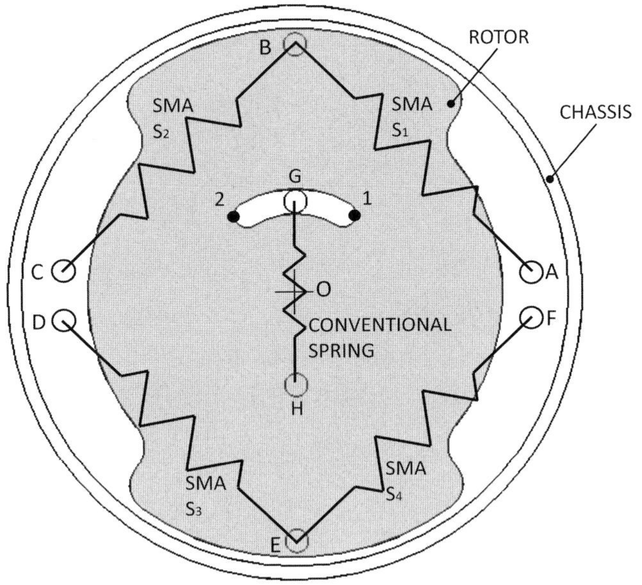

Fig. 3 Final concept for the basic actuator module

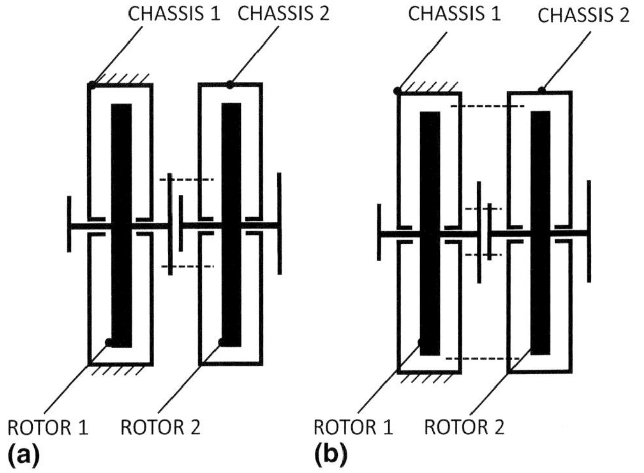

Fig. 4 Angle assembly (a) and Torque assembly (b) of two basic modules

were performed. Finally, a quantitative scoring matrix was used to choose the best concept.

Figures 3 and 4 display the final concept that was chosen for further development. The system is fully modular. The basic modules of Fig. 3 can be assembled to increase the angular stroke, as shown in Fig. 4(a), or to increase the output torque, as shown in Fig. 4(b). Proper combination of the right number of elementary modules, arranged according either to Fig. 4(a) or (b), can thus match any requirement in terms of output torque and angular stroke. The basic module is made of an outer chassis (white ring in Fig. 3), an inner rotor (grey area in Fig. 3) and an arbitrary number, n, of spring units, joining the rotor to the chassis. In Fig. 3 just two units are represented for clarity, one unit formed by SMA springs S_1 and S_2, and the other formed by SMA springs S_3 and S_4. Each SMA spring is linked to the chassis (points A, C, D and F, respectively) and to the rotor (points B and E). Only one spring of each elastic unit is warmed at a time: for example, if S_1 and S_3 are activated, S_2 and S_4 are not activated and vice versa. When S_1 and S_3 are activated, a clockwise motion occurs. On the contrary, when S_2 and S_4 are activated, an anticlockwise motion takes place.

In addition to the SMA spring units, the basic module contains also a conventional spring (GH in Fig. 3). The spring is linked to the chassis at point G and to the rotor at point H. This spring is introduced to reduce the torque ripple of the actuator. In fact, when SMA springs S_1 and S_3 are activated starting from the extreme position 1 of Fig. 3, they are fully extended, and for this reason they supply a high value of output torque, which decrease reaching its minimum in the extreme position 2. The conventional spring acts against S_1 and S_3 when they are fully extended, reducing the peak of the output torque, and helps S_1 and S_3 when they are near the extreme position 2. The same reasoning applies to springs S_2 and S_4.

Warming of the SMA springs is achieved by means of Joule effect, when a suitable current is supplied to them. To this aim, an electronic board on each module receives the commands (see Fig. 2) and applies a voltage to the SMA springs to actuate them properly. An external board controls all the modules of a multistage actuator, deciding the proper commands to provide to each one. This control system allows both the on/off and the proportional control of the actuator. Cooling of the SMA springs is achieved by means of natural convention.

3. Modelling and Numerical Optimization

This section describes the mathematical modelling of the basic module of Fig. 3, which represents the heart of the actuator. As a first step, a thermo-electromechanical model was developed, which was able to predict the behavior of the basic module. Then, the thermo-electromechanical model was optimized numerically to minimize the electrical consumption of the module, subject to the mechanical and dynamic constrains of the HOQ. Finally, based on the optimized configuration, the conventional spring that smoothes out the output torque is calculated.

3.1 Thermo-Electromechanical Model

The model is divided into three parts: mechanical model, thermal model and electrical model.

3.1.1 Mechanical Model. The mechanical model of the basic module is shown in Fig. 5, with only one spring unit displayed for reasons of clarity. Two opposite SMA traction springs S_1 and S_2 act on a cursor B. The cursor B is rigidly linked with the centre O, which is the output shaft of the actuator. For this reason, B can only rotate around O maintaining constant the arm b and its generic polar position is (b, α). The other extremities of S_1 and S_2 are fixed on the chassis of the actuator, points A and C, respectively at a polar position of (a, α_P) and $(a, -\alpha_P)$. Thus, the angle $2\alpha_P$ is proportional to the total pre-stretch imparted to the system. The active springs S_1 and S_2 are made of the same shape memory material and are geometrically identical. So, when S_1 and S_2 are at the same temperature and no external moment is applied, B remains on the polar position of Fig. 5(a) $(b, 0)$.

An external dissipative moment T_D, which is always opposite to the velocity of the cursor, acts against the cursor. So, when the cursor moves clockwise, T_D is anticlockwise and vice versa. For a correct design of the system, the module must supply an output moment always higher than T_D.

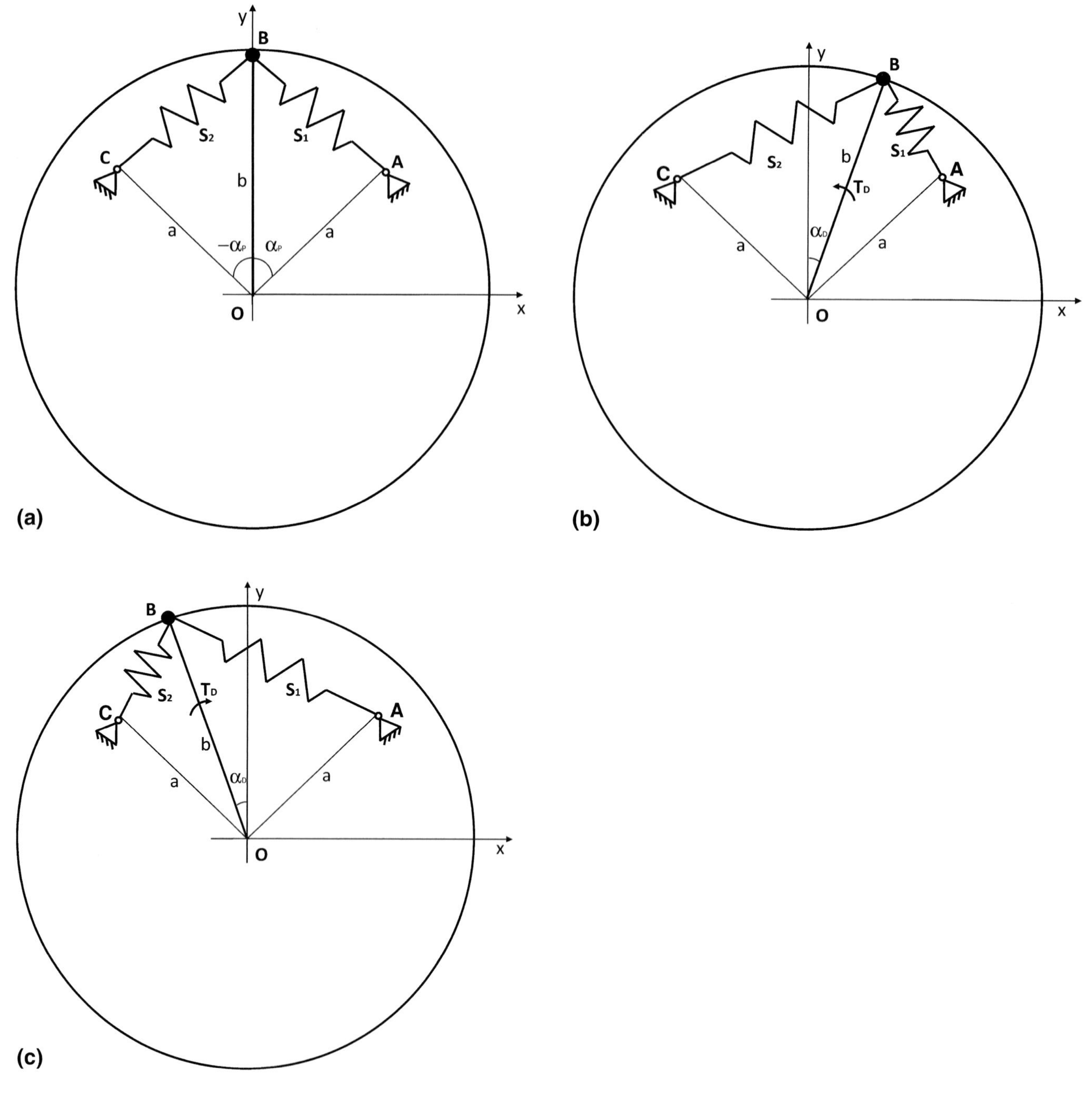

Fig. 5 Mechanical scheme of the basic module: (a) all springs cool, (b) S_1 "hot", (c) S_2 "hot"

When S_1 is warmed up over the austenite finish temperature, A_F, maintaining S_2 below the martensite finish temperature, M_F, the cursor moves clockwise reaching the position (b, α_D), as shown in Fig. 5(b). For the symmetry of the problem, when S_2 is warmed up over A_F and S_1 is cooled below M_F, the cursor moves anticlockwise reaching the position (b, $-\alpha_D$), as shown in Fig. 5(c). As a result, the total angular stroke of the actuator is $\alpha_{tot} = 2\alpha_D$.

Each spring in Fig. 5 is modelled as a helical traction spring exhibiting a linear force-deflection relationship. Although the real stress-strain response of the material can be far from linear in the cold state, assumption of linear behavior is acceptable when only the end positions of the actuator are of interest (Ref 12). Figure 6 shows the characteristic lines of the alloy, identified by austenite shear modulus G_A at high temperature and martensite modulus G_M at low temperature.

An important dimensionless parameter, s_1, can be defined as the ratio between the hot stiffness, K_{SH}, and the cold stiffness, K_{SC}, of each SMA spring. If the geometric changes related to the deformations of the springs do not influence the elastic constant value, s_1 can also be expressed by the ratio between the austenite and martensite elastic moduli of the spring material as follows:

$$s_1 = \frac{K_{SH}}{K_{SC}} = \frac{G_A}{G_M} \qquad \text{(Eq 1)}$$

From Fig. 5(a), by requesting equilibrium of the rotor under the moments about axis O and applying simple geometrical relationships, the expression of the output moment can be calculated as a function of the angular coordinate α of point B. From this expression, by substituting $\alpha = \alpha_D$ and considering S_1 in a hot state (stiffness $K_{SH} = s_1 K_{SC}$) and S_2 in a cold state (stiffness K_{SC}), the value of the minimum output moment, T_{min}, can be determined as

$$T_{min} = nabK_{SC}\left(\frac{s_1 \sin(\alpha_P - \alpha_D) f_{min}}{(b - a\cos(\alpha_P - \alpha_D))\sqrt{1 + \frac{a^2 \sin^2(\alpha_P - \alpha_D)}{(b - a\cos(\alpha_P - \alpha_D))^2}}} - \frac{\sin(\alpha_D + \alpha_P) f_{max}}{(b - a\cos(\alpha_D + \alpha_P))\sqrt{1 + \frac{a^2 \sin^2(\alpha_D + \alpha_P)}{(b - a\cos(\alpha_D + \alpha_P))^2}}}\right) \quad \text{(Eq 2)}$$

where f_{max} and f_{min} are the maximum deflection and the minimum deflection, respectively. Using geometric considerations and standard machine design formulas for helically coiled traction springs (Ref 13, 14), these quantities can be written as follows:

$$f_{max} = \left(\sqrt{a^2 + b^2 - 2ab\cos(\alpha_D + \alpha_P)} - L_{0S}\right) \quad \text{(Eq 3)}$$

$$f_{min} = \left(\sqrt{a^2 + b^2 - 2ab\cos(\alpha_P - \alpha_D)} - L_{0S}\right) \quad \text{(Eq 4)}$$

$$L_{0S} = N_S d_S + C_S d_S \quad \text{(Eq 5)}$$

where N_S is the coil number, d_S is the wire diameter, C_S is the spring index of the active springs and L_{0S} is the free length of each spring. All the geometric nonlinearities of the problem were considered in Eq 2-4. Equation 5 takes into account even the size of the spring ends.

Finally, using standard formulas of machine design (Ref 13, 14), the shear strain, $\gamma_{max\ s}$, of the SMA spring can be written as follows:

$$\gamma_{max\ s} = \frac{G_M f_{max}}{\pi d_S N_S C_S^2} \quad \text{(Eq 6)}$$

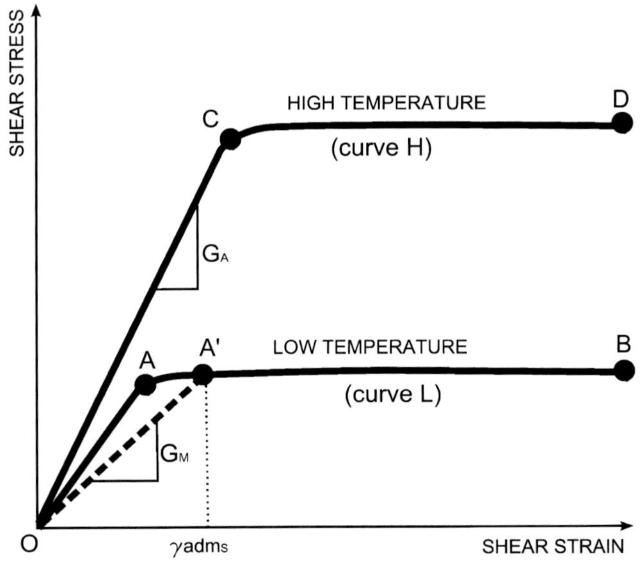

Fig. 6 Simplified shear stress-shear strain behavior of SMAs at different temperatures

3.1.2 Thermal Model. The actuation cycle was assumed like the one shown schematically in Fig. 7. An applied peak voltage, V_P, ensures that the temperature of the SMA spring increases to T_P, just above A_F, in a rising time, t_R. Then, the applied voltage falls to a lower maintenance value, V_M, that ensures a temperature, T_M, slightly above M_S. Due to the hysteresis of the SMAs, the SMA spring remains in austenitic state, thus its shear modulus is still G_A. This approach helps to reduce, at the same time, both the dissipated energy, E_D, and the cooling time t_{cool}. Cooling of the spring is performed by turning off the voltage. With no voltage applied, the temperature decreases from $T_M \approx M_S$ to M_F in the least time that allows the complete transformation from austenite to martensite to be achieved. The value of the activation time, t_{ON} = 30 s, was chosen.

The expressions of the rising time, t_R, and the cooling time, t_{cool}, can be obtained using the simplified model proposed by Reynaerts and Van Brussel (Ref 3). The model starts from the consideration that in a polycrystalline SMA

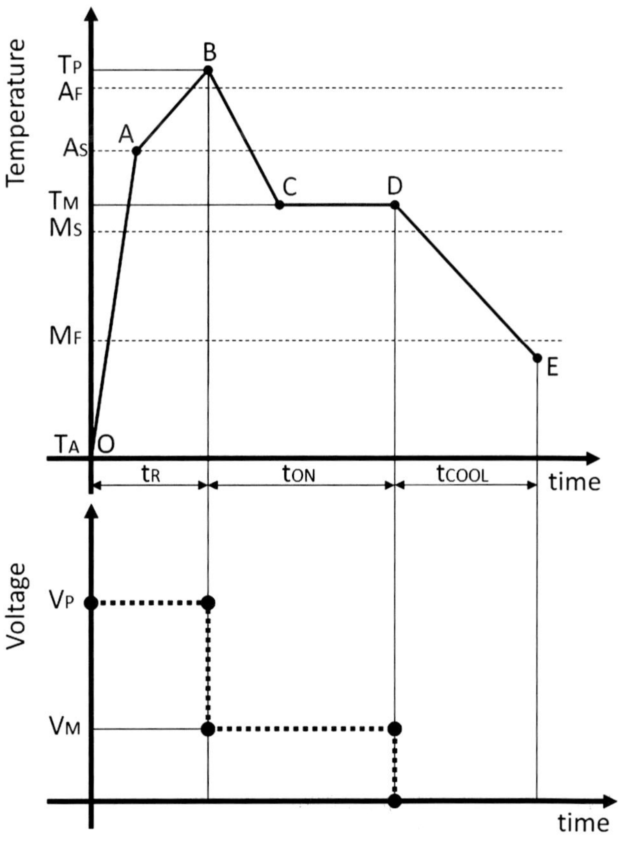

Fig. 7 Actuation cycle

the transformation temperatures, during both heating and cooling, are not precise and well-identified values as it occurs for a single crystal. This is because the transformation temperatures of all the aggregated crystals are not equal but spread over a finite range. When measured for a polycrystal, the martensite to austenite transformation begins at A_S (austenite start) and terminates at A_F (austenite finish). Likewise, the austenite to martensite transformation begins at M_S (martensite start) and terminates at M_F (martensite finish). Throughout both temperature ranges, whenever heat is supplied to or taken away from the polycrystal, the temperature of the SMA varies as it does (though with a different rate) when its state is either entirely austenitic or entirely martensitic. Outside the transformation ranges, the thermal response of the SMA to a net heat input is completely defined by its true specific heats of the material (c_A for austenite and c_M for martensite). Within the transformation ranges, the temperature response is affected not only by the true specific heat (which falls between c_A and c_M) but also by the transformation enthalpies (latent heats) characterizing the polycrystal as a whole during heating (X_{MA}) and during cooling (X_{AM}). Assuming that enthalpies X_{MA} and X_{AM} are absorbed or freed uniformly during the transformations, the thermal effect of those enthalpies can be described by introducing two fictitious specific heats defined as the transformation enthalpies divided by the corresponding transformation temperature interval. By approximating the true specific heat as the mean between c_A and c_M, the global equivalent specific heats throughout austenite to martensite transformation, c_{AM}, and martensite to austenite transformation, c_{MA}, can be written as

$$c_{AM} = \frac{c_A + c_M}{2} + \frac{X_{AM}}{M_S - M_F} \qquad \text{(Eq 7)}$$

$$c_{MA} = \frac{c_A + c_M}{2} + \frac{X_{MA}}{A_F - A_S} \qquad \text{(Eq 8)}$$

If the assumption of uniform enthalpy absorption or release is not met, relationships (7) and (8) give only average values of the equivalent specific heats. Still, these values provide good predictions of the thermal response of the material whenever (as is the case in this paper) complete transformations are considered.

By virtue of the above approach, the energy balance of the SMA spring at any temperature T is given by the following differential equation (Ref 3)

$$mc\frac{\partial T}{\partial t} + hS(T - T_a) = P \qquad \text{(Eq 9)}$$

where c is the (equivalent) specific heat at temperature T, m is the mass, S is the total surface of the SMA spring, h is the total convective heat transfer coefficient and P is the electrical power entering the spring.

With proper choice of c and by applying the appropriate boundary conditions, Eq 9 gives by integration the thermal transient of the SMA spring between any two selected temperatures. From this transient, the heating or the cooling time required to bring the spring from one temperature to the other is easily found.

Consider, for example, the cooling phase of the spring, ranging from initial temperature $T = T_M \approx M_S$ and final temperature $T \approx M_F$ (line DE in Fig. 7). Integrating (9) between M_S and M_F with $c = c_{AM}$ (austenite to martensite transformation) and $P = 0$ (no power supply) gives the cooling time, t_{cool}, as

$$t_{cool} = \frac{mc_{AM}}{hS}\left\{\ln\left(\frac{M_S - T_a}{M_F - T_a}\right)\right\} \qquad \text{(Eq 10)}$$

For the heating phase (line OAB in Fig. 7), the procedure is similar but integration of (9) is most conveniently divided into two parts. The first part (line OA) starts at $T = T_a$ and terminates at $T = T_S$ during which the (true) specific heat is $c = c_M$. The second part (line AB) starts at $T = T_S$ and terminates at $T = T_P \approx A_F$ during which the (equivalent) specific heat is $c = c_{MA}$. Integrating (9) over these two temperature ranges with power supply $P = P_P$ throughout gives the rising time, t_R, as

$$t_R = \frac{mc_M}{hS}\ln\left\{\frac{-P_P}{hS(A_S - T_a) - P_P}\right\} + \frac{mc_{MA}}{hS}\ln\left[\frac{hS(A_S - T_a) - P_P}{hS(A_F - T_a) - P_P}\right] \qquad \text{(Eq 11)}$$

The peak and the maintenance temperatures, respectively T_P and T_M, can be obtained by considering a steady state ($\partial T/\partial t = 0$) in expression (9).

$$T_P = \frac{P_P}{hS} + T_a \qquad \text{(Eq 12)}$$

$$T_M = \frac{P_M}{hS} + T_a \qquad \text{(Eq 13)}$$

In all the above equations, the surface S, the mass m and the convective coefficient h are supplied by the following expressions (ξ = mass density of the SMA)

$$S = \pi^2 C_S N_S d_S^2 \qquad \text{(Eq 14)}$$

$$m = \xi\frac{\pi^2 C_S N_S d_S^3}{4} \qquad \text{(Eq 15)}$$

$$h = (61.3 - 17.5d_S - 2.7C_S - 0.33N_S + 0.93d_S C_S + 0.08d_S N_S) \quad \mathrm{W\,m^{-2}} \qquad \text{(Eq 16)}$$

Expression (16) for the total convective heat transfer coefficient, h, was derived using the numerical procedure reported in Appendix 1.

3.1.3 Electrical Model. It was assumed that each SMA spring is warmed by means of Joule effect. For this reason the simple relationships of a resistor warmed by an applied electrical tension were used, combined with the expressions of the geometry of the SMA springs.

The expressions of the peak electric power, P_P, and the maintenance electric power, P_M, can be written as follows.

$$P_P = \frac{V_P^2}{R} \qquad \text{(Eq 17)}$$

$$P_M = \frac{V_M^2}{R} \qquad \text{(Eq 18)}$$

In Eq 17 and 18, the resistance, R, of the SMA, can be expressed as

$$R = 4\rho\frac{C_S N_S}{d_S} \qquad \text{(Eq 19)}$$

where ρ is the SMA resistivity, calculated as the average between the austenitic and the martensitic state. Equation 19 is a simplified representation of the resistance, which is actually not a constant but a mixture of martensite and austenite contributions that change continuously during heating and cooling. Although this variation is very critical for control applications (Ref 15), it can be regarded as negligible for the optimization and the simulation purposes of this paper.

Finally, in terms of electric powers, P_P and P_M, and actuation times, t_R and t_{ON} (= 30 s), the total energy dissipated in an actuation cycle, E_D, can be expressed as

$$E_D = P_P t_R + P_M T_{ON} \quad \text{(Eq 20)}$$

By combining relationship (20) with Eq 17-19 and 9-16, the total dissipated energy can be cast as a function of the vector of design variables $\mathbf{X} = \{n, N_S, d_S, C_S, a, b, \alpha_P, V_P, V_M\}$.

3.2 Numerical Optimization

A nonlinear numerical optimization was performed to find an optimal configuration of the basic module of Fig. 3 that minimizes the total energy dissipated in the thermal cycle of Fig. 7. Formally, the optimization problem was stated by requesting that vector, $\mathbf{X}$, of design variables

$$\mathbf{X} = \{n, N_S, d_S, C_S, a, b, \alpha_P, V_P, V_M\} \quad \text{(Eq 21)}$$

must be found so as

$$E_D(\mathbf{X}) = \min \quad \text{(Eq 22)}$$

subject to the following mechanical, thermal and electric constrains:

Mechanical Constraints

$$T_{\min} \geq T_D \quad \text{(Eq 23)}$$

$$n \leq \frac{\pi}{1.1\alpha_p} \quad \text{(Eq 24)}$$

$$a \leq \frac{0.5D_{adm}}{1.2} \quad \text{(Eq 25)}$$

$$b \leq \frac{0.5D_{adm}}{1.2} \quad \text{(Eq 26)}$$

$$\gamma_{\max s} \leq \gamma_{adm\,s} \quad \text{(Eq 27)}$$

Thermal Constraints

$$T_P \geq A_F \quad \text{(Eq 28)}$$

$$T_M \geq M_S \quad \text{(Eq 29)}$$

$$t_{cool} \leq t_{adm} \quad \text{(Eq 30)}$$

Electric Constraints

$$V_P \leq V_{adm} \quad \text{(Eq 31)}$$

$$V_{ON} \leq V_{adm} \quad \text{(Eq 32)}$$

$$P_P \leq P_{adm} \quad \text{(Eq 33)}$$

in which $D_{adm} = 150$ mm, $\gamma_{adm\,s} = 0.038$ rad, $T_D = 1$ Nm, $t_{adm} = 120$ s, $V_{adm} = 24$ V, $P_{adm} = 250$ W, $A_F = 90$ °C, $A_S = 80$ °C, $M_S = 70$ °C, $M_F = 50$ °C. The values of the transformation temperatures and the shear strain were chosen referring to NiTiNOL.

Constraints (23) to (33) have the following physical meaning. Constraint (23) requires that the minimum value of the output torque must overcome the external dissipative torque. Constraint (24) requires that the angular space taken up by the total number of spring units must be contained in the round angle. Constraints (25) and (26) require that the attachment points of the active springs must fall within the admissible outer diameter of the chassis. Constraint (27) requires that the material of the SMA springs must not be overstretched in torsion. Constraint (28) requires that the austenitic transformation must be completed at the end of the heating up ramp while constraint (29) requires that the martensitic transformation must be prevented throughout the maintenance time (see Fig. 7). Constraint (30) requires that the cooling time must not exceed an accepted value. No constraints were imposed on the rising time, t_R, because it is not critical in comparison with the cooling time. Constraints (31) and (32) require that, by heating up and during temperature maintenance, the applied voltage must not exceed an upper limit imposed by security and practical reasons. Finally, constraint (33) requires that the maximum electric power must not exceed an admissible threshold.

Solving this optimization problem, the minimum value of the objective function $E_D^* = 711$ J was achieved for the following optimal embodiment of the basic module.

$$\mathbf{X}^* = \left\{ \begin{array}{l} n = 2 \\ N_S = 13.43 \\ d_S = 1.42 \text{ mm} \\ C_S = 7 \\ a = 54 \text{ mm} \\ b = 54 \text{ mm} \\ \alpha_P = 1.57 \text{ rad} \\ V_P = 5.2 \text{ V} \\ V_{ON} = 0.7 \text{ V} \end{array} \right\} \quad \text{(Eq 34)}$$

3.3 Conventional Spring Design

The basic module of Fig. 3 supplies an output torque that is strongly dependent on the angular position. For instance, when the angular position is $-\alpha_D$ (Fig. 5) and spring S_1 is hot, there is a maximum, $T_{\max}$, in the output torque. The torque decreases gradually upon rotation of the actuator and reaches a minimum value, $T_{\min}$, when the angular position is α_D, where spring S_1 achieves its lowest extension. This behavior is shown in Fig. 8.

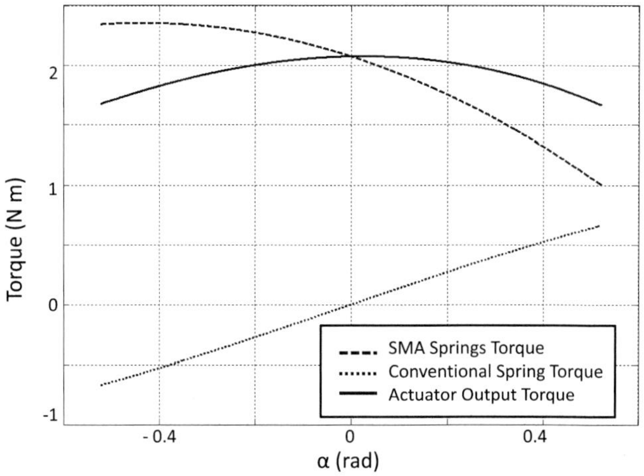

Fig. 8 Diagram of output torque as a function of rotation angle

A conventional spring was thus added to overcome this problem, as shown in Fig. 3 and explained in Sect 2.3. This spring supplies a torque opposite to the one supplied by the hot spring when the hot spring extension is high and a torque that helps the hot spring when the hot spring reaches low extensions. For instance, when S_1 is hot, the traditional spring acts against S_1 for $\alpha = [-\alpha_D \div 0]$, reducing the peak of the output torque, and helps S_1 for $\alpha = [0 \div \alpha_D]$, increasing the minimum output torque. Figure 8 shows that the conventional spring was designed so that the output torque in $\alpha = \alpha_D$ was equal to the output torque in $\alpha = -\alpha_D$, starting from the results of the numerical optimization. To obtain this effect, the spring was designed to ensure a torque which is a half of the difference between the maximum torque, T_{max}, and the minimum torque, T_{min}.

The final result is a reduction in the maximum output torque and an increase in the nominal torque of the actuator (Fig. 8) with a beneficial equalization of the torque profile over the total stroke.

Standard Machine Design formulas (Ref 13, 14) were considered for the embodiment of the spring. The spring is made of steel, with an elastic modulus $E_C = 210000$ MPa, a Poisson ratio $\nu = 0.3$ and an admissible shear stress $\tau_{adm} = 400$ MPa. A numerical optimization was performed to minimize the pitch diameter of the conventional spring. Referring to Fig. 3, the distance between point G, fixed to the chassis, and point H, fixed to the rotor, was chosen of 40 mm so as not to determine interferences with the other parts of the basic stage. A final spring resulted with wire diameter $d_T = 0.85$ mm, pitch diameter $D_T = 4.3$ mm and $N_T = 9.5$ active coils.

4. Detail Design and Virtual Testing

The detail design of the rotary actuator was performed using a 3D CAD system (SolidWorks) in two steps. In the first step, the basic stage was modelled alone. In the second step, an assortment of multiple-stage actuators was obtained by assembling several basic modules in different ways. Figure 9 presents the result of the first step, with Fig. 9(a) showing a front sectioned view and Fig. 9(b) displaying a rear sectioned view of the module. The circular chassis contains the rotor which is connected to a front disk and to a rear disk. The SMA springs link the chassis to the rotor and provide the actuation of the device. The torque-balancing conventional spring is seen in Fig. 9(a), with one end connected to the chassis (through a circular slot in the rotor) and the other end connected to the rotor. The entire mechanism is protected by a closing lid fitted to the chassis. Several patterns of magnetic sockets are arranged on the rear disk, on the front disk, on the lid and on the rear face of the chassis. Movable steel pins can be inserted into the sockets to provide the double function of connecting adjacent stages in the axial direction and transmitting the torque and the rotation between each other. Insertion of the pins in a particular pattern of sockets determines whether the assembly of two modules will produce the sum of angles (as in Fig. 4a) or the sum of torques (as in Fig. 4b), with the same basic module serving both arrangements. Sliding electric contacts ensure the electric supply of all the stages by allowing the relative rotation of adjacent stages. The supply peak voltage is 5.2 V, the maintenance voltage is 0.7 V and a mean power of 23 W is required.

The final module of Fig. 9 has a diameter of 150 mm and a thickness of 32.5 mm. The reciprocating angular stroke amounts to 60° with a nominal (minimum) output torque of 1.6 Nm and a maximum torque of 2.1 Nm (achieved in the middle position of the angular stroke). The torque density is about 2.8 Nm/dm^3.

The second step of the detail design was devoted to checking the architecture of multistage actuators. To this aim, two basic modules as in Fig. 9 were coupled in angle mode (Fig. 4a) and in torque mode (Fig. 4b) by acting on the proper set of pins (see above). Both configurations were tested to verify the complete functionality of the two-stage actuators in terms of mechanical coupling (correspondence of connecting pins) and electrical connectivity (compliance of siding contacts) through the entire angular stroke.

After the geometric design and the architectural check described earlier, the basic module of Fig. 9 was virtually tested using the kineto-dynamic facility (CosmosMotion) provided by the SolidWorks modeller. The analysis was aimed at verifying that the output torque and the angular stroke of the device

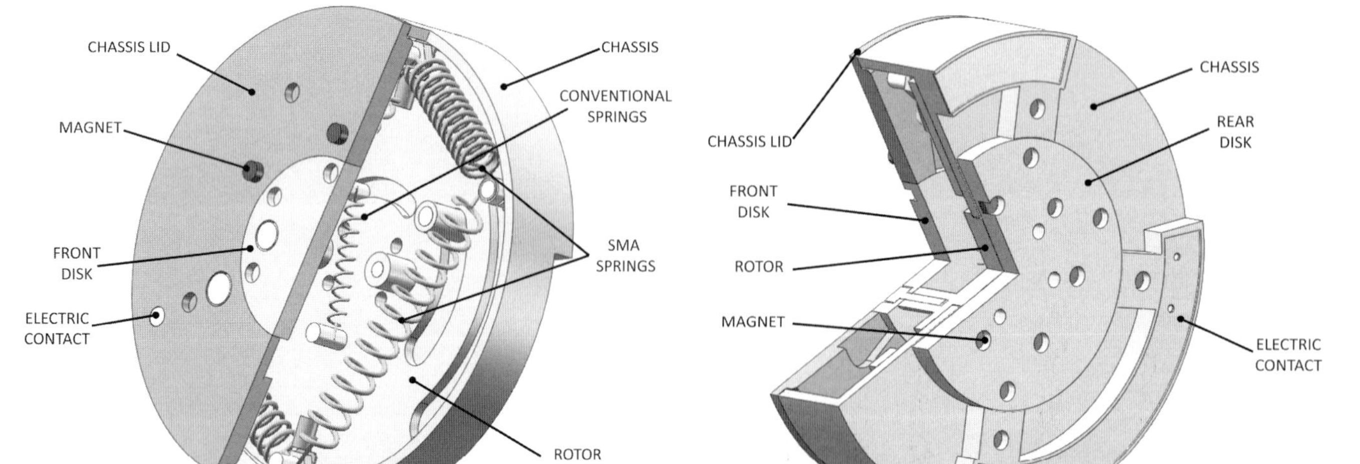

Fig. 9 Detail design of rotary actuator: (a) front view; (b) rear view

would actually meet the target values. The results of the simulations, which are omitted for brevity, fully confirmed the expectations.

5. Discussion

The discussion basically addresses three aspects: the assessment of the actuator performances in comparison with the Target Specifications and with the merits of conventional rotary actuators; the appraisal of the strong and the weak points of the actuator; the suggestion of further work for future developments.

The target specifications related to maximum dimension, thickness and angular stroke of the SMA actuator are fully achieved. Moreover, the nominal torque of the actuator is nearly reached (1.6 Nm instead of 2 Nm). Considering that the actuator thickness is a half of the target specification and that the system is fully modular, the nominal torque target specification can be assumed fully achieved by the assembly of two basic modules coupled together. The designed actuator features a torque density of 2.8 Nm/dm^3 and the target specification was 4 Nm/dm^3. For this reason, this target specification is missed. Likewise also the power consumption specification is not reached with an actual power consumption of 23 W more than doubling the target specification of 10 W. All the other specifications are reached.

The comparison of the SMA rotary actuator with the conventional competitors shows that the proposed SMA design has fair performances in terms of output torque and dimensions. Its performances are acceptable in terms of angular stroke and electric consumption, although their values are worse if compared with conventional actuators. On one hand, the proposed device offers excellent performances in terms of angular stroke, nominal torque, total size and power consumption, if compared with other rotary SMA actuators (Ref 6-9). Furthermore, the proposed methodology is an example of rational and complete design of SMA actuators. The procedure takes into account all the critical aspects that must be considered to design efficient SMA actuators. On the other hand, further work is needed to improve the performance of the present architecture in comparison with conventional competitors. In particular, a mechanical structure must be found that increases the efficiency of the actuator. Although favorable for the achievement of significant strokes, the use of helically coiled SMA springs has limitations in terms of torque generation and can hardly compete with standard actuators. It must be observed, however, that the proposed modular constructions and the use of a conventional balancing spring are very good concepts to enhance angle and torque output. The main conclusion of the paper is that these two key ideas should be retained in the design of future rotary SMA actuators.

6. Conclusions

A shape memory rotary actuator was designed and virtually tested to maximize torque and angular stroke, while limiting overall size and electric consumption. The design process was performed in three steps: conceptual design, modelling and numerical optimization, and detail design and virtual testing. The conceptual design complies with the QFD approach and its output is a global concept that is fully modular: proper combination of the right number of elementary modules can match any requirement in terms of output torque and angular stroke. The basic module is made of shape memory spring units, joining a rotor to a chassis. In addition to these units, the basic module also contains a conventional spring that reduces the torque ripple of the actuator. A thermo-electromechanical model of the basic module was developed and used to perform a numerical optimization, aimed at minimizing the electrical consumption of the actuator. Virtual testing on a 3D digital model confirmed that the output torque and the angular stroke of the device actually meet the target specifications. Although the performances of the present architecture compare unfavorably with conventional competitors, the overall behavior is superior to other shape memory rotary actuators. This improvement is mainly attributable to the proposed modular constructions and to the use of a torque balancing spring: two key ideas that are strongly recommended for the design of future shape memory rotary devices.

Appendix 1

This appendix describes the computational fluid dynamic (CFD) campaign used to determine the total convective heat exchange coefficient, h, of the shape memory spring enclosed in the actuator. For this configuration, the technical literature presents no useful data.

The analyses were performed on the idealized system shown in Fig. 10, with the spring described as a horizontal tube and the enclosing volume described as a horizontal prismatic box with square cross section. The tube represents the envelope of the spring with completely closed coils (mean diameter $D_S = C_S d_S$, wall thickness d_S, axial length $L_S = N_S d_S$). This idealization simplify the CFD problem and ensures conservative results of the heat transfer coefficient with respect to a (stretched) spring with free coils.

The temperature of the tube was set at 373 K (100 °C) and the temperature of the walls of the box was fixed at 293 K (20 °C). The volume of the box was assumed filled with still air at an initial uniform temperature of 293 K (20 °C). In each analysis, the total heat transfer calculated by the software Cosmos Flow Works was divided by the total surface of the

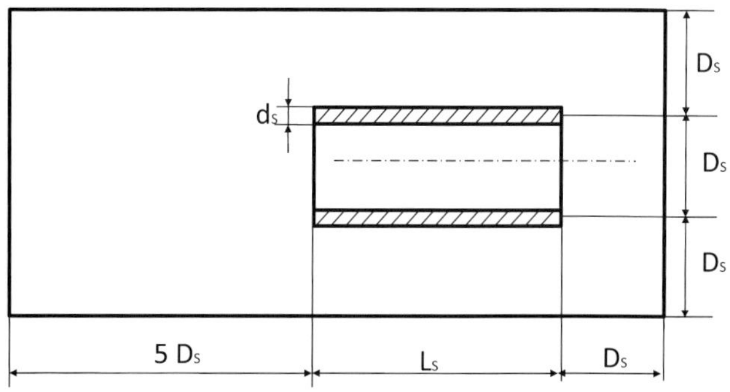

Fig. 10 Geometry investigated in the CFD campaign

tube to obtain the heat transfer coefficient, h, for that particular geometry.

The design of experiments (DOE) approach was adopted to explore the space of all possible spring configurations. Input variables of the DOE were the design parameters of the spring (wire diameter, d_S, spring index, C_S, and coil number, N_S). Two levels were chosen for each variable, and in particular $d_S = 0.25$ and 3 mm, $C_S = 5$ and 12, and $N_S = 12$ and 40.

The values of h calculated for all the configurations were interpolated using statistical instruments to give the polynomial approximation of Eq 15.

References

1. M. Colli, E. Dragoni, and N. Bellato, *System Design of a Shape Memory Alloy Actuator for Automotive Tumble Flap*, SAE Powertrain & Fluid Systems Conference and Exhibition, 2006
2. W. Huang, On the Selection of Shape Memory Alloys for Actuators, *Mater. Des.*, 2002, **23**, p 11–19
3. D. Reynaerts and H. Van Brussel, Design Aspects of Shape Memory Actuators, *Mechatronics*, 1998, **8**, p 635–656
4. A. Lu, D. Grant, and V. Hayward, *Design and Comparison of High Strain Shape Memory Alloys Actuators*, Proceedings 1997 IEEE International Conference on Robotics and Automation
5. E.A. Khidir, N.A. Mohamed, M.J.M. Nor, and M.M. Mustafa, A New Concept of a Linear Smart Actuator, *Sens. Actuators A*, 2007, **135**, p 244–249
6. J. Abadie, N. Chaillette, and C. Lexcellent, An Integrated Shape Memory Alloy Micro-Actuator Controlled by Thermoelectric Effect, *Sens. Actuators A*, 2002, **99**, p 297–303
7. K.J. Gabriel, W.S.N. Trimmer, and J.A. Walzer, A Micro-Rotary Actuator Using Shape Memory Alloys, *Sens. Actuators A*, 1988, **15**, p 95–102
8. K. Kuribayashi, Millimeter-Sized Joint Actuator Using a Shape Memory Alloy, *Sens. Actuators A*, 1989, **20**, p 57–64
9. D. Grant and V. Hayward, *Design of a Shape Memory Alloy Actuator with High Strain and Variable Structure Control*, Proceedings 1995 IEEE International Conference on Robotics and Automation
10. K.T. Ulrich and S.D. Eppinger, *Product Design and Development*, McGraw-Hill, Columbus, OH, 2008
11. D.G. Ullman, *The Mechanical Design Process*, McGraw-Hill, Columbus, OH, 1997
12. I. Spinella and E. Dragoni, Design Equations for Binary Shape Memory Actuators Under Dissipative Forces, *J. Mech. Eng. Sci.*, 2009, **223**(C3), p 531–543
13. J.E. Shigley, C.R. Mischke, and T.H. Brown, *Standard Handbook of Machine Design*, McGraw-Hill, Columbus, OH, 2004
14. G. Niemann, *Machine Elements: Design and Calculation in Mechanical Engineering*, Springer, Berlin, 1978
15. H. Meier and C. Oelschlaeger, Numerical Thermomechanical Modeling of Shape Memory Alloy Wires, *Mater. Sci. Eng. A*, 2004, **378**, p 484–489

JMEPEG (2009) 18:649–654
DOI: 10.1007/s11665-009-9409-7

SMA Numerical Modeling Versus Experimental Results: Parameter Identification and Model Prediction Capabilities

Ferdinando Auricchio, Alberto Coda, Alessandro Reali, and Marco Urbano

(Submitted October 14, 2008; in revised form January 17, 2009)

In this work, we briefly review the one-dimensional version of a well-known phenomenological shape memory alloy (SMA) constitutive model able to represent the main macroscopic SMA macroscopic behaviors (i.e., superelasticity and shape-memory effect). We then show how to identify the needed parameters from experimental results and, in particular, from strain-temperature tests. We finally use the obtained material parameters to test the prediction properties of the model, comparing numerical results with some experiments (different from those used for the identification), and we discuss model capabilities and further required enhancements.

Keywords experimental results, numerical modeling, parameter identification, shape memory alloys

1. Introduction

The research toward an exhaustive modeling of the macroscopic behavior of shape memory alloys (SMAs) has been widely growing in last years because of the increasing employment of such smart materials in a large number of applications in many fields of engineering.

Thus, many models able to reproduce SMA main macroscopic effects have been recently proposed in the literature in one or in three dimensions (see, e.g., Ref 1-8).

The purpose of such works is to give reliable models which could be used by engineers who need to perform accurate simulations to design and study the response of SMA structures or components. To reach this goal, the proposed models have to be interfaced with analysis programs which are commonly used in engineering offices during the design process of a real device; accordingly, a good model should also be robust and flexible, possibly including both superelasticity and shape-memory effect, without becoming too complex. Also, the problem of the physical interpretation of the model parameters is an important issue, as a clear and effective parameter identification procedure is the key for the employment of a constitutive model by engineers in real-life simulations.

Among the different models available in the recent literature, the one introduced in Ref 9 and improved in Ref 10 seems to be very attractive. Developed within the theory of irreversible thermodynamics, this model is in fact able to describe both superelasticity and shape-memory effect (see Fig. 1), and the corresponding solution algorithm is simple and robust as it is based on a plasticity-like return map procedure. Its robustness makes the model particularly suitable for implementation within finite element codes, allowing the simulation of the behavior of complex SMA devices.

In the following, we focus on the 1D model obtained by the reduction of the 3D model described in Ref 10. We briefly show and comment the basic equations, as well as the physical meaning and a possible identification strategy for the needed material parameters. We then show the application of the proposed procedure in order to simulate the behavior of a SMA wire, comparing our numerical results with experimental tests.

2. Basic Model Equations

The model equations arise from a free energy density function Ψ defined as

$$\Psi(\varepsilon, \varepsilon^{tr}, T) = \frac{1}{2}E(\varepsilon - \varepsilon^{tr})^2 + \tau_m(T)|\varepsilon^{tr}| + \frac{1}{2}h(\varepsilon^{tr})^2 + I_{\varepsilon_L}(\varepsilon^{tr}),$$

where the physical meaning of the different material parameters are discussed in the next section. To satisfy the transformation strain constraint $|\varepsilon^{tr}| \le \varepsilon_L$, the indicator function $I_{\varepsilon_L}(\varepsilon^{tr})$ is used, which is defined as

$$I_{\varepsilon_L}(\varepsilon^{tr}) = \begin{cases} 0 & \text{if } |\varepsilon^{tr}| \le \varepsilon_L, \\ +\infty & \text{otherwise.} \end{cases}$$

This article is an invited paper selected from presentations at Shape Memory and Superelastic Technologies 2008, held September 21-25, 2008, in Stresa, Italy, and has been expanded from the original presentation.

Ferdinando Auricchio and **Alessandro Reali**, Dipartimento di Meccanica Strutturale, Università degli Studi di Pavia, Pavia, Italy, European Centre for Training and Research in Earthquake Engineering (EUCENTRE), Pavia, Italy, and Istituto di Matematica Applicata e Tecnologie Informatiche (IMATI), CNR, Pavia, Italy; and **Alberto Coda** and **Marco Urbano**, SAES Getters Group, Lainate, MI, Italy. Contact e-mail: auricchio@unipv.it.

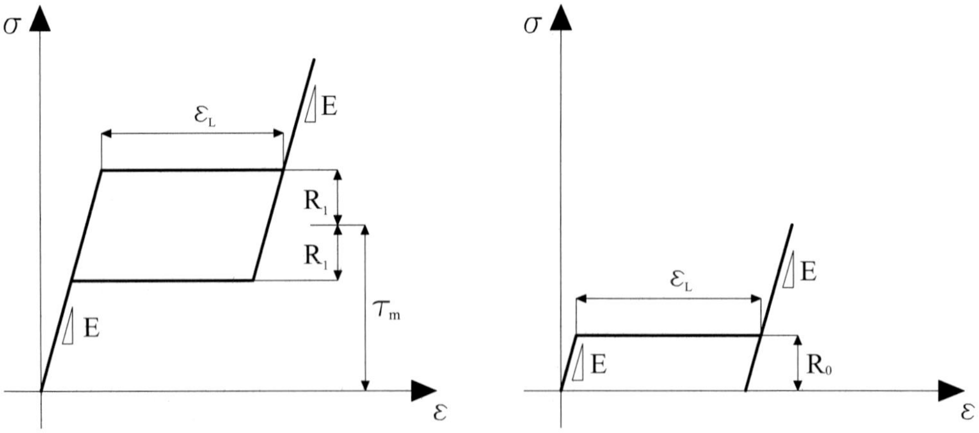

Fig. 1 Superelastic effect (*left*) and shape-memory effect (*right*). Graphical representation for the material parameters required by the constitutive model

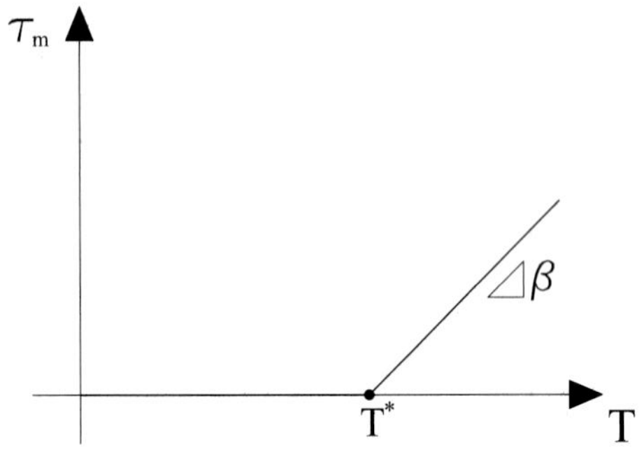

Fig. 2 Temperature dependence for τ_m

Moreover, the dependence of τ_m on temperature (depicted in Fig. 2) is expressed as

$$\tau_m(T) = \beta(T - T^*)^+,$$

where the positive part function is defined as $(a)^+ = (a + |a|)/2$. The constitutive equations for the model are then classically obtained as follows:

$$\begin{cases} \sigma = \frac{\partial \Psi}{\partial \varepsilon} = E(\varepsilon - \varepsilon^{tr}), \\ X = -\frac{\partial \Psi}{\partial \varepsilon^{tr}} = \sigma - \tau_m(T)\frac{\varepsilon^{tr}}{|\varepsilon^{tr}|} - h\varepsilon^{tr} - \gamma\frac{\varepsilon^{tr}}{|\varepsilon^{tr}|}, \end{cases}$$

where X is the thermodynamic stress-like quantity associated to the transformation strain ε^{tr}. The variable γ results from the (sub)differentiation of the indicator function and fulfills

$$\begin{cases} \gamma = 0 \ \text{ if } |\varepsilon^{tr}| < \varepsilon_L, \\ \gamma \geq 0 \ \text{ if } |\varepsilon^{tr}| = \varepsilon_L. \end{cases}$$

To control the evolution of the internal variable ε^{tr}, a limit function F is introduced and is defined as

$$F(X) = |X| - R(T),$$

where the radius of the elastic domain $R(T)$ controls the width of the hysteresis loops (its dependence on temperature is illustrated in Fig. 3).

Considering then an associative framework, the flow rule for the internal variable takes the form

$$\dot{\varepsilon}^{tr} = \dot{\zeta}\frac{\partial F}{\partial X} = \dot{\zeta}\frac{X}{|X|}.$$

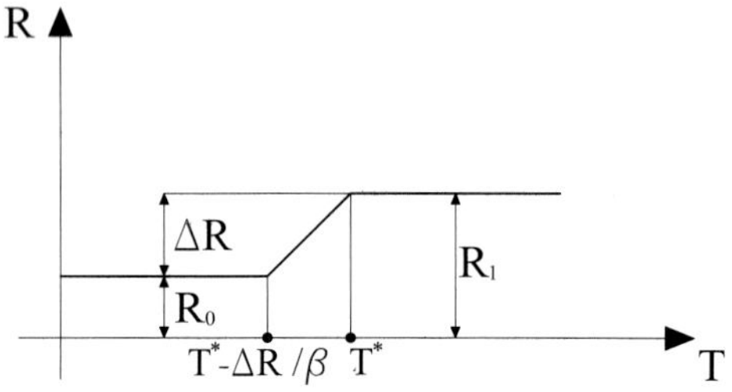

Fig. 3 Temperature dependence for the elastic domain radius *R*

The model is finally completed by the classical Kuhn-Tucker conditions

$$\begin{cases} \dot{\zeta} \geq 0, \\ F \leq 0, \\ \dot{\zeta}F = 0. \end{cases}$$

3. Model Input Material Parameters

As seen from the equations introduced in the previous section, the 1D model requires eight material parameters indicated as E, ε_L, h, R_0, R_1, β, T^*, and δ.

A brief description along with some comments on such parameters follows:

- E: elastic modulus.
  In this model, the same elastic modulus for austenite and martensite is assumed. Different moduli for the two phases will be accounted for in future developments.
- ε_L: maximum transformation strain.
  This material constant measures the maximum strain obtainable through alignment of the martensite variants.
- h: stress-strain slope measure during transformation.
  This material constant is important in measuring the slope of the stress-strain diagram during the transformation, in fact $\frac{\partial \sigma}{\partial \varepsilon} = \frac{Eh}{E+h}$ (h can be interpreted as a standard linear kinematic hardening parameter, as in classical elasto-plastic models). Note that $\frac{\partial \sigma}{\partial \varepsilon} = \frac{Eh}{E+h} \simeq h$ when $E \gg h$. In particular, for $h = 0$ the stress-strain diagram shows a flat plateau (as in Fig. 1).

- R_0: elastic domain radius at low temperatures.
- R_1: elastic domain radius at high temperatures.
  To match the different radii with temperature, we assume a dependence of R on T as described in Fig. 3. In particular, R_0 is equal to half of the hysteresis loop width at low temperatures. R_1, instead, is equal to half of the hysteresis loop width at high temperatures, i.e. for $T > T^*$.
- β: slope of τ_m (see Section 2 and Fig. 1) with respect to temperature, as represented in Fig. 2.
- T^*: reference temperature, important in defining the dependence of τ_m on temperature (see again Fig. 2).

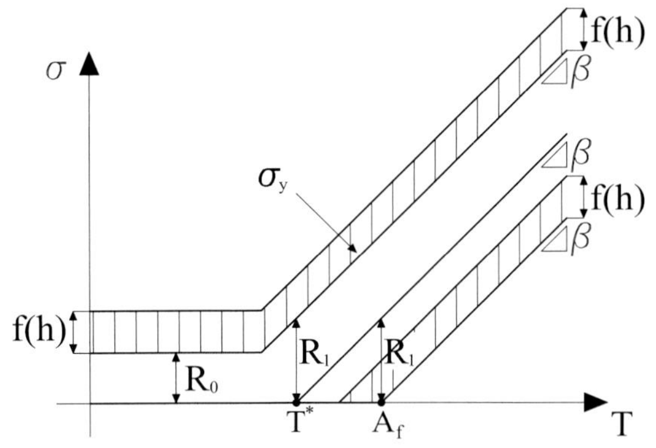

Fig. 4 Temperature dependence for the transformation stress σ_y. $f(h) = \frac{Eh}{E+h}\varepsilon_L$, and, when $E \gg h, f(h) \simeq h\varepsilon_L$

Moreover, β and T^* are also important in expressing the dependence on temperature of the transformation stress $\sigma_y = \tau_m + R$ (see Fig. 4). In particular, β represents the slope of σ_y with respect to temperature.

- δ: regularization parameter.
  To guarantee numerical solvability, a regularization is performed on the norm of the transformation strain as follows:

$$\overline{|\varepsilon^{tr}|} = \sqrt{(\varepsilon^{tr})^2 + \delta} - \sqrt{\delta},$$

where δ is a small (e.g., $\sim 10^{-8}$) user-defined regularization parameter.

4. Model Response to Simple Test Situations

The most common situations that can be encountered in standard tests at constant temperature or stress are here considered and explained, also through graphical representations (see Fig. 5-7).

In particular, in Fig. 5, a superelastic test at a constant temperature $T = T_{SE} > T^*$ is described. The test starts at $\sigma = 0$ and a loading phase is followed by an unloading phase up again to $\sigma = 0$. No residual strain is observed.

In Fig. 6, instead, a shape-memory test at a constant temperature $T = T_{SME} < T^*$ is described. The test starts at $\sigma = 0$ and a loading phase is followed by an unloading phase up again to $\sigma = 0$. In this case, a residual strain is observed (that can be, however, recovered by heating the material).

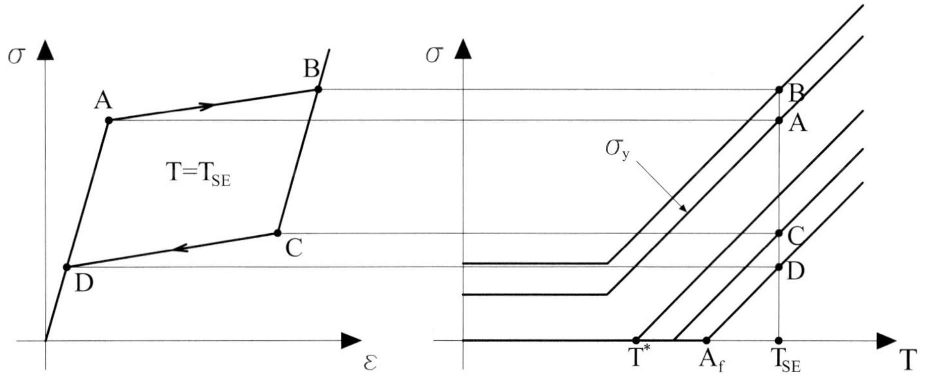

Fig. 5 Superelastic test (at a constant temperature $T = T_{SE}$): stress-strain diagram (*left*) and stress-temperature diagram (*right*)

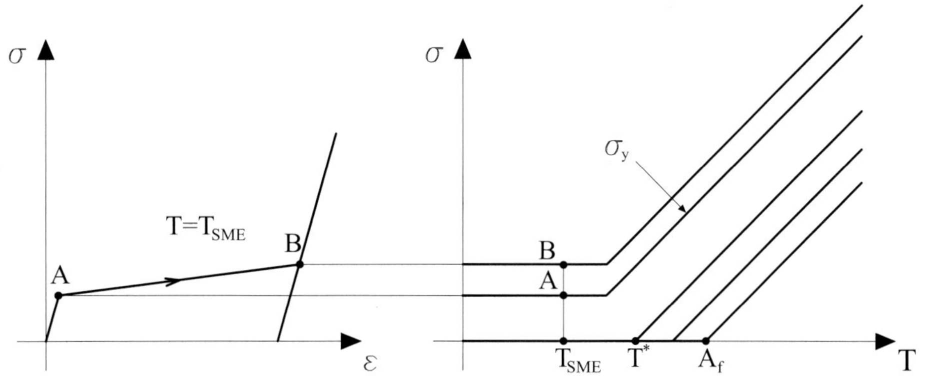

Fig. 6 Shape-memory test (at a constant temperature $T = T_{SME}$): stress-strain diagram (*left*) and stress-temperature diagram (*right*)

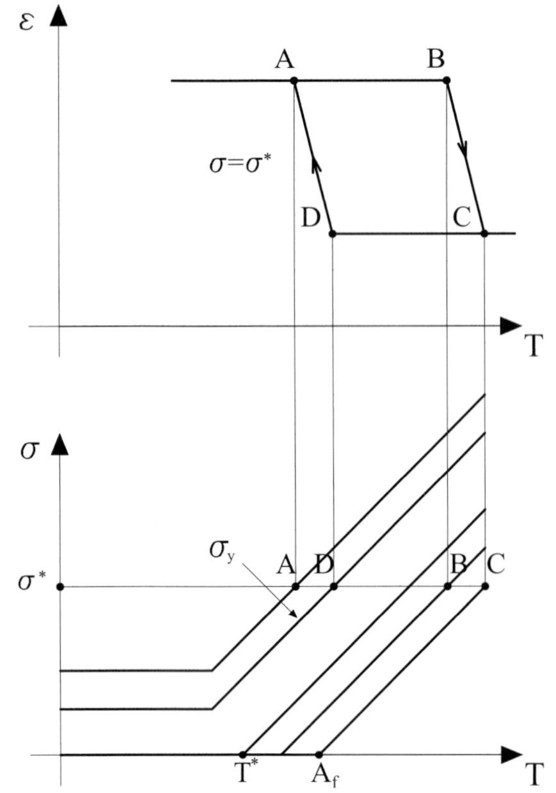

Fig. 7 $\varepsilon - T$ test (at a constant stress $\sigma = \sigma^*$): strain-temperature diagram (*top*) and stress-temperature diagram (*bottom*)

Finally, in Fig. 7, a strain-temperature test at a constant stress $\sigma = \sigma^*$ is described. The test starts at a low temperature (with the material at a stress level such that only martensite is present) and a heating phase is followed by a cooling phase up to the initial temperature.

5. Parameter Identification from a ε–*T* Test

Given two strain-temperature curves obtained at two constant stresses $\sigma^1 > \sigma^2$, as represented in Fig. 8, it is possible to identify in a simple way for all the needed material parameters.

In fact, as highlighted in the figure, ε_L is immediately obtained as the difference between the maximum and the minimum strain. E and β can be computed, using their definition, respectively, as

$$E = \frac{\Delta\sigma}{\Delta\varepsilon} \quad \text{and} \quad \beta = \frac{\Delta\sigma}{\Delta T},$$

with $\Delta\varepsilon$, $\Delta T = T^1_{yM} - T^2_{yM}$ and $\Delta\sigma = \sigma^1 - \sigma^2$ as indicated in the figure.

Then, it is possible to compute R_1 knowing that the hysteresis width is $2R_1/\beta$ (see again Fig. 8). The equation $T^1_{yA} - T^1_{yM} = 2R_1/\beta$ is obtained subtracting the two following equations (deriving from $F = |X| - R = 0$)

$$\begin{cases} \sigma - \beta(T_{yM} - T^*) = R_1, \\ \sigma - \beta(T_{yA} - T^*) = R_1. \end{cases}$$

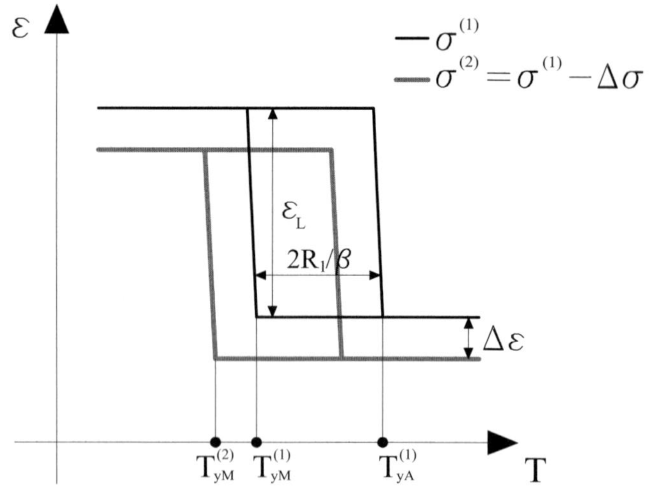

Fig. 8 Two strain-temperature curves at two constant stresses $\sigma^1 > \sigma^2$

Moreover, starting from $\frac{\partial\sigma}{\partial\varepsilon} = \frac{Eh}{E+h}$ and considering the dependence of the stress on temperature $\frac{\partial\sigma}{\partial\varepsilon} = \frac{\partial\sigma}{\partial T}\frac{\partial T}{\partial\varepsilon} = \beta\frac{\partial T}{\partial\varepsilon}$, with simple computations it can be found that

$$h = \frac{1}{\frac{1}{\beta}\frac{\partial\varepsilon}{\partial T} - \frac{1}{E}}.$$

Finally, to compute T^*, the following expression, directly derived from

$\sigma^1 - \beta(T^1_{yM} - T^*) = R_1$, can be used

$$T^* = T^1_{yM} + \frac{R_1 - \sigma^1}{\beta}.$$

Alternatively, T^* could be computed from

$$T^* = A_f - \frac{R_1}{\beta},$$

as illustrated in Fig. 4. In this case, the austenite finishing temperature A_f is needed (and it may be obtained, e.g., from a differential scanning calorimetry, DSC).

6. An Example of Parameter Identification From a ε–*T* Test: Numerical Versus Experimental Results

In this section, an example of parameter identification (and numerical simulation) from a ε-*T* test is illustrated. We remark that we choose to use this kind of experimental results, instead of classical σ-ε curves, because the ε-*T* response is very significant for SMA producers, being one of the most important characteristic of SMA for actuators.

The experimental data (provided by the SAES Getters Group) shown in Fig. 9 and 10 have been obtained testing 0.2 mm NiTi wires at different temperatures, keeping the stress constant (in these examples, strain-temperature cycles at 100, 150, and 200 MPa are considered). It is to be remarked that here, instead of the strain ε, the "stroke" (referred to zero) is considered, i.e. for each "ε-*T*" curve the actual ordinate is $-(\varepsilon - \varepsilon_{min})$.

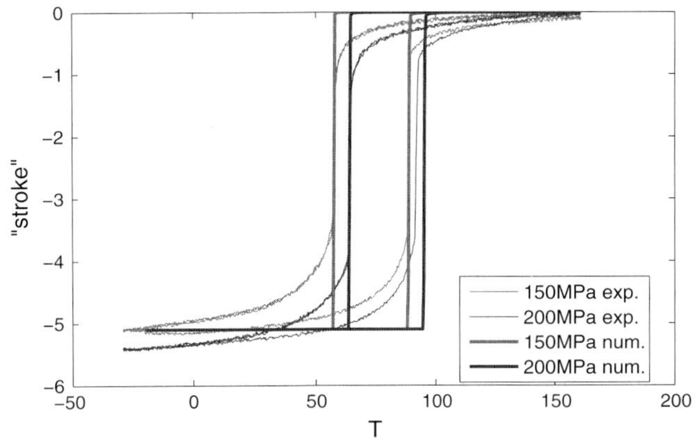

Fig. 9 Comparison of experimentally and numerically obtained curves relative to 0.2 mm wires at 150 and 200 MPa; the numerical curve at 150 MPa is fully fitted, while the one at 200 MPa is partially fitted and partially predicted

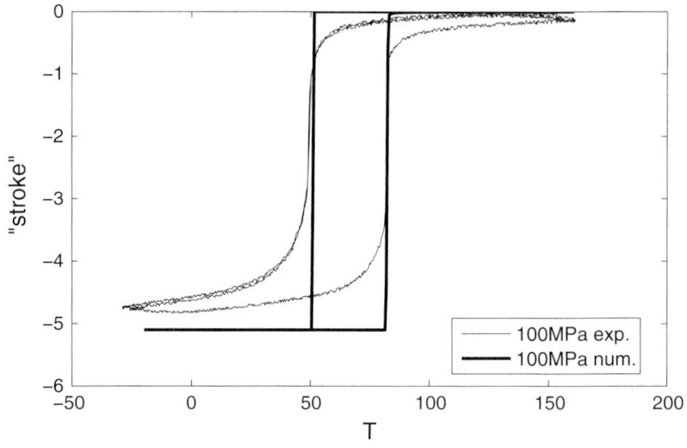

Fig. 10 Comparison of experimentally and numerically obtained curves relative to 0.2 mm wires at 100 MPa; the numerical curve is fully predicted

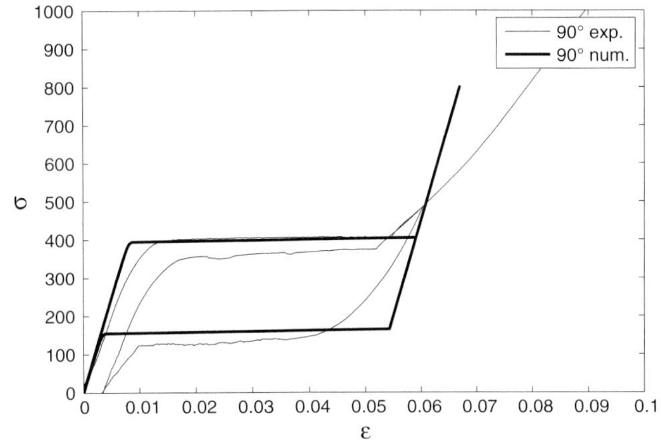

Fig. 11 Stress-strain tests at the temperature of 90 °C; the numerical curve is fully predicted (i.e., no curve fitting at all)

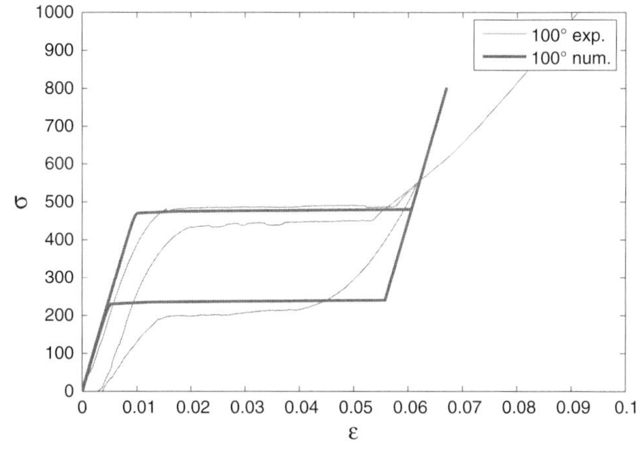

Fig. 12 Stress-strain tests at the temperature of 100 °C; the numerical curve is fully predicted (i.e., no curve fitting at all)

Performing the parameter identification procedure discussed in the previous section, the following set of material parameters has been obtained:

$$
\begin{aligned}
E &= 50000 \text{ MPa} \\
\varepsilon_L &= 5.1\% \\
h &= 100\,\text{MPa} \\
R_0 &= 40\,\text{MPa} \\
R_1 &= 120\,\text{MPa} \\
\beta &= 7.7\,\text{MPa}/^\circ\text{C} \\
T^* &= 54\,^\circ\text{C} \\
\delta &= 10^{-8}
\end{aligned}
$$

For the identification procedure, the 150 MPa curve has been used to compute ε_L and T^*, and, for the other parameters, the 150 and 200 MPa curves have been employed. Moreover, R_0 has been assumed to be smaller than 50 MPa (i.e., 40 MPa) as it has been experimentally observed that a hysteresis is present even at 50 MPa.

With these parameters, numerical curves have been produced and those for the 150 and 200 MPa cases are reported in Fig. 9 in order to compare them with the experimental ones from which the parameters have been identified. Moreover, Fig. 10 shows that a good agreement is obtained also for the 100 MPa curve, which has not been used for the identification procedure.

It is also to be remarked that in all these results an important role is played by the fact that in the model considered, at the moment, there is no distinction between the elastic moduli for austenite and martensite. This, in fact, makes constant the gap between the maximum and the minimum strain, while in the experimental results significant differences are observed.

Finally, some stress-strain tests at temperatures of 90, 100, 110, and 120 °C have been performed using the parameters computed above, and they have been compared with experimental data. The results are reported in Fig. 11-14, where a good agreement is observed in all cases.

7. Conclusions and Future Developments

In this work, we have considered the 1D version of the phenomenological SMA constitutive model presented in Ref 10

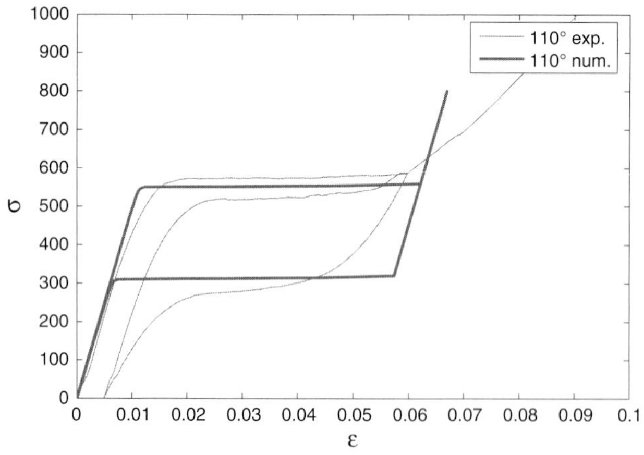

Fig. 13 Stress-strain tests at the temperature of 110 °C; the numerical curve is fully predicted (i.e., no curve fitting at all)

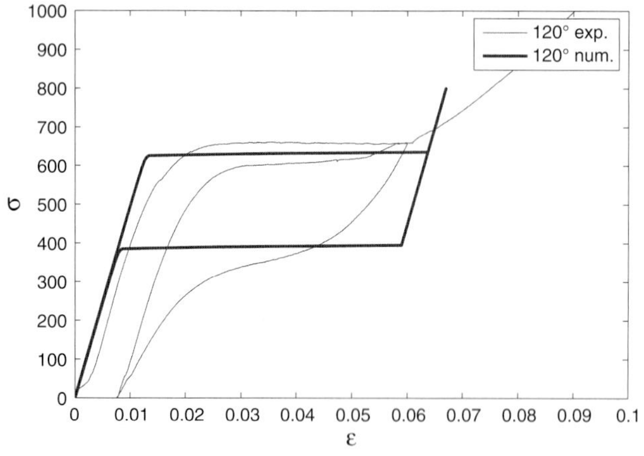

Fig. 14 Stress-strain tests at the temperature of 120 °C; the numerical curve is fully predicted (i.e., no curve fitting at all)

and, after a review of the model including a description of all the material parameters, we have shown a strategy to identify those parameters from experimental results and, in particular, from strain-temperature tests. We have then used such a procedure to obtain the material parameters for a real SMA wire in order to test the prediction properties of the model, comparing numerical results with some experiments (different from those used for identification) and obtaining a good agreement.

Finally, we remark that interesting extensions of the model, in order to have an even more accurate response, could be the implementation of:

- elastic properties depending on the transformation strain level (see Ref 11);
- plasticity and training effects (see Ref 12, 13);
- tension-compression asymmetries (see Ref 11); and
- fully coupled electric-thermo-mechanical analyses (see Ref 14).

References

1. S. Govindjee and C. Miehe, A Multi-Variant Martensitic Phase Transformation Model: Formulation and Numerical Implementation, *Comput. Methods Appl. Mech. Eng.*, 2001, **191**, p 215–238
2. D. Helm and P. Haupt, Shape Memory Behaviour: Modelling within Continuum Thermomechanics, *Int. J. Solids Struct.*, 2003, **40**, p 827–849
3. D.C. Lagoudas, P.B. Entchev, P. Popov, E. Patoor, L.C. Brinson, and X. Gao, Shape Memory Alloys, Part II: Modeling of Polycrystals, *Mech. Mater.*, 2006, **38**, p 391–429
4. P. Popov and D.C. Lagoudas, A 3-D Constitutive Model for Shape Memory Alloys Incorporating Pseudoelasticity and Detwinning of Self-Accommodated Martensite, *Int. J. Plasticity*, 2007, **23**, p 1679–1720
5. V.I. Levitas, Thermomechanical Theory of Martensitic Phase Transformations in Inelastic Materials, *Int. J. Solids Struct.*, 1998, **35**, p 889–940
6. B. Peultier, T. Ben Zineb, and E. Patoor, Macroscopic Constitutive Law for SMA: Application to Structure Analysis by FEM, *Mater. Sci. Eng. A*, 2006, **438–440**, p 454–458
7. B. Raniecki and Ch. Lexcellent, R_L Models of Pseudoelasticity and Their Specification for Some Shape-Memory Solids, *Eur. J. Mech. A Solids*, 1994, **13**, p 21–50
8. S. Reese and D. Christ, Finite Deformation Pseudo-Elasticity of Shape Memory Alloys—Constitutive Modeling and Finite Element Implementation, *Int. J. Plasticity*, 2008, **24**, p 455–482
9. A.C. Souza, E.N. Mamiya, and N. Zouain, Three-Dimensional Model for Solids Undergoing Stress-Induced Phase Transformations, *Eur. J. Mech. A: Solids*, 1998, **17**, p 789–806
10. F. Auricchio and L. Petrini, A Three-Dimensional Model Describing Stress-Temperature Induced Solid Phase Transformations: Solution Algorithm and Boundary Value Problems, *Int. J. Numer. Methods Eng.*, 2004, **61**, p 807–836
11. F. Auricchio, A. Reali, and U. Stefanelli, A Macroscopic 1D Model for Shape Memory Alloys Including Asymmetric Behaviors and Transformation-Dependent Elastic Properties, *Comput. Methods Appl. Mech. Eng.*, 2009, **198**, p 1631–1637
12. F. Auricchio and A. Reali, A Phenomenological One-Dimensional Model Describing Stress-Induced Solid Phase Transformation with Permanent Inelasticity, *Mech. Adv. Mater. Struct.*, 2007, **14**, p 43–55
13. F. Auricchio, A. Reali, and U. Stefanelli, A Three-Dimensional Model Describing Stress-Induced Solid Phase Transformation with Permanent Inelasticity, *Int. J. Plasticity*, 2007, **23**, p 207–226
14. F. Auricchio and L. Petrini, A Three-Dimensional Model Describing Stress-Temperature Induced Solid Phase Transformations: Thermomechanical Coupling and Hybrid Composite Applications, *Int. J. Numer. Methods Eng.*, 2004, **61**, p 716–737

JMEPEG (2009) 18:655–663
DOI: 10.1007/s11665-009-9375-0

1059-9495/$19.00

Study of a Nitinol Stent Deployed into Anatomically Accurate Artery Geometry and Subjected to Realistic Service Loading

Nuno Rebelo, Rachel Fu, and Michael Lawrenchuk

(*Submitted September 12, 2008; in revised form December 4, 2008*)

Realistic simulation of a stent in service loading is necessary for the design of Nitinol stents. This type of analysis is complicated by several issues: artery geometry and material properties are largely unknown and may vary from person to person; the loading on the stent is due to interactions between pulsatile blood flow, the artery, and stent. This study addresses the first issue by obtaining realistic artery geometry from magnetic resonance imaging (MRI) scans of a patient in the fetal and supine positions, and comparing deformation of the stent deployed into the two different artery configurations. The second issue is addressed by investigating two different methods of applying service loading. The first method applies sinusoidal pressure waves directly to the artery wall. The second method considers the effect of blood flow within the artery by performing a fluid-structure interaction (FSI) simulation. Simulation of expansion, annealing, crimping, and deployment of a Nitinol stent into the two artery configurations was performed using Abaqus. Following the deployment, sinusoidal pressure pulses, emulating systolic and diastolic blood pressures, were applied to the inner surface of the artery. To simulate the effect of blood within the artery, the coupled Eulerian-Lagrangian method in Abaqus was used. The blood was modeled as a partially filled Eulerian domain, whose material boundary contacted the Lagrangian artery. Transient inlet velocity and outlet pressure conditions were applied to the Eulerian domain to simulate blood flow. Since fatigue life calculations are important for the design of Nitinol stents, the maximum principal strains at the diastolic pressure troughs and the difference between the maximum principal strains at diastolic and systolic pressure were investigated for five cycles. Different maximum principal strains resulted for the different artery geometries (fetal or supine), while the difference between maximum principal strains was similar. For the simulations including the effect of blood flow, the maximum principal strains at the diastolic pressure troughs were somewhat higher than the simulations in which pressure was directly applied to the artery. In conclusion, this study established and compared various methods of improving the service loading in the simulation of Nitinol stents, specifically including anatomically correct artery geometry and the effect of blood flow.

Keywords biomaterials, modeling processes, titanium

1. Background/Objectives

The objective of this study is to demonstrate finite element analysis (FEA) methodologies which can be used to simulate the service loading on a Nitinol self-expanding stent with a more realistic approach than typical methods. The cyclic loading on a stent deployed into an artery is due to complex interactions between the pulsatile blood flow, the deformable artery, and the stent. In general, the loading is modeled using an expanding and contracting rigid cylinder in contact with the inner and outer surfaces of the stent (Ref 1, 2). This approximation does not account for the nonuniform deformation in the artery due to the pressure pulses, which is the main cause of the loading experienced by the stent. For this study, the stent was first deployed into a deformable artery, whose geometry was determined from MRI scans of a patient in both the supine and fetal positions. The interior surface of the artery was then subjected to systolic and diastolic pressure wave loading and the strains in the stent were compared for the two artery configurations. Finally, to examine the effect of blood flow within the supine artery, a coupled Eulerian-Lagrangian (CEL) analysis was performed in which blood contained within the artery was subjected to transient inlet velocity and outlet pressure conditions. The deformations and strains from this FSI analysis were then compared with those of the purely structural analysis.

This article is an invited paper selected from presentations at Shape Memory and Superelastic Technologies 2008, held September 21-25, 2008, in Stressa, Italy, and has been expanded from the original presentation.

Nuno Rebelo and **Rachel Fu**, DS Simulia Corp, SIMULIA Western Region, 39221 Paseo Padre Parkway, Suite F, Fremont, CA 94538; and **Michael Lawrenchuk**, Materialise, 6111 Jackson Rd., Ann Arbor, MI 48103. Contact e-mail: Nuno.Rebelo@3ds.com.

2. Methodology

The RESIStent consortium, working to improve the durability of Nitinol stents (Ref 3), produced a wealth of medical images of patients' superior femoral arteries (SFA). We have used a sample of these in order to create a realistic environment into which a stent was deployed. Mimics software (Materialise, Ann Arbor, MI) was used to obtain the artery geometry utilized in this study. The artery geometry was isolated from the MRI data through semi-automated segmentation and surface modeling techniques. A gray value threshold was used to differentiate the artery from surrounding tissues, and some further editing was applied to create the initial artery geometry. The Mimics FEA Module was then used to optimize the surface mesh and facilitate the link with Abaqus (SIMULIA, Providence, RI). The artery model was then exported as a surface mesh to Abaqus. The procedure was performed for MRI scans of a patient in both the fetal and supine positions. Figure 1 shows the contrast MRI images in both positions as well as the corresponding FEA geometry generated. Because the simulations performed in this study pertain to stent deployment, only a small section of the artery was included in the FEA models, as shown in Fig. 1.

Abaqus was used as the FEA solver in this study. The stent design was laser-cut from a Nitinol tube with an outer diameter of 1.47 mm and a thickness of 0.216 mm. The stent mesh consisted of 114,561 linear hexahedral reduced integration elements. The material model for Nitinol behavior, built into Abaqus and based on the Auricchio and Taylor (Ref 4, 5) model was used. It included different behavior in tension and compression, different moduli for austenite and martensite, and transformation plateaus determined at body temperature. The artery was modeled with triangular shell elements of thickness and linear material properties as specified by Tittelbaugh et al. (Ref 6), in which the Young's modulus of the artery was calculated based on a compliance of 2.2%. All six degrees of freedom of the ends of the artery segments were modeled as fixed, and translational degrees of freedom of two points on one end of the stent were constrained to prevent rigid body motion.

Prior to any service loadings imposed on the stent, it was necessary to simulate the stent processing steps as well as the stent deployment into the artery. The first processing step modeled was the stent expansion from the as-cut configuration using a rigid surface to a specified inner diameter of 5.54 mm for the straight artery section (supine) and 6.56 mm for the bent artery section (fetal). The deviation from the rated stent expansion for this design was contrived in order for the stent to self-expand beyond the artery section. It should be noted that in principle, a stent would not be deployed into a bent SFA. A bent configuration would rather be obtained by bending both artery and stent after implantation into a straight configuration. Since the mechanical description of such deformations is not well characterized at this time, we have opted to deploy into a geometrically well-characterized bent configuration. Although not realistic, the procedure may yield interesting information about what happens when a self-expanding stent is deployed into a curved artery. The annealing step was then modeled by removing the stent elements and adding them back into the model in a strain free manner. The final processing step modeled was the crimping of the stent using a rigid surface to an outer diameter of 3.30 mm. The crimped stent was then released into either the bent or straight artery sections through removal of contact with the rigid surface. The deformation at each of these steps is shown in Fig. 2. These simulation steps were performed in Abaqus/Standard as static analyses.

It is well known that average blood pressure fluctuates between a systolic value of 120 mmHg (15,999 Pa) and a diastolic value of 80 mmHg (10,666 Pa) with a mean value of 100 mmHg (13,332 Pa). Since the artery geometry was obtained in vivo, it corresponds to a deformation state which accounts for the mean internal pressure of 100 mmHg. Therefore, to simulate systolic and diastolic fluctuations, a sinusoidal pressure wave fluctuating between 20 mmHg

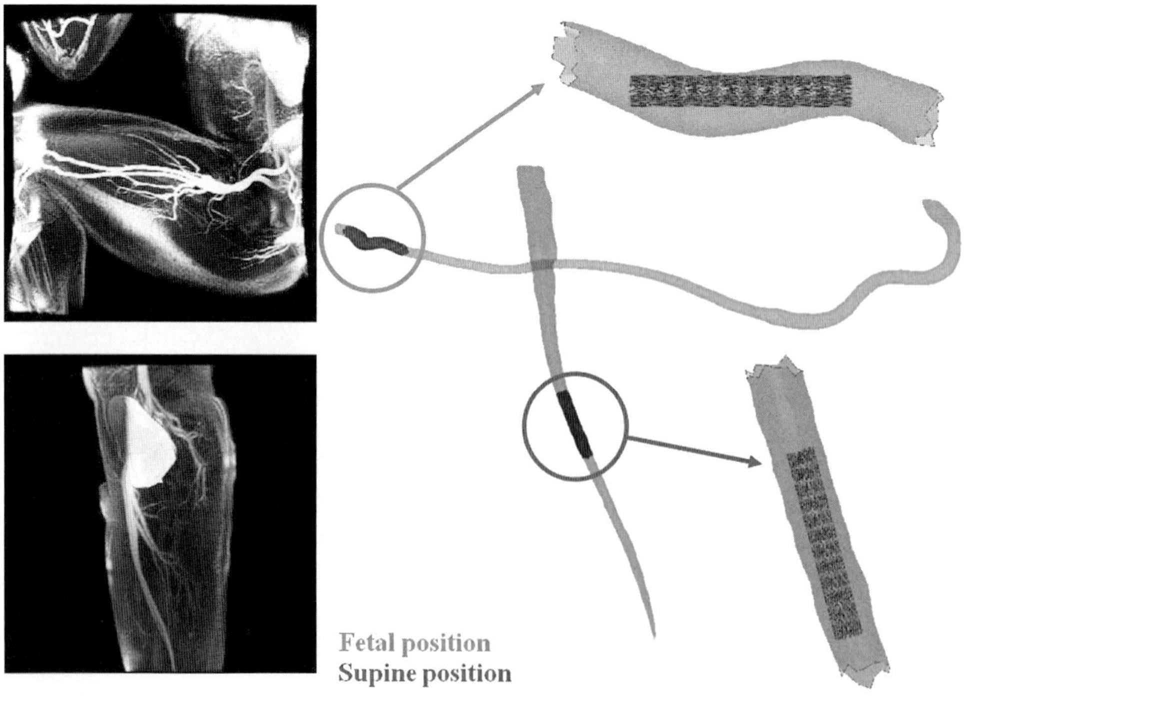

Fig. 1 Femoral artery geometry including the MRI scan in fetal (*upper*) and supine (*lower*) positions and the FEA geometry segments modeled (stent prior to deployment shown)

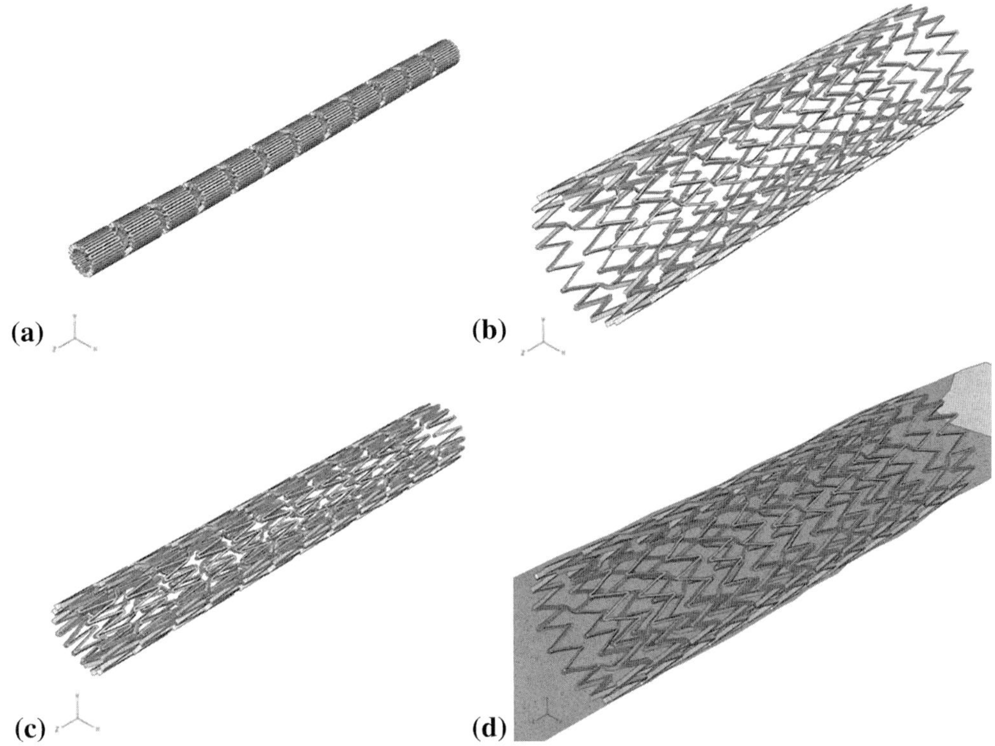

Fig. 2 Analysis steps prior to service loading (a) laser-cut stent, (b) expanded, (c), crimped, and (d) deployed into artery

(2666 Pa) and −20 mmHg (−2666 Pa) was applied to the inner artery surface at a frequency of 1.2 Hz, and five cycles were modeled. Tittelbaugh et al. (Ref 6) showed that for this frequency, dynamic effects are negligible, so the pressure loading was also modeled as a static procedure.

For the next step in realistic service loading, we wished to investigate the effect of the blood flow within the artery, as it is the contact pressure exerted by the fluid which is actually causing the structural loading. Abaqus/Explicit contains an Eulerian capability to complement and interact with the traditional Lagrangian analysis. In a Lagrangian analysis, nodes are fixed within the material and the elements deform as the material deforms; in an Eulerian analysis, nodes are fixed in space and material flows through elements that do not deform (Ref 7). Since Eulerian elements may be partially filled with or completely void of material, the Eulerian material boundary is computed during each time increment using a volume of fluid formulation. Eulerian-Lagrangian contact allows Eulerian material to interact with Lagrangian elements, enabling simulation of FSI.

The main advantage of the CEL method is that FSI analyses can be performed within the same model using a single analysis code. Other methods of FSI generally involve transferring loads and boundary conditions between simultaneous structural and computational fluid mechanics models, requiring multiple analysis codes as well as a means of information transfer between the two codes.

Because the Eulerian capability has been implemented in Abaqus/Explicit, the CEL method is most suitable for high-speed dynamic events. However, because the period of pulsatile blood flow is approximately 1 s, various modeling techniques were required to obtain a solution from the Explicit solver within a reasonable amount of solution time. The integration operator in the explicit procedure is conditionally stable, so incrementation is controlled by a stable time increment calculated by the solver.

Figure 3 shows the model setup of the coupled FSI model. The stress state and deformed configuration of the stent released into the straight artery was imported into a model containing the definition of the Eulerian domain. The Eulerian domain consisted of 6804 elements and was initially specified as partially filled with material boundaries at the artery wall. The transient inlet velocity condition specified was determined from Fourier coefficients derived by Steinman et al. (Ref 8). The transient outlet pressure condition applied was as determined by Tittelbaugh et al. (Ref 6) from modeling a long length of vessel with the inlet velocity condition and a zero pressure outlet condition. The outlet pressure approximately fluctuates between positive and negative 20 mmHg (2666 Pa). The inlet velocity and outlet pressure profiles are also shown in Fig. 3. Note that because the elements in the Eulerian domain are larger than the elements in the stent, contact between the stent and the blood was not included. It was assumed that the driving load on the stent is the artery motion. Mesh refinement of the Eulerian domain would be required to capture the flow details between the struts of the stent; however, such an analysis would become considerably more expensive in computational terms.

Blood was modeled using the linear Us-Up Hugoniot form equation of state (EOS) material model. This type of material model determines the pressure as a function of density and specific energy. Blood density was specified to be the same as water (983 kg/m^3) and additionally, EOS shear viscosity was specified to be approximately three times that of water (0.003 Pa s). The speed of sound in blood was calculated to

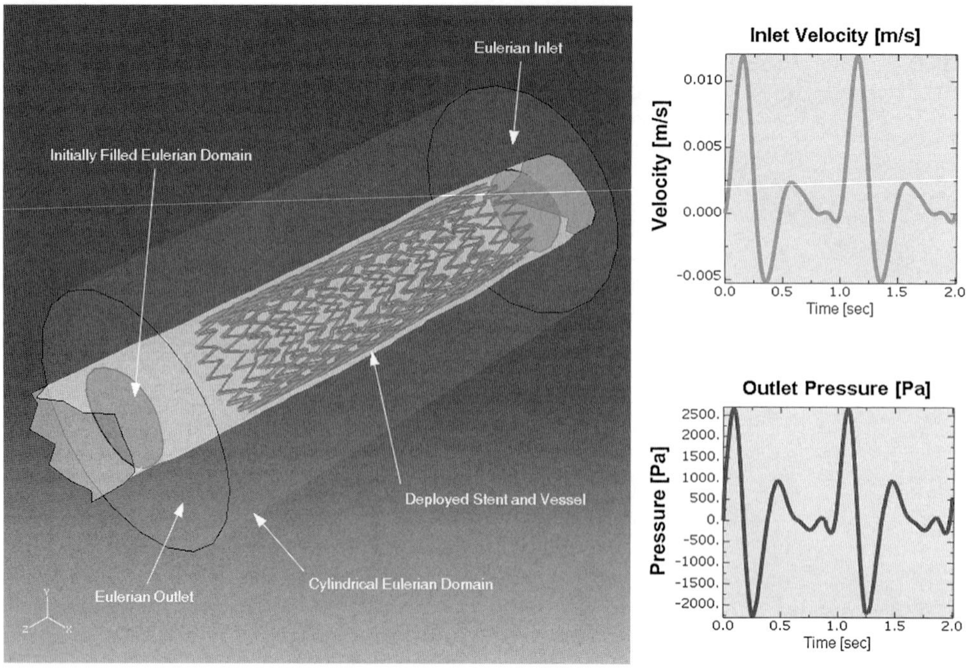

Fig. 3 Coupled Eulerian-Lagrangian model setup and transient inlet velocity and outlet pressure profiles

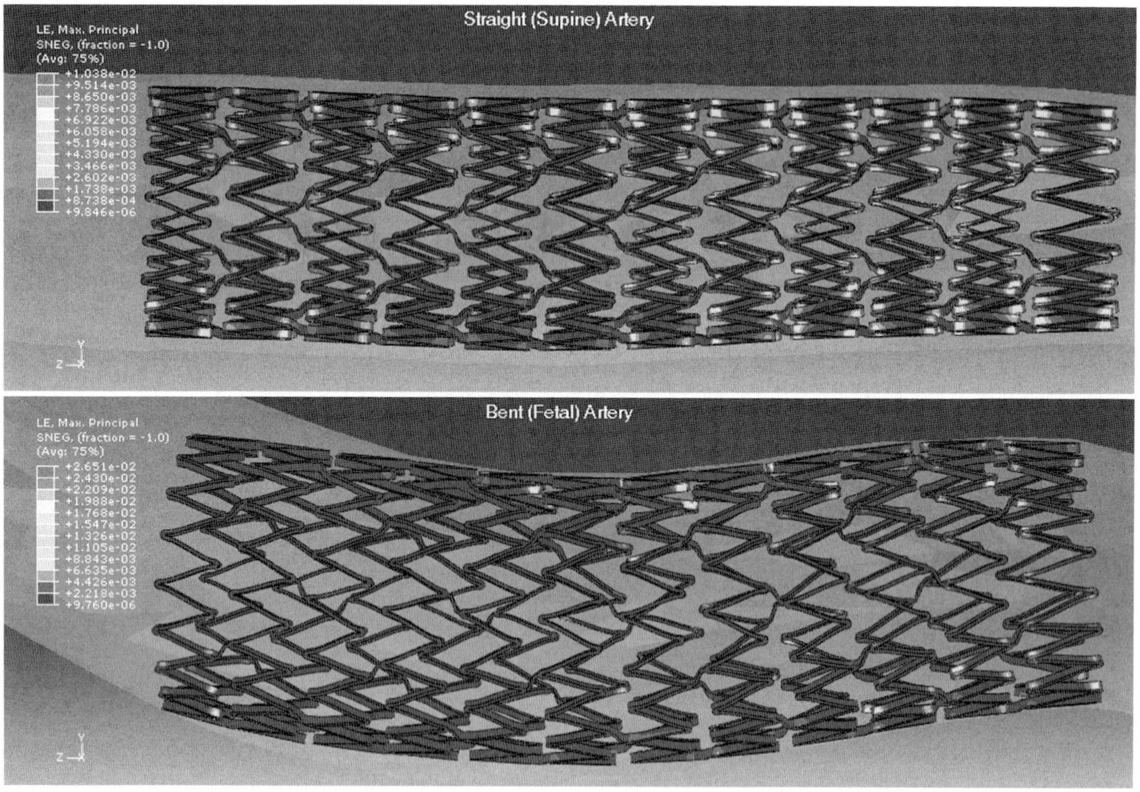

Fig. 4 Maximum principal strains of the fully deployed stent in the straight and the bent artery sections

be 1450 m/s. For EOS materials, the stable time increment is inversely proportional to the speed of sound in the material, and for nearly incompressible fluids such as blood, the stable time increment can be very small. Decreasing the speed of sound increases the stable time increment, but also increases the compressibility of the fluid.

3. Results

The stent configuration after deployment into both the straight and bent artery sections is shown in Fig. 4. The stent deployed into the bent artery section does not have enough flexibility to make full contact with the artery wall at all locations, and so may be less effective than a stent deployed into a straight artery section. Figure 4 also shows that the peak maximum principal strain value in the fully deployed stent is more than double in the bent artery section.

The maximum principal strains of the stent subjected to cyclic loading are generally of interest for fatigue life calculations. For the sinusoidal pressure loading applied to the artery, the maximum principal strain in the stent will occur at the diastolic pressure trough, and these results are shown for the straight and bent artery sections on the left side of Fig. 5 and 6, respectively. Since the stent generally takes the first cycle to settle, only results from the second to fifth cycles are shown. The change in maximum principal strains between the diastolic and systolic pressure peaks and troughs is also required for fatigue life calculations, and are also shown on the right side of Fig. 5 and 6. The cyclic evolution of results is due to the nonlinear material behavior and to contact conditions as well. No changes in material behavior properties with cycling were taken into account in this study.

As expected, for both the bent and straight artery sections, the maximum principal strains decrease with cycling until approximately steady values are achieved. The peak maximum principal strains in the stent deployed into the bent artery section are approximately three times higher than in the straight section, highlighting the importance of utilizing accurate artery geometry of the location of deployment. The higher maximum principal strains in the stent deployed into the bent artery section may indicate that the contrived expansion radius may have been sized incorrectly.

For the stent in both the bent and straight artery sections, the delta maximum principal strains are similar for all cycles modeled, and are also similar between the straight and bent artery segments. This is reasonable, since the same pressure fluctuations were applied to the bent and straight arteries.

As previously mentioned, because the CEL capability is available in Abaqus/Explicit, approximations were made to obtain a solution in a reasonable amount of analysis time. First, the wave speed of blood was specified to be one order of magnitude lower than actual, to increase the stable time increment. Second, the period of the inlet velocity and outlet pressure cycles was decreased from 1 to 0.01 s. Simplified CEL analyses of the blood and artery without the stent were run to investigate the validity of these assumptions.

Because decreasing the wave speed of blood increases the compressibility, the main consequence for this application would be a decrease in contact pressure between the blood and artery. A simplified CEL analysis in which the fluid inlet and outlet were pressurized to 5333 Pa was compared with a structural analysis in which a pressure of 5333 Pa was directly applied to the artery. Figure 7 shows that the artery deformations are very similar for these two cases, indicating that the lower wave speed does not significantly affect contact pressures between the blood and the artery. Figure 7 also shows that the maximum increase in density in the blood is below 0.5%, so the fluid behavior is still close to incompressible, even with the reduced wave speed.

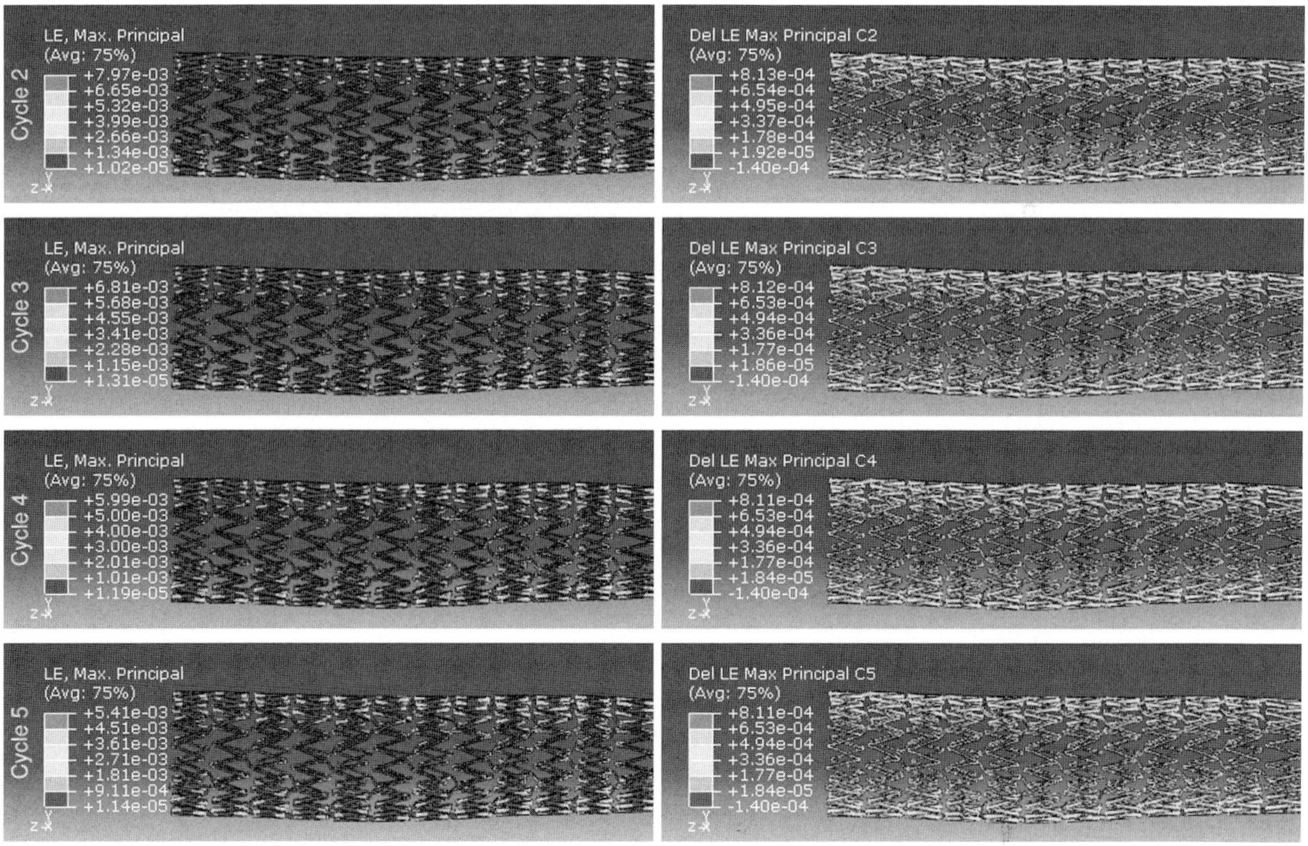

Fig. 5 Maximum principal strains at diastolic pressure (*left*) and delta maximum principal strains between diastolic and systolic (*right*) in the straight artery segment for the second to fifth cycles

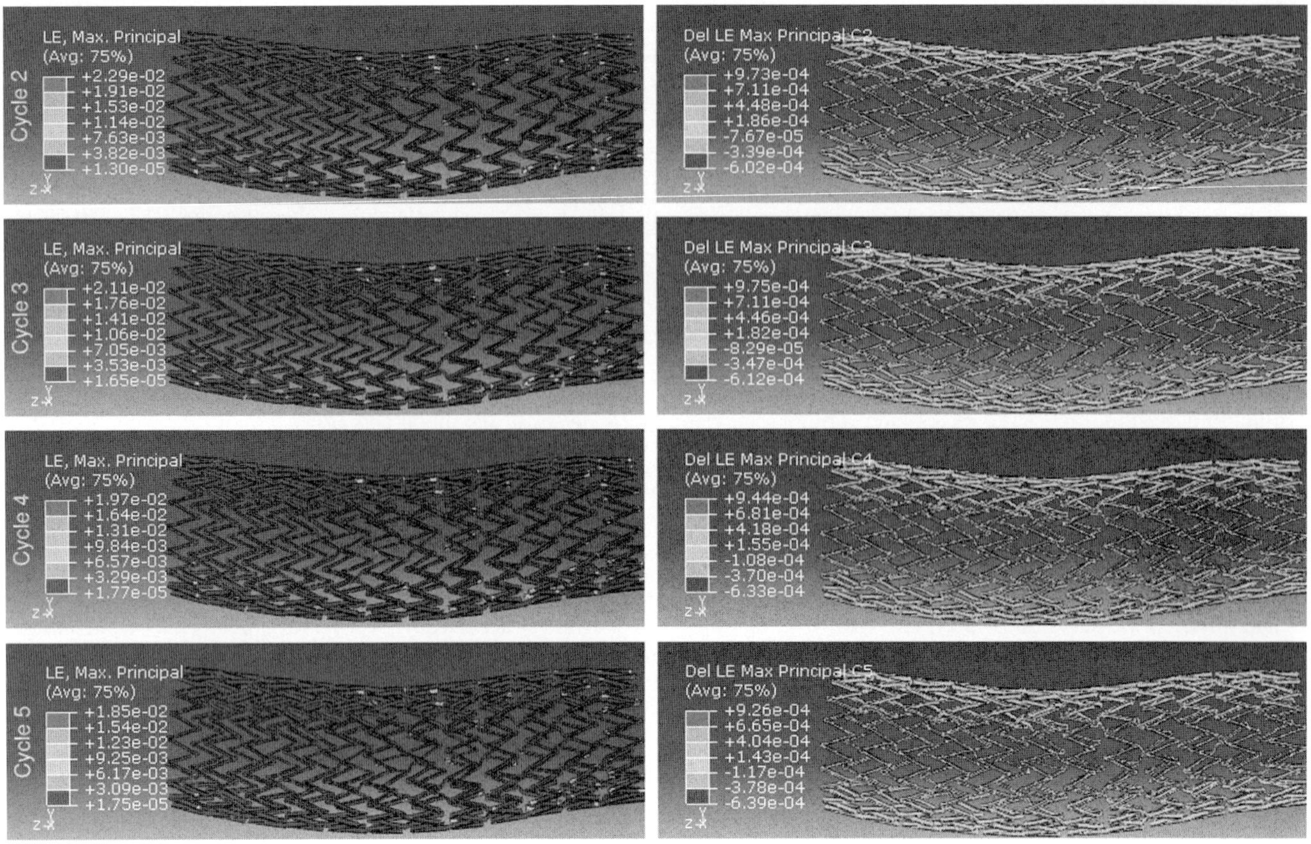

Fig. 6 Maximum principal strains at diastolic pressure (*left*) and delta maximum principal strains between diastolic and systolic (*right*) in the bent artery segment for the second to fifth cycles

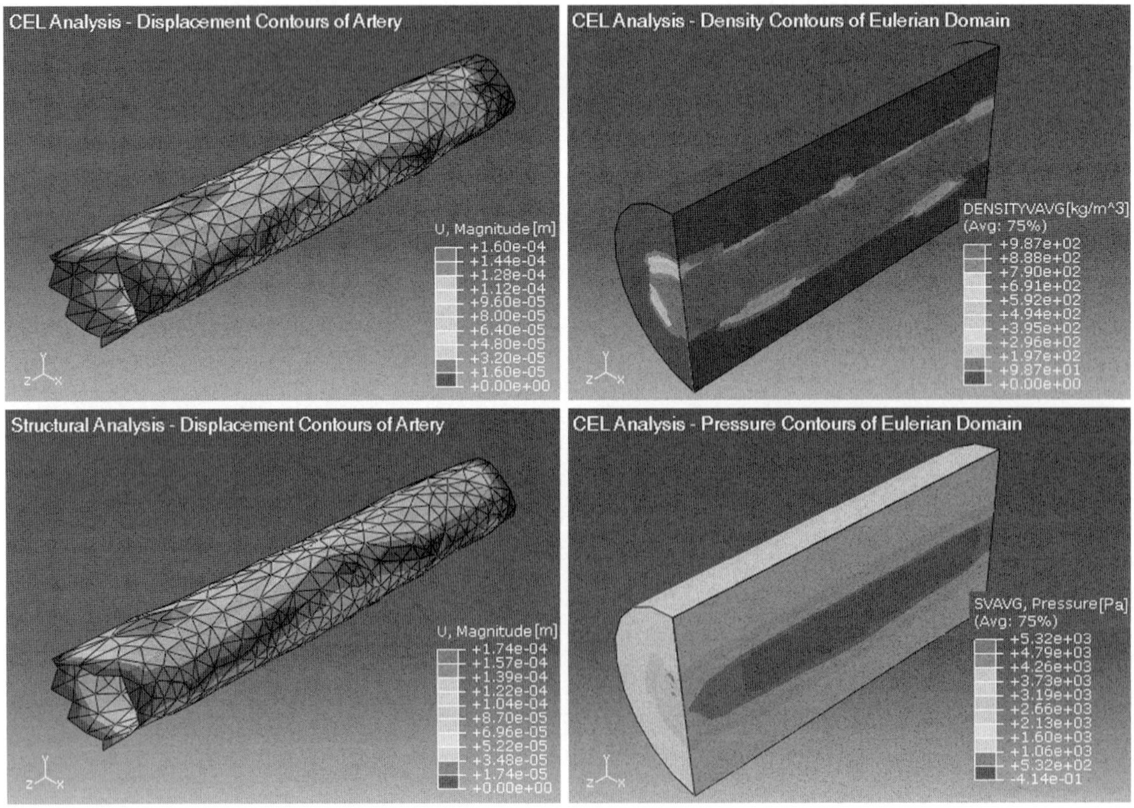

Fig. 7 Deformation contours of the artery for pressure directly applied to the artery and for the simplified CEL model; density and pressure contours for the simplified CEL model

The kinetic energy was compared with the internal energy for a simplified CEL analysis with the inlet velocity and outlet pressure cycles accelerated to reduce computation time. Figure 8 shows that the ratio of kinetic energy (ALLKE) to internal energy (ALLIE) in the model fluctuates about an average of approximately 3.5%, indicating that the higher frequency of the inlet velocity and outlet pressure cycles still yields reasonably quasi-static results.

Figure 9 shows the transient velocity vectors and pressure contours of a mid-plane cut through the Eulerian domain for the fifth cycle of the full CEL analysis. The outline of the artery is also shown to indicate the Eulerian material boundary location. These results show a parabolic flow distribution as expected; however, the outlet pressure condition generates a backflow at the systolic peak. This may indicate that the phase difference between the periodic velocity and pressure

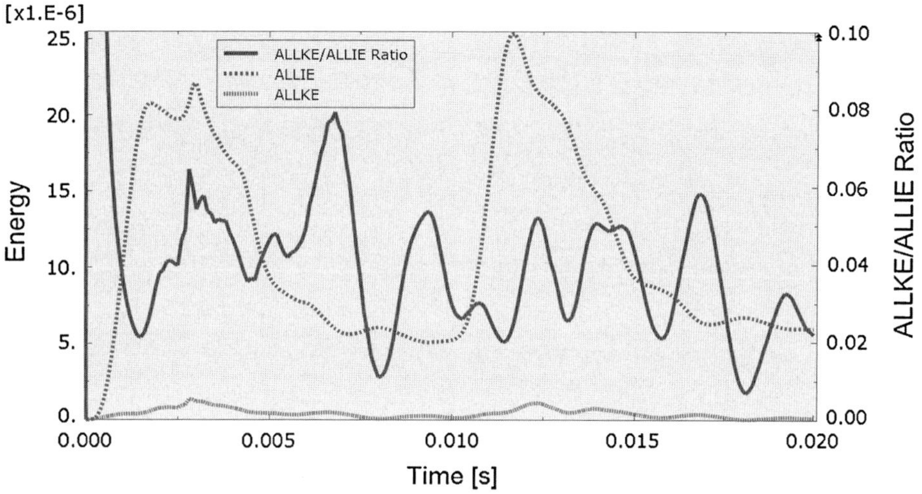

Fig. 8 Kinetic (ALLKE) and internal (ALLIE) energy results for the simplified CEL model

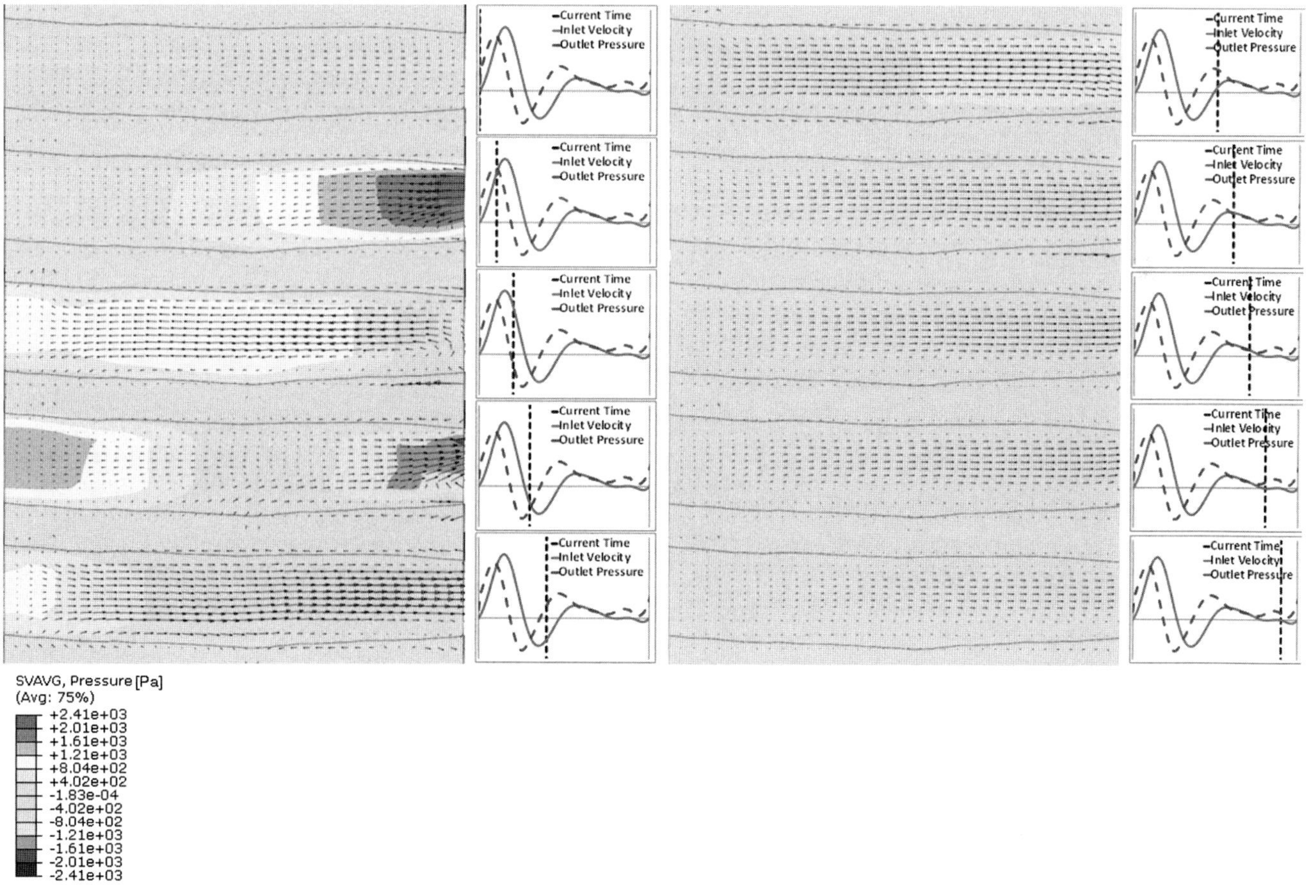

Fig. 9 Pressure contours and velocity vectors of blood in the full CEL model during the fifth cycle

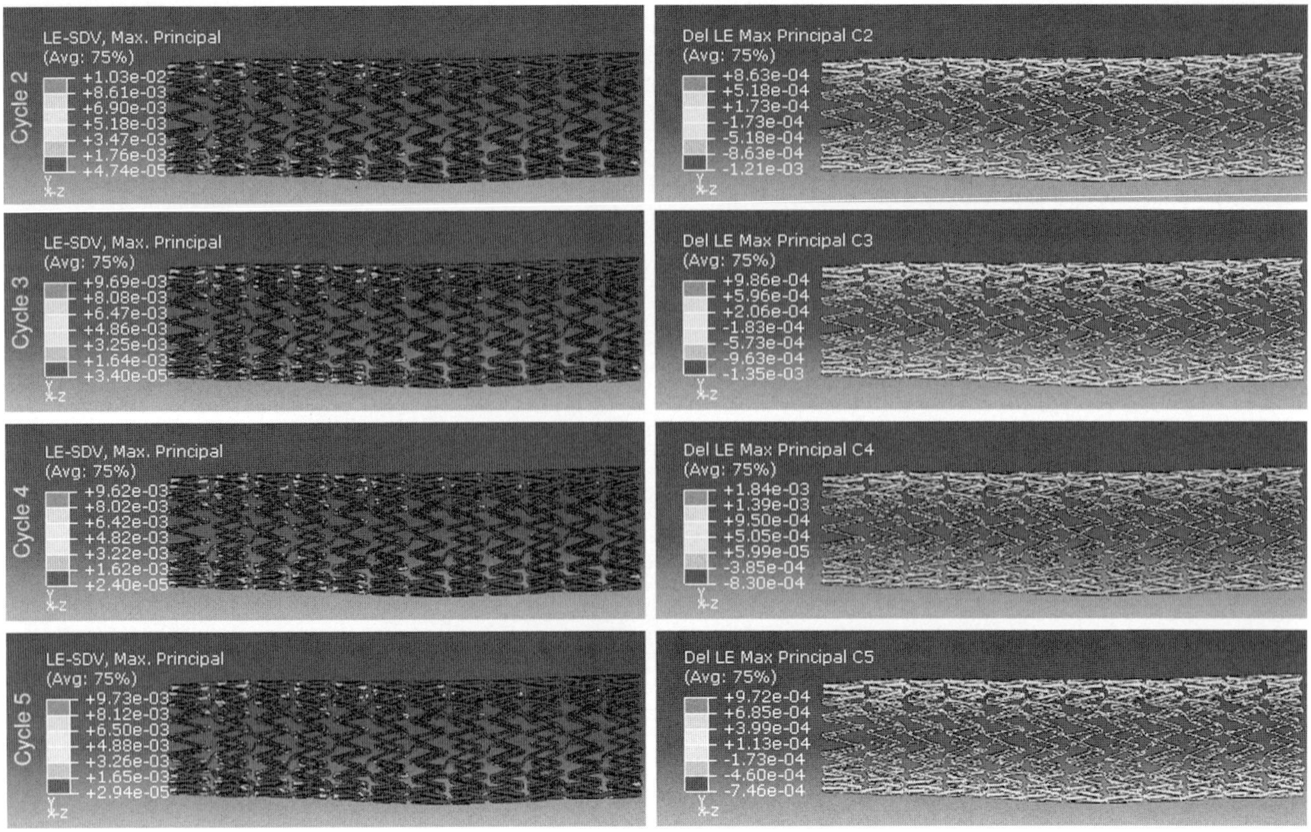

Fig. 10 Maximum principal strains at diastolic pressure (*left*) and delta maximum principal strains between diastolic and systolic (*right*) from the full CEL model for the second to fifth cycles

conditions may not be appropriate for the length of artery modeled in this study.

For the full CEL model, the systolic peak and diastolic trough are defined by the peak and trough of the transient outlet pressure condition. Figure 10 shows that the maximum principal strains in the stent at the diastolic troughs are somewhat higher for the CEL model than for the model in which pressure was directly applied to the artery. Differences are reasonable, and attributable to the fact that in the CEL model the pressure distribution is not uniform along the length of the artery. However, the difference in maximum principal strain between the systolic peak and diastolic trough is similar to the case with the pressure waves directly applied to the artery.

4. Discussion

Loading on the stent in the femoral artery is due to the pulsatile nature of blood flow, which can fluctuate about a mean pressure of 100 mmHg (13,332 Pa) with an amplitude of 20 mmHg (2666 Pa). Since the artery geometry was obtained in vivo, its deformation accounts for the mean pressure and only the pressure fluctuations need to be applied. Therefore, sinusoidal pressure waves were applied to the inner surface of the artery, and the resulting deformation of the stent was observed. The stent will experience maximum strains at the diastolic pressure trough, and typically the difference in maximum principal strains between the systolic peak and diastolic trough is used for fatigue life calculations. After five cycles, the largest maximum principal strain converged to approximately 0.5% in the stent deployed into the supine (straight) artery section and approximately 1.9% in the stent deployed into the fetal (bent) artery section. However, the maximum difference in maximum principal strains between systolic and diastolic is similar between 0.08% and 0.09%.

The CEL capability in Abaqus/Explicit was investigated to include the effects of blood flow within the supine artery on the stent deformation. Loading and boundary conditions become even more difficult to ascertain. Transient inlet velocity and outlet pressure conditions were specified on the Eulerian domain. Since Abaqus/Explicit is ideally suited for high-speed dynamic events, approximations were made to obtain results in a reasonable amount of time: reducing the wave speed of the fluid and decreasing the period of the cyclic velocity and pressure loading. Using a simplified CEL model which included only the blood and the artery, these assumptions were shown to yield reasonable results.

For the CEL model with the stent deployed into the supine artery, the peak maximum principal strain converged to approximately 0.97% in five cycles, which is slightly higher than the model with the pressure directly applied to the artery. The peak difference in maximum principal strains between systolic and diastolic is approximately 0.1%, which is similar to the values obtained with the pressure directly applied to the artery.

The original intent of investigating the stent deployed into the bent or fetal artery section was to ascertain the stent deformation changes when the patient shifts from the supine to the fetal position. The method in this study of deploying the stent into the bent artery section is only an approximation of this event, and future work would involve determining a more accurate method to simulate the shift.

For the CEL model in this study, backflow at the outlet during the systolic pressure peak indicated that the phase difference between the transient inlet velocity and outlet pressure may not be appropriate for the length of artery modeled. A partially blocked outlet to induce back pressure or an experimentally determined pressure condition could be investigated in future work. Also, the Eulerian mesh could be refined so that contact between the blood and stent could be included. Still, this study has demonstrated a method of FSI analysis which includes the effects of blood flow in simulating service loading of a stent deployed into an anatomically correct artery geometry.

References

1. N. Rebelo and M. Perry, Finite Element Analysis for the Design of Nitinol Medical Devices, *Min. Invas. Ther. Allied Technol.*, 2000, **9**(2), p 75–80
2. N. Rebelo, S. Prabhu, C. Feezor, and A. Denison, Deployment of a Self-Expanding Stent in an Artery, *Proceedings of the International Conference on Shape Memory and Superelastic Technologies* (Kurhaus Baden-Baden, Baden-Baden, Germany), 2004
3. RESIStent Program: SRI International, Stanford University and a Consortium of Stent Manufacturers, 2004-2006, private communication
4. F. Auricchio and R. Taylor, Shape-Memory Alloys: Modeling and Numerical Simulations of the Finite-Strain Superelastic Behavior, *Comput. Meth. Appl. Mech. Eng.*, 1996, **143**, p 175–194
5. F. Auricchio and R. Taylor, Shape-Memory Alloys: Macromodeling and Numerical Simulations of the Superelastic Behavior, *Comput. Meth. Appl. Mech. Eng.*, 1997, **146**, p 281–312
6. E.M. Tittelbaugh, R. Fu, and S. Sett, Coupling FEA to CFD to Investigate the Effects of Pulsatile Blood Flow on the Dilatation of Artery Walls, *Proceedings of the NAFEMS World Congress* (Vancouver, BC, Canada), 2007
7. Dassault Systemes SIMULIA Corp., *Abaqus User's Manual 6.8*, Providence, Rhode Island, 2008
8. D. Steinman, B. Vinh, C. Ethier, M. Ojha, R. Cobbold, and K. Johnston, A Numerical Simulation of Flow in a Two-Dimensional End-to-Side Anastomosis Model, *J. Biomed. Eng.*, 1993, **115**, p 112–118

JMEPEG (2009) 18:664–671
DOI: 10.1007/s11665-009-9384-z

1059-9495/$19.00

Carbon Fiber Reinforced Smart Laminates with Embedded SMA Actuators—Part I: Embedding Techniques and Interface Analysis

P. Bettini, M. Riva, G. Sala, L. Di Landro, A. Airoldi, and J. Cucco

(Submitted October 21, 2008; in revised form December 19, 2008)

Up to now one of the main limits for a large use of shape memory alloys (SMA)-based smart composite structures in the aerospace industry is the lack of useful numerical tools for design. Moreover, technological aspects still need a more detailed investigation. This paper shows how to overcome issues regarding embedding of NiTiNOL wires in carbon fibre/epoxy laminates. A crucial aspect of those structures is related to the load transfer capabilities between the SMA actuators and the host material during their activation. Embedding techniques developed for taking into account problems like thermal and electrical compatibility between actuators and host material and passive/active invasivity are reported in this paper. Simple smart laminates with several actuators were manufactured, tested, and deeply analyzed. In order to characterize the interface in the real operative conditions, pull-out tests were conducted on NiTiNOL wires embedded in composite fiber laminates. The results were compared to standard experiments on wires embedded in pure epoxy resin blocks.

Keywords embedding technologies, interface analysis, NiTiNOL, pull-out, smart composites structures

1. Introduction

Materials for primary structures in aerospace applications must have high mechanical properties, lightness, and durability. It is important for these structures to exhibit good behavior in terms of impact resistance, environmental endurance, fire and corrosion resistance. Composite materials meet well such requirements and nowadays are widely used in many engineering fields.

On the other hand, a large number of new materials and technologies (i.e. micro-electronic devices, shape memory alloys (SMA)-based micro-actuators, piezo-ceramics (PZT), electro-active polymers, optical fibers (OF), etc.) allow to imagine extensive diffusion of smart materials in many aerospace and ground applications. Their capabilities in terms of shape and vibration control of large space structures, acoustic control for noise reduction in civil aircraft, health monitoring and in-site structure identification are well assessed and promising results are also obtained in many other applications.

In the last years, several research activities in the manufacturing processes field have been devoted to develop smart structures that could combine high mechanical efficiency, due to composite host materials, with good functional properties of the embedded sensors and actuators (Ref 1-3).

The embedding of these smart devices into the structures could give some advantages compared to their bonding onto the outer skin. For example, they allow the actuation in locations hardly accessible from outside because of shape constraints; the protection of the actuation system inside host structure from all environmental effects, which may reduce its performance, can be also achieved. In addition, active structures could be applied even in those cases where a clean and smooth surface is required (i.e. aerodynamic surfaces) (Ref 4).

Abbreviations	
A_s, A_f	austenite start and finish temperature
CFRP	carbon fiber reinforced plastic
CFRSL	carbon fiber reinforced smart laminates
CTE	coefficient of thermal expansion
DOF	degree of freedom
FE	finite element
GFRP	glass fiber reinforced plastic
HF	hydrofluoric acid
M_s, M_f	martensite start and finish temperature
OWSME	one way shape memory effect
OF	optical fiber
PZT	piezo-ceramic
QP	quick-pack
SMA	shape memory alloys
TWSME	two way shape memory effect
T_g	glass transition temperature
UD	uni-directional laminate

This article is an invited paper selected from presentations at Shape Memory and Superelastic Technologies 2008, held September 21-25, 2008, in Stresa, Italy, and has been expanded from the original presentation.

P. Bettini, M. Riva, G. Sala, L. Di Landro, A. Airoldi, and **J. Cucco**, Department of Politecnico di Milano, Aerospace Engineering, Milan, Italy. Contact e-mail: riva@aero.polimi.it.

The embedding technique of these components within the load-carrying structure is still a leading edge issue. Research activities oriented both to the development of manufacturing techniques, to the characterization of embedded sensors and to the estimation of actuators invasivity (active and passive) on the smart structure performance are of valuable importance.

Moreover, most of the applications are strongly dependent on the availability of useful numerical tools. In fact, all the design activities regarding structures are nowadays related to the adoption of numerical models able to predict the efficiency and performance of the systems (Ref 5). The peculiarity of the smart composite structures underlines this problem due to the joined complexity of the active materials behavior (non-linearity) with the composite material modelling (anisotropy, damage mechanisms).

The activity of the Smart Composite Laboratory at Politecnico di Milano is dedicated to the development of both technological processes and numerical tools for Smart Carbon Laminates with embedded sensors and actuators. Some of the results are presented in this paper. In a further paper, modelling and numerical analysis of embedded SMA will be presented. This activity is focused on the embedding of Shape Memory wires, selected taking advantage of their functional properties in terms of low activation speed combined with high loading capability. In fact, these characteristics are useful for most of the aerospace applications (airfoil morphing, shape control, etc.). Shape Memory effect of these actuators can be exploited in two different ways. In the first one, wires are embedded in a detwinned martensite phase and the activation of the structure is obtained using the one way shape memory effect (OWSME) due to the reverse transition induced by heating (from detwinned martensite to the austenite phase); then the structure (and the embedded wires too) come back to the initial form through its stiffness (host material remain in the elastic range) when the wires are cooled and the forward transition is induced. In the second one, wires can be embedded both in austenite or in detwinned martensite phase considering two way shape memory effect (TWSME). A structure can change its form by heating or cooling so that reverse or forward transition, respectively, occurs (Ref 6, 7).

In any case, the coupling between SMA and host material poses a number of critical aspects which have to be taken into account during the embedding process. This paper presents the development of technological process and the investigation of the interface between active SMA and host materials in case of carbon fiber reinforced plastic (CFRP) laminates. A first aspect is related to the curing temperatures of the thermoset epoxy matrices used in aeronautical composites; often these are higher than the transition temperatures of standard achievable actuators so that both reverse and forward transition occur during the curing process. A second aspect is related to the activation of the wires. Joule heating requires electrical insulation between wires and CFRP host material. Finally, a good adhesion of the resin around the actuators has to be obtained to guarantee high levels of load transfer capability.

The objective of this work, together with what will be presented in a further paper oriented to the development of useful numerical tools for SMA-based micro-actuators, is the definition of a design strategy that takes into account the computational costs and the level of accuracy required by a smart structure for a preliminary prediction of the overall performance of the system.

2. Technological Aspects

2.1 Materials Choice

Ideal smart structures should exhibit high mechanical performance and good functional properties keeping at the same time very low invasivity of the embedded actuators. If such a design concept is considered, a trade-off between actuators dimensions and the host material stiffness is always required. For instance, in case of uni-directional laminates, wires SMA actuators with diameter comparable to the single ply thickness (125 μm) should be embedded with the same orientation as reinforcing fibers to minimize their invasivity. On the other hand, this configuration usually leads to a restrained activating power considering the high stiffness of this host material (due to very high content of fibers). To improve the activation efficiency while keeping a low invasivity, new technologies are being developed oriented toward the increasing of the volume fraction of SMAs in the laminate by substituting part of the reinforcing fibers with thin NiTiNOL wires at a single ply level.

The material choice is also dependent on the thermo-mechanical compatibility between the actuators and standard epoxy resin used for CFRP in aeronautical field. In particular, SMAs with A_f lower than the resin T_g have to be used to avoid a reduction of the stiffness of resin and interface; at the same time, A_s must be higher than normal operating condition temperatures of the smart structure in order to avoid undesirable activations. In addition, it is useful to have M_f greater than room temperature to avoid the use of a cooling system (i.e. Peltier) for the complete forward transition (Ref 8).

In order to maximize the activation effects, a host material with limited stiffness and actuators with relatively high diameters were thus chosen for this work. A previous research on embedded OF showed that their passive invasivity over the host material mechanical properties was negligible in static conditions (Ref 9). Considering the similarity between OF and SMA wires in terms of dimensional and morphological characteristics, the invasivity of thin NiTiNOL wires was assumed negligible. As a consequence, attention has to be paid only to the thermal loads during the activation that must keep safe the resin around wires. NiTiNOL wires with 381 μm diameter (Dynalloy) and fabric prepreg of 120 μm of thickness (Seal) were then selected. Their properties are summarized in Table 1 and 2, respectively.

2.1.1 Characterization of the Materials. A preliminary thermo-mechanical characterization of the SMA actuators was done to determine the mechanical properties, the transition temperatures and their stress dependency (Ref 10, 11). Results showed a multistep forward transition due to the presence of the

Table 1 Characteristics of SMA actuators as-received (transitions evaluated by DSC)

Material type	NiTiNOL $Ni_{52}Ti_{48}$
Shape	Wire
Dimensions	0.3181 mm
A_s-A_f	41.23-69.60 °C
M_s-M_f	19.67-3.91 °C
R_s-R_f	72.95-39.98 °C
E^A	68 GPa
E^M	21 GPa
E^R	14 GPa

Table 2 Characteristics of the host material

Material type	Carbon fabric prepreg SEAL CC90 ET443
Cured thickness ply	0.12 mm
Curing temperature	125 °C
T_g	127 °C
$E_{xx} = E_{yy}$	56.55 GPa
ν_{yx}	0.0514
G_{xy}	4.043 GPa

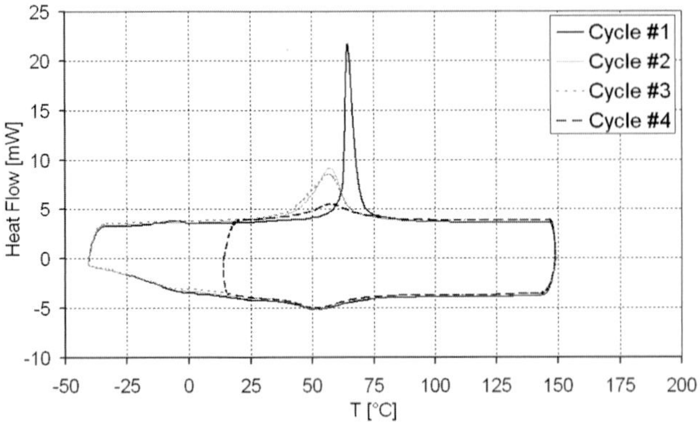

Fig. 1 DSC analysis of the Dynalloy SMA actuators: determination of transition temperatures without stress applied by using heat flow curves (Exo Up)

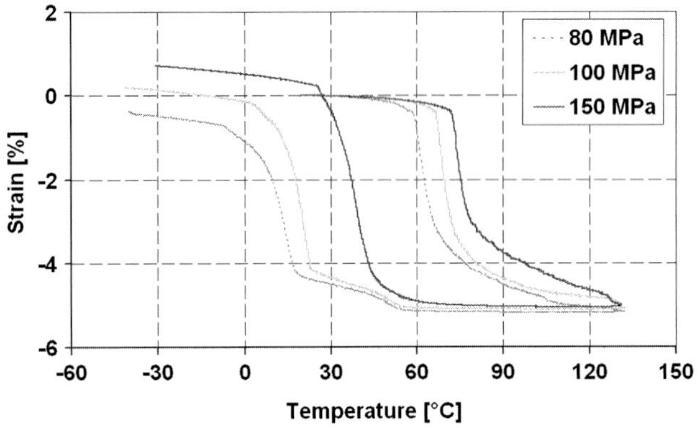

Fig. 2 Strain vs. temperature curves of SMA trained wire detwinned at 4% at different loading levels

R-phase that can be removed by an annealing treatment (Ref 12). DSC analyses were performed and the transition temperatures were evaluated from the curves adopting the method suggested by ASTM standard (Table 1) (Ref 13). Figure 1 shows the results obtained with the SMA wire as-received. The superposition of all the scanning cycles (after the first one) confirms the good quality of the analyses. The scanning n°4 (dashed line) underlines that A_s and A_f do not change even if the reverse transition start in the R-phase condition.

Typical strain versus temperature curves with different applied stress obtained by using a Dynamic Mechanical Analyzer are reported in Fig. 2. It can be observed that R-phase disappears when high stresses are imposed. Stress dependence of the Dynalloy SMA transition temperatures experimentally obtained is summarized in Fig. 3. Young

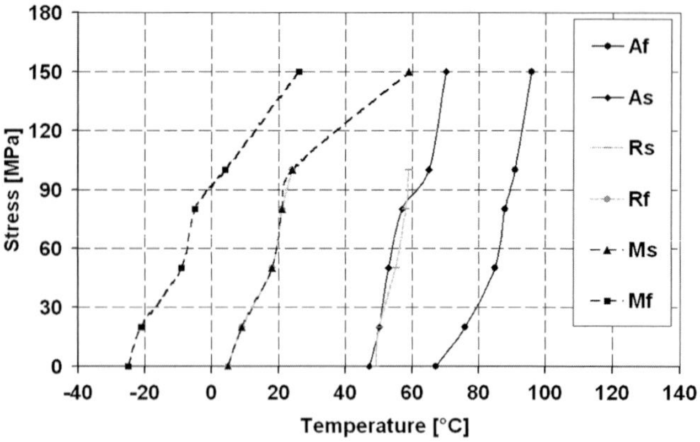

Fig. 3 Stress dependence of the Dynalloy SMA transition temperatures obtained using a Dynamo Mechanical Analyzer

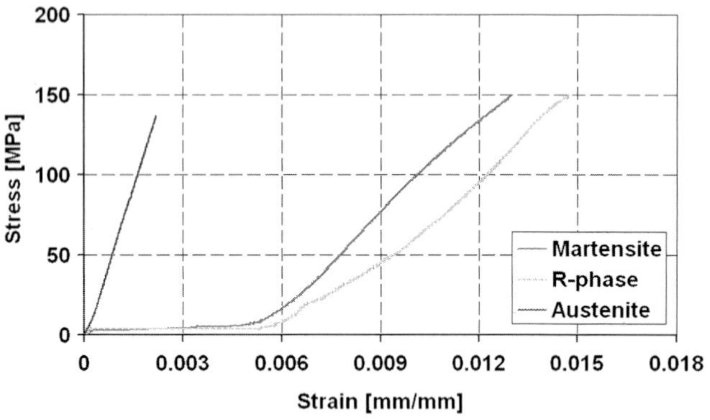

Fig. 4 Stress vs. strain curves of the SMA wires obtained in the three different phases with a Dynamo Mechanical analyzer

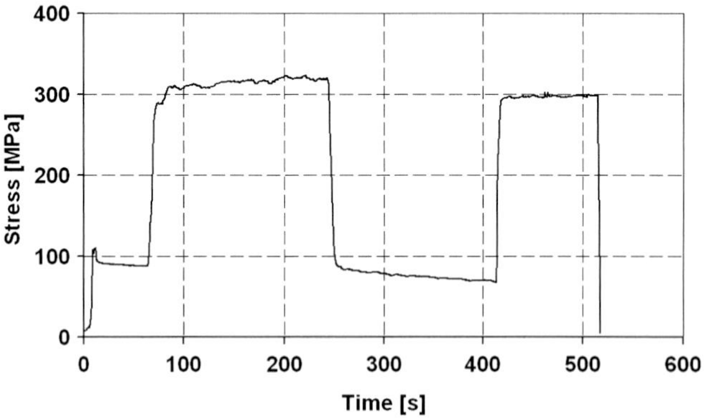

Fig. 5 Iso-length test on wire detwinned at 4%: SMA exhibits about 300 MPa of recovery stress

moduli in the three different phases were evaluated through tensile tests (Table 1). The obtained stress-strain curves (Fig. 4) are limited to 150 MPa corresponding to the load cell limit of the loading device (18 N). To evaluate SMAs recovery stress, iso-length tests were also performed on 4% pre-strained wires by using an electromechanical dynamometer machine (INSTRON). The test implies two cycles of turning on and switching off the power supply for Joule activation to verify the repeatability of the performance. Results are shown in Fig. 5.

Finally, tensile specimens according ASTM D3039 were manufactured and tested in order to obtain the mechanical characteristics of the host material (Table 2). A DSC analysis was also performed to evaluate the cured resin glass transition temperature (Ref 14).

2.2 Embedding Technique

In order to manufacture a smart panel based on OWSME using the chosen materials, SMA wires have to be embedded in their martensite detwinned phase. A direct free embedding in CFRP is not possible due to the transition induced by the resin curing cycle.

One of the first methods developed to overcome this limit consisted in using special sheaths where the wires could pass into after the resin curing stage; the wires were then externally connected to the panel to give them load transfer capabilities (Ref 2, 15). This solution presents more drawbacks than benefits: very high invasivity, presence of an external device for the wires-structure connection and consequent less compact laminate with lower aerodynamic capacities when used as aeronautical shell. The technique developed in this research consists in a direct embedding of SMA wires with constraints that prevent the shrinkage due to the reverse transition. Doing so, a transient stress is induced on the constrained wires when the reverse transition occurs during the heating step of the resin curing cycle. After the polymer crosslinking, this state-of-stress gets off when the SMA comes back in martensite phase during the final cooling step of the cycle. At the end, the adhesion between embedded wires and the host material matrix gives the mutual load transfer capability without using external devices. If such an embedding technique is applied, the actuators efficiency strongly depends on the quality of this bonding. For this reason, the interface analysis, which is a crucial task, is required.

At the end of the process the wires are exiting the final laminate. If the activation is given by thermal convection or radiation (using hot air flow or electrical resistances), the edges of the panel can be trimmed removing the resin squeeze-out and the wire ends. On the other hand, if Joule-effect activation is used, the wire tips must be connected to the power supply system thus limitating the possibility of profiling the laminate. A special mold with elastomeric dams was designed and manufactured so that good quality edges can be guaranteed. In addition, dedicated connectors have simplified the activation system allowing the wires to be cut.

Figure 6 shows some details of the described technique. They are related to the manufacturing of a smart panel embedding six SMA. The Joule-activation in itself also requires an electrical insulation to avoid shorts between the host material and the embedded wires. This problem was solved adopting a glass fiber reinforced plastic (GFRP) insulating sub-laminate (named Quick-Pack (QP)) originally developed to embed different sensors or actuators (Ref 4, 9).

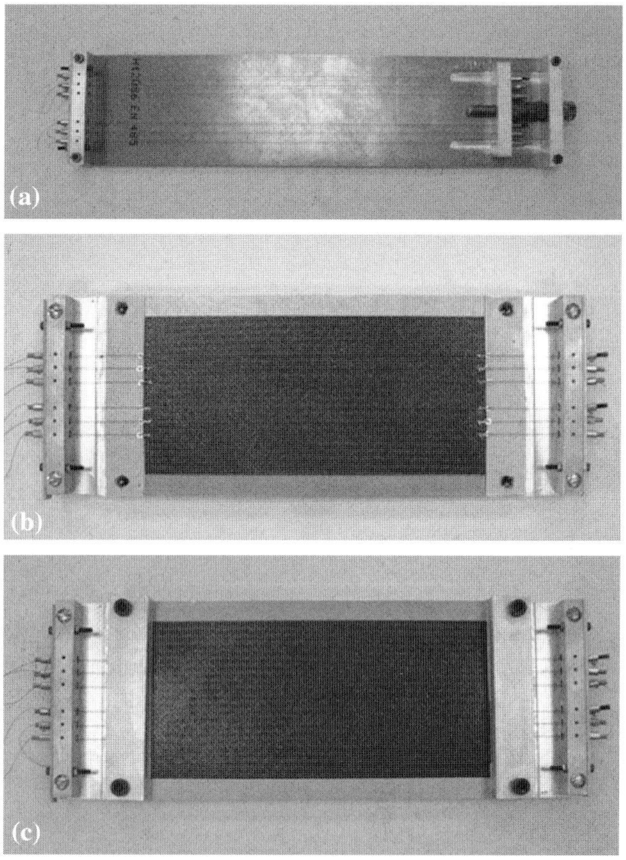

Fig. 6 Direct constraint embedding technique for SMA detwinned wires: pre-strain tool (a), constraining mechanism, crimp and elastomeric dams adopted during the lamination phase (b) and the whole system ready to be cured (c)

2.3 Quick-Pack

Quick-Pack is a thin laminate made of two fabric plies of glass fibers and epoxy matrix originally developed to protect OF sensors and PZT actuators during their embedding in a host material. With this preliminary process the very fragile OF or PZT result easier to handle. The result is a powerful system ready to be embedded (Ref 4, 9).

A preliminary validation of this QP technique for SMA wires was performed by accuracy and efficiency tests. Fatigue tests were included in a previous research activity on OFs (Ref 9). The porting of this technology for SMA actuators was done using NiTiNOL wires embedded in a 50 μm of thickness GFRP/epoxy fabric with 66% of resin content of the same type used in the CFRP host material. The choice of the QP prepreg was done in order to minimize its invasivity. On this basis very thin prepreg with high resin percentage were selected so that limited cross section dimension changes and improvement of the adhesion between wires and plies can be obtained.

After the manufacturing of the QP, its final embedding in the host material requires subsequent thermal processes; these do not degrade the polymer characteristics; since should be assumed as post-curing processes which should usually lead to improved stability and mechanical performance. On the other hand, different coefficient of thermal expansion (CTE) between GFRP and CFRP may induce residual stresses inside the laminate. Fatigue tests carried out in the previously mentioned work demonstrated that the presence of this new device does not create any significant degradation of the final structure (Ref 9).

The QP tips are a further critical issue because they must exit the CFRP panel to avoid shorts. For this reason, special elastomeric pads were used during the final curing to improve layers compaction.

The carbon fiber reinforced smart laminates (CFRSL) based on NiTiNOL micro-actuators obtained at the end of the whole process (Fig. 7) was used for experiments and numerical

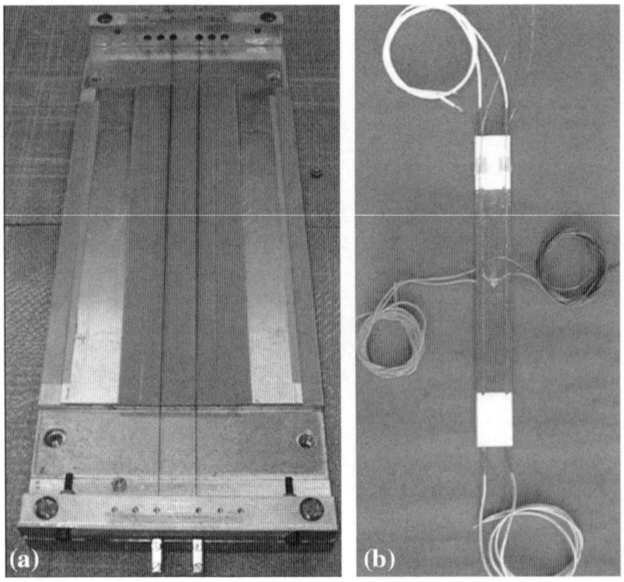

Fig. 7 Quick-Pack manufacturing phase (a) and final CFRSL specimen (b)

Table 3 Characteristics of manufactured CFRSL

Specimen ID	003
Dimensions	260 × 40 × 1.27 mm
Host material	Carbon fabric prepreg SEAL CC90 ET443
Cured thickness ply	0.12 mm
Lamination sequence	$[90°/(0°)_2/90/45°/-45°]_s$
Actuators material	Dynalloy wires of 0.381 mm diameters
Number of actuators	2

correlations that will be presented in a further paper. The main characteristics and dimensions of this panel are summarized in Table 3.

3. Interface Analysis

The efficiency of a smart structure is strongly dependent on the interface between the sensors/actuators and the host materials. Their adhesion is an essential requirement to ensure congruence and high load transfer capability. The same considerations have always driven the advance of composite materials and several methods have been developed to study the interface between reinforced fibers and matrix (like microtension, fragmentation, microcompression, pull-out tests) (Ref 16).

These methods can be independently used to evaluate the maximum interfacial shear stress (IFSS) that gives the interface failure and to understand its failure mechanisms. Pull-out test was selected for this work due to its simple set-up and specimens manufacturing phase (Ref 17).

Specimens are made up of small resin blocks with a single fiber embedded for a length (L) that must be less than a critical value (L_c); the critical length is defined as the embedded length that gives a pull-out force equal to the fiber failure one. In such a way debonding occurs instead of fiber failure (Fig. 8a). The average value of maximum shear stress (τ_{IFSS}) is obtained from

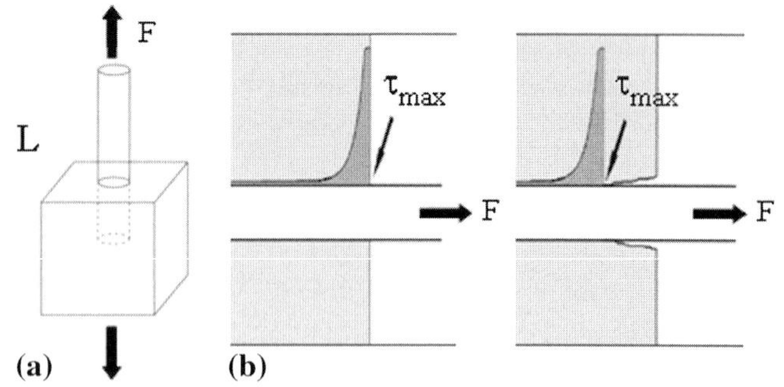

Fig. 8 Standard pull-out test to evaluate the IFSS between fiber and resin of the composite materials: specimen scheme (a) and failure mechanism (b)

the tests by using an equilibrium relationship based on a simplified stress distribution at the interface (Ref 17):

$$\tau_{IFSS} = \frac{F_{max}}{\pi \cdot d \cdot L}$$

where F_{max} is the maximum recorded force corresponding to the debonding and d is the fiber diameter.

Two observations can be done. First, this formula underestimates the stress levels since it considers a uniform state-of-shear-stress along the interface in the fiber direction. In fact, the failure propagation mechanism implies much different stress conditions as shown in Fig. 8(b). The fracture nucleates and propagates due to a peak of stress localized at the propagation front. Second, the friction forces between resin and fiber are neglected. Their evaluation could be done considering both friction coefficients and pressure at the interface due to technological cycles but they are very difficult to be obtained (Ref 18).

Pull-out tests on standard and non-standard specimens with SMA embedded wires are presented next. A hydrofluoric acid (HF) surface treatment was also performed on the embedded wires to evaluate the possible improvement of the adhesion between resin and wires.

In view of the fact that the presence of reinforcing fibers in the composite laminates as well as the high pressures applied during a typical prepreg curing cycles may strongly influence interface characteristics (Ref 19), a particular kind of non-standard specimens was also manufactured and tested in order to analyze the interface in the real operative conditions of the wires.

3.1 Standard Pull-Out Tests

3.1.1 Specimens Manufacturing and Tests Set-up. Specimens were made with an Araldite LY5052/Aradur 5052 epoxy resin. A curing cycle of 3 h at 100 °C was chosen in order to have the glass transition temperature (130 °C) higher than NiTiNOL A_f. All the wires were embedded in their austenite phase for a length of about 2÷3 mm. During the manufacturing phase a particular attention was used to eliminate voids in the resin and to avoid its rising up along the wires at the exit of the blocks. After a degassing process, the resin was poured in a special aluminum mold equipped with a drilled PTFE plate for the wires crossing. Moreover, the mold had a flexible wall to prevent possible residual stresses due to curing cycles (Fig. 9). At the end of curing all the specimens were cooled down to

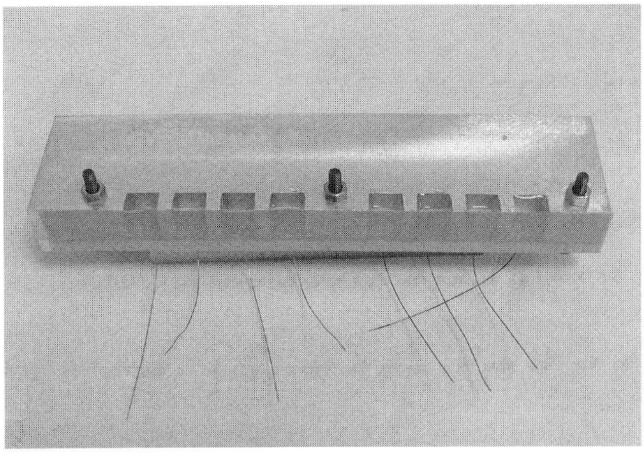

Fig. 9 Special mold for the pull-out specimens manufacturing

Fig. 10 Detailed view of a standard pull-out specimen showing the absence of resin rising

Fig. 11 Pull-out equipment: the tensile tools are clamped and installed in the environmental chamber

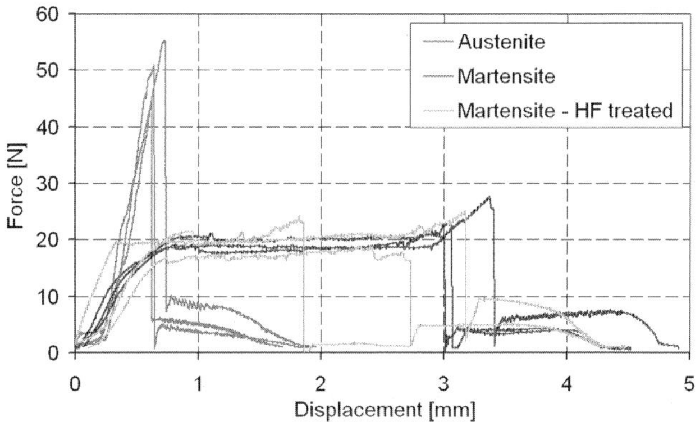

Fig. 12 Pull-out curves of standard specimens in all configurations (austenite, martensite and martensite—HF treated)

ensure martensite phase in the wires. A detailed view of a manufactured specimen shows the absence of the rise up (Fig. 10).

A specific testing equipment was also designed and manufactured to apply the restrained top loading method so that pre-load stresses on the resin blocks due to the test machine clamping can be avoided. The tools were made of CFRP plies covered by two GFRP plies that fulfill the electrical insulation (required in the Joule-activation of the wires adopted for non-standard specimens).

Two set of specimens were tested at low and high temperature to evaluate the IFSS in martensite and in austenite phase, respectively. All the tests were performed using a mono-axial electro-mechanical testing machine (INSTRON 4302) in displacement control mode with crosshead speed of 0.5 mm/min equipped with a 1 KN cell load and a thermal chamber (Fig. 11).

3.1.2 Results of Standard Pull-Out Tests. Results are summarized in Fig. 12 that reports the force versus displacement curves referred to all the different sets (martensite with and without HF surface treatment, austenite). All the pull-out tests led to complete debonding: the load rises until debonding and then a sudden drop occurs; afterward the effects of friction between wire and resin are evidenced by a slightly varying force. In the martensite test, the debonding occurs at higher loads than those leading to stress induced martensite (SIM) underlined by the presence of a plateau. The comparison between the curves referred to HF treated and not treated wires does not show appreciable differences.

On the other hand, in austenite tests, the debonding loads are higher than in martensite ones. This behavior could be related to the thermal mismatch between resin and metal which affects the mechanical constraint on the wires, or to a variation of wire roughness consequent to phase transformation.

Figure 13 shows IFSS obtained from all experimental tests and it underlines the SMA phase influence on the interface. Even if experimental data are limited for a reliable statistical analysis, the low standard deviation obtained permits to evaluate a meaningful average of IFSS reported in Table 4. It can be seen that austenite IFSS (20.05 MPa) is three times larger than the martensite one (7.59 MPa) while the interface is not significantly influenced by the HF treatment (6.67 MPa).

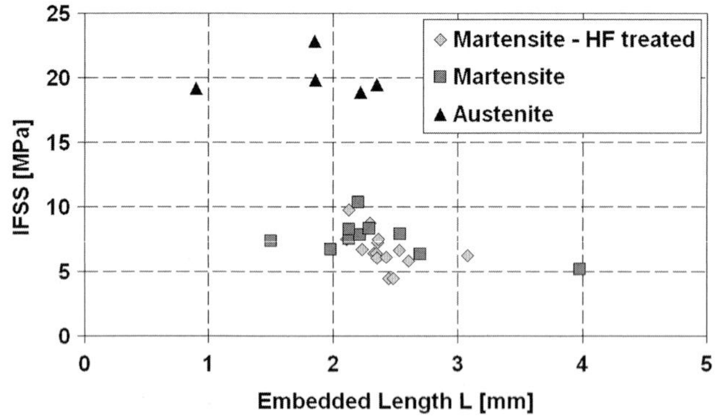

Fig. 13 IFSS obtained in the standard pull-out tests. Comparison between martensite, martensite with hydrofluoric acid surface treatment and austenite phases

Table 4 IFSS obtained from standard pull-out tests

Type of specimens	IFSS, MPa
Martensite—HF treated	
Number of data	15
Average IFSS	6.67
Standard deviation	1.35
Martensite	
Number of data	10
Average IFSS	7.59
Standard deviation	1.32
Austenite	
Number of data	15
Average IFSS	20.05
Standard deviation	1.43

3.2 Non-Standard Tests

3.2.1 Specimens Manufacturing and Tests Set-up. Non-standard pull-out tests were considered in order to evaluate the interface in operating conditions. To achieve this goal, a set of specimens made of a CFRP fabric laminate embedding a single martensite detwinned SMA wire deformed with 4% of strain was designed and manufactured using the technologies previously described in this paper (Fig. 14a). Finally special tabs were bonded at room temperature to ease clamping to the testing machine (Fig. 14b).

The specimens were manufactured in series of two by modifying the tools used for the curing of the smart laminates (Sect 2.2). As a consequence, a loss of repeatability in terms of geometrical and physical characteristics must be taken into account during the analysis of results.

Standard methods for the IFSS determination cannot be adopted in this configuration because of the wires exiting the specimens. Thus, estimation of interface quality can be based only on the overall pull-out force applied to extract the wires. As in previous standard tests, specimens were tested at low and high temperature adopting a thermal chamber (Fig. 14c). A different specimen series was tested by Joule-activation maintaining a fixed crosshead position in order to evaluate the wires loading capacity.

3.2.2 Results of Non-Standard Pull-Out Tests. Results of the tensile tests are reported in Fig. 15: again austenite interface results are stronger than martensite one. In this

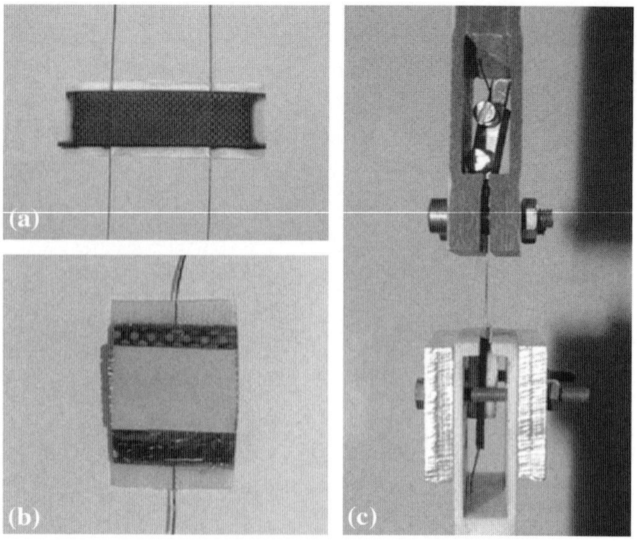

Fig. 14 Non-standard pull-out specimen manufacturing and test set up: specimen after the curing (a), after the gluing of tabs (b) and installed in the test facility (c)

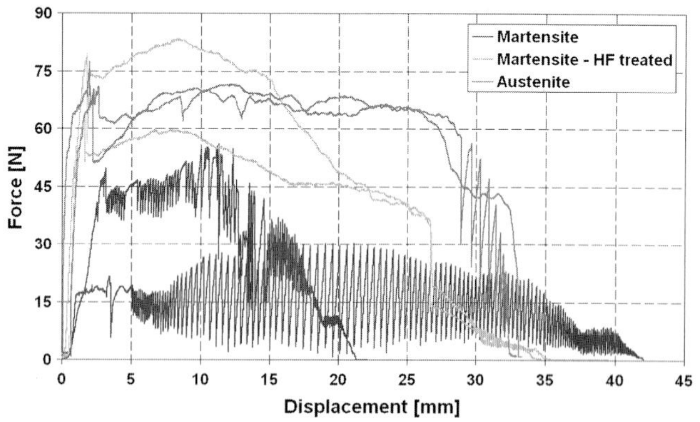

Fig. 15 Pull-out curves of non-standard specimens in all configurations (austenite, martensite and martensite—HF treated)

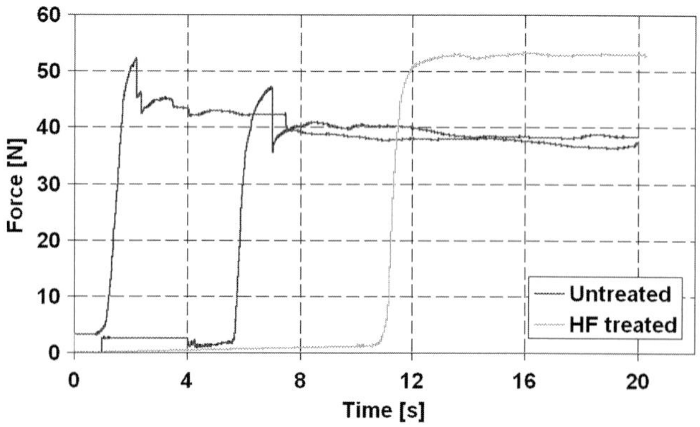

Fig. 16 Joule activation of non-standard specimen: untreated and HydroFluoric treated

situation, the HF treatment seems to effectively improve the bonding performance between actuator and host material. In fact, the martensite-HF treated curves are similar to the austenite ones both up to the debonding peak load and in the following wire slipping out stage. The reason of this behavior could be due to a possible roughness variation or to higher wettability of the austenite and HF-martensite surfaces with respect to the martensite one.

The final activation tests performed (Fig. 16) have confirmed the activation time repeatability in all configurations and again the good stress transfer capability of austenite.

4. Conclusions

The results coming from non-standard pull-out test show the efficiency of the embedding technique. The porting of the technology, originally developed for the embedding of OF, has proven its efficiency also for SMA-based composites. In addition, the direct constraint embedding technique of detwinned martensite wires allows the adoption of advanced composites as host materials. The results in terms of IFSS are encouraging considering that real peak values are much higher than recorded ones (which assumes a uniform stress distribution over the whole embedding length).

Standard pull-out tests have shown that higher load transfer capability is obtained when the wires are in the austenite phase (three times higher than martensite).

These considerations should be kept in mind in any future improvements of this embedding technique. Some preliminary studies showed that QP technique may overcome technological embedding difficulties. With this technique SMA wires can be embedded without constraints because of the detwinning being prevented by the cured QP. This improvement should be a very useful future development in terms of manufacturing and well-characterized actuators. This development must be subjected to a detailed interface analysis devoted to the optimization of materials and to the forming tools upgrade.

The limited invasivity of the developed QP with the composite host material is suggesting the implementation of this technology also in complex shape laminates.

References

1. I. Chopra, Review of State of Art Smart Structures and Integrated Systems, *AIAA J.*, 2002, **40**(11), p 2145–2187
2. S.P. Thompson and J. Loughlan, Adaptive Post Buckling Response of Carbon Fibre Plates Employing SMA Actuators, *Compos. Struct.*, 1997, **38**(1–4), p 667–678
3. Y.J. Zheng, L.S. Cui, and J. Schrooten, Basic Design Guidelines for SMA/Epoxy Smart Composites, *Mater. Sci. Eng. A*, 2005, **390**, p 139–143
4. P. Gaudenzi, M. Olivier, G. Sala, D. Sciacovelli, M. Wheelan, P. Bettini, G. Nosenzo, and A. Tralli, Development of an Active Composite with Embedded Piezoelectric Sensors and Actuators for Structure Actuation and Control, *Proceedings of the 54th International Astronautical*, IAC-03-I.4.03, Bremen, 2003
5. Q.P. Sun and K.C. Hwang, Micromechanics Modeling for the Constitutive Behavior of Polycrystalline Shape Memory Alloys, *J. Mech. Phys. Solids*, 1993, **I**, p 1–17, **II**, p 19–33
6. M. Cho and S. Kim, Structural Morphing Using Two-Way Shape Memory Effect of SMA, *Int. J. Solids Struct.*, 2005, **42**, p 1759–1776
7. M. Langelaar and F. van Keulen, Design Optimization of Shape Memory Alloy Structures, *10th AIAA/ISSMO Multidisciplinary Analysis and Optimization Conference* (Albany, NY), 2004
8. W. Huang, On the Selection of Shape Memory Alloys for Actuators, *Mater. Des.*, 2002, **23**, p 11–19
9. P. Bettini and G. Sala, Preliminary Assessment of Helicopter Rotor Blades Fatigue Endurance Through Embedded F.O. Sensors, *Proceeding of the 24th ICAF Symposium*, May 16–18 (Naples), 2007
10. D.A. Miller and D.C. Lagoudas, Thermo-Mechanical Characterization of NiTiCu and NiTi SMA Actuators: Influence of Plastic Strains, *Smart Mater. Struct.*, 2000, **9**(5), p 640–652
11. J. Khalil-Allafi, G. Eggeler, W. Schmahl, and D. Sheptyakov, Quantitative Phase Analysis in Microstructures which Display Multiple Step Martensitic Transformation in Ni-Rich NiTi Shape Memory Alloys, *Mater. Sci. Eng. A*, 2006, **438–440**, p 593–596
12. Z.G. Wang, X.T. Zu, X.D. Feng, S. Zhu, J.M. Zhou, and L.M. Wang, Annealing-Induced of Transformation Characteristics in TiNi Shape Memory Alloys, *Physica B*, 2004, **353**, p 9–14
13. ASTM F2004-03, Standard Test Method for Transformation Temperatures of Nickel-Titanium Alloys by Thermal Analysis
14. ASTM D3418-03, Standard Test Method for Transition Temperatures of Polymers by Differential Scanning Calorimetry
15. V. Birman, Stability of Functionally Graded Hybrid Composite Plates, *Compos. Eng.*, 1995, **5**(7), p 913–921
16. S.F. Zhandarov and E.V. Pisanova, The Local Bond Strength and Its Determination by Fragmentation and Pull-Out Tests, *Compos. Sci. Technol.*, 1997, **57**, p 957–964
17. C.Y. Yue, H.C. Looi, and M.Y. Quek, Assessment on Fibre-Matrix Adhesion and Interfacial Properties Using the Pull-Out Test, *Int. J. Adhes. Adhes.*, 1995, **15**(2), p 73–80
18. M.R. Piggot, Why Interface Testing by Single-Fibre Methods can be Misleading, *Compos. Sci. Technol.*, 1997, **57**, p 965–974
19. W. Beckert and B. Lauke, Critical Discussion of the Single Fibre Pull-Out Test: Does it Measure Adhesion?, *Compos. Sci. Technol.*, 1997, **57**, p 1689–1706

JMEPEG (2009) 18:672–678
DOI: 10.1007/s11665-009-9417-7

Carbon Fiber-Reinforced Smart Laminates with Embedded SMA Actuators—Part II: Numerical Models and Empirical Correlations

M. Riva, P. Bettini, L. Di Landro, G. Sala, and A. Airoldi

(Submitted October 21, 2008; in revised form February 16, 2009)

Up to now one of the main limitations for a large use of shape memory alloys (SMA)-based smart composite structures in the aerospace industry is the lack of useful numerical tools for design; in addition, some technological aspects still need a more detailed investigation. This article shows numerical modeling approaches adopted for the implementation of SMA constitutive laws in commercial codes such as ABAQUS. Two different approaches were selected. The first one is based on the thermomechanical model proposed by Turner and the other one follows the thermodynamic macromechanical constitutive law developed by Lagoudas. The implementation in ABAQUS code was followed by a procedure to evaluate model parameters and to experimentally validate the reliability of code predictions for specifically designed test situations. This article presents the test campaign carried out for the definition of these parameters and the numerical-experimental correlation for both the models.

Keywords embedded SMA, polymer matrix composites, shape memory alloy, smart structures

1. Introduction

Materials for primary structures in aerospace applications must have high mechanical properties, lightness, and durability. It is important for these structures to exhibit good behavior particularly in terms of impact resistance and environmental endurance; fire and corrosion resistance are further characteristics often required. Composite materials often conform well to such requirements and nowadays are widely used in many engineering fields.

On the other hand, a number of new materials and technologies (i.e., microelectronic devices, shape memory alloys (SMA)-based microactuators, piezo-ceramics, electroactive polymers, optical fibers, etc.) allow to realize structures able to actively respond to specific external inputs. Thus, extensive diffusion of smart materials in many aerospace and ground commercial applications is expected. Their capabilities in terms of shape and vibration control of large space structures, acoustic control for noise reduction in civil aircraft, health monitoring, and in-site structure identification are well assessed and promising results are also obtained in many other applications.

In the last years, several research activities in the manufacturing processes field were devoted to develop smart structures that could combine high mechanical efficiency, due to composite host materials, with good functional properties of embedded sensors and actuators (Ref 1-3).

The embedding of these smart devices into the structures could give some advantages compared to their bonding onto the outer skin. For example, they allow the actuation in locations hardly accessible from outside because of shape constraints; the protection of the actuation system inside the host structure from all environmental effects reducing its performance can be also achieved; in addition, active structures could be applied even in those cases where a clean surface is required (i.e., skin of the aerodynamic surfaces) (Ref 4).

The embedding technique of these components within the load-carrying structure is still a leading-edge application. Moreover, research activities must be oriented both to the development of those manufacturing techniques and to the characterization of embedded sensors and actuators invasivity (active and passive) on the smart structure performance.

Nomenclature

A_s, A_f	austenite start and finish temperature
CTE	coefficient of thermal expansion
CFRSL	carbon fiber-reinforced smart laminates
FE	finite element
M_s, M_f	martensite start and finish temperature
OWSME	one-way shape memory effect
SMA	shape memory alloys
TWSME	two-way shape memory effect
T_g	glass transition temperature

This article is an invited paper selected from presentations at Shape Memory and Superelastic Technologies 2008, held September 21-25, 2008, in Stresa, Italy, and has been expanded from the original presentation.

M. Riva, P. Bettini, L. Di Landro, G. Sala and **A. Airoldi**, Aerospace Engineering Department, Politecnico di Milano, Milan, Italy. Contact e-mail: riva@aero.polimi.it.

Finally, most of the applications are also strongly dependent on the availability of useful numerical tools. In fact, all the design activities regarding structures are nowadays related to the adoption of numerical models able to predict the efficiency and performance of the systems (Ref 5). The peculiarity of smart composite structures underlines this problem due to the higher complexity of the active materials behavior: the typically nonlinear response of these materials joined to the composite material modeling, which must account for laminates anisotropy and complex damage mechanism pose a difficult task.

In view of a development of both technological processes and numerical tools for smart carbon laminates with embedded sensors and actuators, this research is focused on the embedding of shape memory wires, selectively taking advantage of their functional properties in terms of low activation speed combined with high authority, i.e., capacity to effectively transfer their load/deformation to the host structure. In fact, these characteristics are useful for most of the aerospace applications (airfoil morphing, shape control, etc.). Authority of these actuators can be exploited in two different ways. First, the wires are embedded in a detwinned martensite phase and the activation of the structure is obtained using the OWSME due to the reverse transition induced by heating (from detwinned martensite to the austenite phase); then the structure (and the embedded wires too) come back to the initial form through its stiffness (host material remain in the elastic range) when the wires are cooled and the forward transition is induced. Second, the wires can be embedded both in austenite or detwinned martensite phase considering TWSME. A structure can change its form by heating or cooling so that reverse or forward transition, respectively, occurs (Ref 6, 7).

A previous work was oriented to the description of the embedding techniques developed for SMA actuators and the analysis of the interface with the host materials (Ref 8). This article is focused on numerical aspects. The target is the definition of a design strategy that takes into account the computational cost and the level of accuracy required by a smart structure for a preliminary prediction of the overall performance of the system. The implementation of the constitutive laws of the actuators in commercial codes is the crucial point for the modeling of several actuators and is still an open issue regarding shape memory actuators. Several constitutive laws have been developed in the last decade for the prediction of the behavior of SMAs. These laws derive from micromechanical approaches (Ref 5) or are based on macromechanical models (Ref 9-14). Liang and Rogers (Ref 9) or Turner (Ref 12, 14) proposed a description of material behavior on a phenomenological basis; Tanaka and Lagoudas (Ref 10, 11) proposed a constitutive law based on thermodynamical approaches. This work is focused on the implementation and comparison between the relatively simple model (Ref 12, 14) with the more detailed one (Ref 11) to have two different tools depending on the accuracy required by the application design.

2. Design and Modeling of Smart Structure

Smart structures design is strongly dependent on the availability of useful numerical tools for the performance prediction of the whole system. In this article, the attention is focused on two complementary constitutive laws: the first one is the Lagoudas' thermodynamical model and the other is the Turner's thermomechanical model. The work was oriented to implement these constitutive laws in ABAQUS, a commercial code, trying to identify the peculiarity of the two different models to have a criterium for the proper numerical procedure selection.

The finite element analysis time, considered as the sum of the time required to prepare the FE model and the solving time, played an important role in the choice of a modeling technique. The capabilities of the constitutive laws were also taken into account in terms of how well they represent of the real behavior of the modeled SMA (two way shape memory effect, superelasticity, hysteresis modeling, etc.).

2.1 Turner Model

The Turner model assumes a nonlinear thermoelastic behavior defining an effective coefficient of thermal expansion (CTE) starting from the classic total strain definition. The strain applied to a SMA is the sum of mechanical, thermal, and transformation contributes:

$$\varepsilon = \varepsilon_e + \varepsilon_p + \varepsilon_{th} + \varepsilon_m \qquad \text{(Eq 1)}$$

where ε_e, ε_p, and ε_{th} are the elastic, plastic, and thermal strains, respectively and ε_m derives from the phase transition. Considering the loading situations below the elastic limit, the plastic deformation can be dismissed. The elastic strain can be expressed as the ratio between applied stress and a temperature-dependent Young modulus. An equivalent thermal strain $\tilde{\varepsilon}_{th}$ can be introduced that encompasses both transition induced and thermal strains. The model considers that this factor is equal to the simple thermal strain of martensite phase when the temperature is less or equal to A_S. When the temperature exceeds this value, another thermal contribution is introduced, which depends on a temperature-dependent fictitious expansion coefficient CTE $\alpha_E(T)$. This can be written as:

$$\tilde{\varepsilon}_{th} = \begin{cases} \alpha_M(T - T_0) & T \leq A_S \\ \alpha_M(T - T_0) + \alpha_E(T)(T - A_S) & T > A_S \end{cases} \qquad \text{(Eq 2)}$$

By writing Eq 2 ($T > A_S$) in an explicit form with respect to the effective CTE, one obtains:

$$\alpha_E(T) = \frac{\tilde{\varepsilon}_{th} + \alpha_M(T_0 - A_S)}{T - A_S} \quad \text{with} \quad \tilde{\varepsilon}_{th} = \frac{\sigma_{REC}(T)}{E(T)} \qquad \text{(Eq 3)}$$

The $\sigma_{REC}(T)$ is the recovery stress function dependent on temperature. All the parameters can be determined with empirical tests. In fact α_M and A_S are obtained from dilatometric and calorimetric tests (dilatometer and differential scanning calorimeter). Then, $E(T)$ and $\sigma_{REC}(T)$ can be measured through tensile tests at different temperatures (isothermal) (Fig. 1) and constrained activation tests (isolength), respectively. Considering the characteristics of Turner model, it is apparent that this is able to model the recovery phase of OWSME only.

The implementation in commercial codes, like ABAQUS, is quite simple because the software allows the definition of temperature-dependent material properties and, in particular, Young modulus ($E(T)$, Fig. 1) and CTE ($\alpha_E(T)$, Fig. 2). A thermal analysis is performed (temperature is the only degree of freedom) to define the temperature map for each step. This thermal field is then applied to the complete model to define the strain displacement and the stress fields.

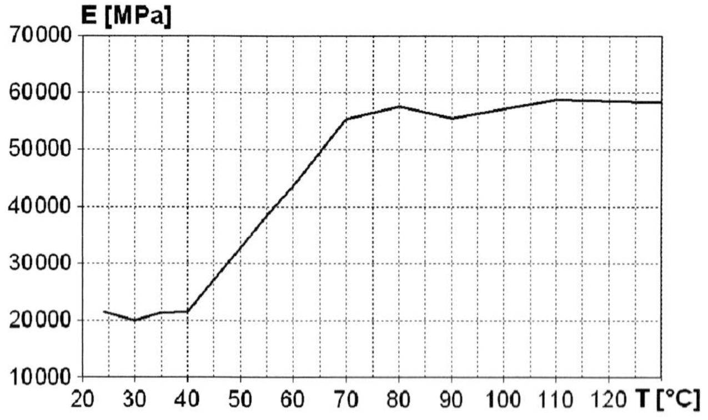

Fig. 1 NiTiNol temperature-dependent Young Modulus

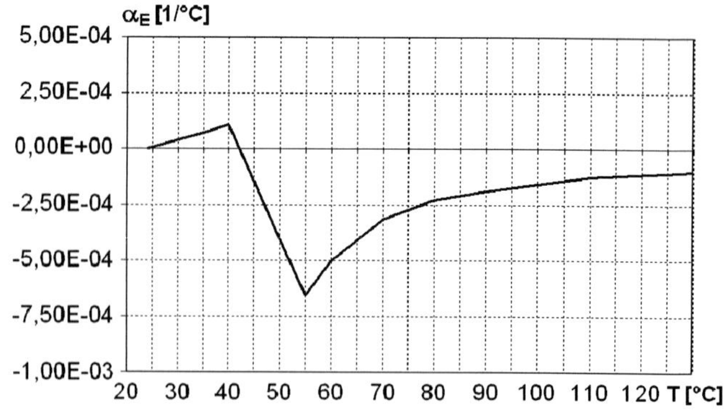

Fig. 2 NiTiNol temperature-dependent CTE

2.2 Lagoudas Model

The Lagoudas model was implemented in a reduced formulation; to take advantage of the actuators shape (wires), it was simplified into a one-dimensional law. The plastic deformation was neglected following the similar reduction imposed in the Turner model. The model is based on five state variables: strain ε, stress σ, temperature T, martensitic fraction ξ, and martensitic strain ε_{m}.

The constitutive equations are:

$$\varepsilon = S\sigma + \alpha(T - T_0) + \varepsilon_{\mathrm{m}}$$
$$S = \frac{1}{E^{\mathrm{A}}} + \xi\left(\frac{1}{E^{\mathrm{M}}} - \frac{1}{E^{\mathrm{A}}}\right) \quad \text{(Eq 4)}$$
$$\alpha = \alpha^{\mathrm{A}} + \xi(\alpha^{\mathrm{M}} - \alpha^{\mathrm{A}})$$

where S is the compliance, α is the CTE, and T_0 is the reference environment temperature; according to Lagoudas model, the evolution of the martensitic transformation is considered similar to a plastic deformation and defines the flow rule in its differential form (Ref 11, 13)

$$\dot{\varepsilon}_{\mathrm{m}} = \Lambda\dot{\xi} \quad \text{where} \quad \Lambda = \begin{cases} H^{\mathrm{cur}}\mathrm{sign}(\Sigma^{\mathrm{eff}}), & \dot{\xi} > 0 \\ \frac{\varepsilon^{\mathrm{m}}_{\mathrm{max}}}{\xi_{\mathrm{max}}}, & \dot{\xi} < 0 \end{cases} \quad \text{(Eq 5)}$$

In this formulation, Λ is the transformation tensor, the expressions $\dot{\xi} > 0$ and $\dot{\xi} < 0$ specify the transition phase direction, H^{cur} is the current maximum deformation strain (dependent on the applied stress), Σ^{eff} is the effective stress, $\varepsilon^{\mathrm{m}}_{\mathrm{max}}$ is the maximal martensite strain ever reached in the process, and ξ_{max} is the correspondent martensitic fraction. Lagoudas defines an empirical formula (Ref 11, 13) for the definition of the Σ^{eff} that is, in fact, the stress that causes the phase transition and depends on the martensitic fraction, the deformation, and some material parameters:

$$\Sigma^{\mathrm{eff}} = \sigma + \left\{-D_3\left[-\ln\left(1 - \xi^{\mathrm{d}}\right)\right]^{\frac{1}{m_1}} + D_2 H \xi^{\mathrm{d}} + D_1\right\}\mathrm{sign}(\varepsilon_{\mathrm{m}})$$

The strain evolution consequent to the martensitic fraction is treated as a plastic deformation, and a function Φ that determines the behavior of the transition is defined as follows:

$$\Phi \leq 0, \quad \Phi\dot{\xi} = 0, \quad \Phi = \begin{cases} \Pi - Y, & \dot{\xi} > 0 \\ -\Pi - Y, & \dot{\xi} < 0 \end{cases} \quad \text{(Eq 6)}$$

where Y is a material constant, which defines the energy dissipation over volume due to a complete transformation. Π is the thermodynamic force which drives the transformations and is representative of the chance that this phenomenon is occurring. It is experimentally defined by a function dependent on characteristic temperatures of the material and other material parameters.

$$\Pi = \Sigma^{\mathrm{eff}}\Lambda + \frac{1}{2}\Delta S\sigma^2 + \sigma\Delta\alpha(T - T_0) + \rho\Delta s_0(T - M^{0\mathrm{S}}) + D_4\xi$$
$$- \rho\Delta c\left[T - T_0 - T\ln\left(\frac{T}{T_0}\right)\right] - D_5[-\ln(1 - \xi)]^{\frac{1}{m_2}} + Y$$

The finite element method implies the solving of the nonlinear system $[K(\{u\})]\{u\} = \{F\}$, where the nonlinearity is due to the dependence of the stiffness matrix ($\{K\}$) on the displacement variable $\{u\}$ and $\{F\}$ is the loading vector. The implementation into the ABAQUS code of this model was done writing a UMAT (User Material) subroutine, which allows the user to define the material behavior. UMAT initializes the state variables with a function called SDVINI (Solution-Dependent INItial Variables) and then evaluates them in each time increment using the Lagoudas formulation. The state variables (σ, ε, T, ξ, ε^{m}) are managed by the STATEV (STATE Variables) parameter that interacts between the subroutine and the main code. A return mapping closest point projection algorithm was adopted for the evaluation of strains and temperatures at each step, which verifies if the transition phase is occurring (Ref 15, 16). The subroutine extracts the state variables from ABAQUS and by applying mechanical or thermal strains, it derives the new stress and the transition function Φ; if the transition is not occurring, the prediction is accepted and the state variables are updated for the next step, otherwise, if the transition is in process, the subroutine considers substeps where imposed strains and temperatures are fixed and a variation of ξ and ε^{m} is applied to evaluate the transition tensor and the influence of the transition phase on the state variables for this step.

2.2.1 Constitutive Law Calibration. To model the NiTiNOL wires, a subroutine was developed taking into account the unidimensional simplified model. This choice also simplifies the calibration of the subroutine because starting from the Lagoudas formulation there are 19 parameters to be obtained through proper experimental measurements. These are listed in Table 1.

Most of these parameters can, in principle, be obtained through thermomechanical tests, whereas back-and-drag-stress-related parameters are obtained by interpolation on the basis of specific experimental tests.

2.3 Comparisons

A useful numerical approach should be flexible and easy to implement, but, at the same time, it has to significantly represent the smart structures. The constitutive models considered here are in this sense complementary. The Turner model is characterized by easy implementation and low CPU time, but it is not able to adequately describe the transition from detwinned martensite to austenite; moreover, it does not take into account the "hysteresis" between the two transitions (back and forward). On the other hand, the Lagoudas law is able to cover those deficiencies, but it requires too long solving time to be used in a design and optimization phase.

In this research, both models were applied. Usually, the procedure implemented requires a preliminary modeling using the faster Turner model to allow operations like optimization and performances prediction. The following step, when necessary, consists in the implementation of the Lagoudas model to have a more detailed description of the performances particularly in terms of deployable load, back and forward transition, and phase transitions.

Table 1 Lagoudas parameters

Martensitic and austenitic physical constants	
E^A, E^M	Young's moduli
ν^A, ν^M	Poisson coefficients
α^A, α^M	Thermal expansion coefficients
$\rho c^A, \rho c^M$	Specific heats
$\rho \Delta S^0$	Specific enthalpy difference
Transition phase parameters	
H	Maximum transition strain
Y	Total hysteresis for a single transition cycle
M^{0S}	Starting martensite temperature
D_1, D_2, D_3, m_1	Back stress-related parameters
D_4, D_5, m_2	Drag stress-related parameters

3. Experimental Validation

The two models were validated by testing dedicated specimens specifically manufactured. The Turner model was used for the experimental/numerical comparison with a carbon fiber-reinforced panel with OWSME NiTiNOL wires embedded, whereas the Lagoudas model was validated with a simple wire of both OWSME and TWSME (Ref 21). NiTiNOL wires from Furukawa Techno Materials with diameter of 0.38 mm were employed.

3.1 Turner Model Validation

Different verification models were performed to select the type of the elements. The comparison underlines how a shell model is precise enough with a high gain in terms of CPU time. Due to this consideration, the final comparison was conducted on a shell-based model, which contains NiTiNOL wires, modeled via proper ABAQUS properties: REBAR consisting in the modeling of a reinforcing inclusion made by a proper material in a shell element (Fig. 3). The thermoelastic model receives as input the thermal load on the wires and derives the temperature behavior of all the nodes.

A carbon fiber-reinforced panel with two embedded NiTiNOL wires was modeled and analyzed using the ABAQUS commercial code. The actual panel was manufactured with 12 plies $[90°/(0°)_2/90°/+45°/-45°]_s$ and its overall dimensions were: 30 × 170 × 1.2 mm. The actuators chosen were OWSME wires (∅ = 0.38 mm) trained using standard heat treatments in accordance with Ref 17. They were embedded between the 2nd and 3rd plies with a 4% imposed strain. All manufacturing techniques and mechanical characteristics are reported in Ref 8. The panel was constrained on one side and it was activated via Joule effect (Fig. 4). The numerical/experimental comparison shows that, after a short transition, the predicted displacement matches (max. error 20 μm) the experimental one (Fig. 5).

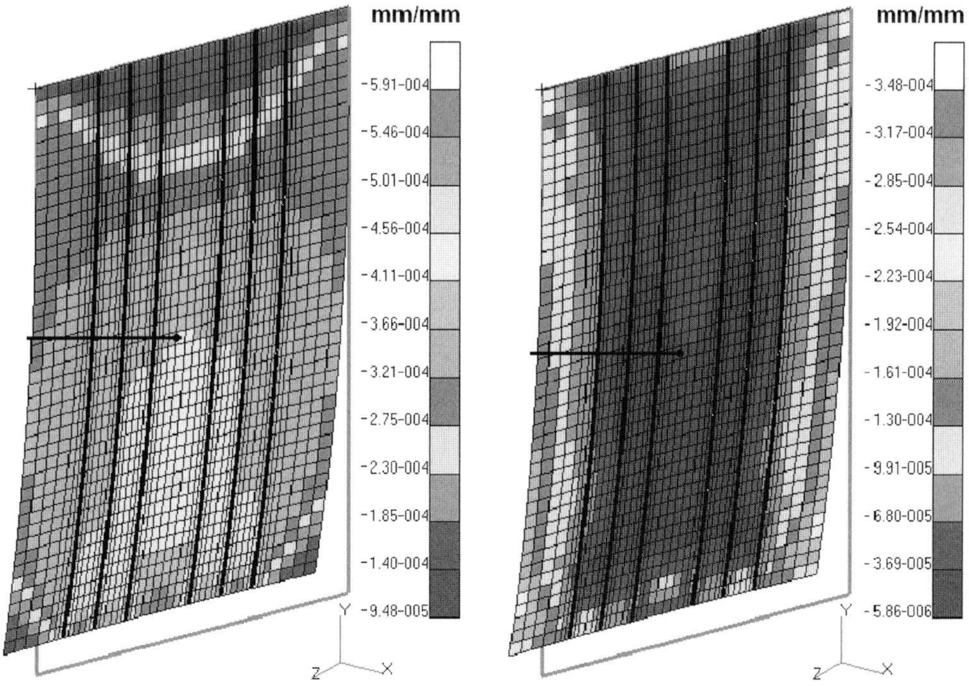

Fig. 3 FE model of the Turner experiment

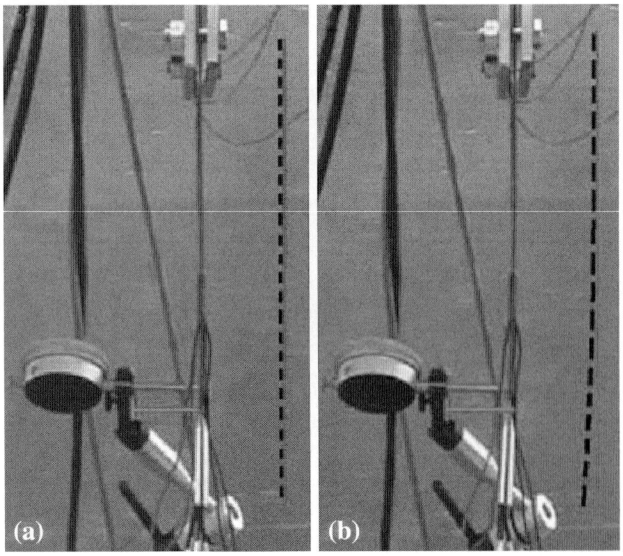

Fig. 4 Turner validation experimental test setup

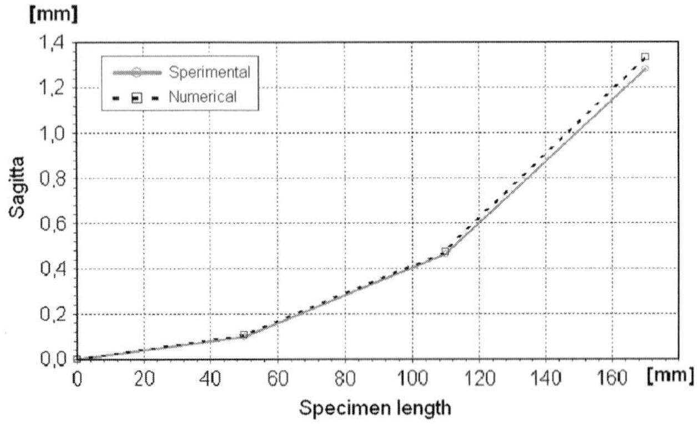

Fig. 5 FE displacement comparison for the Turner model

3.2 Lagoudas Validation

The validation model is very simple; in fact, it was decided to adopt a single ABAQUS TRUSS element, which has two nodes and two linear degrees of freedom per node, to reduce the elevated analysis time due to the subroutine implementation. One of the nodes was constrained keeping the other free. For the applied stress cases, a *DLOAD card was used, which applies axial stress to the element along its main direction. The thermal load was applied in two steps: a cooling down from 393.15 to 273.15 K and a heating to the initial temperature. The first validation was conducted on the Lagoudas experimental test by considering the calibration parameters presented in Ref 11.

The experiment selected for the validation of the subroutine was a simple DMA tensile test. The specimen is heated up to a temperature higher than its A_f and cooled down below the M_f. One tip of the wire is constrained while the other is fixed on a movable part, which is able to induce a constant stress (at a desired level) during the whole test. The strain is recorded and plotted with respect to the temperature, obtaining the typical behavior of Fig. 6. Figure 7 shows that the results derived from Lagoudas model and that implemented for the ABAQUS code

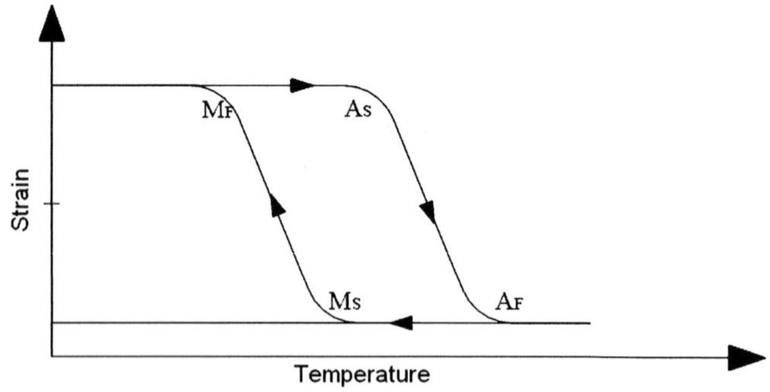

Fig. 6 Typical temperature vs. strain behavior for SMA

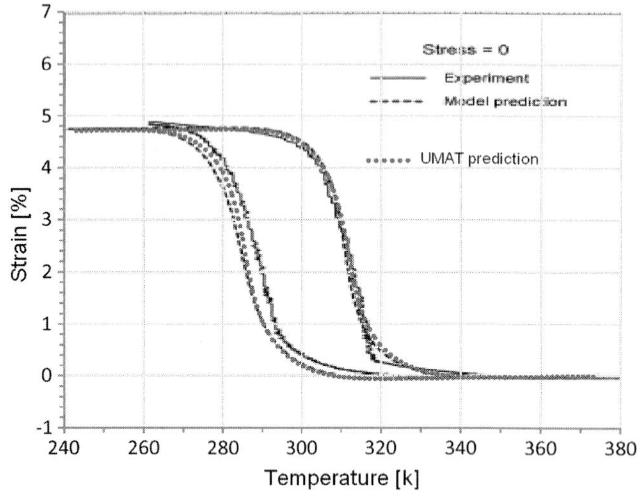

Fig. 7 Comparison between UMAT, Lagoudas, and experimental results

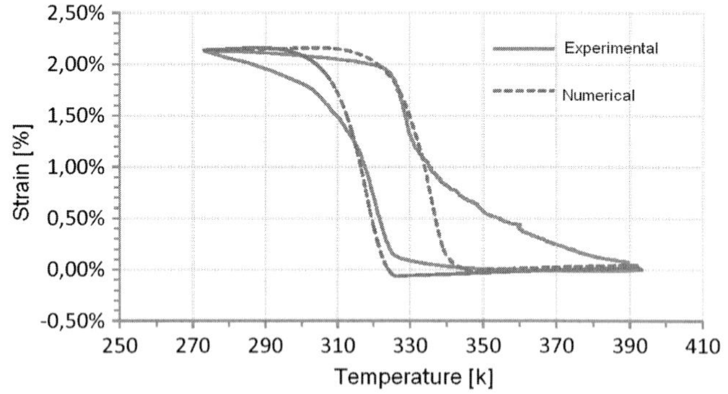

Fig. 8 Correlation of one-way trained wire with 20 MPa applied stress

are almost coincident. Both the routines follow the experimental results very well, either in terms of local predictions (i.e., transition temperatures and maximum strain) or in terms of overall predictions (general temperature and strain behavior).

The OWSME wires were experimentally tested and compared with numerical results considering a constant applied stress during the whole test. Figures 8 and 9 show an overall good correlation. In particular, it can be underlined that the starting temperatures A_s and M_s are reproduced well by the UMAT subroutine; the maximum strain is also predicted with

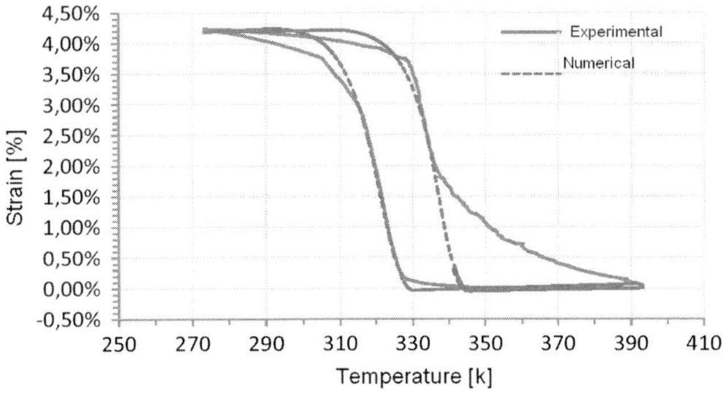

Fig. 9 Correlation of one-way trained wire with 50 MPa applied stress

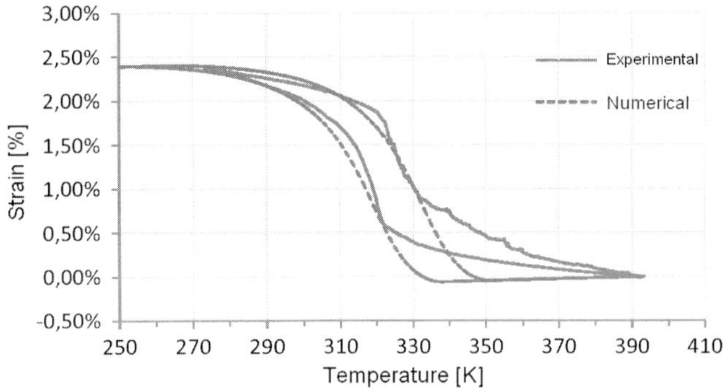

Fig. 10 Correlation of two-way trained wire without applied stress

limited error and finally the first portion of each transition (M to A and A to M) has a good correlation. Some of the mismatch (in particular the final part of the reverse transition) can be the result of previous heat treatments conducted on the same wire used as specimen.

The numerical validation of the TWSME was conducted for both the unloaded (Fig. 10) and loaded specimen tests (Fig. 11). Also in this case, the maximum strain is reproduced well. Somewhat lower agreement is presented in the transition temperatures. This could be induced by some imperfections in the two-way training process, or to the fact that the calibration of the model was done with different wires, although of the same batch, which have been subjected to several heat treatments.

4. Conclusions

The numerical tools developed can help both the performance prediction (Turner's model) and a behavior verification in terms of stress-induced transition level and characteristic temperatures (M_s, M_f, A_s, and A_f).

Considering the results derived from the two different models implemented, a modeling strategy to be applied to several smart structures was setup. Turner model is used for a preliminary performance prediction in terms of maximum displacement and stresses induced by the actuator into the structure. The lightness of this modeling technique guarantees useful optimization of the positioning procedure and dimension choice of the actuators. The Lagoudas model can be used after

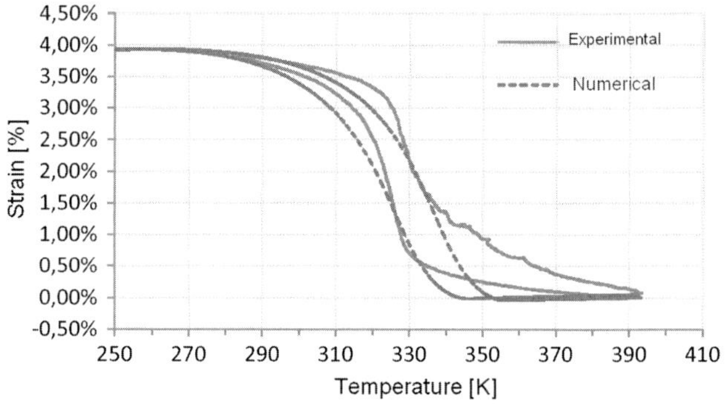

Fig. 11 Correlation of two-way trained wire with 40 MPa applied stress

the optimization process to verify the stress induced from the structure to the actuators and the consequent variation in the transition temperatures of the active materials. In this case, a further loop with Turner model, if the A_f is higher than the glass transition temperature (T_g) of the matrix, may be required.

Future work is in progress to test different host materials and types of SMA actuators. Thus, smart structures modeling and manufacturing taking advantages of the know-how acquired will move from simple specimen panels to more complex smart structures, such as laminates with single or double curvature.

Acknowledgment

The sponsorship of FIRB project RBIP06AWF9 is gratefully acknowledged.

References

1. I. Chopra, Review of State of Art Smart Structures and Integrated Systems, *AIAA J.*, 2002, **40**(11), p 2145–2187
2. S.P. Thompson and J. Loughlan, Adaptive Post Buckling Response of Carbon Fibre Plates Employing SMA Actuators, *Compos. Struct.*, 1997, **38**(1–4), p 667–678
3. Y.J. Zheng, L.S. Cui, and J. Schrooten, Basic Design Guidelines for SMA/Epoxy Smart Composites, *Mater. Sci. Eng. A*, 2005, **390**, p 139–143
4. P. Gaudenzi, M. Olivier, G. Sala, D. Sciacovelli, M. Wheelan, P. Bettini, G. Nosenzo, and A. Tralli, Development of an Active Composite with Embedded Piezoelectric Sensors and Actuators for Structure Actuation and Control, *Proceedings of the 54th International Astronautical*, IAC-03-I.4.03 (Bremen, Germany), 2003
5. Q.P. Sun and K.C. Hwang, Micromechanics Modeling for the Constitutive Behavior of Polycrystalline Shape Memory Alloys, *J. Mech. Phys. Solids*, 1993, **41**(I), p 1–17, (II), p 19–33
6. M. Cho and S. Kim, Structural Morphing Using Two-Way Shape Memory Effect of SMA, *Int. J. Solids Struct.*, 2005, **42**, p 1759–1776
7. M. Langelaar and F. van Keulen, Design Optimization of Shape Memory Alloy Structures, *10th AIAA/ISSMO Multidisciplinary Analysis and Optimization Conference* (Albany, NY)
8. P. Bettini, M. Riva, G. Sala, L. Di Landro, A. Airoldi, and J. Cucco, Carbon Fiber Reinforced Smart Laminates with Embedded SMA Actuators—Part I: Embedding Techniques & Interface Analysis, accepted in the August 2009 SMST Special Issue of *J. Mater. Eng. Perform.* doi:10.1007/s11665-009-9384-z
9. C. Liang and C.A. Rogers, One-Dimensional Thermomechanical Constitutive Relations for Shape Memory Materials, *J. Intell. Mater. Syst. Struct.*, 1990, **1**(2), p 207–234
10. K. Tanaka, F. Nishimura, T. Hayashi, H. Tobushi, and C. Lexcellent, Phenomenological Analysis on Subloops and Cyclic Behavior in Shape Memory Alloys Under Mechanical and/or Thermal Loads, *Mech. Mater.*, 1995, **19**, p 281–292

11. D.C. Lagoudas and Z. Bo, Thermomechanical Modeling of Polycrystalline SMAs Under Cyclic Loading, *Int. J. Eng. Sci.*, 1999, **37**(Part I, II, III, IV), p 1089–1204
12. T.L. Turner, 2001 "Thermomechanical Response of Shape Memory Alloy Hybrid Composites", NASA-2001-tm210656, Langley Research Center, Hampton, VA
13. D.C. Lagoudas and P.B. Entchev, Modeling of Transformation-Induced Plasticity and its Effect on the Behavior of Porous Shape Memory Alloys. Part I: Constitutive Model for Fully Dense SMAs, *Mech. Mater.*, 2004, **36**, p 865–892
14. T.L. Turner, 2001 "Experimental Validation of a Thermoelastic Model for SMA Hybrid Composites", NASA-2001-8spie-tlt24, NASA Langley Research Center, Hampton, VA
15. J.C. Simo and S. Govindjee, Non-Linear B-Stability and Symmetry Preserving Return Mapping Algorithms for Plasticity and Viscoplasticity, *Int. J. Numer. Methods Eng.*, 1991, **31**, p 151–176
16. M.A. Qidwai and D.C. Lagoudas, Numerical Implementation of Shape Memory Alloy Thermomechanical Constitutive Model Using Return Mapping Algorithms, *Int. J. Numer. Methods Eng.*, 2000, **47**, p 1123–1168
17. F. Peterson, "Method of Training NiTiNOL Wire," US Patent 6797083, 28 Sep 2004
18. T.L. Turner and H.D. Patel, "Input Files and Procedures for Analysis of SMA Hybrid Composite Beams in MSC.Nastran and ABAQUS", NASA-TM-2005-213517, NASA Langley Research Center, Hampton, VA
19. E. Patoor, D.C. Lagoudas, P.V. Entchev, L.C. Brinson, and X. Gao, Shape Memory Alloys, Part I: General Properties and Modeling of Single Crystal, *Mech. Mater.*, 2006, **38**, p 391–429
20. L. Colonna Preti, "Microattuatori in leghe a memoria di forma: sperimentazione, implementazione di leggi costitutive e validazione numerico-sperimentale", Master Thesis, Politecnico di Milano, 2006, in Italian
21. A. Fusi, "Utilizzo degli Smart Material nelle vetture di F.1: progetto e sviluppo di un dimostratore tecnologico", Master Thesis, Politecnico di Milano, 2007 (in Italian)

JMEPEG (2009) 18:679–685
DOI: 10.1007/s11665-009-9361-6

Stress-Induced Martensitic Transformation in the Crack Tip Region of a NiTi Alloy

A. Falvo, F. Furgiuele, A. Leonardi, and C. Maletta

(Submitted September 23, 2008)

The evolution of stress-induced martensitic transformation in front of the crack tip in a NiTi alloy is analyzed in this investigation, by two-dimensional finite element simulations of single edge-crack specimens. In particular, the transformation start and finish contours, i.e., the boundaries of the transformation zone, were obtained by using plasticity concepts, and the effects of the temperature were taken into account by using the Clausius-Clapeyron relation. Furthermore, comparisons between numerical and analytical results, obtained by Irwin's modified linear elastic fracture mechanics relations, were carried out. These comparisons show that a good agreement in terms of the martensite start and finish sizes is obtained; moreover, the analytic approach could be able to describe the stress field in the crack tip region outside the phase transformation zone, i.e., in the austenitic region, but a proper equation to estimate the effective crack length should be found. To this aim, further studies should be carried out.

Keywords cracks, finite element analyses, fracture, NiTi alloys, shape memory alloys, stress-induced martensitic transformation

1. Introduction

In the last decades, shape memory alloys (SMAs), and in particular the Nickel Titanium-based ones (NiTi), have received more attention from scientific and engineering communities, owing to their unique characteristics, namely shape memory effect (SME) and superelastic effect (SE) (Ref 1). In particular, these properties allow large recoverable strains or large induced internal forces due to a reversible solid-state phase transformation between austenite and martensite; this transformation can be activated by a temperature variation (TIM, thermally induced martensite) or by the application of external forces (SIM, stress-induced martensite). Due to these interesting features, as well as to their good mechanical performances and biocompatibility, NiTi alloys have seen growing use in the last years in many branches of engineering and medicine. As a direct consequence of this interest, many researchers have studied the thermo-mechanical behavior of SMAs, in terms of both SME and SE (Ref 1), and various numerical models have been developed in order to describe their mechanical and functional behavior (Ref 2). However, a limited number of works have been devoted to the study of the fatigue (Ref 3-6) and, above all, fracture behavior (Ref 7-11) of NiTi alloys. From a materials science point of view, an accurate knowledge of these topics is essential for predicting the functional and structural life of damaged structures as well as their failure modes, with the aim to improve the overall performances of NiTi-based components or structures. The fracture behavior of austenitic NiTi alloys strongly depends on the SIM transformation which occurs in the crack tip region, as a consequence of the high values of local stresses.

In an experimental investigation of fracture mechanics in NiTi alloys (Ref 11), it has been shown that the crack is stress control propagated in line with the direction of the maximum normal stress. In addition, it was observed that the notch geometries and the martensitic transformation play significant roles on the fracture process of SMAs. In the last years, some finite element analyses concerning the martensitic transformation in front of notches or cracks have been carried out (Ref 12, 13); in particular, it has been found that the shape of the martensitic transformation zone at the crack tip is similar to the characteristic shape of plastic zone for metals. This means that modified plasticity concepts can be used to predict the transformation zone in SMAs.

In the present work, two-dimensional (2D) numerical simulations were carried out in order to investigate the effects of the temperature on the stress-strain distributions and on the martensitic transformation zone in front of the crack tip in a single edge-crack (SEC) specimen. Furthermore, comparisons between numerical and analytical results, obtained by Irwin's modified linear elastic fracture mechanics relations, were carried out. These comparisons show that a good agreement in terms of the martensite start and finish sizes is obtained; moreover, the analytic approach could be able to describe the stress field in the crack tip region outside the phase transformation zone, i.e., in the austenitic region, but a proper equation to estimate the effective crack length should be found. To this aim, further studies should be carried out.

This article is an invited paper selected from presentations at Shape Memory and Superelastic Technologies 2008, held September 21-25, 2008, in Stressa, Italy, and has been expanded from the original presentation.

A. Falvo, F. Furgiuele, A. Leonardi, and **C. Maletta**, Department of Mechanical Engineering, University of Calabria, 87036, Arcavacata di Rende, Italy. Contact e-mail: carmine.maletta@unical.it.

2. Constitutive Behavior

The classic simplified isothermal stress-strain curve for an austenitic NiTi-based SMA is represented with the solid piecewise line in Fig. 1. It is worth noting that this curve is relative to a monotonically increasing loading path, while if the load is decreased before failure the material exhibits the typical stress-strain hysteretic behavior, as the unloading path is represented by the dashed lines in Fig. 1. In the figure, the stress-strain relation is approximated with a piecewise linear curve, for sake of simplicity. The figure clearly shows that, when the applied load is increased, the SIM transformation (A $\rightarrow$ M) starts as σ reaches a critical value σ_s^{AM}, and the transformation is completed when σ is equal to σ_f^{AM}, i.e., the martensite fraction ξ_M increases from 0 to 1 and the material exhibits a characteristic deformation namely, ε_L. When unloading, the reverse transformation (M $\rightarrow$ A) occurs in the stress range σ_s^{MA}-σ_f^{MA}, resulting in a stress-strain hysteretic behavior. However, unloading and the corresponding reverse transformation (martensite to austenite) are not taken into account in this investigation since, for fracture purposes, the specimens for toughness measurements are generally loaded to fracture without unloading.

The main thermo-mechanical properties of the material, to be used in the numerical model described in the following section, were experimentally obtained in a previous work of the authors (Ref 14) by carrying out different isothermal tensile tests on dog bone-shaped test specimens in Ni-49 at.% Ti sheet (Type S, Memory Metalle, Germany, 0.75 mm thick) with a nominal austenite finish temperature $A_f = -7$ °C, as shown in Fig. 2.

The test were executed by using a universal testing machine (Instron 8500), equipped with a furnace (MTS 653) and a resistance extensometer with a gauge length of 10 mm. The stress-strain hysteretic behavior of the material was measured by isothermal strain controlled cycles, at a strain rate of 200 $\mu\varepsilon\ s^{-1}$, that were carried out for different values of maximum deformation and at different temperatures. Figure 3(a) shows the evolution of the stress-strain hysteretic behavior of the material in the first 25 mechanical cycles for a fixed value of total strain $\varepsilon_{tot} = 3.5\%$. As well known (Ref 1, 14), the figure clearly shows that the material exhibits a strong modification of the stress-strain response in the first cycles, but the stress-strain loops become stable after 20 cycles and no significant modifications are observed with further increasing the number of mechanical cycles.

Figure 3(b) shows the modification of the isothermal stress-strain hysteresis loops under different values of the temperature for $T > A_f$, in the temperature range between 303 K and 328 K.

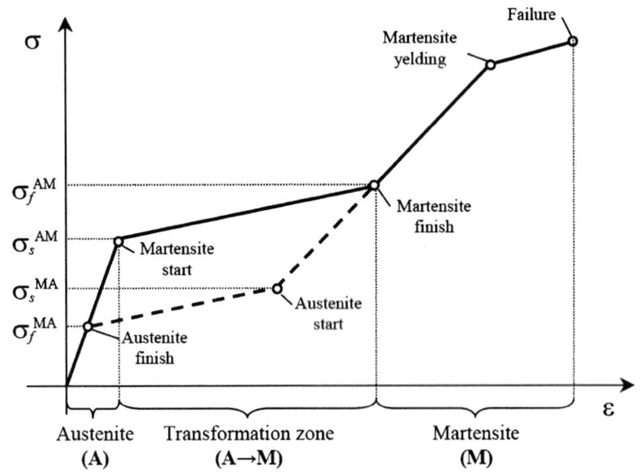

Fig. 1 Typical stress-strain relation of an austenitic NiTi alloy

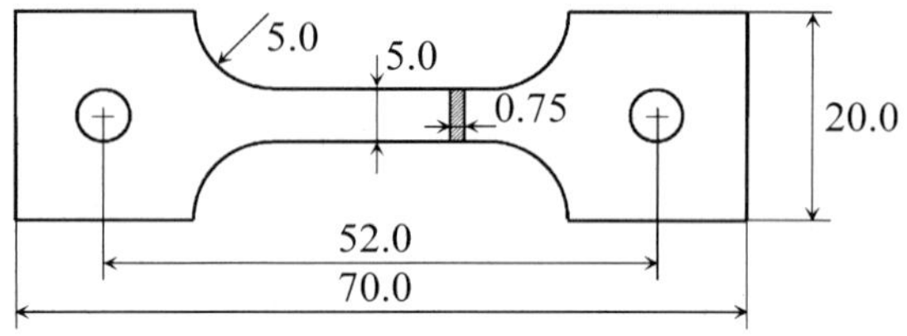

Fig. 2 Test specimen

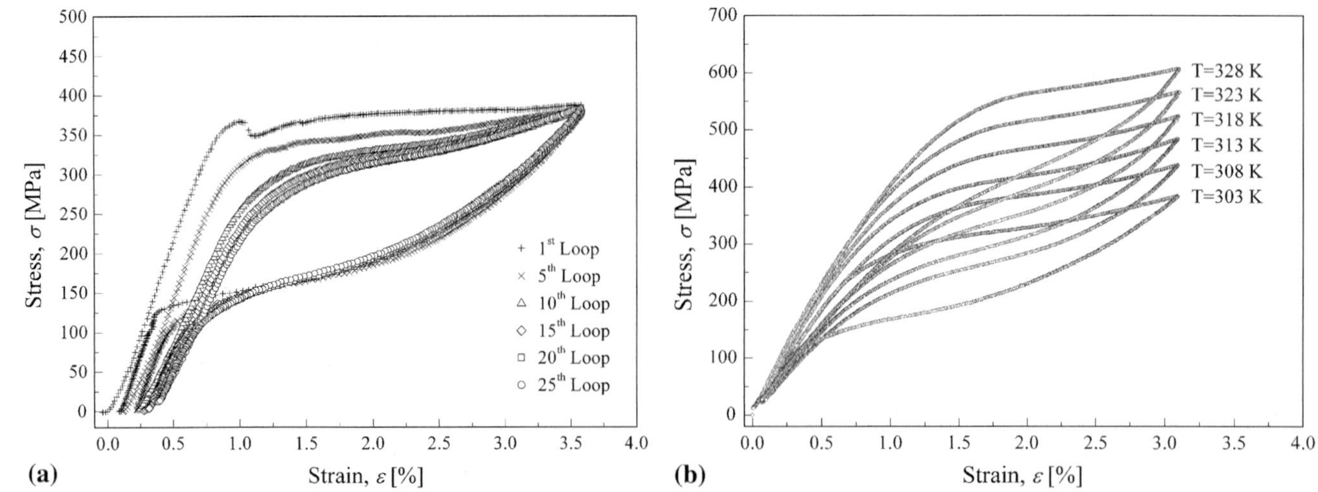

Fig. 3 Isothermal stress-strain hysteretic behavior: (a) evolution in the first 25 cycles at $T = 303$ K; (b) for different values of the temperature (303-328 K)

These results were used to measure the constants of the Clausius-Clapeyron relation:

$$\frac{d\sigma}{dT} = \text{const.} \quad \text{(Eq 1)}$$

which define the change in the stress for the onset of martensite during loading and the stress to obtain austenite during unloading.

Let C^{AM} be the constant governing the austenite-to-martensite transformation, then the critical stress values can be obtained as follows:

$$\begin{cases} \sigma_s^{AM} = \sigma_{s0}^{AM} + C^{AM}(T - T_0) \\ \sigma_f^{AM} = \sigma_{f0}^{AM} + C^{AM}(T - T_0) \end{cases}, \quad \text{(Eq 2)}$$

in which the subscript 0 is referred to the quantities measured at the reference temperature ($T_0 = 303$ K). In Table 1, the values of C^{AM}, σ_{s0}^{AM}, and σ_{f0}^{AM} for the investigated NiTi alloy, together with the Young's moduli and the Poisson's ratios, are reported.

3. Numerical Model

A SEC specimen was analyzed by 2D plane stress FE analyses carried out with the commercial finite element code MSC Marc®. The geometry and the dimensions of the SEC specimen are illustrated in Fig. 4, while Fig. 5 shows the corresponding FE model.

Due to the symmetry of the SEC specimen, only the upper half of the whole specimen was modeled. The model consists of 5629 eight-node bi-quadratic plane stress quadrilateral elements. A nonuniform mesh was introduced by employing a relatively fine mesh in the crack tip region, and a coarse mesh elsewhere, as clearly shown in Fig. 5.

The tensile load is applied on the upper half of the hole. As previously explained, unloading and the corresponding reverse transformation processes, from martensite to austenite (M → A), are not taken into account.

The isothermal stress-strain curves of the NiTi alloy, used in the numerical model, have been obtained for any temperature from data in Table 1, and by applying Eq 2; the resulting piecewise linear-stress relations are shown in Fig. 6, in which ε_L was fixed to 7%. Furthermore, the curves in Fig. 6 were truncated at a maximum stress equal to 1000 MPa, i.e., the yielding of martensite was not modeled, as the size of the transformation zone in the crack tip region is expected to be much greater than the plastic zone of the martensitic structure.

The martensite volume fraction, ξ_M, during the SIM, can be described by the equivalent plastic strain ε_{eq}^p (Ref 13). Indeed, the yielding condition for a conventional material is:

$$\sigma_{eq} = Y_p\left(\varepsilon_{eq}^p\right), \quad \text{(Eq 3)}$$

while the phase transformation condition in a SMA is:

$$\sigma_{eq} = Y_M(\xi_M), \quad \text{(Eq 4)}$$

in which σ_{eq} is the von Mises equivalent stress, while Y_p and Y_M are the plastic and transformation hardening function, respectively. By comparing Eqs 3 and 4, it emerges that the

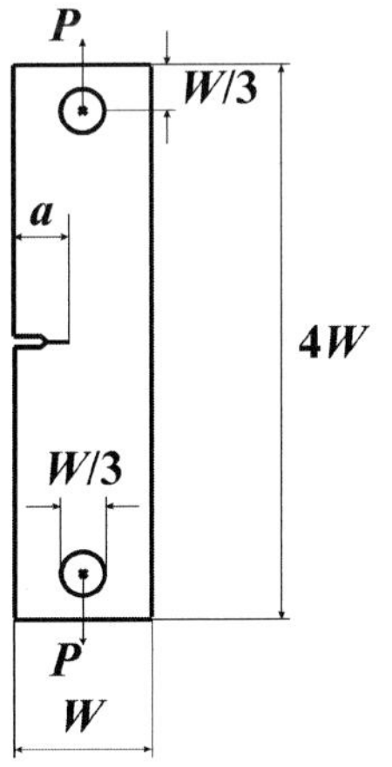

Dimensions [mm]	
Width, W	17,60
Crack length, a	6,80
Thickness, t	1,00

Fig. 4 SEC specimen geometry and dimensions

Table 1 Thermo-mechanical parameters of a Ni-49 at.% Ti alloy

E_A, MPa	ν_A	E_M, MPa	ν_M	σ_{s0}^{AM}, MPa	σ_{f0}^{AM}, MPa	C^{AM}, MPa K^{-1}
39000	0.3	20000	0.3	260	618.4	10.3

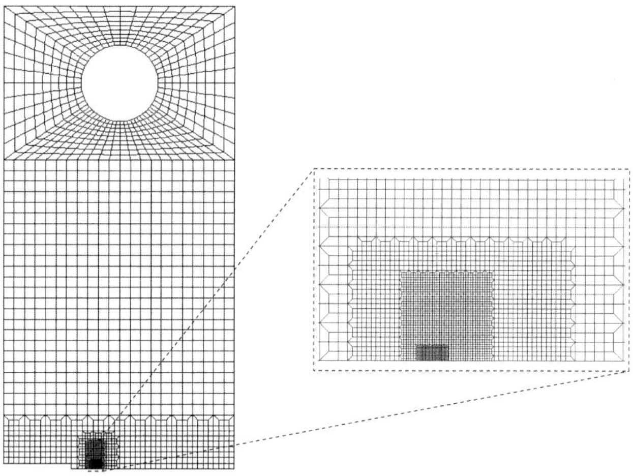

Fig. 5 FE model of the SEC specimen

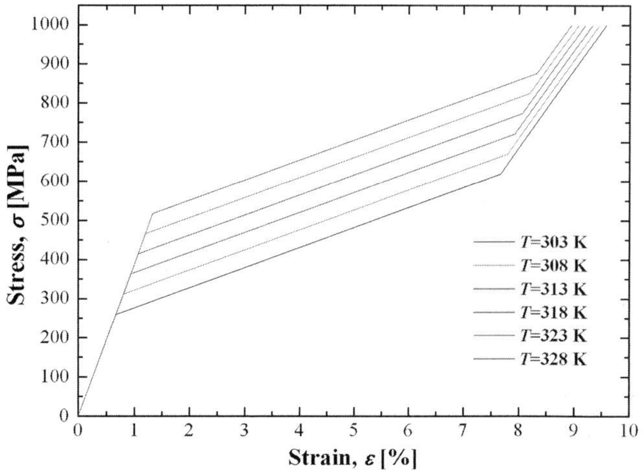

Fig. 6 Piecewise linear stress-strain relation of the NiTi alloy at several temperatures

transformation condition in SMAs is similar to the plastic yield condition in common metals, thus a simple linear relation between ξ_M and ε_{eq}^p can be assumed (Ref 15):

$$\xi_M = C\varepsilon_{eq}^p, \quad \text{(Eq 5)}$$

where C is a constant, which can be easily obtained from the uniaxial stress-strain relation of the SMA. In fact, at the beginning of phase transformation, the martensitic fraction and the equivalent plastic strain are zero, while at its end the martensitic fraction is 1 and the equivalent plastic strain is given by:

$$\varepsilon_{eq}^p = \varepsilon_{eq} - \varepsilon_{eq}^{el} = \varepsilon_{eq} - \frac{\sigma_{eq}}{E_A}, \quad \text{(Eq 6)}$$

in which ε_{eq} and ε_{eq}^{el} are the total and elastic equivalent strain, respectively. For the piecewise linear stress-strain relations used herein, ε_L and C^{AM} are constant, thus the equivalent plastic strain value at the end of the transformation is also constant and equals to 6.08×10^{-2}. By substituting these values into Eq 5, it follows:

$$C = \frac{\xi_M}{\varepsilon_{eq}^p} = \frac{1}{6.08 \times 10^{-2}} \approx 16.44. \quad \text{(Eq 7)}$$

Therefore, the distributions of martensite volume fraction ξ_M in front of the crack tip can be obtained by the distributions of the equivalent plastic strain ε_e^p through the following equation:

$$\xi_M = 16.44\varepsilon_{eq}^p. \quad \text{(Eq 8)}$$

4. Analytical Model

As it is well known, the normal stress σ_y, for $\theta = 0$, given by the linear elastic fracture mechanics (LEFMs) in the case of mode I loading condition is given by:

$$\sigma_y = \frac{K_I}{\sqrt{2\pi r}}, \quad \text{(Eq 9)}$$

in which r is the radial distance from the crack tip and K_I is the mode I stress intensity factor (SIF). For the SEC specimen in Fig. 4 (Ref 16), the SIF can be evaluated as follows:

$$K_I = \frac{P}{t\sqrt{W}}\sqrt{7.59\frac{a}{W} - 32\left(\frac{a}{W}\right)^2 + 117\left(\frac{a}{W}\right)^3}. \quad \text{(Eq 10)}$$

Equation 9 shows that the elastic stress would become very large in the proximity of the crack tip, but it does not occur due to SIM, which causes a different stress distribution; a similar behavior is observed in ductile metals due to the nonlinear effects of the plastic deformations. Irwin (Ref 17) has shown that when yielding occurs at the crack tip in common metals, LEFM may still be applied if an effective crack length is used, i.e., a physical crack length plus an allowance for the extent of the plastic zone:

$$a_{eff} = a + r_y. \quad \text{(Eq 11)}$$

where r_y can be expressed as a function of the size of the plastic zone r_p, as illustrated in Fig. 7; for plane stress conditions is given by:

$$r_y = \frac{r_p}{2} = \frac{1}{2\pi}\left(\frac{K_I}{S_y}\right)^2 \quad \text{(Eq 12)}$$

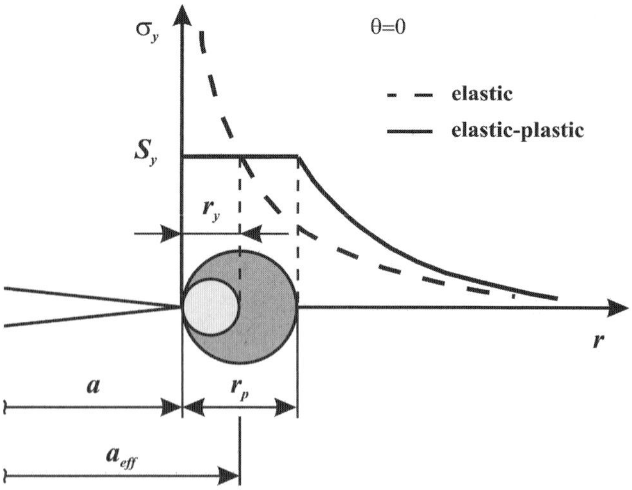

Fig. 7 Stress distribution in the crack tip region of a ductile metal

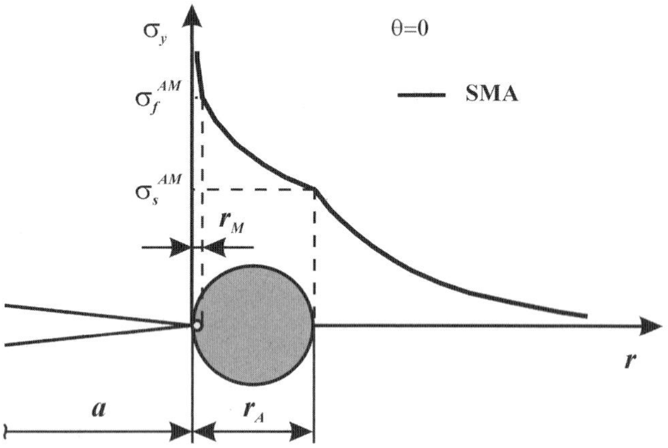

Fig. 8 Stress distribution in the crack tip region of an austenitic NiTi alloy

where S_y represents the yielding strength in ductile metals. Thus, the stress distribution in the crack tip region can be determined by the following equation:

$$\sigma_y = \frac{K_{Ieff}}{\sqrt{2\pi(r - r_y)}} \quad \text{(Eq 13)}$$

where K_{Ieff} is the effective SIF obtained by using $a_{eff} = a + r_y$ in Eq 10.

In Fig. 8, a schematic representation of the stress distribution in the crack tip region of an austenitic NiTi alloy is shown. The comparison between Fig. 7 and 8 shows that Eq 12 cannot be used to predict the size of the transformation zone, as a consequence of the different stress distributions in the crack tip region for ductile metals and NiTi alloys, since the equilibrium conditions are no more satisfied.

However, by assuming a linear evolution of the stress in the transformation zone, i.e., between σ_s^{AM} and σ_f^{AM}, and by neglecting the size of the martensitic zone, as $r_M << r_A$, the equilibrium condition can be rewritten as follows:

$$\frac{\sigma_s^{AM} + \sigma_f^{AM}}{2} r_A = \int_0^{r'_A} \sigma_y^{el} dr \quad \text{(Eq 14)}$$

where σ_y^{el} represents the stress distribution in the crack tip region of a linear elastic material (Eq 9), while r'_A is the extent of the transformation start contour, obtained by applying the condition $\sigma_y^{el} = \sigma_s^{AM}$:

$$r'_A = \frac{1}{2\pi}\left(\frac{K_I}{\sigma_s^{AM}}\right)^2 \quad \text{(Eq 15)}$$

By substituting Eq 9 in Eq 14 and taking into account Eq 15, the extent of the transformation start contour is given by:

$$r_A = \frac{2}{\pi}\frac{K_I^2}{\sigma_s^{AM}\left(\sigma_s^{AM} + \sigma_f^{AM}\right)}. \quad \text{(Eq 16)}$$

It is worth noting that in the case of elastic plastic material ($\sigma_s^{AM} = \sigma_f^{AM} = S_y$), the previous equation gives the extent of the plastic zone in common metals (Eq 12).

Based on the previous consideration, the Irwin correction can be applied to predict the stress distribution in the crack tip region of a NiTi alloy, by using the effective crack length:

$$a_{eff} = a + \frac{r_A}{2}. \quad \text{(Eq 17)}$$

Furthermore, Eq 13 gives the stress distribution in the crack tip region of a SMA by substituting $r_y = r_A/2$.

5. Results and Discussions

5.1 FE Model Validation

The FE model was initially validated by comparing the constitutive relation with the numerical results, in terms of the equivalent stress and strain ($\sigma_{eq} - \varepsilon_{eq}$), obtained in the crack tip region for a tensile load $P = 360$ N, which corresponds to a remote stress $\sigma^\infty = P/(tW) = 20.45$ MPa.

The comparisons between constitutive relation and numerical results were carried out for several values of the temperature, and a good agreement was observed in all cases, as illustrated in Fig. 9 which illustrates the results for $T = 303$ K.

Figure 10 illustrates stress contours for the transformation start ($\sigma_{eq} = \sigma_s^{AM}$) and the transformation finish ($\sigma_{eq} = \sigma_f^{AM}$) conditions, together with the maximum stress $\sigma_{eq} = 1000$ MPa

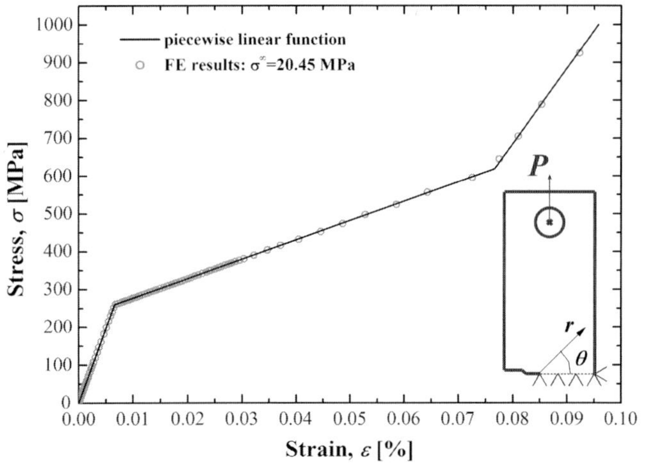

Fig. 9 Piecewise linear constitutive relation vs. FEM results ($T =$ 303 K)

(see Fig. 6); as expected, the figure clearly shows that the size of the transformation zone is much greater than the zone limited by the condition $\sigma_{eq} = 1000$ MPa.

5.2 FE Results

By using Eq 8, the martensite volume fraction, ξ_M, in the crack tip region was computed for several values of the temperature in the range between 303 and 328 K; as an example, Fig. 11 shows the contours with constant ξ_M values at $T = 303$ K. The martensite volume fraction ξ_M in the partial martensite zone decreases with increasing the distance from the crack tip. Similar results have been obtained for all the investigated temperatures.

The transformation zone was analyzed by the distances from the crack tip and for $\theta = 0$ (see Fig. 9), which correspond to the transformation start (r_A) and finish (r_M) condition. In Fig. 12, the trends of r_A and r_M versus temperature are shown. The transformation start size rapidly decreases with increasing the temperature, while the transformation finish size is almost constant (it slightly decreases), resulting in an overall reduction of the transformation zone, which is represented by the gapping area between the curves in Fig. 12.

5.3 Analytical Predictions Versus Numerical Results

In Table 2, the numerical values of the martensite start region for different values of temperature are reported, together with the analytical predictions by LEFM and modified LEFM. The same results are schematically reported in Fig. 13.

The analytic approach underestimates the martensite start size for low values of the temperature while an overestimation is observed increasing the test temperature, giving relative differences of about 10%. These effects are due to the assumption $r_M \ll r_A$, taken to rewrite the equilibrium condition at the crack tip in the NiTi alloy, since the martensite finish radius, r_M, is not negligible respect to r_A with increasing the temperature, as shown in Fig. 12.

In Fig. 14, the numerical normal tension, σ_y, has been plotted versus the radial distance from the crack tip, r, for $T = 303$ K, in the stress-singularity region ($r < 2$ mm) in a bi-logarithmic plot. Increasing the distance from the crack tip, two knees are clearly visible, corresponding to the martensite finish and start, respectively; then, it is possible to identify in the graph the martensitic (M), the transformation (A → M) and the austenitic (A) regions. Using ordinary least square, the equation of the linear region of the austenitic zone was evaluated; in particular, the slope is −0.5, which corresponds to

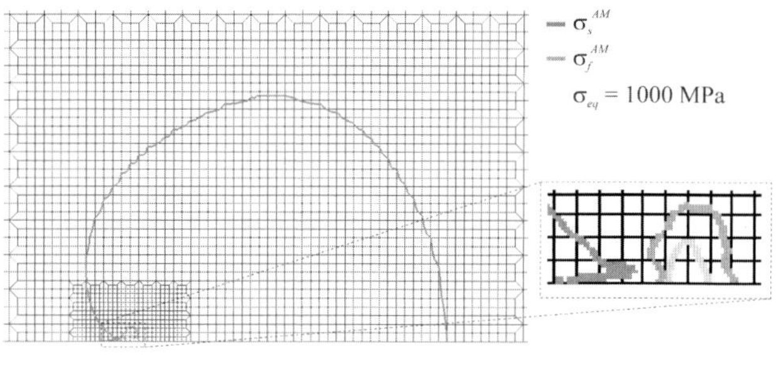

Fig. 10 Stress contours in the crack tip region: martensite start (*red*), martensite finish (*orange*) and $\sigma_{eq} = 1000$ MPa (*yellow*) ($T = 303$ K)

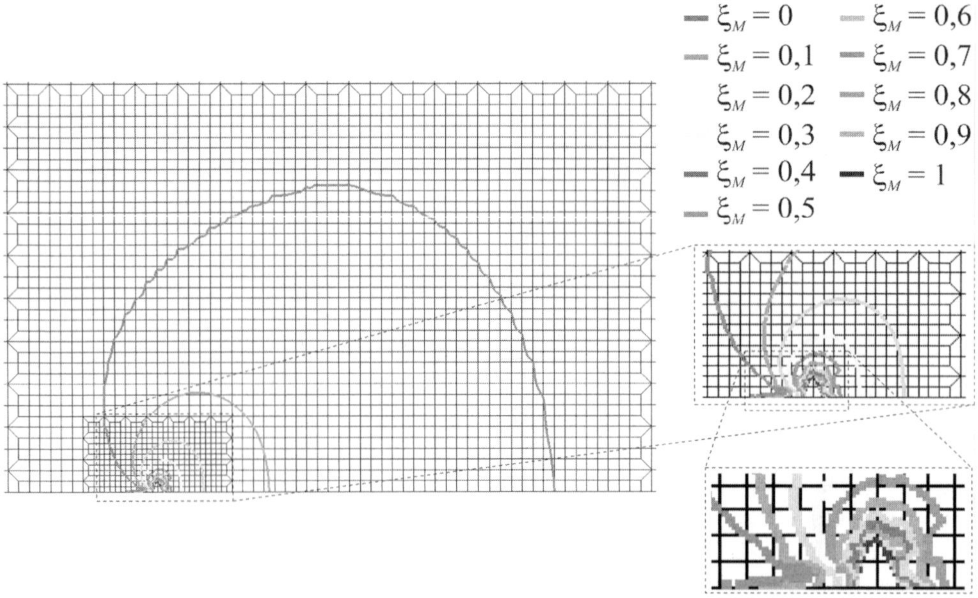

Fig. 11 Contour lines of constant martensite volume fraction ξ_M ($T = 303$ K)

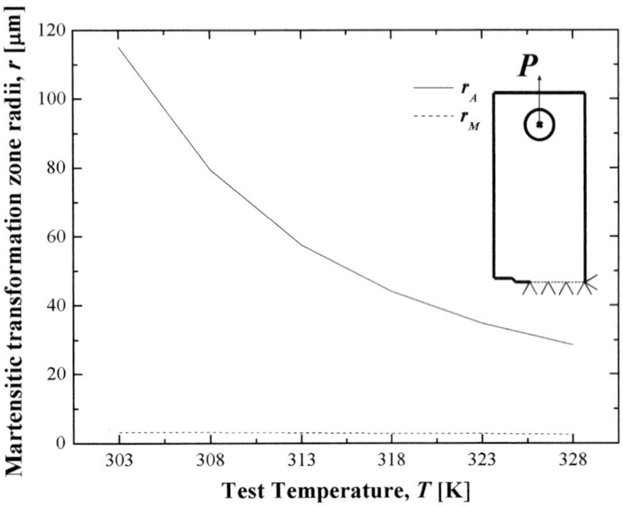

Fig. 12 Martensitic transformation zone vs. temperature, T

Table 2 Martensite start size, r_A, for different temperatures: LEFM, modified LEFM, and numerical results

	r_A, μm		
T, K	LEFM	Modified LEFM	Numeric
303	176.2	102.8	115.1
308	121.4	76.4	79.5
313	88.8	59.1	57.5
318	67.8	47.1	44.1
323	53.5	38.5	34.8
328	43.3	32.1	28.6

the theoretical value of the LEFM. Thus, the Irwin correction (Eq 13) could be used to describe the stress distribution in a SMA for $r > r_A$, i.e., in the austenitic region, but not within the phase transformation zone.

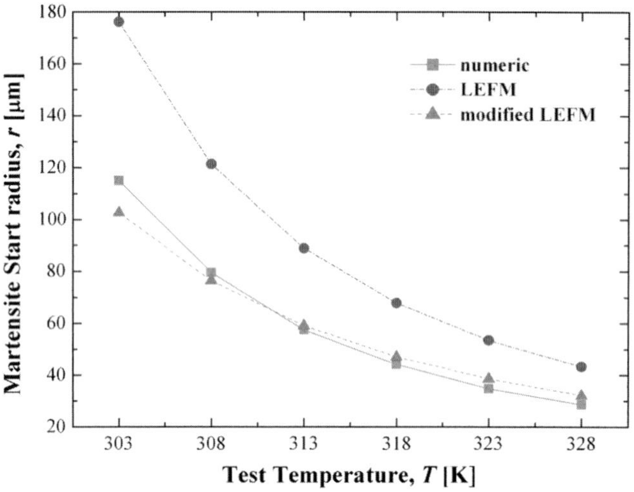

Fig. 13 Martensite start size, r_A, vs. temperature: LEFM, modified LEFM, and numerical results

In Fig. 15, a comparison between the numerical results and analytical predictions in terms of normal stress is presented for the reference temperature, $T = 303$ K. The figure shows similar trends between the curves; in particular, the LEFM overestimates the normal stress σ_y, while satisfactory agreement between the modified LEFM and numerical results has been obtained, with relative difference $<10\%$. Notwithstanding these differences, the results indicate that the modified Irwin correction of LEFM could be used to describe the stress distribution in the austenitic region but a proper equation should be found to estimate the effective crack length. In fact, the gap between the curves in Fig. 15 can be reduced by moving the vertical asymptote of the Irwin curve, i.e., by changing the value of r_y in Eq 13. Further studies should be carried out in order to define a modified relation for the effective crack length, which takes into account the whole stress-strain curve of the material, i.e., by considering the stress

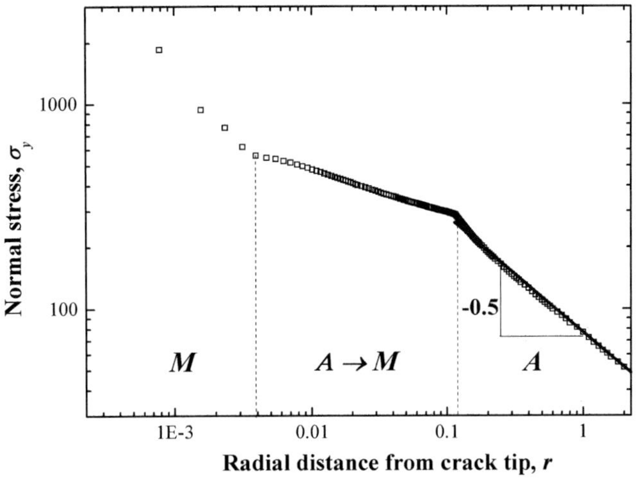

Fig. 14 Normal stress, σ_y, vs. radial distance from the crack tip, r ($T = 303$ K)

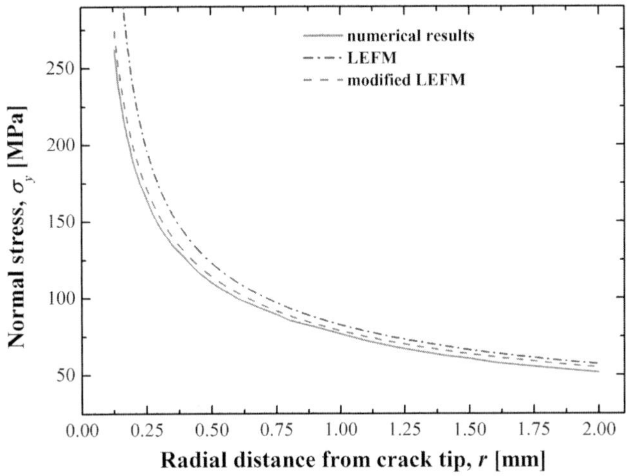

Fig. 15 Normal stress, σ_y, distribution in the austenitic region ($T = 303$ K)

distribution in the transformation zone as well as in the full martensitic region.

6. Conclusions and Perspectives

In the present work, the effects of the temperature on the stress-strain distributions and on the martensitic transformation zone in the crack tip region of a SEC specimen in NiTi alloy were analyzed by 2D numerical simulations. In particular, the martensite start and finish contours were evaluated from the numerical results. It was demonstrated that the distributions of martensite volume fraction, ξ_M, in front of the crack tip can be obtained by the distributions of the equivalent plastic strain. The numerical results, presented herein by contour lines, show that the martensite volume fraction ξ_M in the partial martensite zone gradually decreases with increasing the radial distance from the crack tip. Similar results have been obtained for all the investigated temperatures. Moreover, the numerical results show that the size of the transformation start rapidly decreases with increasing the temperature while the finish one is almost constant, resulting in an overall reduction of the transformation zone with increasing the temperature. Finally, comparisons between numerical and analytical results, obtained by Irwin's modified LEFMs relations, were carried out. These comparisons show that a satisfactory agreement in terms of the martensite start and finish sizes is obtained; moreover, the analytical approach could be able to describe the stress field in the crack tip region outside the phase transformation zone, i.e., in the austenitic region, but a proper equation to estimate the effective crack length should be found. To this aim, further studies should be carried out.

References

1. K. Otsuka and X. Ren, Physical Metallurgy of Ti-Ni-Based Shape Memory Alloys, *Prog. Mat. Sci.*, 2005, **50**, p 511–678
2. A. Paiva and M.A. Savi, An Overview of Constitutive for Shape Memory Alloys, *Math. Prob. Eng.*, 2006, **2006**, Art. no. 56876, p 1–30
3. G. Eggeler, E. Hornbogen, A. Yawnyl, A. Heckmann, and M. Wagner, Structural and Functional Fatigue of NiTi Shape Memory Alloys, *Mater. Sci. Eng. A*, 2004, **378**, p 24–33
4. M. Wagner, T. Sawaguchi, G. Kausträter, D. Höffken, and G. Eggeler, Structural Fatigue of Pseudoelastic NiTi Shape Memory Wires, *Mater. Sci. Eng. A*, 2004, **378**, p 105–109
5. G.B. Rao, J.Q. Wang, E.H. Han, and W. Ke, Study of Residual Stress Accumulation in TiNi Shape Memory Alloy During Fatigue Using EBSD Technique, *Mater. Lett.*, 2006, **60**, p 779–782
6. M. Guiomar de Azevedo Bahia, R. Fonseca Dias, and V. Tadeu Lopes Buono, The Influence of High Amplitude Cyclic Straining on the Behaviour of Superelastic NiTi, *Int. J. Fatig.*, 2006, **28**, p 1087–1091
7. S. Yi and S. Gao, Fracture Toughening Mechanism of Shape Memory Alloys due to Martensite Transformation, *Int. J. Solids Struct.*, 2000, **37**, p 5315–5327
8. S. Yi, S. Gao, and L. Shen, Fracture Toughening Mechanism of Shape Memory Alloys Under Mixed-Mode Loading due to Martensitic Transformation, *Int. J. Solids Struct.*, 2001, **38**, p 4463–4476
9. K. Gall, N. Yang, H. Sehitoglu, and Y.I. Chumlyakov, Fracture of Precipitated NiTi Shape Memory Alloys, *Int. J. Fract.*, 2001, **109**, p 189–207
10. G.M. Loughran, T.W. Shield, and P.H. Leo, Fracture of Shape Memory CuAlNi Single Crystals, *Int. J. Solids Struct.*, 2003, **40**, p 271–294
11. J.H. Chen, W. Sun, and G.Z. Wang, Investigation on the Fracture Behavior of Shape Memory Alloy NiTi, *Metall. Mater. Trans. A*, 2005, **36**, p 941–955
12. G.Z. Wang, Effects of Notch Geometry on Stress-Strain Distribution, Martensite Transformation and Fracture Behavior in Shape Memory Alloy NiTi, *Mat. Sci. Eng. A*, 2006, **434**, p 269–279
13. G.Z. Wang, A Finite Element Analysis of Evolution of Stress-Strain and Martensite Transformation in Front of a Notch in Shape Memory Alloy NiTi, *Mat. Sci. Eng. A*, 2007, **460-461**, p 383–391
14. C. Maletta, A. Falvo, F. Furgiuele, and J.N. Reddy, A Phenomenological Model for Superelasticity in NiTi Alloys, *Smart Mater. Struct.*, 2009, **18**, Art. no. 025005
15. Y. Jung, P. Papadopoulos, and R.O. Ritchie, Constitutive Modelling and Numerical Simulation of Multivariant Phase Transformation in Superelastic Shape-Memory Alloys, *Int. J. Numer. Meth. Eng.*, 2004, **60**, p 429–460
16. P.P. Milella, La MFLE nella pratica sperimentale, *Meccanica della Frattura Lineare Elastica ed Elastoplastica*, Chap. 4, Ansaldo Nucleare Editore, Ed., 1999, Italia, Genova (in italian)
17. G.R. Irwin, Plastic Zone Near a Crack and Fracture Toughness, *Proceedings of 7th Sagamore Ordnance Materials Conference*, 1960 (Syracuse, NY), Syracuse University Press, 1960, p 63–78

JMEPEG (2009) 18:686–690
DOI: 10.1007/s11665-009-9424-8

1059-9495/$19.00

Simulation-Based Design of a Rotatory SMA Drive

Sascha Dilthey and Horst Meier

(Submitted September 15, 2008; in revised form January 28, 2009)

The design and optimization of a rotatory drive powered by shape memory alloy (SMA) actuators is described in this paper. SMA actuators used in technical applications are parameterized by the use of trial-and-error methods, because there is a lack of computer-aided design tools for this active material. A numerical modeling approach was developed to design and optimize the geometry and the load and heating conditions of SMA actuators in a technical system to achieve a good dynamic and a high reliability. The shape memory effect used in most technical systems is the extrinsic two way effect (2WE). This effect can be simulated with the numerical model which was implemented in MATLAB/SIMULINK. The focus of the model is on the activation behavior of the SMA actuator, which defines its rate of heating and cooling. Different load conditions and various actuator geometries and shapes, e.g. wire or spring actuator, are simulated by the calculation of the energetic balance of the whole system. The numerical model can be used to simulate time variant heating currents in order to obtain an optimal system performance. The model was used to design a rotatory SMA-drive system, which is based on the moving concept of a wave drive gear set. In contrast to the conventional system, which is driven by an electric motor, the SMA drive consists of a strain wave gear and SMA wire actuators that are applied circularly to generate a rotatory movement. Special characteristics of this drive system are a high torque density and a high positioning accuracy.

Keywords mechanical testing, modeling processes, superalloys

1. Introduction

Actuators are always series of energy providers and energy transducers. An actuator can be considered a 'new actuator' when its function is essentially based on physical characteristics of new transducer materials [e.g. shape memory alloys (SMAs)] (Ref 1). SMAs show a characteristic property by which the metal remembers a high-temperature shape and reverts to it as a result of a phase transformation between the low temperature phase (martensite, α) and the high temperature phase (austenite, β). This transformation process can be utilized to generate the work load (Ref 2). The characteristic transformation temperatures M_s, M_f, A_s, A_f (where M is martensite, A is austenite, s is start, and f is finish) represent the start and finish temperatures of this phase transformation. The phase change between the two solid phases involves a rearrangement of atoms within the crystal lattice. The fraction of martensite and austenite in the SMA device depends mainly on its actual temperature, its stress condition, and the transformation temperatures which are determined by the chemical alloy composition of the SM material. The results of research presented in this paper are all based on the use of binary nickel titanium (NiTi) as SM alloy.

The fundamental features of shape memory actuators are (i) the high energy density (characterizing the work load to volume ratio) and (ii) the high reversible strain when exploiting the one way effect (1WE). This makes SMAs unique compared to other smart materials that can be used for actuator applications. Cyclic actuator movements of SMAs are often realized by what is referred to as an "extrinsic two way effect" (2WE). The shape memory effect only governs one direction, i.e. the SMA device remembers its shape as it transforms from the low-temperature to the high-temperature phase whereas the backward movement on cooling (which transforms the device back into α) results from an external resetting force, e.g. a biasing spring. Experiments show that extrinsic two-way effects are superior in engineering applications to the so-called intrinsic two-way effect which suffers from being difficult to achieve by thermo-mechanical training, from a lack of stability and from only small exploitable strain amplitudes. In order to use SMAs as switching transducer materials, particularly SMA wires can easily be heated with electric current by the use of their electrical resistance. As the time response of SMA actuators is limited by the velocity in which heat can flow in and out of the material, small devices like thin wires outperform thick section components in this respect.

Although the exceeding properties of SMAs are already known, these materials could not be established in R&D processes so far. A number of problems have limited their success on the market which is partly due to a lack of generally accepted design procedures and difficulties in manufacturing suitable components. In this paper, the design and optimization of an SMA actuator in a technical application based on a numerical simulation of the activation behavior is presented. The simulation is validated experimentally and used exemplary for a SMA powered Rotatory Drive System.

This article is an invited paper selected from presentations at Shape Memory and Superelastic Technologies 2008, held September 21-25, 2008, in Stresa, Italy, and has been expanded from the original presentation.

Sascha Dilthey, Chair of Production Systems, Mechanical Engineering, Universitaetsstr. 150, IB 2/126, Bochum 44780, Germany; and **Horst Meier**, Ruhr-University Bochum, Mechanical Engineering, Universitätsstr. 150, LPS, IB2/126, Bochum 44801, Germany. Contact e-mail: dilthey@lps.rub.de.

2. Numerical Simulation of Shape Memory Actuators

The activation of SMAs, defined as the stroke-time-relation of the actuator within transformation, is solely specified by the velocity of cooling and heating the material. Many factors determine the dynamic behavior of SMA actuators: the chemical alloy composition, the actuator's geometry, the position and width of the hysteresis, the electrical heating current (DC) I and its cycle I(t), the mechanical load and the number of cycles as well as the environmental conditions. Especially with time variable current profiles and also variable load conditions, these parameters cannot or only inadequately be described with simplifying algebraic functions. In order to characterize the activation behavior of a SMA actuator system, all time variant and invariant energy fluxes that pass the system border have to be balanced. This energy balance of the whole system for a straight SMA wire with a bias spring is shown in Fig. 1.

The only energy source during the heating phase is an external electric current that heats the wire by using its ohmic resistance. Regarding the heat transmission for the cooling process of the actuator, three effects are distinguished in thermodynamics: convection, thermal radiation, and thermal conduction (Ref 3). The heat transmission caused by free air convection is about 90% and by heat radiation about 10% of the total emission.

The low influence of heat conduction due to the clamping of the wire can be neglected. Similarly to the consideration of the emitted heat flow, the required mechanical energy, resulting from the returning device and external loads as well as the integral latent heat for transformation from martensite to austenite has to be quantified.

The latent heat for transformation is considered proportional to the derivation of the volumetric martensite fraction ξ. Its value for the applied SMA actuators was specified by a differential scanning calorimetry (DSC) measurement to about 24000 J/kg. According to the first law of thermodynamics, the caloric state equation for the actuator's inner energy can be described for heating and cooling by differential equations.

These non-linear first order differential equations cannot be solved definitely with analytic methods. Therefore, the solution curve for one unknown parameter, as the wire temperature T_w for instance, has to be calculated numerically by using an approximation procedure. Besides the solvability of the problem at all, the main advantage of using a numerical method is the possibility of implementing time-variable parameters that depend on the mixture between martensite and austenite and thus have an impact on the phase transformation. These mixture-dependent parameters are the convective heat-transfer coefficient, the thermal expansion, the elastic elongation, the geometry and the ohmic resistance of the SMA actuator. Thereby the quality of the generated model can be enhanced.

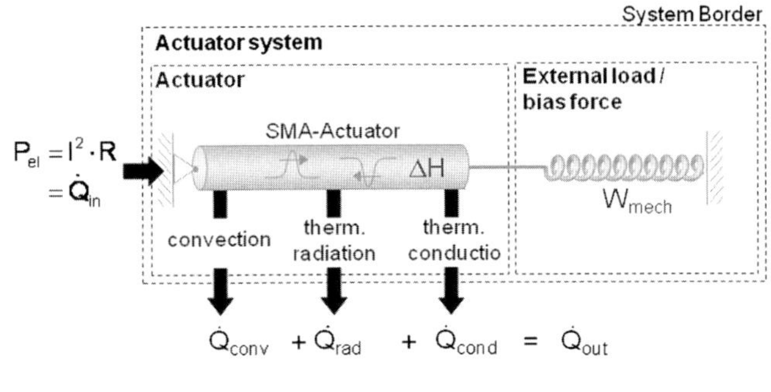

Fig. 1 Energy balance of a SMA actuator system

All these parameters are influenced by the geometric shape of an actuator. The most common SMA actuator shape is a thin wire. For this shape the stress behavior under load conditions can easily be described, as well as the change of the geometry and the ohmic resistance. For some applications SMA spring actuators are used. The behavior of a spring-shaped actuator is different to a wire actuator. The convective heat-transfer coefficient is temperature dependent, on the one hand, but on the other hand, the shape generates interdependencies between the sections of the spring which have a huge influence on the numerical model. A complete uncompress spring behaves like a wire-shaped actuator whereas a fully compressed spring can be described as a hollow cylinder. Between these two states of the spring a linear transformation is assumed. The stress condition differs as well. A wire actuator is loaded by axial stress only, which is constant over the cross section. A spring actuator shows a completely different condition with shear stress and a cross section related increased maximal stress that can be described with a known parameter k.

For the analytic description of the SMA's characteristics during phase transformation, an interconnection with a hysteresis model that has to be parameterized by the input of specific actuator attributes is necessary. For this purpose a simple hysteresis model from Liang and Roger (Ref 4) that segmentally defines the characteristic hysteresis curve of a SMA with trigonometric functions in the different temperature ranges is implemented into the numerical model. With an extension of this hysteresis model it is possible to describe not only the outer hysteresis routes, but also inner hysteresis route (Ref 5). On these routes the actuator is not heated up to or over the temperature where the transformation into austenite is completed or the heating of the actuator is started before the back transformation into the martensitic phase is finished. These activation strategies can be used to increase the dynamic of the actuator system.

3. Numerical Model

The computer-aided computation of the differential equations and the relevant process parameters in order to simulate the time response of a SMA actuator during phase transformation is implemented in MATLAB/SIMULINK. The complete SMA actuator system was built in SIMULINK (Fig. 2) as a block diagram consisting of data sources and sinks, algebraic functions, logic operations and customized subsystems. Important inputs for the SIMULINK model (e.g. length and diameter and number of coils of the actuator, constant of the opposed bias spring(s), environmental conditions) are separately combined in a MATLAB script for easy parameterization. By using virtual switches (upper left) the heating current can be defined (constant value, sine wave, ramp, arbitrary sequence…). The energy sinks (convection, radiation, latent heat for transformation, mechanical energy) are defined in subsystems and added to the model. In the loop shown, the wire's temperature, the mechanical stress, the transformation temperatures depending on the stress and the martensite fraction are calculated. The change in the ohmic resistance of the actuator, as a

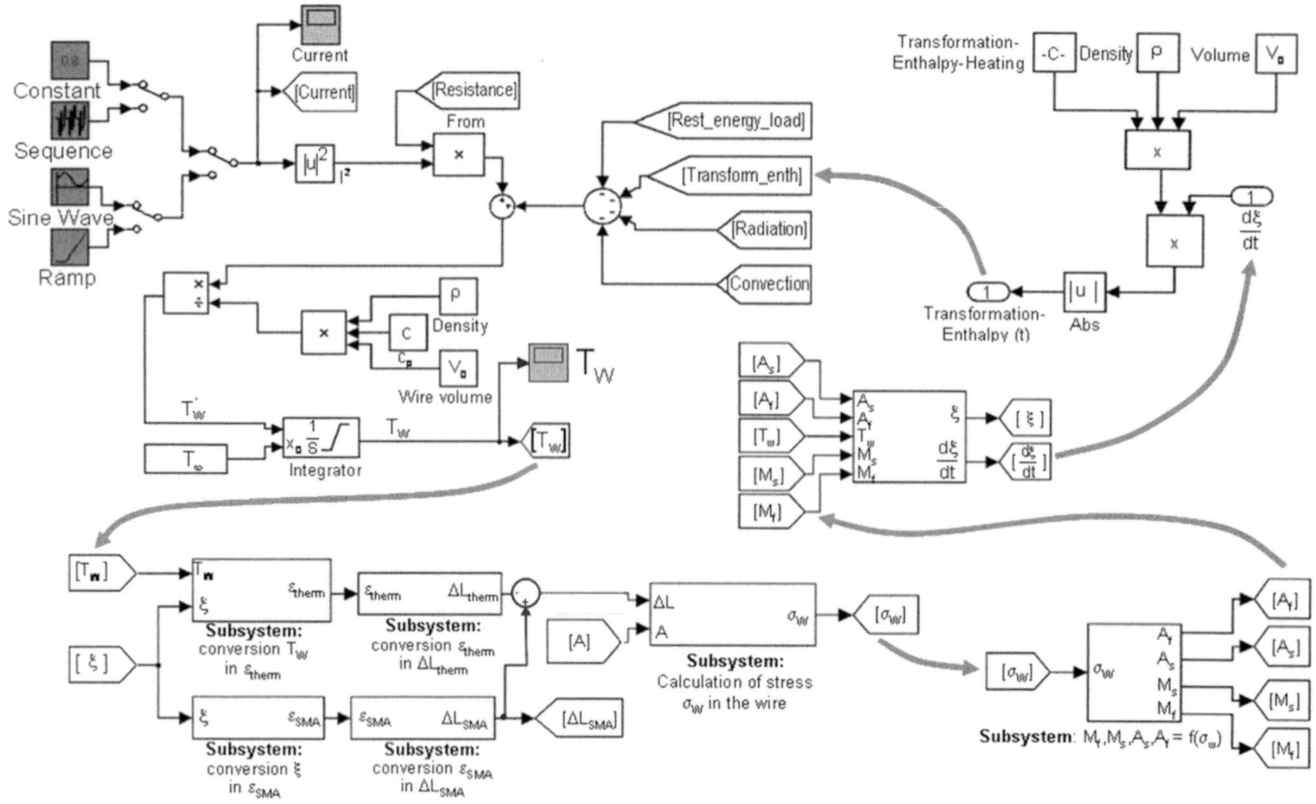

Fig. 2 Simulink flowchart: heating stage

function of the temperature and the geometry of the wire, is calculated in the subsystem *Resistance*. These values—each interdependent—can be visualized by the use of scopes that can be placed at any position in the model.

A comparable block diagram which can be parameterized by the input of the initial values for the start temperature TW_start and the start martensite fraction ξ_start was generated for the cooling phase.

This model was designed to simulate the time behavior of a SMA actuator, depending on the boundary conditions, like the bias force and the heating current. Within the model the heating current, the stress depending transformation temperatures (M_f, M_s, A_s, A_f) and the wire temperature TW are calculated. Using the hysteresis model the current martensite fraction ξ during phase transformation is computed. With the martensite fraction ξ the current stroke of the actuator ΔlW is calculated.

4. Experimental Validation and Simulation Results

In order to validate the SIMULINK model, an SMA actuator is analyzed experimentally and compared numerically. For simulation and experiment, the applied load, the heating current profile, and the boundary conditions were the same.

An experimental setup has been developed, shown in Fig. 3. It was built to analyze the time response of straight SMA actuator wires or spring actuators. The SMA actuator (5) is clamped between a force transducer (6) on the one side and an aerostatic motion slide (3) that is connected with the bias load (1) and a measuring plate for the ultrasonic displacement transducer (7) on the other side. The programmable power

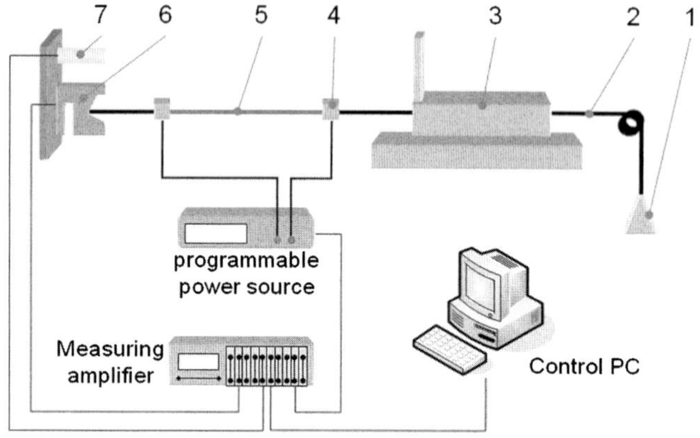

Fig. 3 Experimental setup for a constant load

source enables the generation of a totally arbitrary heating current that is switched over to the actuator with a relay.

Figure 4, showing an uncompleted cycle of only a third of the maximum stroke with the heating stage on the left and the cooling stage on the right, confirms the physically good implementation of the actuator's properties and its behavior in the SIMULINK model exemplary. The used NiTi wire actuator came from Memory Metalle, Allow Dy90, and had a diameter of 0.25 mm with a length of 800 mm. A constant bias force of 5 N was applied and the heating current was set to 0.6 A.

It is evident that the computed stroke-time curves match the measured curves well. Differences between the measured and the simulated process can be detected at the beginning of the transformation and at the end of the cooling phase. These differences may occur as a result of the simplified, but wrong

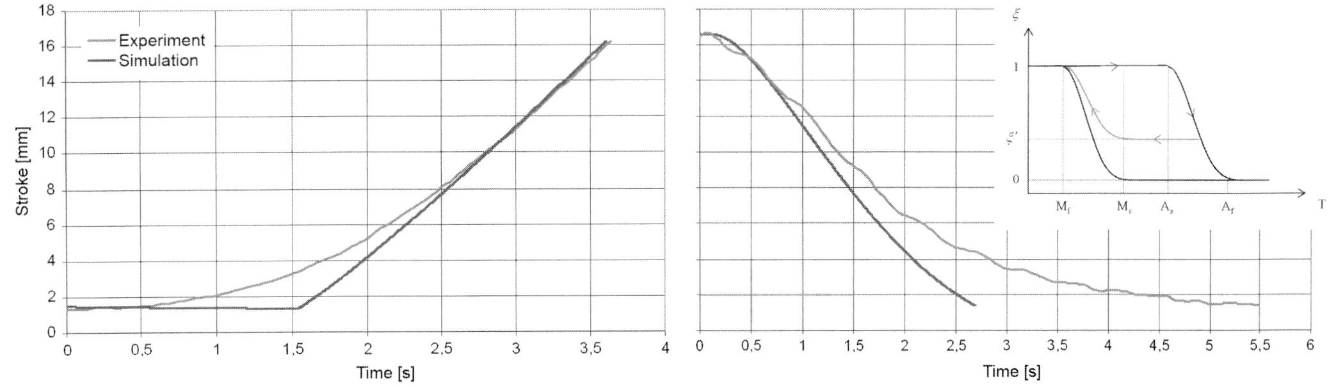

Fig. 4 Simulation vs. experiment: results of an inner hysteresis route

simulation boundary condition of a constant surrounding temperature in the direct actuator proximity and the friction of the direction change mechanism for the bias force occurring mainly during the cooling phase.

Even better results were obtained for a complete transformation with not only constant heating currents, but also ramp or step current curves. To match the loading conditions of the application described in the next section, the experimental setup can be changed and resetting forces can be generated by installing a bias spring or a pneumatic force generator, which can create variable forces by a regulating valve that controls the volume flow rate blown on a piston connected to the aerostatic motion slide.

5. Simulation-Based Design

Based on the results gained by the numerical analysis of the SMA actuator's activation behavior, a rotatory stepping motor powered by SMA actuator wires has been developed. By using the numerical model and its simulation results, the properties of the stepping drive were optimized.

The drive system with straight SMA actuator wires that is presented in this paper is based on a harmonic drive (HD) gear (Ref 6). A conventional HD gear consists of three components: the drive is executed by the Wave Generator, a thin-raced ball bearing that is fitted onto an elliptical plug. The Flexspline is a flexible steel cylinder with external teeth. The Circular Spline is a solid steel ring with internal teeth and has one to two teeth more than the Flexspline. When the elliptical Wave Generator rotates, the pivotable Flexspline shifts within the Circular Spline in a rotatory motion free from slippage. As Flex- and Circular Spline have different numbers of teeth and therefore different circumferences, a relative movement occurs. The gears reduction ratio results from the different number of gear teeth. The functional principle of the HD gear was adopted for the design of a rotatory drive system and enhanced with an actuator component: the Wave Generator is left out and the deformation of the Flexspline is realized by external forces that are provided by eight electrically driven double laid straight SMA actuator wires (see Fig. 5). In this setup, the bias force, which is needed to reset the actuator during the cooling phase, is provided by the elastic properties of the Flexspline which is made of aluminium and the deforming force of the other activated actuators.

With this basic concept the numerical model was used to calculate the optimal diameter of the SMA wires and the

Fig. 5 SMA-powered Rotatory Drive System

constant heating current, which can be applied with two opposing strategies to generate a movement in both clockwise and anticlockwise direction. The optimization process works in an iterative way. A start set of actuator and boundary parameters is computed. With the fast and modular simulation tool a quick change and calculation with a variation of single parameters is possible. Although this iterative optimization method is a fast and inexpensive way to simulate and design a broad range of technical SMA actuator systems, actual research is focused on the development of a design tool, which uses special algorithms to automatically select and parameterize an optimal SMA actuator.

6. Conclusion

Based on an energy balance calculation of the actuator system a numerical model for the activation behavior of SMA actuators (wire and spring shaped) was presented. The model features a simple way to parameterize the actuator system and is therefore well suited for the design and layout of technical solutions with SMA actuators. It includes different environmental conditions, temporally variable heating currents as well as different time variable loads. In addition, inner transformation routes can be simulated. The qualitatively good implementation of the system's properties was proven experimentally. Using the gained results, a stepping drive with straight SMA wires was developed, which was optimized by use of the simulation model.

References

1. H. Janocha, *Adaptronics and Smart Structures—Basics, Materials, Design and Application*, 1999
2. J. Van Humbeeck, Shape Memory Alloys: A Material and a Technology, *Adv. Eng. Mater.*, 2001, **3**(11), p 837–885
3. VDI (The Association of Engineers, Germany), VDI-Waermeatlas, Berechnungsblaetter für den Waermeuebergang, 1997 (in German)
4. C. Liang and C.A. Rogers, One-Dimensional Thermomechanical Constitutive Relations for Shape Memory Materials, *J. Intell. Mater. Syst. Struct.*, 1990, **1**, p 207–234
5. A.V. Srinivasan and D.M. McFarland, *Smart Structures*, Cambridge University Press, Cambridge, 2001
6. L. Oelschlaeger, "Numerische Modellierung des Aktivierungsverhaltens von Formgedaechtnisaktoren am Beispiel eines Schrittantriebes," Ph.D. Thesis, Ruhr-University Bochum, 2004 (in German)

JMEPEG (2009) 18:691–695
DOI: 10.1007/s11665-009-9407-9

1059-9495/$19.00

SmartFlex® NiTi Wires for Shape Memory Actuators

L. Fumagalli, F. Butera, and A. Coda

(Submitted September 15, 2008; in revised form January 29, 2009)

Shape memory alloys (SMAs) are active metallic materials classified nowadays as "smart" or "intelligent" materials. One of their main areas of interest is that of actuators. The use of SMAs in actuators offers the opportunity to develop robust, simple, and lightweight elements that can represent an alternative to electro-magnetic actuators commonly used in several fields of industrial applications, such as automotive, appliances, etc. SAES Getters S.p.A. thanks to its vertically integrated process and to the scientific and quality approach, developed a NiTi-based wires family which can represent a solution for shape memory actuators. In this paper, the mechanical, thermal, and electrical response of these shape memory wires, with diameters ranging from 20 to 500 μm, will be examined and discussed, with particular focus on the design of the actuator. The thermo-mechanical properties have been investigated and measured by several methods. The most common and useful tests for these commercially available wires will be also described.

Keywords advanced characterization, automotive, mechanical testing, nondestructive testing, titanium

1. Introduction

For simple actuations involving, for example, linear movements of a few millimeters or angular movements of just a few degrees, the usage of a motor rotating at thousands of rpm coupled with gears, clutches, and reducers is certainly not the best solution. The use of "smart materials" for actuation represents an excellent technological opportunity for the development of electro-mechanical components. In particular, the typical characteristic of "smart materials" promotes the development of simple, very compact, reliable actuators that can be integrated in the components or in the structures thus transforming these from static to dynamic or, in some cases, adaptive as they can react directly to environmental stimuli. Component designers can therefore use smart materials to simplify products, to add new functions, to upgrade performance, to improve reliability, and to reduce component cost accompanied by a significant reduction in mechanical complexity and size (see Fig. 1).

Shape memory alloys (SMAs) are functional materials with a multitude of different dependent properties. Among them, the equiatomic system NiTi has been established over the years as a standard alloy covering a wide range of application requirements (Ref 1). NiTi shows the best combination of properties, especially in terms of the amount of work output per material volume and the large amount of recoverable strain compared to any other actuator principles (Fig. 1). The obvious simplicity of the mechanical design and the small number of moved parts is amazing for an actuator and make SMAs particularly attractive for microsystems applications.

SMA actuation is based on the one-way effect. In principle, the effect can be used for generating motion and/or force. The required heating power is supplied either indirectly by variation of ambient temperature or directly by electrical heating of the SMA component (Joule effect). Electrically driven actuators carry out specific functions at ambient temperatures below the M_s temperature, which can be controlled via heating power. Further functions of SMA components may be structural or reset functions.

The principle of work production with an external reset force is applied in most SMA actuators (Ref 2). Depending on the choice of the reset mechanism, different trajectories can be obtained in the stress-strain space. Figure 2 shows various reset mechanisms using a SMA tension wire. In case (a), a weight produces a constant force level. Another widely used variant is passive bias spring (b). The case of decreasing reset force for increasing deflection is described in (c). Moreover, the use of a second SMA tension wire with opposite actuation direction leads to an antagonistic mechanism (d). As the reset mechanism determines the stress profile in the SMA component, the typical hysteretic behavior of the phase transformation can be adjusted or modulated by selecting an appropriate mechanism, according to specific requirements of the final application. Definitely, the reverse bias design concept can open new applications for SMAs actuators. The correct usage of a straight wire shape memory actuator is based on the sensible design of the surrounding mechanical structure that has to provide the resetting force for the actuator. Under optimal design condition, the transformation temperatures can be significantly increased and the maximum actuator stroke can be optimized.

Under pure tension or compression load, high forces but relatively small displacements can be reached. Larger displacements are achieved under torsion or bending loads. Wires and springs are the most important shapes of NiTi that can be used into the design of an actuator. Depending on the kind of

This article is an invited paper selected from presentations at Shape Memory and Superelastic Technologies 2008, held September 21-25, 2008, in Stresa, Italy, and has been expanded from the original presentation.

L. Fumagalli, F. Butera and **A. Coda**, Saes Getters S.p.A, Lainate, MI, Italy. Contact e-mail: alberto_coda@saes-group.com.

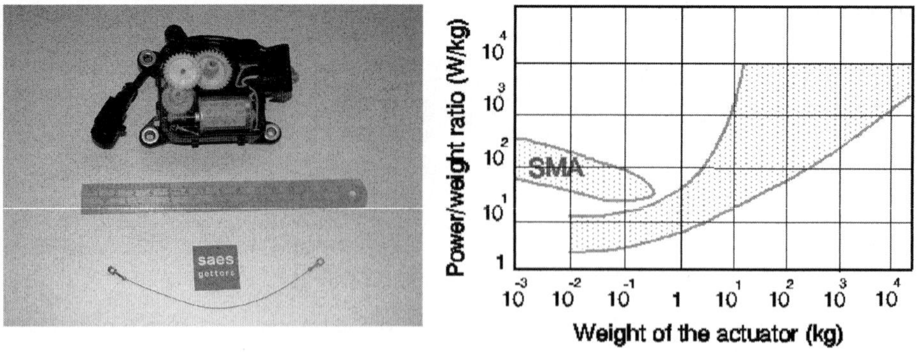

Fig. 1 Comparison between traditional and shape memory actuators in terms of size and power/weight ratio

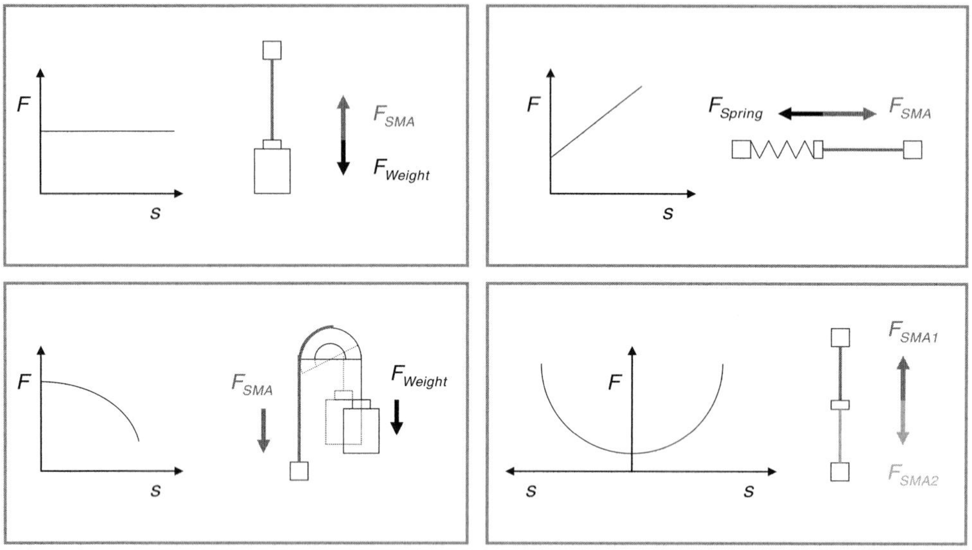

Fig. 2 Some possible reset mechanism for tensile shape memory wire

application, typical distinctions can be made which should be understood as guidelines (Ref 3), as they are not firm rules:

- Straight wire for resistance heated actuators with small displacement (up to 5%),
- Compression or tensile springs for indirect heated actuators with large displacement (more than 5%).

The mechanically stabilized SAES SmartFlex* NiTi (49-51 at.%) actuators show a very sophisticated profile of properties. In this paper, the mechanical, thermal, and electrical response of these shape memory wires, at diameters ranging from 20 to 500 μm, under different working conditions, simulating the actual operating condition in real actuators, will be examined in depth and discussed, in order to direct the design of the actuator so that the functional properties of the material can be completely exploited. The thermo-mechanical properties have been investigated and measured by several methods. The most common and useful tests for these commercially available wires will be also described.

*SmartFlex is a registered trademark of SAES Getters S.p.A., Italy.

2. Discussion

2.1 Hysteresis Test

The knowledge about the transformation and the hysteresis related properties is of importance for the design of shape memory actuators. In this test, the wire is subjected to a constant load and its deformation is measured during a controlled temperature profile in an environmental cell.

In Fig. 3, the test output for Smartflex 76 under constant stress of 150 MPa is displayed. As it can be seen from the graph, some important information can be gathered, such as the maximum stroke (difference in wire length between the cold shape and the hot shape) and the transition temperatures (beginning and end of the direct and reverse transformation) at a given load. Most often, the transition temperatures are determined using the tangent method (Ref 4). In the following curve, the maximum stroke of the wire is around 5%, M_f at 65 °C and A_s at 85 °C.

As it is well known, the applied load is an important parameter that can affect the test: one can run tests with different loads to analyze their influence on the intrinsic characteristics of the wire, such as transformation temperatures and stroke; otherwise a standard load, near to that of usual actuators applications, can be chosen to make a comparison between different wires. In Fig. 4, the transition temperatures measured by hysteresis test on Smartflex 76 wire under different applied loads are reported.

As it can be seen, the Martensite (M)-Austenite (A) transformation temperatures increase by increasing the level of stress. The increase is due to the fact that an applied load during M-A transformation generally supports the transformation itself as one can derive from a modified Clausius-Clapeyron equation (Ref 5):

$$\frac{d\sigma_c}{dT} = -\frac{\Delta H^{A \to M}}{T_0 \cdot \varepsilon}$$

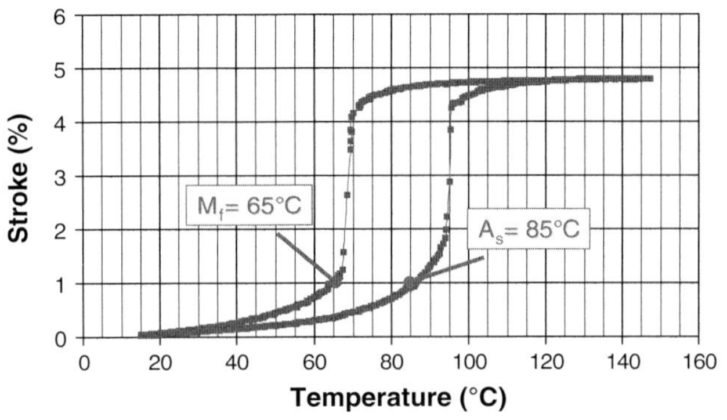

Fig. 3 Constant load test (150 MPa) on Smartflex shape memory wire

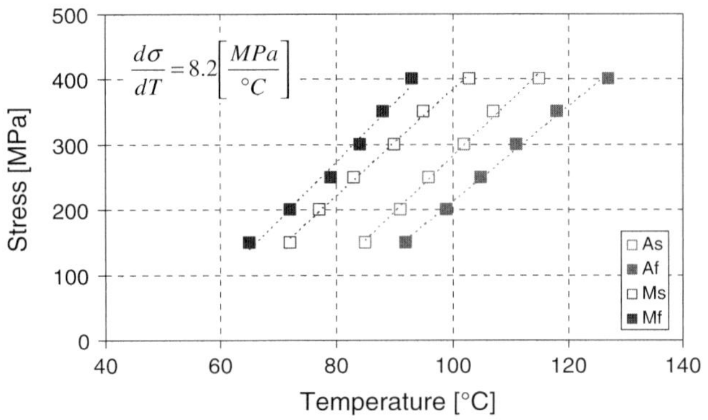

Fig. 4 Transition temperatures behavior of Smartflex shape memory wire under different loading conditions

Typically, in actuators, the wire deformation is kept below the maximum stroke, since the smaller the difference (maximum—working strain), the shorter the wire lifetime is; so the largest recoverable strain is a very important parameter for a SMA wire.

Wire transformation temperatures are of course fundamental parameters to measure, know, and control (Ref 6): in the case of wire actuated by Joule effect, reverse transition temperatures concern the needed actuation current and time. Also, the A-M transition temperatures are very useful to determine the system speed for recovering the cold shape, once the electrical feeding is removed. A limitation for usage of NiTi wire is the relatively low transformation temperature, which makes applications with environmental temperatures above 80 °C fairly critical: if the environmental temperature increases above this value, the SMA wire does not fully recover the initial shape, so that a certain strain deficit accumulates over the number of cycles. Of course this can be healed by just one cycle at lower temperatures, but it certainly can lead to functional problems during application.

The main problem related with the hysteresis test is the duration (a single cycle between 15 and 150 °C at a rate of 1 °C/min lasts more than 4 h). Another problem is the maximum usable length: in a typical hysteresis system only samples of about 100-150 mm can be analyzed. For this reason, Saes Getters S.p.A. has developed and patented (Ref 7) a new characterization method in which the quality control on the total length of the produced wire is possible. This equipment will enable an on-line 100% product quality control to measure and guarantee NiTi wire thermo-mechanical properties.

2.2 Functional Characterization

By using the equipment shown in Fig. 5(a), it is possible to carry out what we call "one-shot" cycles. The wire is hanged with a weight clamped at the bottom end and heated up by Joule effect under a constant load. A position laser sensor is located at the bottom side of the weight. This means that studies of current, applied load or fixed stroke effects on actuation times can be assessed, in order to optimize the working conditions of shape memory wires.

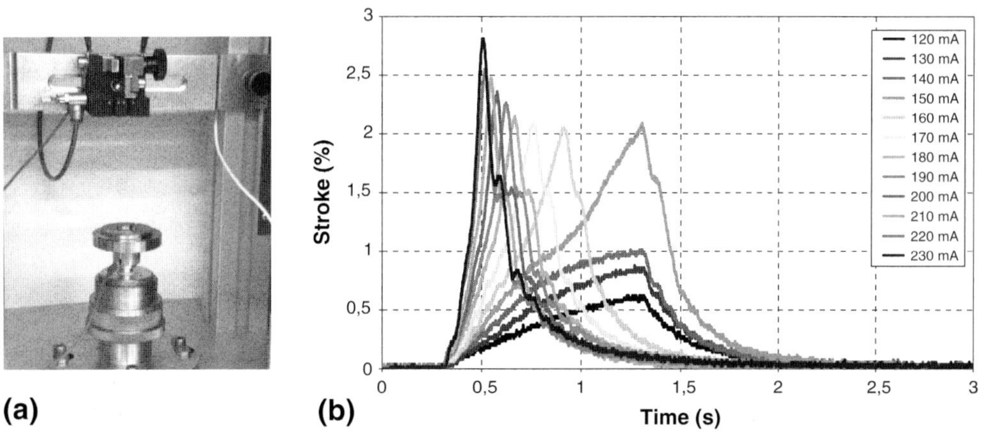

Fig. 5 Experimental set-up for "one-shot" test (a). Actuation times and strokes of Smartflex 76 using different heating currents (b)

For example, in Fig. 5(b) the actuation time and stroke of Smartflex 76 have been measured as a function of the feeding current, by controlling the stroke at 2% during all tests. What can be observed is that for currents <150 mA the wire is not able to reach the fixed stroke. On the other hand, using currents higher than 180 mA, the actuation is so fast that the wire acceleration cannot be exactly controlled by the position sensor, due to its slower response time. These data may be collected varying the wire operating conditions, such as the fixed stroke, so to have a picture of the actuation times using different heating currents. The results on Smartflex 76 shape memory wire have been summarized in Fig. 6. Such a test has been done not only on SmartFlex 76, but also on all the available sizes of SAES shape memory wires. In Fig. 7, the typical actuation conditions (time and current), under a constant stress of

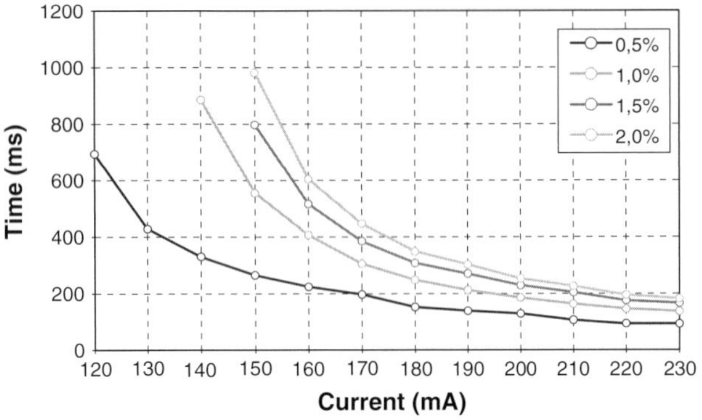

Fig. 6 Smartflex 76 actuation times as a function of different feeding currents and strokes

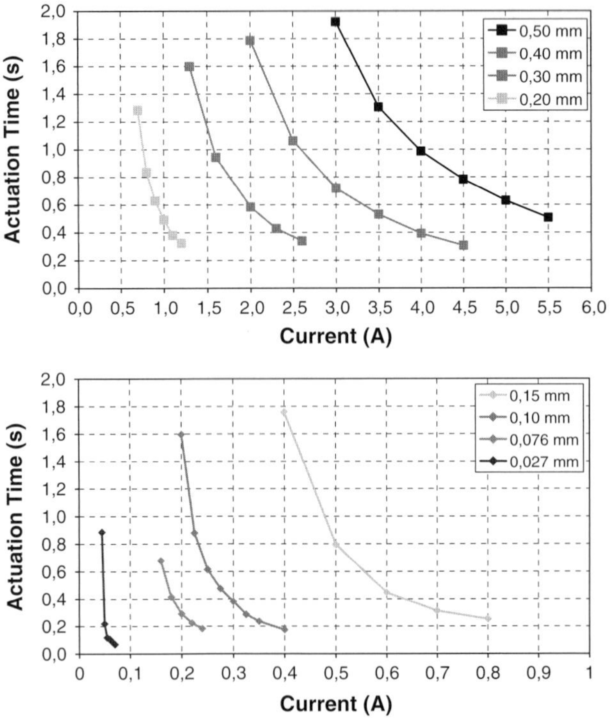

Fig. 7 Typical actuation times and currents for SMA wires from 500 to 27 μm under constant stress and stroke conditions (200 MPa, 3.5% stroke)

200 MPa and a fixed stroke of 3.5%, have been reported for wires from 500 to 27 μm.

As expected, the smaller the wire cross section the lower the actuation time and the electrical power. Particularly interesting is the case of very thin wire SmartFlex 27 whose actuation is so fast that the actuation time seems to be almost constant even varying the heating current. This work is very important since at best it allows exploitation of the material performances and by an appropriate design adapts them to a very wide range of industrial and automotive applications.

2.3 Fatigue Life Test

Beside the maximum stroke and the transition temperatures, a very important feature that defines the wire suitability for a specific application is the shape memory effect stability during cycling and the number of thermal cycles which can accomplish before fracture. Thus, a fatigue life test, in which the wire is subjected to electrical actuation cycles under an applied constant load, is fundamental in the characterization of SMA wires. To do this, the wire is mounted on the equipment shown in Fig. 5(a) and thermally cycled by Joule effect under a constant load. In this test, the heating of the wire is very fast (typically <1 s) and the stroke is fixed so that a position sensor can stop the current passage once this value is reached. This allows to investigate the fatigue wire behavior in a very wide range of operating conditions, being the lifetime strongly dependent upon the heating current, actuation time, applied load, and required stroke.

Different configurations and conditions can be set for the analysis: one can observe the largest stroke setting a constant actuation time high enough to assure the M-A transition to completely take place. Otherwise, it is possible to carry out the test at constant strain (generally closer to the application conditions). In this case, the electrical feeding is stopped when a certain set stroke is reached. One can also choose the electrical feeding conditions (constant I or U). Given the test parameters and configuration, the total number of thermal cycles is a customer's typical specification (on the order of 10^4-10^5 cycles) that must be respected.

Similar to the hysteresis test, the analysis conditions should be chosen close to the ones of the real application, in order to gather useful information about the behavior of the wire during its life. Smartflex 76 shape memory wires have been tested (Fig. 8) by using a current of about 170 mA under a constant stress of 350 MPa and controlling the stroke to 1%. Under these operating conditions, the samples have survived longer

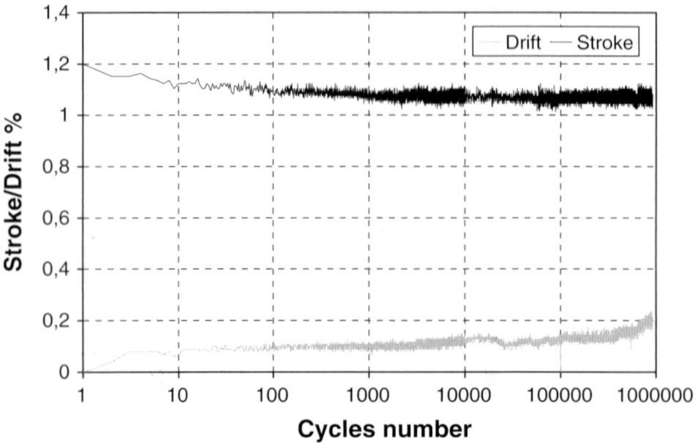

Fig. 8 Fatigue lifetime output (stroke and drift) for Smartflex 76

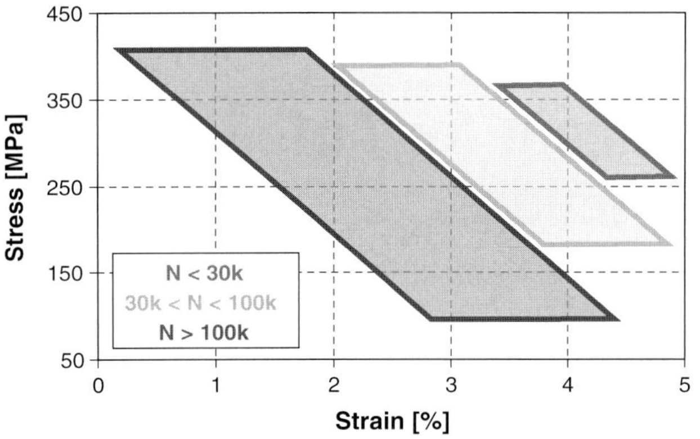

Fig. 9 Fatigue lifetime for Smartflex 76 under different stress-strain conditions

than 10^6 cycles, assuring constant stroke and a drift $<0.2\%$ over the life.

As before mentioned, the dependence of the achievable reversible strain on different stress-strain conditions and the number of cycles is of considerable interest with regard to the service life of SMA actuators. For a deep and complete understanding of the effects of stress and strain levels on the lifetime of SMA wires, a lot of measurements have been carried out. In Fig. 9, for example, three differently colored areas have been identified. In each area, all the experiments have been collected according the utilized stress-strain conditions and the corresponding lifetime. As it can be seen from the graph, increasing both stress and strain, the lifetime of shape memory wires is strongly reduced. On the other hand, by selecting appropriate working conditions, it is possible to obtain very long lifetime and very high reliability from shape memory wires.

3. Conclusions

Shape memory alloys NiTi-based actuators are being used in many different industrial applications. They also have already a great potential as miniaturized sensors and actuators. SAES Getters S.p.A., thanks to its vertically integrated process and to the scientific and quality approach, developed a NiTi-based wires family which can represent a very good solution for shape memory actuators. The trained SAES SmartFlex NiTi actuators show a very sophisticated profile of properties. The performances and the lifetime of SmartFlex wires were tested under different working conditions. They can fulfill the main actuators requirements and are ready for high volume production with a 100% of quality control.

References

1. H. Funakubo, *Shape Memory Alloys*, Gordon and Breach Science Publisher, New York, 1987
2. M. Kohl, *Shape Memory Microactuators*, Springer-Verlag, Berlin, Heidelberg, 2004
3. M. Mertmann, and M. Wuttig, Application and Hysteresis of Different Shape Memory Alloys for Actuators, *Proceedings of ACTUATOR 2004*, 2004, p 72–77
4. H. Luo, E. Abel, J. Hewit, M. Pridham, B. Steele, Z. Wang, and A. Slade, Issues Concerning the Measurement of Transformation Temperatures in NiTi Alloys, *Proc. SPIE*, 2002, **4934**, p 226–233
5. M. Mertmann, S. Kautz, and W. Brown, Design Principles and New Actuator Applications with NiTi Straight Wire Actuators, *Proceedings of ACTUATOR 2002*, 2002, p 85–90
6. K. Otsuka and C. M. Wayman, *Shape Memory Materials*, Cambridge University Press, Cambridge, 1998, p 149–177
7. International Patent Application No. WO2005/106441

JMEPEG (2009) 18:696–705
DOI: 10.1007/s11665-009-9356-3

A Novel SMA-based Concept for Airfoil Structural Morphing

S. Barbarino, R. Pecora, L. Lecce, A. Concilio, S. Ameduri, and E. Calvi

(Submitted September 16, 2008; in revised form November 28, 2008)

The adaptive structures concept is of great interest in the aerospace field because of the several benefits which can be accomplished in the fields including noise reduction, load alleviation, weight reduction, etc., at a level in which they can be considered as compulsory in the design of future aircraft. Improvements in terms of the aerodynamic efficiency, aeroelastic behavior, stability, and manoeuvrability performance have already been proved through many international studies in the past. In the family of the Smart Materials, Shape Memory Alloys (SMA) seem to be a suitable solution for many static applications. Their high structural integrability in conjunction with actuation capabilities and a favorable performance per weight ratio, allows the development of original architectures. In this study, a morphing wing trailing edge concept is presented; morphing ability was introduced with the aim of replacing a conventional flap device. A compliant rib structure was designed, based on SMA actuators exhibiting structural potential (bearing external aerodynamic loads). Numerical results, achieved through a FE approach, are presented in terms of trailing edge induced displacement and morphed shape.

Keywords Aerodynamic surfaces, Morphing, Shape control, SMA

1. Introduction

Because several flight regimes occur during a typical aircraft mission, it is practically impossible to define a single configuration able to maximize aerodynamic efficiency, manoeuvrability, stability, fuel consumption and so on in any circumstance.

Currently, a common design strategy is to identify few dominant flight regimes within a typical mission and define a specific set of parameters like wing body geometry and control surfaces attitude, with respect to which the aircraft performance is maximized. In this way, drawbacks within conditions that fall outside the design range are unavoidable. Solutions characterized by a certain adaptivity level can lead to overcome these limits. Components like ailerons, flaps, equilibrators, together with the related actuation strategies, correspond somehow to the necessity of facing very different states, even dramatically far from the nominal design ones (e.g., take-off and landing) (Ref 1).

To simplify the actuation chain and deform a limited part of the structure, these movable devices are confined to limited regions of the wing; this causes, however, sharp shape modifications, badly affecting the aerodynamic field and, due to the concentration of the external loads around the hinges zone, a significant increase of the structural weight. Classical actuator systems generally lead to a 50% increase in weight relative to a naked wing.

To avoid such problems, different design strategies have been taken into account. An interesting one was implemented within the 3AS (Active Aeroelastic Aircraft Structures) European Project, (Ref 2, 3). By controlling the rigidity of some parts of a wing, the suitable displacement field corresponding to the said optimal configuration for a specific flight regime was attained. This strategy, recalling the classical aeroelastic tailoring philosophy, was applied to an all-movable vertical tail (AMVT), and led to an integrated shaft with a variable stiffness. Instead of using the external loads to deform the aerodynamic surfaces, the so-called 'morphing' strategy may be pursued. In this case, large and distributed deformation of specific parts of the wing body are achieved by generating proper internal forces (Ref 4-6). In order to reach this objective, leading and trailing edges have to be designed to withstand the external loads and undertake dramatic deformations, following the selected flight regime requirements (Ref 7, 8). Large displacement capabilities associated with load bearing capacities is the paradox which the smart or adaptive structure systems have to overcome in order to provide a reliable alternative to classical devices. In this article, the proposed solution to eliminate the above paradox is to refer to actuators that play a structural role, also.

The use of smart materials has a crucial role in this challenging approach. What the authors use in the research

This article is an invited article selected from presentations at Shape Memory and Superelastic Technologies 2008, held September 21-25, 2008, in Stressa, Italy, and has been further expanded from the original presentation.

S. Barbarino, R. Pecora, and **L. Lecce**, Department of Aerospace Engineering, University of Naples "Federico II", Via Claudio 21, Naples 80125, Italy; **A. Concilio** and **S. Ameduri**, Smart Structures Lab., The Italian Aerospace Research Centre (CIRA), Via Maiorise, Capua, 81043 CE, Italy; and **E. Calvi**, ALENIA AERONAUTICA S.p.a., Viale dell'Aeronautica snc, Pomigliano d'Arco, 80038 NA, Italy. Contact e-mail: silvestro.barbarino@gmail.com.

herein presented, are the Shape Memory Alloys, SMA. Their Super Elastic and Shape Memory properties, jointly with their capacity to transmit large loads, make them a privileged material for applications that require both large strains and high actuation forces (hundreds to thousands of Newton for classical commercial elements) (Ref 9).

In this article, the design of an adaptive flap, driven by SMA-based actuators, is illustrated (Ref 10, 11). More in detail, by taking advantage of a FE approach and implementing an SMA descriptive model (Ref 12), the concept of an original rib is dealt with. This structural element is made of different plates that are connected to each other through crossed spring laminas. These are in charge of providing a certain amount of rigidity to the structure, assuring the required preload for the SMA elements and allowing for large relative rotations. SMA elements are wires, located under the springs, which connect two consecutive plates. When they are activated, their contraction leads to a rotation of the rib parts, with consequent camber variations. External aerodynamic loads are considered when the system performance is evaluated; in detail the configuration corresponding to a traditional rotated flap is referred to.

On this architecture, an optimization study is carried out, over certain selected architecture parameters such as the springs, length, their angular position, and location. SMA mechanical properties and geometry are also considered. Results are presented in terms of achievable camber variations and trailing edge displacements.

2. Architecture Working Principle and Modeling Strategy

The considered structure is sketched in Fig. 1. It is made of five bulkily shaped panels, connected in couples by crossed laminas, working as spring elements. The single panel may be assumed to be rigid in the performed study; the set of five constitutes the skeleton of the rib.

The connecting thin plates (beams in a 2D representation, shown in the same figure) exhibit the necessary elastic properties to allow the architecture undergo suitable deflections. Their rigidity is tuned so that deflections related to the external loads are restricted into small fractions of the ones associated with the driving commands.

The proposed actuator system is composed of the union of the active SMA wires and the spring elements. They are thought to show cyclic behavior and to carry out the prescribed external load, i.e., to have a structural purpose that is a function of their characteristic elasticity. These components can be referred to as "bearing actuators." The activation of SMA generates a contraction of the wires that in turn causes, because of the eccentricity with respect to the spring subsystem center of rotation, a relative angular movement of the bulk plates. SMA ribbons and rods are also considered in the following discussion.

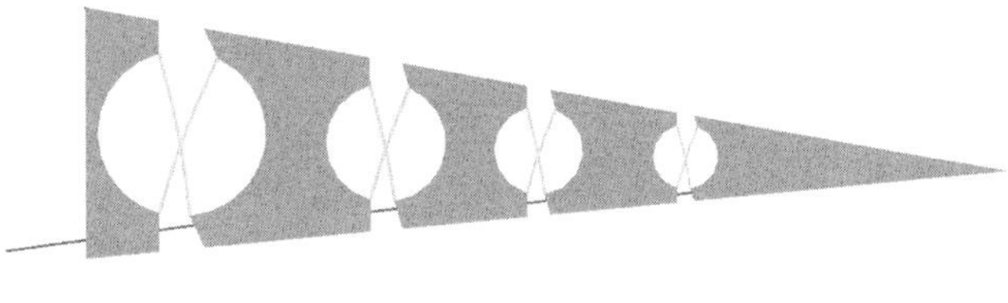

Fig. 1 Rib architecture

The used springs show a nonlinear behavior, because of the translation of the rotation pivot during deflection. This property goes into the direction of achieving in principle higher preloads for SMA wires recovery and larger relative rotations, while assuring the desired rigidity levels. These properties are strongly dependent on some geometric parameters, as shown in the next paragraph; then, by properly adjusting those values, the performance may be improved or optimized.

MSC/Nastran FE models were set up to investigate the behavior of the described structural system, integrated with an external proprietary Matlab routine aimed at simulating the behavior of the SMA elements, implementing the Liang and Rogers' 1D model (Ref 12). The assumptions, this model is based on, are coherent with the working modality of the actuators: no bending actions (the wires are hinged with the rib panels) and 1D behavior of the material (strain recovery is thought to occur only along the axial direction). Liang and Rogers' is an approximation of the former Tanaka's model (Ref 13), substituting exponential with sinusoidal expression in the description of the material composition (austenite versus martensite fraction, ξ_M and ξ_A), function of the temperature, T, and the stress field, σ:

$$\xi = \frac{\xi_M}{2}\cos[a_A(T - A_S) + b_A\sigma] + \frac{\xi_M}{2} \quad \text{Austenite} \rightarrow \text{Marteniste} \qquad \text{(Eq 1)}$$

and

$$\xi = \frac{1 - \xi_A}{2}\cos[a_M(T - M_F) + b_M\sigma] + \frac{1 + \xi_A}{2}$$

Marteniste → Austenite

The values $a_A = \pi/(A_F - A_S)$ and $a_M = \pi/(M_S - M_F)$ are combinations of the material characteristic constants, A_F, A_S, M_F, M_S, namely, austenite final, austenite start, martensite final, and martensite start temperature, respectively. They refer to the beginning or the completion of phase transformation and define when a particular crystal structure appears or when the conversion is complete. Two other parameters appearing in the above formulas are $b_A = -a_A/C_A$ and $b_M = -a_M/C_M$, C_A and C_M being the slopes of the linear relationship between stress and temperature characteristic transformation.

Non-linearity assumption is necessary in order to properly consider the large strains, the characteristic of an SMA application. SMA-active elements are simulated through thermal 1D elements characterized by a negative expansion coefficient (SMA contracts as the temperature increases). Following activation, the structure deforms under the action of the wires. They both undergo a stressed condition. The attained displacement is associated with the computed SMA material stress. This operation is repeated for increasing values of the activation temperature, so that the dot-line curve in Fig. 2 may be sketched. As the force increases, the strain also does and because of the large strains attained, the classical non-linear behavior is shown (smaller strains for an equal increase of the forces). The initial configuration is assumed to have an offset of a roughly 5% strain, while the starting point of the operative function is at 3%, corresponding to the imposed recoverable strain for repeated cyclic applications (Ref 14, 15). In reality, the SMA wires will exhibit growing contractions until it transforms completely into austenite. This is represented by the intersection point of the dot-line curve with the dotted curve that is the graphical representation of the constitutive material law for a generic SMA. The intersection point determines then the working condition of the proposed actuator. This kind of procedure allows reaching the "maximum" recoverable strain. The process may instead be stopped at a certain point, not letting the material undergo the complete phase transformation (partial recovery).

The process herein described is a simplification of the followed strategy. In fact, in order than the process may be cyclic, the activation point shall be moved at a prestressed configuration and, particularly, fixed at the complete martensite state. This assures the capability of recovering the maximum value of strain (complete transformation martensite-austenite is thus accessible). As the activation takes place (austenite start), the working point moves toward higher stress values. The final point is established when the desired strain recovery percentage is attained that is a function of the cycles number to be faced (Ref 14, 15).

From a figurative point of view, increasing temperature makes the SMA material curve to move up in the 2D stress-strain material plane and the working point to shift towards left, over the lower boundary of the SMA material curve (austenite transformation). On cooling the wire, the process is reversed under the action of a recalling force that makes the SMA to transform again into the martensite phase. The working point in this case runs over the upper part of the SMA material curve (martensite transformation). Thus, in the loading phase, the

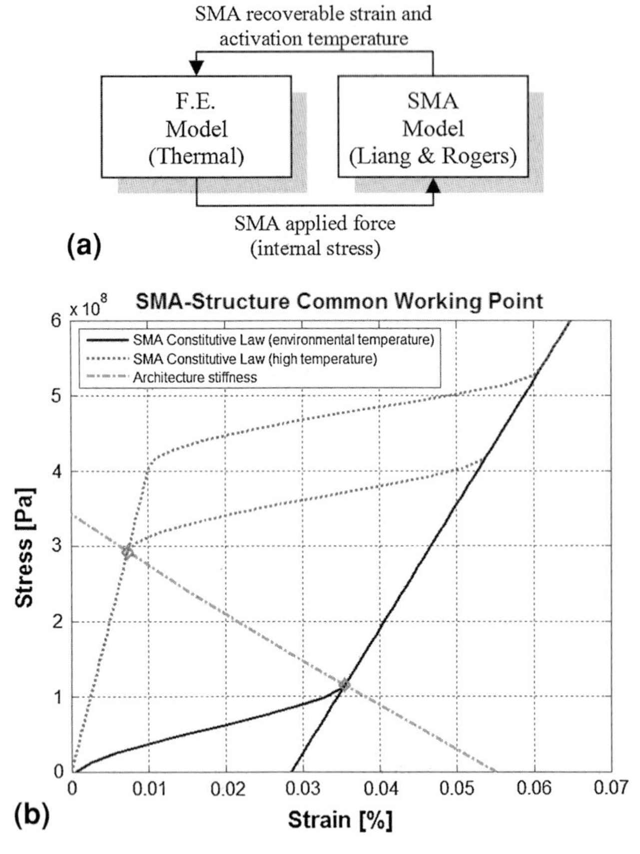

Fig. 2 Simulation scheme adopted for SMA integration within the FE approach

transformation may be said to be temperature driven; and in the recovery phase, stress driven.

3. Aerodynamic Loads

The aerodynamic load computation has been addressed by assuming the RA 18-43NIL1 as wing airfoil; the RA 18-43NIL1 is a typical wing airfoil used in regional transport aircraft. In Fig. 3 the RA 18-43NIL1 geometry is plotted and some related parameters are indicated.

Points O and P are, respectively, called *leading* and *trailing edge* while the segment OP is referred to as the airfoil chord c. The chord divides the airfoil boundary curve in two parts: the *upper camber* (represented by the curve connecting O to P in a clockwise orientation) and the *lower camber* (represented by the curve connecting P to O in a clockwise orientation). The airfoil *mean camber line*, or more synthetically, the airfoil *camber line* is defined as the locus of the centers of circles tangent both to upper and to lower cambers.

Referring to the coordinate system of Fig. 3, the function $z_C = z_C(x)$ describing the camber line is given by $z_C(x) = [z_U(x) + z_L(x)]/2$, where $z_U = z_U(x)$ and $z_L = z_L(x)$ are the functions describing the upper and lower cambers. An airflow characterized by a velocity vector V induces a pressure distribution on the airfoil, along the upper and the lower camber, essentially related to: the shape of the airfoil, the magnitude of V, the air density, ρ, and the airfoil *angle of attack* defined as the angle between the airfoil chord and the direction of V.

Approximating the upper and the lower camber by means of n linear segments delimited by points A_i and A_{i+1} ($i = 1 \ldots n$ 1), the pressure p_i arising on each segment A_iA_{i+1}, can be expressed according to the following equation:

$$p_i = \frac{1}{2}\rho V^2 C_{p,i} = qC_{p,i} \qquad \text{(Eq 2)}$$

where $q = \rho V^2/2$ is called the *dynamic pressure* and $C_{p,i}$ the *local pressure coefficient*. The local pressure coefficient $C_{p,i}$ is a non-dimensional parameter whose value depends on the orientation of the segment A_iA_{i+1} in the airflow and therefore by the airfoil shape and angle of attack. The evaluation of the local pressure coefficients is usually performed by means of specific numerical methods of computational aerodynamics. In this study, they were evaluated by means of a two-dimensional vortex lattice method (VLM) (Ref 16).

A split-flap was considered, hinged at a distance of 0.7c from the airfoil leading edge; the wing airfoil portion between 0.7c and c was assumed to be representative of a flap airfoil and rotated around the hinge point in order to simulate the split-flap down deflection. Deflection angles δ between 2 and 24 deg

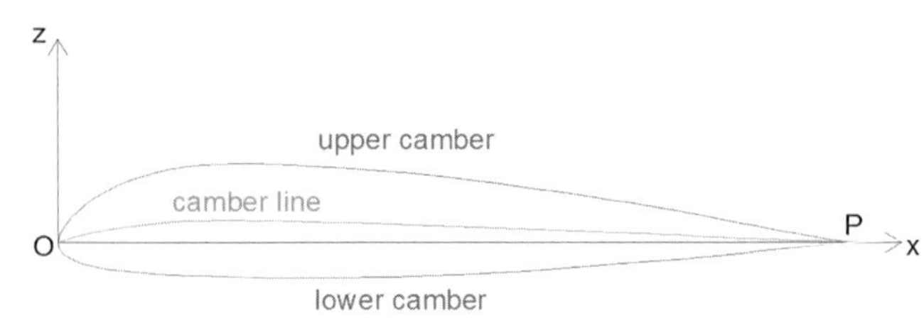

Fig. 3 RA 18-43NIL1 airfoil

were considered; for each step the C_p distribution and the wing airfoil lift coefficient, C_l, were evaluated in correspondence of a 5-deg angle of incidence (angle of attack). The airfoil lift coefficient C_l is defined by the following equation:

$$C_l = \sum_{i=1}^{n} C_{p,i}\left(\underline{n_i} \cdot \underline{\zeta}\right) \quad \text{(Eq 3)}$$

where n is the number of linear segments used to approximate the upper and the lower camber, $C_{p,i}$ is the pressure coefficient related to the i-th segment, n_i is the normal unit vector of the segment contained in the airfoil plane and positively oriented outwards of the airfoil surface, and ζ is the normal unit vector to the airflow velocity vector V. The airfoil lift coefficient, C_l, is a non-dimensional parameter used to characterize the lifting performance of airfoils or, in other terms, the airfoil's capability to generate a force orthogonal to the airflow velocity direction. Such a force, called *lift*, is responsible of aircraft sustentation and is proportional to C_l; higher C_l leads to higher lift. The obtained C_l values for each examined flap deflection are reported in Table 1 ($C_{l(\text{conventional})}$).

Let us consider a two-dimensional Cartesian reference system S_0, defined as follows: origin located at the airfoil leading edge; X-axis directed along the non-flapped chord and oriented towards the trailing edge; Z-axis in the airfoil plane and oriented upward. In correspondence of a flap rotation δ_i, around the hinge point A $[0.7 \cdot c;\ Z_A]$, the trailing edge moves from the point P $[c;\ 0]$ to the point P', defined by the coordinates: $X_{P'} = 0.7 \cdot c + \overline{AP} \cdot \cos(\alpha + \delta_i),\quad Y_{P'} = \overline{AP} \cdot \sin(\alpha + \delta_i) - Z_A$ where $\alpha = \sin^{-1}(0.3 \cdot c/\overline{AP})$. Fixed by the same trailing edge position due to a δ_i flap deflection, a quadratic morphing law was imposed to the camber line related to the flapped airfoil. In detail, the motion of point P to point P' was obtained by morphing the flapped airfoil camber line according to a parabolic arch connecting the point A to the point P', and the tangent in A to the wing airfoil camber line (Fig. 4). Camber line morphing was obviously obtained by means of a congruent morphing of the upper and lower cambers of the flapped airfoil.

According to this procedure, for each of the previously investigated hinged flap rotations, a corresponding morphed flap shape was found and the related global C_l was evaluated. The obtained results are reported in Table 1 ($C_{l(\text{morphed})}$). In Fig. 5, the wing airfoil C_ls related to conventional split and morphed flap are compared for each δ_i. For a given trailing edge down-deflection, the morphed flap solution leads to higher wing airfoil C_l than the conventional split solution.

The morphed flap configuration leading to a wing airfoil $C_l = 2.09$ was considered the most satisfactory from an aerodynamic performance point of view, and was chosen as the target of this. It corresponds to a 12-deg deflection angle and a 160-mm vertical trailing edge displacement.

A morphed flap portion of span $b = 300$ mm was then considered for aerodynamic load evaluation purpose. Let us call, MS_a and MS_b, the two morphed flap airfoil shapes which delimit the flap portion. Considering the generic aerodynamic strip s_i, delimited by two homologues segments A_i,A_{i+1} and B_i,B_{i+1} lying, respectively, on MS_a and MS_b boundaries, according to (2), the aerodynamic load acting normally to s_i was evaluated as

$$F_{n,i} = qS_iC_{p,i} \quad \text{(Eq 4)}$$

where q is the dynamic pressure, S_i is the strip area, given by the product of the span, b, between the segments A_i,A_{i+1} and B_i,B_{i+1}, and length d_i; $C_{p,i}$ is the pressure coefficient, assumed equal for both segments A_i,A_{i+1} and B_i,B_{i+1}. A dynamic pressure of around 1500.00 N/mm^2 was considered, typical for take-off and landing of a regional transport aircraft. Using Eq 4, the aerodynamic load distribution acting on the flap portion was evaluated; this load distribution was then used for a proper sizing of the ribs to be located at the end of the investigated wing portion.

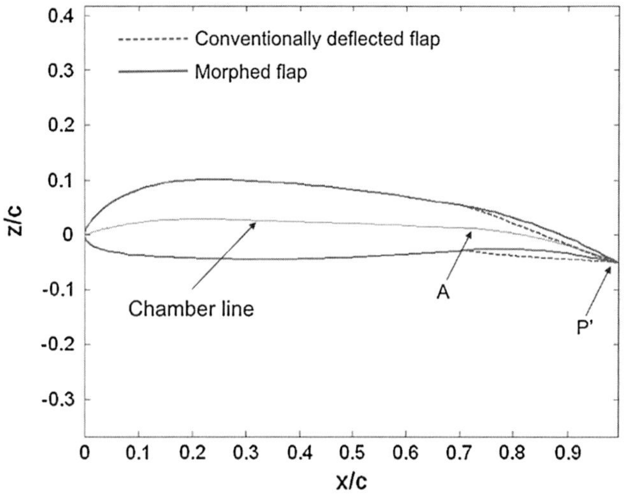

Fig. 4 Wing airfoil shapes corresponding to conventionally deflected flap and morphed flap

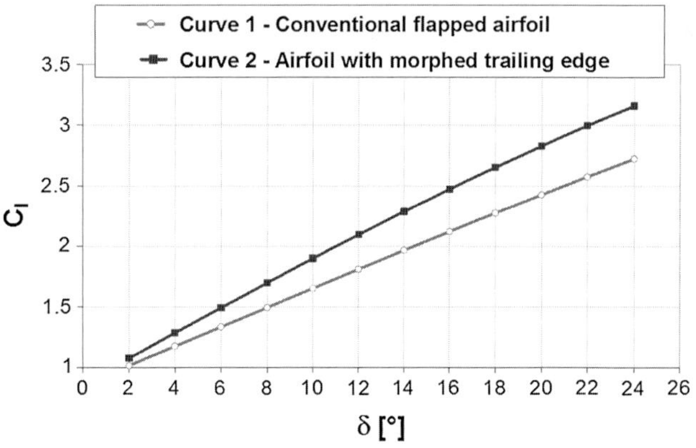

Fig. 5 Wing airfoil C_l comparison

Table 1 Wing airfoil C_l for several flap deflection angles

δ, °	2	4	6	8	10	12	14	16	18	20	22	24
C_l (conventional)	1.0155	1.1760	1.3359	1.4950	1.6532	1.8103	1.9663	2.1209	2.2740	2.4253	2.5749	2.7224
C_l (morphed)	1.0760	1.2866	1.4944	1.6990	1.8996	2.0956	2.2868	2.4727	2.6530	2.8275	2.9962	3.1589

4. Design Criteria of the Elastic Hinge

A parametric study was carried out on the aforementioned architecture. Since the elastic hinges play a fundamental role, the first efforts were focused on the identification of their optimal configuration. Through a FE approach, a single element was studied: its performance was estimated with respect to the cross plates length, d, and their relative angle, γ (Fig. 6). Discrete values for both parameters were assumed: 100, 120, and 140 mm for d; 60, 90, 120, and 150 deg for γ. A total of 12 runs were carried out to exploit all the possible combinations. Values higher than 150 deg were not considered, because of practical integration problems. The spring thin plates were assumed to be made of titanium, to meet both typical aeronautic lightness requirements and high allowable stresses, within the elastic range of the material.

The first phase was aimed at evaluating the performance dependence on the hinge geometry. Laminas, rectangular cross section was considered constant and set to 3×8 mm^2. Dimensions were defined according to design considerations based on the required flexibility and rigidity of the structure, together with the necessity of attaining a sufficient SMA preload and the capability of sustaining sufficient stress levels during actuation. SMA element features were fixed as follows: a wire 280-mm long, with a 0.5-mm diameter, called to recover a max strain of 3% (Ref 14, 15) with respect to its original length (martensite phase); further details can be found in Table 3.

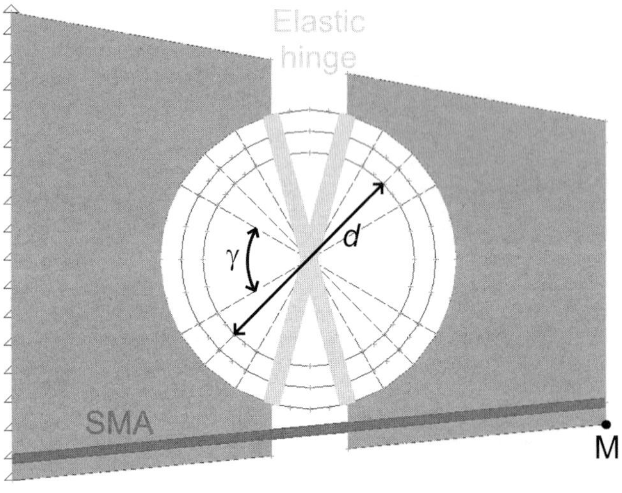

Fig. 6 Single elastic hinge scheme adopted for the parametric study

The results obtained are expressed in terms of hinge rotation and vertical displacement of the trailing edge. They are normalized and presented as histograms in Fig. 7, while actual values are reported in Table 2.

As a result of this first study, the max performance is achieved by maximising both d and γ: in this case, a displacement of 16.48 mm and a rotation of 5.91 deg is obtained ($d = 140$ mm; $\gamma = 150$ deg).

Assuming then the above parameters as the optimal configuration, further investigations were carried out by changing the SMA element characteristics and dimensioning the plate's cross section to withstand SMA preload and actuation requirements.

The active component properties were not made to change as a parametric variation but for modifying its nature, instead. Ribbon and rod forms were tested. The architecture of the actuators was maintained the same in each of the three cited cases, as well as the mechanism by which the load is transferred to the structure (two hinges). In Table 3, the features of the different SMA elements taken into account are reported, basing on experimental tests of material characterization carried out in our laboratory (Ref 17-19); length was set as the original, at 280 mm.

For the sake of clarity, it is useful to explain the exact meaning of the martensite and austenite Young moduli. The behavior of an SMA material is generally nonlinear, having to consider transformation between a phase and another and, contemporaneously having to take into account both the phases. However, when the transformation fully occurs (in a sense or the other), the material behaves as a classical linear one with distinct elastic moduli for the austenite and martensite status, respectively, E_A and E_M. Typically, E_M is one-third of E_A. Given the maximum recoverable strain ε_{max} for a given SMA, maximum recoverable force may be attained as:

$$F_{max} = E_A A \varepsilon_{max} \qquad \text{(Eq 5)}$$

SMA commercial products were considered in the comparison. The use of different SMA materials for the different forms (wires, ribbons, rods) does not allow for an immediate comparison of the results but gives an idea of what can be expected in the different cases.

As already mentioned, preload is necessary to guarantee a full cyclic behavior inside the boundaries of the environmental and the austenite finish temperature (A_f), i.e., to guarantee the working point to run a complete loop over its martensite-austenite-martensite transformation. Because of this necessity, preload requires stiffer elastic elements for stiffer active elements: Table 3 shows the necessary preload for the diverse material typology and cross section. Furthermore, max stress

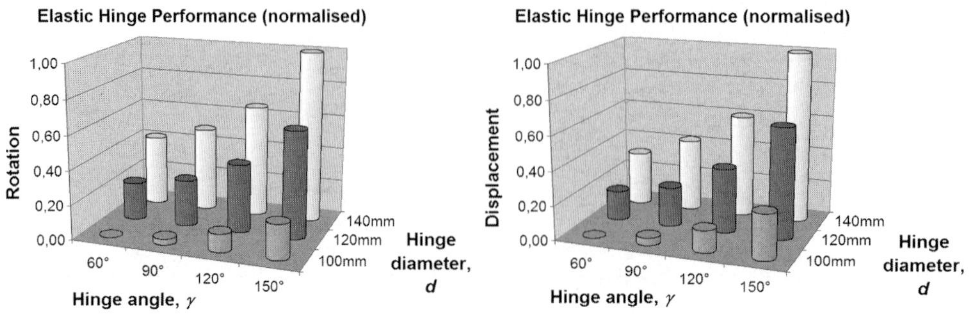

Fig. 7 Normalized performance of the elastic hinge vs. length, d, vs. relative angle

Table 2 **Performance of the elastic hinge vs. length, *d*, vs. relative angle, γ**

	Rotations				Displacements			
d/γ, mm	60°	90°	120°	150°	60°	90°	120°	150°
100	5.7452	5.7509	5.7624	5.7778	0.01615	0.01617	0.01619	0.01624
120	5.7807	5.7887	5.8094	5.8466	0.01621	0.01623	0.01627	0.01636
140	5.8122	5.8248	5.8512	5.9068	0.01626	0.01629	0.01635	0.01648

Table 3 **SMA actuators features (Ref 17-19)**

	Wire	Ribbon	Rod
Cross section dimensions	0.5 mm diameter	1.0 × 10 mm	5.0 mm diameter
Cross section area	0.196 mm^2	10 mm^2	19.63 mm^2
Young's Modulus of Austenite E_A	24.67 GPa	40.18 GPa	28.16 GPa
Young's Modulus of Martensite E_M	18.27 GPa	16.5 GPa	8.78 GPa
Martensite start temperature	35 °C	43.3 °C	10 °C
Martensite finish temperature	12.6 °C	27.5 °C	−20 °C
Austenite start temperature	59.3 °C	47.3 °C	39 °C
Austenite finish temperature	85.9 °C	58.9 °C	82 °C
Stress necessary to achieve 3% strain (preload)	175 MPa	155 MPa	165 MPa
Force necessary to achieve preload	35 N	1550 N	3239 N
Maximum recoverable force (actuation)	~550 N	~8000 N	~15000 N

Table 4 **Results summary of the plate cross section dimensioning**

	Wire	Ribbon	Rod
Elastic hinge cross section dimensions	3.0 × 8.0 mm	8.0 × 20 mm	8.5 × 32 mm
Elastic hinge cross section area	24 mm^2	160 mm^2	272 mm^2
Elastic hinge material	Titanium	Titanium	Titanium
Elastic hinge estimated weight	29.75 g	198.37 g	337.23 g
SMA preload			
SMA force level	35.65 N	1550.08 N	3238.69 N
Elastic hinge max stress	128 MPa	317 MPa	366 MPa
SMA actuation			
SMA force level	79.43 N	3426.84 N	6686.59 N
Elastic hinge max stress	289 MPa	715 MPa	769 MPa
Hinge performance			
Net displacement	13.217 mm	12.788 mm	12.495 mm
Net rotation	5.89 deg	5.67 deg	5.55 deg

following actuation shall be kept inside the titanium elastic range. The elastic hinges were then dimensioned according to a trial and error procedure: an arbitrary cross section was preliminarily assumed and stress conditions were evaluated at preload and activation phases. According to the preliminary outcomes, the cross sections were modified, taking into consideration the inertia moment and aiming at minimizing the total weight and bulk. The final results are summarized in Table 4, for the three configurations. The actuation performance in terms of hinge rotation and trailing edge displacement are finally shown in Fig. 8.

Elastic hinge resulting stress field presents the classical composed normal-bending shape, being caused by an eccentric normal force. Its maximum value is reported in the table and an example of the stress state into the plate components is shown in Fig. 9, where normal stress is shown.

From these results the wire architecture appears to lead to the best actuator performance. This trend is also confirmed when comparing the three configurations in terms of weight and SMA activation temperature (which in turn influences the required energy), estimated through the Liang & Rogers' model-based Matlab routine (Fig. 10).

Although the SMA represents a small amount of the overall weight, related preload and working constraints heavily condition the elastic hinge size. The SMA choice demonstrates once more to be a critical design aspect: an excessive powerful actuator may not lead to better performance while increasing total weight.

5. Design Criteria of the Rib Structure

The higher hinge performance, measured in terms of trailing edge vertical displacement and rotation is thus obtained for the highest values of d and γ in the considered interval and for a configuration wire-based SMA. By following these results, the design of the rib was accomplished.

In order to maximize the architecture features, it is necessary to accommodate the largest number of elastic hinges, each

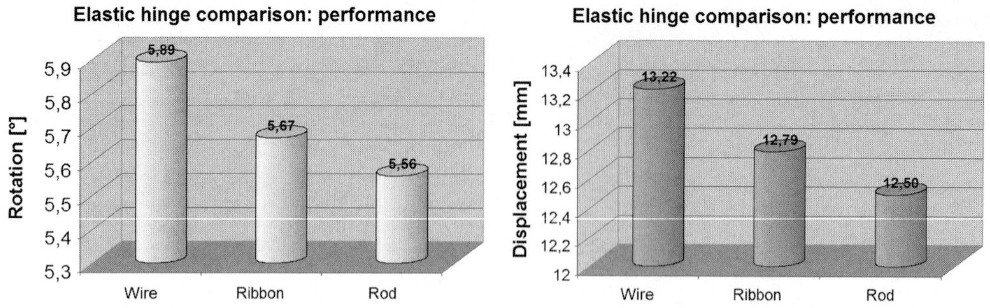

Fig. 8 Performance of the elastic hinge vs. SMA element typology

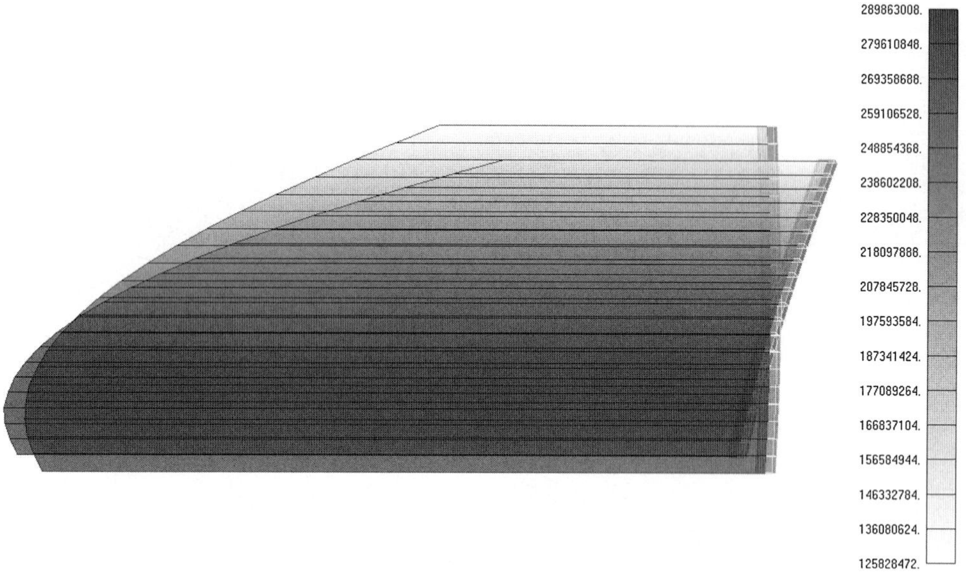

Fig. 9 Stress field distribution into the component plates of a reference hinge (wire solution)

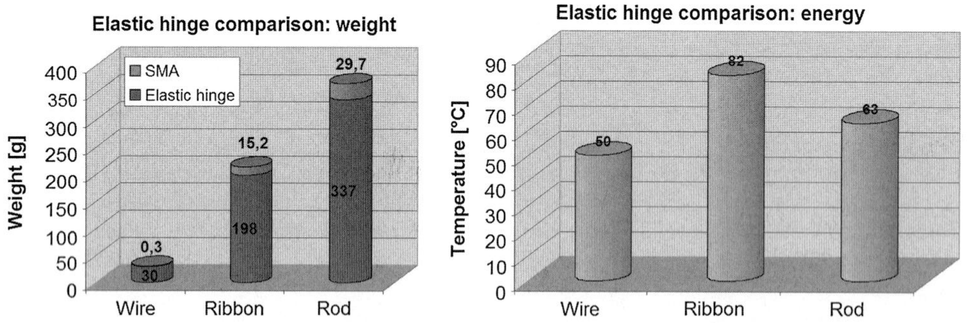

Fig. 10 Weight and SMA activation temperature comparison

characterized by the largest *d* parameter (of course, compatible with the local airfoil thickness) and the defined relative angle γ, in the considered section.

Finally, the actuator length also affects the attainable figures in a direct way (the longer the wire, the higher the performance). In Fig. 11, a sketch of the final rib arrangement is reported, while the design parameters are reported in Table 5.

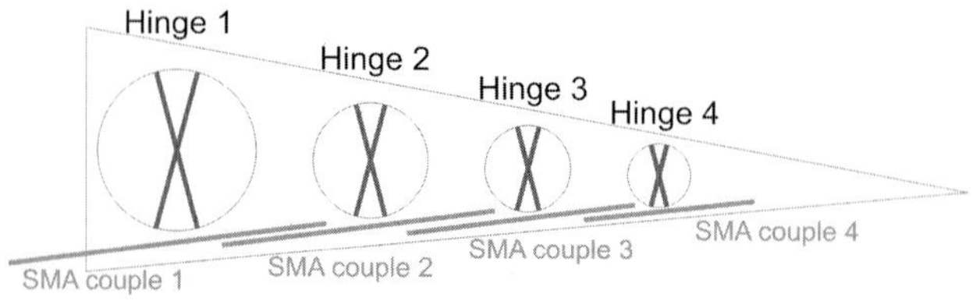

Fig. 11 Elastic hinges disposition within the rib

Table 5 Design parameters for each elastic hinge of the rib

	Hinge 1	Hinge 2	Hinge 3	Hinge 4
Elastic plates length, d	140 mm	100 mm	75 mm	55 mm
Elastic plates cross section	3.0 × 8.0 mm	3.0 × 6.0 mm	2.5 × 6.0 mm	2.0 × 7.0 mm
Elastic plates angle, γ	150 deg	150 deg	150 deg	150 deg
SMA wires length	280 mm	240 mm	200 mm	150 mm

Two SMA wires were considered for each elastic hinge, in order to properly counteract the aerodynamic loads, also being compatible with typical fail-safe design criteria, because of being associated with a certain redundancy level. SMA element splitting together with the geometrical variation of the d parameter, limited by the local airfoil thickness, led to a new sizing of the plate cross sections. Final results are reported in Table 5.

6. Numerical Simulation and Results

Once the final configuration was defined, a FE model was realized; analysis was carried out through MSC/Nastran. 7075 T6 aluminium alloy, 2-mm thick plates were considered for the rib panels. SMA wires were simulated through beam elements with a suitable thermal expansion coefficient (negative). This allowed for simulating the SMA contraction following the activation input; however, the nonlinear numerical analysis was performed by step, until reaching the equilibrium value, shown in Fig. 2(b). SMA beams were hinged at the rib panels in order to avoid any bending stress. In the same way, the titanium elastic hinges were simulated through beam elements.

In order to meet the necessity of attaining cyclic behavior and accessing the maximum recoverable strain, the configuration shall undergo prestress. In the real case, this can be attained by deforming the structure downward and then installing the SMA element. The elastic force tries to restore the initial configuration, deforming the SMA that, in turn, resists this recovery. Now, both the structural elements (elastic hinge) and the SMA wires go through a prestress condition.

The designer's ability is crucial to individualize the structural deformed configuration that is able to generate the stress in the SMA wire that assures equilibrium with the "residual" elastic force at the required "neutral" position (Fig. 12b).

In the numerical simulation, to replay the physical process is difficult enough. Instead the following alternative process may be followed. A non-deformed structure is realized according to the schematic shown in Fig. 12(a). Numerically, this configuration is not loaded (no internal stress) at the level of the crossed laminas or the SMA wires. Starting from this configuration, the rib is brought to the "neutral" condition (Fig. 12b), by activating the SMA via a fictitious negative thermal load. In this condition, both the structure and the wires experience a stress field. In spite of the different theoretical paths, followed to reach this condition, the numerical and the real configuration are conceptually the same. The numerical approach may be said to properly imitate the reality. Finally, a further activation of the SMA wires simulate the action of the active devices aimed at deflecting the proposed flap.

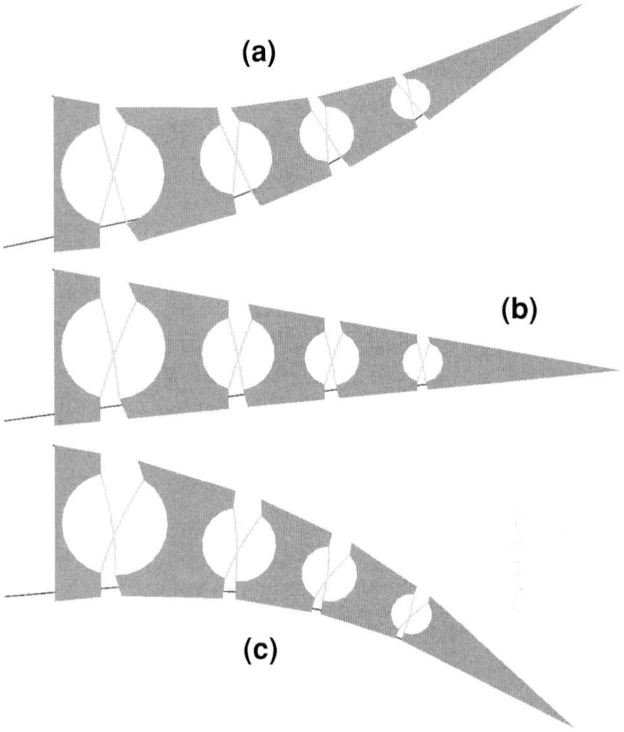

Fig. 12 Adaptive rib reference layouts: unloaded configuration (a); "neutral" position (b); activated rib (c)

Table 6 Preload condition details

	Hinge 1	Hinge 2	Hinge 3	Hinge 4
SMA force level[a]	82.6 N	81.94 N	81.65 N	79.91 N
Elastic hinge max stress	250 MPa	252 MPa	278 MPa	263 MPa
Hinge pre-load angle	10.17 deg	7.32 deg	7.42 deg	6.58 deg

[a]Each hinge has 2 SMA wires as actuator, doubling the necessary preload

Due to the modularity of the referred architecture, intermediate configurations are possible, between neutral and totally extended outline, according to the couples of SMA wires that are activated.

In Table 6, preload details are reported for both the laminas and SMA elements. It can be remarked that the maximum stress level reached in the elastic hinges under activation is always under 700 MPa.

Trailing edge vertical displacement and rotation, with and without aerodynamic loads, are reported in Table 7 for different activation conditions. Figure 13 shows finally the morphed

Table 7 Morphed flap performance: trailing edge displacement and rotation

	Performance without aerodynamic loads		Performance with aerodynamic loads	
	Vertical displacement, mm	Rotation, °	Vertical displacement, mm	Rotation, °
All SMA couples activated	246.7	32.24	223.9 (−9.24%)	29.84 (−7.44%)
Only first couple	84.83	7.01	71.24 (−16.20%)	5.71 (−18.54%)
Only second couple	72.31	7.95	61.2 (−15.30%)	6.83 (−14.08%)
Only third couple	58.33	8.51	49.53 (−15.08%)	7.62 (−10.45%)
Only forth	41.23	8.77	32.37 (−21.48%)	7.88 (−10.15%)

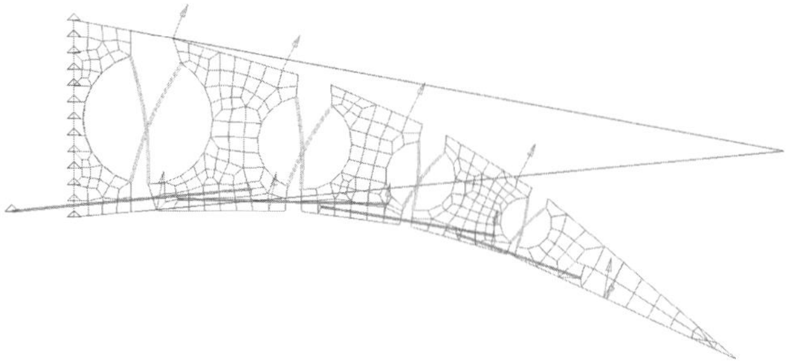

Fig. 13 Rib morphed shape for all SMA wires activated under aerodynamic loads

Table 8 Load conditions for the SMA wires during actuation

	Axial force[a] during actuation (without aerodynamic loads), N	Axial force[a] during actuation (with aerodynamic loads), N
SMA wires for hinge 1	153.27	651.73
SMA wires for hinge 2	183.28	446.09
SMA wires for hinge 3	186.65	310.01
SMA wires for hinge 4	199.21	238.77

[a]Force levels must be divided for the SMA actuators (2 wires for each hinge)

shape of the rib in presence of aerodynamic loads when all the SMA wires are activated.

The rib performance is well over the target of 160-mm trailing edge vertical displacement, estimated to lead to a lift coefficient of 2. Aerodynamic loads affect this result by 9.24% (a 7.44% difference is instead evaluated for rotation). This effect can be easily attributed to the flexibility introduced by the crossed laminas and the major stress level the SMA wires are subjected to (Table 8), however, compatible with their actuation capabilities.

When activating a single couple of SMA wires, vertical displacement and rotation show an opposite behavior; the more the "active" hinge is backward, the more the rotation increases while the vertical displacement reduces.

7. Conclusions and Further Steps

In this article, the design and the optimization of an SMA-based architecture, aimed at producing a morphing flap, have been presented.

The main innovation proposed by this study consists of the specific features of the rib architecture, following the principles of a "smart structure" (integrated, light, compact actuators), which is able both to largely deform itself (morphing) as well as to withstand external loads (bearing actuators).

A multi-body rib is referred to, whose components are linked to each other through elastic hinges made of crossed thin plates. In spite of the large forces produced, all the involved materials work inside the linear range. Geometrically, the configuration has intrinsic nonlinearities, caused by the pivots' translations. This effect has two main consequences: it decreases the effective carried load and increases the net rotation.

In order to maximize the adaptive flap performance in terms of max achievable rotations and vertical displacements, evaluated at the trailing edge, a parametric study was carried out, involving crossed laminas length and their relative angle. The best configuration was found to correspond at the higher values for thin plates length and angle, within the considered intervals.

Another study was then accomplished, by considering three different SMA element typologies: wires, ribbons, and rods. The optimal choice, when performance, compactness, and lightness are considered, was proved to be the wire-based one.

The problem of integrating the optimized elastic hinges within the rib was then faced. In order to obtain the max vertical displacement of the trailing edge, the largest number of these devices was integrated in the spar, considering the compatibility with the local airfoil thickness. Thus, four hinges suitably scaled and characterized by a 150-deg plate's relative angle were figuratively installed in the rib. Each of them presented peculiar characteristics, following the different geometric features, selected in order to guarantee the linear working of the involved materials.

Nonlinear numerical simulations were finally carried out, which can take into account a large number of expected displacements. A dedicated FE model of the rib architecture and the surrounding structure (i.e., skin elements and stiffeners) was realized. MSC/Nastran code was selected for simulations, integrated with an external Matlab routine for replicating the SMA elements behavior. A dedicated procedure was also set up in order to integrate the two programs. External aerodynamic loads were estimated by using a panel method on the morphing airfoil.

The structure behavior with and without aerodynamic loads, under the action of the designed device, was estimated and compared indicating the performance reduction following the presence of the external pressure field. Max displacement proved, however, to stand well above the target of 160 mm, corresponding to a C_L (lift coefficient) of 2.

The sophisticated numerical implementation of the morphing trailing edge herein presented would deserve a wider use. For example, to answer questions linked to the real use of these systems, such as the influence of the SMA actuator geometry on the cooling time, max achievable actuation frequency etc. This would provide new and interesting details in evaluating the active devices performance in view of the integration on actual elements and will be the subject of future investigations.

In the same way, the active rib design and functionality could be further improved. Attention should be paid to the skin elements: sliding panels or more performing aerodynamic solutions like "elastomeric" skins, which can absorb cyclically large deformations could be examined.

Moreover, the possibility of expanding the proposed device working envelop, as its implementation like an aileron (able to undergo both positive and negative camber variations) shall be considered. Such architecture would require for instance SMA wires working in an antagonistic way, i.e., mounted, respectively, over and under the hinges' location. One of the most important additions to the complexity of the implemented model should be the one of pondering over whether to opt for the SMA Shape Memory or the Super Elastic Effect.

References

1. E. Stanewsky, Adaptive Wing and Flow Control Technology, *Prog. Aerospace Sci.*, 2001, **37**, p 583–667 (Elsevier Science Ltd.)
2. S. Kuzmina, G. Amiryants, J. Schweiger, J. Cooper, M. Amprikidis, and O. Sensberg, Review and Outlook on Active and Passive Aeroelastic Design Concept for Future Aircraft, *ICAS 2002 Congress*, September 8-13, Toronto, Canada, ICAS Vol. 432, 2002, p 1–10
3. J. Schweiger, A. Suleman, S. Kuzmina, and V. Chedrik, MDO Concepts for an European Research Project on Active Aeroelastic Structures, *9th AIAA/NASA/ISSMO Symposium on Multidisciplinary Analysis and Optimisation*, CP Atlanta, GA, 2002
4. J. Spillman, The Use of Variable Chamber to Reduce Drag, Weight and Costs of Transport Aircraft, *Aeronaut. J.*, 1992, **96**, p 1–8
5. H.P. Monner, T. Bein, H. Hanselka, and E. Breitbach, *Design Aspects of the Adaptive Wing—The Elastic Trailing Edge and the Local Spoiler Bump*, Royal Aeronautical Society, Multidisciplinary Design and Optimization, London, 1998
6. J.R. Wilson, Active Aeroelastic Wing: A New/Old Twist on Flight, *Aerospace Am.*, 2002, **40**(99), p 34–37
7. I. Chopra, Review of State of Art of Smart Structures and Integrated Systems, *42nd AIAA/ASME/ASCE/AHS/ASC Structures, Structural Dynamics, and Materials Conference*, Seattle, WA, *AIAA J.*, **40**(11), 2002
8. J. Browman, B. Sanders, and T. Weisshaar, Evaluating the Impact of Morphing Technologies on Aircraft Performance, *43rd AIAA/ASME/ ASCE/AHS/ASC Structures, Structural Dynamics, and Materials Conference*, Denver, CO, USA, AIAA Paper 2002-1631, 22-25 April 2002
9. W.J. Buehler, J.V. Gilfrich, and R.C. Wiley, Effect of Low-temperature Phase Changes on the Mechanical Properties of Alloys Near Composition TiNi, *J. Appl. Phys.*, 1963, **34**, p 1475
10. S. Barbarino, S. Ameduri, and R. Pecora, Wing Chamber Control Architectures based on SMA: Numerical Investigation, *International Conference on Smart Materials and Nanotechnology in Engineering (SMN2007)*, 1-4 July 2007, Harbin, China, SPIE No. 6423-077
11. S. Barbarino, R. Pecora, L. Lecce, S. Ameduri, and A. Concilio, Airfoil Morphing Architecture Based on Shape Memory Alloys, *2008 ASME International Conference on Smart Materials, Adaptive Structures & Intelligent Systems (SMASIS08)*, October 28-30, Ellicot City, MD, USA, SMASIS08-480
12. C. Liang and C.A. Rogers, One-Dimensional Thermomechanical Constitutive Relations for Shape Memory Material, *J. Intel. Mater. Syst. Struct.*, 1990, **1**(2), p 207–234
13. K. Tanaka, A Thermomechanical Sketch of Shape Memory Effect: One-Dimensional Tensile Behaviour, *Res. Mechanica*, 1986, **18**(3), p 251–263
14. A.V. Srinivasan and D.M. McFarland, *Smart Structures: Analysis and Design*, Cambridge University Press, ISBN 0-521-65977-9, 1995
15. K. Shimizu and T. Tadaki, *Shape Memory Alloys*, H. Funakubo, Ed., Gordon and Breach Science Publishers, New York, ISSN 0889-860X, 1987, p 1–60
16. A.M. Kuethe and C.Y. Chow, *Foundations of Aerodynamics*, 4th ed., Wiley and Sons, 1986
17. S. Ameduri, P. Caramuta, and F. Lenzi, Experimental Characterisation of an SMA Rod Element for Helicopter Blade Morphing, *Internal Report*, CIRA-CF-08-0185, 2008
18. S. Ameduri, Test report di accettazione per nastri di SMA, *Internal Report* (in Italian), CIRA-TR-06-0215, 2006
19. A. Vigliotti and L. Lecce, Prove su campioni di SMA, *Internal Report* (in Italian), CIRA-TR-06-0346, 2006

JMEPEG (2009) 18:706–710
DOI: 10.1007/s11665-009-9390-1

Shape Memory Hooks Employed in Fasteners

D. Vokoun, D. Majtás, M. Frost, P. Sedlák, and P. Šittner

(Submitted October 17, 2008; in revised form December 9, 2008)

We study shape memory hooks employed in releasable fasteners. The designed fastener in question consists of two identical parts, each one containing an array of hooks made of the shape memory alloy with the pseudoelastic behavior at room temperature. The fastening function consists in interlocking of the hooks of the either part together. The firmness of the connection is tested for various hook geometry and heat treatment of the hooks. A model for the mechanical response of a loaded single shape memory hook is developed and described. Both the model and the experimental data indicate the firmness of the connection increases with increasing hook curvature and hook density. In the study, we also present the fastener manufacture using NiTi wires with a diameter of 150 µm.

Keywords fastener, modeling, NiTi hooks, pseudoelasticity

1. Introduction

Shape memory hooks may be used as fasteners (Fig. 1). In the recent shape memory alloy (SMA) fastener development, Powell et al. (Ref 1) construct an array of SMA hooks while each hook consists of either two kinds of SMAs or a SMA and other shape memory material. The hooks can open on an external stimulus (e.g., heating by electrical current) due to the shape memory effect. The main function of such a fastener is to interconnect the hook array with an array of loops and allow a release of the connection by making the hooks straight upon heating.

In our research, we study releasable fasteners made of SMA hooks utilizing the pseudoelastic effect. Although the pseudoelastic behavior of SMAs has been studied in numerous papers related to the SMAs, there is not much known about the mechanical behavior of pseudoelastic hooks made of NiTi alloys employed in SMA fasteners. The reason is first our application of the pseudoelastic effect is very specific and second, during the fastener releasing by pulling apart of the two hook arrays each individual hook is bent, especially in the curved part causing a complicated nonhomogeneous stress state. Additionally, not each hook is interlocked and the interlocked hooks are not being released at the same moment due to a certain variation in the hook length. Hence, the prediction of the thermo-mechanical behavior based on geometrical and material parameters and SMA hook density is not easy.

If the substrates in Fig. 1 are flexible, our SMA fasteners may serve as a kind of Velcro fasteners. In the Velcro type fastener, the peel-off force is considerably smaller than the pulling force in the normal direction or in any other direction. In our study, we focus on the measurement of the normal pulling force. Therefore, the substrates for attaching SMA hook arrays in our study are rigid. However, flexible substrates may be used in our further study to employ the SMA hooks in the Velcro type fasteners. One may ask what is the advantage of the SMA hooks used as Velcro type fasteners if much cheaper plastic hooks can do the same job. Actually, the advantage is less noise during the hook release when compared to the release action of the typical Velcro fasteners. Furthermore, other advantages are better durability, heat resistance, or functioning in chemically aggressive environments. The noise reduction of SMA hooks is due to damping capacities of SMAs due to ongoing face change transformations.

In the first part of our study, we describe the SMA hook shape setting. Further, we deal with mechanical loading of a single NiTi hook providing with some experimental data. We measure the minimum force necessary for a separation of two interlocked hooks, one of which is NiTi hook and the other is a steel rigid hook. In this kind of measurement, we vary the hook curvature and the NiTi hook heat treatment. In the second part of the work, we introduce a simple 1D model capable of simulating the load-stroke curves during the unhooking process. Finally, in the third part, we introduce the hook array manufacture and some measurements with complete hook arrays where an attention is paid to the hook density while the overall unhooking force in the perpendicular direction is measured.

2. SMA Hook Shape Setting

A cold-drawn NiTi wire with a diameter of 0.5 mm purchased from a commercial company was shape-set with a special fixture (Fig. 2) causing the wire to have the shape of a hook after releasing from the fixture. The fixing device was

This article is an invited paper selected from presentations at Shape Memory and Superelastic Technologies 2008, held September 21-25, 2008, in Stresa, Italy, and has been expanded from the original presentation.

D. Vokoun, D. Majtás, M. Frost, and **P. Šittner**, Institute of Physics ASCR, v.v.i., Na Slovance 2, 182 00 Prague 8, Czech Republic; and **M. Frost** and **P. Sedlák**, Institute of Thermomechanics ASCR, v.v.i., Dolejškova 1402/5, 182 00 Prague 8, Czech Republic. Contact e-mail: vokoun@fzu.cz.

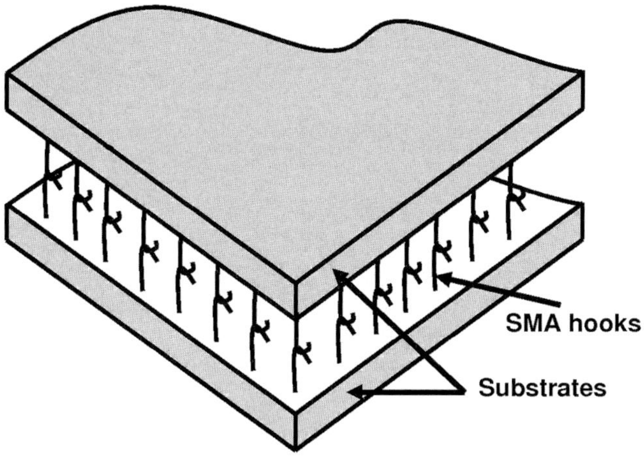

Fig. 1 A scheme of a SMA hook fastener

Fig. 2 A fixture for shape setting of the SMA hooks

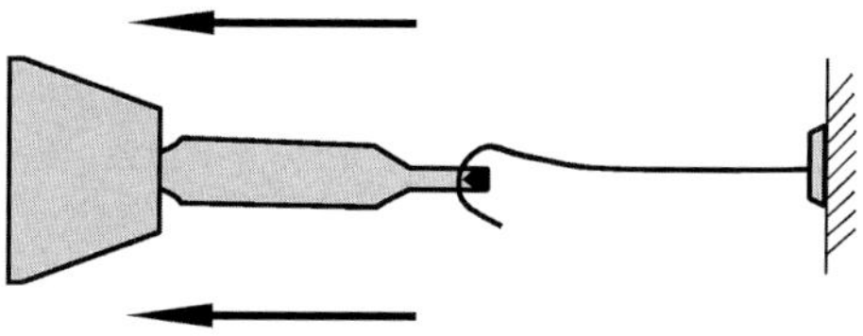

Fig. 3 A scheme of loading of a single SMA hook

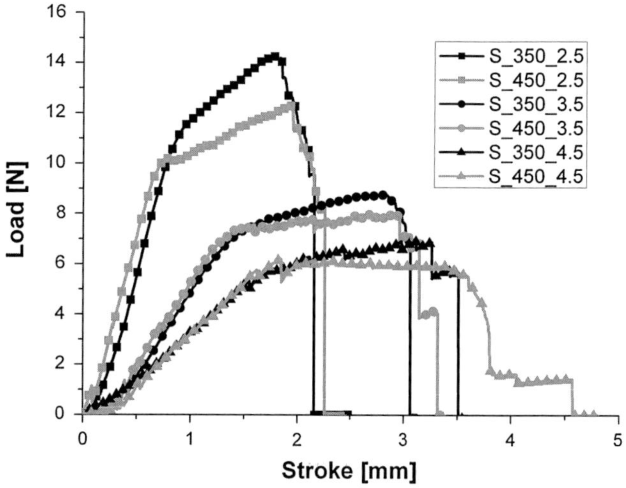

Fig. 4 The diagram of hook load vs. stroke for various samples

Table 1 A list of tested single SMA hook specimens

Sample	Heat treatment, °C	Hook inner diameter at the bent place, mm
S_350_2.5	350	2.5
S_450_2.5	450	2.5
S_350_3.5	350	3.5
S_450_3.5	450	3.5
S_350_4.5	350	4.5
S_450_4.5	450	4.5

designed to vary the NiTi wire hook curvature and to keep the hook prestressed. The shape-setting temperatures were 350 and 450 °C applied to the wire in the shape of a hook for 30 min.

We also prepared smaller SMA hooks using another cold-drawn NiTi wire with a diameter of 0.15 mm. The SMA hook shape setting was different: The wire was wound around a rectangular plate with a thickness of 0.8 mm and annealed at 350 °C for 30 min. The individual hooks were then cut from the shape-set superelastic wire.

Both the wires (diameters 0.15 and 0.5 mm) after the chosen heat treatment were superelastic at room temperatures. While the large hooks were used for single hook loading experiments and modeling, the smaller hooks were further manufactured into SMA hook arrays.

3. SMA Hook Loading Experiments

The single hook loading experiments were carried out with six various hook samples. The samples are listed in Table 1. The experimental setup is shown in Fig. 3. The long axes of both the hooks are co-linear. The hooks were pulled apart along the axis line while measuring the pulling force. We stopped pulling as soon as the unhooking happened. The unhooking event is characterized by a sharp drop of the measured force to the zero value.

The measured load-stroke curves in Fig. 4 can be grouped into three curve couples. Each curve couple corresponds to one of the hook curvature. It means that the curves in one couple contain the data related to both the heat treatments. We can see that the curve course and the maximum force reached strongly depend on the hook curvature. The larger the curvature the larger maximum the force is. In each couple, the maximum force associated with the 350 °C heat treatment is higher than that associated with the 450 °C heat treatment. The reason for that can be seen in Fig. 5: The tensile stress-strain curve for the 350 °C heat treatment has higher upper transformational plateau than that one for the 450 °C heat treatment. Most of the curves in Fig. 4 have a rugged course due to the frictional force acting at the point of the contact between the two hooks (steel and SMA hook). It is worth noting that sample S_350_2.5 after unhooking did not return to its original shape due to caused plastic deformation.

4. Model

The main motivation for creating a mechanical model simulating the SMA hook loading as described in the previous part was the need to determine the minimum force necessary for SMA hook releasing. However, the created model is additionally able to predict the whole load-stroke curves shown in Fig. 4 and load-stroke curves for SMA hooks of any other geometrical or material parameters.

For simulation of the material behavior, we used a pseudoelasticity module of the phenomenological 1D model iRLOOP (Ref 2). This module of the iRLOOP model, which was developed at IT and IP ASCR, v.v.i., in 2007 has the following features:

- keeps the concept of thermodynamic force linking stress and temperature, which drives the transformation processes,
- reliably describes both entire and partial martensite to austenite and reverse transformation processes with partial cycles and internal subloops,
- captures asymmetric behavior in tension and compression in pseudoelasticity,
- is able to be driven by prescribed evolution of strain and temperature of a wire, and
- is able to adapt to particular material by changing several experimentally well-measurable input parameters.

The model introduces two internal state variables: tensile martensite volume fraction, ξ^+, and compressive martensite volume fraction, ξ^-. The value of tensile resp. compressive martensite depends on the driving force for martensitic transformation, ϕ^+ resp. ϕ^- defined as:

$$\phi^+ = \frac{\sigma}{s^+} - (T - A_f) \quad s^+ = \frac{c}{\Lambda^+}$$
$$\phi^- = \frac{\sigma}{s^-} - (T - A_f) \quad s^- = \frac{c}{\Lambda^-} \qquad \text{(Eq 1)}$$

It also depends on the set of ordered pairs of real numbers called return points. In (1), Λ^+ and Λ^- are the transformation strains in tension and compression, and c is the material parameter. The model uses the adaptation of the Duhem-Madelung model of hysteresis described in Ref 3; the envelope functions of the hysteretic loop are approximated by piecewise linear functions which best fit the experimental data.

Since tensile and compressive martensite never appear simultaneously for temperature $T > A_f$ (austenite finish), which is the case of our study, we can consider the evolution of ξ^+ and ξ^- as completely independent and the constitutive equation for the total strain of the material in the simple form (thermal expansion neglected):

$$\varepsilon = \left(1 - (\xi^+ + \xi^-)\right)\frac{\sigma}{E_A} + (\xi^+ + \xi^-)\frac{\sigma}{E_M} + \Lambda^+\xi^+ + \Lambda^-\xi^-, \qquad \text{(Eq 2)}$$

where E_M and E_A denote the Young modulus of martensite and austenite.

In the simulation of the SMA hook loading, the hook was divided to a system of beam elements, each of them consisting of several material slices-truss elements (Fig. 6). We assume

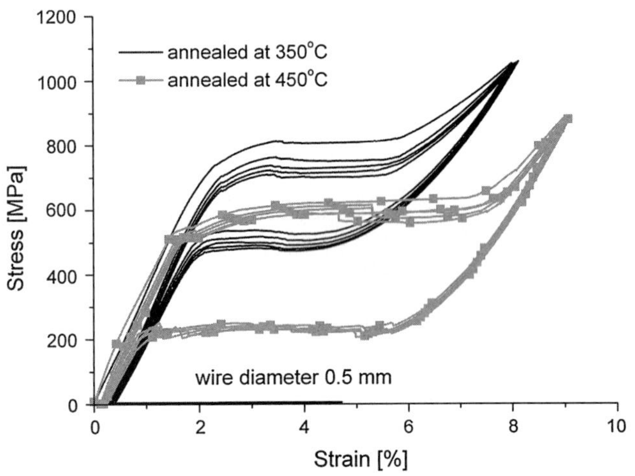

Fig. 5 The diagram of the tensile stress vs. strain dependence measured for the straight NiTi wire samples having the identical heat treatment as the SMA hooks

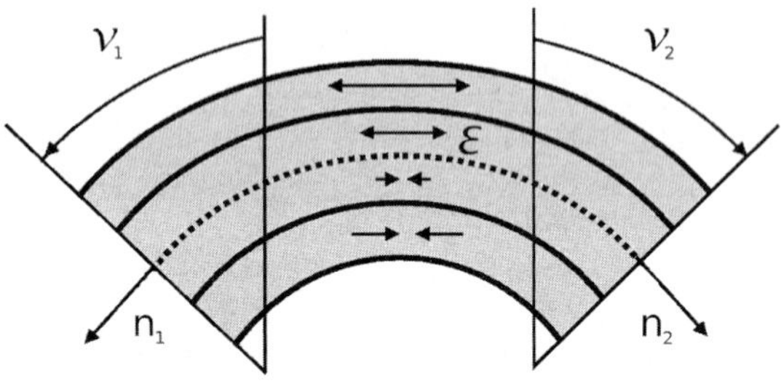

Fig. 6 The beam element consisting of several material slices-truss elements. The normal of cross section of the beam is assumed to be parallel to the beam axis

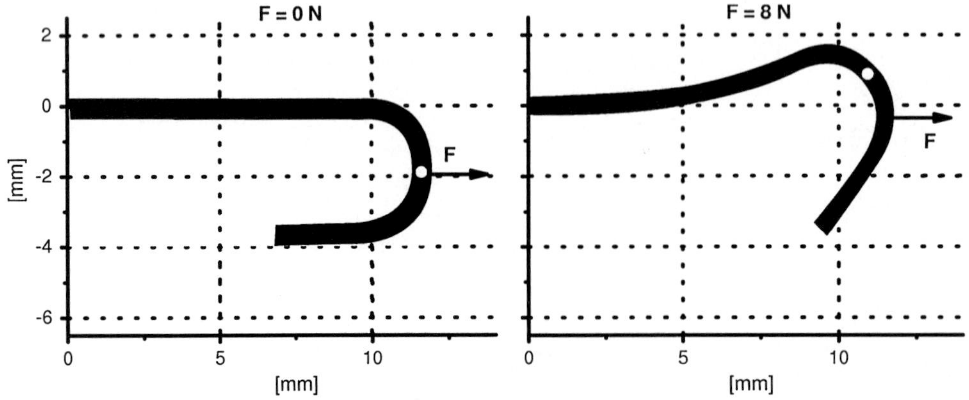

Fig. 7 The simulation of the hook deformation. It is assumed that the force direction remains perpendicular to the beam axis in the loading point

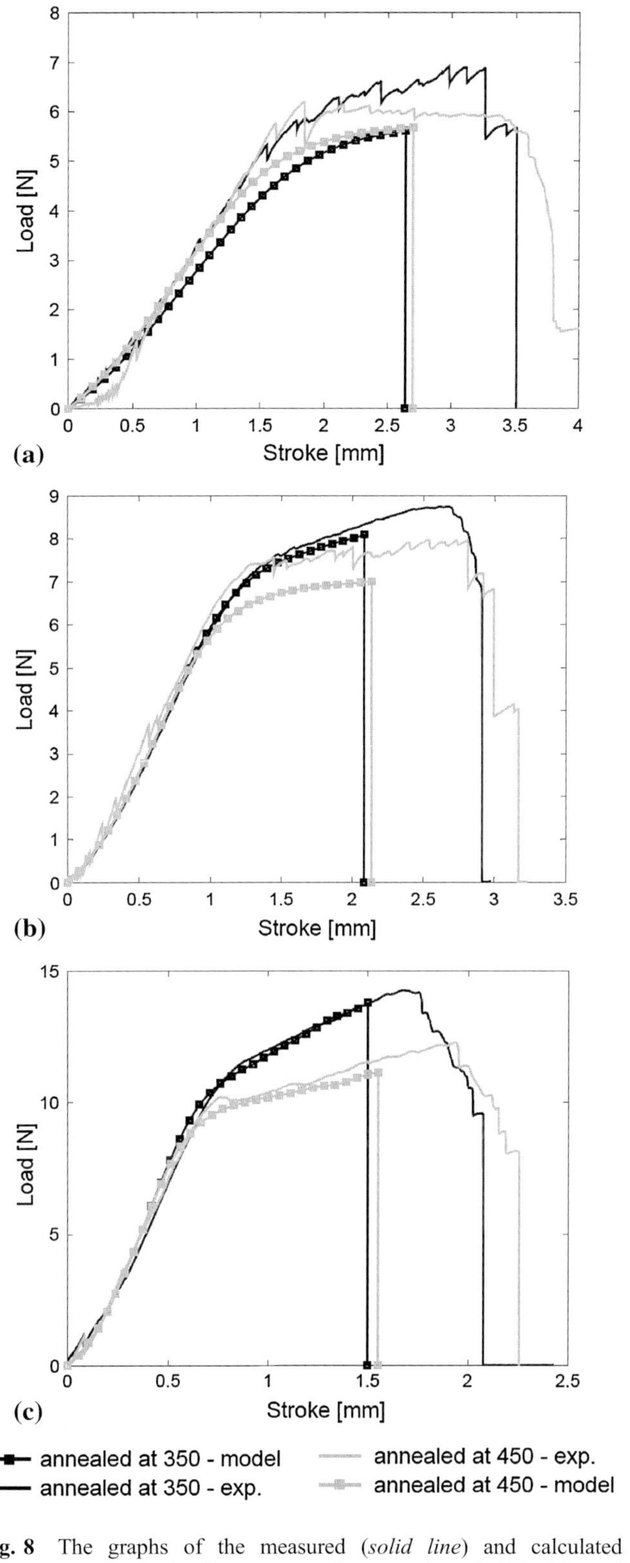

Fig. 8 The graphs of the measured (*solid line*) and calculated (*square markers*) loading curves for samples S_350_4.5 and S_450_4.5 (a), S_450_3.5 and S_350_3.5 (b), and S_350_2.5 and S_450_2.5 (c)

that during bending the normal lines to the cross section areas ($\mathbf{n}_1$, $\mathbf{n}_2$ in Fig. 6) of the beam remain parallel to the beam axis. Furthermore, the superelasticity module of iRLOOP was used for each of the slices.

The boundary conditions are given by the fixation of the straight end of the hook (fixation of displacement and rotation) and by placing the loading force to the point, where the force direction is perpendicular to the beam axis (assuming thus a neglectable friction between hook and its counterpart, Fig. 7). By this way, we have obtained a movement of the loading point within the hook arc in our model similar to that observed in experiment. This loading point movement adds another nonlinearity to the problem in addition to the material and geometrical ones.

Table 2 A comparison of the experimental and calculated maximum pulling forces

Sample	Maximum load—exp., N	Maximum load—model, N	Model inaccuracy, %
S_350_2.5	14.3	13.8	3.4
S_450_2.5	12.3	11.2	9.3
S_350_3.5	8.8	8.1	7.5
S_450_3.5	8.0	7.0	12.2
S_350_4.5	6.9	5.6	18.8
S_450_4.5	6.2	5.7	8.5

Figure 8 shows the measured and calculated loading force versus stroke for all our samples listed in Table 1. The model input data for our simulation were the hook geometry and the tensile stress-strain curves in Fig. 5. There is a fairly good agreement between the calculated and the experimental curves at the beginning of the loading. Also, the prediction of the maximum pulling forces is close to the real respective values (Table 2). There is a certain disagreement in the maximum stroke of the measured and calculated curves. The reason for it is the simplification of the contact condition by neglecting any friction forces.

5. Hook Array, Manufacture, and the Measurement

In order to manufacture the hook arrays, two steel plates with the dimension $22 \times 22 \times 2$ mm^3 and an array of microchannels with a diameter of 0.2 mm laser drilled in the plates were prepared. Two microchannel densities were realized: 121 and 441 holes per one plate evenly arranged in the area of 4 cm^2. A rubber layer with a thickness of 2 mm was deposited on one side of either plate to allow individual inserted hooks to be fixed in each hole. After SMA hook insertion into the plate holes, a thin rubber layer was deposited at the other surface of the steel plate. This thin layer had a function to separate the steel plate from the resin that was deposited on the top of it. After resin curing at temperature 80 °C for 2 h, both the rubber layers and the steel plates were easily removed so that only the array of hooks connected together in the right positions by the cured resin remained. Finally, the straight hook ends on one side of the resin plate were polished away. In this way, the SMA hook array samples were ready for the mechanical experiments. The experimental setup is shown in Fig. 9.

Three series of mechanical tests were carried out. In each series, the interlocked SMA hook arrays were pulled apart along the axis that is perpendicular to both the array planes while force was measured until complete SMA hook release. In the first [third] series of experiments, we used the hook arrays with the low [high] densities (Fig. 10). In the second series, we used the hook arrays with low density on one side and the high density on the other side.

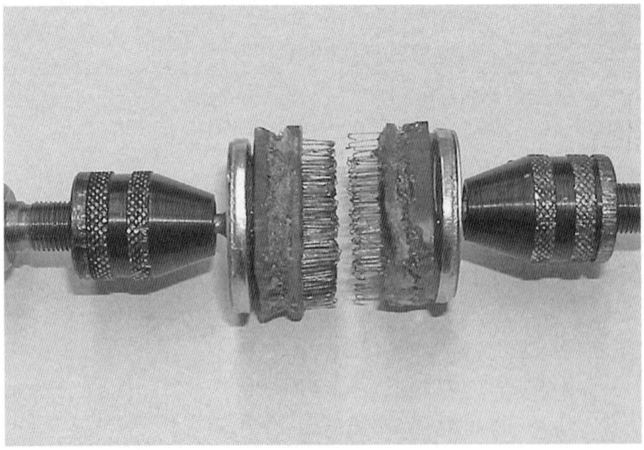

Fig. 9 Loading of SMA hook arrays

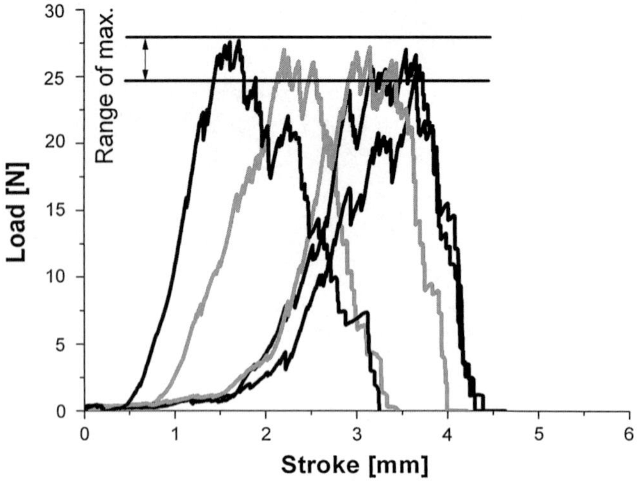

Fig. 10 The diagram of hook load vs. stroke for two SMA hook arrays with the high density of hooks. By the high density, we mean 441 hooks evenly arranged in the area of 4 cm^2

In Fig. 10, we show only the loading curves for the SMA hook arrays with the high density of SMA hooks. Since the number of the interlocked hooks is variable in each experiment and because at a given stroke each hook pair is in different stage of releasing, the loading curves in Fig. 10 are not overlapping and have a completely different character than those of Fig. 4. The maximum force lies inside a certain force interval in each series and the largest maximum force values are attained in the third series using SMA hook arrays with the high density. The mean values of the maximum forces in series 1, 2, and 3 are 5, 10, and 26 N, respectively. Hence, the maximum force may be optimized by changing SMA hook array density. Other force measurements such as measurements of the maximum force in shear loading and measurement of the maximum force during peel off are important for testing our superelastic applications. Therefore, in our further research, we plan to arrange the abovementioned measurements including the noise measurement during peeling off our SMA hook fasteners.

6. Conclusions

(1) The minimum force necessary for the single SMA hook releasing strongly depends on the hook geometry and depends less on the hook heat treatment.
(2) The model predicts the mechanic response well. The only disagreement in the maximum stroke before the unhooking event consists in the simplified contact condition. Based on the experimental and model data, the maximum force is increasing with the decreasing hook diameter. However for the diameters at the curved part of a hook <2.5 mm, plastic deformations develop after hook releasing (the wire diameter considered is 0.5 mm) and the hook does not return to its original shape any more.
(3) A method for making SMA hook array was shown. This method enables us to form precise alignment of hooks and reach predefined hook densities.
(4) The maximum force for normal direction loading of the SMA hook arrays may be optimized by changing the SMA hook array density.

Acknowledgments

D.V., D.M. and M.F. acknowledge support by a Marie Curie International Reintegration Grant within the 6th European Community Framework Programme. This research has also been supported by EC via FP6 integrated project AVALON (NMP2-CT-2005-515813). M.F. and P·S. acknowledge the Ministry of Education, Youth and Sports of the Czech Republic (MSMT) for financial support under Grant No. 1M06031.

References

1. Powell et al., U.S. Patent 7,032,282 B2
2. M. Frost, P. Sedlák, M. Sippola, and P. Šittner, *Smart Mater. Struct.*, in review
3. J.A. DeCastro, K.J. Melcher, R.D. Noebe, and D.J. Gazdosh, *Smart. Mater. Struct.*, 2007, **16**, p 2080–2090

JMEPEG (2009) 18:711–716
DOI: 10.1007/s11665-009-9383-0

1059-9495/$19.00

"One-Module"-Actuators Based on Partial Activation of Shape Memory Components

S. Langbein and E.G. Welp

(Submitted September 18, 2008; in revised form November 29, 2008)

An advantage of shape memory alloys (SMAs) is their potential to generate integrated actuator systems with a shape memory component. This can be accomplished for example by activating the thermal shape memory effect in selected regions of the SMA-component. We refer to this process as partial activation. The purpose of the present study is to find a way to create universal actuators with properties adjustable for various applications solely by partial activation. Thus, an object of investigation is the analysis of properties and capabilities of partial activation. Furthermore this study also implicates the survey of possibilities for partial power supply and electrical contacting. One possibility to use partial activation in integrated systems is given by the agonist-antagonist design. This type of design offers the advantage that a return spring or a mechanical brake for clamping the position without feeding electrical power is not necessary. On the other hand retention force is limited by the martensitic plateau and positioning accuracy by the elastic portion of mechanical stress. To solve these problems with constructive or control-oriented solutions is furthermore an aim of this study. Another approach is to use partial activation for influencing passive superelastic structures like hinges, dampers, or return elements by changing the austenitic plateau stress in integrated systems. To create a multifunctional integrated system, the NiTi-elements presented in this study offer various options since they apply partial activation both for thermal shape memory and for influencing super elasticity.

Keywords actuator, modular system, partial activation, shape memory alloys

1. Introduction

Integrated actuator systems begin to play a more and more important role. Above all in the field of microtechnology, conventional actuators will soon reach their limits and further miniaturizations are barely feasible so that new and particularly integrative solutions need to be developed (Ref 1). Especially in a critical environment like in fluids, in a vacuum or in clean rooms, systems are required that will not become affected by environmental conditions and vice versa which will have no impact on their environment, e.g., by polluting. Due to their considerable volume-specific work capacity, shape memory alloy (SMA) structures are particularly suitable for such kind of applications (Ref 2). In particular in microsystems technology and in micro assembly, there is a high demand for Smart Materials that can be produced in smallest dimensions and that are able to create considerable forces and large displacements.

This article is an invited paper selected from presentations at Shape Memory and Superelastic Technologies 2008, held September 21-25, 2008, in Stresa, Italy, and has been expanded from the original presentation.

S. Langbein and **E.G. Welp**, Institute Product and Service Engineering, Ruhr-University Bochum, Bochum, Germany. Contact e-mails: langbein@lmk.rub.de and welp@lmk.rub.de.

As for the applications, we can differentiate between tools or grippers for the manufacture and assembly of micro components and drives of micro systems. But also beyond microtechnology, the SMA prove of the potential to realize simple actuator systems based on integrated structures resulting in a reduced complexity and thus in a reduction of assembly costs and costs in general.

One ambition of this research project is to offer integrated actuators based on partial activation of shape memory components. While doing so, the complete system is not composed of several standardized components, as in a conventional sense, but a single, monolithic component is made available that can be divided into various, partial activatable regions. Such a SMA-component can be understood as an "one-module" or integrated modular system, with different local operating centers, which can be activated individually.

2. Basis of Partial Activation

Compared to local configuration (Ref 3, 4) with partial activation the functional characteristics result only temporary by the heating of a particular part of the component zone. Therefore the phase change respectively the shape memory effect in operating condition is limited to this local area. A specific thermo-mechanical pre-treatment for the implementation of partial activation is not necessary. In this context general mechanisms used for thermal activation such as self-heating, thermal conduction, or thermal radiation are applicable as heating mechanisms. A valuation which sums up the activation processes is given in Table 1. With partial activation either the one-way effect or the two-way effect, or a

Table 1 Thermal methods of activation

Method of activation	Principle of operation	Advantages	Disadvantages
Heating by inherent resistance	The material is heated during current flow by its electrical inherent resistance	-Single partial activation possible -Heating is well controllable -Movement of the structure not interfered by activation principle -Quick response time	-Relatively high currents necessary -Contact for connection cable required
Heating by radiation	The material is heated by thermal radiation either by a radiation heater or a laser diode	-Contactless, no interference of movement	-High heat loss -Hardly controllable local heating -High complexity
Heating by heating elements	The material is heated by attached heating elements	-Single partial activation possible -Heating is well controllable	-Interference of movement by heating elements -Slower response time
Heating by induction	The material is heated by electromagnetic induction	-Contactless, no wiring effort regarding component -Quick response time	-High effort regarding constructive fitting of inductor coils -High currents necessary -Potential interference of movement by inductor coils

variation of the pseudo-elastic effect is implemented. The variation of the pseudo-elastic effect depends on the temperature sensitivity of the pseudo-elastic parameters, e.g., the pseudo-elastic plateau stress. A coexistence of both effects in one component as with local configuration is only possible, if the corresponding and dimensioned reset area is heated at first, to create superelastic characteristics. The warm-up period of the reset area only ends after a complete resetting of the structure. Depending on the structural design this process requires a local memory imprinting, as the reset area must be set under stress when the actuator area has reached its high-temperature form. The heating of pre-strained structure areas and the corresponding increase of its stiffness can also contribute to the improvement of the structural stability of such structures. As a result there are the following functional variants, which can be generated by partially activatable structures:

- actuator functions with the following characteristics:
  - standard design
  - agonist-antagonist design
  - step-actuator design
  - integrated return spring design
- pivotal functions with different stiffness
- spring functions with different stiffness
- damping functions with different damping characteristics.

3. Experimental Pilot Surveys

The basic aim of the present study was to test the impact of partial activation on stress- and strain capability. Hereby the wire was activated in its full length, in half of its length and in one-fourth of its length. This activation was conducted with stresses of 25 N/mm^2, 50 N/mm^2, 100 N/mm^2, and 200 N/mm^2, to find out the exact impact of the stresses on the respective activation. A wire with the marking H (approx. 49.5 at.% Ni) with two different thermo-mechanical pre-treatments was used. The wire, cold shaped with 30%, was shaped by heat treatment once at 300 °C for 20 min and once at 400 °C for 20 min. After that the wires were electrically

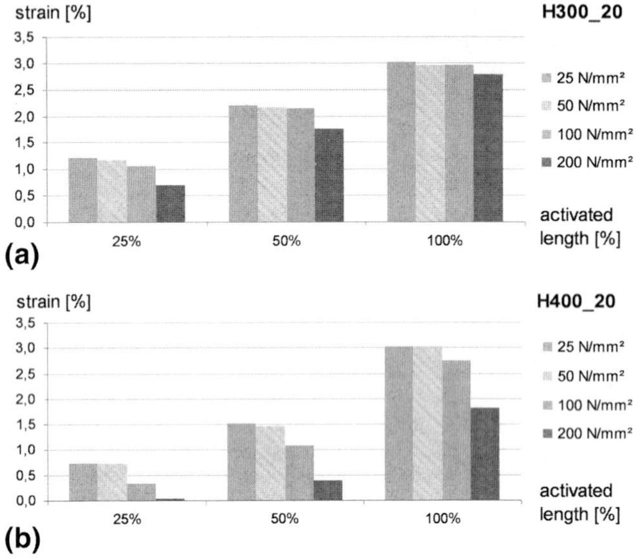

Fig. 1 Maximum strain of two wires with different thermo-mechanical pre-treatments

activated in a test rig by means of their inherent resistance. The contacting clamps were attached to the wires in distances of 25, 50, and 100%, relating to the clamped lengths of the wires. Hereby the displacement respectively the strain of the wire was measured. The results are presented in Fig. 1. In this context the maximum strain of the respective wires is related to the activated length occurring with different stresses.

The results show that the stress prevailing inside the wires can have a considerable impact on the strains of the partially activated wires. This results from the fact that with stresses above the martensitic plateau in combination with partial activation, the inactivated areas of the wire are expanded and that at the same time their strains seem to decline toward the periphery. The varying results of the two wires can be traced back to the fact that the 300-20 wire has a plateau stress of approx. 220 N/mm^2, while the 400-20 wire is at 100 N/mm^2. Therefore the strains of the 300-20 wire remain constant, while the 400-20 wire shows a significant strain loss at its periphery.

Another aspect that can be noticed while comparing both diagrams is that there are differences in strain value of the two wires, even at low stresses and strains. One can see that the 300-20 wire shows a strain of approx. 2%, e.g., at a stress of 25 N/mm^2 and an activated length of 50%, while the 400-20 wire on that condition only shows a strain of 1.5%. The reasons for this are the different A_S-temperatures of the two wires. With the 300-20 wire the temperature is 90 °C, with the 400-20 wire it is, however, 110 °C. This means that the existing thermal conduction within the wires and the low transformation temperature within the 300-20 wire cause a transformation of 50% of the wire's length respectively activate 50% of it. The basic experiments show that with the application of a partial activation in shape memory components, the load on the one hand and the transformation temperatures respectively the control accuracy on the other hand, has a considerable impact on the operating mode of these shape memory components. The outcomes of this are the following basic rules for the application of such components:

- The stress of the shape memory components musst be clearly below the martensitic plateau stress,
- The heating of the areas which are to be activated should not exceed the transformation temperatures too much, to avoid a thermal conduction and with it an expansion of the areas which are to be activated,
- The reset force may not affect the activation cycle.

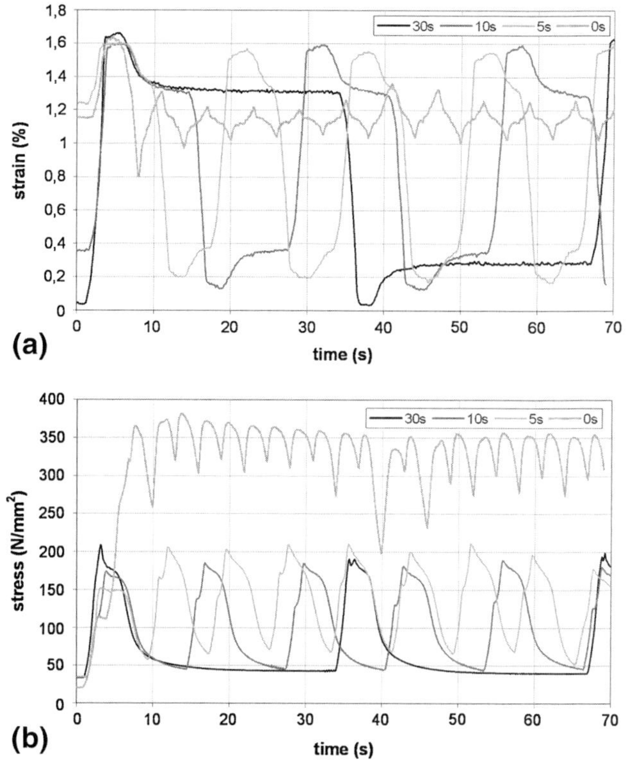

Fig. 2 Agonist-antagonist design wire actuator with different cool-down periods in a specified time frame

4. Experimental Surveys of Agonist-Antagonist Design

The problem of a variable reset force can be solved in an effective way by using the agonist-antagonist design. Another advantage of this construction design is that the intermediate positions can be kept continuously variable without the application of a mechanical construction. The positioning accuracy of this construction design, however, is a problem (Ref 5). Another problem is the regulation regarding the warm-up and cool-down periods. Figure 2 presents the different cool-down periods in a specified time frame. The heating-up period is always 3 sec.

One can observe that a steady-state equilibrium occurs between the two wires both at a cool-down period of 30, 10, and 5 sec. In addition to that, the force decreases to the same value in its currentless condition, whether the cool-down period is 30, 10, or 5 sec. This value is equivalent to the pre-stress of the wires due to assembly. The maximum value reaches a force of 14 N, which corresponds to about 200 N/mm^2. If the wires are warmed up immediately one after another without an adequate cooling phase, the force increases and with it the stress inside the wire increases considerably, too. The measured force will no longer be equivalent to the value of the pre-stress due to assembly. The forces between the wires rise up to 27 N which corresponds to a stress inside the wires of approx. 375 N/mm^2. Since the respective antagonist is not martensitic and for this reason not malleable, the agonist cannot develop its adjustment travel. The adjustment travel remains minimal with a value of only 1 mm. To achieve the same adjustment travel of about 5.5 mm, as in the 10 sec and 5 sec experiments, even without a cooling phase, one would need to extend the warm-up period of the agonist as well as the cool-down period of the antagonist. As the effectively sufficient conductor temperature increases due to the extended warm-up period, this would cause the warm-up period of the antagonist to be even longer. Furthermore the graph of the experiment significantly shows that there is a positioning deviation of approx. 1 mm—almost 20% of the adjustment travel—which occurs in every positioning step. The positioning deviation can be put down to the elastic portions of the martensitic deformation. In conclusion the experiments referring to the agonist-antagonist design show two things:

- There is a considerable positioning deviation, which can be reduced in technical applications by a positioning regulation.
- A precise regulation for the warm-up and cool-down periods needs to be made, so that the two wires do not work against each other while being in austenitic condition. This leads to high stress values inside the wire which can have a negative effect on the fatigue behavior.

The constructive solution of both regulating difficulties is a challenge for further research.

5. Development of Partial Activated Structures

5.1 Selection of Activation Method

There are various methods to activate the shape memory structures, which have been already mentioned in the previous paragraph. The presented activation methods are all in principle applicable for a partial activation of the structures. The advantages and disadvantages, given in Table 1, need to be considered to select the best possible operation and to meet the favoured requirements.

5.2 Selection of Connection Method

The easiest operation for partial activation is the heating by electrical inherent resistance. To employ this activation method, the connection between the actuators respectively the structures and the connection cable are necessary. There are different ways to establish this contact. First, there is the option to attach the connection cable to the shape memory structure by connection techniques such as substance-bonding, friction-locking, or form-locking. Second, there is the option to connect the structures via sliding contacts or similar. Within these two fields innumerable technical solutions exist already; one has to select the most adequate one. In general crimp or welded connections are suited for contact purposes.

5.3 Conceptual Design of Structures and Selection of Construction Design

In general three types of construction designs can be distinguished, the differential construction design, the partial integral construction design, and the complete integral construction design. If one wants to implement the complete integral construction design and with it the "one-module"-module system, without using the agonist-antagonist construction design, it is an advantage to combine the partial activation with the local configuration as described in (Ref 3). Pseudo-elastic pivotal areas or when needed reset elements can only be integrated into the structure by local configuration. With pure local configurations a contact with the actuator area is not required because the entire component can be heated.

Tables 2 and 3 also show a systematic development of actuator structures, which are applicable for partial activation. Furthermore the conceptions take into consideration the combination with local configurations. For the development, selected basic structures were organized by suitable classification criteria and recorded in classification schemes. The presented structures are only an extract from an optionally expandable selection of solutions. Further solutions result from, e.g., a variation of the position of actuating elements and reset areas. Due to this variation other transmission ratios, adjustment forces, and travels can be implemented. An extension would also be achieved by connecting similar or separate structures or structure elements to complex or potential spatial structures.

The classification schemes which depict the structure synthesis are allied as follows: The structures are divided into return spring design (Table 2) and agonist-antagonist design (Table 3). Within the tables the structures are divided into gripper structures and movement structures. Two different basic structures are described each time. The rows are ordered, consecutively from top to bottom, and show the development of the integration potential. In row 1 the construction design is constructed completely differential, i.e., the carrier consists of steel or plastic, the resetting force is achieved by a steel spring or a second SMA-component. The shape memory effect only inherits the actuator functions. Such a construction design can only be found in SMA-based applications. Rows 2, 3, and 4 (rows 2 and 3 in Table 3) show partially integrated structures, in which the function integration develops via different parameters: In row 2 carrier and return spring are put together as one component (only by return spring design). Such an actuator is shown in (Ref 6). In row 3 (row 2 in Table 3) the carrier is manufactured from an SMA and connected with

Table 2 Systematic development of partially activatable actuator structures (return spring design)

partial activatable SMA-structures (return spring design)									
design	parts	material		integrated parts	gripper-structures		actuator-structures		
		SMA	no SMA		1	2	1	2
differential	carrier		X	-				
	return spring		X	-				
	SMA-component	X		-				
partial integrated	carrier		X	X				
	return spring		X	X				
	SMA-component	X		-				
	carrier	X		X				
	return spring		X	-				
	SMA-component	X		X				
	carrier		X	-				
	return spring	X		X				
	SMA-component	X		X				
integrated	carrier	X		X				
	return spring	X		X				
	SMA-component	X		X				

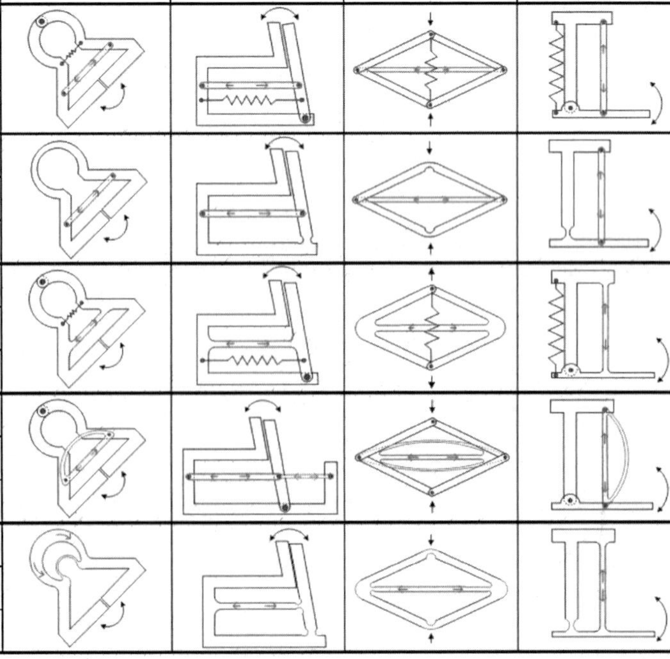

the shape memory actuator element to a single component. The return spring or the second SMA-component is separated. Row 4 (row 3 in Table 3) shows an assembly made of a carrier and a local set up of shape memory structure, which integrates actuating elements and reset elements. The solutions as given in row 5 (row 4 in Table 3) have a completely integrated, monolithic construction design and meet the criteria of a "one-module"-module system. All functional and structural characteristics are combined in a single component. Thereby the solutions given in row 5 (Table 2) consist of an SMA with an in situ configuration of actuator areas and pseudo-elastic flexure hinges or of an SMA were the reset area is heated at first to perform structure deformation and resetting. The solution in row 6 has a similar assembly, differing only because here the reset element results from the agonist-antagonist principle.

5.4 Construction of a Demonstrator

Figure 3 shows the developed demonstrator. The system consists of four different components: bottom plate, bracket, lever arm, and the actuator wire. The actuator wire is conducted via a bore hole in the bottom plate to the corresponding bore hole in the lever arm. There the wire is fastened while the lever arm is in a horizontal position. Then the wire is connected via a required bore hole in the lever arm again. The bore hole depends on the required transmission ratio. The wire is fastened

Table 3 Systematic development of partially activatable actuator structures (agonist-antagonist design)

partial activatable SMA-structures (agonist-antagonist design)								
design	parts	material		integrated parts	gripper-structures		actuator-structures	
		SMA	no SMA		1	2	1	2
differential	carrier		X	-				
	SMA-component1	X		-				
	SMA-component2	X		-				
partial integrated	carrier	X		X				
	SMA-component1	X		X				
	SMA-component2	X		-				
	carrier		X	-				
	SMA-component1	X		X				
	SMA-component2	X		X				
integrated	carrier	X		X				
	SMA-component1	X		X				
	SMA-component2	X		X				

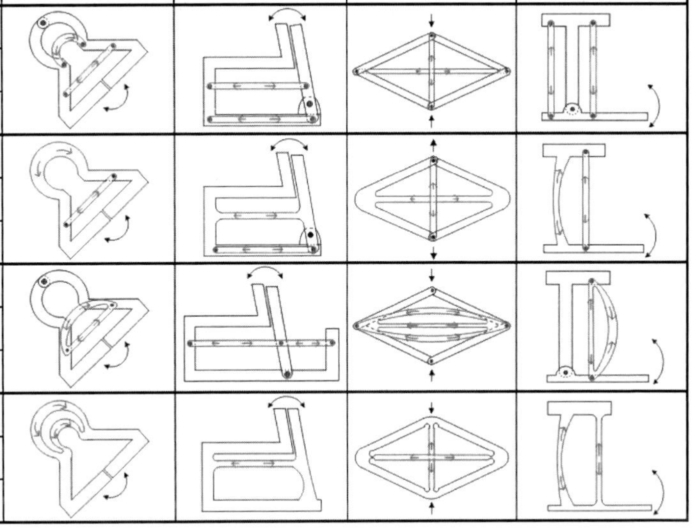

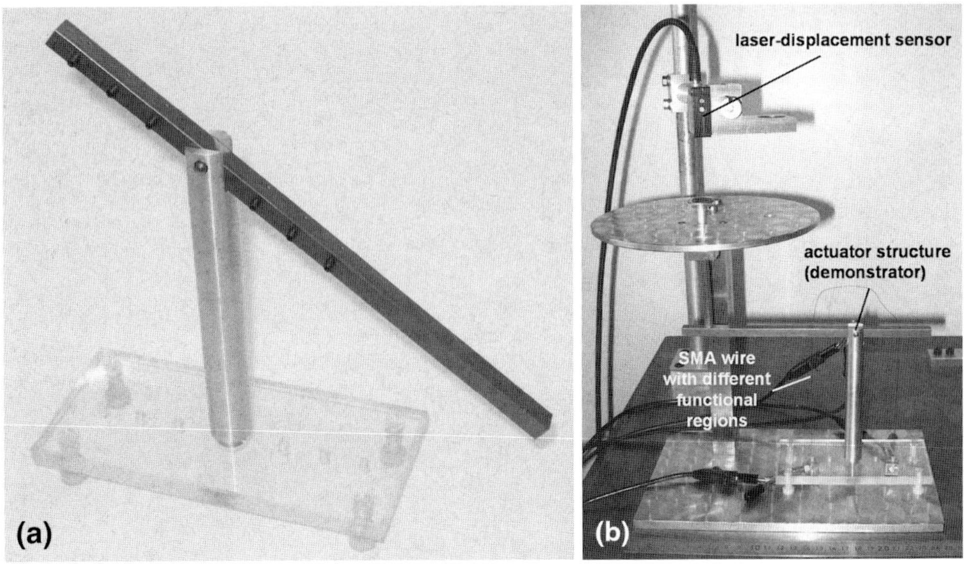

Fig. 3 Demonstrator and experimental setup

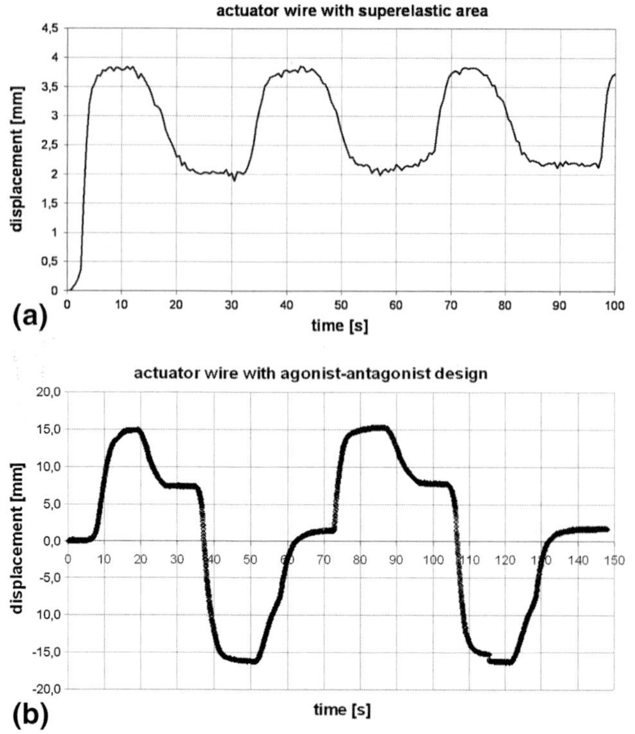

Fig. 4 Experimental results

for a second time and after that it is connected via the corresponding bore hole in the bottom plate again and it is clamped there. The actuator is now able to perform movements according to the agonist-antagonist principle.

The experimental results, generated by means of this demonstrator are presented in Fig. 4. On the one hand typical results of the agonist-antagonist design are displayed, but on the other hand Fig. 4 also shows results where the actuator wire was set up locally and therefore has superelastic characteristics on one side in order to initiate independent resetting.

6. Summary/Conclusions

The present research shows that it is feasible to generate "one-module"-modular system structures. These structures have the potential to increase standardization and the potential to reduce the number of components. Due to the manifold types of construction and movement these structures are particularly suitable to meet complex movement tasks. In micro technology as well, the idea of partial activation offers many possibilities. The problem of local heating and its associated contact could be solved by conductor paths integrated on the board, being built together with the actuator. For this reason partially controllable actuator arrays could be generated in a simple way. Furthermore the basic experiments have identified, the problems occurring with the application of partial activation and the agonist-antagonist design and how these problems can be solved in technical systems.

References

1. H. Janocha, *Adaptronics and Smart Structures-Basics, Materials, Design and Applications*, Springer-Verlag, Berlin, Heidelberg, New York, 1999
2. Å. Burman, E. Møster, and P. Abrahamsson, *On the Influence of Functional Materials on Engineering Design, Research in Engineering Design*, Springer-Verlag, London, 2000
3. S. Langbein and E.G. Welp, *Generation of Smart Structures on the Basis of In Situ Configuration of Shape Memory Alloys, Advances in Science and Technology*, Vol 59, Trans Tech Publications, Switzerland, 2008, p 184–189
4. Y. Bellouard, T. Lehnert, J.-E. Bidaux, T. Sidler, R. Cavel, and R. Gotthardt, Local Annealing of Complex Mechanical Devices: A New Approach for Developing Mono-Lithic Micro-Devices, *Mater. Sci. Eng.*, **A273–275**, p 795–798, Elsevier Science, 1999
5. M. Kohl, *Entwicklung von Mikroaktoren aus Formgedächtnislegierungen*, Dissertation, Forschungszentrum Karlsruhe, 2002
6. S. Langbein, E.G. Welp, and J. Sohn, Development of a Variable and Integratively Structured SMA-Actuator for Multifunctional Application, *Proceedings of the 11th International Conference on New Actuators*, Bremen, 2008

JMEPEG (2009) 18:717–724
DOI: 10.1007/s11665-009-9487-6

Electrically Actuated Antiglare Rear-View Mirror Based on a Shape Memory Alloy Actuator

T. Luchetti, A. Zanella, M. Biasiotto, and A. Saccagno

(Submitted September 15, 2008; in revised form April 3, 2009)

This article focuses on the experience of Centro Ricerche FIAT (CRF) regarding the development of shape memory alloy (SMA) actuators, and addressed some new design approaches which have been defined. Specific characteristics of shape memory materials, such as the efficiency of the transformation, have oriented the design of actuators toward occasionally used devices. The antiglare manual mechanism, incorporated in the internal rear-view mirror of a car, fits this new approach well. An antiglare rear-view mirror is a system capable of detecting a glare situation during night-time driving in order to automatically switch the mirror plane so as not to distract the driver. The low forces required, together with the silent, bistable movement are suitable for the use of a SMA actuator in this application. In the first part of the paper, the conceptual design is illustrated and a preliminary overview of the working principle is provided together with a series of considerations regarding the kinematics and the layout of electronic sensors in order to realize a fully controlled mechatronic prototype. Before concluding, the description of the realization of a working prototype is presented. The prototype of the EAGLE (Electrically Actuated antiGLare rEar-view mirror) system has provided experimental confirmation that such a device can satisfy fatigue and functional test requirements, thus offering the opportunity to spread the use of SMA devices in the automotive field.

Keywords automotive antiglare mirrors, device re-engineering, shape memory alloy

1. Introduction

Exploiting compactness and silent operation of shape memory actuators, a new internal rear-view mirror with an automatic antiglare mechanism has been developed.

To date, automatic antiglare internal mirrors have been made of electro-chromic materials only. Electro-chromic rear-view mirrors produce a variation in the reflectance in case of a dazzling light. These materials have a response time of several seconds after the glare condition has occurred, meaning that the device is not very effective in terms of protecting against the disturbance to the eyes of the driver. On the other hand, relatively expensive materials are needed and the darkening produced on the mirror surface to eliminate the glare is often considered to be a negative aspect.

An improved solution from the ergonomic point of view therefore is offered by the mechanical movement of the reflecting prismatic surface. Traditionally, manufacturers have equipped vehicles with prismatic overhead rear-view mirrors which function by using a mirror backed triangular piece of glass attached to a mechanical pivot. During normal operation, collimated incident light from the rear windshield is reflected off the mirror. Upon angular adjustment, the prism reflects light off its frontal surface at night, creating a dimmer reflection (Ref 1).

The beneficial features of the manual actuator include low cost and silent operation, but consumers are generally averse to manual actuation in today's vehicles; as a consequence, most car drivers simply do not use such a function.

Furthermore, the need to re-engineer this component derives from the safety perspective since with the manual device drivers must take one hand off the driving wheel for a couple of seconds in order to rotate the mirror. Often car buyers are interested in technology features if offered at a relatively low price which also helps to distinguish the specific product. Following this reasoning, car designers work at continually developing at new technologies to perform a particular function but at lower cost, and potentially with a higher level of quality, functionality and reliability.

In recent years, mirror activation by electro-magnetic actuators have been attempted but these have not been particularly successful due to the size of the actuators and actuation noise which can distract the driver. Clearly inside a car every distraction to the driver is potentially dangerous. For this reason, the antiglare system exists and the manual or the electro-chromic technology is used.

The device with shape memory actuator overcomes these restrictions with a very fast actuation (from the normal to antiglare position in less than a second) that is also silent and gradual without mechanical jerking.

This article is an invited paper selected from presentations at Shape Memory and Superelastic Technologies 2008, held September 21-25, 2008, in Stresa, Italy, and has been expanded from the original presentation.

T. Luchetti, A. Zanella, and **M. Biasiotto**, Centro Ricerche Fiat S.c.p.a, Torino, Italy; and **A. Saccagno**, Ficomirrors, Ficosa International, Barcelona, Spain. Contact e-mail: tommaso.luchetti@crf.it.

This paper intends to demonstrate that shape memory alloys (SMAs) actuators represent an appropriate and feasible solution for providing a low cost and reliable automatic device.

The main features of shape memory actuators are the high density power and silent operation. The first feature is essential to enable a simple and low weight device, whereas low noise in the cabin of a car is fundamental to avoid driver distractions.

Previous experience in Centro Ricerche FIAT (CRF) gave rise to the realization of a prototype, the main aim of which was to produce a mirror rotation through the shape memory wire contraction during the glaring time. This system was composed of a direct connection between the SMA wire and mirror. Thus, during glare, the wire has to be heated continuously to maintain the required twisted position. As a consequence, a series of problems related to activation and deactivation time, power consuming and fatigue lifetime were encountered.

One aim of this paper is to describe the criteria adopted in the design of a new electrically actuated antiglare rear-view mirror.

2. Component Requirements

For the development of the device, the automotive specifications and the experience on mirrors of the development partner Ficosa International were taken into account.

The typical requirements of automotive components are very demanding both in terms of performance and reliability (Ref 2). Each automotive component maker must ensure these specifications are respected before supplying the component to the car maker. The component checking process is very stringent and often makes the introduction of new devices in the car difficult, particularly in the case of new, emerging technologies. However, it also stimulates development of new materials capable of meeting the specifications required.

Figure 1 shows the main specifications that an automotive component must comply with. The most critical points for development of new actuators based on shape memory materials are as follows:

- operating temperatures,
- fatigue and number of cycles, and
- long-term stability of mechanical characteristics.

As for the SMA's operating temperature, it must be stressed that the SMAs on the market based on Ni-Ti are characterized by martensite-austenite transition temperatures (A_s—austenite start) of up to 99 °C. However, due to the high-level hysteresis involved (usually above 20 °C), the austenite-martensite transformation temperature (M_f—martensite finish) is below 78 °C (Ref 3). These materials can be used in actuators that normally operate below 78 °C, such as those that operate only during the night or those located in points of the body where the temperature does not rise above M_f. Evidently it is necessary to ensure that the actuator is not damaged at higher temperatures and, at the same time, is not actuated automatically due to a high ambient temperature. This problem is related to the transformation temperatures and restricts the potential field of application of SMA alloys in the car. Over recent years considerable effort has been dedicated to resolve these issues; nevertheless, it is still appropriate to shift the entire shape memory transformation cycle to above 100 °C.

In order to exploit the characteristics of SMA devices and their miniaturization potential, there is a tendency to integrate the actuator within the mechanics or even inside the structure of the component. Therefore, the component can be considered to become "smart", meaning also that the functional performance must be assessed at a system level. In some cases, this prevents the use of standard tests, thus requiring the development of "ad hoc" procedures in order to characterize the "smart components" (Ref 4).

Activation time	< 1 s
Operating voltage	10 ÷ 16 VDC
Operating temperature	-40 ÷ +85°C
Numbers of cycles	60000
Work in harsh environment	\
Low cost	\
Vibration	50 ÷ 2000 Hz, up to 10g

Fig. 1 Typical requirements of automotive components

3. Conceptual Design

The technical specification of the device are

- bi-stable mechanism,
- silent operation,
- approximately 8° mirror rotation,
- fatigue life up to 60,000 cycles, and
- low cost.

The design process started with the study of kinematics. Attempting to merge compactness with simplicity, a bi-stable mechanism similar to those present in pens was designed, the idea being to use a SMA wire to push a mechanism that forces the mirror to assume alternately the normal position and the rotated position. This movement allows the prismatic mirror to reflect the undesired light on to the headliner and consequently perform the antiglare function.

The new device comprises three main components:

- a SMA wire,
- a piston, and
- a bracket to support the mirror.

The SMA wire forces the piston to move vertically; on the piston, a pin connected to a hinged bracket slides within a slot into the piston. The bracket is rigidly connected to the prismatic mirror. Each wire contraction causes the piston to move and forces the bracket to rotate alternately following the slot path. The shape of the slot is designed with two engaging points corresponding to the stable positions of minimum potential: one of these positions is reached and maintained during the glare phase, while the other is maintained during normal operation. The mechanism is designed to provide a separation between the wire position and the mirror position, allowing the wire to cool down to be ready for the successive actuation. Figure 2 shows the first sketches made to visualize the mechanism working principle.

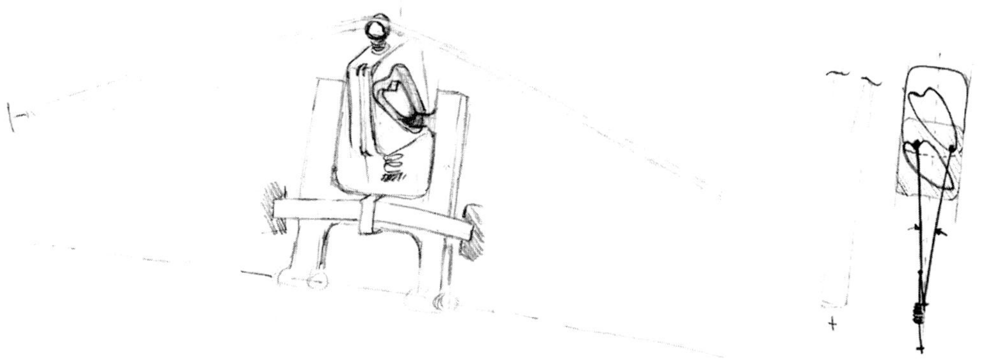

Fig. 2 First draft of the mechanism showing the main working principle

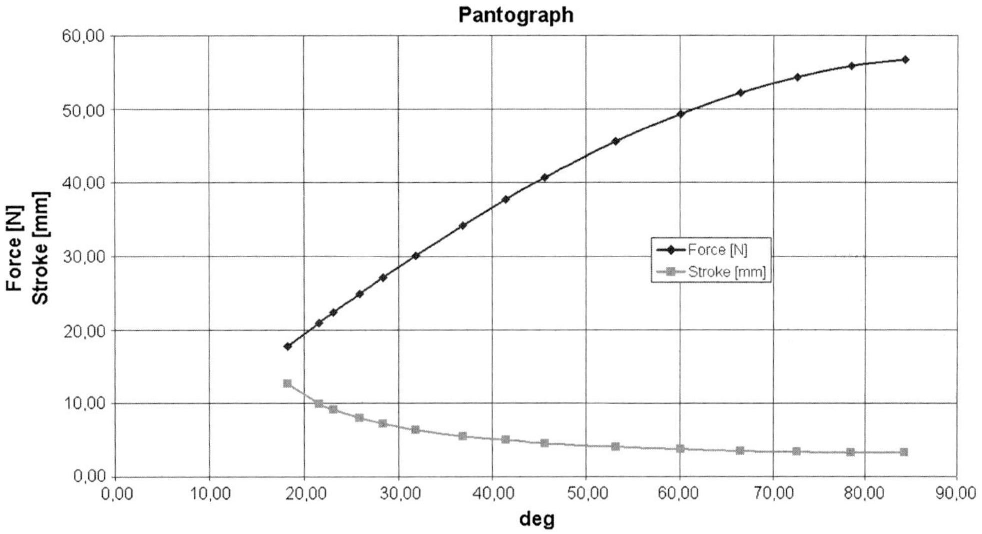

Fig. 3 Wire response simulation; vertical stroke and force vs. 0.5 mm diameter SMA wire starting position angle

The mechanism has been designed in order to ensure no error possibility during actuation, to guarantee the stability of the mirror from vibration and to be silent during activation and deactivation. The result is a simple, thin, wire-integrated component that provides designers with styling freedom.

4. Wire Engineering

At the heart of the system is the interaction between the wire and the slot profile, and a significant proportion of the design effort focused on minimizing the stress over the wire during actuation.

The objective here is to ensure a predetermined displacement so as to guarantee the appropriate relative movement between the slot and the pin (which is rigidly connected to the bracket); a two-way shape memory effect stabilized wire was used to perform this task (Ref 5).

Experimental data provided an indication of the nominal load to which the wire could be stressed, in order to achieve a fatigue life of over 500,000 cycles: this value is about 150 MPa. Correspondingly, it is appropriate not to exceed this value systematically since the hysteresis curve switches to a higher temperature if the load is increased (Ref 6). With the increase of the transformation temperatures, various drawbacks can arise such as the increase of necessary actuation energy, the increase of final temperature and, as a consequence, the increase of required heating and cooling times.

Additionally, the wire was positioned like a pantograph to enhance the stroke available, causing the axial stress to increase. Therefore, one aim of the design process was to arrange the device in such a way as to make the wire work in the range in which the shape memory wire is safe.

Tensile force and available stroke depend on the angle that the wire forms when mounted in the device (pantograph angle); using the diagram shown in Fig. 3, it is possible to choose the pantograph angle that enables the right balance between stroke and force needed. The diagram illustrates the force and stroke values at different angles of the pantograph placed wire. The values are calculated assuming a 150 MPa tensile strength on a 0.5-mm-diameter wire. The angle chosen is 24° and the wire length is 200 mm. The wire contracts 8 mm linearly and the available useful displacement for the actuation is 10 mm (2 mm over the nominal 4% contraction). That range provides a safety margin during actuation. At the same time, the theoretical force the wire can exert is 23 N (@ 150 MPa).

The wire shown in Fig. 4 has a silicone covering (Ref 7). Since the transformation is completely controlled by temperature, the wire section is affected by high thermal gradients. Besides the interface between the crimp and the wire with the free air is a breaking affected point, therefore the wire covering uniforms the conductivity coefficient on the heat exchanging surface and reduces the thermal stress on the material itself.

Furthermore, the silicone cover causes the thermal conductivity to increase and the wire cooling time to decrease. This means that the whole antiglare movement can be very fast and the activation frequency can be raised.

5. Mechanism Engineering

One of the first aspects investigated regarding the mechanism was the layout of the components incorporated in the device.

For reasons of symmetry the piston was placed in the middle of the mirror, the 200 mm SMA wire was laid behind the mirror plane as presented in Fig. 5 and 6.

The SMA actuator moves the piston and simultaneously compresses the two reaction springs which have a dual role:

- To bring the system always to a minimum-potential engaged position in order to keep the bracket (mirror) blocked.

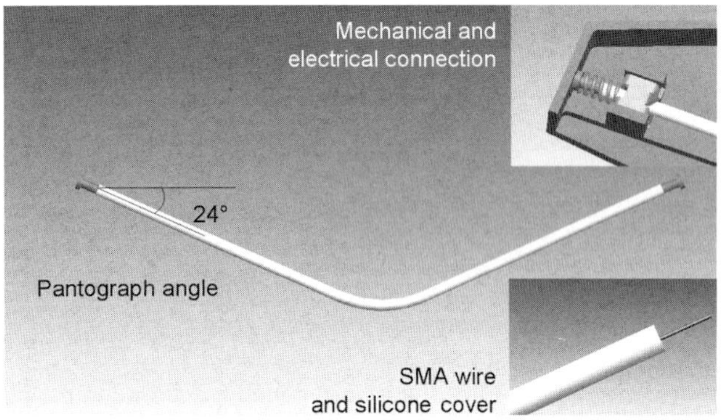

Fig. 4 Shape memory wire final layout with mechanical crimps and silicone covering

- To stress the wire during the cooling time in order to accelerate the shape recovery.

Two constant load springs were mounted to transfer a constant load on the SMA wire during the actuation. Correspondingly, considering force losses due to friction to be a nominally constant value, it was possible to forecast the loads acting on wire section and consequently simplify the device design.

During normal operation of the device, accidents may occur such as mechanism blockage or poor stroke end alignment. The system is protected from misuse by the insertion of a compression spring mounted axially and in series with the wire. In this way, the connection between the wire and the main structure collapses when the wire tension exceeds a predetermined value of about 33 N, the aim being to lengthen the life of the wire loading mechanism by preventing it from breaking prematurely.

A large part of the work conducted was dedicated to the design of the slot path in order to convert wire work (in terms of forces and stroke) into an ergonomic movement which minimizing friction losses. In the design of the slot profile, it is important to imagine how the bracket needed to move. When glare occurs, the mirror should rotate 8° from the starting position and then must be blocked in that position. The releasing actuation should disengage the mirror and bring it to the original position.

The question was how to make this rotation starting from the SMA wire contraction, enabling smooth movement of the mirror and consequently an "ergonomic" behavior of the device. Indeed, the asymmetric "heart" shape and the slot slope are the results of kinematic and ergonomic considerations, considerable time having been spent on its design.

To better understand the reasons behind arriving at this particular shape, it is useful to note that

- the vertical extension is 8 mm which corresponds to the available stroke provided by the wire,
- the width is defined by the 8° mirror rotation, and
- a mechanical stroke end-stop is useful to limit piston stroke and relax the wire in both stable positions.

The diameter of the pin was selected by considering the force acting on it and so as to avoid the second stable position

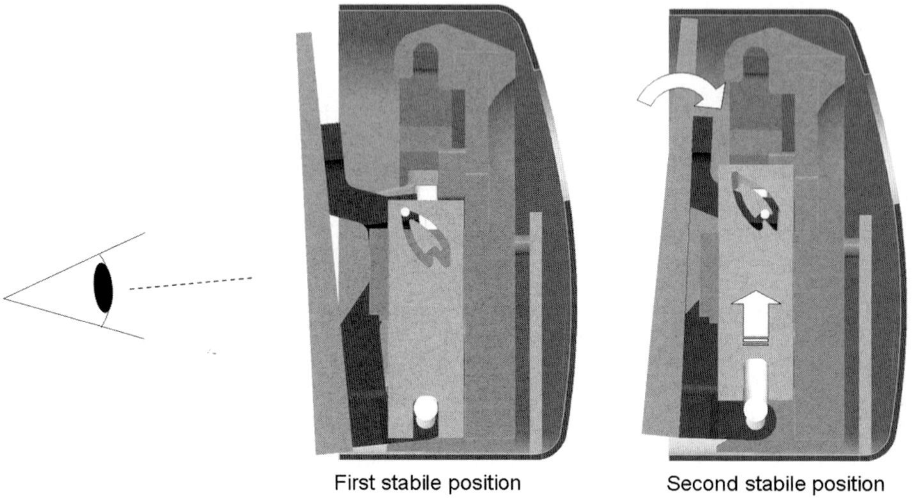

Fig. 5 Cross section showing the sloth path and the pin connected with the handle

bypass. At the same time, the slope of the ramp determines the loads that the wire has to sustain during actuation.

During actuation, an appropriate level of force must act to ensure proper operation of the bi-stable mechanism in every condition of use. This force must act in a direction during the engaging run, and in the reverse sense during the disengaging phase. This is obtained by a plastic flexure inspired by the conventional solution of the manual actuated rear-view mirror connected to a handle.

The flexure upper hinge, the bracket lower hinge and pin axis had to be aligned in the vertical plane. That plane defines the bracket instability position (Fig. 6). When the pin axis crosses that plane, the pin is pushed toward the opposite arm of the heart-shaped slot. The force produced by the flexure forces the pin to follow the right ramp. Consequently, the alternation of the engaging states is guaranteed since there is no possibility for the pin to come back on the same side again. The instability zone is relatively wide due to friction, and so the theoretic slot was modified respectively.

Kinematics and dynamics were simulated in Pro/Engineering Mechanism application, the results enabling appropriate operation of the system to be predicted.

6. System Logic and Electronic Board Layout

To understand the logic behind the development of the system, lets consider driving on a highway at night during a relatively long journey—the driver is already feeling tired and eyes are starting to be overstressed. A car behind approaches with incorrect projector lights leveling resulting in glare.

The system has been designed to detect such situation using two photodiodes mounted on the electronic drive board; these sensors trigger the start of the actuation sequence. The SMA wire is then heated and the mirror rotates.

A hall-effect position sensor detects the moment when the system has reached the point at which the bi-stable device changes its stable equilibrium reference; then the wire is de-energized but the mirror is maintained mechanically in this position. When glare stops (e.g. the car behind overtakes), the system activates the wire a second time, so the mechanism becomes unclamped and the mirror rotates back to the starting position.

The anti-glare mirror movement is silent, rapid (taking less than 400 ms) and does not distract or interfere in any way with the attention of the driver (Fig. 7). In order to avoid continual

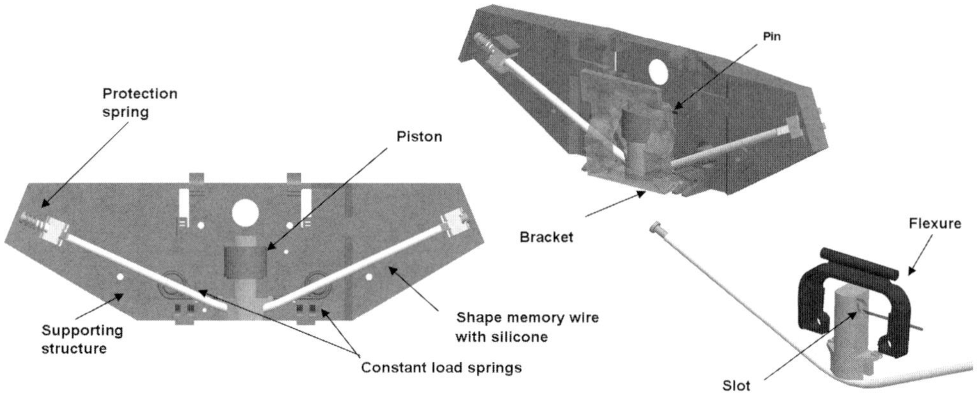

Fig. 6 CAD of the device and description of the components

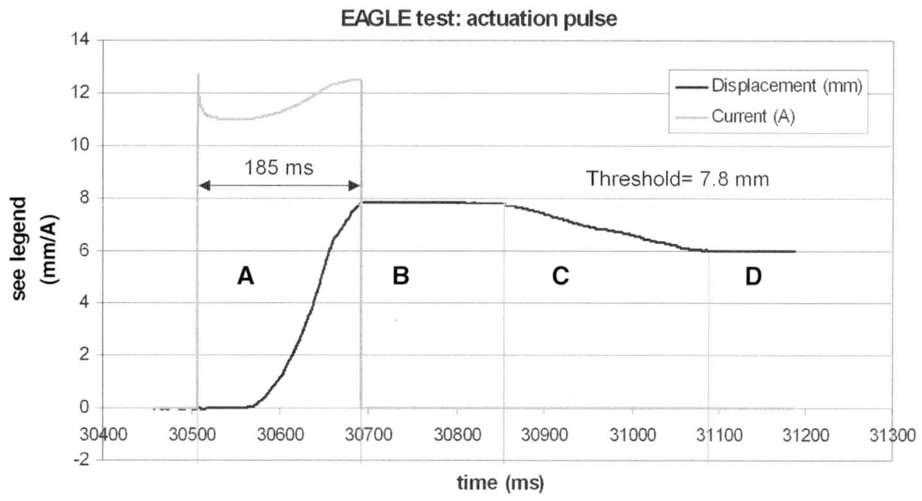

Fig. 7 Typical activation pulse and wire activation response time

operations of the system, which may cause an annoying flickering effect, a filtering time between a wire activation and the following is enforced (i.e. no actuations are allowed for a determined time after an actuation has occurred). In the current prototype, it was decided to pause the actuations for 20 s (Fig. 8).

The wire current absorption is regulated by means of a PWM controlled switch to optimize the behavior of the wire. The PWM frequency is about 100 Hz, sufficiently far from the typical acoustic spectrum. Any kind of current control can be used as well.

The heating time cannot be predetermined since it depends on the starting temperature of the wire, and the cabin of a car is a highly variable environment. For this reason, it is not possible to operate with a constant duration pulse and thus effective heating duration is defined by a position sensor that cuts the power supply. A hall sensor was chosen as position sensor in order to be noncontact, being placed onto the electronic board and allowing an easy mounting process.

Three different uniquely identifiable voltage values occur for the three different positions of the piston.

- Saturation for 1st stable position.
- <1 V for stroke end position.
- 3.2 V for 2nd stable position.

7. Testing the Wire

An initial fatigue test was performed on the sole actuating wire in the same geometrical configuration of the whole device. According to specifications, 60,000 actuations must be guaranteed. The wire was set up on the base with 24° angle and loaded with 2.34 kg (corresponding to 150 MPa). A potentiometer is mounted on the testing device to enable control similar to the expected one. A blocking device was mounted in order the simulate the behavior of the definitive

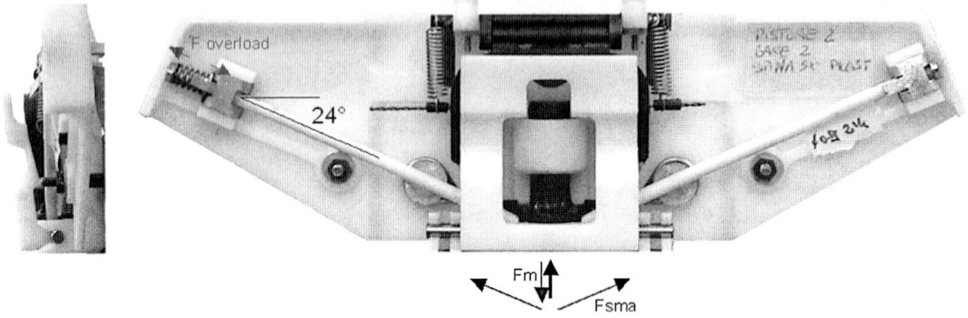

Fig. 8 Final prototype

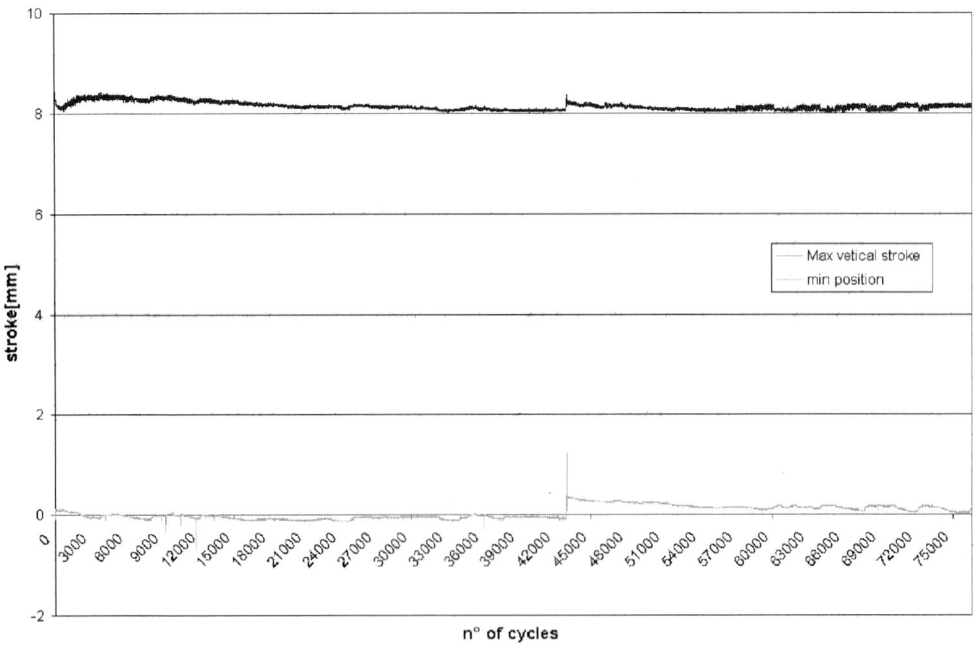

Fig. 9 Fatigue test on Dynalloy SMA wire mounted in the same geometrical configuration of the whole device (pantograph angle: 24°, wire loaded with 2.34 kg, wire diameter: 0.5 mm)

mechanism. During heating and cooling, maximum and minimum position, maximum current, applied voltage and heating time were recorded. The system cuts the power as soon as the stroke reaches 8 mm; the recorded overshoot can be attributed to inertial "bouncing" of the system which is not constrained. The electrical power is applied by a voltage generator at 10 V. The voltage in this case is lower than the automotive reference (used in the test of the whole system) since the absence of mechanical frictions would make the actuation too fast. A constant cooling time between subsequent pulses was applied.

A first test has been performed on the Dynalloy SMA wire (0.5 mm diameter). The results are shown in the graph below, and are positive with more than 75.000 actuations (Fig. 9).

Results demonstrate that the wire can withstand the fatigue requirements under the specified working conditions.

A second fatigue test has then been performed on the SAES Getters SmartFlex wire (0.5 mm diameter), with the same set of parameters. Again results demonstrate high fatigue resistance of the wire under the specified working conditions.

8. Complete Device Testing

The EAGLE (Electrically Actuated antiGLare rEar-view mirror) prototype was produced and assembled in order to perform a complete experimental activity. A fatigue test was performed on the prototype provided with a SMA wire controlled externally.

The forces acting on the piston and on the wire are summarized in the table shown in Fig. 10.

The device was mounted in standard working configuration without the mirror, and a potentiometer was installed and maintained in contact with the piston by a weak spring. The potentiometer was used to simulate the hall sensor and identify the position of the piston. All the tests were performed at room temperature (between 20 and 26 °C).

The target for the tests was to exceed 60.000 cycles (30,000 complete cycles of activation and deactivation).

A fatigue test was performed on the EAGLE mechanism mounted with a SMA wire controlled externally. The control was such that the current was interrupted as soon as a certain threshold was exceeded. The current itself was applied by a short circuit on 13 V power supply. As a further protection a "maximum duration time" was defined after which the current was interrupted even if the mechanical threshold had not been reached.

The diagram shown in Fig. 11 represents the overall behavior of the characteristic parameters during the fatigue test. The system achieved 73,455 actuations before the test was been interrupted before failure having already exceeded expectations.

F_m min	18.2 N	Minimum Piston load
F_m max	25.2 N	Maximum Piston load
F_{sma} min	22.4 N	SMA wire load
F_{sma} max	38.7 N	SMA wire load
σ_{min}	114 Mpa	On wire section
σ_{max}	197 Mpa	On wire section
$F_{protection}$	33.4 N	Protection spring in action

Fig. 10 The table summarizes the forces and the stresses acting on the wire and on the piston

9. Conclusions

This paper demonstrates that SMAs represent a good solution to create automatic devices inside vehicles. Typical automotive specifications have been presented and commented with reference to the main features of shape memory materials.

The example of an innovative design of an internal antiglare rear view mirror has been presented starting with the state-of-the-art of antiglare rear-view mirrors and continuing with shape memory wire characteristics and the target device function. The wire was used as a pantograph to enhance stroke output and was covered by a silicone cover to accelerate cooling. Electronic control ensures an adaptive power input according to the environment conditions, resulting in a long fatigue life for the device.

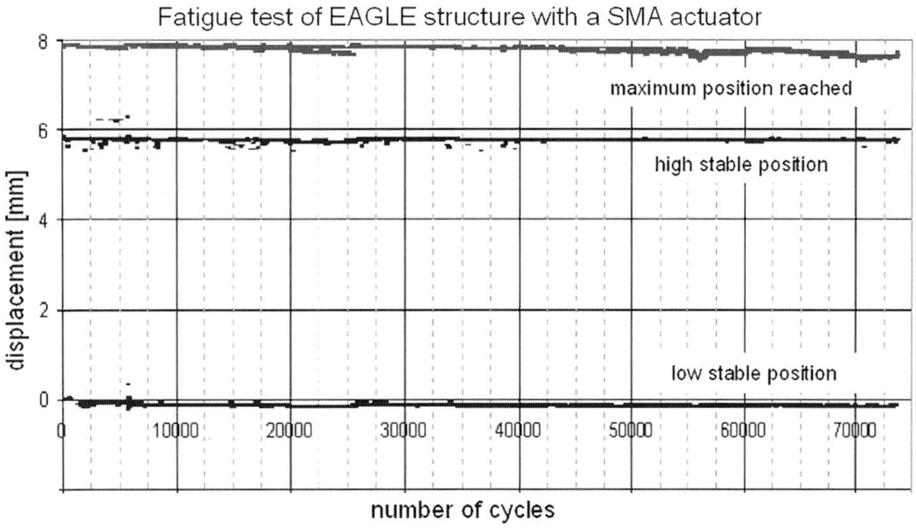

Fig. 11 Complete device fatigue test results with SAES GETTERS "SmartFlex" wire; during the test it has been measured the piston displacement and it has been registered its position in real time

The mechanism was designed with a bi-stable kinematics in order to protect the wire from thermal and mechanical overstress. The mechanism model was simulated with CAD tools in order to verify the kinematics. This activity was conducted directly with the supplier and produced a series of working prototypes ready to be tested on vehicles.

The testing activity, conducted to evaluate the prototype system, demonstrated that the mechanism is ready to respect automotive requirements.

This work also underlines some of the stringent design criteria that engineers should face when designing an innovative automotive device using shape memory materials. Such considerations will also help suppliers in understanding the specific issues and the complexity of the shape memory material engineering process.

References

1. J.B. Gross, P. Donatelli, F.A. Pertl, and J.E. Smith, Mechanically Dimmable Mirrors, Center for Industrial Research Applications, West Virginia University, SAE International, 2006
2. SAES GETTERS. www.saesgetters.com
3. D.J. Leo et al., Vehicular Applications of Smart Material Systems, *SPIE Proceedings*, 1998 Shape Memory
4. F. Butera, Shape Memory Actuators for Automotive Applications, *Shape Memory Alloys. Advances in Modelling and Applications*, F. Auricchio, L. Faravelli, G. Magonette, and V.Torra, Ed., CIMNE, Barcelona, 2001, p 405–425
5. H. Funakubo, *Shape Memory Alloys*, Gordon and Breach Science Publishers, New York, 1984
6. H. Funakubo, *Shape Memory Alloys*, Gordon and Breach Science Publishers, New York, 1987
7. CRF Patents. US6835083B1; EP1460284A2

JMEPEG (2009) 18:725–728
DOI: 10.1007/s11665-009-9492-9

Investigation on the Hysteretic Behavior of NiTi Shape Memory Wires Actuated Under Quasi-Equilibrium and Dynamic Conditions

A. Coda, M. Urbano, L. Fumagalli, and F. Butera

(Submitted September 17, 2008; in revised form March 20, 2009)

The functional characterization of SMAs for actuation is typically performed by measuring the specimen deformation under constant load during a controlled thermal cycling across transformation temperatures. Under dynamic actuation, transformation temperatures different from those measured in quasi-equilibrium conditions have been observed. The aim of this work is to better investigate and understand these phenomena. Direct and indirect heating of shape memory wires under several loading conditions are examined in detail. According to the experimental results, the hypothesis is to consider the observed differences as an effect of the thermal cycling rate on the internal friction. However, the presented data seem do not fully confirm this idea. Further experiments will be carried out in order to directly measure the internal friction of the material under the same working conditions.

Keywords advanced characterization, nondestructive testing, titanium

1. Introduction

Despite SMA for actuation are more and more utilized in industrial applications, no standard testing specifications have been prescribed so far. Functional characterization is typically negotiated according to customer needs and manufacturer characterization facilities.

As far as wires are concerned, thermal cycling under constant load turns out to be a fundamental functional test (Ref 1). Preferably the test is carried out at several load values, for example at the operating load and at lower and higher loads. The test consists in loading a SMA wire with a constant tensile force and recording its length while increasing and decreasing its temperature across the transformation temperatures. The test, also referred to as *hysteresis test*, provides information about the strain, transformation temperatures and load dependence of transformation temperatures.

For practical reasons, the test is typically carried out in a climatic chamber by recording chamber temperature together with the wire length. Temperature rates are in this case very slow as thermal equilibrium must be ensured. As test conditions are different from those of most wire applications, where very fast wire heating is obtained by Joule effect, questions may arise about the suitability of this procedure.

In this article we investigate the effect of the temperature rate and heating method on the hysteresis test: in particular, thermal cycling tests carried out in climatic chamber are compared to those performed by SMAq, a patented instrument (Ref 2) where wire is heated by current flow. Besides, pulsed actuations are performed in climatic chamber at increasing chamber temperature, providing fast actuation information of the austenite-martensite (A-M) transformation. With the help of literature fundamental research results, physical interpretation of experiments will be provided.

2. Experimental Results

Hysteresis tests were performed on a hanging wire with a weight hooked at its bottom end in a climatic chamber. The chamber temperature was cycled across the transformation temperatures. The tests were carried on in a Weiss WK11-600 climatic chamber, performing thermal cycles between 15 and 170 °C at different temperature rates (0.3, 1, 5 and 10 °C/min); the temperature was measured by a PT100 thermocouple. Inside the chamber, four samples of 0.2 mm diameter SAES SmartFlex® wires (Ni49Ti51 at.%), 150 mm long, were constrained at the top end to a steel structure, so that four independent tests could be performed at the same time for a better reliability. A constant weight of 4.7 N (150 MPa) was applied to the bottom end of each wires and a position sensor measured the displacement (shortening) of the wire during the cycle. In order to compensate for the thermal deformation of the structure, the raw data were corrected by subtracting the displacement of a molybdenum wire, which has a low coefficient of thermal expansion. The acquisition of the signal was made by means of a SAES Field Interface, with the acquisition time set at 20 s. Both sensors were calibrated: the

This article is an invited paper selected from presentations at Shape Memory and Superelastic Technologies 2008, held September 21-25, 2008, in Stresa, Italy, and has been expanded from the original presentation.

A. Coda, M. Urbano, L. Fumagalli, and **F. Butera**, SAES Getters S.p.A., 20020 Lainate (MI), Italy. Contact e-mail: alberto_coda@saes-group.com.

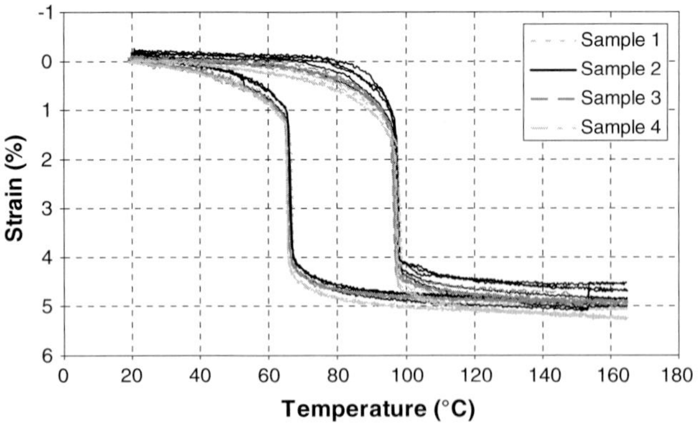

Fig. 1 Hysteresis curves of samples 1-4 at 150 MPa and 1 °C/min temperature rate

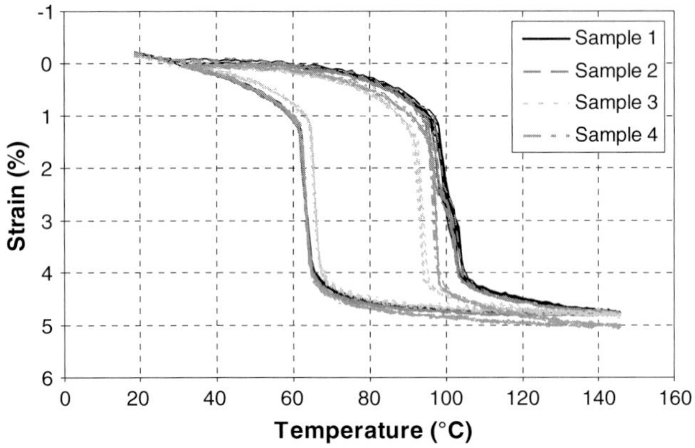

Fig. 2 Hysteresis curves of samples 1-4 at 150 MPa and 5 °C/min temperature rate

PT100 in an oven with a reference thermocouple, at different temperatures, while the displacement sensor with standard blocks of certified thickness at room temperature.

In Fig. 1 the tests output for a 0.2 mm SmartFlex® wire under constant stress of 150 MPa are displayed. The temperature rate is 1 °C/min: this is the highest temperature rate providing sufficient measurement reproducibility. As shown in Fig. 2, in fact, at higher temperature rates (5 °C/min), the differences of the four samples hysteresis curves indicate bad temperature homogeneity inside the chamber. Curves obtained at the rate of 1 °C/min will be considered as reference later in the discussion.

For comparison with the previous tests, thermal cycling at constant load is performed by means of the instrument SMAq whose hardware and functionalities can be found in Ref 2. Here a brief instrument description is presented.

The most important component of SMAq is a linear motor with a peak force of 430 N and a continuous force of 220 N. The motor is equipped with an absolute linear optical encoder (Heidenhain LC) with a resolution of ±2 μm over a wire length up to 240 mm and controlled by the driver Control Technique Unidrive SP, which features a position repeatability of ±1 μm. SMA wires having a maximum diameter of 0.5 mm are horizontally restrained between a first clamp located on the moving element of the linear motor and a second clamp located on a tension load cell (Laumas SA) featuring an accuracy of ±0.02% and a repeatability of 0.01%. The linear motor can be closed-loop controlled through a feedback from the load cell or through a feedback from the linear optical encoder with an overall frequency response of 20 kHz. An infrared camera (Jenoptik VarioCAM), with a resolution of 320 × 240 pixels and a frequency response of 50 Hz over the spectral range 7.5-14 μm, allows for non-contact measurements of the wire temperature between −40 °C and +1200 °C.

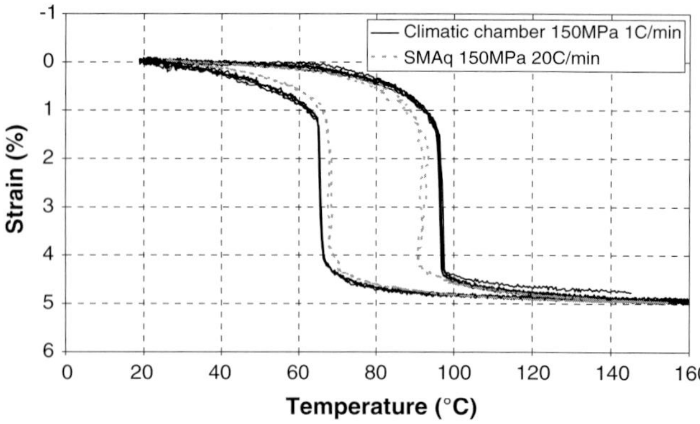

Fig. 3 Comparison between climatic chamber and SMAq thermal cycling at 150 MPa

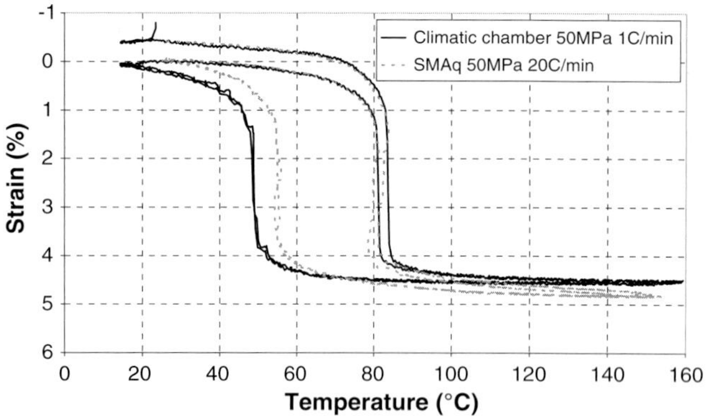

Fig. 4 Comparison between climatic chamber and SMAq thermal cycling at 50 MPa

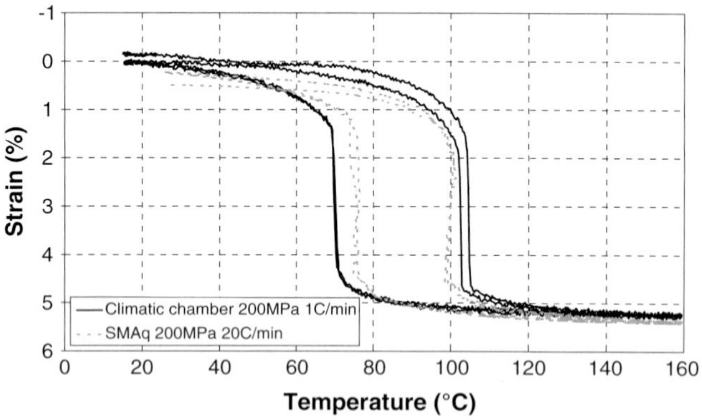

Fig. 5 Comparison between climatic chamber and SMAq thermal cycling at 200 MPa

Comparison of climatic chamber and SMAq curves at different loads are reported in Fig. 3-5: the SMAq curves hysteresis width is systematically lower than that measured in the climatic chamber.

3. Discussion

The small discrepancies between the results obtained by the different measuring systems must be ascribed to the different heating methods. Two possible explanations have been initially guessed. As a first hypothesis, it was supposed that the differences observed in the hysteresis curves could be related to the heating conditions which, in case of fast actuation, may generate inhomogeneity in the temperature profile of the wire during thermal cycling, especially on SMAq where the ambient is at room temperature and the wire is heated at a temperature above A_f. In order to assess the core temperature versus periphery temperature difference during the SMAq cycling, a thermal 2D simulation has been carried out. The material parameters reported in Table 1 have been utilized for the time dependent computation.

The free convection film coefficient of a horizontal infinite cylinder has been calculated according to Ref 3. In Fig. 6 simulation results are shown: the temperature difference during the whole time history remains below 0.05 °C. According to the calculation, the temperature inside the wire is practically homogeneous.

The second and more reasonable hypothesis is to consider the observed differences as an effect of the thermal cycling rate on the internal friction. In the case of the hysteresis test carried out in a climatic chamber, the conventional ramp up and down is very slow (1 °C/min). On the other hand, by Joule effect on SMAq the heating and cooling rates are higher (more than 20 °C/min). As measured by Yoshida et al. (Ref 4) several internal friction peaks have been observed during the phase transformation. The behavior of these peaks is, however, very complicated since it depends on experimental parameters such as the frequency, the amplitude and the heating or cooling rate, as well as on the composition of the sample and the history of the heat treatment (Ref 5-7). Sugimoto et al. (Ref 8) found the relation between the appearance of the internal friction peaks and Ni/Ti ratio. According to them when the ratio is smaller than 1 the transformation to martensite occurs at a single stage and the spectrum shows very low internal friction in the parent phase and very high in the martensite. In this case a sharp peak is typically observed just below the M_s and A_s transition temperatures. Associated with the internal friction peak a remarkable depression of the rigidity is observed. Also, the cycling rate (frequency) plays an important role on internal friction. By decreasing frequency the internal friction typically increases.

On the basis of these considerations we may suppose that the martensitic transformation at very low rates, like in a climatic chamber, can increase the internal friction. This effect is more evident under heating and cooling at the beginning of the direct and reverse transformation, confirming the Yoshida observations. In fact, looking at the typical hysteresis curve at 1 °C/min in the climatic chamber, it is evident that the hysteresis is larger and the course of the martensitic transformation is much steeper than that collected on SMAq by faster heating and cooling.

In order to confirm the effect of heating rate on the properties of the transformation, a special test has been carried out in the climatic chamber. The same setup described before (climatic chamber, position sensor, PT100 thermocouple) was used. In this case, the chamber temperature was set at different constant values, ranging from 25 to 85 °C, and the wire was actuated by an electrical current with an external Delta Elektronika power unit (0-30 V, 0-5 A). The current pulse duration was set to 3 s for all actuations; the current was varied according to the chamber temperature in order to achieve a 160 °C peak temperature. The displacement was recorded with a National Instrument logger that allows a very fast acquisition (in this case 0.01 s). In Fig. 7, the sequence of actuation cycles at increasing room temperature is shown.

By plotting the asymptotic displacement of the wire at the cooling cycle versus testing chamber temperature, a cooling

Table 1 Material parameters for time dependent computation

Parameter	Value
Wire diameter	0.2 mm
A_s	90 °C
A_f	110 °C
M_s	70 °C
M_f	50 °C
Resistivity	90 μΩ cm
C_p	400 J/kg/K
Density	6450 kg/m^3
Thermal conductivity martensite	8.6 W/mK
Thermal conductivity austenite	18 W/mK
Emissivity	0.3

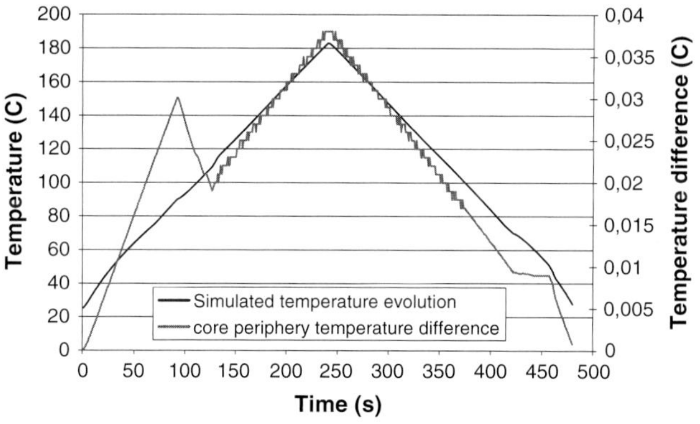

Fig. 6 Core temperature vs. periphery temperature simulation

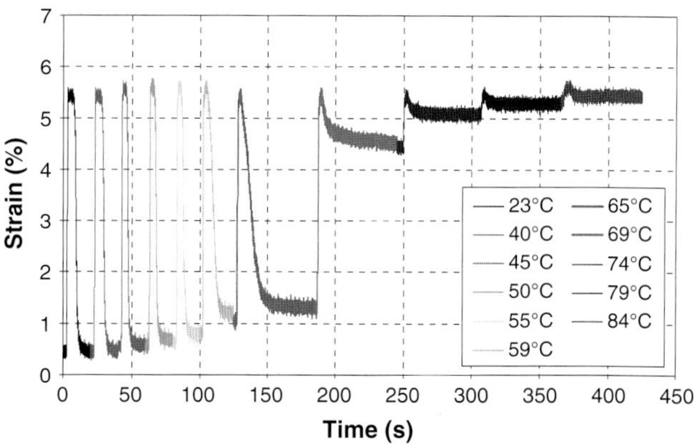

Fig. 7 Wire deformation at increasing ambient temperature

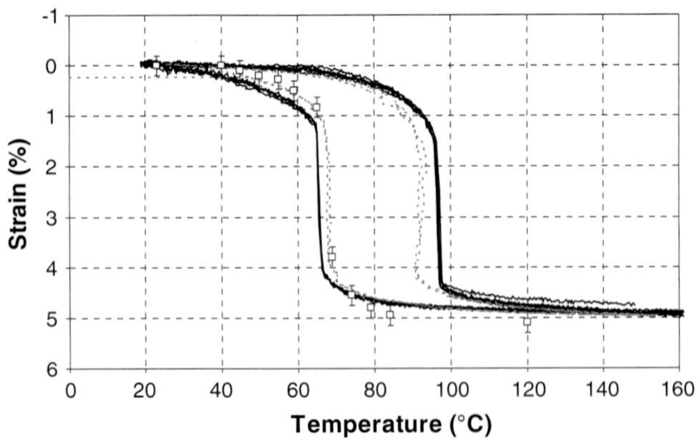

Fig. 8 The cooling branch of hysteresis reconstructed by collecting asymptotic deformation of fast actuations vs. SMAq and climatic chamber results

branch of the hysteresis curve can be constructed and compared to SMAq and climatic chamber, as shown in Fig. 8. The collected points do not overlap the climatic chamber curve, but seem to better match the SMAq hysteresis curve, thus confirming the above discussed hypothesis. However, further experiments should be carried out in order to definitely prove the temperature rate dependence of internal friction.

4. Conclusions

Thermal cycling at constant load, also known as hysteresis test, is a fundamental functional characterization of SMA wires for actuation. In this paper, two different experimental approaches for measuring the hysteresis curve of SMA wires have been compared and discussed. The first one has been carried out in a climatic chamber by submitting the material to a very slow thermal cycling across the transformation temperatures, so that quasi-equilibrium conditions may be attained. The second consists in a faster cycling of the material by Joule effect on SMAq.

On the basis of the experimental results, slight differences in the hysteresis width have been measured. In particular, we observed that in the case of high heating rates, the hysteresis curve becomes narrower with a reduction of A_s and an increase of M_s transformation temperatures. These results have been confirmed by performing pulsed actuations in climatic chamber at increasing chamber (ambient) temperatures. This behavior could be related to the internal frictions generated during the martensitic transformation. As measured by Yoshida et al. the internal friction in NiTi is strongly affected by the composition, thermo-mechanical history and thermal cycling rates. At very low rates, like in a climatic chamber, the internal friction can increase. This effect is more evident under heating and cooling at the beginning of the direct and reverse transformation, confirming the Yoshida observations. This possible hypothesis is not fully supported by the presented data. Further experiments must be carried out in order to directly measure the internal friction of the material.

References

1. K. Otsuka and C.M. Wayman, *Shape Memory Materials*, Cambridge University Press, Cambridge, 1998, p 149
2. M. Urbano, A. Coda, and R. Giannantonio, SMAq: A Novel Integrated Instrument for the Characterization SMA Wires, *SMST-06 Proceedings*, 2006, p 177
3. F.P. Incropera and D.P. De Witt, *Fundamental of Heat and Mass Transfer*, Wiley, New York, 1996
4. I. Yoshida, T. Ono, and M. Asai, Internal Friction of Ti-Ni Alloys, *J. Alloys Compd.*, 2000, **310**, p 339
5. G.J. Wasilewski, Compressive Strength of Ni-Al, *Trans. Metall. Soc. AIME*, 1965, **233**, p 169
6. R.R. Hasiguti and K. Iwasaki, Internal Friction and Related Properties of the TiNi Intermetallic Compound, *J. Appl. Phys.*, 1968, **39**, p 2182
7. O. Mercier, K.N. Melton, and Y. de Preville, Low-Frequency Internal Friction Peaks Associated with the Martensitic Phase Transformation of NiTi, *Acta Metall.*, 1979, **27**, p 1467
8. K. Sugimoto, K. Kamei, T. Sugimoto, and T. Sodeoka, Effect of Nickel and Aluminum on Internal Friction in Ti-Ni and Ti-Ni-Al Alloys, *Proc. Int. Conf. Martens. Transf.*, 1986, p 729

JMEPEG (2009) 18:729–737
DOI: 10.1007/s11665-009-9372-3

1059-9495/$19.00

Feasibility Investigation of Superelastic Effect Devices for Seismic Isolation Applications

G. Attanasi, F. Auricchio, and G.L. Fenves

(Submitted October 2, 2008; in revised form December 10, 2008)

The objective of this work is to investigate the feasibility of a new seismic isolation device concept, in which the restoring force is given by the superelastic effect of shape memory alloys. Seismic isolation is an option for passive protection of structure when an earthquake occurs, because it modifies the structural global response and improves performance. Dynamic responses of proposed innovative SMA isolation system and of traditional bearing device are compared through dynamic time history analyses. Results show that the SMA system is effective in reducing force and displacement demands, dissipating the input seismic energy and limiting the residual displacements.

Keywords seismic isolation, shape memory alloys, superelastic effect

1. Introduction to Seismic Isolation

When *seismic isolation* is mentioned, we refer to a technology which mitigates the earthquake effects on buildings and on their vulnerable contents.

1.1 Base Concept of Seismic Isolation

The base concept in seismic isolation is to protect the structure from the damaging effects of an earthquake by introducing a support isolating the building from the shaking ground. In the literal sense, the structure (a building, a bridge, or a piece of equipment) is separated from its foundations. After all, in an earthquake, the ground moves and this ground movement causes most of the damages to the structure. So, the principle is simple: to separate the structure from the ground so that the ground will move but the building will not. In practice, a full separation of the structure from its foundations is impossible, and large relative horizontal displacements have to be avoided either during the earthquakes or when other horizontal loads, such as wind, are present. Hence, the common solution is to use a layer, usually between foundation and superstructure, which is more flexible than the other structural elements and is able to transmit the vertical load undergoing lateral displacements without critical damage.

This article is an invited paper selected from presentations at Shape Memory and Superelastic Technologies 2008, held September 21-25, 2008, in Stresa, Italy, and has been expanded from the original presentation.

G. Attanasi, European School for Advanced Studies in Reduction of Seismic Risk, ROSE School, IUSS Pavia, Pavia, Italy; **F. Auricchio**, Department of Structural Mechanics, Università degli Studi di Pavia, Pavia, Italy; and **G.L. Fenves**, Department of Civil and Environmental Engineering, University of California, Berkeley, CA. Contact e-mails: gattanasi@roseschool.it, auricchio@unipv.it, fenves@berkeley.edu.

1.2 Effects of Seismic Isolation

Advantages in seismic isolation are evident. First of all, the level of damage is more safely controlled and confined to generally well-replaceable spots; second, an isolation system is usually a nonlinear system which damps and reduces the action demand of the global structure and also limits the transmittable force to the superstructure. However, design of isolated structures has some particular concerns. Practical isolation systems must balance between the extent of force limitation and the acceptable relative displacements across the isolation system during earthquakes. For this purpose, additional damping sources can be useful to reduce both the forces transmitted and the deformations within the structure.

1.3 Evaluation of Seismic Isolation Effectiveness

An important dynamic property of every structure is the period of the first mode of vibration (*fundamental period*). If other parameters are constant, this is a measure of the structural global stiffness. For example, a short masonry building is very stiff, and it is characterized by a short period of vibration, while a frame system skyscraper is more flexible and the period of vibration is longer. In seismic design, the structural period computation leads to the estimation of the seismic load on the buildings. The design spectra summarize the dynamic response of an elastic single degree of freedom (SDOF) system subjected to a ground motion. Seismic design demand is provided in terms of acceleration and displacement as a function of the fundamental period, as shown in Fig. 1.

Nonlinear systems are typically less demanding than the elastic spectra values because the inelastic hysteresis provides some energy dissipation. Hence, if the system is characterized by relevant nonlinear inelastic response, in the design procedure then the elastic spectrum is reduced by a coefficient representing the *hysteretic damping* effect. For example, if an additional damping component acts and the reduction coefficient is larger, then the hysteresis loop is "fatter."

Given that the isolation layer is more flexible than the rest of the structure, the fundamental period of isolated structures is longer than that of nonisolated structures, changing either the displacement or the acceleration demand. As shown in Fig. 1,

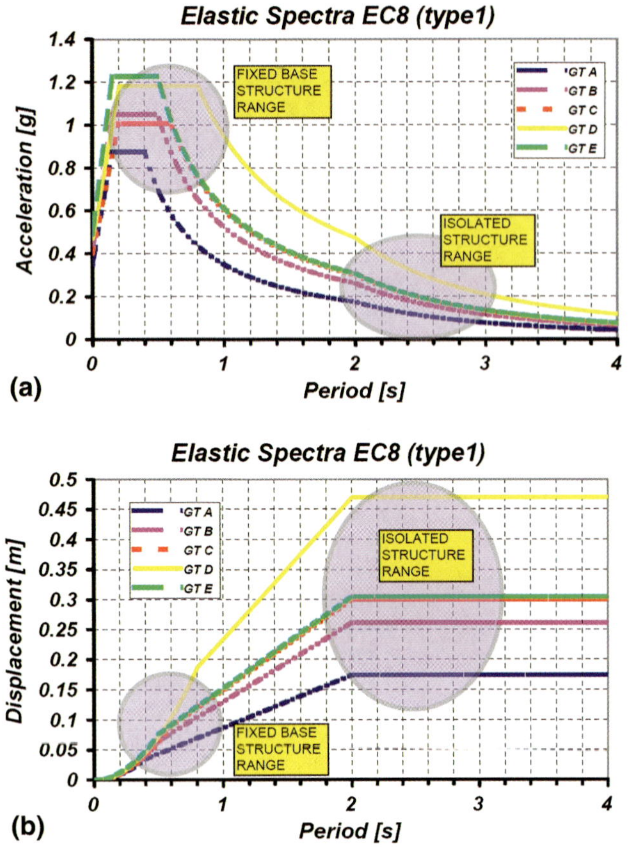

Fig. 1 Elastic design spectra in terms of acceleration (a) and displacement (b) as defined in the Eurocode 8; the different lines refer to different site soil properties; fixed base, and isolated structure domain are shown

usually an isolated structure experiences a smaller acceleration than a fixed base structure and a larger displacement, which in any case is occurring mainly at the isolation level, limiting damage in the superstructure. An extensive and exhaustive description of this topic can be found in Ref 1-4.

2. Superelastic Isolation Bearing Device Design

The effectiveness of a hypothetical superelastic isolation device in order to reach the structural design goals is investigated and compared with the traditional isolation device response. In this context, the term superelastic isolation device refers to a bearing system characterized by a nonlinear horizontal force-displacement relation which can be described by a flag-shaped hysteresis.

It is assumed that a bearing system providing a horizontal base shear as a function of displacement such as the one shown in Fig. 2 can be manufactured. At this first stage of work, the device is defined just in terms of hysteresis rule.

The key parameters characterizing the nonlinear behavior of the device in Fig. 2 are:

- K, αK, r_1K, r_2K are the initial stiffness, the final stiffness, the second stiffness in loading, and the second stiffness in unloading, respectively;
- V_d, u_d, V_y, and u_y are the design shear and displacement, the yielding shear and displacement, respectively; V_{max} is the maximum lateral force which the device can withstand without breaking; and
- βV_y is the shear force difference between the level of force at which the first transformation (when it is loaded) occurs and the level of force at which the second transformation (when it is unloaded) occurs; therefore, β is a parameter that measures the dissipation capability of the flag-shaped hysteresis.

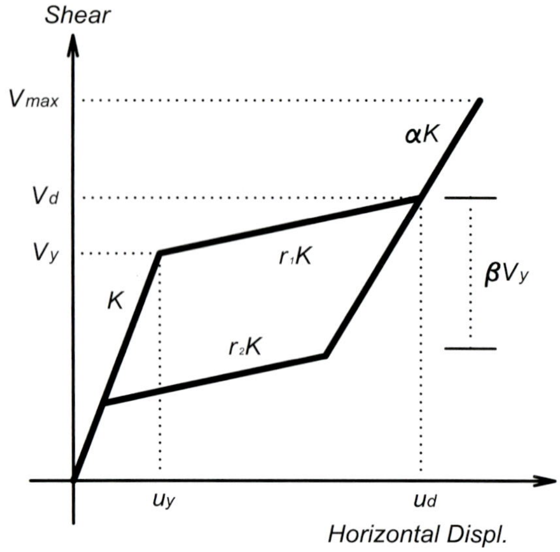

Fig. 2 Parameters for the shear-displacement model of the superelastic isolation device

The device behaves in the same way in both tension and compression, which implies that the force-displacement relation is symmetric with respect to the origin in the first and third quadrant. The flag-shaped relation is attractive for an isolator device because, in theory, residual displacements are zero at the end of a seismic event. A few of the actual isolation systems avoid the critical failure of the structure, but even if well designed, need to be repaired after the earthquake to bring the system back to the original position with respect to the foundations. On the contrary, an eventual superelastic device is re-centering (Ref 5).

2.1 Reference Isolation Device

The superelastic isolation device response is compared with respect to an actual isolation bearing. This is a lead rubber bearing (LRB), which is a bearing composed of a rubber cylinder with a lead core in the middle. The design philosophy of this device is to have the elastic rubber component to transmit the vertical load collaborating for the horizontal stiffness and the lead component that provides a large initial stiffness to resist the nonseismic horizontal load and yields when the design earthquake occurs, providing large hysteretic dissipation.

The device is experimentally characterized (Ref 6). It is produced by *AGOM International srl* (www.agom.it), and the nominal parameters used for the design purpose are listed in Table 1.

Results of the experimental tests carried out on the bearings are shown in Fig. 3 compared with the elastoplastic hysteresis parameters used in the model. Elastoplastic model is clearly an

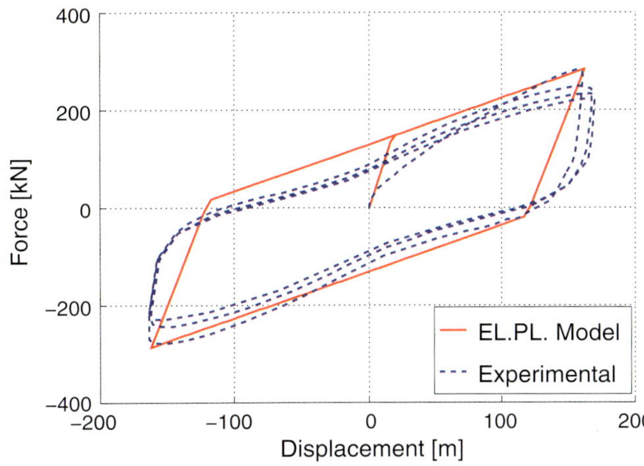

Fig. 3 Force-deformation relations for LRB 500: comparison of experimental test and elastoplastic model (experimental data and model parameters have been provided by *AGOM International srl*)

Table 1 Nominal design properties of reference LRB diameter 500 mm (a) and elastoplastic model parameters (b) [courtesy AGOM International srl]

(a) LRB 500		
Diameter		500 mm
Effective horizontal stiffness		1.62 kN/mm
Seismic combination vertical load		1653 kN
Seismic design displacement		162 mm
Hysteretic damping ratio		28%
(b) LRB 500		
Yielding shear	V_y	147 kN
Design shear	V_d	262 kN
Yielding displacement	u_y	17.5 mm
Design displacement	u_d	162 mm
Initial stiffness	k	8.4 kN/mm
Second stiffness	rk	0.8 kN/mm

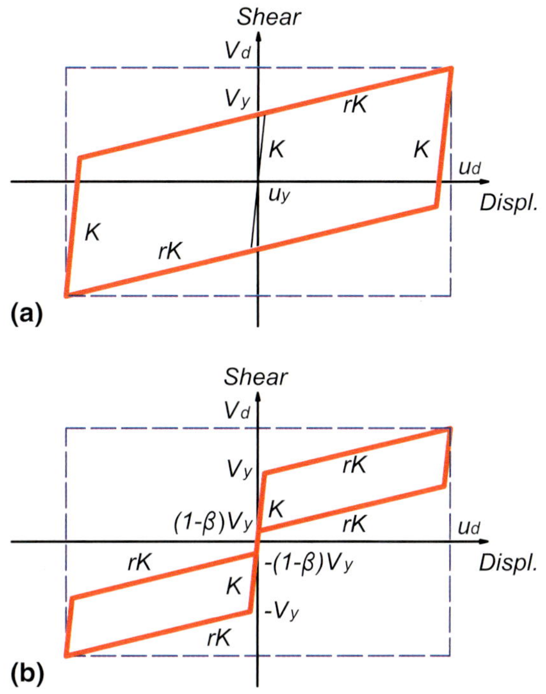

Fig. 4 Hysteretic comparison between LRB elastoplastic model (a) and superelastic flag-shaped model (b)

Table 2 Flag-shaped model parameters for SMA bearing equivalent to LRB diameter 500 mm (design parameters are the same as those referred to Table 1(b) and as shown in Fig. 2)

	SMA eq. LRB500	
Yielding shear	V_y	147
Design shear	V_d	262 kN
Yielding displacement	u_y	17.5 mm
Design displacement	u_d	162 mm
Initial stiffness	k	8.4 kN/mm
Second stiffness	rk	0.8 kN/mm

approximation of the real behavior of the isolator. In particular, the comparison with the experimental results shows that the elastoplastic model does not estimate the initial stiffness or the degradation well. However, for the representation of the general characteristics of the devices, the adopted model is acceptable, since it is exactly the same as the experimental results in terms of secant stiffness at the design displacement, and since it gives a good estimation of the hysteretic energy dissipated and of the residual displacements.

2.2 Equivalent Superelastic Isolation Device

The idealized design of an equivalent SMA isolator is based on the properties of the LRB. Starting from the elastoplastic model of the actual device (described in Table 1), an equivalent flag-shaped model related to the hysteresis in Fig. 2 is defined. The SMA device force-displacement relation parameters are summarized in Table 2.

With respect to model in Fig. 2, it is considered that $r_1 = r_2 = r$, which means that the device has the same stiffness for loading and unloading in the flag-shaped plateau, and it is assumed that the final hardening occurs far away from the area of displacement of interest. The dissipation capability of the flag-shaped hysteresis is a parameter for the investigation.

The two systems, the LRB and the SMA, are characterized by the same initial and second stiffness and the same yielding point. The difference between the two systems is only in the hysteresis: the LRB has an elastoplastic force-displacement relation whereas the superelastic device shows a flag-shaped relation, as shown in Fig. 4.

2.3 Estimation of the Isolation Device Performance

The seismic displacement and the force demand are investigated by considering the same structural system with two different isolation bearings: one with the isolation layer being given by LRB devices (Fig. 4a), and the other by superelastic devices (Fig. 4b). For the seismic design purpose, the fundamental period is computed through the secant stiffness to the design displacement is used. Provided that u_d is the same in both the LRB and in the superelastic models and given that the initial and second stiffness values are also the same, the secant stiffness to the design displacement in our example is the same, and the spectrum in Fig. 1 is achieved with the same period value.

Nevertheless, the hysteresis relation produces important differences in the response evaluation. As mentioned previously, the isolation system is more effective in reducing the seismic input if the hysteresis loop is fatter. The reduction is particularly important especially for the displacement demand, as shown in Fig. 1, where it is critical to accommodate the displacement for long-period structures. Usually in seismic design, the hysteretic damping is computed using the expression (Ref 3):

$$\xi_{\mathrm{hyst}} = \frac{2A_1}{\pi A_2}$$

in which A_1 is the area of the hysteresis skeleton curve and A_2 is the area of the rectangle enveloping the hysteresis relation. Referring to Fig. 4, A_1 is the area of the portion defined by the continuous line of the force-displacement relation; and A_2 is the area inside the lined rectangle enveloping the relation. Given the properties of the considered systems, the hysteretic damping given by an elastoplastic force-displacement relation is larger than the one given by a flag-shaped relation. In case of a flag-shaped system with the maximum dissipation capability (meaning $\beta = 1$), the hysteretic damping in the superelastic device would be exactly one-half with respect the LRB device. If the ductility is defined, which is a measure of the nonlinearity of the system given by the ratio between the yielding displacement u_y and the maximum design displacement u_d as (Ref 4):

$$\mu = \frac{u_d}{u_y}$$

The hysteretic damping of the each of two systems is plotted and compared in Fig. 5.

Based on the hysteretic damping, some nonlinear equations have been created to estimate the reduction in terms of the demand both in displacement and in force. They are different as a function of the formulation, but the expression proposed by Ref 7 is

$$R_\xi = \left(\frac{10}{5 + \xi_{\mathrm{hyst}}}\right)^\alpha$$

where R_ξ is the reduction factor, and α is 0.5 if the earthquake is a far-field event, and 0.25 if a near-field event is considered (Ref 4).

For design purposes, a reduction factor is used to multiply the spectra shown in Fig. 1 to obtain the relative inelastic spectra (Ref 8). For example, a reduction factor 0.8 means that the system is designed for a force and a displacement equal to the 80% of the elastic demands given by the spectra, while a reduction factor of 0.4 means that the system is designed only for the 40% of the elastic demands. The reduction factor is shown as a function of the ductility in Fig. 6, which refers to the hysteretic damping as reported in Fig. 5 and takes into account both the near-fault and the far-field expressions.

The comparison in Fig. 6 shows that according to the hysteretic damping theory, the flag-shaped device is always significantly more demanding than the LRB. Therefore, although in theory the flag-shaped hysteresis isolation system is attractive because it is characterized by zero residual displacements, the results here indicate that the structures have to resist a larger force and are subjected to larger displacement with respect to the actual LRB.

3. Feasibility Investigation on the Superelastic Isolation Bearing Seismic Behavior

The equivalent viscous damping approach is a model for seismic design based on the principle that a nonlinear system can be described using a linear system based on the secant stiffness to the design displacement and on an additional viscous damping component (Ref 4 and 8). According to this approach, the flag-shaped isolation system is always more

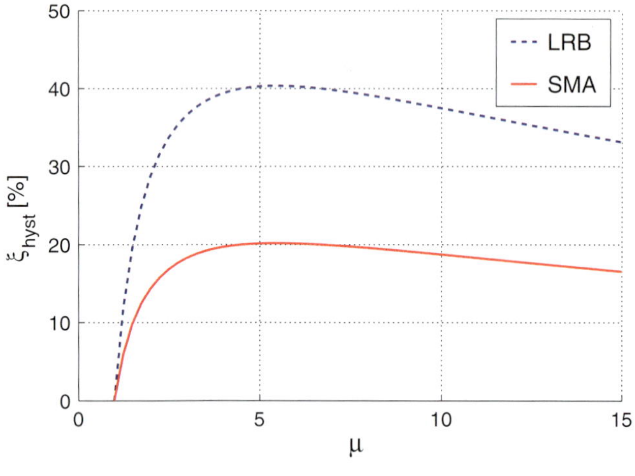

Fig. 5 Hysteretic area-based damping comparison as a function of ductility for superelastic maximum dissipation capability ($\beta = 1$)

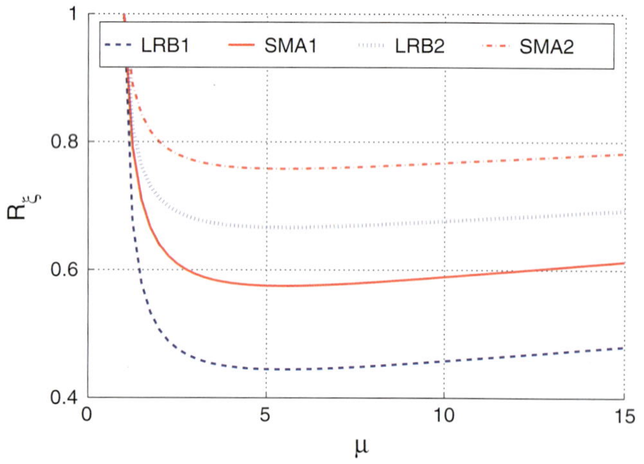

Fig. 6 Force and displacement reduction factors: LRB1 and LRB2 are, respectively, the far-field and the near-field reduction factors computed from LRB hysteretic damping; SMA1 and SMA2 are, respectively, the far-field and the near-field reduction factors computed from SMA hysteretic damping

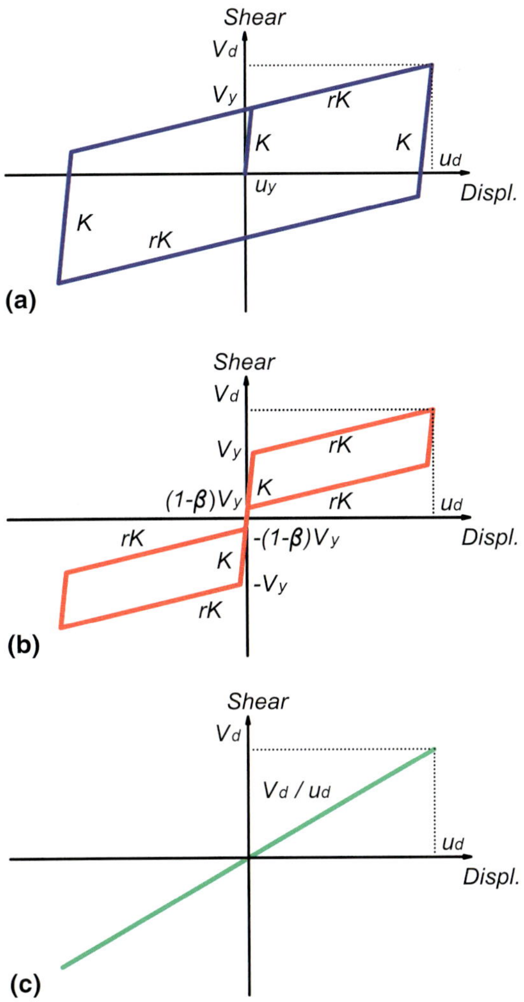

Fig. 7 Base shear-horizontal displacement model for LRB (a), superelastic (b), and equivalent linear elastic with secant stiffness (c) isolation device

demanding than an elastoplastic model. In this section, the effective response in terms of the different isolation systems subjected to ground accelerations is investigated, evaluating time history analysis results.

3.1 Effective Seismic Response Evaluation of Different Hysteresis Isolation Devices

The investigation is performed considering a SDOF system, in which all the deformation is assumed to occur at the isolation level, with a rigid superstructure affecting the system only as an additional mass. Therefore, the force-displacement relation of the global system is given just by the isolation layer contribution. Fourteen earthquake ground motions are considered and the responses of the following force-displacement relations are compared:

- *Elastoplastic model* (Fig. 7a). It represents the real LRB device, and the parameters used are those reported in Table 1(b);
- *Flag-shaped model* (Fig. 7b). It reproduces the shear-horizontal displacement relation of the superelastic device characterized by the design properties reported in Table 2. The model targets a device that performs similar to the real LRB system in the sense of equivalent shear and displacement capacity and initial and second stiffness;
- *Linear elastic model* (Fig. 7c). Given the design displacement, u_d and the design shear, V_d, which are the same as those in the previous models, the equivalent linear system was developed using parameters reported in Table 1(a).

The elastic damping component is modeled in proportion to the initial stiffness so that its contribution to the three relations is the same provided that the displacement history is also the same (Ref 6). Elastic damping ratio of 5% in the linear elastic system and of 2.2% in the elastoplastic and flag-shaped systems are used.

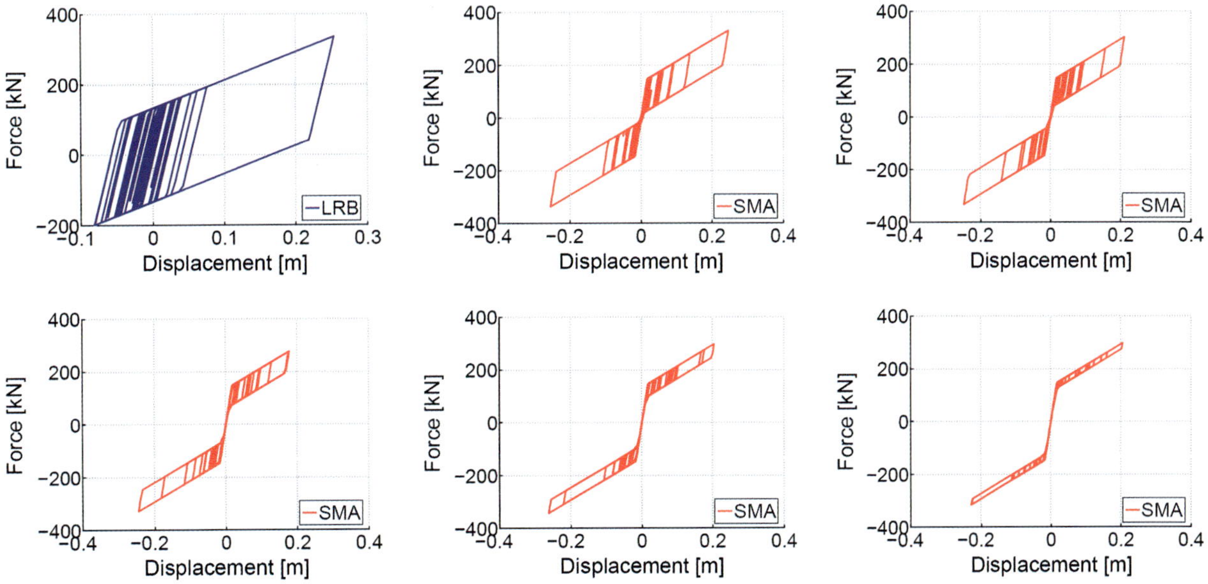

Fig. 8 Elastoplastic and flag-shaped hysteresis considering different dissipation parameters β = 0.95; 0.75; 0.55; 0.35; 0.15 used in the analyses (from *Tabas* (FN) ground motion record analysis)

Table 3 **Ground motion records considered in the analysis**

Earthquake	Date	Location	Comp.	PGA, g	PGV, m/s	PGD, m	Scaling factor	PGA, g	PGV, m/s	PGD, m
Tabas	16 Sep 78	Tabas stat.	FN	0.900	1.100	0.513	0.6257	0.563	0.689	0.321
Tabas	16 Sep 78	Tabas stat.	FP	0.977	1.058	0.752	0.6382	0.624	0.675	0.480
Erzinican	13 Mar 92	Meteor. stat.	FN	0.432	1.192	0.423	0.6672	0.288	0.795	0.282
Erzinican	13 Mar 92	Meteor. stat.	FP	0.457	0.581	0.295	1.2406	0.567	0.721	0.365
Landers	28 Giu 92	Lucerne	FN	0.713	1.360	2.298	0.6161	0.439	0.838	1.416
Northridge	17 Jan 94	Olive View	FN	0.732	1.222	0.310	0.5847	0.428	0.714	0.181
Kobe	16 Jan 95	Tato	FP	0.424	0.637	0.233	0.7109	0.302	0.453	0.165

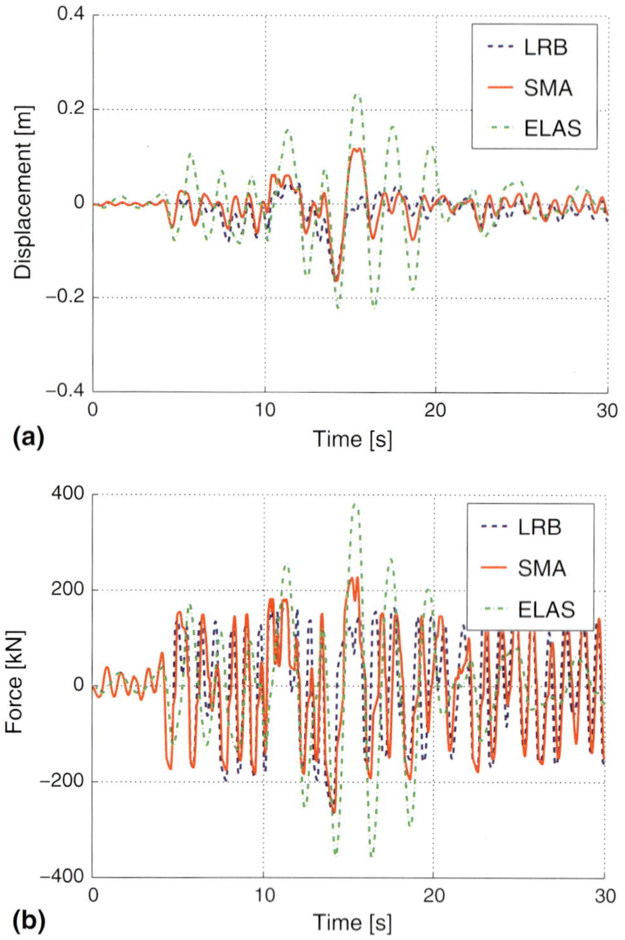

Fig. 9 Comparison between the responses of the three models in terms of displacement (a) and force (b) versus time considering Tabas (FP) record

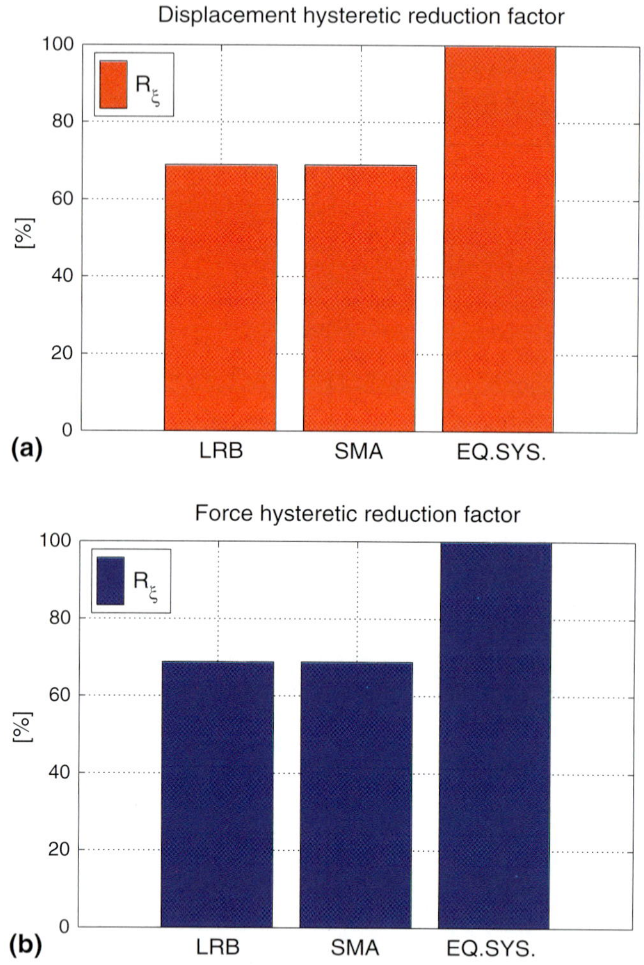

Fig. 10 Maximum displacement (a) and force (b) reduction factor in the three systems considering Tabas (FP) record

Even if in theory a high dissipation flag-shaped hysteresis is supposed to be the best for achieving the structural goals, for the sake of investigation, the effect of different dissipation coefficient hysteresis responses is studied. Five cases are studied, with the dissipation parameter (as defined in Fig. 2) spanning between $\beta = 0.95$ and 0.15. All the considered SMA hysteresis models, including the one from elastoplastic models used for comparison, are shown in Fig. 8.

3.2 Ground Motions for the Time History Analyses

To include in the analyses the seismic input variability, several ground motions are considered. The base for evaluating the demand is the original design spectra of the actual LRB isolator, which is the Eurocode 8 type 1 PGA = 0.25 g soil C (Ref 7), as shown in Fig. 1. Being given the properties of isolated systems, the corner period, which is the period value at which the displacement spectra reach the plateau, has been changed from 2 to 4 s. This induced a more realistic and severe condition in record displacement demand.

The first set of comparison is composed of seven artificial generated ground motions compatible with the design spectra. Seven natural scaled ground motions are also used. The used natural ground motion and the scaling factors are shown in Table 3.

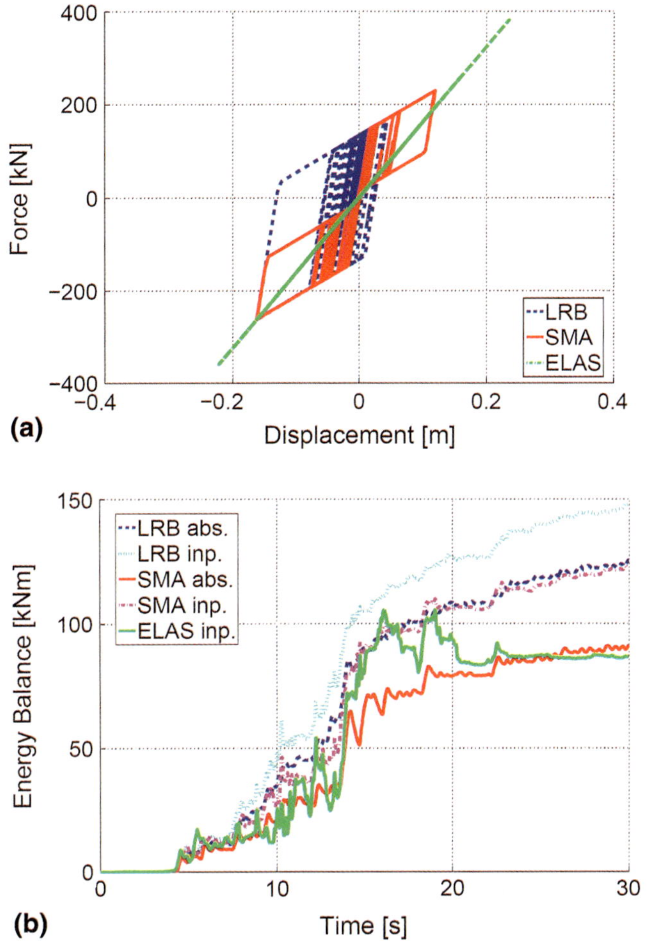

Fig. 11 Comparison of the hysteresis loops (a) and energy balance (b) of the three systems in Tabas (FP) record

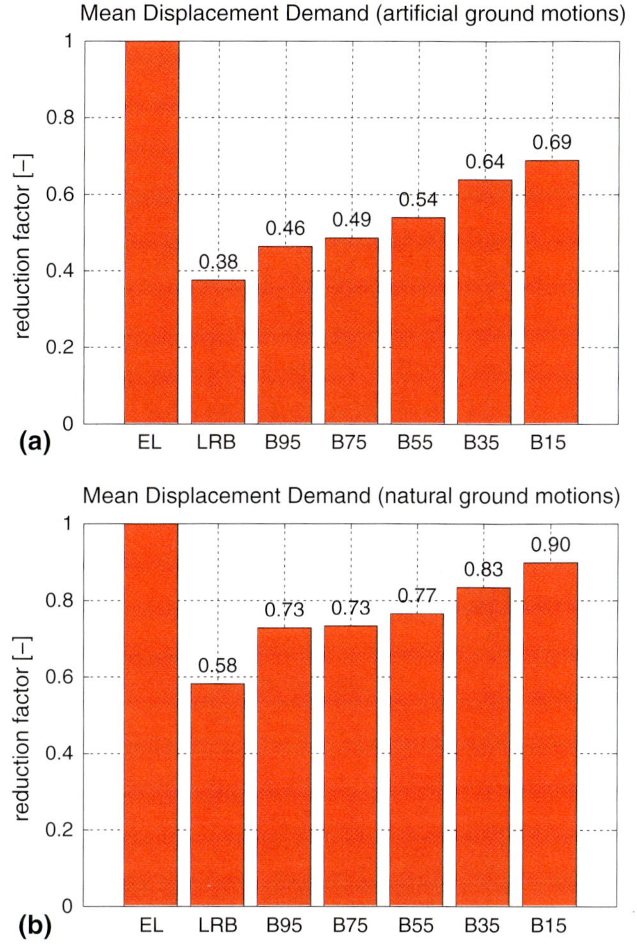

Fig. 12 Displacement mean reduction factor for artificial ground motion (a) and near-fault ground motion (b) considering linear elastic system (EL) and hysteresis in Fig. 8 (LRB, B95, B75, B55, B35, and B15)

3.3 Time History Analysis Results

The responses of flag-shaped, elastoplastic, and linear elastic systems are computed considering force and displacement time history demand and system energy balance using Ruaumoko, a nonlinear time history finite element analysis program (Ref 9). Figure 9 shows an example of the system response for a flag-shaped model with β value equal to 0.95. The record considered is Tabas (FP). Figure 9(a) shows the history of displacement of the three systems and Fig. 9(b) reports the history of base shear. Using the results shown in Fig. 9, the maximum displacement and shear demand in the linear elastic system with respect to the envelopes in the LRB and SMA systems can be compared. Figure 10 shows the maximum displacement and force demands normalized with respect the maxima in linear elastic system, and Fig. 11 shows the comparison between hysteresis loops, and of energy balance at different times.

Figure 10 shows the exact reduction factor due to hysteresis computed in time histories. The demand of the nonlinear system is smaller than one of the linear systems as expected. What was not anticipated is that the differences between elastoplastic (LRB) and superelastic flag-shaped model (SMA) are very small, even though a significantly vast difference was expected given the differences in the hysteretic area (Fig. 11a).

The similarity of the results shown in Fig. 10 would be clearer if the energy balance comparison of the systems as shown in Fig. 11(b) is considered. In this set of comparisons, the history of the system input energy in the three models is shown, which is due to the work done by the ground motion on the system. The system responses are different, even if the ground motion is the same for the three different input energy histories (SMA, LRB, and ELAS).

The input energy is converted by the system into components of kinetic, damping, and strain energy. The most interesting component is the strain energy, reported in Fig. 11(b) as *absorbed* energy, which is integrated with time due to the work done by the reaction force and the lateral displacement of the isolator. The absorbed energy is composed of an elastic and an inelastic contribution, and the comparison focuses on the final inelastic component which is the value at the end of the ground motion, as after some free vibration most of the elastic strain energy is recovered. It is trivial to point out that there is no inelastic strain energy at the end of the history in the linear elastic system. Looking at Fig. 11(b), it turns out that even if the energies in LRB and SMA are different, the ratio of input energies dissipated by inelastic strain given by the hysteresis is very similar.

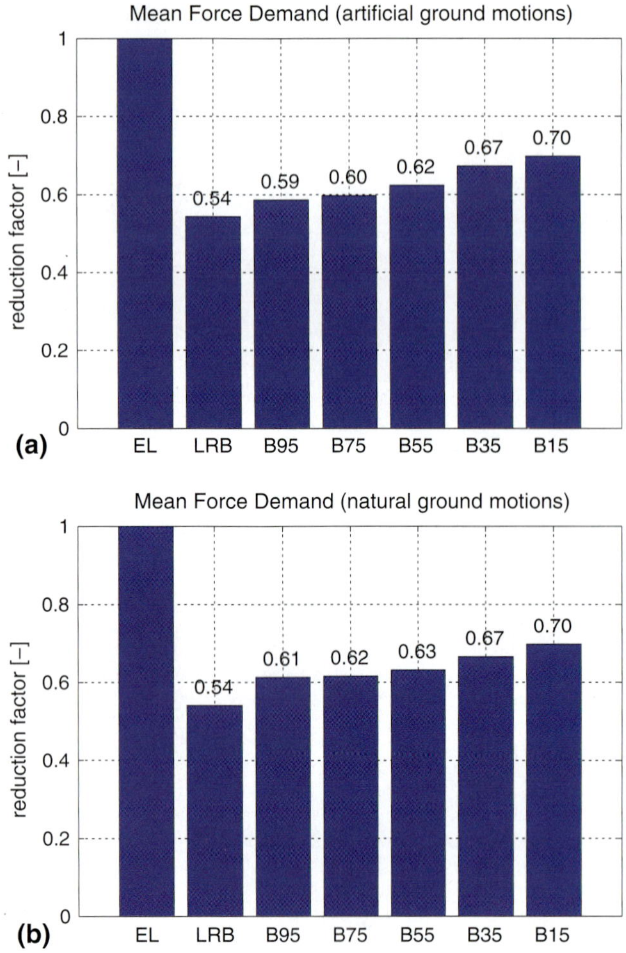

Fig. 13 Mean force reduction factor for artificial ground motion (a) and near-fault ground motion (b) considering linear elastic system (EL) and hysteresis in Fig. 8 (LRB, B95, B75, B55, B35, and B15)

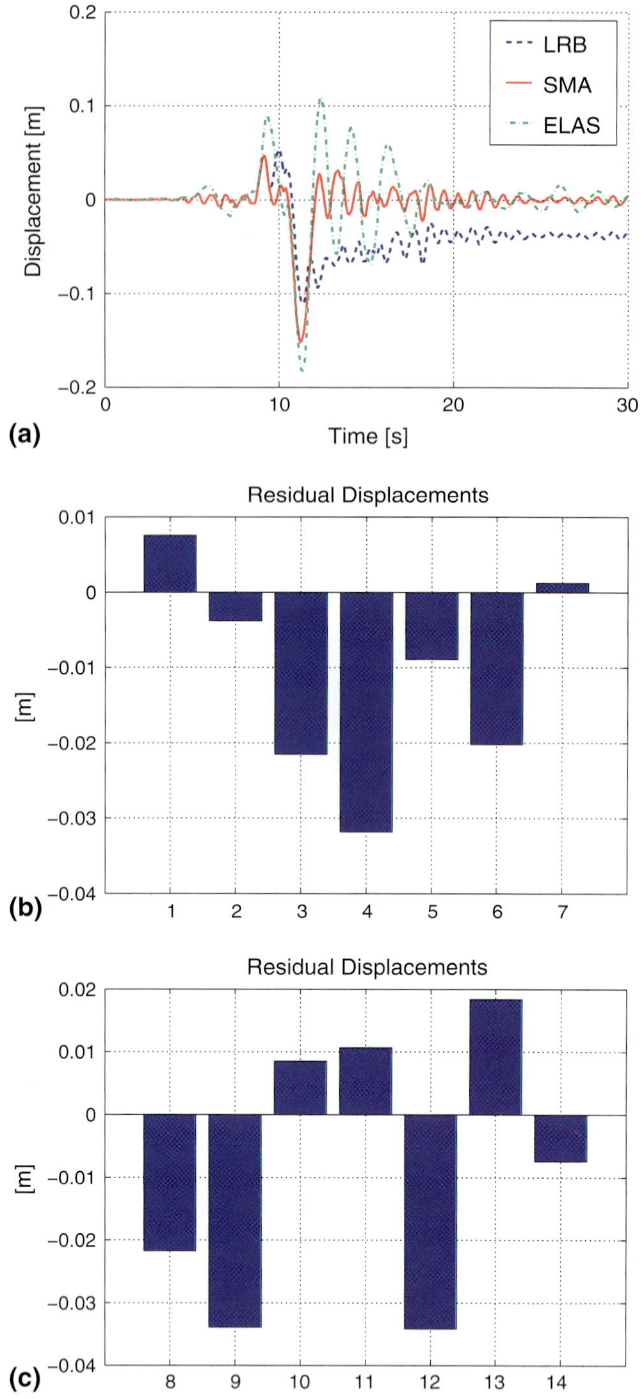

Fig. 14 Displacement-time response from an analysis of the three systems (a) and LRB model residual displacements in artificial (b) and near-fault (c) ground motion

The general comparison of all the ground motion results is shown in Fig. 12 and 13, where mean reduction factors in terms of displacements (Fig. 12) and the forces (Fig. 13) for artificial and natural ground motion records are reported. The comparison is similar to the one shown in Fig. 10, but this time the linear elastic response is compared with the six hystereses found in Fig. 8, which shows the elastoplastic model and the five different dissipation capabilities of the flag-shaped SMA models.

The first and most important result from the comparison is that the differences between the LRB elastoplastic model and superelastic flag-shaped model are small when compared with the linear elastic displacement and force demand for medium-to-high flag-shaped models' dissipation capabilities. Moreover, even by changing the dissipation factor significantly and getting a flag-shaped hysteresis very narrow, the response would not be very different to the most dissipating case. Even though this does not mean the differences are negligible, but these differences turn out to be smaller than the ones expected from those when considering an area-based approach for reduction factor computation.

The response of an idealized SDOF isolation system based on flag-shaped hysteresis considering different dissipations is quite good and comparable to the actual lead rubber response at least for a dissipation value in the range $\beta = 50\%$.

Another issue for isolation bearings is the residual displacement. This problem strongly affects the elastoplastic model, especially with a pulse-like near-fault seismic input. Besides, the SMA bearing induces zero residual displacements. An example of displacement-time graph is plotted in Fig. 14, and it shows that the LRB system is characterized by the residuals in all the analyses.

4. Conclusions

The feasibility of superelastic technology application to seismic isolation devices is investigated in this article. According to the area-based hysteretic damping approach, the behavior of a flag-shaped device is expected to be significantly more demanding in terms of base shear and displacement than that of an actual device.

The evaluation of the response data from time history analyses has been considered as the most suitable method to study the problem in detail. The behavior of a model representing a conventional LRB device, the behavior of a superelastic technology device, and of an equivalent linear elastic model have been compared. Results show that the overall behavior of the isolation system characterized by the flag-shaped hysteresis with a medium-to-high dissipation capability is close to the response of an elastoplastic isolation system. In particular, the displacement and force demands of LRB and SMA normalized with respect to the linear elastic secant model response are quite similar for large SMA dissipation capability. Besides, the differences between the ratios of absorbed energy and input energy are small.

Based on the investigation results, the SMA is possible to be used in seismic isolation and can lead to several advantages. SMA devices have good recentering capability. Moreover, superelastic devices are characterized by energy dissipation and numerical investigations have demonstrated that it is comparable, for its influence on the response, to the dissipation of actual highly dissipating devices.

References

1. R.I. Skinner, W.H. Robinson, and G.H. McVerry, *Seismic Isolation*, Wiley, Chichester, UK, 1993
2. F. Naeim and J.M. Kelly, *Design of Seismic Isolated Structures. From Theory to Practice*, Wiley, New York, 1999
3. C. Christopoulos and A. Filiatrault, *Principles of Passive Supplemental Damping and Seismic Isolation*, IUSS Press, Pavia, Italy, 2006
4. M.J.N. Priestley, G.M. Calvi, and M.J. Kowalsky, *Displacement-Based Seismic Design of Structures*, IUSS Press, Pavia, Italy, 2007
5. R. DesRoches and B. Smith, Shape Memory Alloys in Seismic Resistant Design and Retrofit: A Critical Review of Their Potential and Limitations, *J. Earthquake Eng.*, 2003, **7**, p 1–15
6. G. Attanasi, "Feasibility Assessment of Innovative Isolation Bearing System with Shape Memory Alloys," Master thesis, ROSE School, University of Pavia, Italy, 2008
7. Eurocode 8. *Design of Structures for Earthquake Resistance, Part I: General Rules, Seismic Actions and Rules for Buildings*. pre-ENV 1998-1
8. D.N. Grant, C.A. Blandon, and M.J.N. Priestley, *Modelling Inelastic Response in Direct Displacement Based Design*, IUSS Press, Pavia, Italy, 2005
9. A.J. Carr, *Ruaumoko Manual Volume 4: User Guide to Associated Programs*, Department of Civil Engineering, University of Canterbury, New Zealand, 2007

JMEPEG (2009) 18:738–745
DOI: 10.1007/s11665-009-9442-6

1059-9495/$19.00

Damping in Civil Engineering Using SMA. The Fatigue Behavior and Stability of CuAlBe and NiTi Alloys

V. Torra, A. Isalgue, C. Auguet, G. Carreras, F.C. Lovey, H. Soul, and P. Terriault

(*Submitted September 10, 2008; in revised form February 27, 2009*)

Two types of application in damping of structures by SMA in Civil Engineering are considered. The first one is related to the reduction of the damage produced by earthquakes. The second one is concerned with the increase of the lifetime of the stayed cables in bridges. The analyses of the experimental conditions required for each application are different: Several years or decades without any activity (excepted the summer-winter room temperature parasitic effects) followed by one or two minutes of oscillations under the earthquake affects, or near 100000 oscillations per day with pauses of several hours or days in the damping of stayed cables in bridges. This article analyzes the fatigue behavior of the CuAlBe alloy (appropriate for earthquakes) and of the NiTi alloy. Measurements of the damping of stayed cables indicate that the oscillation amplitude could be reduced up to one-third by using a NiTi wire as a damper device.

Keywords advanced characterization, failure analysis, heat treating, mechanical testing

1. Introduction

Shape Memory Alloys (SMAs) show particular properties associated to their martensitic transformation between metastable phases. The pseudoelasticity and thermoelasticity are effects related to their phase transition, used in sensors and actuators. The transformation often referred as a first-order phase transition with latent heat, shows hysteresis provided by different irreversible phenomena, i.e., when the interface overcomes internal barriers associated to dislocations (or arrays of dislocations, grain boundaries, and precipitates) (Ref 1).

The hysteretic behavior in this solid-solid transformation can be used in damping applications via the passive action of a structural component built in SMA by the conversion of mechanical energy to heat without hydraulic devices and/or external active or semi-active control (Ref 2). The active control requires expensive structures and external power supply that suppress the oscillations. The semi-active control requires sensors and actuators (i.e., as magneto-rheological fluids), appropriate computing systems and, also, a power supply without stops (ensuring a quick run for the 365/365 days) useful in skyscraper, but usually difficult in free surrounding structures (i.e., bridges in the wilderness) under the direct action of external storms, which can generate, by electric sparks (lightning) in a thundercloud, spontaneous power failures. High-rise buildings are also vulnerable to hurricane-force winds, particularly at the higher levels, since wind speed tends to increase with height (Ref 3).

Many papers and reviews (Ref 4, 5) have been on the application of SMA's as damping devices in civil engineering, but they are difficult to apply because the behavior of the alloys is not carefully analyzed. In particular, the summer-winter temperature effects, self-heating, or residual strains requirements for the given applications are rarely well established. The use of SMA dampers requires a deep knowledge of SMA properties and their coherence with the application with clear explicit and/or implicit requirements (Ref 6). For instance, in earthquake engineering, the available experimental conditions shows that, after several years or decades at rest, i.e., without any oscillation, the damper is required to show an excellent behavior in one or two minutes of destructive activity (i.e., for structural frequencies around 1 Hz close or under 200 oscillations). The long duration at rest conditions supposes that the SMA does not show relevant diffusive effects. In particular, the mechanical analysis of the energetic conversion, from mechanical energy to heat, suggests the advantages of prestressed SMA (Ref 7). The use of Cu-based alloys after several years of prestressing can modify the thermomechanical behavior of the alloy via the martensite stabilization producing negative results during (and after) the earthquake. However, the prestressing is mechanically advantageous. It is also suggested the use of clamps, in the damper, reducing the length of the SMA wire in the "compressive part". Also, other diffusion effects need to be studied and quantified for these applications after extremely larger initial pauses (Ref 8, 9).

The Fig. 1 shows one of the targets in Civil Engineering: the damping of stayed cables in bridges. The Fig. 1(a) shows an outline of the Iroise bridge with stayed cables situated in the Elorn River (near Brest, France). Damping the cables'

This article is an invited paper selected from presentations at Shape Memory and Superelastic Technologies 2008, held September 21-25, 2008, in Stresa, Italy, and has been expanded from the original presentation.

V. Torra, A. Isalgue, C. Auguet, and **G. Carreras**, CIRG-DFA-ETSECCPB, Polytechnical University of Catalonia, Gran Capita s/n, CAMPUS NORD B-4, 08034 Barcelona, Catalonia, Spain; **F.C. Lovey** and **H. Soul**, Dívisión Metales, Centro Atomico Bariloche and Instituto Balseiro, Avenida Bustillo 9500, Bariloche 8400, Argentina; and **P. Terriault**, Mechanical Engineering, Ecole de technologie superieure, 1100, Notre-Dame Street West, Montreal, QC H3C 1K3, Canada. Contact e-mail: vtorra@fa.upc.edu.

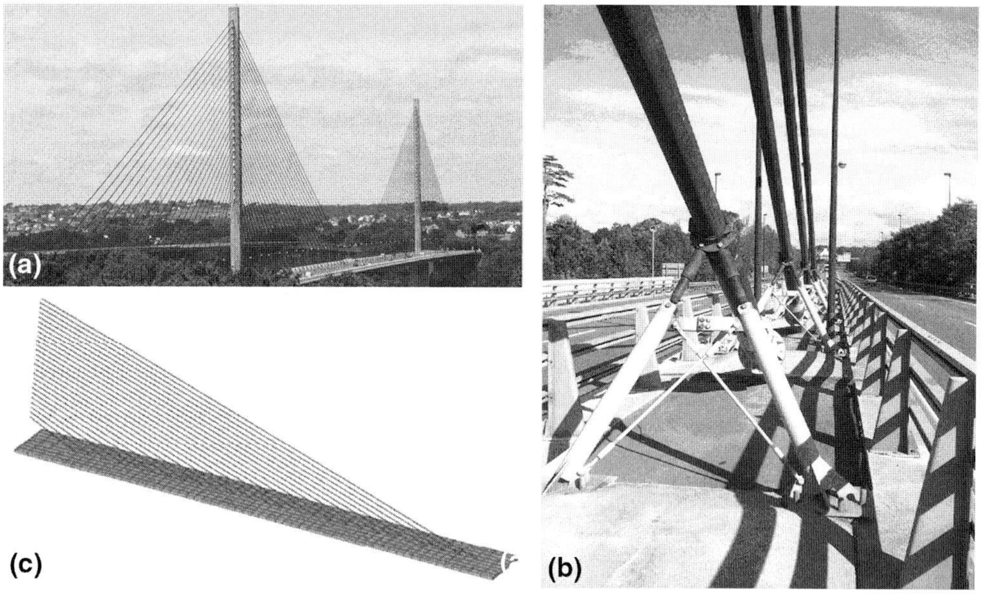

Fig. 1 (a) Highway bridge ("pont de l'Iroise" between Brest and Plougastel, France), platform: 200 + 400 + 200 m (length) 2 + 2 lanes. (b) Link of the cables to platform with two short hydraulic dampers. (c) Outline of the basic structure for simulation (one quarter of the bridge)

oscillation induced by wind and rain is of great technological importance. The Fig. 1(b) outlines the fixation of the cables to the platform (freeway with 2 + 2 lanes). The Fig. 1(c) shows a schema of the representative model used with dampers situated between stayed cables and platform.

In damping of earthquakes only, several minutes of working are required for one earthquake and replicas acting in a family house. The SMA dampers are situated in the diagonals of selected porticos. The situation is completely different when the target is the damping of stayed cables in bridges under the wind or rain effects. The cable frequencies are, for instance, close to 1-2 Hz (or more) and a large number of oscillations (100000) are expected for each day of work (with strong wind as in big storms). At this level, the damper can work "immediately" (or a few days) after installation without relevant initial quiet conditions (several years or decades) as in dampers used for family houses in earthquakes. In this situation, the initial time effects are extremely reduced and the useful working conditions are mainly controlled by the fatigue life. These conditions suggest that the prestressed material is probably a tool for an increased work. The more adequate analysis of these dampers relates to a deep study of the fatigue life and, eventually, positive ideas toward a practical increasing of the fatigue life of the material and the effect of pauses between series of oscillations, i.e., related to calm in the external wind/rain. In these applications (damping of seismic action effects or damping of stayed cable vibration) the SMA material is not allowed to increase significantly its length at working (Ref 10). However, all of the SMAs show some increase of length in cycling. This phenomenon always requires a particular study and, possibly, its reduction or suppression. Once again, the prestressing of the damper can compensate the progressive deformation (SMA creep) associated to working cycles. However, as said above, in a Cu-based alloy, the progressive appearance of stabilized martensite without retransformation destroys gradually the pseudoelastic effect and the applicability of the alloy for a large number of cycles. Furthermore, for a NiTi alloy, the dynamic effects of stress and temperature and pauses between series of cycles cannot be avoided, and therefore requires larger experimental analysis. This study, with an experimental approach, is centered on the fatigue-life analysis for a Cu-based alloy (i.e., CuAlBe) and for a NiTi alloy and, also, the aging effects of both alloys. The CuAlBe alloy is used after an appropriate heat treatment reducing the SMA creep associated to cycles, ensuring that the damper length remains "invariant" in service. Furthermore, the SMA creep in NiTi needs to be minimized to ensure a sequence of a high number of cycles without relevant increase of the length of the SMA component.

2. Experimental

The experimental analysis is focused on two SMA alloys. The CuAlBe alloy was furnished first by the research center of Trefimetaux (in 2003 and 2004) and later by NIMESIS, France (in 2005-2007) with similar original casts. For the cast AH140, the reference data are: Ms = 255 K; Mf = 226 K; As = 253 K; Af = 275 K with the following chemical composition in mass percent: Al = 11.8; Be = 0.5; Cu = 87.7. The wire from the cast AH140 of CuAlBe shows a Clausius-Clapeyron coefficient near 2.2 MPa/K (Ref 11). Several lengths of wires are used in the testing, with diameter of 3.4 mm. The CuAlBe alloy in the "as furnished state" made by direct extrusion of the wires is a mixture of phases $\alpha + \beta$ with easy progressive and permanent residual deformation that permits plastic deformation without pseudoelastic behavior. Practical application of the alloy requires a homogenization (or betatization) process. In this case, the following thermal treatment was used: heating to temperatures of 1123 K or of 1093 K, for a duration of 10 or 40 min, ensuring that the beta (bcc) phase was reached. After this annealing, the specimen was quenched in water at room temperature (i.e., 293 K). The mean grain size was 0.58 mm for 40 min at 1123 K. Further, long aging of 1 month at 373 K, and eventually, also, one additional month (or more) at 333 K were found useful to improve the stability of the material (Ref 12) as indicated in the Fig. 5 (see below). This aging process induces a slow decrease in the CuAlBe Ms temperature.

The NiTi alloys, in the pseudoelastic state, were ordered from the SAES Getters Co. (old furnisher: Special Metals, New Hartford, NY). These alloys normally show an excellent reproducible pseudoelastic behavior from one to other sets of samples for different furnishers, and we used the wires as arrived. Two diameters were used (0.5 and 2.46 mm). In the wires of 0.5 and in 2.46 mm of diameter, the alloy was furnished in black oxide and in light oxide surfaces, respectively, with nominal As temperatures of 248/247 K and 243 K, respectively. The nominal compositions are, respectively, Ni 55.95 wt.% and 55.92 wt.% with balance in Ti. The microstructure of the NiTi wire was studied by transmission electron microscopy. A Philips CM200 LaB6 microscope was used. The crystallographic orientation of the grains is textured along the ⟨111⟩ direction (Fig. 2, left). The weakness of the 001 and the 200 type reflections indicates that the wire is textured along the ⟨111⟩ direction. The microstructure of NiTi appears in the form of small grains and subgrains with a minimum size of about 80 nm (Fig. 2, right). In addition, some precipitates of TiC and Ni3Ti2 (not visible in the figure) were observed.

The main part of measurements was performed using two "Universal Testing Machines", one is an INSTRON, model 5567 with temperature chamber 3119-005 operating between 200 and 373 K with one load cell of 30 kN and, the other is a MTS 810 with load cell of 100 kN. Eventually, the MTS uses a home-made furnace for lower temperatures (i.e., up to 373 K).

2.1 Fatigue Analysis (CuAlBe and NiTi)

The Fig. 3 shows the fracture-life for CuAlBe (left) and for NiTi (as furnished) (right). The results for CuAlBe are highly dependent on the heat treatment. The measurements indicate that true strains (when the permanent deformation or SMA

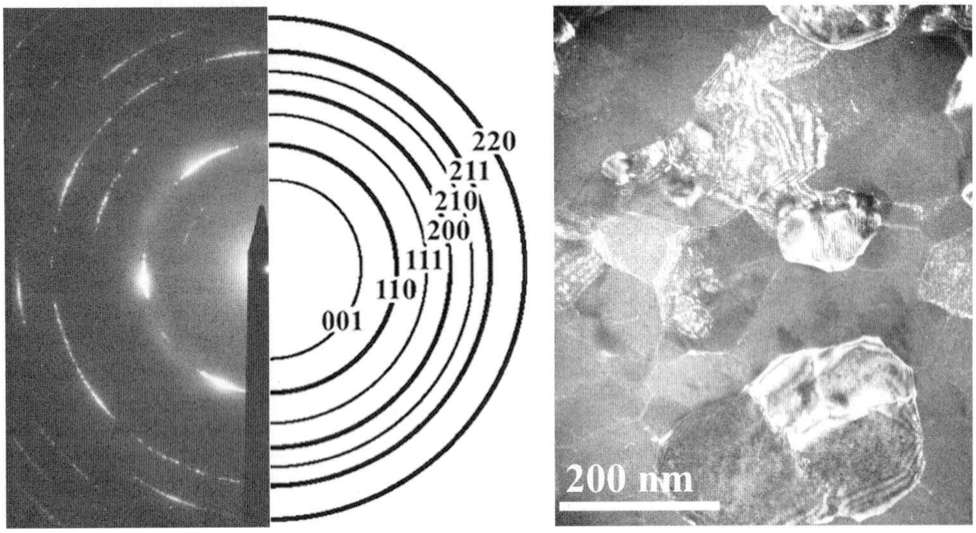

Fig. 2 Left: Electron diffraction pattern of the NiTi wire along the wire axis. Right: Dark field image taken with a small part of the 110-type ring at left

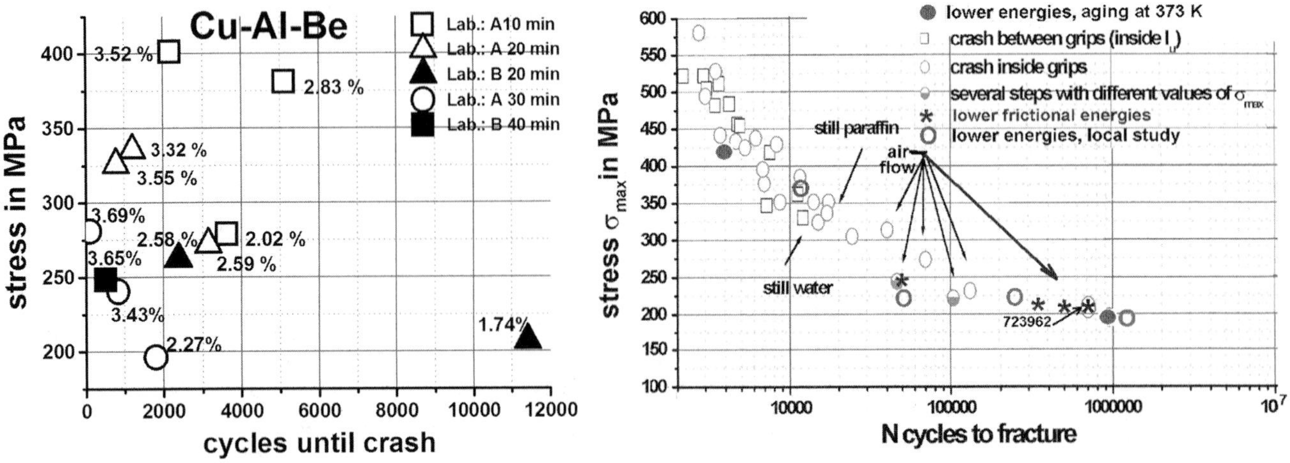

Fig. 3 Left: Fatigue and fracture behavior for CuAlBe. Wires with 3.4 mm of diameter and cycling at 0.5 Hz are used. Right: Fatigue and fracture for NiTi wire of 2.46 mm of diameter. The measurements on CuAlBe relate different length in time for the betatization process (10, 20, 30, 40 min realized in two laboratories A [Barcelona] and B [Bariloche])

creep is subtracted) ranging from 2% to 3% and homogenization at 10 min at 1093 K produces wires which support more than 2000 working cycles, appropriate for earthquake dampers. On the other hand, the fatigue life for NiTi in cycles with only tensile stress (without bending or compression) is a function of the deformation (or the stress). Improved results (overcoming 500000 cycles) relates to lower stresses close to 200 MPa associated with lower deformation or reduced hysteresis area. In NiTi, no significant differences for rupture in grips or in free part of the sample are visualized in Fig. 3, right. Some tests using water or paraffin baths were used to diminish the overheating of the sample, but in reduced deformation, the self-heating is practically irrelevant.

The Basquin law establishes a phenomenological relation between the stress and the fatigue life via the number of cycles to failure, N_f. The law states that

$$\sigma_{\text{MAX}} = A + B^* \left(\frac{1}{(N_f)^a} \right)$$

A set of measurements is presented in Fig. 4, left. In the appropriate coordinates, the experimental measurements fit, approximately, with the Basquin law. The value of the calculated exponent (a) is close to 0.22, which is clearly greater that the classical value (situated between 1/8 and 1/16). In fact, the fatigue measurements are associated to non-classical phenomena: a martensitic transformation. The Fig. 4, right, shows a comparison with results of fatigue in the Ref 13 and 14.

2.2 Aging: Time and Temperature Effects (CuAlBe and NiTi)

The main part of the analysis is done with electrical resistance and partially with stress-strain measurements in comparison with calorimetric measurements extracted from references. The Fig. 5 shows results from the CuAlBe alloy. On the left, the shape of the resistance measurements near the transformation corresponding to three different times (a, b, and c) are presented. When the aging temperatures are similar, after sufficient time, the resistance cycles are practically coincident (a and b). From the difference between a and c cycles the Ms changes after temperature changes can be obtained. The Fig. 5, right, shows the "recoverable" path induced by the aging. The Ms is a well-determined value associated to the aging

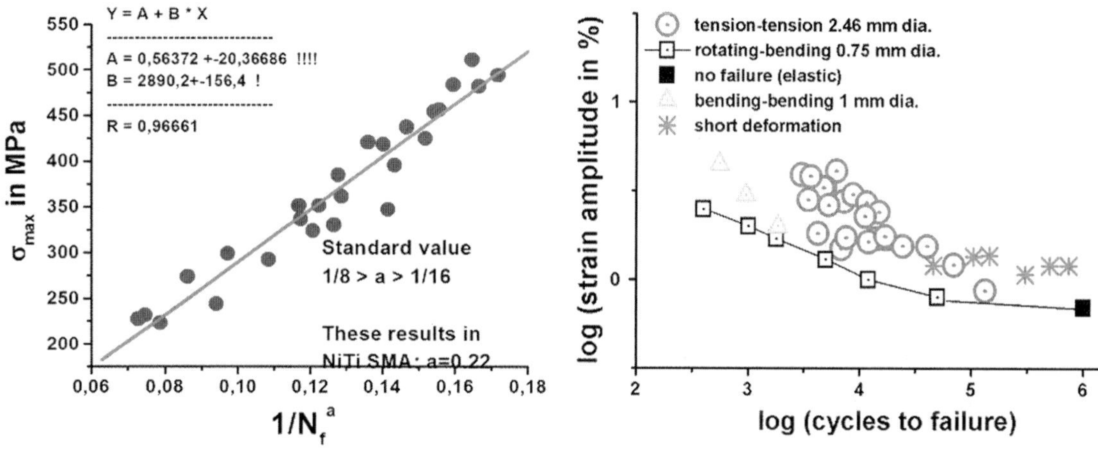

Fig. 4 Experimental data and fit via the Basquin Law (left). Right: comparison with data existing in the Ref 13 and 14 with more complex stresses

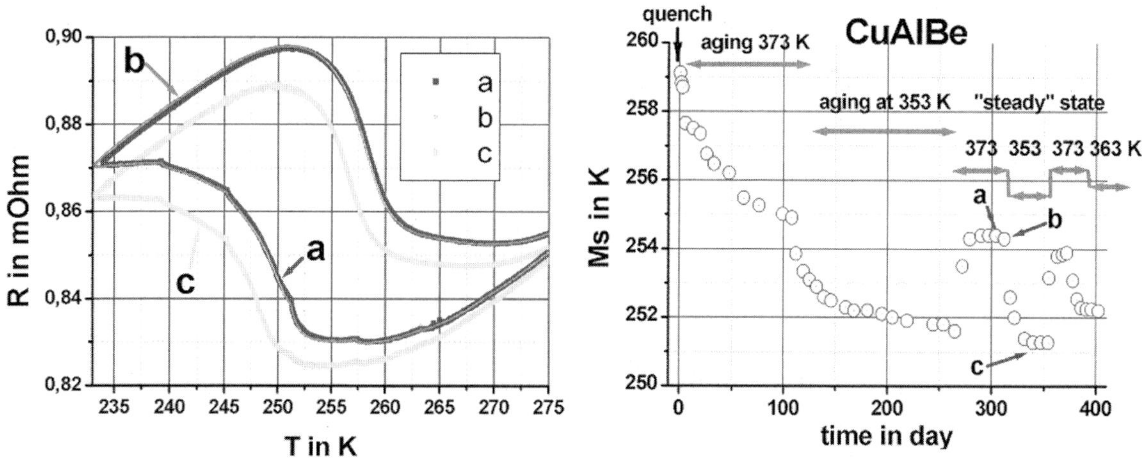

Fig. 5 Hysteresis cycle evolution due to aging determined from measurements of resistance against temperature and time. Left: Hysteresis cycle in resistance against temperature (after aging at 373 K (a and b), and at 353 K (c)). Right: Ms evolution against time after quench and further aging at different aging temperature (373, 353, and 363 K). The points (a), (b), and (c) correspond to the cycles in the left part of the figure

temperature in "steady" conditions. The Fig. 6 shows the resistance measurements associated to the R-phase transformation in NiTi. The evolution of the Rs against the aging temperature (under 800 days at 373, 388, and 398 K) is clearly visualized. The state reached after ageing does not reverse when lowering the temperature from 388 or 398 to 373 K. As a matter of fact, the Rs value increases monotonically.

The resistance measurements (as the calorimetric measurements, (Ref 9)) furnish the effect of aging on the Rs temperature. The effect of aging in NiTi is clearly different from that of the CuAlBe aging. In the Cu-based alloys, the transformation temperature (Ms) is a function of the aging temperature (after sufficient time, for steady state). In fact, the measurements indicate that the Ms tracks the room temperature evolution (Ref 8) as shown in the right part of the Fig. 5. In other words, the aging temperature induces "recoverable" effects in Ms. In NiTi, the aging effect on the R-phase transformation is much slower, but is clearly irreversible.

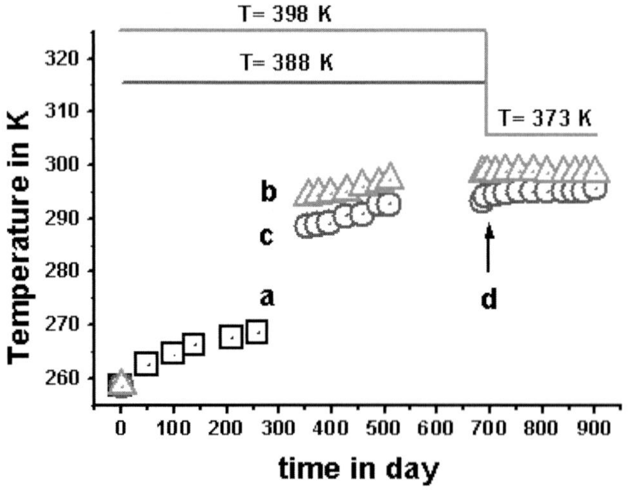

Fig. 6 Evolution on aging of the Rs temperature in NiTi alloy. Aging temperatures: 398 K for b curve, 388 K for c curve, and 373 K for a curve. After the point d, the higher temperatures were reduced to 373 K

2.3 NiTi Aging Effects on Loading

We measured the stress effect in the coexistence zone for different times at room temperature for eventual comparison with the Cu-based alloys (Ref 8). Furthermore, we measured the step obtained in the coexistence zone when a pause is imposed in an intermediate point of the "transformation plateau" (wires of 0.5 mm of diameter were used), and a subsequent cycle is performed, as indicated in Fig. 7, left. We show two consecutive measurements of the stress-strain hysteresis cycles at room temperature: the first one corresponds to the cycle with the pause in the coexistence zone for 27 days, and the second one is the immediate following cycle after the first one was completed. In Fig. 7, right, the part of the stress step is amplified to see that the global increase, including the stress reduction in the martensitic part of the sample, reaches 35 MPa, after one month of pause. The value is associated to two contributions: one part (20 MPa) is due to an increase of the transformation stress in the part of the sample that remained in parent phase. This means that the martensite initial temperature of Ms has decreased, and is more difficult to transform to martensite. The second one, 15 MPa, corresponds to a decrease of the necessary stress to maintain the transformation in the existent martensite phase. The global increase is then

$$\Delta\sigma_{\mathrm{tot}} = \Delta\sigma_{\beta} + |\Delta\sigma_{\mathrm{m}}|$$

We found that the variation of the necessary stress to continue the transformation when the pause is applied for one month in the coexistence zone increases with time and can reach a few tens of MPa. It is a minor effect for one month, but for longer durations (10, 20 years in prestressing for the application of SMA in earthquakes), it requires further deep study, or always avoiding the prestress.

Aging at 373 K for two months induces relevant changes in the force-deformation cycles in comparison with samples used "as furnished". The effect of aging induces a relevant reduction (35%) in the transformation stress after 100 cycles at 0.01 Hz and an increase of the hysteresis area. The experimental observations are clearly observed in Fig. 8.

The monotonic effect on Rs (Fig. 6) seems probably induced by a minor effect of precipitates associated to the lengthy aging. In Fig. 8, a decrease of the critical stress after

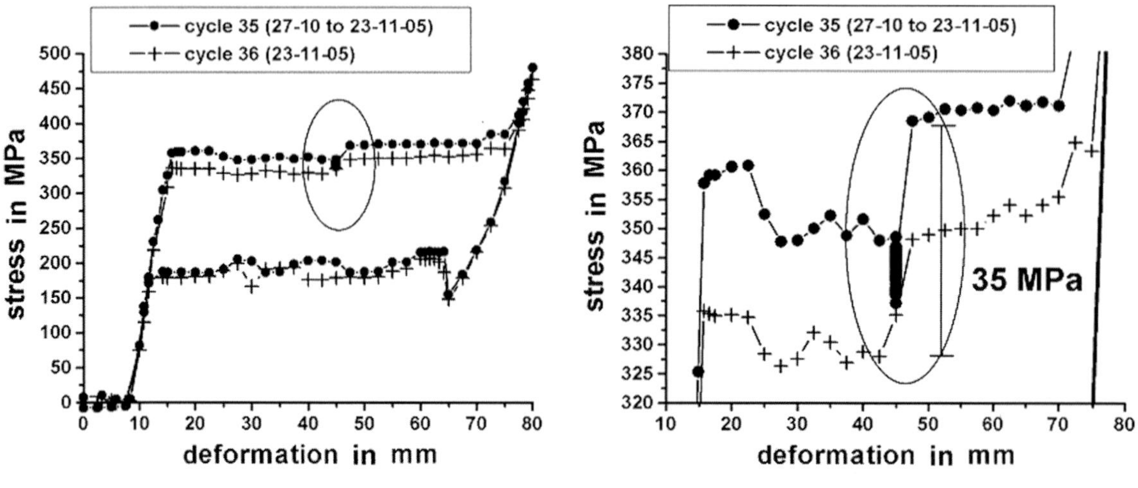

Fig. 7 Left: Two consecutive hysteresis cycles at room temperature (293 K), the first one obtained by applying a prestress in the coexistence zone for 27 days, and the immediate following one. Right: Global increase of 35 MPa ($\Delta\sigma_{\mathrm{tot}}$) to be able to continue the cycle

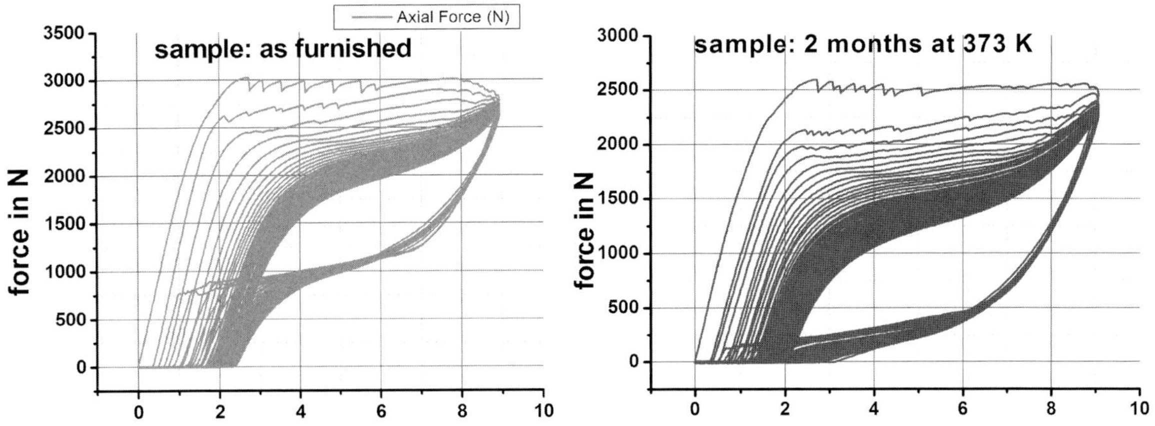

Fig. 8 Hysteresis cycles for NiTi of 2.46 mm of diameter, 100 cycles at 0.01 Hz (useful samples length close to 106.5 mm). Left: Sample as furnished, Right: sample after two months at 373 K

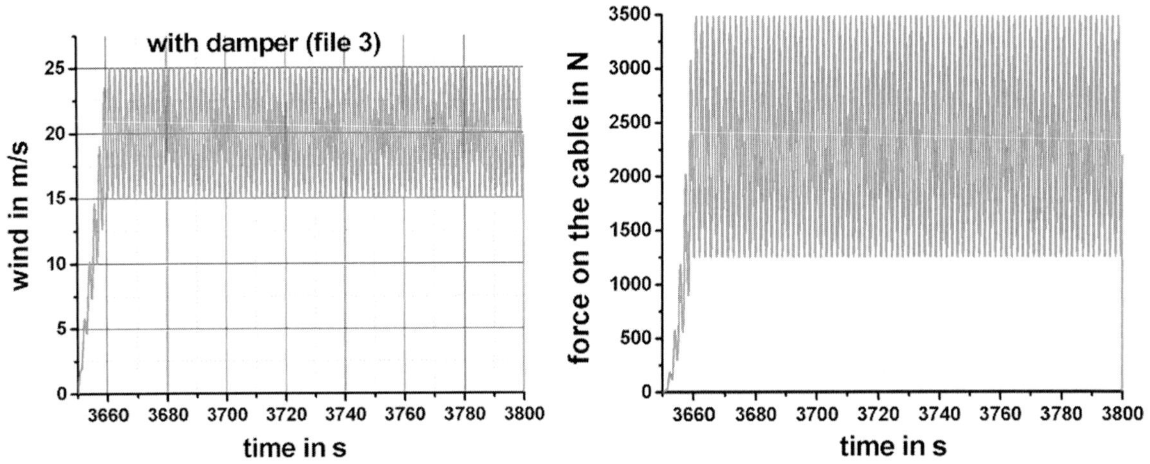

Fig. 9 Left: Periodic wind action on the cable in m/s. Right: Force on the cable in *N*

100 cycles is also observed for the aged samples (two months at 373 K). The effect seems coherent with the Rs extracted from resistance measurements. See, for instance, similar effects induced by the precipitates in CuZnAl in Ref 15.

The results obtained form stress-strain representation shows some differences in comparison with calorimetric or resistance measurements of the aged samples (Ref 9). The differences between resistance against temperature and force-deformation can be associated to the different characteristics of the used thermodynamic forces: (1) the temperature is one scalar thermodynamic force that does not induce relevant shape change in the transformation cycles and (2) the force induces relevant shape change by the oriented martensite variants and the privileged transformation direction.

2.4 The SMA as Dampers

The positive effects of CuAlBe in dampers for a family house are described in Ref 16. The damping actions of NiTi for the stayed cables in the Iroise bridges and in the ELSA cable 1 facility are partly investigated in this article. Using the ANSYS code, several simulations were performed using appropriate proprietary routines for the SMA behavior. The analysis is centered in only one cable of the bridge. A sinusoidal wind is simulated (Fig. 9, left) acting on the cable with one periodical force to induce oscillations (Fig. 9, right).

The Fig. 10 includes the amplitude evolution in free and damped cable (left) and the kinetic energy in the cable free and damped. It also includes the energy absorbed by the damper and transformed to heat. The left part of Fig. 10, left, shows the displacement for free cable, with progressive increase of amplitude. The right part of Fig. 10, left, visualizes the amplitude reduction and some beats when NiTi SMA damper is introduced in simulation. The Fig. 10, right, shows the energies in the cable. At left, progressive increase of kinetic energy in the undamped cable; right curve: kinetic energy under the combined effects of wind and damper. The line "mechanical to heat" shows the monotonic growth of the energy absorbed by the damper.

One experimental analysis of effectiveness of the NiTi SMA is realized using the cable 1 in the ELSA-JRC facility. The cross-sectional area of the steel inside the cable is 7×10^{-4} m^2, the tension on the cable is 250 kN, and the free oscillation frequency is 1.8 Hz. Since the cable is covered by wax, spontaneous damping of oscillations appears before the SMA device is connected. In this case, a minimal scaling in external actions and in SMA wires is necessary to account for the differences in, for instance, length and cross section. The selected Iroise cable has an external diameter, length, and mass

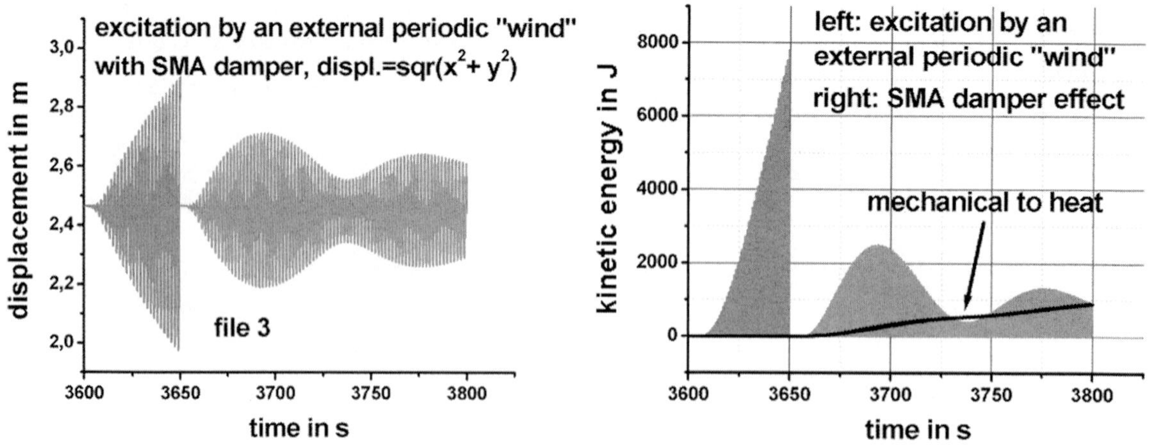

Fig. 10 ANSYS simulation: Effects on the Iroise cable. Cable without damper (outline bottom) and cable with SMA damper (outline top) figures displaced in time in order to compare. Left: Oscillation amplitude (at left) for free cable with progressively increasing amplitude. The right part shows the amplitude reduction, and some beats when NiTi SMA damper is introduced in simulation. Right: Energies in the cable. The continuous line indicate by the arrow corresponds to dissipated energy by the SMA damper

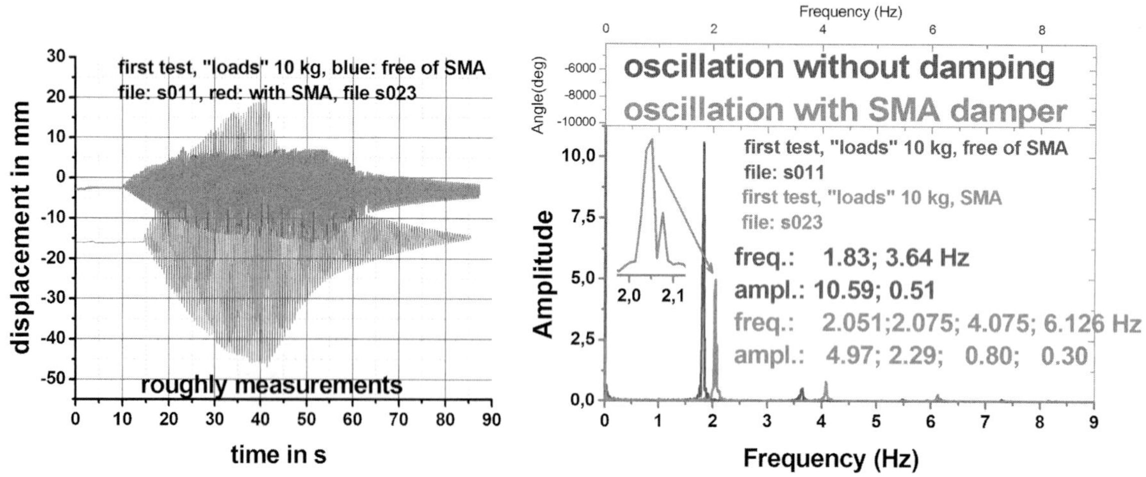

Fig. 11 Effects on the ELSA cable when a periodic excitation (amplitude close to 100 N) associated to resonant frequency is used. Left: Oscillations when there is no SMA (blue). With SMA (red), oscillations appear with reduced amplitude (near 1/3 of the amplitude oscillations in free conditions). Right: Frequency analysis of the experimental data

of 18 cm, 172 m, and 79.6 kg/m, respectively, and is subjected to 5.2 MN of tension. The ELSA cable has around 8 cm of external diameter, 45 m of length, and the mass 9 kg/m, and is subjected to 250 kN of tension. Two types of actions are realized on the ELSA cable, periodic force (near 100 N) to increase the amplitude without, and with only one SMA wire, Fig. 11 (left, displacement, and right, frequency analysis). The SMA wire (length 414 cm), is previously "trained" via a mechanical treatment of 100 cycles at 0.01 Hz corresponding to cycles as shown in Fig. 8, left. The damper is situated perpendicular to the cable at ¼ of their length. The frequency analysis shows that the oscillations occur at resonance frequencies, and then the decrease in amplitude is directly related to an increased damping. Their action induces minor changes in the oscillation frequency as shown in the Fig. 11, right. The frequency analysis using the Fast Fourier Transform of displacements against time shows that without the SMA, the main frequency is always close to 1.8 Hz, and that with SMA damper, the main frequency increases by a 10% and splits in two close frequencies near 2.05 Hz inducing eventual beats in the oscillations (as shows the Fig. 10). The results determined by simulation are similar to those of the experimental measurements.

3. Remarks and Conclusions

This study analyzed the fatigue behavior of a CuAlBe SMA alloy, showing that while it is appropriate for damping in Civil Engineering against the effects of earthquakes, it is not well suited in the actual state of the art, for continuous, long fatigue life. However, some experiments about the performance of a NiTi alloy show that this alloy could indeed be suitable for their eventual application to damping of stayed cables.

For one or two minutes of induced oscillations at 1 Hz, the fatigue life for CuAlBe is sufficient for damping the earthquakes effect; the action of strong winds and rain associated to strong storms (several days) requires around 500000 cycles of work, and the NiTi is more appropriate for this job.

Recoverable against nonrecoverable long-time aging effects in CuAlBe and in NiTi can be observed using, for instance, electrical resistance-temperature cycles. The effect on Ms (CuAlBe) and on Rs (NiTi) can be associated, respectively, to effects of vacancy concentration (tracking between the Ms and the external room temperature) and minor precipitation. The different behavior of the alloys motivates different choices for different applications, as earthquake effects mitigation (large waiting durations, few cycles of activity), or damping of oscillations induced by wind in cables (nearly immediate action, long-fatigue life needed).

The analysis of the simulation of damper behavior in stayed cable shows satisfactory behavior: the relevant reduction of amplitude to less than one half. The experimental observations indicate an increase of cable damping, and are coherent with the calculated values. In order to optimize the action of SMA damper, deeper study of the frequencies changes with the damper loads seems necessary.

Acknowledgments

Our thanks are due to Dr J. Pelegrina of CONICET-CAB and his group (Bariloche, Argentina) for their measurements and comments about the resistance and calorimetric measurements in NiTi alloy. Cooperation between CIRG (UPC) and ETS (Montreal, Quebec, CA) and with CAB-IB (University of Cuyo, Argentina) is supported by Quebec government and CNEA and, earlier, by DURSI (Gen. Catalonia). V.T. gratefully acknowledges the technical support from CAB-IB. The research is carried out within the framework of ESF EUROCORES S3T Project Number 05-S3T-FP014-SMARTeR and the associated "complementary action" of MEC-Spain. Also, the Project Number C22/06 from M. Fomento is gratefully acknowledged. The support provided by Dr. D. Tirelli and B. Zapico of ELSA (JRC-EU, Ispra, Italy) for the damping measurements in the cable 1 are decisive for the study of NiTi effects.

References

1. K. Otsuka and C.M. Waymann, Ed., *Shape Memory Materials*, Cambridge Univ. Press, Cambridge, 1998
2. V. Torra, A. Isalgue, and F.C. Lovey, Microstructure and Thermodynamics of the Martensitic Transformation, *Can. Metall. Quart.*, 2000, **39**, p 207–216
3. H. Janocha, Ed., *Adaptronics and Smart Structures*, Springer-Verlag, Berlin, 1999, p 334–369
4. J. Salichs, Z. Hou, and M. Noori, Vibration Suppression of Structures Using Passive Shape Memory Alloy Energy Dissipation Devices, *J. Intell. Mater. Syst. Struct.*, 2001, **12**, p 671–680
5. J.C. Wilson, M. Eeri, and M.J. Wesolowsky, Shape Memory Alloys for Seismic Response Modification: A State of the Art Review, *Earthquake Spectra*, 2005, **21**, p 569–601
6. C. Auguet, A. Isalgue, F.C. Lovey, F. Martorell, and V. Torra, Metastable Effects on Martensitic Transformation in SMA Part IV: Thermomechanical Properties of CuAlBe and NiTi Observations for Dampers in Family Houses, *J. Thermal Anal. Calorim.*, 2007, **88**, p 537–548
7. V. Torra, A. Isalgue, C. Auguet, F.C. Lovey, J.L. Pelegrina, and P. Terriault, Pre-stressed NiTi: Effects of the Thermodynamic Forces and Time, *Proceedings of the International Conference on Shape Memory and Superelastic Technologies* (Tsukuba, Japan), S. Miyazaki, Ed., ASM International, OH, 2007, p 195–203. ISBN: 978-0-87170-722-2
8. V. Torra, J.L. Pelegrina, A. Isalgue, and F.C. Lovey, Metastable Effects on Martensitic Transformation in SMA (I): Recoverable Effects by the Action of Thermodynamic Forces in Parent Phase, *J. Thermal Anal. Calorim.*, 2005, **81**, p 131–135
9. C. Auguet, A. Isalgue, F.C. Lovey, J.L. Pelegrina, S. Ruiz, and V. Torra, Metastable Effects on Martensitic Transformation in SMA. Part III. Tentative Temperature Effects in a NiTi Alloy, *J. Thermal Anal. Calorim.*, 2007, **89**(2), p 537–542
10. A. Isalgué, F.C. Lovey, P. Terriault, F. Martorell, R.M. Torra, and V. Torra, SMA for Dampers in Civil Engineering, *Mater. Trans.*, 2006, **47**, p 682–690
11. A. Isalgue, V. Torra, A. Yawny, and F.C. Lovey, Metastable Effects on Martensitic Transformation in SMA. Part VI. The Clausius-Clapeyron Relationship, *J. Thermal Anal. Calorim.*, 2008, **91**(3), p 991–998
12. A. Isalgue, J. Fernandez, V. Torra, and F.C. Lovey, Conditioning Treatments of CuAlBe SMA for Dampers, *Mater. Sci. Eng. A*, 2006, **438–440**, p 1085–1088
13. H. Tobushi, T. Hachisuka, S. Yamada, and P.-H. Lin, Rotating-bending Fatigue of a NiTi Shape Memory Alloy Wire, *Mech. Mater.*, 1997, **26**, p 35–42
14. M.G. de Azevedo, R. Fonseca, and V.T. Lopes, The Influence of High Amplitude Cycling Straining on the Behaviour of Superelastic NiTi, *Int. J. Fatigue*, 2006, **28**, p 1087–1091
15. F.C. Lovey and V. Torra, Shape Memory in Cu-based Alloys: Phenomenological Behavior at the Mesoscale Level and Interaction of Martensitic Transformation with Structural Defects in Cu-Zn-Al, *Prog. Mater. Sci.*, 1999, **44**(3), p 189–289
16. V. Torra, A. Isalgue, F. Martorell, P. Terriault, and F.C. Lovey, Built in Dampers for Family Homes via SMA: An ANSYS Computation Scheme Based on Mesoscopic and Microscopic Experimental Analyses, *Eng. Struct.*, 2007, **29**, p 1889–1902

JMEPEG (2009) 18:746–753
DOI: 10.1007/s11665-009-9433-7

Shape Memory Alloy Tension/Compression Device for Seismic Retrofit of Buildings

Matthew Speicher, Darel E. Hodgson, Reginald DesRoches, and Roberto T. Leon

(Submitted September 16, 2008; in revised form February 20, 2009)

A tension/compression device is developed for applications as bracing elements in buildings. The device is designed to allow Nitinol forms, such as helical springs or Belleville washers, to be used in compression. The device allows both overall extension (tension) and compression while subjecting the Nitinol to an optimum deformation mode. It is possible, due to the versatility of the design, to adjust the force and stroke of the device without changing the overall configuration. This new device is subjected to a cyclic loading protocol that tests the Nitinol element's ability to recover large deformations. The effect of different Nitinol configurations and a cyclic loading history are evaluated in the study. The results show that Nitinol helical springs produce good recentering and damping behavior while Nitinol Belleville washers show good potential to form the basis for a Nitinol damping device.

Keywords Belleville washer, damper, NiTi, Nitinol, seismic retrofit, SMA

1. Introduction

Numerous researchers have investigated the use of recentering systems to improve the performance of structures during seismic events. The 2008 Sichuan Earthquake in China was a reminder of how a large portion of today's infrastructure is vulnerable to damage from earthquakes. One strategy to improve performance of buildings is to create systems that have the ability to recenter. In comparison to traditional passively damped systems, recentering systems have been shown to have at least equivalent performance (Ref 1, 2). Recentering systems can be broken down into two categories: (1) post-tensioned (PT) based systems and (2) shape memory alloy (SMA) based systems. PT systems use cables or elements to provide a restoring force during and after a loading event. Research on these systems first began in concrete buildings (Ref 3) and then was expanded to steel framed structures (Ref 4-7). These systems are often combined with supplemental damping to improve their performance.

SMA-based systems take advantage of the material's unique ability to recover large strains due to a solid-solid phase transformation and dissipate energy because of the resulting internal friction. For civil engineering, NiTi SMAs, otherwise know as Nitinol, have become the most popular SMA due to its favorable properties including hysteretic damping, large strain recovery, and excellent corrosion resistance. From a micro-scale to a macro-scale, Nitinol shows great promise for its use in seismic resisting systems (Ref 8). Nitinol SMA is the material which is evaluated in this paper.

A variety of research initiatives have looked at SMAs' viability as a structural recentering and damping element for both bridges and buildings (Ref 9). Dolce et al. (Ref 10) devised several types of recentering systems using SMA wires. They tested some devices which produced predicable behavior with good energy dissipation and excellent recentering. However, none of these devices have become widely used. In addition, other SMA-based systems have been investigated without wide implementation such as a self-centering friction damped brace by Zhu and Zhang (Ref 11) and beam-column connections by Ocel et al. (Ref 12), Penar (Ref 13), and Sepulveda et al. (Ref 14).

The inherent advantage of using an SMA-based system is that it can be designed to have (1) the ductility and energy dissipation to prevent collapse and (2) the ability to significantly reduce the residual deformations after an earthquake occurs. This research presents the development and initial testing of a new device that facilitates the use of two forms of Nitinol SMA: helical springs and Belleville washers. Whether the device is in tension or compression, the SMA elements are compressed. In addition, the device can use a wide array of shapes as the active components, which have a tremendous range of possible force and deformation characteristics.

2. Tension/Compression Device Description

The tension/compression device is a cylindrical-shaped damper that provides the ability to test a variety of SMA

This article is an invited paper selected from presentations at Shape Memory and Superelastic Technologies 2008, held September 21-25, 2008, in Stresa, Italy, and has been expanded from the original presentation.

Matthew Speicher, Reginald DesRoches, and **Roberto T. Leon**, School of Civil Engineering, Georgia Institute of Technology, 790 Atlantic Drive, Atlanta, GA 30332-0355; and **Darel E. Hodgson**, Nitinol Technology, Inc., 740-D Sierra Vista Ave., Mountain View, CA 94043. Contact e-mails: matthew.speicher@ce.gatech.edu, darel@nititech.com, reginald.desroches@ce.gatech.edu, and roberto.leon@ce.gatech.edu.

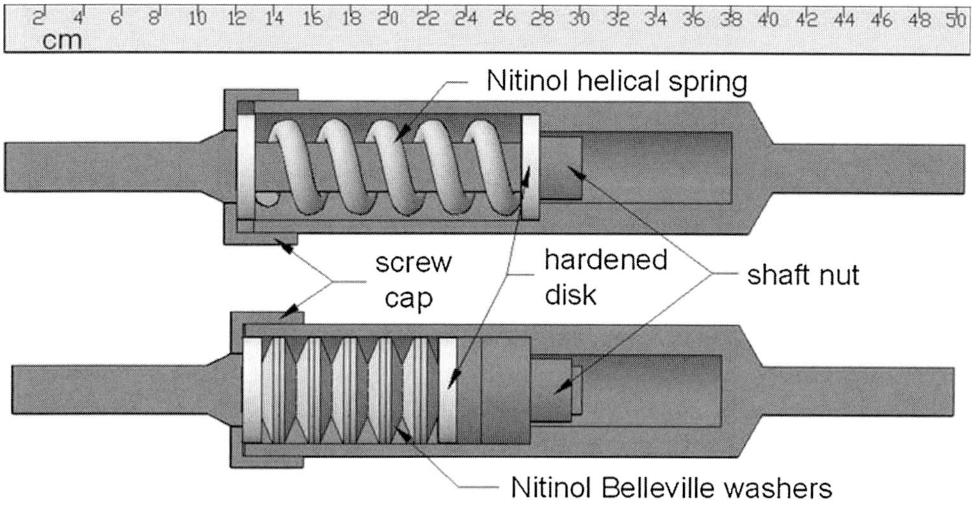

Fig. 1 Internal view of tension/compression device

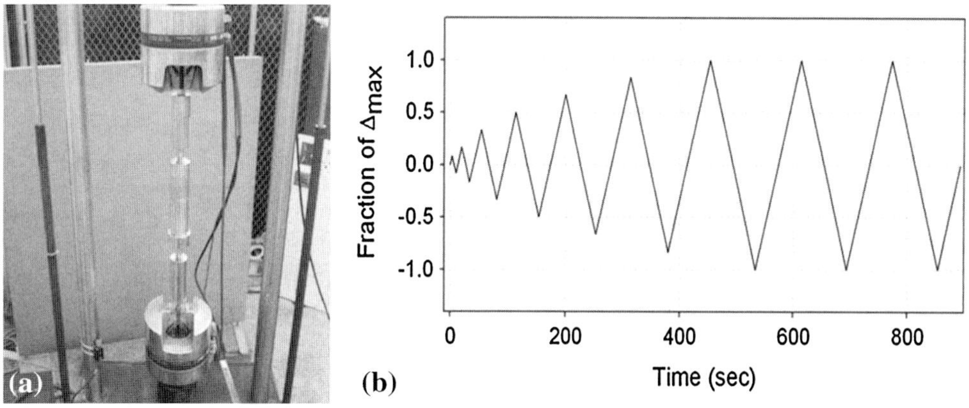

Fig. 2 (a) Test machine setup and (b) example loading protocol

elements. The body and the shaft of the device are made out of standard 304 stainless steel cylinders. Drawings of the device are shown in Fig. 1. The device is approximately 50 cm in length and 6.4 cm in diameter. The stroke capacity of the damper is dependant on the active element that is inserted, with values in the range of 2-5 cm for the devices presented in this paper. The damper is fitted with either helical springs or Belleville washers for the testing. Different strength and stiffness properties can be easily obtained by using different combinations of these elements. Two different springs and three different arrangements of Belleville washer are investigated.

3. Experimental Setup

An experimental scheme is devised to assess the viability of the SMA-based device using a 250 kN MTS Universal Testing Machine shown in Fig. 2(a). The MTS machine is fitted with hydraulic vee-notched wedge grips which can accommodate a rod diameter up to 1.9 cm. Stainless steel coupler elements are used to transfer the force from the 2.54 cm diameter rods of the damper to 1.9 cm diameter rods gripped by the MTS machine. The MTS machine is run by a Teststar controller running Testware software. The tests are conducted in displacement control using the built-in LVDT attached to the bottom grip.

A far-field loading protocol modeled after the protocol used in the SAC Steel Project (Ref 15) was selected for the experiments. The loading protocol, shown in Fig. 2(b), uses the testing machine's stroke as the deformation parameter. For each specimen arrangement, the maximum deformation was calculated and then 90% of this value was set as the target test maximum to prevent overloading of the device. A quasi-static loading rate for the testing was set at 0.127 cm/s to eliminate dynamic effects. Additionally, all experiments were carried out under ambient temperatures in the range of 26-28 °C. Self heating was not recorded because of the difficulty in monitoring the heat in the confined device and it was not expected to greatly impact the tests due to the quasi-static protocol implemented.

The tension/compression device was fabricated to facilitate the ability to implement different types of active elements.

In this paper six different tests are presented. The tests performed are as follows:

A. Cyclic protocol using the damper loaded with a hollow NiTi spring.
B. Cyclic protocol using the damper loaded with a solid NiTi spring.
C. Monotonic protocol compressing individual NiTi Belleville washers.
D. Cyclic protocol using the damper loaded with a single stack of 10 NiTi Belleville washers.
E. Cyclic protocol using the damper loaded with a double stack of 12 NiTi Belleville washers.
F. Cyclic protocol using the damper loaded with a triple stack of 12 NiTi Belleville washers.

4. Active Elements

Three different types of active elements are used for testing: a hollow spring, a solid spring, and a set of Belleville washers. The hollow NiTi spring has a diameter of 3.81 cm, an initial length of 14.20 cm, and a deformation capacity of 7.95 cm. The hollow spring was made from tubing with an outside and inside diameter of 12.5 and 9.5 mm, respectively. The solid SMA spring has a diameter of 3.81 cm, an initial length of 12.98 cm, and a deformation capacity of 4.57 cm. The solid spring was made from solid stock with a 12.5 mm diameter. The springs were made from Nitinol Alloy 508 (50.8% at.% Nickel). Both springs were made by heating small sequential sections of the stock with a small torch just to the point of initial softening (very low red heat, ~650-700 °C). The softened metal was then bent around a mandrel to produce a helix. The finished coils were given a final heat treatment to achieve uniform properties and good superelasticity.

The NiTi Belleville washers are 5.50 cm wide, 0.64 cm tall, and 0.305 cm thick. They were cut with a waterjet from a hot-rolled sheet of standard Nitinol Alloy 508. The conical shape was made by using a 30° angled cone. This yielded washers with a 26-27° angle. The same set of 12 washers is used for all the testing in this paper excluding washer 3 in the individual tests (test C). Washer 3 was from a set given a different heat treatment that yielded a higher transformation temperature resulting in reduced superelasticity. Additionally, a set of pretests were performed on these washers which is not presented in this paper.

Fig. 3 Nitinol spring loaded on center shaft

5. Results of Helical Spring Tests

5.1 Hollow Helical Spring

Test A was performed using the hollow NiTi spring as the active element. The spring was inserted into the device and the shaft nut was tightened to give the spring 4.14 cm of precompression (Fig. 3). This was done to increase the initial stiffness of the device and to give a good flag-shaped hysteresis. The precompressed length of the spring was 10.06 cm and the fully compressed length was estimated at 5.83 cm. This resulted in a maximum device stroke of 4.23 cm to bring the spring loops into contact with each other. For the tests, 90% of this value (3.81 cm) was used as the test target maximum deformation. Using the previously described protocol, the resulting force-deformation plot was obtained (Fig. 4).

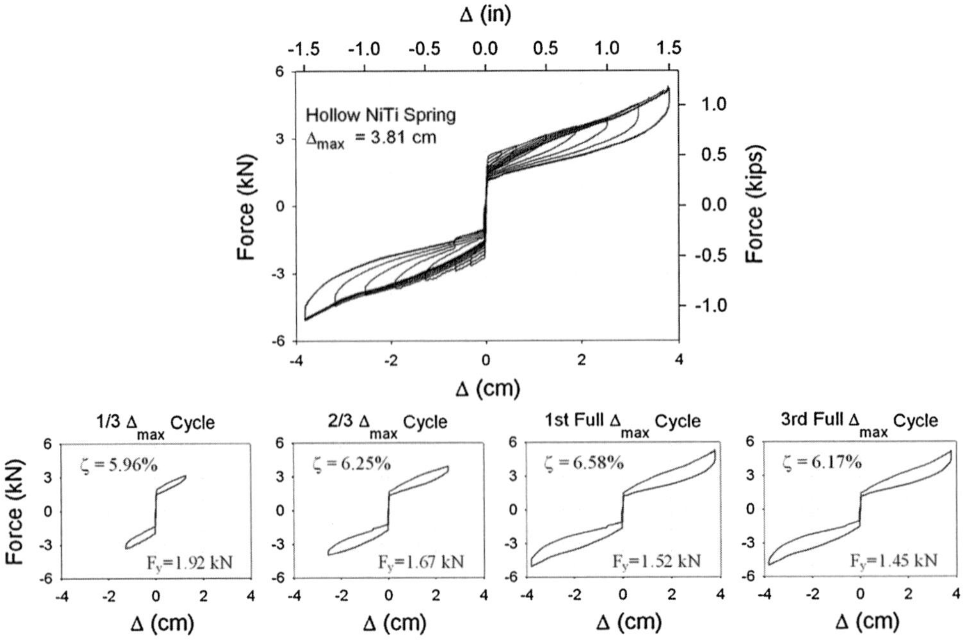

Fig. 4 Force-deformation response of hollow Nitinol spring in device (test A)

The 1/3, 2/3, and Full Δ_{max} cycles are highlighted to show the progressive behavior. The precompressed helical spring provided high initial stiffness and recentering over the entire cyclic protocol.

5.2 Solid Helical Spring

Test B was performed using the solid NiTi spring as the active element. Again the spring was inserted into the device and the shaft nut was tightened to give the spring 1.15 cm of precompression. The spring was only precompressed this amount because of how the spring fit on the shaft. The precompressed length of the spring was 11.83 cm and the fully compressed length was estimated at 6.75 cm. This resulted in a maximum device stroke of 5.08 cm; in which 90% of this value (4.57 cm) was used for testing. The cyclic loading protocol was applied and a similar quantitative behavior as the hollow spring was observed. The force-deformation curves are shown in Fig. 5, with the 1/3, 2/3, and Full Δ_{max} cycles individually displayed.

6. Discussion of Helical Spring Tests

The force-deformation relationships for both tests A and B demonstrate good hysteretic damping, limited strength degradation, and excellent repeatability. In assessing the performance of a damper, the equivalent viscous damping and the consistency of force levels are two important quantities. The equivalent viscous damping is defined as:

$$\zeta = \frac{E_D}{4\pi E_{so}} \qquad \text{(Eq 1)}$$

where E_D is the energy dissipated in one cycle and E_{so} is the energy absorbed by an equivalent linear elastic system loaded to the same maximum force and displacement levels as used in E_D. In structural engineering, values of ζ around 5-10% are typical. For this paper, the yield force is defined as the force level in which the overall device breaks from its initial stiffness; which is the "kinked" point in the bilinear loading response. Also, the yield force is from the positive (tension) stroke of the device. Ideally this should not matter but in practice there was some difference between the positive and negative behavior. However, there was reasonable overall anti-symmetry observed.

The superelasticity combined with precompression of the spring resulted in a slender flag-shaped hysteric loop that accounted for ζ ranging from 6 to 11%, with the majority of the values between 6 and 7% for the hollow spring and 7 and 9% for the solid spring. In Fig. 6(a), the trend of ζ is graphically presented. In comparison to the solid spring, the hollow spring tended to have more stable levels of damping over the cyclic range.

When interpreting the behavior of the force levels in tests A and B, it should be noted that the precompression given to the hollow spring was almost four times that given to the solid spring. As a result, the hollow spring setup had a higher yield force than the solid spring, which at first glance is counterintuitive. However, the stiffness of the sloped loading plateau is clearly larger for the solid spring in comparison to the hollow spring. The yield forces for the hollow and solid springs are shown versus the fraction of Δ_{max} in Fig. 6(b). Both springs have the same trend; the yield force decreases as the loading cycles increase. The hollow spring yield force decreases faster than that of its solid counterpart. This can be contributed to the hollow spring having a larger precompression. When the cyclic loading is applied, the hollow spring is pushed further into its superelastic range which results in an increase in residual deformations due to accumulation of dislocations in the microstructure. This accumulation presets stress in the material causing a reduction in the yield force (Ref 16).

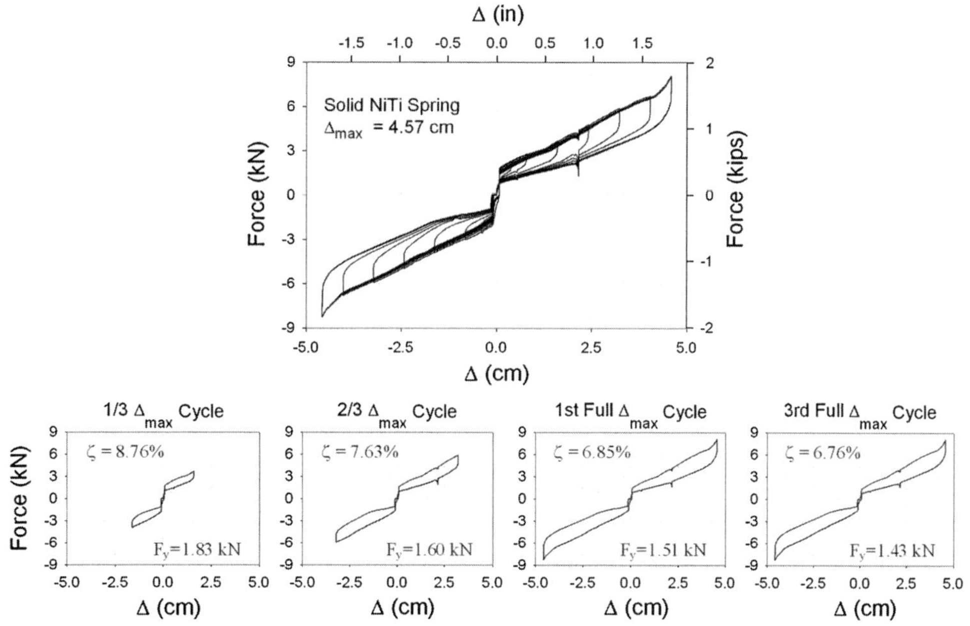

Fig. 5 Force-deformation response of the solid Nitinol spring in device (test B)

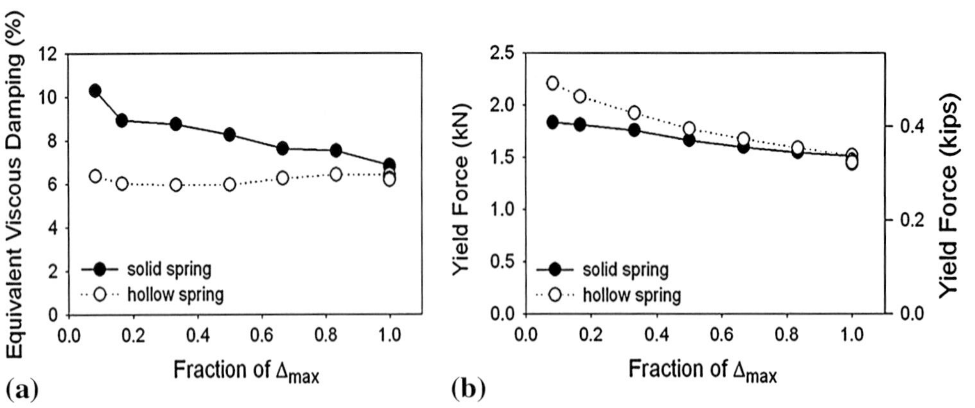

Fig. 6 (a) The damper's equivalent viscous damping and (b) yield forces over a range of deformations

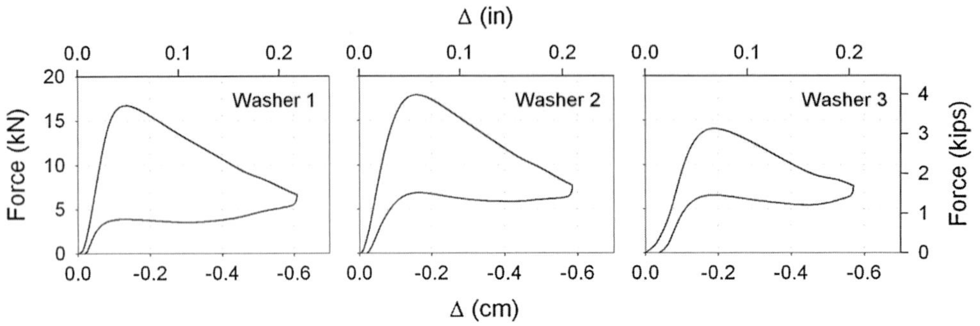

Fig. 7 Response of single NiTi Belleville washers under compression

7. Results of NiTi Belleville Washer Tests

7.1 Individual

In order to investigate the unknown behavior of the NiTi Belleville washers, three washers were randomly selected for compression testing (test C). These tests were performed in the MTS machine. The SMA Belleville washer was placed between two 1.27 cm hardened disks and compressed until just short of flat. The resulting duck-head-shaped force-deformation curves are shown in Fig. 7. The washers have good initial stiffness and strength, but at a deformation of approximately 0.15 cm the force carrying capacity peaks and there is a significant drop off until the imposed deformation is released at 0.605 cm. Washer 3 had the lowest strength and the highest residual deformation.

7.2 Single Stacked

Test D was conducted on a washer stack with the most flexible configuration of 10 singly stacked washers as shown in Fig. 8. Flat steel washers of 0.2 cm thickness were inserted between each SMA washer in an attempt to prevent the washers from inverting or buckling as seen in preliminary tests. As stated in the specimen description section, the SMA washers used in this test were not virgin; they had been used in preliminary tests.

The initial length and the test target maximum deformation of the washer configuration were 10.41 and 5.08 cm, respectively. The shaft nut was hand tightened. The resulting cyclic force-deformation curves are shown in Fig. 9. The washer stack gave semi-sporadic looking loading and unloading plateau tracks. The loading paths appeared to be repeatable and consistent. Some residual deformations were observed, especially as the cyclic loading got into larger deformations. After the test was complete and the damper was dissembled, it was observed that one of the washers had inverted. To recover its original shape, the washer was hit with a wooden mallet against a hard surface. The washer promptly snapped back.

Fig. 8 Washer configuration for test D

7.3 Double Stacked

Test E was done using a double-stacked 12 washer configuration (Fig. 10), giving increased stiffness and strength in comparison to the single-stacked test. In this test, flat washers were not added in order to see if the double stack would naturally prevent individual washers from inverting. Additional stainless steel cylinders were added as fillers between the shaft nut and the hardened disk. Again the shaft nut was hand tightened and the cylinder was slid into position.

The initial length and test target maximum deformation of the washer configuration were 7.91 and 3.30 cm, respectively. The resulting force-deformation is shown in Fig. 11. A semi-sporadic

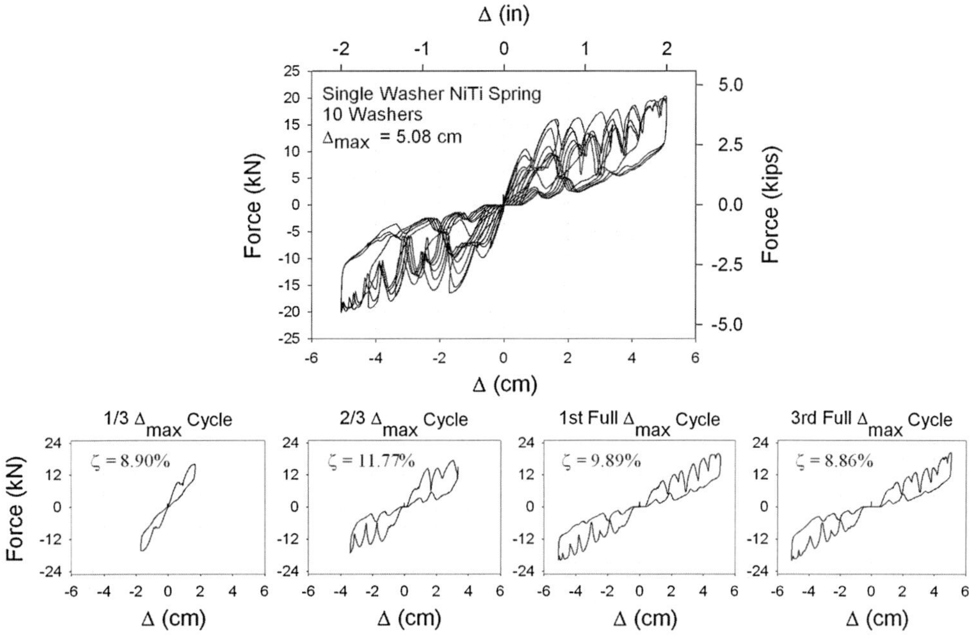

Fig. 9 Force-deformation response of single-stacked washer configuration (test D)

Fig. 10 Washer configuration for test E

response was observed, but this time fewer humps were noticed. After the test was complete and the damper was dissembled, one washer was found to be inverted. This washer was not the same one which inverted in the single-stacked test (test D).

7.4 Triple Stacked

The final test, test F, was done using a triple-stacked 12 washer configuration (Fig. 12), giving even more stiffness and strength in comparison to the double-stacked test. This time flat washers were again added to try to prevent the inverting

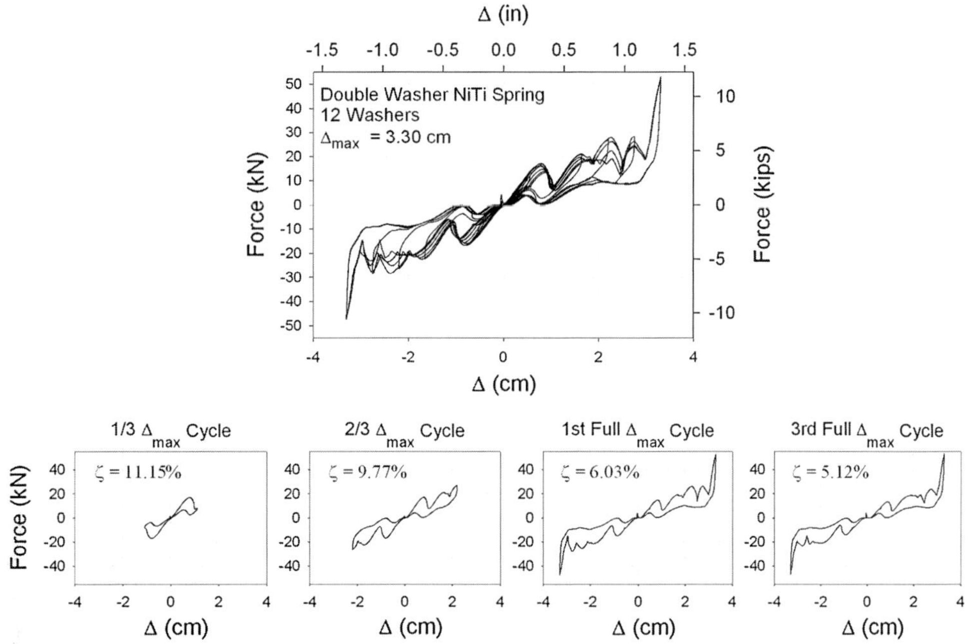

Fig. 11 Force-deformation response of double-stacked washer configuration (test E)

observed in the previous test. Additional stainless steel cylinders were added as fillers between the shaft nut and the hardened disk. Again the shaft nut was hand tightened and the cylinder was slid into position.

The initial length and test target maximum deformation of the washer configuration were 6.76 and 2.03 cm, respectively. The resulting force-deformation is shown in Fig. 13. This time a less sporadic load path was observed. The force-deformation curve had two distinct humps with a minor intermediate hump. After the test was complete and the damper was dissembled, no washers were found to be inverted.

8. Discussion of NiTi Belleville Washer Tests

8.1 Individual

As presented in the results section, the trend for all three individual washers were the same but there were varying levels of strength, stiffness, and residual deformations. The large initial stiffness followed by a peaking and falling off of load carrying capacity can be explained when looking at the geometry of the washer. A handbook (Ref 17) produced by Schnorr Corporation thoroughly details the behavior of Belleville washers made from Hookian elastic materials.

Fig. 12 Washer configuration for test F

Washers with similar geometry to those used in these tests show a response characteristic similar to Fig. 7, excluding the superelastic recovery seen during the release of the deformation. The effect of the superelasticity of the washer is not fully understood and more investigation is needed. However, when the washers were loaded to the flat position, some had the tendency to pass through a region of bifurcation and buckle into another stable configuration. After recovering the shape with a wooden mallet, no residual deformations were observed. Generally in structural engineering it is desired to have materials that retain load carrying capacity when subjected to large deformations, which is not the case here.

8.2 Single, Double, and Triple Stacked

The behavior of the single-stacked configuration was governed by the tendency of individual washers losing their load carrying capacity as they are deformed beyond a certain limit. This is the direct result of the duck-head-shaped behavior observed in the individual washer tests. Since each SMA Belleville washer can be assumed to have different peak strengths, the response of the single-stacked washer configuration is governed by the weakest link. As soon as the weakest washer is deformed to approximately 0.15 cm, this washer begins to loose its strength and thus take on the deformations of the other washers. This is reflected in the device's force-deformations humps. Once the displacement is increased enough to flatten the weakest washer, the remaining washers begin to pick up the force. This cycle is repeated until the entire group of washers has flattened or device deformation is decreased. Upon the release of the imposed deformation, the flattened Belleville washers spring back causing the force in the device to increase at moments even while the deformation is decreasing (see unloading path for Fig. 9, 11, and 13). The residual deformations of the setup begin to accumulate under the 2/3 Δ_{max} cycle. Upon the completion of the third Full Δ_{max} cycle, the residual deformation can be attributed to one

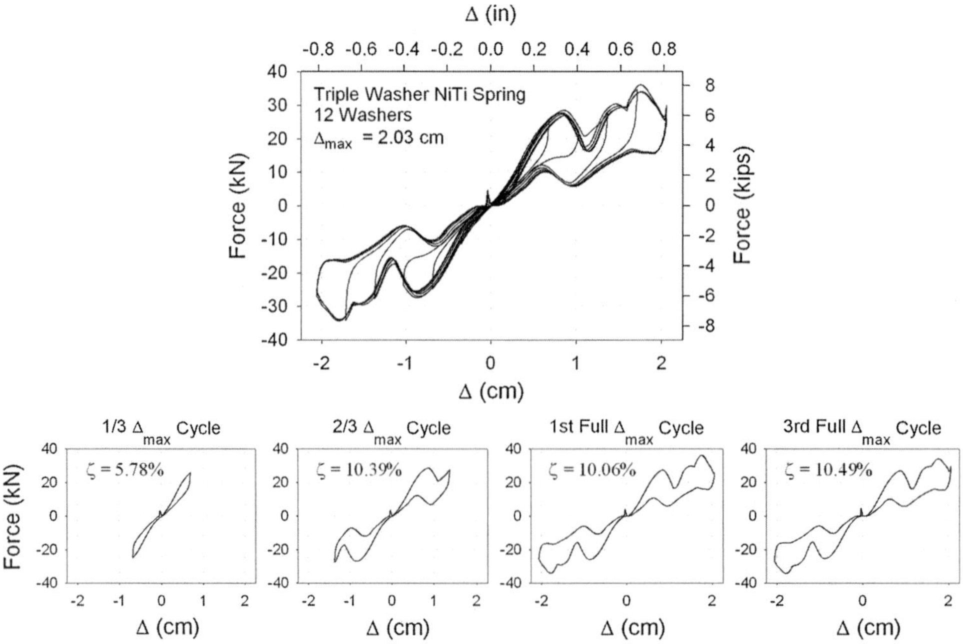

Fig. 13 Force-deformation response of triple-stacked washer configuration (test F)

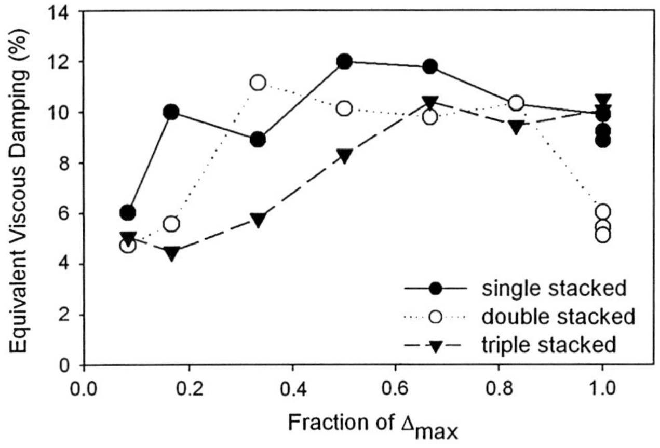

Fig. 14 Comparison of equivalent viscous damping ratios for the stacked washer tests

Belleville washer completely inverting as noted in the results section.

The double- and triple-stacked configurations had progressively less sporadic loading paths. Beyond the initial softness observed due to some of the washers settling into their positions, the stiffness increased as more washers were nested together. The initial stiffness, defined at a point half way up the first hump, was 1.74, 2.45, and 3.93 kN/mm for the single-, double-, and triple-stacked configurations, respectively. By adding more washers to each nest, fewer humps were able to form because there were fewer nests to allow this to occur. The drawback to the added stiffness and reduced hump count is the reduced deformation capacity. This could be improved by adding more nested washers to the stack, assuming one stays within the limitations of the device.

An important quantity to discuss is the equivalent viscous damping, ζ, of the stacked washer configurations. In Fig. 14, the ζ is compared for each washer test. The ζ ranged from just over 4% to just under 13%. The E_{so} was calculated by using the maximum deformation and the maximum force of each respective cycle (not necessarily the same point). The damping values were dependent on the location each deformation cycle fell in the humped response. The triple-stacked configuration showed increasing damping for increasing deformation. It was expected that for the more washers nested together, the more damping would be observed from the friction action between the nested surfaces. However, this was not clearly observed in the results.

9. Conclusions

Two SMA tension/compression devices were developed and tested with good initial results. In particular, the responses of the NiTi helical spring setups were promising because of the good superelasticity (as noted by the recentering), damping, and repeatability. More research needs to be done on the helical springs to determine if full scale load and stiffness levels can be achieved. The NiTi Belleville washers gave results that call for further development. Work needs to be done to develop washers that have improved individual force-deformation characteristics. These improvements should include creating washers that do not have the tendency to invert and also do not have the reduction in load carrying capacity as deformation is increased. Overall, the tension/compression device provides a convenient platform for implementing different forms of SMAs.

Acknowledgment

The authors thank Johnson Matthey for donating the stock NiTi materials for these experiments.

References

1. G.S. Cheok, W.C. Stone, and H.S. Lew, Seismic Performance Behavior of Precast Concrete Beam-Column Joints, *Proceedings of the Symposium on Structural Engineering in Natural Hazards Mitigation*, April 19–21, 1993 (Irvine, CA), p 83–88
2. J.M. Ricles, R. Sause, M.M. Garlock, and C. Zhao, Posttensioned Seismic-Resistant Connections for Steel Frames, *J. Struct. Eng.*, 2001, **127**(2), p 113–121
3. C. Christopoulos and A. Filiatrault, Post-Tensioned Energy Dissipating (PTED) Steel Frames for Seismic Regions, *2003 ASCE/SEI Structures Congress and Exposition: Engineering Smarter* (Seattle, WA), 2003, p 155–157
4. D.K. Nims, P.J. Richter, and R.E. Bachman, Use of the Energy Dissipating Restraint for Seismic Hazard Mitigation, *Earthq. Spectra*, 1993, **9**(3), p 467
5. J.M. Ricles, R. Sause, S.W. Peng, and L.W. Lu, Experimental Evaluation of Earthquake Resistant Posttensioned Steel Connections, *J. Struct. Eng.*, 2002, **128**(7), p 850–859
6. C. Christopoulos and A. Filiatrault, Seismic Response of Post-Tensioned Energy Dissipating Moment Resisting Steel Frames, *12th European Conference on Earthquake Engineering* (Como, Italy), 2002
7. C. Christopoulos, R. Tremblay, H.-J. Kim, and M. Lacerte, Self-Centering Energy Dissipative Bracing System for the Seismic Resistance of Structures: Development and Validation, *J. Struct. Eng.*, 2008, **134**(1), p 96–107
8. J. McCormick, J. Tyber, R. DesRoches, K. Gall, and H.J. Maier, Structural Engineering with NiTi. II: Mechanical Behavior and Scaling, *J. Eng. Mech.*, 2007, **133**(9), p 1019–1029
9. J.C. Wilson and M.J. Wesolowsky, Shape Memory Alloys for Seismic Response Modification: A State-of-the-Art Review, *Earthq. Spectra*, 2005, **21**(2), p 569–601
10. M. Dolce, D. Cardone, and R. Marnetto, Implementation and Testing of Passive Control Devices Based on Shape Memory Alloys, *Earthq. Eng. Struct. Dyn.*, 2000, **29**(7), p 945–968
11. S. Zhu and Y. Zhang, Seismic Analysis of Concentrically Braced Frame Systems with Self-Centering Friction Damping Braces, *J. Struct. Eng.*, 2008, **134**(1), p 121–131
12. J. Ocel, R. DesRoches, R.T. Leon, W.G. Hess, R. Krumme, J.R. Hayes, and S. Sweeney, Steel Beam-Column Connections Using Shape Memory Alloys, *J. Struct. Eng.*, 2004, **130**(5), p 732–740
13. B. Penar, "Recentering Beam-Column Connections Using Shape Memory Alloys," Masters Thesis, Georgia Institute of Technology, Atlanta, GA, 2004
14. J. Sepulveda, R. Boroschek, R. Herrera, O. Moroni, and M. Sarrazin, Steel Beam-Column Connection Using Copper-Based Shape Memory Alloy Dampers, *J. Constr. Steel Res.*, 2008, **64**(4), p 429–435
15. H. Krawinkler, A. Gupta, R. Medina, and N. Luco, Loading Histories for Seimsic Performance Testing of SMRF Components and Assemblies, *SAC/BD-00/10*, SAC Joint Venture, 2000
16. J.M. Gong, H. Tobushi, K. Takata, K. Okumura, and M. Endo, Cyclic Superelastic Deformation of TiNi Shape-Memory Alloy, *Mater. Sci. Forum*, 2000, **394–395**, p 245–248
17. E. Fromm and W. Kleiner. *Handbook for Disc Springs*, Schnorr Corporation, Hela Werbung, Hellbronn, 2003. http://www.schnorr.com/docs/Handbook.pdf

JMEPEG (2009) 18:754–759
DOI: 10.1007/s11665-009-9404-z

1059-9495/$19.00

The Effect of Exposure to Simulated Body Fluids on Breakdown Potentials

Clifford P. Warner

(*Submitted September 12, 2008*)

Current methods to evaluate the corrosion resistance of small medical implants are typically based on short-term in vitro tests. While the duration of these tests is kept to a minimum to make it feasible to evaluate a large number of samples in a reasonable time, these methods do not account for the long-term changes that can occur in the oxides of metals exposed to biological fluids. Given other electrochemical changes to these materials with time in solution, it is a reasonable question to consider whether breakdown potentials are a fundamental parameter of a material and surface condition. Data on 316L and nitinol wire samples, and nitinol stents, show that breakdown potentials increase with time in solution up to 28 days. The difference between the breakdown potential and rest potentials either increased or exhibited no change.

Keywords biomaterials, corrosion testing, material selection

1. Introduction

Evaluation of the corrosion resistance of medical implants is a complicated subject. Modern materials utilized in devices, such as 316L stainless steel, MP35N, and nitinol, are inherently resistant to uniform corrosion. Assessment of corrosion resistance then moves to concerns such as pitting, crevice, galvanic and fretting corrosion. Part of the complication in characterizing the corrosion resistance of a device is the inherent stochastic nature of pitting corrosion; the other is the complexity of modern devices.

There are a number of standards that can be used to evaluate devices, such as ASTM F2129 (Ref 1), F746 (Ref 2), and G71 (Ref 3). However, minimum criteria have not been defined for some of these standards, particularly for ASTM F2129. One of the challenges in determining minimum criteria is that these methods are admittedly harsh. They are designed to make materials fail, so a failure in the test is not necessarily indicative that a device will fail in vivo. All of these tests were designed to enable a researcher to rank materials, rather than provide in vivo simulation for design verification testing. That leaves one with the need for in vivo failure data. Given a lack of real failure data in the literature to base criteria, there is significant ambiguity for manufacturers and testing laboratories in determining whether their devices are adequately corrosion-resistant based on these short-term electrochemical tests.

In this vacuum, criteria were proposed specifying an absolute breakdown potential (E_b) of 600 mV (all voltages versus SCE) for good performance and 300 mV as a minimum acceptable breakdown potential (Ref 4, 5). These criteria were suggested for cases where there were no corroborating data, such as explants or animal trial evaluations. The basis of these criteria was an adequate safety margin to the maximum values for in vivo open circuit potentials of a number of implant materials published in the literature (Ref 6, 7).

Discussion of these criteria by Pound (Ref 8) suggests that a fixed minimum potential may be too simplistic. Pound suggests that $E_b - E_r$ is a better predictor of the resistance to pitting corrosion because it takes into consideration the gap between the breakdown potential and the rest potential. This view is supported by the work of Wilde (Ref 9) and Silverman (Ref 10). Wilde showed that the difference between the breakdown potential and repassivation potential, and not absolute E_b, was correlated to mass loss due to crevice corrosion in environmental exposure testing.

One factor missing from the discussion of criteria in the literature is whether E_b is a constant or if it varies with immersion time. Several studies (Ref 6, 11-14) have demonstrated that the oxides of titanium and nitinol undergo considerable change with time in solution. The oxide dome structure on 316L stainless steel (Ref 11), titanium (Ref 12), and nitinol (Ref 13) have been shown to undergo significant morphological changes with time in solution. Other studies demonstrated electrical properties of the oxides, such as Er (Ref 6) and the polarization resistance (Ref 14), change over long periods of time. Given these observations, it is a reasonable question to ask whether E_b and $E_b - E_r$ are material constants or are dependant on time in solution. This paper will evaluate whether E_b and $E_b - E_r$ change with the length of time a sample is held in solution. This will be evaluated for 316LVM stainless steel wires, nitinol wires, and for nitinol stents.

2. Materials and Methods

All experiments were performed on commercially available 316LVM stainless steel and nitinol wires. The 316LVM

This article is an invited paper selected from presentations at Shape Memory and Superelastic Technologies 2008, held September 21-25, 2008, in Stresa, Italy, and has been expanded from the original presentation.

Clifford P. Warner, W.L. Gore & Associates, 501 Vieve's Way, Elkton, MD 21921. Contact e-mail: cwarner@wlgore.com.

wire was acquired in a spring temper condition with an as-drawn surface condition and a 0.254 mm nominal diameter. Two lots of nitinol wire with a nominal composition of 56.0 wt.% Ni and balance Ti were acquired. One lot was acquired in a straight annealed condition with a black oxide surface condition and was used for testing with straight wire samples with no further processing. Sections of a second lot of nitinol wire were electropolished, wound on mandrels, and then heat-treated in air to form stents. The aim of the processing was to produce a stent with a thermally grown oxide on the surface. Both nitinol surface conditions were created with the intention that the samples would exhibit pitting corrosion in the cyclic potentiodynamic polarization (CPP) test.

All cyclic potentiodynamic experiments followed ASTM F2129 with the exception of changes to the open circuit hold time, and a reverse scan rate of 10 mV/s. Electrochemical experiments were performed with a Gamry Instruments PCI-4/300™ potentiostat and Gamry Framework™ software. Tests were performed in a standard 1 L corrosion cell utilizing a saturated calomel electrode (SCE) for a reference electrode and graphite rods for counter electrodes. All tests were performed in phosphate buffered saline at pH 7.4, 37 °C and under nitrogen purge. A forward scan rate of 1.0 mV/s, a reverse scan rate of 10 mV/s, and an initial potential equal to the rest potential were used for all experiments. Temperature was maintained with a hot water bath.

In the first series of experiments with straight wire samples, two open circuit potential (OCP) hold times were used: 1 and 66 h. Up to 30 samples were tested per combination of material and OCP hold technique, with one notable exception being the nitinol wire samples held for 66 h where 8 samples were tested due to a limited supply of that lot of material.

Stent samples were tested to the ASTM F2129 test method, again varying the OCP hold time. The first 15 samples followed ASTM F2129 using a 1 h OCP hold. The other half used a modified protocol where the samples were held for 672 h (28 days) at OCP before the CPP portion of the test. A Gamry Instruments ECM8™ multiplexer was used to record the open circuit potential of 5 samples at a time in order to test 15 samples with this protocol. Each sample was tested in an independent corrosion cell with its own reference electrode and counter electrode. Once the OCP hold was completed, the CPP portion of the experiment was run sequentially, one cell at a time.

Statistical analyses of breakdown potentials (E_b) and the difference between the breakdown potential and the rest potential ($E_b - E_r$) were performed using JMP™ version 7.0.1. Statistical evaluation was performed using the Wilcoxon/Kruskal-Wallis test with normal approximation, a non-parametric test, with a 95% confidence level used for statistical significance in all analyses.

3. Results

The changes in rest potentials and breakdown potentials for stainless steel samples are shown as a function of OCP hold time in Fig. 1. Rest potentials increased from a mean of −81 to 33 mV and breakdown potentials increased from 613 to 961 mV. Mean and standard deviations for E_r, E_b, and $E_b - E_r$ for all materials are listed in Table 1. While E_r and E_b values increased with increasing hold time, the change in E_b was greater than for E_r as can be seen by the increase in $E_b - E_r$ with increasing OCP hold time in Fig. 3. The mean $E_b - E_r$ increased from 694 mV with a 1 h OCP hold to 928 mV at 66 h. Increases to E_r, E_b, and $E_b - E_r$ for stainless steel samples were all statistically significant as can be seen by the Wilcoxon p-values listed in Table 2.

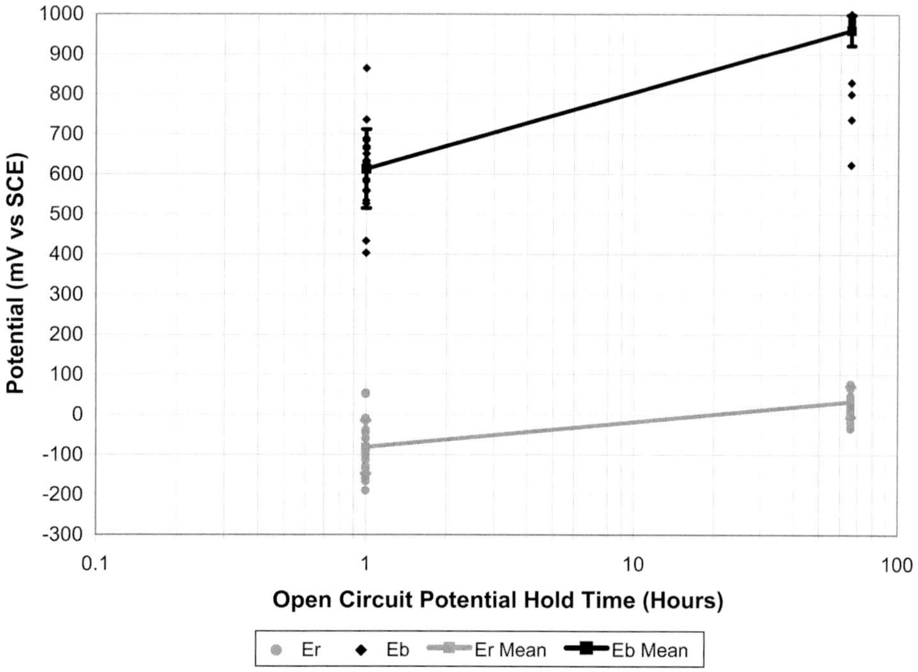

Fig. 1 Rest potentials and breakdown potentials as a function of open circuit potential hold time for stainless steel straight wire samples. Means are connected by the lines

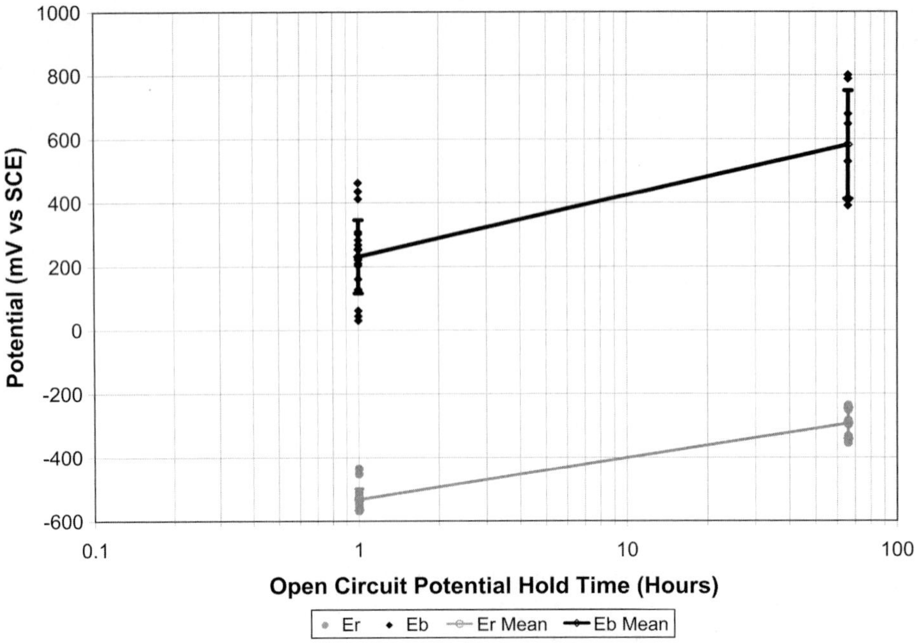

Fig. 2 Rest potentials and breakdown potentials as a function of open circuit potential hold time for nitinol straight wire samples. Means are connected by the lines

Table 1 Mean and standard deviation for E_r, E_b, and $E_b - E_r$ values for stainless steel and nitinol straight wire samples

			E_r, mV		E_b, mV		$E_b - E_r$, mV	
Material	**OCP hold time, h**	**Number of samples**	**Mean**	**SD**	**Mean**	**SD**	**Mean**	**SD**
316 SS	1	21	−81	66	613	99	694	127
	66	30	33	38	961	38	928	89
Nitinol	1	22	−532	34	231	115	763	115
	66	8	−294	48	582	170	876	131
Nitinol stent	1	15	−106	23	446	148	553	150
	672	13	36	21	680	149	644	152

Table 2 p-Values for statistical comparisons of E_r, E_b, and $E_b - E_r$ vs. OCP hold time for 316 stainless steel and nitinol wire samples

Material	Parameter	Wilcoxon p-value
316 SS	E_r	<0.0001
	E_b	<0.0001
	$E_b - E_r$	<0.0001
Nitinol	E_r	<0.0001
	E_b	0.0002
	$E_b - E_r$	0.039
Nitinol stent	E_r	<0.0001
	E_b	0.0013
	$E_b - E_r$	0.1405

Nitinol straight wire samples performed similarly to the stainless steel samples. E_r and E_b increased with increasing OCP hold time as can be seen in Fig. 2. Mean rest potentials increased from −532 to −294 mV, while mean breakdown potentials increased from 231 to 582 mV. $E_b - E_r$ values also increased with increasing OCP hold time for the straight wire samples, as shown in Fig. 3. The change in $E_b - E_r$ was not as large as observed with the stainless steel samples, but the mean still increased from 763 to 876 mV. The changes in E_r, E_b, and $E_b - E_r$ were all statistically significant as can be seen by the Wilcoxon p-values shown in Table 2.

Rest potentials and breakdown potentials for the stent frames are shown in Fig. 4. As with the wire samples, breakdown potentials increased with increased open circuit potential hold in solution, in this case 672 h (28 days). Two samples in the 672 h hold period developed bubbles in the luggin probe which led to problems in the CPP test resulting in a set of 13 data points for the longer OCP hold. Figure 5 shows the open circuit potential as a function of time for one of the stent samples. Most of the changes in OCP occurred in the first 4 to 6 days with little subsequent change in the remaining 22 days of exposure. No significant transients were observed in the OCP traces, indicating that no metastable pitting occurred during the OCP hold. Mean rest potentials increased from −106 mV at 1 h to 36 mV at 672 h, and mean breakdown potentials increased from 446 to 680 mV. Increases in rest and breakdown potentials for the stents were statistically significant as shown by the p-values in Table 2. $E_b - E_r$ values are shown in Fig. 6, with the mean $E_b - E_r$ value increasing from 553 to 644 mV. However, the increase was not statistically significant

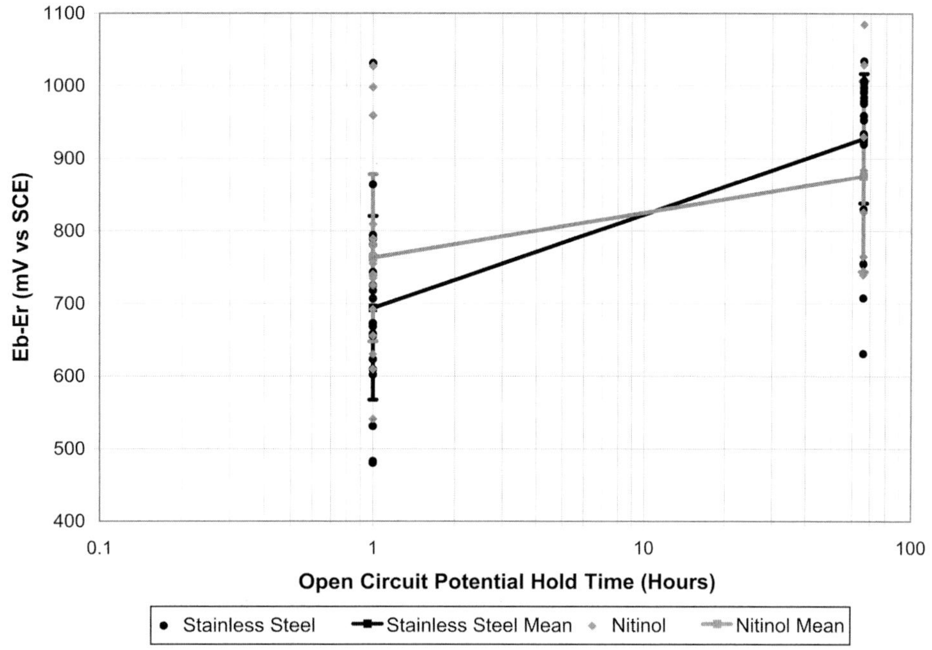

Fig. 3 $E_b - E_r$ as a function of open circuit potential hold time for stainless steel and nitinol straight wire samples. Means are connected by the lines

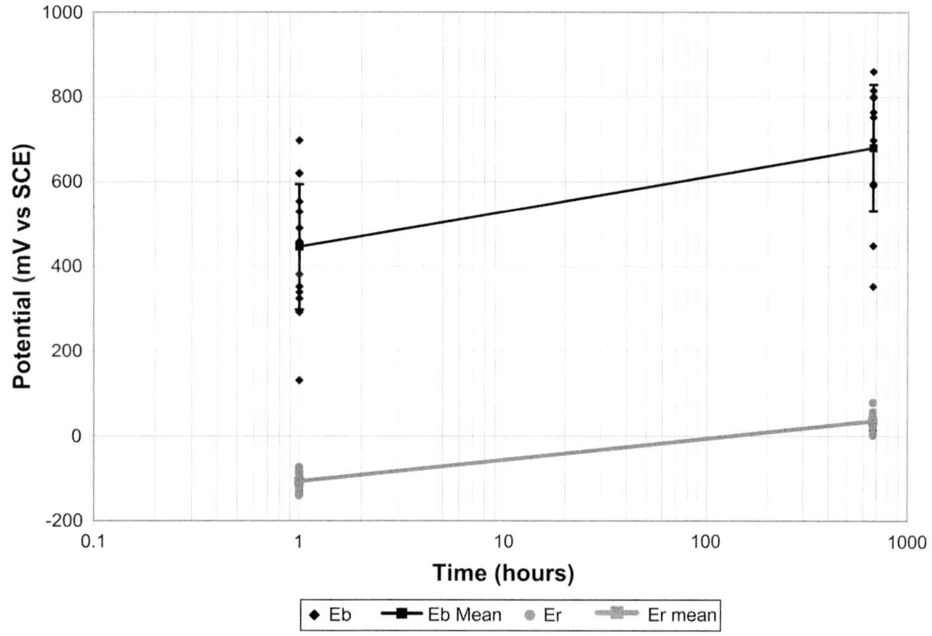

Fig. 4 Rest potentials and breakdown potentials as function of open circuit potential hold time for nitinol stents. Means are connected by the lines

as shown by the Wilcoxon p-value of 0.1344. Typical cyclic polarization curves for 1 h and 28 day OCP hold samples are shown in Fig. 7. The 28 day hold samples exhibited a decrease in current density in the passive region of between 1 and 1.5 orders of magnitude from that of the 1 h OCP hold samples. Additionally, it should be noted that none of the longer hold samples exhibited a repassivation potential on the reverse curve, while 13 of 15 of the 1 h hold samples did repassivate.

4. Discussion

Data from both wire samples and stents show that E_r and E_b values increased with increasing open circuit potential hold time for all the materials evaluated in this study. The increase in E_b values with time is an interesting and new finding. $E_b - E_r$ values either stayed constant or increased with increasing open circuit hold time. What is most significant

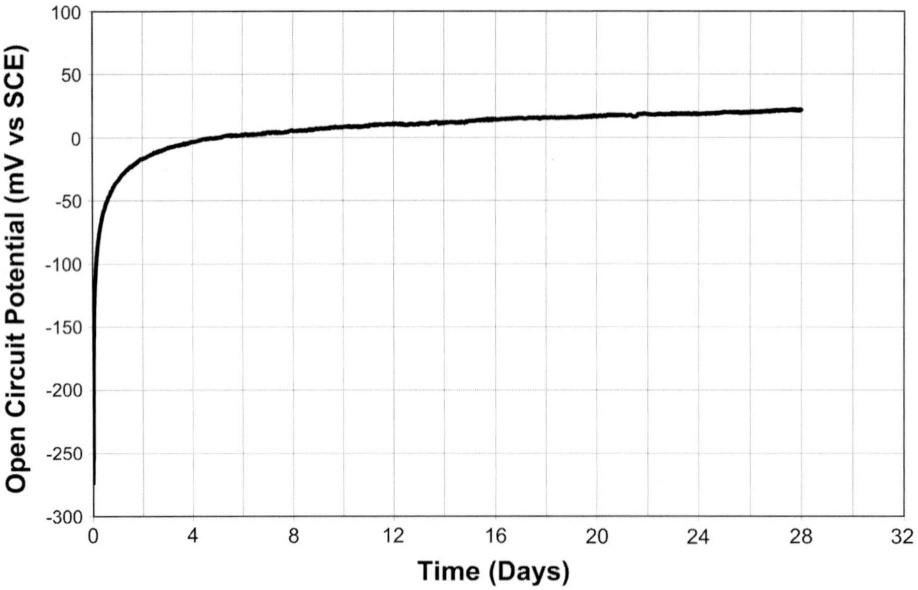

Fig. 5 Open circuit potential data from one representative sample showing increasing potential in the first 4 days with very little subsequent change

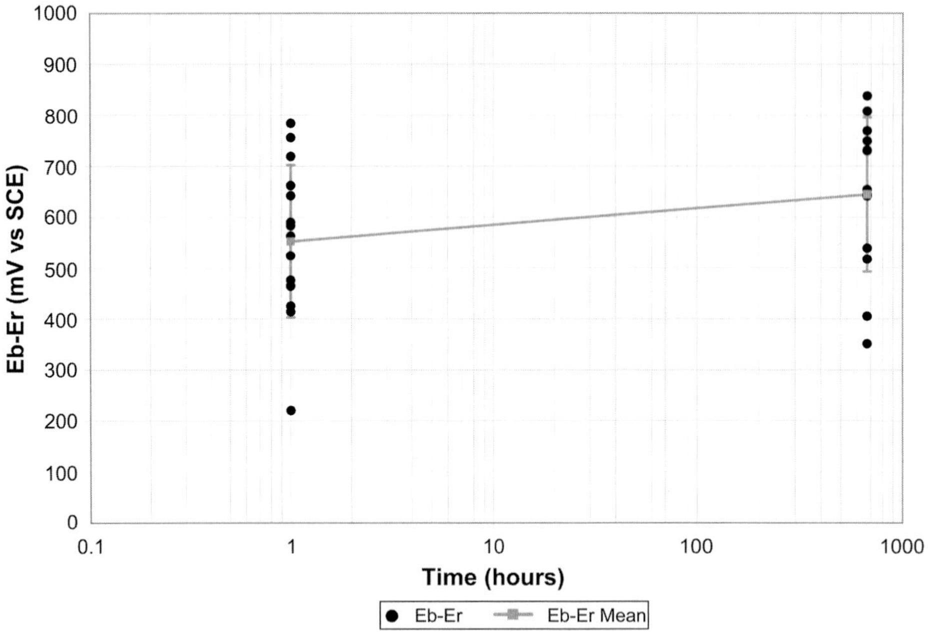

Fig. 6 $E_b - E_r$ as a function of open circuit potential hold time for nitinol stents. Means are connected by the line

here is that $E_b - E_r$ did not decrease with increasing time in solution.

Beyond the changes to E_r and E_b, there were additional changes observed in the CPP curves. The difference in current density in the passive region suggests a greater polarization resistance with time in solution. This corresponds well with the increased polarization resistance found in F138 and nitinol observed by Rondelli (Ref 14) through EIS studies with immersions of up to 1 week.

The results here have interesting implications to the discussion on criteria for ASTM F2129. Referring back to the stent data in Fig. 4, the mean breakdown potential of 446 mV is clearly below the 600 mV criterion proposed for good material and some of the values fall below the 300 mV minimum criterion suggested by Corbett (Ref 4). However, with increasing time in solution, E_bs climbed to a mean of 680 mV and a minimum value of around 400 mV showing acceptable performance by Corbett's suggested criteria. Data from longer immersions show that Corbett's criteria would provide a conservative evaluation of the corrosion resistance of these devices. In contrast, $E_b - E_r$ values taken at 1 h are consistent with longer term data, showing that $E_b - E_r$ may be a better

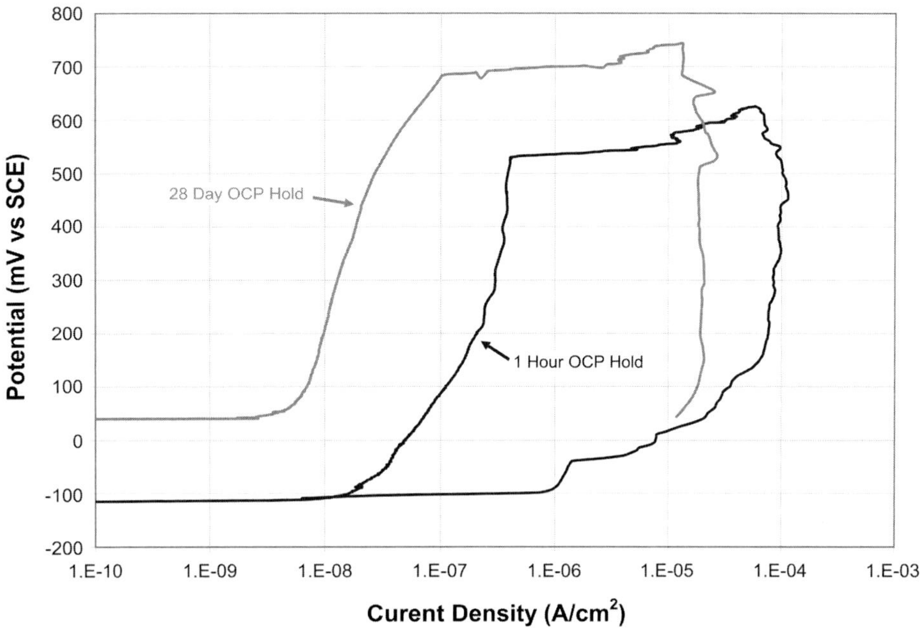

Fig. 7 Typical cyclic potentiodynamic polarization curves for 1 h and 28 day OCP hold times on nitinol stents

parameter due to its consistency. Others (Ref 8-10) have suggested that $E_b - E_r$ is a more relevant parameter for the prediction of corrosion performance from in vitro test data. Overall, this suggests that $E_b - E_r$ is a better parameter for evaluation of the resistance to pitting corrosion of implant devices than a fixed value of E_b.

5. Conclusions

- Rest potentials and breakdown potentials increased with increasing open circuit potential hold time. Changes were statistically significant for 316L stainless steel wires, nitinol wires, and nitinol stents.
- Statistically significant increases were observed for $E_b - E_r$ with increasing open circuit potential hold times for 316L stainless steel wires and nitinol wires.
- No statistically significant change was observed to $E_b - E_r$ with increasing open circuit potential hold times up to 28 days for nitinol stents.

References

1. Standard Test Method for Conducting Cyclic Potentiodynamic Polarization Measurements to Determine the Corrosion Susceptability of Small Implant Devices, F2129-06, ASTM International, 2006
2. Standard Test Method for Pitting and Crevice Corrosion of Metallic Surgical Implant Materials, F746-04, ASTM International, 2004
3. Standard Guide for Conducting and Evaluating Galvanic Corrosion Tests in Electrolytes, G71-83, ASTM International, 2003
4. R.A. Corbett, Laboratory Corrosion Testing of Medical Implants, *Proceedings of the Materials and Processes for Medical Devices Conference*, S. Shrivastava, Ed., Sept 8-10, 2003 (Anaheim), ASM International, 2004, p 133–138
5. S.N. Rosenbloom and R.A. Corbett, An Assessment of ASTM F2129 Test Results Comparing Nitinol to Other Implant Alloys, *Proceedings of the International Conference on Shape Memory and Superelastic Technologies*, B. Berg, M.R. Mitchell, and J. Proft, Eds., May 7-11, 2006 (Pacific Grove, CA), ASM International, 2008, p 243–251
6. T.P. Hoar and D.C. Mears, Corrosion-Resistant Alloys in Chloride Solutions: Materials for Surgical Implants, *Proc. R. Soc. Lond. A*, 1966, **294**, p 486–510
7. C.C. Shih, S.J. Lin, K.H. Chung, Y.L. Chen, and Y.Y. Su, Increased Corrosion Resistance of Stent Materials by Converting Current Surface Film of Polycrystalline Oxide into Amorphous Oxide, *J. Biomed. Mater. Res.*, 2000, **52**, p 323–332
8. B.G. Pound, Susceptibility of Nitinol to Localized Corrosion, *J. Biomed. Mater. Res.*, 2006, **77A**, p 185–191
9. B.E. Wilde and E. Williams, The Use of Current/Voltage Curves for the Study of Localized Corrosion and Passivity Breakdown on Stainless Steels in Chloride Media, *Electrochem. Acta*, 1971, **16**, p 1971–1985
10. D.C. Silverman, Tutorial on Cyclic Potentiodynamic Polarization Technique, paper no. 299, *Corrosion* 1998 (Houston, TX), NACE International, p 1179–1225 (2003)
11. N. Istephanous, Z. Bai, J.L. Gilbert, K. Rohly, A. Belu, I. Trausch, and D. Untereker, Oxide Films on Metallic Biomaterials: Myths, Facts and Opportunities, *Mater. Sci. Forum*, 2003, **426–432**, p 3157–3164
12. J.P. Bearinger, C.A. Orme, and J.L. Gilbert, In Situ Imaging and Impedance Measurements of Titanium Surfaces Using AFM and SPIS, *Biomaterials*, 2003, **24**, p 1837–1852
13. J. Bai and J.L. Gilbert, In Situ Electrochemical Atomic Force Microscopy Study of Morphology of Oxide Film on Etched NITI Alloy: Effects of Hydration and Potential, World Congress of Biomaterials, Sydney, 2004
14. G. Rondelli, P. Torricelli, M. Fini, L. Rimondini, and R. Giardino, In Vitro Corrosion Study by EIS of an Equiatomic NiTi Alloy and Implant Quality AISI Stainless Steel, *J. Biomed. Mater. Res. B Appl. Biomater.*, 2006, **79B**, p 320–324

JMEPEG (2009) 18:760–764
DOI: 10.1007/s11665-009-9435-5

1059-9495/$19.00

A Comparative Biocompatibility Analysis of Ternary Nitinol Alloys

Waseem Haider, Norman Munroe, Chandan Pulletikurthi, Puneet K. Singh Gill, and Sushma Amruthaluri

(Submitted October 9, 2008; in revised form March 10, 2009)

Nitinol alloys are rapidly being utilized as the material of choice in a variety of applications in the medical industry. It has been used for self-expanding stents, graft support systems, and various other devices for minimally invasive interventional and endoscopic procedures. However, the biocompatibility of this alloy remains a concern to many practitioners in the industry due to nickel sensitivity experienced by many patients. In recent times, several new Nitinol alloys have been introduced with the addition of a ternary element. Nevertheless, there is still a dearth of information concerning the biocompatibility and corrosion resistance of these alloys. This study compared the biocompatibility of two ternary Nitinol alloys prepared by powder metallurgy (PM) and arc melting (AM) and critically assessed the influence of the ternary element. ASTM F 2129-08 cyclic polarization in vitro corrosion tests were conducted to evaluate the corrosion resistance in phosphate buffered saline (PBS). The growth of endothelial cells on NiTi was examined using optical microscopy.

Keywords biocompatibility, endothelial cells, nitinol, sintering, stents

1. Introduction

Shape Memory Alloys (SMAs) are a group of metallic materials that demonstrate the ability to return to some previously defined shape or size when subjected to an appropriate thermal or mechanical procedure. In general, these materials can be plastically deformed at some relatively low temperature, and on exposure to some higher temperature, return to their original shapes prior to the deformation. Materials that exhibit shape memory only on heating are referred to as having a one-way shape memory. Some materials also undergo a change in shape on re-cooling. These materials have a two-way shape memory.

Nitinol alloys are rapidly becoming the material of choice for self-expanding stents, graft support systems, filters, baskets, and various other devices for minimally invasive interventional and endoscopic procedures due to their excellent superelasticity, radiopacity and shape memory properties (Ref 1-6).

In this study, Nitinol alloys have been prepared using the powder metallurgy (PM) and arc melting (AM) methods. Tantalum (Ta) and Copper (Cu) have been added to NiTi as ternary elements. Tantalum imparts radiopacity, thermal stability, and control over transformation temperatures (Ref 3), whereas copper increases corrosion resistance and martensitic transformation temperature, and prevents Ti_3Ni_4 precipitation (Ref 4).

The susceptibility to corrosion of Nitinol alloys was evaluated by conducting Cyclic Polarization tests in accordance with ASTM F 2129-08 (Ref 7).

This article is an invited paper selected from presentations at Shape Memory and Superelastic Technologies 2008, held September 21-25, 2008, in Stresa, Italy, and has been expanded from the original presentation.

Waseem Haider, Norman Munroe, Chandan Pulletikurthi, Puneet K. Singh Gill, and **Sushma Amruthaluri**, Applied Research Center, Florida International University, 10555 West Flagler Street, Miami, FL 33174. Contact e-mail: whaid001@fiu.edu.

2. Materials

2.1 Metal Powders

All the metal powders were supplied by Atlantic Equipment Engineers (AEE), Bergenfield, NJ, USA. The size and purity of powders are shown below:

- Nickel, 4-8 μm (99.9%)
- Titanium, <20 μm (99.7%)
- Tantalum, 1-5 μm (99.8%)
- Copper, 1-5 μm (99.7%)

Ni, Ti, Ta, and Cu powders were used as starting materials for the production of three different Nitinol alloys by PM, i.e., NiTi, NiTiTa, and NiTiCu.

NiTi, NiTiTa, and NiTiCu pellets prepared by AM method were obtained from National Institute of Standards and Technology (NIST) and were used for comparison.

2.2 Reagent

Phosphate buffered saline (PBS), a reagent grade chemical conforming to the specifications of the Committee on Analytical Reagents of the American Chemical Society was used as the standard test solution.

Table 1 Alloy composition (in at.%)

Ni, at.%	Ti, at.%	Ta or Cu, at.%
51	49	0
48.45	46.55	5

3. Methodology

3.1 Mixing

Pure metal powders were weighed in the appropriate proportions as shown in Table 1. These powders were mixed in a glove box that was first evacuated and then purged with argon to maintain a stable and inert atmosphere in order to minimize oxidation due to the reactive nature of very fine particles.

3.2 Milling

A high-energy ball mill (SPEX 8000), equipped with a hardened steel vial (Ø 2¼ in. × 3 in.) as the milling container and stainless steel balls as the grinding medium, was used to alloy the metal powders mechanically. A milling speed of 1200 rpm, milling duration of 60 min, and a ball-to-powder ratio (BPR) of 10:1 were utilized for effective milling of the powders. The milled powder was collected and stored in an airtight container within the glove box for future usage (Ref 3).

3.3 Pelletizing

A stainless steel die was specially designed and manufactured (Fig. 1) for the consolidation of the milled powders into pellets. The powder was uni-axially cold pressed at 15,000 lbs force (66723 N) with a manual pellet press and held at that pressure for about 30 min to form the pellet. The pellet was then carefully ejected out of the die and stored in the glove box in separate labeled plastic boxes under argon.

3.4 Sintering

Pellets were sintered at 900 °C under argon, utilizing a ramp time of 25 min and a hold time of 2.5 h. Afterward, pellets were allowed to cool down to room temperature. These pellets were used to perform corrosion and biocompatibility studies.

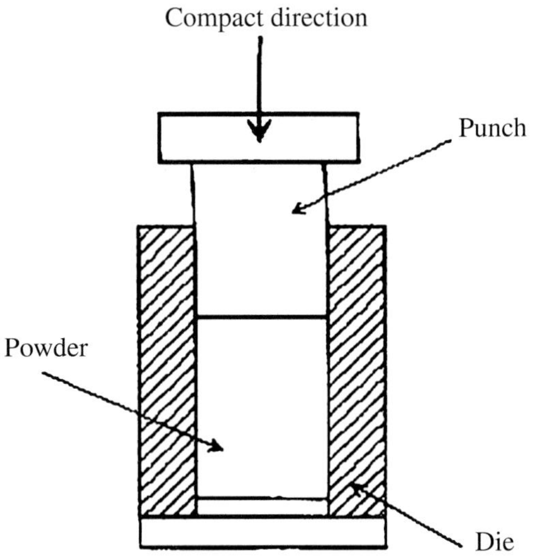

Fig. 1 Pelletizing die

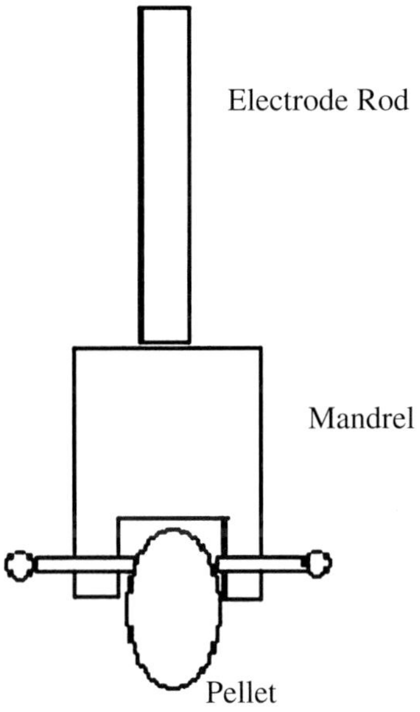

Fig. 2 The holding mandrel

4. Experimental Procedure

4.1 Comparative Corrosion Studies

All the pellets were polished with a series of 200, 320, and 600 grit SiC paper. The pellets were then degreased ultrasonically with acetone, rinsed in distilled water, and air dried.

The corrosion cell was first cleaned with deionized water, rinsed with PBS solution, and filled with approximately 70 mL of PBS. The cell with PBS solution was brought up to 37 °C by placing it in a controlled temperature water bath, and the solution was purged with ultra high purity nitrogen for 30 min prior to immersion of the pellet (Fig. 2). A saturated calomel electrode was used as the reference electrode, and it was inserted into a Luggin Capillary. In order to increase the accuracy of the corrosion data, the surface area of the pellet in contact with PBS was carefully calculated. The cyclic polarization option was then selected on a GAMRY® Instrument Framework Software with a scan rate of 1 mV/s over a potential range between −0.5 and 2.2 V_{SCE}.

4.2 Endothelial Cell Growth

Growth of endothelial cells on the NiTi alloys was assessed using the ISO 10993 protocols for biological evaluation of medical devices (Ref 8) by using an Olympus IX81 microscope. Cells were first cultured using F-12K medium in a 75 cm^2 cell culture flask. When the cells reached confluency, they were trypsinized, centrifuged, and resuspended in culture media for cell counting and cell seeding. A solution of 500 μL of cell culture medium (seeding density 2×10^5 endothelial cells) was transferred to each NiTi alloy located in a 24-well plate. The well plate was then placed in an incubator for 72 h at 37 °C having 5% CO_2.

5. Results and Discussion

Table 2-7 show various corrosion parameters that were determined at 37 ± 1 °C for three types of Nitinol alloys prepared by two different methods using PBS as the electrolyte. Cyclic polarization curves for PM and AM Nitinol alloys are shown in Fig. 3 and 4, respectively.

The true measure of a material's pitting corrosion resistance is usually associated with the breakdown potential, E_b, as well as the gap, $E_b - E_r$. It appeared that NiTiTa prepared by both PM and AM had the greatest corrosion resistance among the three alloys as indicated by the highest E_b. NiTiCu alloy exhibited a corrosion resistance value lying those of between NiTiTa and NiTi.

According to Otsuka et al. (Ref 9-11), Ta and Cu are known to impart corrosion resistance as well as biocompatibility to NiTi.

The PM method is known to avoid problems associated with casting, such as segregation or extensive grain growth. Moreover, it can allow an exact control of the chemical composition (Ref 12). This may explain the difference in corrosion resistance obtained between AM and PM alloys, with the latter proving to be more resistant to pitting corrosion.

Attempts were made to grow endothelial cells on each alloy by adopting the ISO 10993 series, a set of standards for evaluating the biocompatibility of implant devices, prior to a clinical study, but only the result of cell growth on AM NiTi have been obtained till date.

Table 2 Average corrosion parameters obtained for PM NiTi

Corrosion parameters	Sample 1	Sample 2	Sample 3	Sample 4	Sample 5	Average
E_b, V	1.121	1.118	1.168	1.170	1.169	1.149
E_r, V	−0.005	−0.277	−0.367	−0.351	−0.140	−0.228
E_p, V	1.024	1.033	1.094	1.128	1.125	1.080
$E_b - E_r$, V	1.126	1.395	1.535	1.521	1.309	1.377

Table 3 Average corrosion parameters obtained for PM NiTiTa

Corrosion parameters	Sample 1	Sample 2	Sample 3	Sample 4	Sample 5	Average
E_b, V	1.200	1.144	1.561	1.184	1.112	1.240
E_r, V	−0.403	−0.459	−0.372	−0.418	−0.527	−0.435
E_p, V	...	1.027	1.513	1.035	1.029	1.151
$E_b - E_r$, V	1.603	1.603	1.933	1.602	1.639	1.675

Table 4 Average corrosion parameters obtained for PM NiTiCu

Corrosion parameters	Sample 1	Sample 2	Sample 3	Sample 4	Sample 5	Average
E_b, V	1.099	1.354	1.137	1.186	1.170	1.189
E_r, V	−0.478	−0.477	−0.498	−0.365	−0.488	−0.461
E_p, V	0.991	1.275	1.021	...	1.099	1.096
$E_b - E_r$, V	1.577	1.831	1.635	1.551	1.658	1.650

Table 5 Average corrosion parameters obtained for AM NiTi

Corrosion parameters	Sample 1	Sample 2	Sample 3	Sample 4	Sample 5	Average
E_b, V	1.082	1.060	1.070	1.016	1.110	1.067
E_r, V	−0.492	−0.458	−0.508	−0.541	−0.536	−0.507
E_p, V	1.010	1.026	1.007	0.891	1.076	1.002
$E_b - E_r$, V	1.574	1.518	1.578	1.557	1.646	1.574

Table 6 Average corrosion parameters obtained for AM NiTiTa

Corrosion parameters	Sample 1	Sample 2	Sample 3	Sample 4	Sample 5	Average
E_b, V	1.101	1.102	1.132	1.123	1.131	1.117
E_r, V	−0.565	−0.485	−0.592	−0.501	−0.451	−0.518
E_p, V	0.996	1.070	1.078	1.029	1.044	1.043
$E_b - E_r$, V	1.666	1.587	1.724	1.624	1.582	1.635

Table 7 Average corrosion parameters obtained for AM NiTiCu

Corrosion parameters	Sample 1	Sample 2	Sample 3	Sample 4	Sample 5	Average
E_b, V	1.116	1.080	1.123	1.016	1.110	1.089
E_r, V	−0.531	−0.398	−0.474	−0.541	−0.536	−0.496
E_p, V	1.101	1.016	1.041	0.891	1.076	1.025
$E_b - E_r$, V	1.647	1.478	1.597	1.557	1.646	1.585

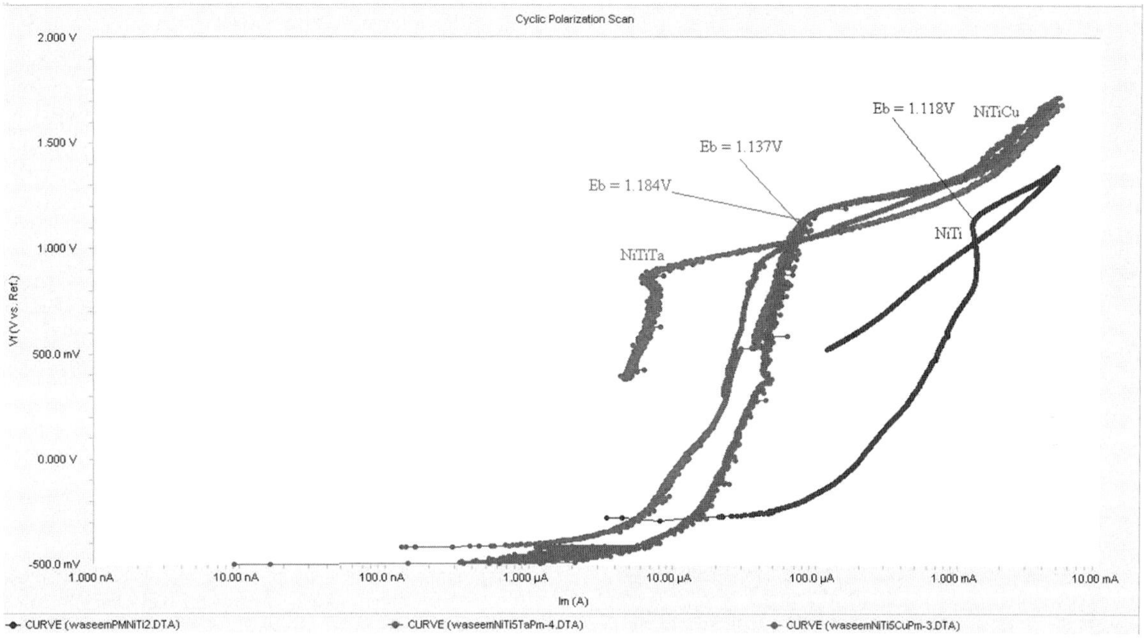

Fig. 3 Cyclic polarization curves of PM Nitinol alloys

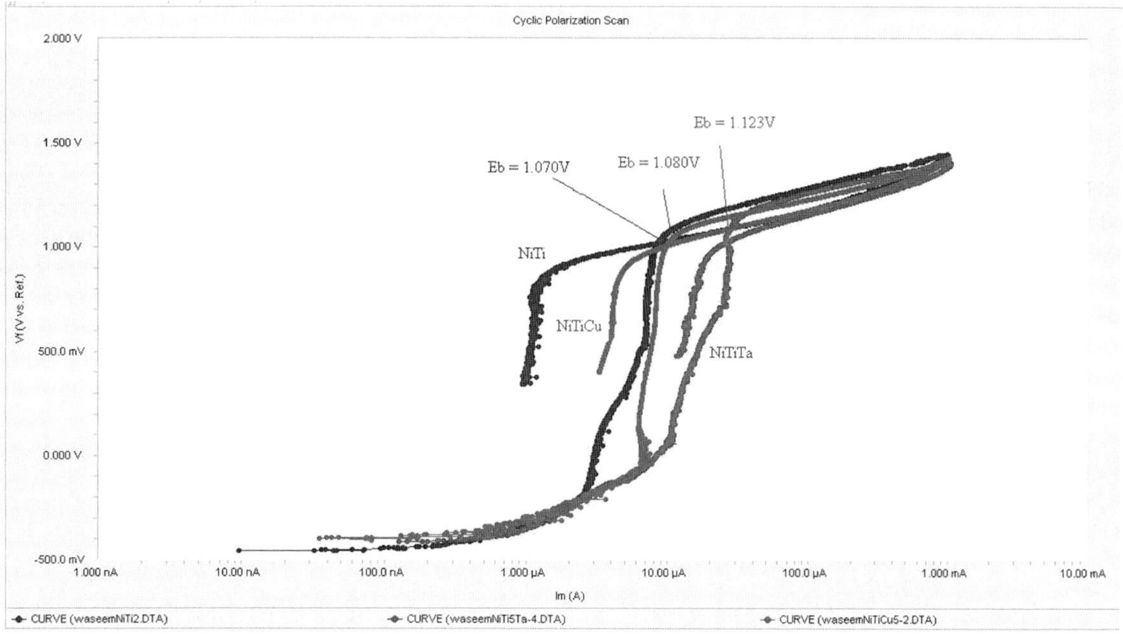

Fig. 4 Cyclic polarization curves of AM Nitinol alloys

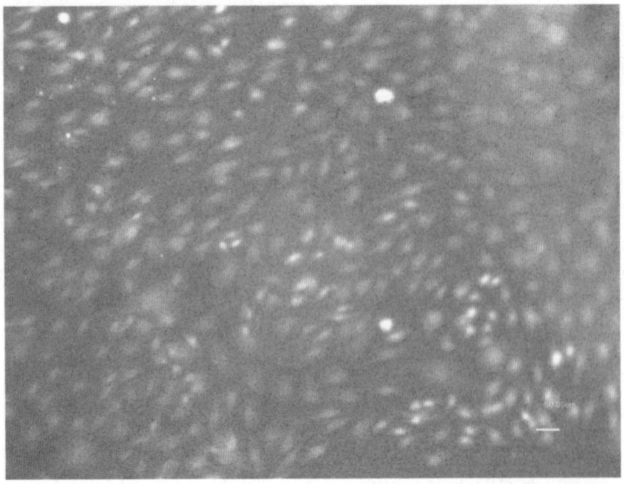

Fig. 5 Endothelial cell growth on AM NiTi (10×)

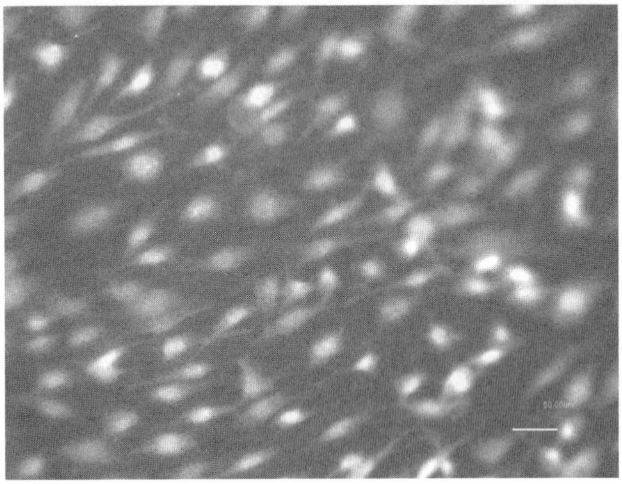

Fig. 6 Endothelial cell growth on AM NiTi (20×)

Figures 5 and 6 show the photomicrographs of endothelial cells on NiTi alloy after 72 h. The cells appeared to be healthy, and the growth was prolific.

6. Conclusions

Cyclic polarization corrosion tests performed on NiTi, NiTiTa, and NiTiCu alloys revealed that the breakdown potential increased as Cu or Ta was added as the ternary element, with NiTiTa being the most resistant to pitting corrosion. Based on the results of cyclic polarization tests, PM alloys appeared to be more resistant to pitting corrosion as compared with AM alloys. The growth and viability of endothelial cells on NiTi were prolific. However, the effect of the ternary elements Ta and Cu on endothelial cell growth will be the focus of future work.

Acknowledgment

This project was supported by Award Number SC3GM084816 from the National Institute of General Medical Sciences. The content is solely the responsibility of the authors and does not necessarily represent the official views of the National Institute of General Medical Sciences or the National Institutes of Health.

References

1. S. Shabalovakaya, Physicochemical and Biological Aspects of Nitinol as a Biomaterial, *Int. Mater. Rev.*, 2001, **46**(4), p 230–250
2. J. Ryhanen, E. Niemi, W. Serlo, E. Niemela, P. Sandvick, H. Pernu, and T. Salo, Biocompatibility of Nickel-Titanium Shape Memory Metal and Its Corrosion Behavior in Human Cell Cultures, *J. Biomed. Mater. Res.*, 1997, **35**(4), p 451–457
3. S. Kanchibbolta and N. Munroe, Amorphization in Ni-Ti-Ta System Through Mechanical Alloying, *J. Mater. Sci.*, 2005, **40**(18), p 5003–5006
4. T. Goryczka and J.V. Humbeeck, NiTiCu Shape Memory Alloy Produced by Powder Technology, *J. Alloys Compd.*, 2008, **456**, p 194–200
5. N.D. Munroe, W. Haider, K.H. Wu, and A. Datye, Corrosion Behavior of Electropolished Implant Alloys, *Proceedings of the International Conference on Shape Memory and Superelastic Technologies*, December 3-5, 2007, Tsukuba, Japan, p 307–314
6. N.D. Munroe, W. Haider, K.H. Wu, and A. Datye, Corrosion Behavior of Cardiovascular Stent Materials, *Proceedings of the International Conference on Shape Memory and Superelastic Technologies*, December 3-5, 2007, Tsukuba, Japan, p 299–306
7. Standard Test Method for Conducting Cyclic Potentiodynamic Polarization Measurements to Determine the Corrosion Susceptibility of Small Implant Devices, *ASTM F 2129-08, Annual Book of ASTM Standards*
8. V.P. Anand, Biocompatibility Safety Assessment of Medical Devices: FDA/ISO & Japanese Guidelines, *MD & DI*, January 2000, p 206–216
9. K. Otsuka and T. Kakeshita, Science and Technology of Shape-memory Alloys: New Developments, *MRS Bulletin*, February 2002, p 91–100
10. J.L. Ma, Master thesis, Florida International University, 1999
11. S. Kanchibhotla, Master thesis, Florida International University, 2004
12. M. Bram, A. Ahmad-Khanlou, A. Heckmann, B. Fuchs, H.P. Buchkremer, and D. Stover, Powder Metallurgical Fabrication Processes for NiTi Shape Memory Alloy Parts, *Mater. Sci. Eng. A*, 2002, **337**, p 254–263

JMEPEG (2009) 18:765–767
DOI: 10.1007/s11665-009-9454-2

Enhanced Biocompatibility of Porous Nitinol

Norman Munroe, Chandan Pulletikurthi, and Waseem Haider

(Submitted November 12, 2008; in revised form March 25, 2009)

Porous Nitinol (PNT) has found vast applications in the medical industry as interbody fusion devices, synthetic bone grafts, etc. However, the tendency of the PNT to corrode is anticipated to be greater as compared to solid nitinol since there is a larger surface area in contact with body fluids. In such cases, surface preparation is known to play a major role in a material's biocompatibility. In an effort to check the effect of surface treatments on the in vitro corrosion properties of PNT, in this investigation, they were subjected to different surface treatments such as boiling in water, dry heating, and passivation. The localized corrosion resistance of alloys before and after each treatment was evaluated in phosphate buffer saline solution (PBS) using cyclic polarization tests in accordance with ASTM F 2129-08.

Keywords biocompatibility, corrosion, cyclic polarization, passivation, porous NiTi

1. Introduction

Porous Nitinol (PNT) is utilized in a wide variety of medical applications such as bone grafts, intervertebral fusion devices, dental implants, tissue repair, and bone reconstruction due to good osteoconductive properties (Ref 1-3), open porosity (65 ± 5%), high strength, and super elastic behavior. Corrosion of Nitinol implants not only leads to failure of the device but also may result in allergic, toxic, and carcinogenic effects in patients (Ref 4, 5). Although Nitinol normally has a 1:1 ratio of titanium and nickel, the phase responsible for the shape memory effect, NiTi, occurs at compositions between 49.5 and 53 at.% Ni. Nevertheless, researchers (Ref 6) have shown that the Ti:Ni ratio varies as much as one order of magnitude as a result of surface treatment. In an effort to understand the effect of surface treatments on the localized corrosion resistance of PNT, a combination of four methods of surface treatment were utilized, namely boiling in distilled water at 132 °C for 30 min, dry heating in a continuous supply of air at 132 °C for 1 h, etching in acid mixture for 1 min, and passivation in 20% concentrated nitric acid for 20 min. The susceptibility to corrosion of surface-treated PNT was evaluated by conducting cyclic polarization tests in accordance with ASTM F 2129-08 (Ref 7). This is the currently preferred in vitro method for assessing the corrosion resistance of implantable medical devices, where the difference between the break down potential, E_b, and rest potential, E_r, is used as a measure of the resistance to localized corrosion. Five samples were subjected to each of the aforementioned surface treatments and the average values of E_b and E_r were evaluated.

This article is an invited paper selected from presentations at Shape Memory and Superelastic Technologies 2008, held September 21-25, 2008, in Stresa, Italy, and has been expanded from the original presentation.

Norman Munroe, Chandan Pulletikurthi, and **Waseem Haider**, Applied Research Center, Florida International University, 10555 West Flagler Street, Miami, FL 33174. Contact e-mail: whaid001@fiu.edu.

2. Materials

2.1 Test Articles

PNT specimens (Actipore) used in this investigation were obtained from Biorthex Inc., Montreal, QC, Canada. The material was prepared by the self-propagating high-temperature synthesis (SHS) method with the following physical properties: 230 ± 130 μm pores, 65 ± 5% porosity, Ø = 12.5 ± 0.25 mm, and L = 30.0 ± 0.1 mm. Several discs of thickness 2.5 ± 0.25 mm were cut with a high-speed saw.

2.2 Reagents

Phosphate buffer saline solution (PBS) of pH 7.2, a reagent grade chemical conforming to the specifications of the Committee on Analytical Reagents of the American Chemical Society was used as the standard test solution.

1HF + $4HNO_3$ + H_2O acid mixture was used as the etchant for the PNT samples 20% conc. HNO_3 was used as the passivation solution.

3. Experimental Methods

3.1 Specimen Preparation

The following surface treatments were adopted:

(a) PNT samples were initially etched (Ref 8) in an acid mixture comprising of 1HF + $4HNO_3$ + $5H_2O$ at room temperature with continuous ultrasonic stirring for 1 min; (b) Etched samples were boiled in distilled water (Ref 9) at 132 °C for 30 min. These samples were then passivated by immersing them in 20% conc. HNO_3 at 80 °C for 20 min (Ref 10); (c) Untreated PNT samples were first dry heated (Ref 9) at 132 °C in a tubular furnace with a continuously supply of air for 1 h and then passivated; (d) Untreated PNT was first water boiled and then passivated.

3.2 Cyclic Polarization Tests

In vitro cyclic polarization tests were conducted (Ref 11) in accordance with the ASTM F 2129-08 (Ref 7) standard on as-received PNT and surface-treated PNTs. The Gammry corrosion cell was first cleaned with deionized water, rinsed with PBS solution, and filled with approximately 50 mL of PBS. The cell with contents was brought up to 37 °C by placing it in a controlled temperature water bath. The PBS solution was purged with ultrahigh-purity nitrogen for 30 min prior to immersion of the pellet. The pellet was immersed with nitrogen purging for an additional 30 min before starting the cyclic polarization tests. A saturated calomel electrode was used as the reference electrode and it was inserted into a Luggin Capillary. The surface area of the pellet in contact with PBS was carefully calculated to increase the accuracy of the corrosion parameters. The cyclic polarization option was selected on a Salartron SI 1287 Electrochemical Interface with a scan rate of 0.167 mV/s over a potential range between −0.5 and 1.5 V versus a saturated calomel electrode (SCE).

4. Results

Cyclic polarization tests were conducted and the average corrosion parameters for five surface treatments each utilizing five samples are summarized in Table 1.

5. Discussion

Table 1 shows the average corrosion parameters for the treated and untreated PNTs. The PNT that was etched, water boiled, and passivated was found to have superior resistance to pitting corrosion in terms of the $E_b - E_r$ (1.5005), whereas the dry-heated and passivated PNT showed the least corrosion resistance (1.1708). The higher resistance to pitting corrosion exhibited by the former etched, water boiled, and passivated PNT was attributed to selective etching of nickel oxides from the surface of the PNT (Ref 12), so that after water boiling and passivation, a higher proportion of titanium oxide was produced at the surface (Ref 6). The least corrosion resistance exhibited by the latter was due to the oxidation of Nitinol at elevated temperature, which leads to the formation of nickel oxide species that are more prone to corrosion (Ref 6). There was no significant difference in corrosion resistance between the untreated (as-received) PNT and those subjected to water boiling with passivation and etching. The highest break down potential (E_b 1.1010 V) was observed for the etched, water boiled, and passivated sample. The etched sample had the lowest value (E_b 0.9507 V). The highest protection potential, E_P, was obtained for the untreated sample, followed by the etched, water boiled, and passivated sample.

The cyclic polarization curve shown in Fig. 1 represents the general corrosion behavior of PNTs in this investigation. Figure 2 shows a comparison of cyclic polarization curves and

Table 1 Average corrosion parameters obtained for treated and untreated PNTs

Corrosion parameters	Untreated PNT	Etched, water boiled, and passivated PNT	Dry-heated and passivated PNT	Water boiled and passivated PNT	Etched PNT
E_b, V	1.0841	1.1010	1.0546	1.0587	0.9507
E_r, V	−0.2272	−0.3280	−0.1162	−0.1942	−0.3995
E_p, V	1.0536	1.0472	0.9810	0.9800	0.9139
$E_b - E_r$, V	1.3113	1.5005	1.1708	1.2530	1.2787

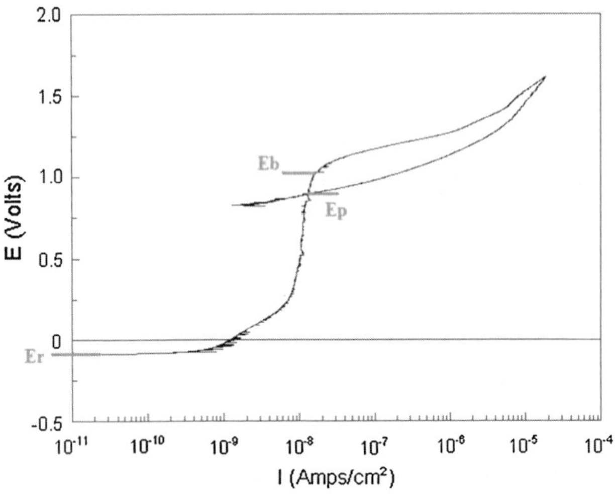

Fig. 1 Cyclic polarization curve of an etched PNT sample showing E_b, break down potential, E_p, protection potential, and E_r, rest or open circuit potential

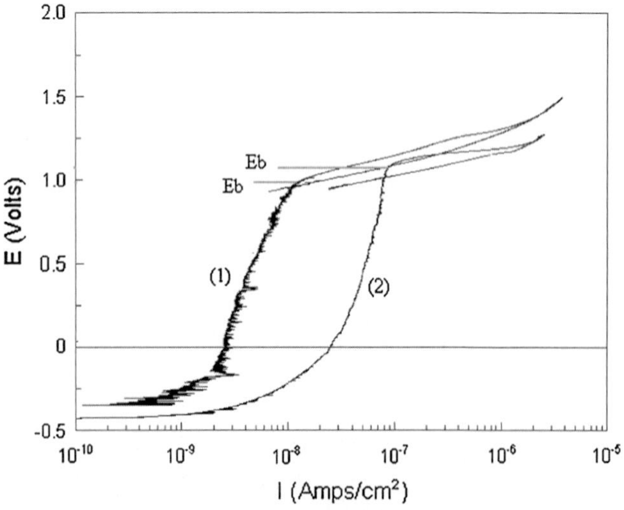

Fig. 2 A comparison of cyclic polarization curves and break down potential (E_b) values for (1) untreated and (2) etched + water boiled + passivated PNTs

Table 2 Summary of test results

	Untreated sample			
	E_b	E_r	E_p	$E_b - E_r$
Mean	1.0841	−0.2272	1.0536	1.3113
SD	0.06229	0.09259	0.0800	0.0654
Max	1.1520	−0.1346	1.1320	1.4050
Min	1.0100	−0.3200	0.9550	1.2168

	Etch only				Dry heated + Passivated			
	E_b	E_r	E_p	$E_b - E_r$	E_b	E_r	E_p	$E_b - E_r$
Mean	0.9507	−0.3280	0.9139	1.2787	1.0546	−0.1162	0.9810	1.1708
SD	0.0154	0.0917	0.0765	0.0964	0.0244	0.0538	0.0446	0.0467
Max	0.9200	−0.1734	0.9400	1.3813	1.0700	0.2107	1.0350	1.2457
Min	0.0961	−0.4213	0.8970	1.1284	1.018	0.0561	0.9100	1.1347

	Etched + water boiled + passivated				Water boiled + Passivated			
	E_b	E_r	E_p	$E_b - E_r$	E_b	E_r	E_p	$E_b - E_r$
Mean	1.1010	−0.3280	1.0472	1.5005	1.0587	−0.1942	0.9800	1.2530
SD	0.0164	0.0381	0.0398	0.0446	0.0375	0.0222	0.0112	0.0264
Max	1.1100	−0.4335	1.1040	1.5400	1.0700	−0.1627	0.9950	1.2752
Min	1.0750	−0.3573	1.0150	1.4960	1.0450	−0.2252	0.9650	1.2077

break down potential (E_b) values for untreated and etched, water boiled, and passivated PNTs (Table 2).

6. Conclusion

In vitro corrosion tests (ASTM F 2129-08) in PBS solution were conducted on surface-treated PNT to assess its localized corrosion resistance. Since Nitinol oxidation at elevated temperature leads to the formation of different oxide species, it is more prone to corrosion. This was verified in the case of dry-heated and passivated PNT samples, which revealed the least resistance to pitting corrosion than all tested samples. The etched, water boiled, and passivated sample produced the highest resistance to pitting corrosion. This was attributed to selective etching of nickel oxides from the surface of the PNT and the formation of a higher proportion of titanium oxide after water boiling and passivation.

Acknowledgment

The project described was supported by Award Number SC3GM084816 from the National Institute of General Medical Sciences. The content is solely the responsibility of the authors and does not necessarily represent the official views of the National Institute of General Medical Sciences or the National Institutes of Health.

References

1. M. Assad, P. Jarzem, M.A. Leroux, and C.H. Rivard, Porous Nitinol for Intervertebral Fusion: A Histomorphometry and Radiological Study in Sheep, *J. Biomed. Mater. Res. B Appl. Biomater.*, 2003, **64**(2), p 107–120
2. M.H. Wu, Fabrication of Nitinol Materials and Components, *Proceedings of the International Conference on Shape Memory and Superelastic Technologies* (Kunming, China), 2001, p 285–292
3. R.A. Ayers, S.J. Simske, T.A. Bateman, A. Petkus, R.L.C. Sachdeva, and V.E. Gyunter, Effect of Nitinol Implants Porosity on Cranial Bone in Growth and Apposition After 6 weeks, *J. Biomed. Mater. Res. B Appl. Biomater.*, 1999, **45**(1), p 42–47
4. G. Riepe, C. Heintz, N. Chakfe, M. Morlock, W.G. Fengles, and H. Imig, Degradation of Stentor Device After Implantation in Human Beings, *Proceedings of the International Conference on Shape Memory and Superelastic Technologies* (Pacific Groove, CA), 2000, p 279–283
5. Y.H. Li, G.B. Rao, L.J. Rong, Y.Y. Li, and W. Ke, Effect of Pores on Corrosion Characteristics of Porous NiTi Alloy in Simulated Body Fluid, *Mater. Sci. Eng. A*, 2003, **363**, p 356–359
6. S.A. Shabalovskaya, Physicochemical and Biological Aspects of Nitinol as a Biomaterial, *Int. Mater. Rev.*, 2001, **46**(4), p 230–250
7. "Standard Test Method for Conducting Cyclic Potentiodynamic Polarization Measurements to Determine the Corrosion Susceptibility of Small Implant Devices," ASTM F 2129-08, *Annual Book of ASTM Standards*
8. S. Trigwell, R.D. Hayden, K.F. Nelson, and G. Selvaduray, Effects of Surface Treatment on the Surface Chemistry of NiTi Alloy for Biomedical Applications, *Surf. Interface Anal.*, 1998, **26**, p 483–489
9. B. Thierry, M. Tabrizian, O. Savadogo, and L'H. Yahia, Effects of Sterilization Processes on NiTi Alloy: Surface Characterization, *J. Biomed. Mater. Res. A*, 2000, **49**(1), p 685–693
10. B. O'Brien, W.M. Carroll, and M.J. Kelly, Passivation of Nitinol Wire for Vascular Implants—A Demonstration of the Benefits, *Biomaterials*, 2002, **23**(8), p 1739–1748
11. N. Munroe, W. Haider, A. Datye, and K.H. Wu, Corrosion Behavior of Electropolished Implants Alloys, *Proceedings of the International Conference on Shape Memory and Superelastic Technologies*, Dec 3-5, 2007 (Tsukuba, Japan), 2007, p 307–314
12. S. Shabalovskaya, G. Rondelli, J. Anderegg, B. Simpson, and S. Budko, Effect of Chemical Etching and Aging in Boiling Water on the Corrosion Resistance of Nitinol Wires with Black Oxide Resulting from Manufacturing Process, *J. Biomed. Mater. Res. B Appl. Biomater.*, 2003, **66**, p 331–340

JMEPEG (2009) 18:768–780
DOI: 10.1007/s11665-009-9420-z

1059-9495/$19.00

Acceptance Criteria for Corrosion Resistance of Medical Devices: Statistical Analysis of Nitinol Pitting in In Vivo Environments

Lawrence E. Eiselstein, Duane Steffey, Andrew Nissan, Nigel Corlett, Roberto Dugnani, Esra Kus, and Sarah G. Stewart

(*Submitted October 24, 2008; in revised form February 23, 2009*)

ASTM F 2129 test method nor the FDA provides any guidance as to what constitutes an acceptance criterion for the corrosion resistance of implantable medical devices. Neither provide any guidance on how many samples to test or how to handle censored data, i.e. datasets where there are only a few tests that breakdown. The development of both a statistically valid acceptance criterion for corrosion resistance and a method of evaluation would be of significant benefit to the medical device community. This study of 420 nitinol cyclic polarization tests, which builds on previous research that was presented at SMST 2007, investigates the effect of long-term exposure to simulated in vivo environments with differing degrees of aeration. This was accomplished by pre-exposing electropolished (EP) nitinol to phosphate buffered saline (PBS) at 37 °C that had been sparged with either ultra high purity nitrogen or laboratory air. Immersion times ranged from 1 h up to 30 days. A total of 290 EP samples were tested in order to obtain a reasonable number of samples with breakdown, i.e. pitted. In addition, a total of 130 mechanical polished (MP) samples were also analyzed. This data allow us to test our statistical model that was presented at SMST 2007. This model takes into account the probability of breakdown per unit of exposed surface area and, if breakdown occurs, predicts the probability that $E_b - E_r$ is greater than some threshold value. Aerated PBS environments were found to have a large influence on the margin of safety against pitting in vivo. Statistical methods for treating highly right censored pitting data are presented.

Keywords ASTM F 2129, biomaterials, corrosion testing, electrochemistry, nitinol, pitting, quality assurance

1. Introduction

Long-term reliability of implantable medical devices requires testing to assure low corrosion and metal ion release rates. Corrosion can reduce the structural integrity of the device and the release of metal ions has been associated with adverse immunological response (Ref 1-3). ASTM F 2129 is a test method used to assess the corrosion susceptibility of small metallic implants using cyclic potentiodynamic polarization (Ref 4). The United States Food and Drug Administration (FDA) typically requires medical-device manufacturers to provide data that will demonstrate that their device will have sufficient corrosion resistance during the expected in vivo service life which could be greater than 10 years. Frequently, medical-device manufacturers use ASTM F 2129 to satisfy this FDA requirement.

This article is an invited paper selected from presentations at Shape Memory and Superelastic Technologies 2008, held September 21-25, 2008, in Stresa, Italy, and has been expanded from the original presentation.

Lawrence E. Eiselstein, Duane Steffey, Andrew Nissan, Nigel Corlett, Roberto Dugnani, Esra Kus, and **Sarah G. Stewart**, Exponent, Menlo Park, CA. Contact e-mail: eisel@exponent.com.

However, neither the ASTM F 2129 test method nor the FDA provides any guidance as to what constitutes an acceptance criterion for the polarization results or how many samples should be tested. ASTM F 2129 suggests comparison with a device already approved and on the market in the United States that is without any postmarket approval incidents of corrosion failures. This approach can, however, be quite expensive as it requires testing twice the number of experimental units (i.e., number of candidate devices and predicate devices); moreover, it presumes the existence of a suitable predicate device, i.e. a device that is implanted in a similar location in the body and manufactured from similar material and that the medical-device company has accurate knowledge regarding the predicate device's in vivo corrosion resistance.

As discussed in our earlier paper on nitinol corrosion, NiTi appears to have very good corrosion resistance based on many patient-years of in vivo experience (Ref 5). Nevertheless, the development of both a statistically valid acceptance criterion for corrosion resistance and a method of evaluation would be of significant benefit to the medical-device community. In addition, there is a need for a protocol by which a medical-device manufacturer can periodically check the corrosion resistance of an approved medical-device to assure that the manufacturing process is still within specification and devices are being manufactured with adequate corrosion resistance, i.e. a need for a manufacturing lot acceptance criterion. This study is an attempt to establish such a methodology and is a continuation of work that was first reported at SMST 2007 in Japan (Ref 5).

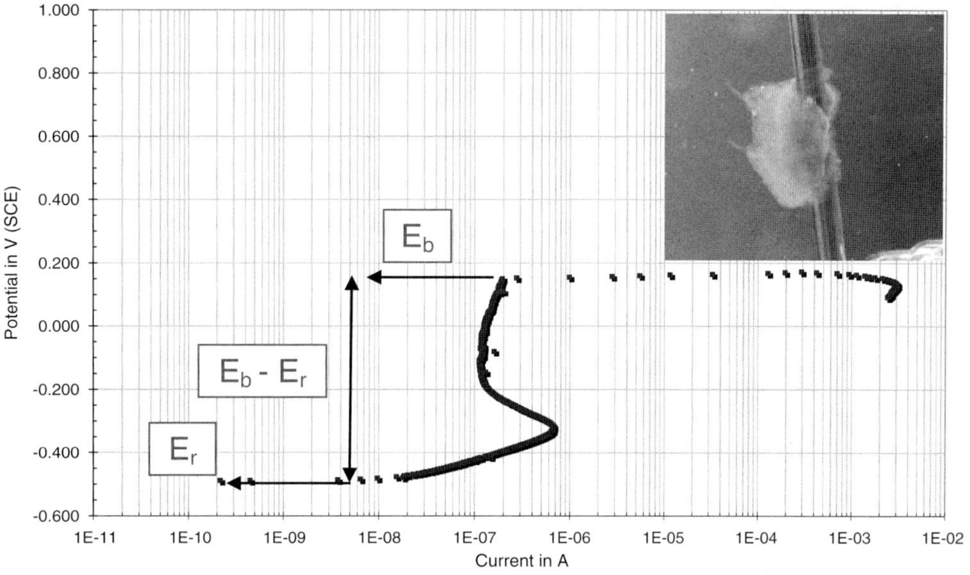

Fig. 1 Definition of electrochemical parameters E_b, E_r, and $E_b - E_r$

Three general methodologies exist as possibilities for an acceptance criterion. The first is that the pitting or breakdown potential (E_b), a measure of resistance to localized corrosion in a specific environment, of an implantable device should have similar or better corrosion resistance as determined by ASTM F 2129 testing than approved devices currently on the market with no known corrosion problems. The electrochemical parameters discussed in this paper are shown in Fig. 1. These exemplar devices should ideally be manufactured from similar materials and be used in the same general location in the body as proposed for the current device to assure a similar in vivo exposure environment, i.e. exposure to blood, bile, urine, saliva, etc. As mentioned above, it is sometimes difficult, and expensive, to find such exemplars and there is little guidance on how many devices should be tested to confirm equivalent or better corrosion resistance. The second methodology that has been proposed is that the breakdown potential (E_b) of the device should be greater than some threshold value, independent of the material used for the implant (Ref 6-8). These implant alloys can have drastically different pitting resistance and it is somewhat arbitrary that a minimum E_b should be specified independent of the material and environment. It is generally agreed that if a NiTi medical-device can be made reliably with an E_b of greater than 0.600 V with respect to a saturated calomel electrode (SCE), then there is little doubt that these materials will be resistant to corrosion in vivo unless other conditions like fretting, galvanic effects, or crevice corrosion take place in vivo. The reality, however, is that many medical devices will not be able to achieve such high values due to material selection, material finish, design, and manufacturing limitations. The third methodology, which we further develop in this paper, is to use the difference between the breakdown potential (E_b) and the open-circuit or rest potential (E_r). This gap ($E_b - E_r$) is considered to be a better parameter to evaluate medical-device margins against pitting in vivo as has previously been discussed by Pound (Ref 9). The basic principle behind this approach is that if the in vivo E_r is much lower than the pitting potential the device will have a large margin of safety against pitting or correspondingly, if $E_b - E_r$ is nearly zero or negative then the device will most likely pit when exposed to that environment. This is illustrated in Fig. 2. This $E_b - E_r$ parameter has been used frequently in the past to minimize the risk of pitting in many industries (Ref 9-13).

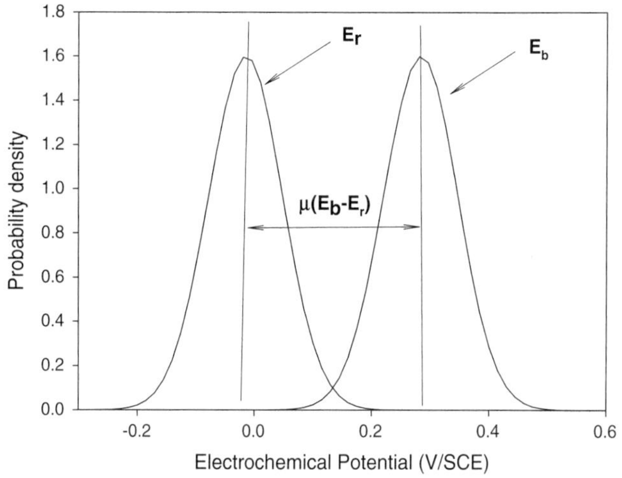

Fig. 2 Schematic illustrating the margin of safety against pitting

One problem with the $E_b - E_r$ criteria, as discussed by Rosenbloom and Corbett, is that there is little information regarding E_r for in vivo conditions (Ref 8). Another issue is how E_b, E_r, and ultimately $E_b - E_r$ vary during long-term in vivo exposures. The current ASTM F 2129 Standard states that devices are to be tested in their finished state. However, it is unlikely that the device, after exposure to in vivo conditions for several months will have the same resistance to pitting corrosion or even the same rest potential as a result of the surface film modification. Finally, there seems to be a wide variation in E_b values for NiTi depending on surface finish. Some NiTi:

1. will always pit below the minimum potential of 800 mV specified in ASTM F 2129;

2. will never undergo pitting below 800 mV; and
3. will infrequently breakdown.

Normal quality-control methodologies can be used for samples that always pit or never pit, but case 3, samples that only infrequently breakdown (datasets that contain censored data), are more difficult to treat statistically.

In an attempt to answer some of these questions, and to investigate whether long-term in vivo exposure might affect the margin of safety against localized corrosion, we report how long-term immersion in aerated and nonaerated phosphate buffered saline (PBS) will affect E_b, E_r, and ultimately $E_b - E_r$ on a statistically significant number of samples with two different surface conditions: mechanically polished (MP) and electropolished (EP).

Experimental data were collected on the breakdown potential (E_b) of replicate surrogate devices (i.e., wire samples) and on the variation in rest potential (E_r) with immersion time for individual units. Using these data, we conducted simulation studies to investigate the implications for device performance—specifically, the probability that the rest potential would exceed breakdown potential during the life of the device. The simulation results provided statistical support in defining a threshold value that would represent an acceptable margin between breakdown and rest potential ($E_b - E_r$), as well as a correspondingly low probability of corrosion-induced failure during the lifetime of the device.

We then implemented a tolerance interval approach for determining, from standard testing of a candidate device, whether there is sufficient confidence that a high proportion of replicate units will attain the specified threshold value. Finally, we developed guidance for medical-device manufacturers' quality-assurance programs in addressing the key practical question of how to determine the number of replicate units required for testing in order to establish an acceptable lower bound on the value of $E_b - E_r$ for manufactured devices. Developing this lower bound can be quite challenging for NiTi materials with good surface treatments since the vast majority of samples do not exhibit breakdown. We suggest a methodology to treat this type of highly censored data, i.e. few breakdowns with a large number of censored data, i.e. values greater than 800 mV SCE.

2. Materials and Methods

This investigation evaluated the pitting resistance of NiTi wire in two as-received surface conditions: MP and EP. The surface preparation of the material is proprietary to the material supplier and was not provided. The EP and MP NiTi used in the experiments came from different material lots.

The NiTi properties and surface characterization for the MP specimens that were purchased were described 55.9%Ni content superelastic wire with an A_f temperature in the 5-15 °C temperature range with a bright finish described as mechanical polishing that was followed by an argon anneal. The EP specimens that were purchased were described as superelastic straight wire that conforms to ASTM F 2063 Specifications (Ref 14). The EP materials had a nominal composition of 55.8 wt.% Ni. This material had an A_f temperature in the 5-18 °C range. Both the MP and EP wire had a diameter of 1.45 mm (0.057 in).

Auger analysis revealed surface oxide thicknesses of 28.6 ± 3.7 and 5.72 ± 0.79 nm for MP and EP wire, respectively (Ref 15). These film thicknesses were based on four measurements from two different locations on two different MP and EP NiTi wires. The thickness measurement was based on the equivalent sputter rate of SiO_2 under identical conditions used for the NiTi. Oxide thickness measurements were calculated as the thickness where the oxygen intensity dropped to half of its maximum within the profile. Both oxide films on the MP and EP had a 2 wt.% or less Ni concentration at the surface (Ref 15). All electrochemical testing was done with 2.5-cm long wire. Both cut ends of the wire were coated with silicone sealant which resulted in a surface area of 1.15 cm^2.

Figure 3 shows a representation of the size distribution of inclusions from longitudinal and transverse cross sections of the wire, and is similar for both materials used. The longitudinal sections indicate that the inclusions are present as stringers that broke up during the drawing process and are relatively strong and brittle compared to the surrounding matrix, as indicated by the fracturing and voids formed around these inclusions. We were unable to classify the inclusions with energy dispersive spectroscopy (EDS) but were able to observe that they are titanium rich. Infrequently, these inclusions intersect the wire surface as is shown in Fig. 4 for an EP specimen. Locating surface inclusions on the MP wire was much more difficult due to the rough surface and thick oxide layer.

All pre-test immersions and electrochemical testing were performed in PBS (8.0 g L^{-1} NaCl, 0.2 g L^{-1} KCl, 1.15 g L^{-1} $Na_2H_2PO_4$, and 0.2 g L^{-1} KH_2PO_4) at 37 ± 1 °C. The pH of the solutions was 7.4 ± 0.1 for all immersions and electrochemical testing. When samples were pre-exposed to aerated PBS, these solutions were aerated with laboratory air (79% N_2, 21% O_2, and 0.038% CO_2). For deaerated tests, the PBS was sparged with ultrahigh-purity nitrogen for 30 mins prior to testing and then continuously throughout the test. The use of deaerated and aerated solutions was chosen to represent the widest range of oxidizing conditions to which an implantable medical-device might be subjected. Obviously, the deaerated PBS has a lower redox potential than would be expected to be found in vivo, and the aerated PBS is expected to provide higher redox potentials than conditions in vivo, except for perhaps implants in the airways or devices that are subjected to peroxides from an immunological response.

All tests were performed using Gamry computer-controlled potentiostats and Gamry DC-105 software. Long-term potential monitoring was performed in accordance with ISO 16429:2004 and the cyclic polarization tests were performed in accordance with ASTM F 2129-06 (Ref 4, 16). All potentials were measured with respect to a SCE. The potentiodynamic scan rate used was 0.167 mV/s and the potential increased until breakdown occurred or 1.1 V (SCE) was reached. All E_r values reported are the corrosion potentials observed at the end of the pre-test immersions, i.e. in aerated or deaerated solutions, not the E_r measured in the deaerated PBS used to determine E_b.

3. Results

We tested a total of 420 nitinol test specimens in order to get sufficient information regarding the probability of pitting in 130 mechanical polished and 290 EP NiTi samples. The pre-test

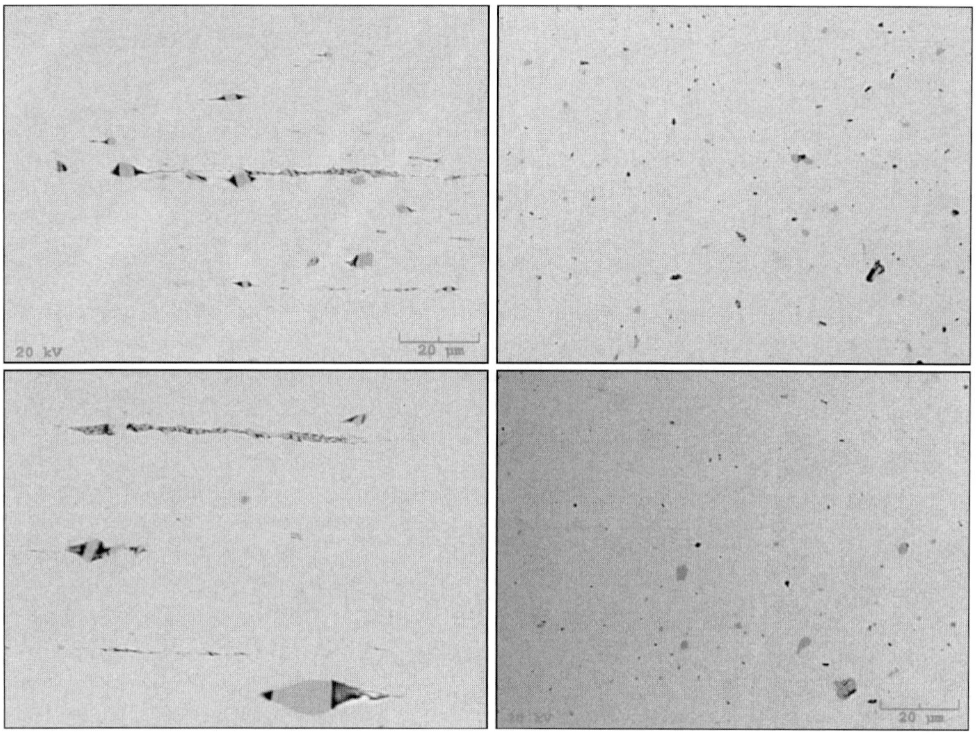

Fig. 3 Inclusions from metallurgical cross sections. Longitudinal (*upper left*) and transverse (*upper right*) of EP wire. The corresponding longitudinal and transverse metallographic sections from the MP wire are shown at the *bottom left* and *right*

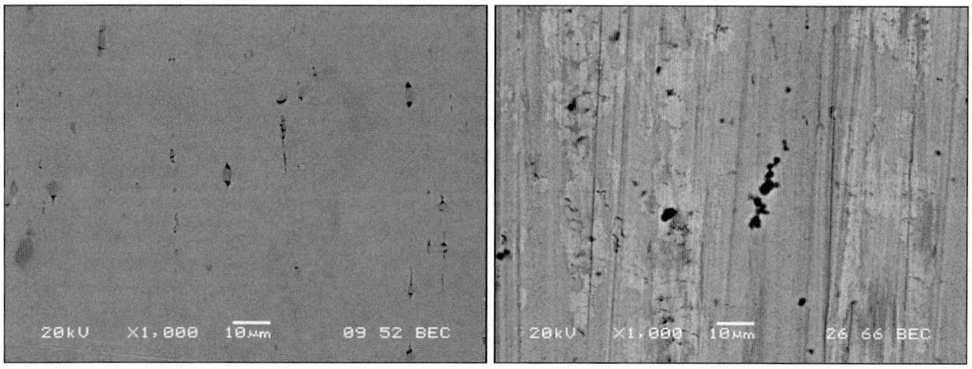

Fig. 4 Inclusions intersecting the surface. EP (*left*) and MP (*right*) in as-received condition

exposure times to the deaerated and aerated environments were 1, 12, 60, 156, and 720 h for the MP surface finish and 1, 12, and 60 h for the EP samples. Then, 3-99 replicate specimens were tested at each pre-test immersion time. The results from our testing are summarized in Fig. 5-7 and Table 1. The mean and standard deviation for E_b for EP samples can only be calculated for the samples that pitted and therefore underestimate the true resistance of the material to pitting. As can be seen from Table 1, there are very clear differences between the corrosion resistance of the MP and EP samples. All of the MP samples exhibited pitting below 0.8 V (SCE) whereas only 13% of the EP samples pitted. Figure 5 illustrates that increasing oxygen concentration in the PBS increases E_r and that E_r also tends to increase with increasing pre-test exposure time in both deaerated and aerated PBS solutions. E_r increases less rapidly with increasing pre-test exposure time for the EP surface condition. E_b for either the MP or EP surface condition is about the same and unaffected by oxygen concentration in the PBS as shown in Fig. 6. However, E_b does increase with increasing pre-test exposure time. The safety margin against pitting corrosion, $E_b - E_r$, is lower in aerated environments primarily from the effect of oxygen on raising E_r (Fig. 7).

4. Discussion

All the MP samples pitted, whereas only about 13% of the EP samples pitted indicating a significant effect of surface treatment. The dissolved oxygen concentration also has a very large influence, as expected on the open-circuit potential (E_r). The MP wire had an oxide film about five times as thick as the EP samples and had significantly lower corrosion resistance as seen in Table 1. Clarke et al. showed that NiTi with a thicker

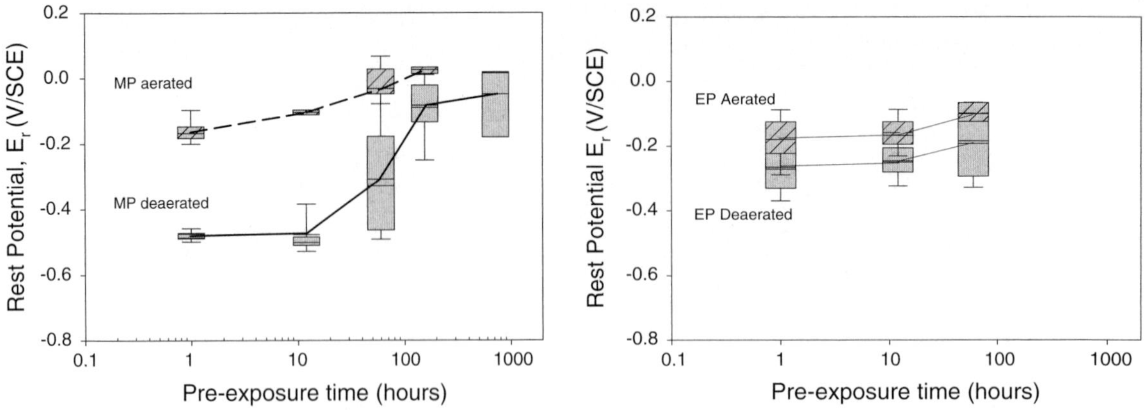

Fig. 5 Effect of pre-test immersion on rest potential E_r

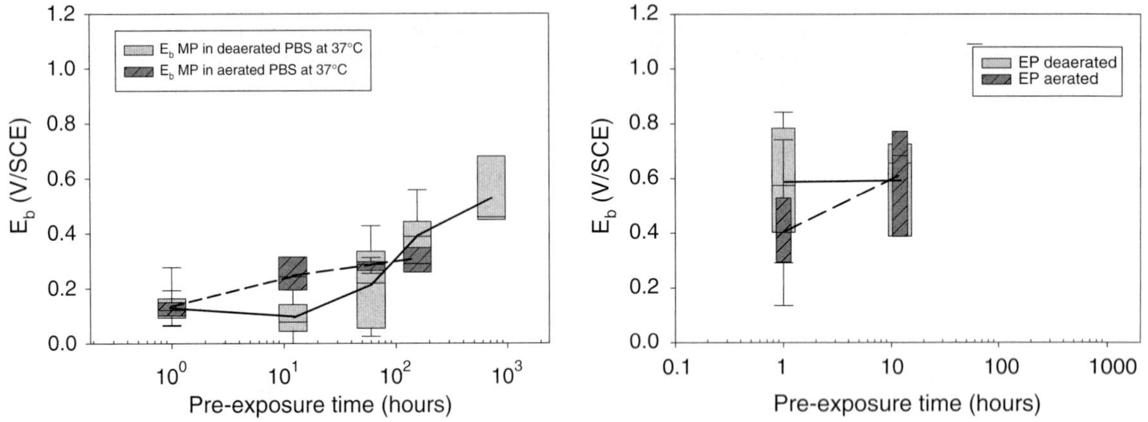

Fig. 6 Effect of pre-test immersion on break down potential E_b

Table 1 Summary of test results

NiTi	Gas	Time, h	No. RPT (a)	E_r Mean	E_r σ	No. CPT (b)	No. pitted	% Pitted	E_b Mean	E_b σ	$E_b - E_r$ Mean	$E_b - E_r$ σ	No. incomplete tests (c)
MP	N_2	1	31	−0.472	0.062	30	30	100	0.130	0.047	0.602	0.078	1
MP	Air	1	13	−0.160	0.035	13	13	100	0.138	0.067	0.298	0.076	0
MP	N_2	12	27	−0.476	0.110	27	27	100	0.095	0.081	0.571	0.095	0
MP	Air	12	6	−0.104	0.009	6	6	100	0.246	0.065	0.350	0.069	0
MP	N_2	60	16	−0.307	0.161	16	16	100	0.212	0.158	0.520	0.153	0
MP	Air	60	13	−0.015	0.049	11	11	100	0.283	0.018	0.310	0.038	2
MP	N_2	156	15	−0.090	0.082	15	15	100	0.393	0.098	0.482	0.168	0
MP	Air	156	6	0.024	0.011	6	6	100	0.306	0.059	0.281	0.062	0
MP	N_2	720	3	−0.049	0.113	3	3	100	0.530	0.131	0.578	0.129	0
MP	Air	720	0	...	...	...	...	...	...	...	...	...	...
EP	N_2	1	98	−0.265	0.082	97	17	18	0.585	0.195	0.832	0.201	1
EP	Air	1	58	−0.180	0.069	57	10	18	0.407	0.184	0.584	0.216	1
EP	N_2	12	58	−0.245	0.060	56	4	7	0.590	0.193	0.769	0.198	2
EP	Air	12	57	−0.159	0.059	55	7	13	0.609	0.259	0.763	0.267	2
EP	N_2	60	13	−0.192	0.093	12	0	0	...	...	...	...	1
EP	Air	60	6	−0.099	0.045	6	1	17	1.090	...	1.199	...	0
EP	N_2	156	0	...	...	...	...	...	...	...	...	...	...
EP	Air	156	0	...	...	...	...	...	...	...	...	...	...

(a) No. RPT: Number of rest potential tests; not all samples had valid cyclic polarization tests
(b) No. CPT: Number of cyclic polarization tests; not all samples had valid cyclic polarization tests
(c) Not all samples had valid cyclic polarization tests

oxide layer has a lower E_b value and a greater nickel release in both long-term and short-term experiments (Ref 1). Clarke et al. also found that after long-term immersion tests breakdown potentials increased for samples that exhibited lower values initially (Ref 1).

The effect of long-term pre-exposures of MP NiTi to 37 °C PBS clearly increases E_b; however, E_b increases faster for pre-test exposures to aerated solutions compared to deaerated solutions. This can be seen by comparing E_b values for various pre-test exposure immersion times for MP NiTi in Table 1 and Fig. 6. Clarke et al. found a similar trend in E_b for NiTi with a thick oxide (340 nm) exposed to 0.9 wt.% saline for up to 4 months although they do not report how E_r values were affected (Ref 1). Our data show that both E_r and E_b increased with increasing pre-immersion duration in deaerated PBS (Fig. 5 and 6), but the trends were not equitable as $E_b - E_r$ decreased with increasing immersion duration because E_r increased faster than E_b.

The pre-test exposure of MP samples to aerated PBS had a more dramatic effect than the longest exposure to deaerated PBS with the $E_b - E_r$ values being about 300 mV lower which is primarily due to the very large increase of E_r when exposed to aerated PBS.

The statistical distribution of E_b for stainless steels has previously been studied by Shibata and others (Ref 17-19). Shibata has shown that a normal distribution adequately describes the distribution of E_b for AISI 304 stainless steel tested in 3.5 wt.% NaCl solutions at 30 °C (Ref 17, 18). Nevertheless, Kowaka notes that pitting potentials are actually the lowest breakdown potential that is found only at a small area on a sample under test, and as such, a "weakest link" statistical distribution such as a Gumbel or Weibull distribution is generally expected to fit such data (Ref 18). Kowaka showed that the Weibull distribution also provided a good fit to Shibata's stainless steel E_b data while at the same time being a "weakest link" type of statistical distribution. The E_b data from the MP and EP specimens exposed to deaerated PBS are shown in Fig. 8. Both datasets are well described by a three-parameter Weibull distribution.

5. MP Modeling

Regression analysis of $E_b - E_r$ for MP NiTi data was performed to quantify the effects of aeration and immersion time. The effect of pre-test immersion exposure time was modeled as linear in the logarithm scale and separate regression analyses were performed on the aerated and deaerated data because of the greater variability of the

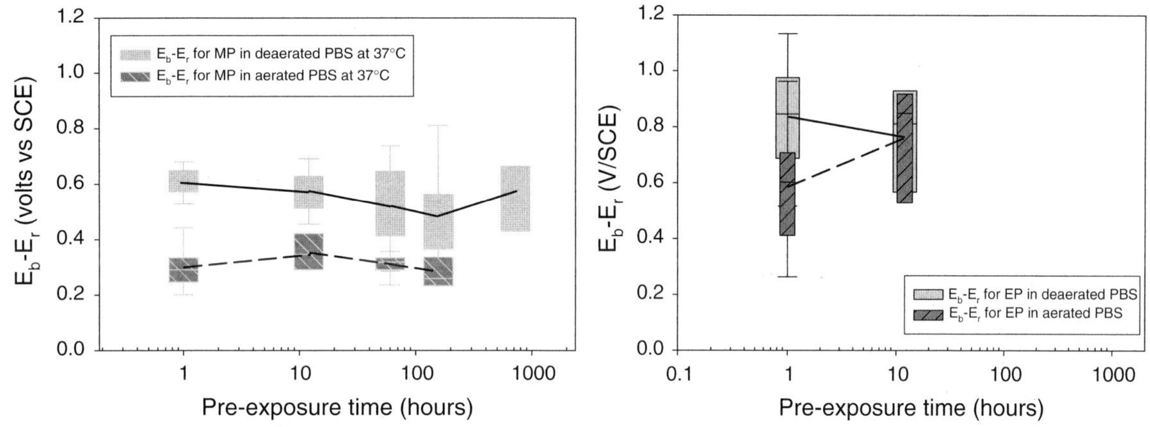

Fig. 7 Effect of pre-test immersion on $E_b - E_r$

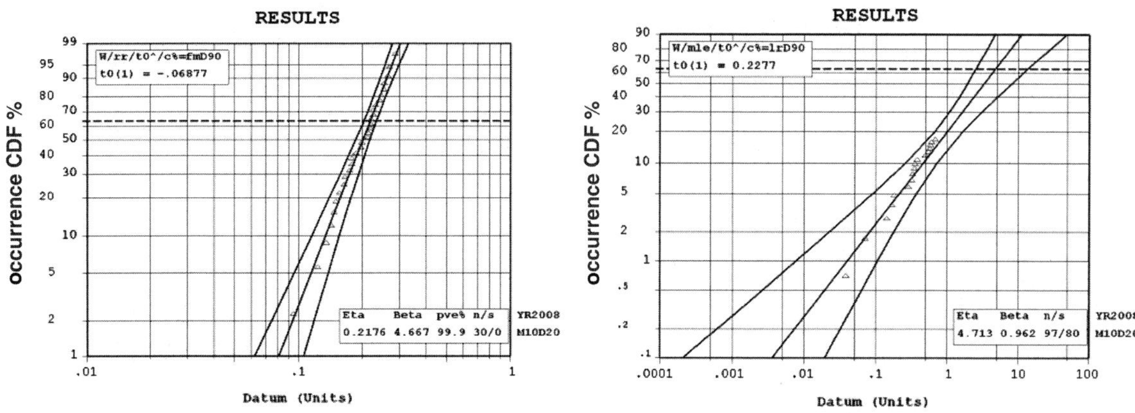

Fig. 8 Three-parameter Weibull distribution of E_b for MP (*left*) and EP (*right*) in deaerated PBS for 1 h

Table 2 Regression analysis results of $E_b - E_r$ for MP NiTi

Predictor	Coefficient	SE coefficient	T	P
Deaerated				
Constant	0.6037	0.01962	30.77	<0.001
ln, t	−0.01838	0.006081	−3.02	0.003
S = 0.1179	R^2 = 9.3%		R^2 (adjusted) = 8.3%	
Aerated				
Constant	0.3083	0.0172	17.92	<0.001
ln, t	−0.00037	0.00533	−0.07	0.945
S = 0.0649	R^2 = 0.0%		R^2 (adjusted) = 0.0%	

deaerated data. The fitted regression equation for the deaerated data is as follows:

$$E_b - E_r(\text{V/SCE}) = 0.604 - 0.0184\ln[t]$$

Ninety-one data points were used in this regression and the ANOVA results are reported in Table 2.

Similarly for the aerated data the fitted regression equation is as follows:

$$E_b - E_r(\text{V/SCE}) = 0.308 - 0.0053\ln[t]$$

Thirty-six data points were used in this regression and the ANOVA results are reported in Table 2.

Pre-test exposure to aerated environments has a large effect on the corrosion margin of safety primarily due to the large effect of aeration on E_r as can be seen by the constant term dropping from 604 to 308 mV. Table 2 shows that the ln(t) factor is statistically significant for deaerated samples but not for aerated samples. However, the low R^2 value for the deaerated samples indicates the presence of large unexplained variability. We suspect that this large variation is a result of the inherent randomness of the pitting process. The average $E_b - E_r$ value for the longest duration pre-test immersion (720 h) appears to violate this decreasing trend, but the average is calculated from only three observations and does not alter the statistical significance of the trend.

For deaerated samples, the baseline for $E_b - E_r$ is defined as material subjected to pre-test exposure for 1 h. The regression equation for deaerated samples can therefore be expressed as:

$$E_b - E_r(\text{V/SCE}) = \text{baseline} + \text{pre-test immersion duration effect}$$

where baseline = 604 mV (SCE); this is the $E_b - E_r$ from the model for NiTi in the as-received condition (i.e., pre-test exposure to deaerated PBS for 1 h).

Pre-test immersion duration term = −0.0184ln[t]; the model predicts a decrease in $E_b - E_r$ as the pre-exposure immersion time increases:

12 h = −30 mV
60 h = −96 mV
156 h = −119 mV
720 h = −121 mV
10-years = −209 mV

Sparging with air provides dissolved oxygen levels that are higher than those generally found in vivo and therefore the aeration effect here is exaggerated over that which would be expected when NiTi is exposed to arterial blood. Nevertheless, an aerated PBS solution better represents the in vivo dissolved oxygen concentration than a deaerated solution. The value provided above of −209 mV for the expected decrease in $E_b - E_r$ after 10-years exposure to deaerated solutions is an extreme extrapolation of the data and therefore may not accurately predict actual in vivo performance. However, if testing is done as we suggest, i.e. pre-exposure to aerated solutions, the inherent variability in $E_b - E_r$ is decreased and there is no need to account for the time the implant will be in service.

The probability that an MP sample will have a negative $E_b - E_r$ can be calculated using the probability distribution statistics shown in Table 3 where we have combined both the 1 and 12 h* and 60 and 156 h** deaerated data since there was no statistically significant difference between the 1 and 12 h or the 60 and 156 h data; however, the difference between the 1 and 12 h and the 60 and 156 h data was statistically significant.† We also calculated the probability that $E_b - E_r$ will be negative for MP NiTi for the combined 60 and 156 h immersions in deaerated PBS and the probability of negative $E_b - E_r$ for MP NiTi exposed to aerated PBS for all our aerated data (1-156 h data combined) since $E_b - E_r$ versus pre-test exposure time is relative insensitive to exposure duration in aerated environments.

Table 3 shows the results using several different types of distributions. A normal distribution is used since it has been previously used to describe E_b data for AISI 304 stainless steel in 3.5 wt.% saline (Ref 17-19). A Gumbel extreme value distribution has also been used since it theoretically should model the distribution of smallest values for multiple measurements taken in different locations of constant area (Ref 18). Among a set of candidate models that included the two- and three-parameter Weibull and lognormal distributions, the Gumbel extreme value distribution was found to be the best fit to the $E_b - E_r$ data for MP samples pre-exposed to deaerated PBS for 1 and 12 h (Fig. 9). The superiority of the Gumbel versus the normal model is reflected in the larger likelihood value of 65.9 in Table 3. This finding is also reflected in the more reasonable (higher) pitting probability predicted using the Gumbel distribution (6.43×10^{-5} versus 5.14×10^{-12}). A three-parameter Weibull distribution was found to best characterize the longer-term deaerated immersion data (60-156 h) with an estimated pitting probability of zero versus a 6.66×10^{-4} probability if a normal distribution is assumed.

The three-parameter Weibull distribution was also found to best characterize the aerated $E_b - E_r$ data. This is not surprising since the two- and three-parameter Weibull distributions are typically among the probability models providing the best fit to minimum extreme value distributions such as E_b. The three-parameter Weibull model gives an estimated probability of pitting in vivo of zero whenever the threshold parameter—i.e. the minimum observable value of $E_b - E_r$—is estimated to be positive. For example, the $E_b - E_r$ threshold is estimated at 149 mV (SCE) for the samples exposed to aerated solution and at 224 mV (SCE) for the deaerated exposures. Although both distributions would predict zero probability of pitting in vivo, the threshold values suggest that the specimens exposed to aerated PBS have a lower margin of safety against pitting. The fit of three-parameter Weibull distributions to $E_b - E_r$ data for short-term and long-term exposure to deaerated PBS are shown in Fig. 9. Similarly, Fig. 10 shows the three-parameter Weibull fit to the aerated $E_b - E_r$ data (all exposure times included).

*Mann-Whitney Rank Sum Test (P = 0.101).
**Mann-Whitney Rank Sum Test (P = 0.385).
†Mann-Whitney Rank Sum Test (P = 0.001).

Table 3 Estimated probabilities that $E_b - E_r < 0$ for MP NiTi

Solution	Distribution type	Immersion time, h	Point estimate (P_E)	95% Upper bound (P_U)	Log-likelihood
Deaerated	Normal	1-12	5.14×10^{-12}	4.96×10^{-9}	58.8
Deaerated	Gumbel (extreme value)	1-12	6.43×10^{-5}	3.60×10^{-4}	65.9
Deaerated	Normal	60-156	6.66×10^{-4}	6.66×10^{-3}	13.6
Deaerated	3-parameter Weibull	60-156	0 (for $E_b - E_r < 224$ mV)	0	16.0
Aerated	Normal	1-156	5.57×10^{-7}	5.07×10^{-5}	48.4
Aerated	3-parameter Weibull	1-156	0 (for $E_b - E_r < 149$ mV)	0	49.5

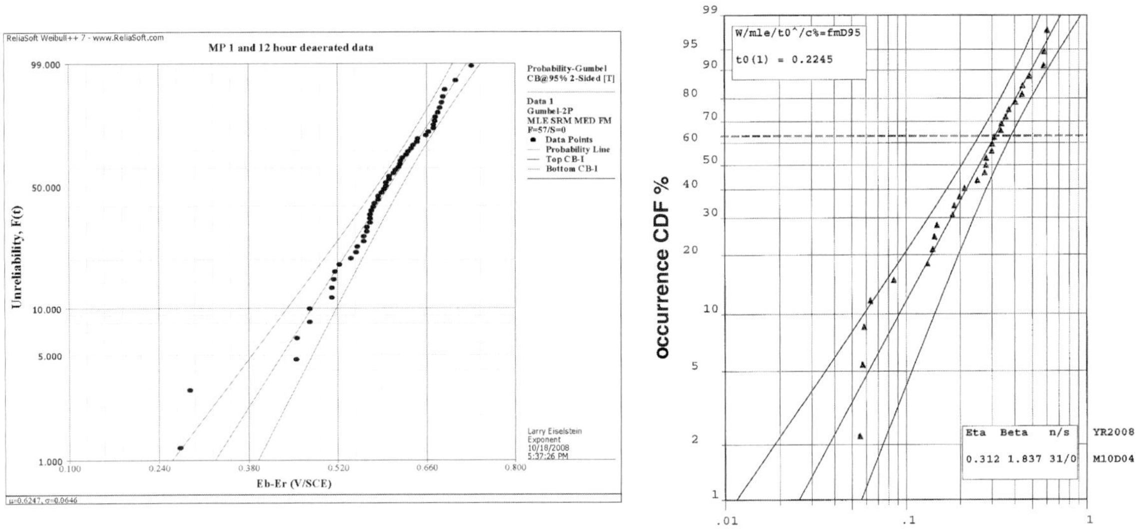

Fig. 9 Distribution of $E_p - E_r$ with confidence bands for MP samples immersed in deaerated PBS for 1 and 12 h as Gumbel (*left*) and immersed in deaerated PBS for 60-156 h as three-parameter Weibull (*right*)

The entries in Table 3 clearly demonstrate that estimated failure probabilities depend very sensitively on the type of probability distribution used to model the observed variation in $E_b - E_r$. When a typical amount of testing is performed, several models may fit the available data and lead to greatly differing inferences regarding extreme values. The normal distribution fits the observed test data reasonably well but, in all cases studied, is not the best distribution to describe the distribution of $E_b - E_r$ data. Specifically, Table 3 shows that, if the normal model is used to calculate the probability of pitting in vivo, it leads the counter intuitive conclusion that pitting is less likely in aerated versus deaerated solutions. This result can be explained by the short tails of the normal distribution and the fact that, although $E_b - E_r$, on average, is greater under deaerated conditions, the associated variability about the mean is also much greater than that under aerated conditions. The Weibull model provides a marginally better fit to the data but effectively precludes the possibility that E_r could exceed E_b. Use of the Gumbel extreme value distribution can be justified by physical considerations and produces conservative estimates of lower-tail probabilities, but the fit of the model to observed data is poor in two of the three cases studied.

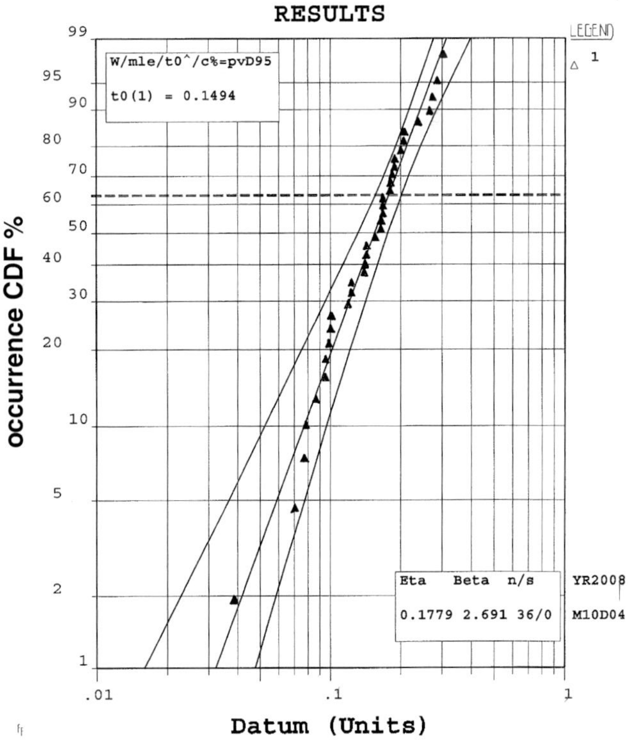

Fig. 10 Distribution of $E_p - E_r$ with confidence bands for MP samples immersed in aerated PBS for 1-156 h (three-parameter Weibull)

6. EP Modeling

A total of 290 EP samples were tested (169 in deaerated PBS and 121 in aerated PBS). Analysis of variance of the EP rest potential data (see Table 4) indicates that E_r is significantly affected by both the amount of dissolved oxygen in the PBS and the length of time of immersion. The effect of dissolved oxygen on E_r is the combined effect of the dissolved on solution redox potential and the effect of long-term exposure on the passive film (morphology, composition, and thickness).

Of the 290 EP samples tested, only 39 (13%) pitted below 0.800 V (SCE) the minimum voltage specified in ASTM F 2129 at which the cyclic polarization test can be stopped. Table 5 shows how probability of pitting of EP samples varies with pre-test exposure time and oxygen concentration in the PBS. The results from Fisher's exact test (Table 5) and binary logistic regression analysis using the censored data versus the oxygen and time variable (Table 6) showed that the differences observed were not significant indicating that neither oxygen concentration or duration of immersion in PBS significantly affect the percentage of EP samples that pit below 1.1 V (SCE).

Modeling of the EP test data is more challenging than for the MP test data as most of the data are censored, i.e. most of the samples did not break down. One way to treat such data is through the use of nonparametric statistical methods. However such methods require a large number of samples. For instance, to specify a 99% reliability value with 95% confidence, 299 samples need to be tested. The number of specimens required for a nonparametric approach is much greater than a parametric approach (Ref 20).

One parametric approach is to fit the censored $E_b - E_r$ values to a standard statistical distribution using a methodology that can handle the censoring. We have done this for the aerated and deaerated EP specimens as shown in Fig. 11. These data were best fit by a three-parameter Weibull distribution for both the aerated and deaerated exposures. The threshold parameter (gamma), below which the probability of pitting is zero, is estimated at 512 mV for the deaerated and 286 mV for the aerated exposures. These results suggest that aerated environments are more prone to produce pitting than deaerated solutions, but neither solution is likely to result in breakdown in vivo since the estimated probability of failure is zero for both.

The prediction of a zero failure probability from a fitted three-parameter Weibull or LogNormal distribution, for instance, does not seem physically reasonable, since there is presumably some probability of pitting no matter how small that actual probability. Further, we believe that single-population models lack physical basis in this context, and our investigation has shown that other

Table 4 Analysis of variance of E_r data from EP samples

General linear model: E_r vs. oxygen concentration and time

Factor	Type	Levels	Values
Oxygen	Fixed	2	Deaerated, aerated
Time	Fixed	3	1, 12, 60 h

Analysis of variance for E_r using adjusted SS for tests

Source	DF	Seq SS	Adj. SS	Adj. MS	*F*	*P*
Oxygen	*1*	*0.531*	*0.230*	*0.230*	*45.55*	*0.000*
Time	*2*	*0.106*	*0.0974*	*0.0487*	*9.65*	*0.000*
Oxygen * Time	2	0.00024	0.00024	0.00012	0.02	0.976
Error	284	1.434	1.434	0.00505		
Total	289	2.071				
$S = 0.0710$	$R^2 = 30.8\%$			R^2 (adj) = 29.6%		

Italics indicate model terms that are statistically significant at 0.05 level

Table 5 EP NiTi—Fisher's exact test indicates probability of breakdown does not vary significantly with oxygen concentration or duration of immersion time in PBS

Oxygen concentration	Immersion time, h	Number of tests	Number pitted	Percent pitted
Deaerated	1	97	17	18
	12	56	4	7
	60	12	0	0
Aerated	1	57	10	18
	12	55	7	13
	60	6	1	17

Fisher's exact test: *P* values of 1.0 and 0.265 for oxygen concentration and time, respectively

Table 6 EP NiTi—binary logistic regression confirms probability of breakdown does not vary significantly with oxygen concentration or duration of immersion time in PBS

Binary logistic regression: censored vs. oxygen, time

Variable	Value	Count
Censored	1	252 (event)
	0	39
	Total	291

Logistic regression table

Predictor	Coefficient	SE coefficient	*Z*	*P*	Odds ratio	95% CI lower	95% CI upper
Constant	1.40	0.296	4.73	0.000			
Aeration	−0.282	0.352	−0.80	0.423	0.75	0.38	1.5
Time							
12 h	0.712	0.385	1.85	0.064	2.04	0.96	4.33
60 h	1.305	1.05	1.24	0.214	3.69	0.47	28.9

Log-likelihood = −111.9. Test that all slopes are zero: $G = 5.41$, DF = 3, *P* value = 0.144

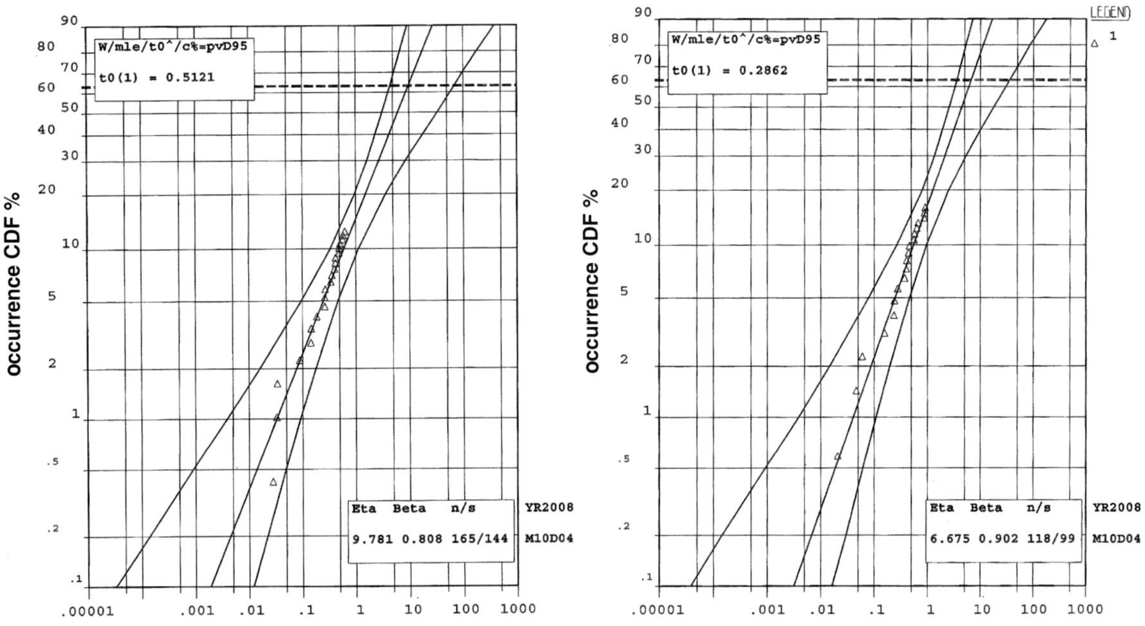

Fig. 11 Censored treatment of $E_b - E_r$ data from EP samples immersed in deaerated (left-two-parameter exponential) and aerated (censored three-parameter Weibull distribution) PBS for 1, 12, and 60 h

probability distributions provide an adequate fit to the observed $E_b - E_r$ data.

To overcome some of the limitations of the parametric approach we have previously proposed (Ref 5), we derive a physically based, mixture model, using a Poisson distribution to describe the fraction of samples that will likely break down. This was done by modeling the occurrence of surface flaws, and then using a normal distribution to model the distribution of breakdown potentials for flawed specimens. The Poisson distribution has been used previously to characterize the number of pits expected (Ref 18). The Poisson probability function ($p(k) = \exp^{-\lambda} \lambda^k / k!$) actually gives the probability of finding a specimen with k flaws that will pit before 0.8 V is reached. In computing this probability, $\lambda = \theta a$ is proportional to the area (a) of the surface being evaluated (1.15 cm^2) and θ is the expected number of occurrences per unit area.

Our proposed approach to an acceptance criterion is to estimate the probability of corrosion (Π) based on the probability that the sample will pit (Π_1) and, if the specimen will pit, the probability (Π_2) that $E_b - E_r$ is negative:

$$\Pi = \Pi_1 \times \Pi_2,$$

where $\Pi_1 = 1 - \exp(-\lambda)$ is the probability that any individual sample will possess one or more flaws, and will therefore pit, and $\Pi_2 = F(0)$, where $F(\cdot)$ is the cumulative distribution function that describes the variation in $E_b - E_r$. We consider alternative parametric choices for $F(\cdot)$ by considering their suitability for and goodness-of-fit to the values of $E_b - E_r$ observed for the pitted specimens. Using the normal distribution for $F(\cdot)$,

$$\Pi_2 = \left(\frac{1}{2}\right)\left[1 + \mathrm{Erf}\left(\frac{\delta - \mu}{\sigma\sqrt{2}}\right)\right]$$

is the probability that $E_b - E_r < 0$ if the specimen pits. Here, δ is an arbitrary extra margin of safety above $E_b - E_r$ which for this analysis we have assumed to be zero.

Other choices for $F(\cdot)$ can be considered; for example, using the three-parameter Weibull distribution,

$$\Pi_2 = 1 - \exp[-(-\gamma/\alpha)^{\beta}],$$

where α and β are the (positive) scale and shape parameters, respectively, and γ is the (negative) threshold parameter. Alternatively, one could develop probability models for E_b and E_r separately. Although we do not pursue such an approach in this paper, some justification can be found in the observed lack of correlation ($r = 0.024$, $P = 0.882$) between the E_b and E_r values for pitted specimens.

The probability (Π) that $E_b - E_r$ is negative can be approximated by:

$$\Pi \approx P_E = P_1 \times P_2$$

where P_1 is the observed proportion of breakdowns and P_2 is computed by substituting data-derived estimates of μ and σ in the expression for Π_2, as shown in Fig. 12 for the deaerated exposures and Fig. 13 for the aerated exposures. The next step is then to determine the upper bound P_U for Π. The lot of material can then be accepted or rejected if P_U is less than a specified limit on the probability for pitting and subsequent breakdown.

For our data, 39 out of 290 specimens broke down. With λ representing the expected number of flaws per 1.15 cm^2 of surface area, this rate can be estimated by

$$\lambda = -\ln(1 - P_1) = -\ln(1 - 39/290) = 0.144$$

or about one flaw per seven specimens.

Restricting attention to the 39 pitted specimens, we can use analysis of variance to see if E_b is affected by oxygen concentration or duration of exposure to PBS. The result of this analysis is shown in Table 7 and confirms that E_b is not significantly affected by oxygen concentration but is influenced by the duration of exposure. Similarly, an analysis of variance of $E_b - E_r$ (see Table 8) indicates that the margin of safety

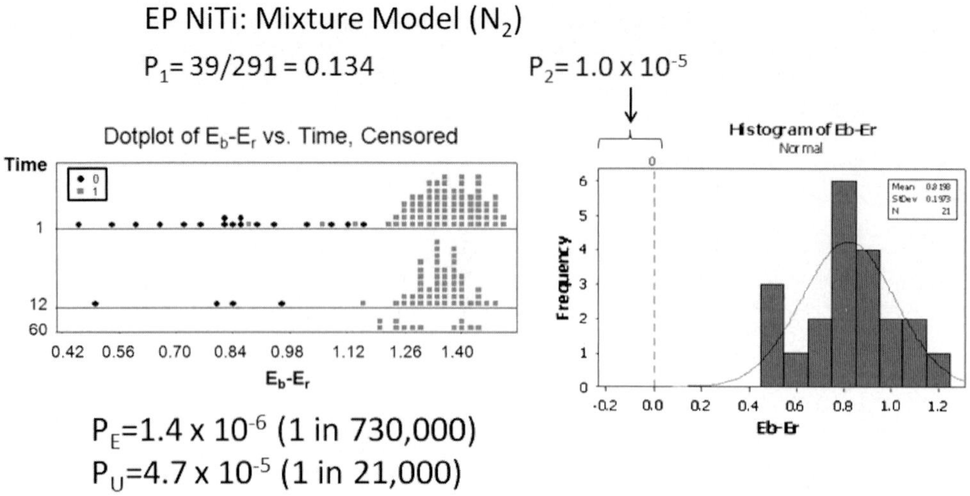

Fig. 12 EP mixed probability model for deaerated PBS

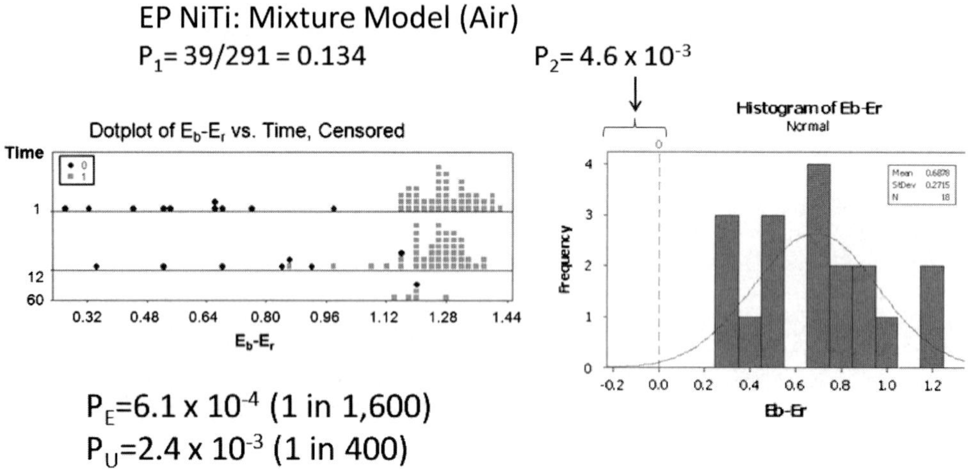

Fig. 13 EP mixed probability model for aerated PBS

Table 7 Analysis of variance of E_b data from EP samples

General linear model: E_b vs. oxygen concentration and time

Factor	Type	Levels	Values
Oxygen	Fixed	2	Deaerated, aerated
Time	Fixed	3	1, 12, 60 h

Analysis of variance for E_b using adjusted SS for tests

Source	DF	Seq SS	Adj. SS	Adj. MS	F	P
Oxygen	1	0.0381	0.131	0.131	3.04	0.090
Time	*2*	*0.437*	*0.437*	*0.219*	*5.10*	*0.011*
Error	35	1.50	1.50	0.0429		
Total	38	1.98				
$S = 0.207$	$R^2 = 24.1\%$			R^2 (adj) = 17.6%		

Italics indicate model terms that are statistically significant at 0.05 level

Table 8 Analysis of variance of $E_b - E_r$ data from EP samples

General linear model: $E_b - E_r$ vs. oxygen concentration and time

Factor	Type	Levels	Values
Oxygen	Fixed	2	Deaerated, aerated
Time	Fixed	3	1, 12, 60 h

Analysis of variance for $E_b - E_r$ using adjusted SS for tests

Source	DF	Seq SS	Adj. SS	Adj. MS	F	P
Oxygen	*1*	*0.169*	*0.280*	*0.280*	*5.72*	*0.022*
Time	2	0.315	0.315	0.158	3.21	0.052
Error	35	1.72	1.72	0.0490		
Total	38	2.20				
$S = 0.221$	$R^2 = 22.0\%$			R^2 (adj) = 15.3%		

Italics indicate model terms that are statistically significant at 0.05 level against pitting $E_b - E_r$ is not significantly affected by duration of exposure but is significantly affected by the dissolved oxygen concentration in the PBS. Thus, from this analysis we see that for EP samples:

1. E_r is greater for aerated solutions
2. E_b is similar for aerated and deaerated solutions
3. Both E_b and E_r increase with exposure time
4. $E_b - E_r$ is lower in aerated solutions
5. $E_b - E_r$ does not vary significantly with immersion time for aerated or deaerated solutions

Because aeration significantly affected $E_b - E_r$ and immersion time did not, estimates of P_2 are developed by fitting probability distributions to the 21 pitted specimens tested under deaerated conditions and to the 18 pitted specimens tested under aerated conditions. Compared to a single-population approach, analysis using this physically based mixture model gives a more realistic result of 1 in 1600 for the probability of pitting for EP exposed to aerated PBS versus 1 in 730,000 for specimens exposed to deaerated solutions. This can be compared to the probability for pitting of MP samples of 1 in 1500 in deaerated PBS (60-156 h) and 1 in 1,800,000 for MP exposed to aerated PBS (1-156 h).

It appears that the MP samples have a very high surface density of pit-initiation sites and that the breakdown potentials for these sites are reasonably well characterized by a normal distribution of voltages with a small standard deviation. The breakdown is most likely associated with defects in the oxide coating from mechanical polishing. The EP samples, however, appear to have a relatively low surface-defect density that results in pitting only once in 10 specimens. For the samples that pit, the distribution of the breakdown potentials are slightly higher than those observed for the MP samples. It is possible that once the defective oxide coating has been removed by electropolishing, the only defects that remain are where inclusions and cracks surrounding the inclusions intersect the surface. Shabalovskaya et al. investigated typical inclusions in NiTi and found that the inclusions are generally titanium rich as we observed (Ref 21, 22). They indicate that these titanium rich inclusions are generally Ti_2NiO, TiO_2, TiC, or oxidized TiC. The probability of inclusions, associated crevices, and voids intersecting the surface is, hopefully, a rare event and may explain why only one in seven of the EP samples failed. The intersection of inclusions and voids and cracks surrounding such inclusions has been used in the past to characterize the probability of fatigue failure (Ref 23) and we believe a similar process can be used to explain the pitting probabilities for NiTi once the more damaging/frequent oxide defects have been removed by electropolishing.

7. Conclusion

1. $E_b - E_r$ as determined from the current version of ASTM F 2129 which does not include pre-test exposures to aerated solutions and can over estimate the margin of safety of NiTi to pitting corrosion in vivo. This indicates that current acceptance testing of medical device implants (such as ASTM F 2129) should be modified or extended to include at least several days pre-test exposure of implant material to aerated physiological solutions prior to cyclic polarization testing using the E_r values at the end of the pre-test immersion and the E_b from the testing in deaerated PBS.
2. Acceptance criterion should be based on a calculation of the probability that the device will pit in vivo is sufficiently small. This can be done by cyclic polarization testing a sufficient number of devices that have been pre-exposed to aerated PBS for sufficient times. This data should then be used to calculate the probability of pitting in vivo, i.e. the probability that $E_b - E_r$ is negative.
3. The margin of safety ($E_b - E_r$) of MP NiTi decreased significantly after exposure to deaerated PBS at 37 °C. Modeling showed that with a pre-test exposure of 1 h E_b E_r decreased approximately 296 mV when exposed to aerated rather than deaerated solutions. This change is primarily a result of the increase in E_r rather than from a change in E_b. If testing is done with pre-exposure to aerated solutions, the inherent variability in $E_b - E_r$ is decreased and there is no need to account for the time the implant will be in service.
4. To achieve a high degree of confidence that the probability of in vivo pitting is acceptably small, the amount of testing required will depend on the population variation in breakdown and rest potentials. Small-scale pilot testing should be performed when this information cannot be obtained from other sources.
5. The pitting resistance ($E_b - E_r$) of EP NiTi is higher than MP NiTi. Only five of the EP NiTi samples broke down whereas all of the MP samples broke down. Using only the samples that broke down, the average $E_b - E_r$ was 0.774 V SCE for the EP NiTi and 0.586 V SCE for the MP NiTi. This, however, does not take into account the fact that over 90% of the EP samples did not break down.
6. MP NiTi breakdown potentials can be treated with a normal quality-assurance type of statistics and sampling theory. $E_b - E_r$ distributions can be used to predict the probability that $E_b - E_r$ from the sample will be negative or some pre-determined threshold value. The use of a threshold value can be used to account for the effect of drift in E_b and E_r with long-term immersion, or testing can be done on samples subjected to long-term immersion in aerated solutions to verify that $E_b - E_r$ has stabilized.
7. Quality-assurance testing and statistical analysis of EP NiTi, and other medical-device materials, with breakdown potentials frequently greater than 0.8 V SCE, is much more challenging. Here we propose a combined model in which the censored data (no break down data) is used to model the probability that any given sample may, or may not break down as a Poisson process, then model the breakdown potentials of the samples that have failed as a normal distribution.
8. The probability that NiTi will pit in vivo ($E_b - E_r < 0$) was estimated as:

   (a) 1 in 15,500 for MP exposed to deaerated PBS for 1 and 12 h (Gumbel)
   (b) 0 for MP exposed to aerated for 1-156 h (three-parameter Weibull)
   (c) 1 in 730,000 for EP exposed to deaerated for 1 h (mixture model)
   (d) 1 in 1,600 for EP exposed to aerated solutions for 1 h (mixture model)

9. The estimated probability of pitting in vivo is higher for EP samples pre-exposed to aerated PBS than for deaerated PBS primarily as a result of the increase in dissolved oxygen raising E_r.

References

1. B. Clarke, et al., Influence of Nitinol Wire Surface Treatment on Oxide Thickness and Composition and Its Subsequent Effect on Corrosion Resistance and Nickel Ion Release, *J. Biomed. Mater. Res. A*, 2006, **79**(1), p 61–70
2. L.E. Eiselstein, D.M. Proctor, and T.C. Flowers, Trivalent and Hexavalent Chromium Issues in Medical Implants, *Mater. Sci. Forum*, 2007, **539–543**, p 698–703
3. J.J. Jacobs, J.L. Gilbert, and R.M. Urban, Current Concepts Review - Corrosion of Metal Orthopaedic Implants, *J. Bone Joint Surg.*, 1998, **80**, p 268–282
4. "Standard Test Method for Conducting Cyclic Potentiodynamic Polarization Measurements to Determine the Corrosion Susceptibility of Small Implant Devices," *ASTM F 2129-06, ASTM International Standard F 2129-06*, Vol. 13.01, ASTM International, West Conshohocken, PA, 2006
5. L. Eiselstein, et al., Toward an Acceptance Criterion for the Corrosion Resistance of Medical Devices: A Statistical Study of the Pitting Susceptibility of Nitinol, *Proceedings of SMST-2007 the International Conference on Shape Memory and Superelastic Technologies*, 2007, ASM International, Tsukuba City, Japan
6. R.A. Corbett, Laboratory Corrosion Testing of Medical Implants, *Proceedings of Materials and Processes for Medical Devices Conference*, S. Shrivastava, Ed., ASM International, Materials Park, OH, 2004, p 166-171
7. R.A. Corbett and S.N. Rosenbloom, An Assessment of ASTM F 2129 Test Results Comparing Nitinol to Other Implant Alloys, *SMST 2006: Proceedings of the International Conference on Shape Memory and Superelastic Technologies*, 2006
8. S.N. Rosenbloom and R. Corbett, "An Assessment of ASTM F 2129 Electrochemical Testing of Small Medical Implants—Lessons Learned," Paper no. 07674, NACE 2007, Houston, TX, 2007
9. B.G. Pound, Susceptibility of Nitinol to Localized Corrosion, *J. Biomed. Mater. Res. A*, 2006, **77**(1), p 185–191
10. G. Frankel, Pitting Corrosion, *ASM Handbook*, Vol 13A, S.D. Cramer and B.S. Covino Jr., Eds. ASM International, Materials Park, OH, 2003, p 236
11. J.R. Scully and M.J. Kelly, Methods for Determining Aqueous Corrosion Reaction Rates, *ASM Handbook*, Vol 13A, S.D. Cramer and B.S. Covino Jr., Eds., ASM International, Materials Park, OH, 2003, p 68
12. D. Silverman, "Tutorial on Cyclic Potentiodynamic Polarization Technique," Paper no 299, CORROSION/98, NACE International, Houston, TX, 1998
13. D. Silverman, Practical Corrosion Prediction Using Electrochemical Techniques, in *Uhlig's Corrosion Handbook*, 2nd ed., R.W. Revie, Ed., Wiley, New York, 2000, p 1179–1225
14. "Standard Specification for Wrought Nickel-Titanium Shape Memory Alloys for Medical Devices and Surgical Implants," ASTM F 2063-05, Vol. 13.01, ASTM International, West Conshohocken, PA, 2005
15. B.J. Burrow and J. Moskito, "Auger Electron Spectroscopy (AES) Surface Analysis Report," Job Number: C07S6006, Evans Analytical Group, Sunnyvale, CA 94086, USA, 2007
16. "Implants for Surgery—Measurements of Open-Circuit Potential to Assess Corrosion Behaviour of Metallic Implantable Materials and Medical Devices Over Extended Time Periods," ISO 16429-04, Geneva, Switzerland. 2004
17. T. Shibata and W.R. Whitney, Award Lecture: Statistical and Stochastic Approaches to Localized Corrosio, *Corrosion*, 1996, **52**(11), p 813–830
18. M. Kowaka, *Introduction to Life Prediction of Industrial Plant Materials*, Allerton Press, Inc., New York, 1994
19. T. Shibata, Corrosion Probability and Statistical Evaluation of Corrosion Data, in *Uhlig's Corrosion Handbook*, R.W. Revie, Ed., Wiley, New York, 2000
20. W.J. Conover, *Practical Nonparametric Statistics*, 2nd ed., Wiley Series in Probability and Mathematical Statistics, Wiley, New York, 1980
21. S. Shabalovskaya, J. Anderegg, and J.V. Humbeeck, Recent Observations of Particulates in Nitinol, *Mater. Sci. Eng. A*, 2008, **481–482**, p 431–436
22. S. Shabalovskaya, et al., The Effect of Surface Particulates on the Corrosion Resistance of Nitinol Wire, *SMST-2003: The International Conference on Shape Memory and Superelastic Technologies*, 2003, p 399–408
23. J. Schaffer, A Probabilistic Approach to Modeling Microstructural Variability and Fatigue Behavior in ASTM F562 Medical Grade Wire, *Presented May 19th, 2006 at the 9th International Fatigue Congress*, Atlanta, USA. 2006

JMEPEG (2009) 18:781–786
DOI: 10.1007/s11665-009-9431-9

1059-9495/$19.00

Analysis and Evaluation of the Dynamic Performance of SMA Actuators for Prosthetic Hand Design

Kevin T. O'Toole, Mark M. McGrath, and Eugene Coyle

(Submitted September 12, 2008; in revised form March 13, 2009)

It is widely acknowledged within the biomedical engineering community that shape memory alloys (SMAs) exhibit great potential for application in the actuation of upper limb prosthesis designs. These lightweight actuators are particularly suitable for prosthetic hand solutions. A four-fingered, 12 degree-of-freedom prosthetic hand has been developed featuring SMA bundle actuators embedded within the palmar structure. Joule heating of the SMA bundle actuators generates sufficient torque at the fingers to allow a wide range of everyday tasks to be carried out. Transient characterization of SMA bundles has shown that performance/response during heating and cooling differs substantially. Natural convection is insufficient to provide for adequate cooling during elongation of the actuators. An experimental test-bed has been developed to facilitate analysis of the heat transfer characteristics of the appropriately sized SMA bundle actuators for use within the prosthetic hand design. Various modes of heat sinking are evaluated so that the most effective wire-cooling solution can be ascertained. SMA bundles of varying size will be used so that a generalized model of the SMA displacement performance under natural and forced cooling conditions can be obtained. The optimum cooling solution will be implemented onto the mechanical hand framework in future work. These results, coupled with phenomenological models of SMA behavior, will be used in the development of an effective control strategy for this application in future work.

Keywords advanced characterization, mechanical testing, superalloys

1. Introduction

Shape memory alloys (SMAs) have the potential to revolutionize the field of biomedical engineering, in particular, lower arm prosthesis design. Currently, most lower arm prostheses are driven by electromechanical DC motors (Ref 1-5). DC motors inhibit aspects of prosthesis design owing to their size and weight. As a result, traditional intelligent prosthetics tend to be bulky, heavy, and noisy. Consequently, designers have been focusing their attention on various alternative actuation techniques. SMAs exhibit favorable properties such as low weight, quiet operation, low cost, and high force to weight ratios. However, SMAs exhibit a significant hysteresis during the heating and cooling cycles which can pose problems in design, operation, and control.

A four-fingered hand prosthesis featuring embedded SMA bundle actuators has been designed by this group (Fig. 1) (Ref 6) requiring 0.66 W/150 μm SMA for maximum safe contraction rate. Within the prosthetic design, the bundle actuators are appropriately sized to give the force requirement at the various phalanges of the human hand. The goal of this work is to optimize the transient performance of the SMA actuators so that they can deliver a response comparable to that

This article is an invited paper selected from presentations at Shape Memory and Superelastic Technologies 2008, held September 21-25, 2008, in Stresa, Italy, and has been expanded from the original presentation.

Kevin T. O'Toole and **Mark M. McGrath**, School of Manufacturing and Design Engineering, Dublin Institute of Technology, Dublin, Ireland; and **Eugene Coyle**, School of Electrical Engineering Systems, Dublin Institute of Technology, Dublin, Ireland. Contact e-mail: kevin.otoole@dit.ie.

Nomenclature	
$\dot{E}_{in}$	rate of energy transfer into a volume, W
$\dot{E}_{g}$	rate of energy generation, W
$\dot{E}_{out}$	rate of energy transfer out of a volume, W
$\dot{E}_{st}$	rate of increase of energy stored in a volume, W
V	electrical voltage, V
I	electrical current, mA
R	electrical resistance of the wire, Ω
M	mass, kg
ξ	martensitic volume
x	displacement of SMA wire/bundle, mm
τ	time constant, s
t_{h90}	90% heating time, s
T	temperature, °C
ΔT	temperature difference ($T_w - T_\infty$), °C
A	surface area of SMA wire, m^2
h	heat transfer coefficient, W/m^2 K
k	thermal conductivity, W/m K
H	latent heat, kJ/kg
T_w	wire surface temperature, °C
T_∞	ambient temperature, °C
f	frictional force, N
c_p	specific heat at constant pressure, J/kg K
t_{c90}	90% cooling time, s
P	electrical power, W

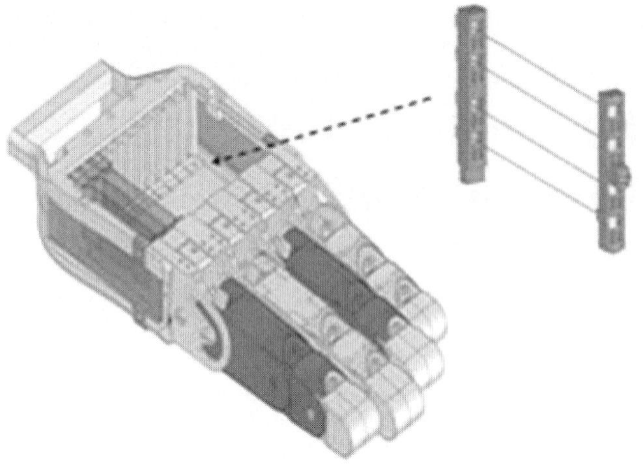

Fig. 1 SMA actuated hand prototype and SMA bundle

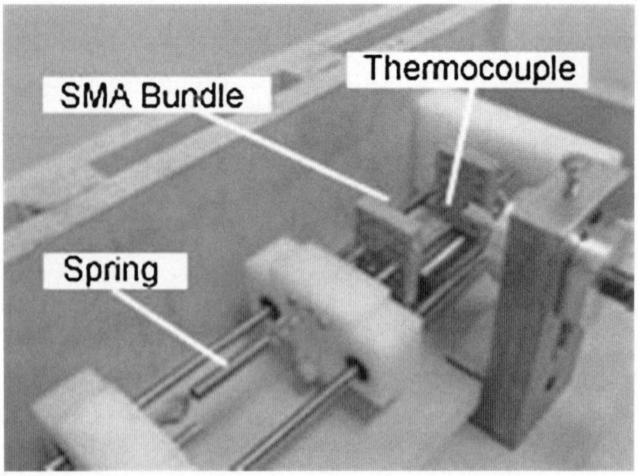

Fig. 2 SMA layout in test rig (*h*-value setup)

Table 1 SMA characterization: six stages

1	Strain characterization of SMA wire and SMA bundle actuator	Completed in previous work (Ref 7)
2	Heat transfer analysis of SMA wire and SMA bundle actuator	Current work
3	Analysis of SMA performance during heating/cooling cycle	Current work
4	Evaluation of forced cooling methods	Current work
5	Dynamic response of SMA bundle actuator during cyclic testing	Preliminary work completed
6	Fatigue testing of SMA bundle actuators	Future Work

of a human muscle performing everyday tasks. Six critical steps for full characterization of SMA wire bundles for use in prosthesis design are outlined in Table 1. This paper focuses on steps 2-4, which examine the relationship between heat transfer and strain response in SMAs. These characteristics are intrinsically linked, and by maximizing the heat transfer to and from the SMA, the strain response can be optimized. This is further illustrated in Section 4 of this work where the strain response is directly compared with the thermal response.

This paper is laid out as follows: Section 2 describes the design and manufacture of the test rig employed in this work. Section 3 analyses the heat transfer to/from single SMA wires and SMA bundle actuators. Section 4 explores the complex relationship among SMA wire temperature, strain response, and force generation in SMA bundle actuators. Section 5 outlines the testing of a suitable forced cooling technique with a view to minimizing the hysteresis associated with SMAs. Some conclusions as well as future work direction are outlined in Section 6.

2. Test Bed Development

A test rig was developed to facilitate testing of both single SMA wires and various SMA bundle sizes (Fig. 2). A cooling system module, developed as a modular attachment to the rig, provides for forced cooling (Fig. 7). The rig can be used in an open or enclosed state depending upon the nature and objectives of each individual experiment. Bundles of different sizes and springs of different stiffness (to provide the relaxation force required) can be inserted to cater for a range of experimental testing.

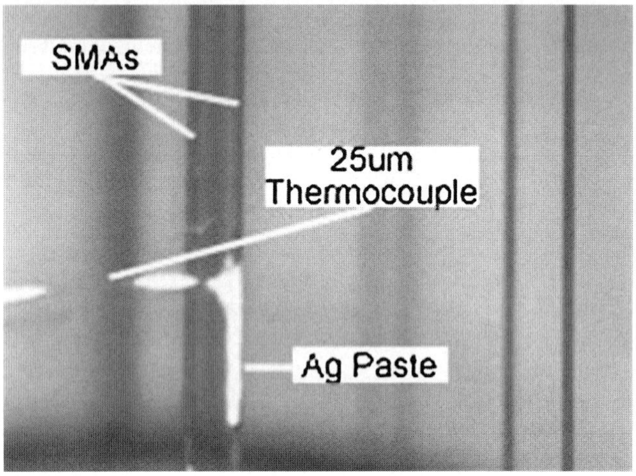

Fig. 3 Sensory system layout

Various sensors are used in the real-time detection of the changing physical phenomena during the experiments. Two K-type micro-thermocouples (25 μm) are used to measure wire temperature and ambient air temperature, respectively (Fig. 3). A Solartron LVDT is employed in the measurement of the rate of contraction and expansion of the SMA wire/bundle under thermal excitation. External noise is filtered using a fourth-order Butterworth filter. Current and voltage readings are also recorded. A NI SCC-68 DAQ is used to condition the outputs from each individual sensor. The results are displayed in NI LabVIEW and compiled in Matlab.

3. SMA Heat Transfer

3.1 Conservation of Energy and Heat Transfer Analysis

The heating and cooling cycles of SMA actuators follow the law of conservation of energy for a control volume. This law is defined as:

$$\dot{E}_{in} + \dot{E}_{g} + \dot{E}_{out} = \dot{E}_{st} \quad \text{(Eq 1)}$$

Szykowny et al. stated that the energy dissipation takes place mostly in the form of heat convection (Ref 8). Previous groups (Ref 9, 10) have shown that the SMA wire heat transfer model can be stated as:

$$VI - h(T)A[T_w - T_\infty] - m\Delta H\frac{d\xi}{dt} - f\frac{dx}{dt} = mC_p\frac{dT}{dt} \quad \text{(Eq 2)}$$

An additional term has been added to the LHS of (2) to account for the power required to overcome the damping effect of the test rig during transient SMA straining activity. By comparing (1) with (2), it can be concluded that during the transient stage of the SMA contraction:

$$\dot{E}_{in} = VI \quad \text{(Eq 3)}$$

$$\dot{E}_g = 0 \quad \text{(Eq 4)}$$

$$\dot{E}_{out} = h(T)A[T_w - T_\infty] + m\Delta H\frac{d\xi}{dt} + f\frac{dx}{dt} \quad \text{(Eq 5)}$$

$$\dot{E}_{st} = mC_p\frac{dT}{dt} \quad \text{(Eq 6)}$$

When the steady-state stage of the cycle is reached, the heat transfer equation can be reduced to

$$\dot{E}_{in} = VI = hA[T_w - T_\infty] \quad \text{(Eq 7)}$$

The transient performance must be optimized to maximize the dynamic performance of the SMA bundle actuator. Previous work by this group has shown that the heating and cooling time responses of SMA bundles differ significantly (Ref 7). Measures must be taken to move toward comparable performance during each stage of the cycle. This can be accomplished by improving the rate of cooling (Section 5).

The heat transfer coefficient, h, is an important consideration when completing the characterization of single SMA wires and SMA wire bundles. This is particularly important in prosthesis design as the SMA bundles are located in an enclosed environment (within the hand). The additional heat convected into this environment by the SMA bundles must be accounted for and removed so that adequate control of actuation can be achieved. Szykowny et al. have suggested that the value of h for a single SMA wire decreases with increasing temperature in an open environment (Ref 8). They have also demonstrated that the calculation of h for a single SMA wire using established analytical methods will yield large error margins suggesting that these methods are unsuitable. The current work aims to establish a correlation between h for a single SMA wire and a bundle of SMA wires placed in a parallel arrangement (Fig. 2 and 4).

3.2 Determination of Heat Transfer Coefficient for a SMA Bundle

The initial testing focused on establishing h for a single SMA wire. This was carried out with the test rig fully sealed to maintain a steady ambient temperature of approximately 22 °C. The wire surface temperature and ambient temperature were continually recorded. Upon completion, the 8-wire SMA bundle (Fig. 2) was loaded into the rig and testing was repeated. Wire surface temperature was taken at one point on the SMA wire located at the top left of the bundle. Ambient temperature was recorded 200 mm from the SMA bundle,

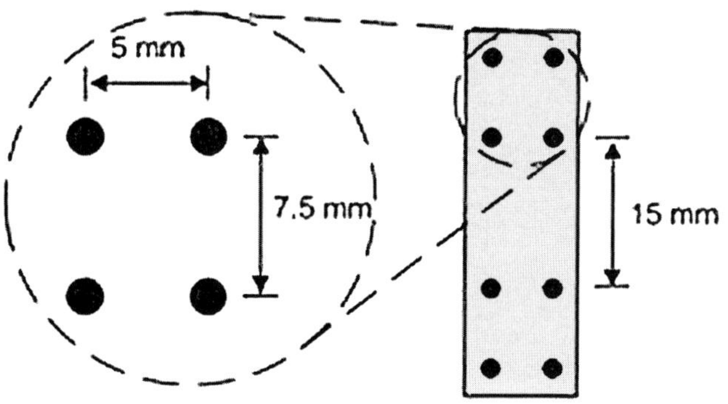

Fig. 4 SMA bundle dimensions

which is well outside the expected thermal boundary region of the SMA bundle. The h value can then be determined from Eq 8.

$$h = \frac{VI}{A[T_w - T_\infty]} \quad \text{(Eq 8)}$$

3.3 Results

The h value was found to increase slightly in a linear fashion as input power is increased. The h values established for the single wire and wire bundle are similar in the SMA active region (200-400 mA) with values in the range of 200-250 W/m^2 K, and followed the linear relationship shown in (9).

$$h = 11.6P + 180 \quad \text{(Eq 9)}$$

The results indicate that the thermal boundary regions of each individual SMA wire within the bundle do not interact substantially with the bundle dimensions used in this work. The force generated remains eight times that of a single SMA wire. The results demonstrate that the thermal analysis of SMA bundles can be carried out by considering the parallel wires to be lined up in series. This approach can only be used if the physical distance between adjacent wires is greater than the thermal boundary layer thickness for each SMA wire in the bundle.

4. Relationship Between Thermal Response and Strain Response

4.1 Fixed Bundle Size—Varying Current

It is important to note that as the SMAs undergo transformation from the martensitic state to the austenitic state, the electrical resistance changes due to the differing properties of martensite and austenite. This change is relatively small and as a result, the electrical resistance of an SMA bundle with a fixed number of wires can be considered constant during the following analysis. Experimental results suggest that

$$\frac{d(\Delta T)}{dt} = f(P) \quad \text{(Eq 10)}$$

Figure 5 shows the temperature response of the 8-wire SMA bundle. It can be seen that the time taken to reach the initial

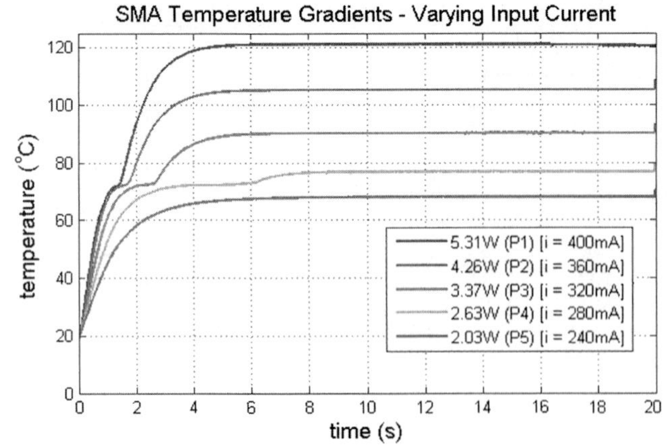

Fig. 5 SMA temperature gradients

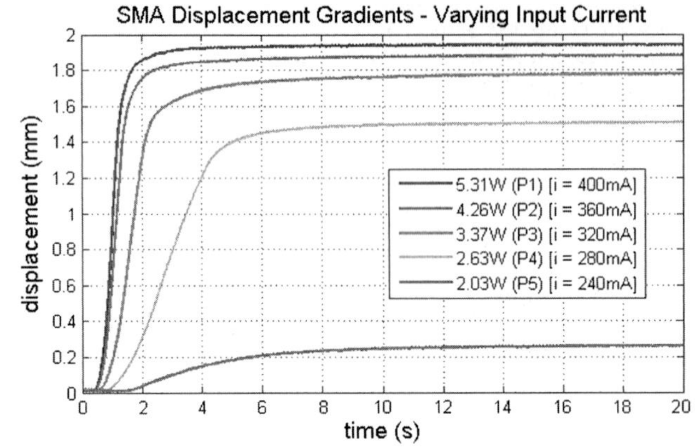

Fig. 6 SMA displacement gradients

point of transformation decreases as energizing current increases. However, by considering the total temperature profile as an approximated first-order response, it is shown that the time constant value for the temperature rise, $\tau(T)$, remains constant for changing values of *P*.

$$\tau_{P1}(T) = \tau_{P2}(T) \cdots\cdots = \tau_{P5}(T) \quad \text{(Eq 11)}$$

In the case outlined in Fig. 5, $P1 > P2 \cdots > P5$, so it can be stated that

$$\frac{d(\Delta T_1)}{dt} > \frac{d(\Delta T_2)}{dt} \cdots\cdots > \frac{d(\Delta T_5)}{dt} \quad \text{(Eq 12)}$$

Since the SMA wires are thermally activated, the rate of contraction and expansion is a function of the rate of heating.

$$\frac{dx}{dt} = f\left(\frac{d(\Delta T)}{dt}\right) \quad \text{(Eq 13)}$$

Therefore, it can be stated that for different levels of power, the 90% heating response time for contraction, t_{h90} varies. This is outlined in Fig. 6 where

$$t_{h90}(P1) < t_{h90}(P2) \cdots\cdots < t_{h90}(P5) \quad \text{(Eq 14)}$$

Since the contraction and expansion of the SMA wires will be used in a novel prosthetic hand design in the generation of grip and release forces via a mechanical framework (Ref 6), generated force, *F*, must also be considered. Analysis of the test bed shows that *F* is a time dependent function of *x* and it is envisaged that

$$\frac{dF}{dt} = f\left(\frac{d(\Delta T)}{dt}\right) \quad \text{(Eq 15)}$$

Equation 15 demonstrates that the higher the rate of heat transfer to/from the SMA bundle within the prosthetic hand, the faster the contraction time and/or the force generation time.

4.2 Fixed Current—Varying Bundle Size

With *I* remaining constant, from (7) it can be stated that:

$$\Delta T \propto R \quad \text{(Eq 16)}$$

It is clear (Eq 16) that the greater the number of SMA wires per bundle, the greater the overall resistance to current. This holds true as the wires are connected electrically in series, while they are mounted mechanically in parallel. This results in a faster heating rate (and requires a greater input voltage). Therefore, it can be concluded that the greater the number of parallel SMA wires in a bundle, the faster the contraction time. In order for a prosthetic device to give a consistent output from each individual phalanx of the hand, differing rates of contraction are unsuitable. As a result, $\dot{E}_{\text{in}}$ must be suitably controlled to give consistent and efficient performance. From Ohm's law, it can be seen that by developing a controller that varies the voltage, thereby allowing the current value to remain constant, heating rates can be regulated in SMA bundles across various bundle sizes.

5. Forced Cooling of SMA Bundles

Previous work has shown that the cooling time of the SMA bundles from a strain perspective is significantly larger than that during the heating phase. This suggests that investigation into methods which improve the cooling time of the SMAs is warranted, particularly in prosthetic actuation applications. Forced cooling can be implemented using a variety of techniques including conduction (heat sinking, Ref 11), convection (forced air cooling and liquid cooling, Ref 12), and radiation (Peltier cooling, Ref 13). The nature of a prosthetic device dictates that many forced cooling techniques are rendered unsuitable for use. A successful cooling system for such a device must be cheap, portable, reusable, and safe. These criteria eliminate convection and radiation cooling strategies, leaving conductive cooling in the form of metallic heat sinks, applied at suitable timed intervals, as the most suitable strategy.

Heat sinks were manufactured from three materials, aluminum ($mc_p = 18.3$ J/K), Brass ($mc_p = 13$ J/K), and Copper ($mc_p = 12.4$ J/K). Each heat sink was individually tested using the cooling module. During the cooling stage, rate of conduction is relatively constant owing to the relatively high heat capacities of the heat sinks. Heat transfer to the sinks follows Fourier's law (17) with convective heat loss being ignored in this analysis.

$$\dot{E}_{\text{in}} = -kA\left(\frac{d(\Delta T)}{dx}\right) \quad \text{(Eq 17)}$$

5.1 Experimental Testing

5.1.1 Test 1. Determination of optimal heat-sinking material—Individual tests were completed using a different heat-sinking material each time. Testing was carried out in a

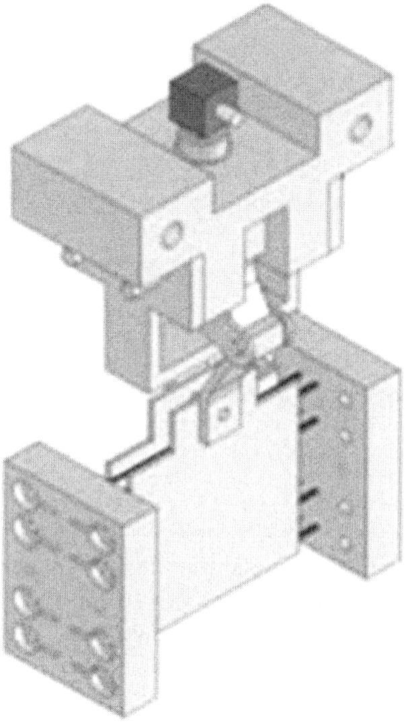

Fig. 7 Cooling module attachment (closed)

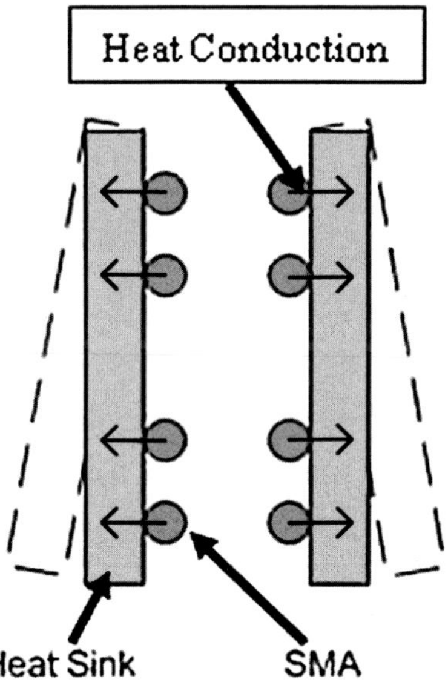

Fig. 8 Heat conduction using heat sink

cyclic manner with the first half of the cycle being the heating period. Heat is delivered to the SMA via Joule heating until maximum contraction has occurred. Upon reaching this value, the current is switched off while a heat sink is simultaneously applied to the SMA wires to initiate the cooling phase (Fig. 7 and 8). The heat sink acts to conduct heat away, thus improving the rate of cooling. A test without heat sinks (natural cooling) was also carried out and used as an experimental control.

5.1.2 Test 2. Assessment of chosen heat sink with variable input power—The most suitable heat sink was tested further by heating the SMA bundles using different energizing currents. The heat sinks were then applied to cool the bundle. This helps in establishing the cooling times for various SMA steady-state temperatures.

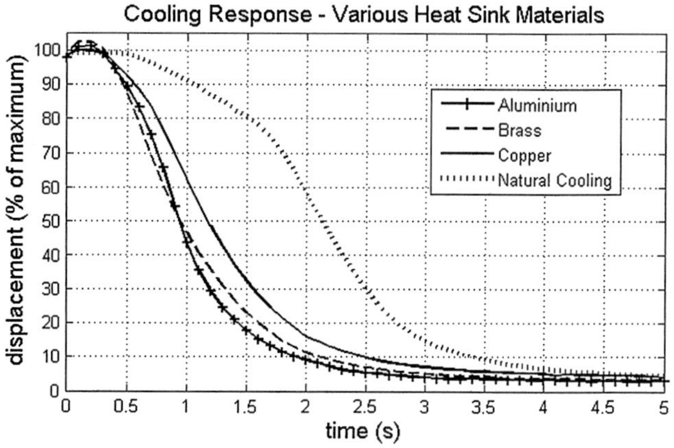

Fig. 9 Cooling response due to heat sinks

5.1.3 Test 3. Analysis of complete cycle time—Cooling of SMAs using the most suitable heat sink was compared to cooling via natural convection. The time to complete a full actuation cycle for the SMA bundle subjected to forced cooling was measured and compared to the complete cycle time of the SMA bundle undergoing natural cooling.

5.2 Results

The results from Test 1 show that the cooling time is reduced significantly through the use of heat sinks (Fig. 9). Aluminum proved to be the most beneficial, reducing the 90% cooling time, t_{c90}, to approximately 1.9 s, compared to 3.4 s for natural cooling. Brass exhibited a t_{c90} value of 2.2 s and Copper produced a t_{c90} value of 2.45 s. In addition to aluminum proving to be the most effective heat sink, it also exhibits additional favorable properties such as low weight and low cost. This makes it suitable for prosthetic applications. Test 2 was carried out using currents up to a maximum of 400 mA, which is rated as 100% of the recommended current for a 150 μm Nitinol wire by the manufacturers (Ref 14). It is demonstrated that the 90% cooling time has a maximum value of 1.9 s with the application of aluminum heat sinks (assuming excess heat is dissipated from the heat sinks) and decreases as energizing currents decrease. The following relationship is observed:

$$I = 157.5(t_{c90})^2 - 9.2(t_{c90}) + 220 \quad \text{(Eq 18)}$$

Potapov and Da Silva (Ref 10) stated that cooling time places limitations on the maximum possible frequency of actuation. To increase the frequency of actuation, cooling times must be decreased. Heating rate is a function of energizing current and SMA wire resistance, and as such cannot be increased further with fixed actuator parameters. The results of a cycle time analysis are illustrated in Fig. 10. The results show that the frequency of actuation can be significantly improved using heat sinks during the cooling phase. The heating phase remains unaffected. It is shown that the use of aluminum heat sinks permits full actuation cycles at 0.3 Hz, in comparison to natural cooling, where actuation cycles are observed at a maximum rate of 0.18 Hz.

It is predicted that during continuous dynamic cycles, heat is intermittently conducted regularly from the SMA wires to the heat-sinking material, and will result in an increase of temperature of the heat sink over time. This will lead to the

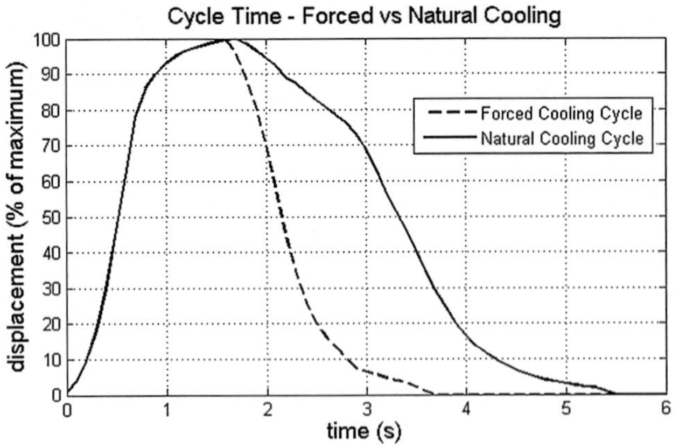

Fig. 10 Forced (Al) and natural cooling

overall temperature difference (ΔT) diminishing, decreasing the capacity of the heat sink to conduct heat away from the SMAs. This will be the focus of future work by this group.

6. Conclusions and Future Work

An analysis of heat transfer and the relationship among temperature, displacement, force generation, and electrical power of single SMAs and SMA bundles has been completed. The conductive forced cooling technique of heat sinking was tested with the SMA bundle in an attempt to improve the dynamic performance of the actuator.

SMAs are thermally activated devices. It is known that a small thermal boundary region encompasses each individual wire. When SMA wires are placed in a parallel bundle formation, the degree of interaction, if any, depends on the closeness of adjacent wires. The results for the eight wire bundle described within this work demonstrate that there is no interaction between the wires as the convective heat transfer coefficient remains relatively unchanged for a single SMA wire and a bundle of SMA wires at steady-state conditions. It can, therefore, be concluded that during the analysis of heat transfer during the transient portion of the heating and cooling cycle, a bundle of parallel SMA wires can be represented as a single SMA wire, with a length equal to the total length of all the parallel SMAs added together. Increasing the number of wires will offset losses in force due to any mechanical inefficiency in the prototype but will result in higher power requirements. Forced cooling via metallic heat sinks of SMAs has been shown to significantly improve the transient response during the cooling (expansion) stage.

Immediate future work will focus on dynamic analysis of the heat sink under long-term cyclic loading. The long-term focus will be on the development of a suitable controller for the effective control of the SMA bundle in the rig described, using the values and relationships developed herein.

References

1. H. Huang, L. Jiang, D.W. Zhao, J.D. Zhao, H.G. Cai, H. Liu, P. Meusel, B. Willberg, and G. Hirzinger, The Development on a New Biomechatronic Prosthetic Hand Based on Under-Actuated Mechanism, *Proceedings of IROS 2006*, Beijing, China, 2006, p 3791–3796
2. L.-R. Lin and H.-P. Huang, NTU Hand: A New Design of Dexterous Hands, *J. Mech. Des. Trans. ASME*, 1998, **120**, p 282–292
3. J. Butterfass, G. Hirzinger, S. Knoch, and H. Liu, DLR's Multisensory Articulated Hand, *IEEE International Conference on Robotics and Automation*, 1998, p 2081–2086
4. C.S. Lovchik and M.A. Diftler, The Robonaut Hand: A Dexterous Robot Hand for Space, *IEEE International Conference on Robotics and Automation*, 1999, p 907–912
5. J.L. Pons, E. Rocon, R. Ceres, D. Reynaerts, B. Saro, S. Levin, and W. Van Moorleghem, The MANUS-HAND Dextrous Robotics Upper Limb Prosthesis: Mechanical and Manipulation Aspects, *Autonom. Robots*, 2004, **16**, p 143–163
6. K.T. O'Toole and M.M. McGrath, Mechanical Design and Theoretical Analysis of a Four Fingered Prosthetic Hand Incorporating Embedded SMA Bundle Actuators, *Int. J. Math. Phys. Eng. Sci.*, 2008, **1**, p 83–90
7. K.T. O'Toole, M.M. McGrath, and D.W. Hatchett, Transient Characterisation and Analysis of Shape Memory Alloy Wire Bundles for the Actuation of Finger Joints in Prosthesis Design, *Mechanika*, 2007, **6**, p 65–69
8. S. Szykowny and M. H. Elahinia, Heat Transfer Analysis of Shape Memory Alloy Actuators, *Proceedings of 2006 ASME International Mechanical Engineering Congress & Exposition (ICEME 2006)*, Chicago, IL, United States, 2006, p. 8
9. A. Bhattacharyya, D.C. Lagoudas, Y. Wang, and V.K. Kinra, On the Role of Thermoelectric Heat Transfer in the Design of SMA Actuators: Theoretical Modeling and Experiment, *Smart Mater. Struct.*, 1995, **4**, p 252–263
10. P.L. Potapov and E.P. Da Silva, Time Response of Shape Memory Alloy Actuators, *J. Intell. Mater. Syst. Struct.*, 2000, **11**, p 125–134
11. R.A. Russell and R.B. Gorbet, Improving the Response of SMA Actuators, *Proceedings of the 1995 IEEE International Conference on Robotics and Automation*, Nagoya, Japan, 1995, p 2299–2304
12. S.A. Mascaro and H.H. Asada, Wet Shape Memory Alloy Actuators for Active Vasculated Robotic Flesh, *Proceedings of the 2003 IEEE International Conference on Robotics and Automation, ICRA 2003*, September 14–19, 2003, Taipei, Taiwan, 2003, p 282–287
13. J. Abadie, N. Chaillet, and C. Lexcellent, An Integrated Shape Memory Alloy Micro-Actuator Controlled by Thermoelectric Effect, *Sensors Actuators A Phys.*, 2002, **99**, p 297–303
14. Dynalloy, Flexinol, Available in http://www.dynalloy.com. Accessed 21 July 2008

JMEPEG (2009) 18:787–792
DOI: 10.1007/s11665-009-9408-8

1059-9495/$19.00

Nitinol Embolic Protection Filters: Design Investigation by Finite Element Analysis

Michele Conti, Matthieu De Beule, Peter Mortier, Denis Van Loo, Pascal Verdonck, Frank Vermassen, Patrick Segers, Ferdinando Auricchio, and Benedict Verhegghe

(Submitted October 16, 2008; in revised form February 17, 2009)

The widespread acceptance of carotid artery stenting (CAS) to treat carotid artery stenosis and its effectiveness compared with surgical counterpart, carotid endarterectomy (CEA), is still a matter of debate. Transient or permanent neurological deficits may develop in patients undergoing CAS due to distal embolization or hemodynamic changes. Design, development, and usage of embolic protection devices (EPDs), such as embolic protection filters, appear to have a significant impact on the success of CAS. Unfortunately, some drawbacks, such as filtering failure, inability to cross tortuous high-grade stenoses, malpositioning and vessel injury, still remain and require design improvement. Currently, many different designs of such devices are available on the rapidly growing dedicated market. In spite of such a growing commercial interest, there is a significant need for design tools as well as for careful engineering investigations and design analyses of such nitinol devices. The present study aims to investigate the embolic protection filter design by finite element analysis. We first developed a parametrical computer-aided design model of an embolic filter based on micro-CT scans of the Angioguard™ XP (Cordis Endovascular, FL) EPD by means of the open source pyFormex software. Subsequently, we used the finite element method to simulate the deployment of the nitinol filter as it exits the delivery sheath. Comparison of the simulations with micro-CT images of the real device exiting the catheter showed excellent correspondence with our simulations. Finally, we evaluated circumferential basket-vessel wall apposition of a 4 mm size filter in a straight vessel of different sizes and shape. We conclude that the proposed methodology offers a useful tool to evaluate and to compare current or new designs of EPDs. Further simulations will investigate vessel wall apposition in a realistic tortuous anatomy.

Keywords Angioguard™, carotid artery stenting, embolic protection device, finite element analysis, nitinol, wall apposition

1. Introduction

Cardiovascular disease (CVD) is the main cause of death in Western countries. Each year CVD causes over 4.35 million deaths in Europe (accounting for 49% of all deaths in 2005) (Ref 1).

CVD is often related to atherosclerosis, a degeneration of the vessel wall which has the main consequence of narrowing the vessel lumen, i.e. a stenosis, leading to blood flow reduction or blockage.

This article is an invited paper selected from presentations at Shape Memory and Superelastic Technologies 2008, held September 21-25, 2008, in Stresa, Italy, and has been expanded from the original presentation.

Michele Conti and **Ferdinando Auricchio**, Universitá degli Studi di Pavia, Pavia, Italy; **Matthieu De Beule, Peter Mortier, Pascal Verdonck, Patrick Segers,** and **Benedict Verhegghe**, IBiTech, Institute of Biomedical Technology, Ghent University, Ghent, Belgium; **Denis Van Loo**, UGCT, Center for X-ray Tomography, Ghent University, Ghent, Belgium; and **Frank Vermassen**, Department of Vascular Surgery, Ghent University Hospital, Ghent, Belgium. Contact e-mail: michele.conti@unipv.it.

At present, the deployment of an intravascular stent has become a common and widely used minimally invasive treatment for stenotic arteries, including mainly coronary arteries but also peripheral stenotic vessels such as the carotid arteries.

In recent years, carotid artery stenting (CAS) has emerged as a less invasive treatment as compared to the conventional surgical approach, carotid endarterectomy (CEA) (Ref 2), but the efficacy of CAS relative to surgery is still a matter of debate (Ref 3).

One major concern related to CAS is the possible distal embolization potentially leading to stroke or other severe neurological complications. Embolization is due mainly to the plaque debris and thrombi generated during the stenosis dilatation and stent apposition. Consequently, embolic protection devices (EPDs) have been developed to capture such released debris and appear to have a significant impact on the success of CAS (Ref 4-8).

EPDs can be classified in the following three categories (Ref 9):

- distal balloon occlusion;
- proximal balloon occlusion;
- embolic protection filters (EPFs).

The balloon occlusion systems block the flow in the internal carotid artery (ICA) and emboli are aspirated before balloon deflation and catheter removal, whereas EPFs maintain blood flow while emboli are captured.

Such a class of filters most often consists of a metallic basket-like structure coated with a membrane made of polymeric material containing numerous pores. Filters are usually mounted on a 0.014 in. (0.036 mm) guidewire, generally 30 mm proximal to a flexible tip and are delivered through a very small profile catheter (<3 French, i.e. 1 mm). During a CAS procedure, before the lesion dilatation and the stent apposition, the filter is opened in the ICA lumen distal to the target lesion, by withdrawing the delivery sheath. At the end of the CAS procedure, a retrieval catheter is advanced onto the guidewire to capture and remove the filter (Ref 10).

EPFs have both advantages and disadvantages with respect to balloon occlusion procedures. On one hand, EPFs have the advantage of preserving cerebral flow throughout the procedure while capturing embolic debris allowing distal perfusion and angiography during the CAS procedure. On the other hand, filters have disadvantages ranging from difficulty in navigating severely stenosed or tortuous vessels to possible incorrect filter apposition against the vessel wall (Ref 11, 12). Currently, several embolic filter designs are available on a fast growing dedicated market (Ref 11).

Because of the relevant differences in EPF designs, it is likely that there are situations in which individual EPFs may be better suited to avoid adverse events during CAS and, consequently, the investigation and the comparison of the performance of different EPFs can play a key role in the improvement of filtering outcomes and potentially on the complete CAS procedure.

Engineering investigation and computer models have shown the capability to provide useful information to understand the mechanics of cardiovascular devices, e.g. stents, and to improve device design (Ref 13-16). To the best of our knowledge, the available comparative investigations about advantages or roles that specific EPDs may have in CAS are mainly based on in vitro, ex vivo, or clinical studies (Ref 12, 17-20).

The purpose of this study is to use finite element analysis (FEA) in the design evaluation of basket-like embolic protection filters. In particular, we investigated the effects of sizing on circumferential basket-vessel wall apposition of a widely adopted embolic filter, i.e. the Angioguard™ XP (Cordis Endovascular, FL).

In order to achieve this purpose, we organized the study as follows:

- Using a script-based procedure, we created a 3D finite element model of the Angioguard™ XP taking into account several geometrical features (such as number of struts, strut diameter, filter length, filter diameter, etc.).
- To validate the model, we qualitatively compared the numerical prediction of a free filter expansion, i.e. as it exits out of the delivery sheath, with a micro-CT scan image of the real device deployment.
- To evaluate the circumferential basket-vessel wall apposition, we simulated the deployment of a 4 mm size Angioguard™ XP in a vessel having different sizes and shape.

2. Materials and Methods

2.1 Model Geometry

The Angioguard™ XP consists of a 0.014 in. (0.036 mm) guidewire with a nitinol basket frame enveloped by a porous

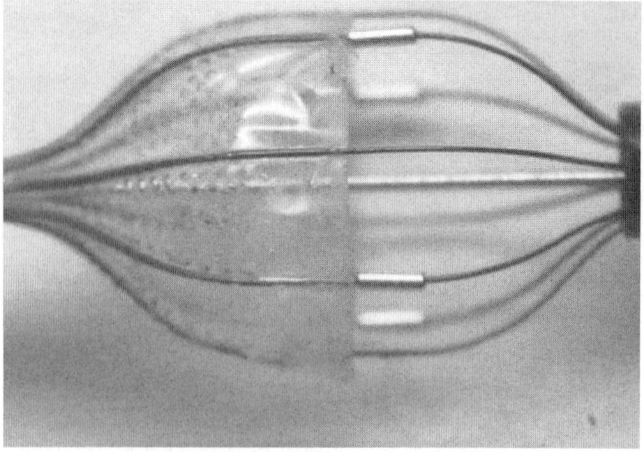

Fig. 1 The Angioguard™ XP embolic protection device

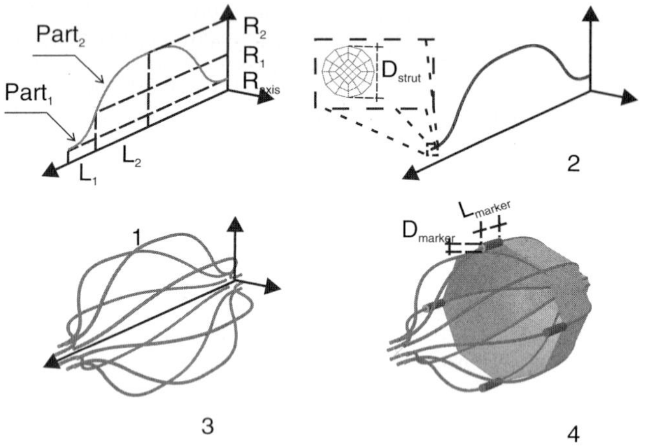

Fig. 2 Example of filter model generation in four steps

polymeric membrane (Fig. 1). The current Angioguard™ XP basket sizes range from 4 to 8 mm in diameter in order to deploy within vessels with diameters ranging from 3.0 to 7.5 mm (Ref 11, 21).

We used the in-house developed pyFormex software (version 0.6.5) (Ref 22) as a pre-processing tool to build a geometrical and finite element model of the Angioguard™ XP.

The model includes the following parts:

- filter struts defining the basket frame;
- membrane;
- markers.

The mesh generation procedure was defined by four essential steps (Fig. 2):

1. creation of path reproducing the shape of one filter strut;
2. sweeping of planar section along the defined path in order to create one 3D filter strut;
3. replication and rotation of the filter strut around the filter axis in order to create the whole basket frame;
4. definition of the whole model assembling the filter struts with the membrane and markers.

From a geometrical point of view, the model is defined by several parameters as depicted in Fig. 2 and reported in Table 1.

The pyFormex script-based mesh generation procedure allows the creation of a finite element model in a quick way facilitating parametrical finite element analyses; moreover, pyFormex generates a finite element mesh that can be easily imported into several commercial finite element solvers. In this study, we used ABAQUS/EXPLICIT 6.8 (Abaqus Inc., Providence, RI).

In order to assess accurately the value of the defined geometrical parameters, a micro-CT scan of a real device in open configuration was performed and the image was analyzed by Materialise miniMagics V12.0.5.1 (see Table 1).

2.2 Material Properties

The filter struts defining the basket frame are manufactured from nitinol, a nickel-titanium-based alloy showing the so-called superelastic effect (SE). A typical superelastic nitinol behavior is shown in Fig. 3 illustrating the ability to undergo large deformations (up to 6-8%) during loading and to fully recover such deformations during unloading. Clearly, the filter exploits this superelastic effect during both the delivery and retrieval phase, allowing the device to switch from a closed configuration to an open configuration and vice versa without experiencing plastic deformation.

In our simulations, nitinol superelasticity is modeled as a predefined user material in the ABAQUS/EXPLICIT 6.8 solver.

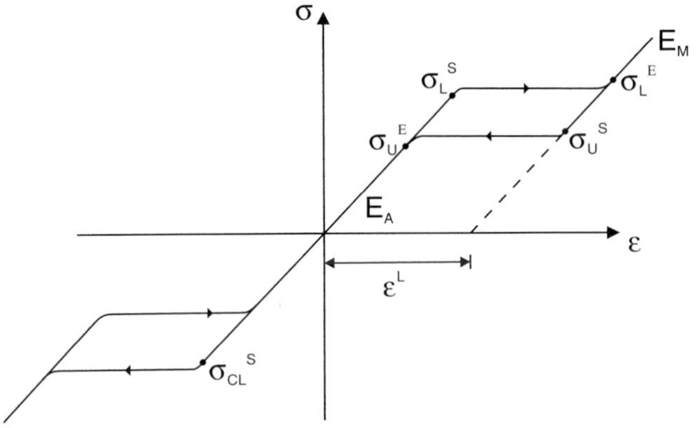

Fig. 3 Typical superelastic nitinol behavior

Table 1 Geometrical parameters of 4.0 mm filter model

Symbol	Description	Value
N_{struts}	Number of struts	8
R_{axis}	Distance of strut centerline from filter axis	0.3 mm
R_1	Filter radius (part 1)	0.68 mm
R_2	Filter radius (part 2)	1.02 mm
$R_{strut} = (R_{axis} + R_1 + R_2)$	Filter radius	2 mm
L_1	Filter length (part 1)	1.49 mm
L_2	Filter length (part 2)	2.76 mm
$L_{strut} = (2 * (L_1 + L_2))$	Filter length	8.5 mm
D_{strut}	Strut diameter	0.08 mm
$\text{Membrane}_{coverage}$	Percentage of filter coverage	50%
Membrane_{th}	Membrane thickness	0.02 mm
L_{marker}	Marker length	0.8 mm
D_{marker}	Marker diameter	0.22 mm

The polymeric porous membrane covering the filter frame was assumed to be manufactured in a Nylon-based material and having 0.02 mm thickness (Ref 14). The holes in the membrane were neglected.

The markers, attached to the nitinol frame to enhance filter radiopacity, were assumed to be manufactured by pure tantalum (Ref 23).

Table 2 reports the adopted material properties employed in the simulations.

2.3 Filter-Free Expansion

To validate the developed finite element model, we simulate the filter-free expansion, i.e. as it exits out of the delivery sheath. Consequently, the numerical analysis of this free expansion of filter is a non-linear problem involving large deformation and contact problems in the numerical analysis.

For this purpose, large deformation analyses were performed using the ABAQUS/EXPLICIT solver, particularly quasi-static procedures were used assuming that inertia forces do not dominate/change the analysis. Kinetic energy was monitored to ensure that the ratio of kinetic energy to internal energy remains less than 10%. Moreover, a mass scaling strategy was used to reduce computational cost.

All finite element simulations were carried out on an in-house built low-cost, high-performance computing cluster (Ref 24).

The delivery sheath was assumed to be a straight rigid cylinder having a diameter equal to 1.06 mm and a length of 15 mm.

Table 2 Material properties used in the simulations

Nitinol (Ref 28)		
E_A	Austenite Young's modulus	35877 MPa
ν_A	Austenite Poisson's ratio	0.33
E_M	Martensite Young's modulus	24462 MPa
ν_M	Martensite Poisson's ratio	0.33
ε^L	Transformation strain	0.0555
$(d\sigma/dT)_L$	Loading temperature derivative of stress	0 MPa/°C
$\sigma^S{}_L$	Loading start of transformation stress	489 MPa
$\sigma^E{}_L$	Loading end of transformation stress	572 MPa
T_0	Temperature	22 °C
$(d\sigma/dT)_U$	Unloading temperature derivative of stress	0 MPa/°C
$\sigma^S{}_U$	Unloading start of transformation stress	230 MPa
$\sigma^E{}_U$	Unloading end of transformation stress	147 MPa
$\sigma^E{}_{CL}$	Start of transformation stress (compression)	...
ρ_{strut}	Density (a)	6.5 g/cm^3
Nylon (Ref 14)		
E_{membr}	Modulus of elasticity	920 N/mm^2
ν_{membr}	Poisson's ratio	0.4
ρ_{membr}	Density	1.1 g/cm^3
Tantalum (Ref 23)		
E_{marker}	Modulus of elasticity	179000 MPa
ν_{marker}	Poisson's ratio	0.35
ρ_{marker}	Density	16.6 g/cm^3

(a) http://www.shape-memory-alloys.com/data_nitinol.htm

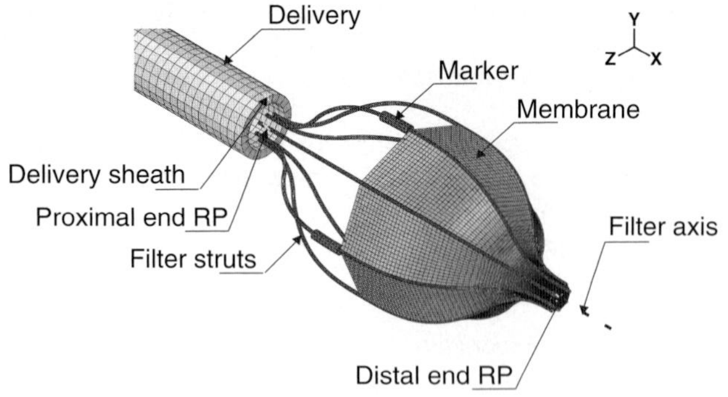

Fig. 4 Finite element model of Angioguard™ XP and part of the delivery sheath

The parts defining filter/catheter systems (see Fig. 4) were modelled as follows:

- *filter struts*: 33,792 three-dimensional 8-node brick 'reduced-integration' elements (C3D8R);
- *membrane*: 8320 three-dimensional 4-node quadrilateral membrane elements with reduced integration (M3D4R);
- *markers*: 1296 three-dimensional 8-node brick 'reduced-integration' elements (C3D8R);
- *catheter*: 3806 three-dimensional 4-node bilinear quadrilateral rigid elements (R3D4).

Mesh tie constraints of type surface-to-surface were applied to tie the markers and the membrane to the filter struts.

Furthermore, the filter frame struts are tied to marker bands defining the relative movement of the filter ends with respect to the guidewire (Ref 21), particularly the proximal marker band is fixed to the guidewire while the distal marker band is free to slip along the guidewire. Therefore, mesh tie constraints, type rigid body, are imposed between the nodes lying on each filter frame end and the predefined reference points. Boundary conditions are then applied to the reference points (RPs) to realistically simulate the movement of the marker bands. In particular, at the proximal end RP, all the degrees of freedom were constrained; at the distal end RP, only displacement along filter axis and rotations were allowed.

The analysis strategy assumes that the filter frame deformation during deployment is comparable to the deformation during the filter frame insertion into the delivery sheath. Consequently, a progressive rigid translation along the filter axis is imposed to the delivery sheath in order to induce the necessary filter deformation to switch from the open configuration (filter out of delivery sheath) to the closed configuration (filter within the delivery sheath).

A frictionless general contact algorithm has been used in order to handle the interactions between the filter frame and the deployment sheath. In this case, the membrane elements were excluded from the contact strategy in order to simplify the simulations assuming that it has a minor role in the overall filter deployment/retrieval mechanics.

2.4 Circumferential Basket-Vessel Wall Apposition

In order to evaluate the circumferential basket-vessel wall apposition, we carried out the simulations of a filter deploying in a straight vessel modelled as a rigid cylinder having different sizes and shape. In particular, two different scenarios were investigated: (i) filter expansion in three circular vessels having respectively 3.0, 3.25, and 3.5 mm diameter as suggested by the manufacturer for a 4 mm size filter (Ref 21); (ii) filter expansion in two oval vessels having both 1.5 mm major semi axis and respectively an ovality* of 0.85 and 0.75.

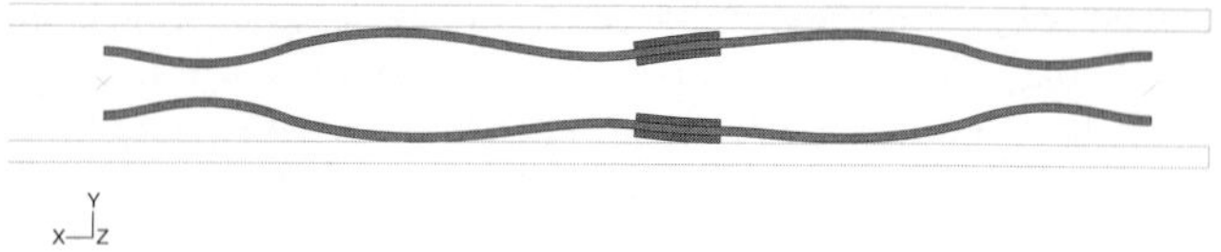

Fig. 5 Longitudinal cut view of the filter frame in the delivery sheath

Preliminary simulations showed that, in the analyzed case, final filter/vessel configuration is not influenced by the deploying mode and consequently the interaction between filter and sheath has not been taken into account.

We defined a two-step simulation strategy as follows: (i) filter diameter reduction is obtained by elongating the basket frame by appropriate axial displacement boundary conditions to the distal end RP, only self contact of membrane was included in this step; (ii) previous boundary conditions are deactivated allowing the filter re-expansion and a global contact algorithm was included allowing the filter/vessel interaction.

The effect of blood flow pressure acting on the filter membrane was included in the analyses applying a uniform pressure of 8.80 mmHg on the inner surface of the membrane (Ref 25).

Circumferential basket-vessel wall apposition was evaluated in terms of gap between the vessel lumen and the circumferential area covered by the filter.

In order to keep the computational cost acceptable and to minimize the impact of the membrane mesh on the gap measurement, a final membrane mesh of M3D4R 24960 elements was chosen by a preliminary mesh sensibility analysis about filter deployment simulation in the 3 mm circular vessel (a finer membrane mesh—29,120 elements—results in a 5.4% gap measurement differing 6% compared to the 5.7% gap obtained with the coarser mesh—24,960 elements).

3. Results

3.1 Filter-Free Expansion

The simulation of the filter-free expansion showed that the basket-frame configuration experiences a severe change. The radial compression of the filter frame is accomplished by an axial elongation bringing the diameter to a final value of 0.96 mm when the filter is completely inserted into the sheath as shown in Fig 5. Moreover, the filter frame experiences a radial deformation that is not uniform along the filter length, resulting in non-uniform adhesion between the strut and the delivery sheath wall.

Comparing the previously described adopted mesh to a finer one (i.e. 67,584, 33,024, and 2,592 elements respectively for filter struts, membrane, and markers) showed a negligible divergence (i.e. difference in maximum axial elongation less than 1%).

*Ovality being defined as the change in cross section roundness and defined as $O = b/a$, where a is cross section semi-major axis length and b is cross section semi-minor axis length.

Moreover, Fig. 6 shows a good qualitative agreement between the numerical results and the micro-CT image of the partially deployed Angioguard™ XP.

3.2 Circumferential Basket-Vessel Wall Apposition

Simulations of the filter deployment in the straight vessel showed that the filter is not able to completely cover the vessel lumen and gaps are present between the vessel wall and the membrane (see Fig. 7), confirming the experimental results recently reported by Sieworek et al. (Ref 26).

As reported in Table 3, the vessel size seems to have a minor impact on the circumferential basket-vessel wall apposition which is in contrast to the impact of the vessel shape. In fact, an increase in the vessel ovality dramatically influences the gap while a change in the vessel diameter does not significantly modify the gap area.

Consequently, the simulations of filter deployment in the straight rigid vessel suggest that, in this case, filter malapposition is caused by inability of the filter struts to accomplish the vessel asymmetry.

4. Limitations

The filter-free expansion simulation showed qualitative agreement with micro-CT images but limitations pertaining to the frictionless contact between the filter and the delivery sheath are present.

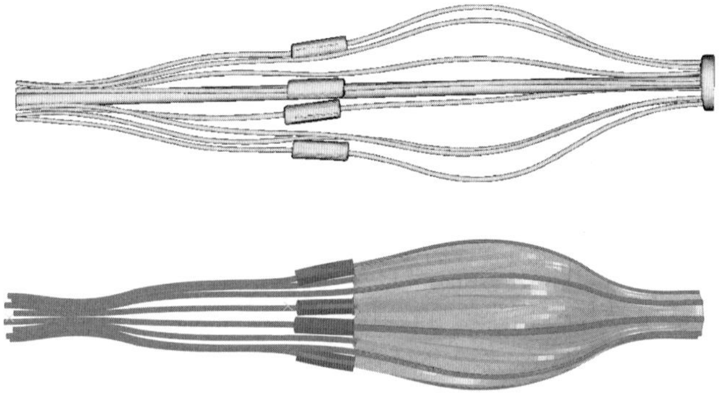

Fig. 6 Partially deployed filter: microCT image (top panel); numerical simulation (bottom panel)

Clearly, the assumption of a uniform pressure distribution along the filter membrane should be improved using more accurate and realistic values achieved, for example, by experiments and computational fluid dynamics analyses.

Material properties of the filter components should also be derived from experimental data to define the mechanical properties more accurately.

The major limitation of this study is the rigid body assumption for the vessel. In reality, the radial outward force of the filter is likely to reshape the vessel and this effect can consequently influence the filter-vessel wall apposition. However, modeling filter apposition in a realistic vessel (i.e. internal carotid artery) introduces substantial computational challenges, such as the incorporation of the variability of the constitution and mechanical properties of the diseased arterial wall, and should be accomplished by adequate experimental validation. This challenge is considered to be beyond the scope of this preliminary investigation.

5. Conclusions

In this study, we proposed and validated a finite element model in order to investigate the basket-like design of an embolic protection filter device. In particular, the circumferential filter/vessel wall apposition was numerically evaluated confirming the inability of the filter to completely adapt to asymmetric vessels.

Clearly, this study needs to be considered as a preliminary proof of concept of the use of finite element based modeling

Table 3 Numerical results of filter/vessel wall apposition

	Filter size, mm—Ovality				
	3.5-0.0	3.25-0.0	3.0-0.0	3.0-0.85	3.0-0.75
Gap (% of vessel lumen)	5.7	4.7	5.7	8.7	14.7

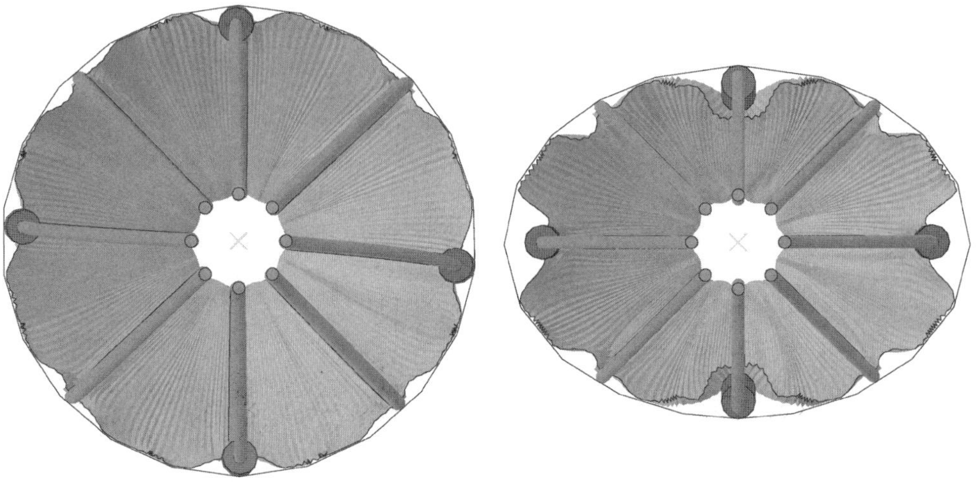

Fig. 7 A 4 mm size filter deployed in a 3 mm circular vessel (on the left) and in 3 mm vessel having 0.75 ovality (on the right)

strategies to investigate and understand the mechanics of embolic protection filters. The proposed model is a base to further investigate the impact of design parameters (filter length, diameter, number of struts, etc.) on filter mechanics (flexibility, radial strength, etc.) and the filter apposition in a more realistic tortuous anatomy taking also into account vessel wall material properties.

Finally, we finally conclude that the proposed methodology could be useful to evaluate and to compare current or new designs of EPDs in the early design phase as recently recommended by the FDA (Ref 27).

References

1. S. Petersen, V. Peto, M. Rayner, J. Leal, R. Luengo-Fernandez, and A. Gray, European Cardiovascular Disease Statistics, BHF London, 2005
2. M.H. Wholey, N. Al-Mubarek, and M.H. Wholey, Updated Review of the Global Carotid Artery Stent Registery, *Catheter. Cardiovasc. Interv.*, 2003, **60**, p 259–266
3. A.J. Furlan, Carotid-Artery Stenting—Case Open or Closed? *N. Engl. J. Med.*, 2006, **355**(16), p 1726–1729
4. N. Al-Mubarak, G.S. Roubin, J.J. Vitek, S.S. Iyer, G. New, and M.B. Leon, Effect of the Distal Balloon Protection System on Microembolization During Carotid Stenting, *Circulation*, 2001, **104**, p 1999–2002
5. J.S. Yadav, M.H. Wholey, R.E. Juntz, P. Fayad, B.T. Katzen, G.J. Mishkel, T.K. Bajwa, P. Whitlow, N.E. Strickman, M.R. Jaff, J.J. Popma, D.B. Snead, D.E. Cutlip, B.G. Firth, and K. Ouriel, Protected Carotid Artery Stenting Versus Endarterectomy in High-Risk Patients, *N. Engl. J. Med.*, 2004, **351**, p 1493–1501
6. P.P. Goodney, M.L. Schermerhorn, and R.J. Powell, Current Status of Carotid Artery Stenting, *J. Vasc. Surg.*, 2006, **43**, p 406–411
7. CaRESS Steering Committee, Carotid Revascularization Using Endarterectomy or Stenting Systems (CaRESS) Phase I Clinical Trial: 1-year Results, *J. Vasc. Surg.*, 2005, **42**, p 213–219
8. M.H. Yen, D.S. Lee, S. Kapadia, R. Sachar, D.L. Bhatt, C.T. Bajzer, and J.S. Yadav, Symptomatic Patients Have Similar Outcomes Compared with Asymptomatic Patents After Carotid Artery Stenting with Emboli Protection, *Am. J. Cardiol.*, 2005, **95**, p 297–300
9. M.K. Eskandari, S.F. Najjar, J.S. Matsumura, M.R. Kibbe, and M.D. Morasch, Technical Limitations of Carotid Filter Embolic Protection Devices, *Ann. Vasc. Surg.*, 2007, **21**, p 403–407
10. F. Fanelli, M. Bezzi, E. Boatta, and R. Passariello, Techniques in Cerebral Protection, *Eur. J. Radiol.*, 2006, **60**, p 26–36
11. K. Kasirajan, P.A. Schneider, and K.C. Kent, Filter Devices for Cerebral Protection During Carotid Angioplasty and Stenting, *J. Endovasc. Ther.*, 2003, **10**, p 1039–1045
12. G.M. Siewiorek, M.H. Wholey, and E.A. Finol, In Vitro Performance Assessment of Distal Protection Devices for Carotid Artery Stenting: Effect of Physiological Anatomy on Vascular Resistance, *J. Endovasc. Ther.*, 2007, **14**(5), p 712–724
13. F. Auricchio, M. Di Loreto, and E. Sacco, Finite Element Analysis of a Stenotic Artery Revascularization Through Stent Insertion, *Comput. Methods Biomech. Biomed. Eng.*, 2000, **0**, p 1–15
14. M. De Beule, P. Mortier, S.G. Carlier, B. Verhegghe, R. Van Impe, and P. Verdonck, Realistic Finite Element-Based Stent Design: The Impact of Balloon Folding, *J. Biomech.*, 2008, **41**(2), p 383–389
15. C. Lally, F. Dolan, and P. Prendergast, Cardiovascular Stent Design and Vessel Stresses: A Finite Element Analysis, *J. Biomech.*, 2005, **38**(8), p 1574–1581
16. P. Mortier, M. De Beule, S.G. Carlier, R. Van Impe, B. Verhegghe, and P. Verdonck, Numerical Study of the Uniformity of Balloon-Expandable Stent Deployment, *J. Biomech. Eng.*, 2008, **130**(2), p 021018
17. S. Muller-Hulsbeck, T. Jahnke, C. Liess, C. Glass, F. Paulsen, J. Grimm, and M. Heller, In Vitro Comparison of Four Cerebral Protection Filters for Preventing Human Plaque Embolization During Carotid Interventions, *J. Endovasc. Ther.*, 2002, **9**(6), p 793–802
18. S. Muller-Hulsbeck, T. Jahnke, C. Liess, C. Glass, J. Grimm, and M. Heller, Comparison of Various Cerebral Protection Devices Used for Carotid Artery Stent Placement: An In Vitro Experiment, *J. Vasc. Interv. Radiol.*, 2003, **14**(5), p 613–620
19. S. Muller-Hulsbeck, P. Stolzmann, C. Liess, J. Hedderich, F. Paulsen, T. Jahnke, and M. Heller, Vessel Wall Damage Caused by Cerebral Protection Devices: Ex Vivo Evaluation in Porcine Carotid Arteries, *Radiology*, 2005, **235**(2), p 454–460
20. R.J. Powell, C. Alessi, B. Nolan, E. Rzucidlo, M. Fillinger, D. Walsh, M. Wyers, R. Zwolak, and J.L. Cronenwett, Comparison of Embolization Protection Device Specific Technical Difficulties during Carotid Artery Stenting, *J. Vasc. Surg.*, 2006, **44**, p 56–61
21. Instructions for use Angioguard XP™ Emboli Capture Guidewire system. www.cordislabeling.com/pdf/24408544_3.pdf. Accessed 16 March 2009
22. http://pyFormex.berlios.de. Accessed 16 March 2009
23. http://www.cabot-corp.com. Accessed 16 March 2009
24. http://bumps.ugent.be/bumper. Accessed 16 March 2009
25. J.M. Hendriks, J.D. Zindler, A. van der Lugt, P. Pattynama, M. van Sambeek, J. Bosch, and L. van Dijk, Embolic Protection Filters for Carotid Artery Stenting: Differences in Flow Obstruction Depending on Filter Construction, *J. Endovasc. Ther.*, 2006, **13**(1), p 47–50
26. G.M. Sieworek, M.K. Eskandari, and E.A. Finol, The Angioguard™ Embolic Protection Device, *Expert Rev. Med. Devices*, 2008, **5**(3), p 287–296
27. Guidance for Industry and FDA Staff, Coronary and Carotid Embolic Protection Devices—Premarket Notification [510(k)] Submissions. http://www.fda.gov/cdrh/ode/guidance/1658.html. Accessed 16 March 2009
28. A.R. Pelton, J. DiCello, and S. Miyazaki, Optimisation of Processing and Properties of Medical Grade Nitinol Wire, *Minim. Invasive Ther. Allied Technol.*, 2000, **9**(1), p 107–118

JMEPEG (2009) 18:793–796
DOI: 10.1007/s11665-009-9488-5

1059-9495/$19.00

Static and Cyclic Load-Deflection Characteristics of NiTi Orthodontic Archwires Using Modified Bending Tests

Mahmoud Nili Ahmadabadi, Tahereh Shahhoseini, Mohamad Habibi-Parsa, Maryam Haj-Fathalian, Tahereh Hoseinzadeh-Nik, and Hananeh Ghadirian

(Submitted September 15, 2008; in revised form April 6, 2009)

Near-equiatomic nickel-titanium (nitinol) has the ability to return to a former shape when subjected to an appropriate thermomechanical procedure. One of the most successful applications of nitinol is orthodontic archwire. One of the suitable characteristics of these wires is superelasticity, a phenomenon that allows better-tolerated loading conditions during clinical therapy. Superelastic nitinol wires deliver clinically desired light continuous force enabling effective tooth movement with minimal damage for periodontal tissues. In this research, a special three-point bending fixture was invented and designed to determine the superelastic property in simulated clinical conditions, where the wire samples were held in the fixture similar to an oral cavity. In this experimental study, the load-deflection characteristics of superelastic NiTi commercial wires were studied through three-point bending test. The superelastic behavior was investigated by focusing on bending time, temperature, and number of cycles which affects the energy dissipating capacity. Experimental results show that the NiTi archwires are well suited for cyclic load-unload dental applications. Results show reduction in superelastic property for used archwires after long-time static bending.

Keywords biomaterials, intermetallics, nonferrous metals

1. Introduction

In the early 1960s, Bohler and co-workers discovered the shape-memory effect in an equiatomic NiTi alloy with generic trade name nitinol (Ref 1).

Nickel-titanium (NiTi) orthodontic wires were introduced in orthodontics by Andreasen and Hilleman (Ref 2). NiTi orthodontic wires exhibit a lower modulus of elasticity, which is particularly useful during the early stages of fixed appliance therapy. Burstone first described the application of a superelastic NiTi wire: low deactivation forces were used to create a "physiologic" bone response (Ref 3).

Burstone and Goldberg observed beneficial characteristics such as low modulus of elasticity combined with a wide elastic working range (Ref 3, 4) that allowed wires to sustain large elastic deflections due to the very high springback quality. These wires enabled the orthodontic profession to achieve ideal tooth movement.

NiTi archwires have gained acceptance by orthodontists as initial alignment wires. Most of the information about the behavior of orthodontic wires is based on mechanical laboratory three-point bending tests that study load-deflection characteristics. Superelasticity is characterized by a load-deflection curve with a plateau during unloading. Results of three-point bending tests are load-deflection plots. These tests offer reproducibility which has facilitated comparison between studies (Ref 5-7).

Variations in model design have been shown to affect load-deflection plots. The load-deflection curves of NiTi wires alter with modifying of the test model (Ref 8-10). Modified three-point bending test which simulates wire force on the teeth in the oral conditions has more accurate results than ordinary three-point bending test. This test method provides more detailed information regarding orthodontic applications.

It is known that the NiTi alloy wire undergoes a phase transformation from an austenitic to a martensitic phase as the load increases during the loading process. Metallurgical studies have attributed these characteristics to a reversible phase transformation from the body-centered cubic structure to the monoclinic structure of Ni-Ti when the stress reaches a certain level during deformation (Ref 11).

The increasing amount of energy stored inside the NiTi wire during this process is consumed during the unloading process as the transformation is reversed, and the martensite structure reverts to austenite. Superelasticity is exhibited by unloading plateau in the reverse transformation from the martensitic phase to the austenitic phase and with low load-deflection ratios. These wires show nearly constant forces in the unloading process, a desirable physiological property for orthodontic tooth movement (Ref 12).

This article is an invited paper selected from presentations at Shape Memory and Superelastic Technologies 2008, held September 21-25, 2008, in Stresa, Italy, and has been expanded from the original presentation.

Mahmoud Nili Ahmadabadi, Tahereh Shahhoseini, Mohamad Habibi-Parsa, and **Maryam Haj-Fathalian**, School of Metallurgy and Materials Engineering, University of Tehran, Tehran, Iran; and **Tahereh Hoseinzadeh-Nik** and **Hananeh Ghadirian**, School of Dentistry, Tehran University of Medical Sciences, Tehran, Iran. Contact e-mail: nili@ut.ac.ir.

According to clinical applications, unloading occurs exactly after loading; therefore, no rest period or delay should occur while loading and unloading is carried out during experimental procedure. In addition, in the oral conditions, loading is being carried out in a short period while the unloading is kept in a long period. In the clinical situation, archwires are kept deflected for a long period of time after being engaged in their brackets (Ref 13).

The purpose of this study is to investigate the load-deflection characteristics of superelastic NiTi wires with a new model design through modified bending tests in cyclic condition and when unloading was delayed for a long period of time.

2. Experiment

In this research, a new three-point bending fixture was invented and designed to determine the superelastic property in clinical conditions, where the wire samples were held in the fixture similar to an oral cavity. By means of this instrument, the three-point bending test simulates wire force on the teeth in the oral configuration.

The lower section of the fixture is a rail fabricated from steel with a special movable base and a curved canal assembled over the rail. The upper section was designed to simulate teeth arrangement and curvature: a stainless steel disk (316L, Ø80 mm, h10 mm) with 12 rods (316L, Ø5 mm, h10 mm) were welded to the points representing the centers of teeth on the disk (Ref 14). The center points of teeth were located on the medium upper standard arc. Distances between the center points of the teeth (interbracket distance) were similar to the Wilkinson model (Ref 8). A fillet face was machined on the rod surface parallel to the standard arc. Brackets were fixed on the flat face of rod using superglue and orthodontic wire attached to the fixed appliance (Fig. 1). The modified three-point bending tests were conducted by the instrument as shown in Fig. 2 and 3. By rotating the steel disk over the fixture, different locations of teeth with different wire curvatures were selected for dynamic loading. The position of tooth 5 (left) was selected and forced to 2 mm deflection (Fig. 1). The fixture was extended for holding 10 wire samples in a dynamic three-point bending test. Consequently, two stainless steel disks were selected and longer steel rods (Ø5 mm, h150 mm) were welded to the points representing center of teeth on the disks. Ten series of brackets were fixed on the flat face of the rods (Fig. 3). This fixture was developed for static bending in a saliva bath at a defined temperature. The saliva was composed of KCl (1.21 g/L), NaCl (0.4 g/L), NaH_2PO_4 (0.78 g/L), and urea (1 g/L). The pH of saliva bath was kept at 7 by means of NaOH. Tests were carried out on the Ø0.016 inch Trueflex (Orthotechnology) archwires with a Zwick universal testing machine fitted with 2.5 kN load cell and at a speed of 1 mm/min.

Fig. 2 Three-point bending fixture

Fig. 3 Extended fixture

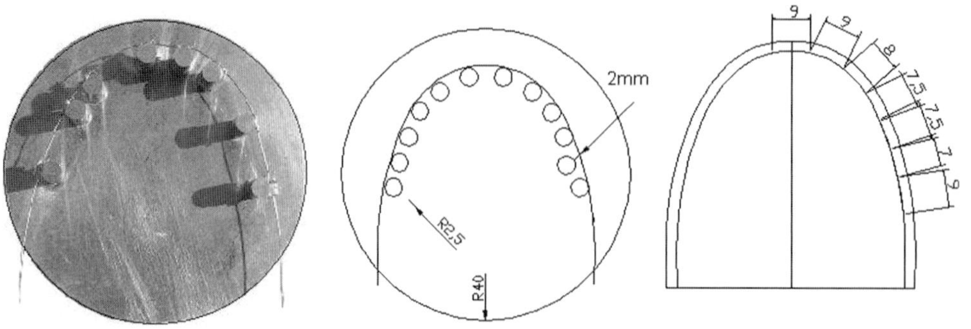

Fig. 1 The points of teeth on the medium standard arc on the upper section of test instrument

Archwires were subjected to bending tests up to 2 mm deflection. The three-point bending tests were conducted in different conditions:

1. Cyclic bending test: As-received arch wire subjected to cyclic bending test at 22 °C (20 cycles).
2. Three-point bending tests: These tests were carried out on the new wire samples at 22 °C (tests were repeated five times with specimen wires changed for every test).
3. Static deflection: Two groups of 15 wires each were subjected to the static deflection for 30, 60, and 90 days in fixture, in an incubator at 37 °C (5 wires for each holding time). Group 1 held in saliva bath and group 2 held dry. The saliva bath was topped up during the long-term studies for the group 1. The amount of deflection in static bending was 2 mm.
4. Dynamic bending: After static deflection, three-point bending tests were carried out on the wires. To maintain similar conditions as for the new wires, dynamic three-point bending tests at this step were performed at 22 °C. All tests were repeated five times in both conditions with specimen wires being changed for every test.

Cyclic bending test was performed by means of the single wire fixture. Other tests were carried out by the extended fixture. Load-deflection graphs show load-unload values (N) versus deflection (mm). Comparisons were made in the following aspects:

- Change in force level and hysteresis range during cycling at loading and unloading sequences.
- Load-deflection curve characteristics between the tests for the static long-time bending and for those with immediate unloading.
- The effect of time on the load-deflection characteristics of orthodontic wires when they were unloaded after 30, 60, and 90 days delay before unloading.
- Force levels and amounts of hysteresis between static bending tests in saliva bath and dry atmosphere.

3. Results

Figure 4 shows cyclic load-deflection plots for as-received archwire at 22 °C. This figure indicates that the stress plateaus associated with loading and unloading decrease with increasing cycle number. Test results indicate wires produced a low slope force at unloading process (lower plateau) at 22 °C. Lower plateau indicates deactivation forces enabling effective tooth movement. Average of force in lower plateau is about 1.5 N (Fig. 4). At 22 °C during increasing number of cycles, level of forces at both load and unload plateau slightly decreased. At this temperature, the hysteresis range is near to constant with increasing cycle number (Ref 14).

During the loading process, NiTi archwire undergoes a phase transformation from austenite to a martensite as the load increases. The increasing amount of energy stored inside the NiTi wire during this process is consumed during the unloading process as the transformation is reversed, and the martensite structure reverts to austenite (Ref 13).

The hysteresis range is equal to the difference between the amount of energy consumed by the reverse transformation and the amount of energy stored in the transformation. The range of the hysteresis observed after cyclic deformation showed energy dissipation. This is accompanied by considerable softening. Softening phenomenon is due to the build-up of residual stresses which may arise from the formation of interstitial dislocations (Ref 15).

Martensitic transformation is a displacive transformation which dislocations accelerated this type of transformation. In the first cycle of bending, a few dislocations remain. In the latter cycles, these dislocations decrease nucleation energy for martensitic transformation. Dislocations work as autocatalyst by aggregation in the austenite-martensite interface and assist nucleation of new martensite blade. Eventually, softening phenomenon occurs due to dislocation-assisted transformation.

Figure 5 represents load-deflection plot for as-received Orthotechnology archwires (mean value of five samples). Figure 6 shows load-deflection plots for archwires after static bending in body temperature. Results show decrease in plateau force levels associated with loading and unloading after 30 days static deflection (Fig. 6) in comparison with new wires (Fig. 5). After a long-time bending, the hysteresis was increased. Also in comparison with new wires (Fig. 5), the force level in loading increased but decreased in the unloading sequence (Fig. 6). This phenomenon has been discussed in relation to stress relaxation. It is a decrease of internal stress over time due to localized flow of solid metal under strain (Ref 13). According to Fig. 6, loading and unloading plateau

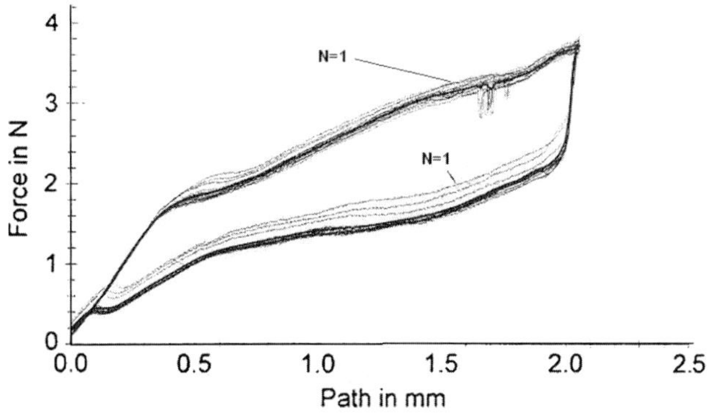

Fig. 4 Cyclic load-deflection plots for as-received archwire at 22 °C

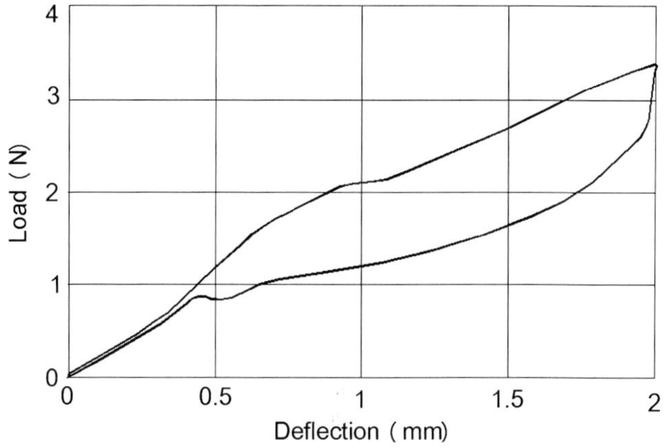

Fig. 5 Load-deflection plots for new archwires (mean value of five samples)

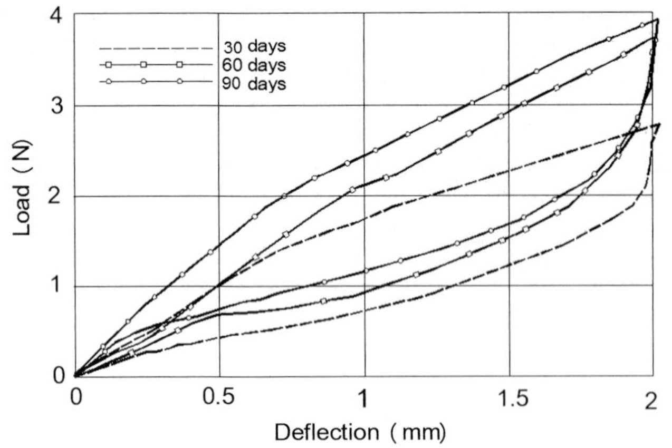

Fig. 6 Load-deflection plots for archwires after 30, 60, and 90 days static bend in body temperature (each curve represents mean value of five samples)

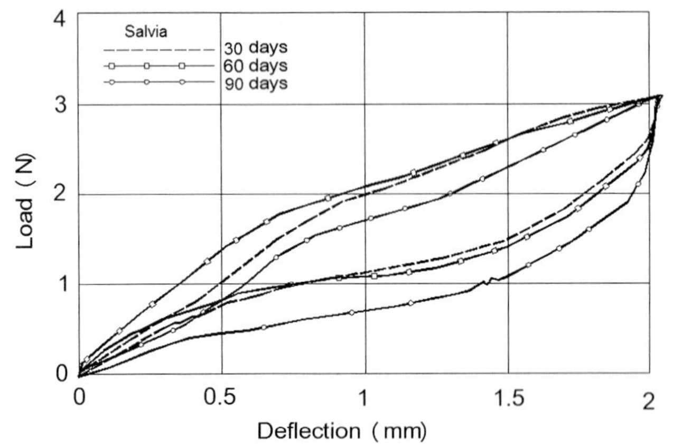

Fig. 7 Load-deflection plots for archwires after 30, 60, and 90 days static bend in body temperature at saliva bath (each curve represents mean value of five samples)

level increase as time in static bend mode increased; however, the magnitude of the increase in load across the loading plateau is greater than the magnitude of the increase in load across the unloading plateau as static bend time increase.

Figure 7 indicates load-deflection plots for archwires after static bending in body temperature in saliva bath. According to Fig. 7, the stress plateaus associated with loading and unloading decrease with increasing static bending in saliva bath. Force levels of both loading and unloading plateaus decrease by increasing holding time in saliva. The hysteresis range is near to constant with increasing holding time.

Usually, hysteresis phenomenon does not occur in the elastic limit. After long-time static bending, the unloading curve runs below the loading curve (under 0.8 N load and 0.4 mm deflection). In other words, the elastic limit is vanished (Fig. 6 and 7). This behavior is probably due to dislocations pile up which prevents reverse transformation after long-time bending.

4. Conclusions

Modified load-deflection tests agreed with tension-compression results and simulated load conditions in an oral cavity. At 22 °C, during cyclic load-unload with increasing number of cycles, level of forces at both load and unload plateau slightly decreased. The range of the hysteresis observed during cyclic load-unload showed energy dissipation. Load relaxation was observed in the wires subjected to long-term static deflection at 37 °C. Load relaxation increased in saliva bath environment and decreased in dry environment as static deflection time increased. The increase of load relaxation in salvia is supposedly due to some interactions between salvia and wires. Hysteresis range in dry wires showed increase after long-term deflection while this range is near to constant with increasing holding time in saliva at 37 °C.

References

1. G. Kaufman and I. Mayo, The *Metal* with a *Memory*, American Heritage.com: Invention & Technology Magazine, Fall 1993, Vol 9, Issue 2
2. G.F. Andreasen and T.B. Hilleman, An Evaluation of 55 Cobalt Substituted Nitinol Wire for Use in Orthodontics, *J. Am. Dent. Assoc.*, 1971, **82**(6), p 1373–1375
3. C.J. Burstone, Variable-Modulus Orthodontics, *Am. J. Orthod.*, 1981, **80**, p 1–16
4. C.J. Burstone and A.J. Goldberg, Beta Titanium: A New Orthodontic Alloy, *Am. J. Orthod.*, 1980, **77**, p 121–132
5. K. Watanabe, Studies on New Superelastic NiTi Orthodontic Wire, Part 1. Tensile and Bending Test, *Shika Rikogaku Zasshi*, 1982, **23**(61), p 47–57
6. F. Miura, M. Mogi, and Y. Ohura, The Superelastic Property of the Japanese NiTi Alloy Wire for Use in Orthodontics, *Am. J. Orthod.*, 1986, **90**(1), p 1–10
7. S.E. Khier and W.A. Brantley, Bending Properties of Superelastic, Nonsuperelastic Nickel-Titanium Orthodontic Wires, *Am. J. Orthod.*, 1991, **99**(4), p 310–318
8. P.D. Wilkinson and A. James, Load-Deflection Characteristics of Superelastic Nickel-Titanium Orthodontic Wires, *Am. J. Orthod.*, 2002, **121**(5), p 483–495
9. F. Parvizi and W.P. Rock, The Load/Deflection Characteristics of Thermally Activated Orthodontic Arch Wires, *Eur. J. Orthod.*, 2003, **25**, p 417–421
10. E.F. Harris, S.M. Newman, and J.A. Nicholson, Nitinol Archwire in a Simulated Oral Environment: Changes in Mechanical Properties, *Am J. Orthod. Dentofacial Orthop.*, 1998, **93**(4), p 508–513
11. K. Otsuka and X. Ren, Physical Metallurgy of Ti–Ni-Based Shape Memory Alloys, *Prog. Mater. Sci.*, 2005, **50**, p 511–678
12. M. Nili-Ahmadabadi and T. Shahhoseini, *Introduction to Biomaterials*, 1st ed., Chap. 3, Biomaterials Research Center, University of Tehran, 2007, p 50–91
13. A. Hanyuda, S. Nagasaka, and T. Yoshida, Long-Term Time Effect on Load–Deflection Characteristics of Orthodontic Wires, *Orthod. Waves*, 2006, **65**(4), p 155–160
14. M. Nili-Ahmadabadi, T. Shahhoseini, M. Haj-Fathalian, and M.H. Parsa, Cyclic Load-Deflection Characteristics of NiTi Orthodontic Arch Wires at Different Temperatures, *Proceedings of APCMP2008*, June 15-20, 2008 (Guilin, China), 2008, p 214–219
15. M. Es-Souni, M. Es-Souni, and H.F. Brandies, On the Transformation Behavior, Mechanical Properties and Biocompatibility of Two NiTi-Based Shape Memory Alloys, *Biomaterials*, 2001, **22**(15), p 2153–2161

JMEPEG (2009) 18:797–804
DOI: 10.1007/s11665-009-9426-6

1059-9495/$19.00

Stress-Shielding Effect of Nitinol Swan-Like Memory Compressive Connector on Fracture Healing of Upper Limb

Q.G. Fu, X.W. Liu, S.G. Xu, M. Li, and C.C. Zhang

(*Submitted September 18, 2008; in revised form March 1, 2009*)

In this article, the stress-shielding effect of a Nitinol swan-like memory compressive connector (SMC) is evaluated. Patients with fracture healing of an upper limb after SMC internal fixation or stainless steel plate fixation were randomly selected and observed comparatively. With the informed consent of the SMC group, minimal cortical bone under the swan-body and swan-neck was harvested; and in the steel plate fixation group, minimal cortical bone under the steel plate and opposite side to the steel plate was also harvested for observation. Main outcome measurements were taken such as osteocyte morphology, Harversian canal histological observation under light microscope; radiographic observation of fracture healing, and computed tomography quantitative scanning of cortical bone. As a conclusion, SMC has a lesser stress-shielding effect to fixed bone than steel plate. Finally, the mechanism of the lesser stress-shielding effect of SMC is discussed.

Keywords fracture healing, internal fracture fixation, nitinol, shape memory alloy, stress-shielding effect

1. Introduction

In 1963, Buehler and Wang (Ref 1) from the American Naval Weapon Research Institute reported on the apparent shape memory effect of Nitinol that aroused widespread interest from relevant scholars. Nitinol, with the effect of shape memory and its superiority in wear and corrosion resistance (Ref 2), has been extensively applied in the medical field (Ref 3) and is considered to be a rare "bio-memory material." Biocompatibility studies have shown NiTi to be a safe implant material (Ref 4, 5). The Chinese scholar, Zhang, has dedicated himself to the study and application of shape memory alloy in the medical field for many years. After intensive research on the physical, chemical, and mechanical nature of Nitinol, and taking into consideration the anatomical characteristics of the long bone, he broke through the traditional concept of internal fixation on fracture and invented the Nitinol swan-like memory compressive connector (SMC) (Ref 6, see Fig. 1a, b, Chinese Patent No. ZL99113873.2) in 1986. By taking advantage of the "thermo-elastic martensite-type" characteristic of Nitinol, the SMC forms the relatively ideal local mechanical environment necessary for the bio-reaction of the fracture union at the fracture end of the long bone. In addition, the SMC enables persistent and initial axial pressure to be exerted on the fracture end, thereby maintaining a long-term and appropriate initial memory load at the fracture end and preventing the occurrence of osteoporosis and re-fracture (Ref 7, 8).

After fixation of the long bone fracture of the upper limb by either dynamic compressive plate or interlocking intramedullary nail, complications often occur, such as bone nonunion, delayed bone union, or breakage of internal fixator in different proportions. The plate or interlocking nail is typically made of stainless steel, whose rigidity will hinder the process of bone reconstruction, leading to the loss of bone mass and the disorder of bone architecture in the fixed bone segment. Moreover, after the removal of the internal fixator, re-fracture is very likely to occur (Ref 9-11). Clinical treatment of long bone fracture of the upper limb by application of SMC has indicated that under the fixation of SMC, the fracture segment of the fractured bones is connected directly by the lamellar bone, avoiding local external callus and osteoporosis of the cortical bone (Ref 6). In order to further confirm such a bone-healing phenomenon and to clarify its nature, our experiment applied SMC to fix long bone fractures of the upper limb in adults, and after bone-healing, the internal fixators were removed to facilitate x-ray imaging, quantitative scanning computed tomography (CT) (Ref 12, 13) of the cortical bone, and histological observation. Finally, by comparing a SMC fixation group with a stainless steel plate fixation group, the stress-shielding effect of SMC on the fixed segment of the fractured bone was evaluated and is discussed.

2. Subjects and Methods

2.1 Design

Controlled observation experiment.

This article is an invited paper selected from presentations at Shape Memory and Superelastic Technologies 2008, held September 21-25, 2008, in Stresa, Italy, and has been expanded from the original presentation.

Q.G. Fu, X.W. Liu, S.G. Xu, M. Li, and **C.C. Zhang**, Department of Orthopedics, Changhai Hospital of Second Military Medical University, Shanghai 200433, China. Contact e-mail: zhangchuncai@vip.sina.com.

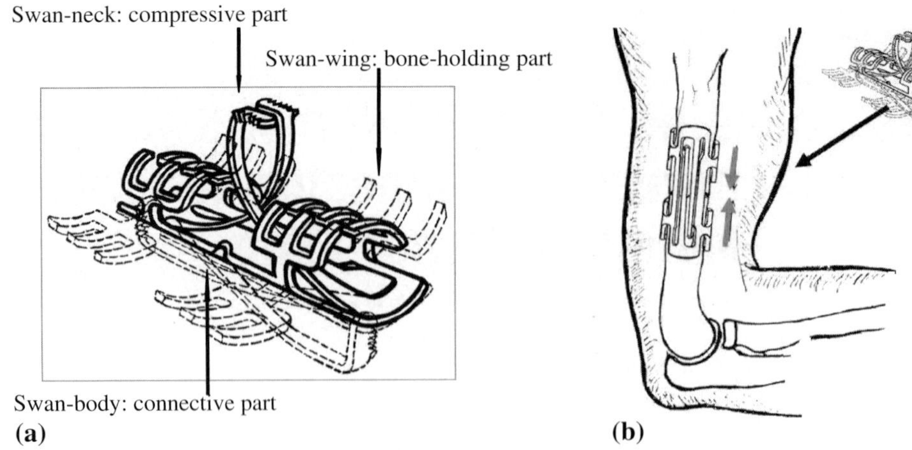

Fig. 1 (a) Schematic diagram of clinical internal fixation of human long bone fracture by SMC. (b) Schematic diagram of SMC fixating shaft of humerus, the *arrow* indicate the direction of pressure

2.2 Time and Setting

The experiment was performed at the Department of Orthopedics, Changhai Hospital of Second Military Medical University in Shanghai from March 2002 to August 2004.

2.3 Participants

Patients with complete healing of bone fracture of the upper limb after SMC internal fixation or stainless steel plate internal fixation were randomly selected. There were 20 cases in the SMC group, including 14 males and 6 females, aged 18-53 years (average = 28.3 years). Their distribution of fracture sites were: eight cases of humerus fracture, one case of ulna fracture, two cases of radius fracture, and nine cases of both ulna and radius fracture. Their time duration from the fracture to the treatment of internal fixation was 3.92 ± 1.47 days. Their time duration from surgery to internal fixator removal was 11.2 ± 3.68 months. There were 20 cases in the stainless steel plate group, including 16 males and 4 females, aged 19-46 years (average = 27.5 years). Their distribution of fracture sites was: 7 cases of humerus fracture, 2 cases of ulna fracture, 1 case of radius fracture, and 10 cases of both ulna and radius fracture. Their time duration from the fracture to the treatment of internal fixation was 4.16 ± 1.52 days. Their time duration from surgery to internal fixator removal was 12.6 ± 4.10 months. All of above fractures were fresh in the upper limbs, received the first internal fixation and achieved complete healing. Both the SMC and the stainless steel plate placed within the body conformed to national standards. According to statistical analysis, there were no significant differences in ages, time duration from the fracture to the treatment of internal fixation, and time duration from surgery to removal of internal fixator between the two groups.

2.4 Internal Fixation Device

2.4.1 The Structure and Functional Mechanism of SMC. The SMC was invented by Zhang et al. (Ref 6) from the Department of Orthopedics, Changhai Hospital of Second Military Medical University in Shanghai, and produced by Ximai Biological Medical Equipment Corporation (Chinese registration number of the medical equipment: 2002, No. 3040024), complying with the national standards. According to the anatomical and biomechanical characteristics of the bone shaft of the upper limb, a plate with the width of 1.5-2.5 mm, containing 50-53% of nickel (the other component being titanium) is made with three parts: connective part (a swan-body), compressive part (two swan-necks), and bone-holding part (four swan-wings). The inner diameter of the SMC is 6-23 mm, and the diameter-length ratio is 1:6. It has a one-way orientation and the shape recovery temperature is (33 ± 2) °C.

Before implantation, the SMC was placed in the 0-4 °C ice water for cooling, because the Nitinol is formable at lower temperatures (martensitic). The four swan-wings were opened with a needle forceps, so the spreading distance could be larger than the diameter of the fixed bone. The two swan-necks were pulled straight toward the dorsal side of the swan-body. The midpoint of the swan-body was aimed at the fracture site, and two holes were drilled on the fractured bone at the same surface with that of tail of the spread swan-neck—both proximal and distal to the fracture site. After reduction of the fracture, the SMC was inserted, and the tail of the spread swan-neck was then inserted into the holes. Afterward, water at 40-50 °C was used to warm the SMC in order to exert its mechanical functions. The effect of shape memory of the swan-wings was produced by the patient's body temperature (~37 °C), and the swan-wings embraced the bone shaft as a ring. Meanwhile, influenced by the tail of swan-neck, the two swan-necks exerted persistent axial pressure on the fracture segment—thus coordinating three-dimensionally to fix the bone shaft (Ref 6).

2.4.2 The Stainless Steel Plate. These were imported from other countries but all conformed to national standards.

3. Methods

3.1 Collection of Samples

Patients with long bone fracture after SMC internal fixation or stainless steel plate internal fixation were selected, and a follow-up showed that the patients had achieved bone-healing. With the patient's permission, a very small quantity of the cortical bone and a small quantity of the normal bone, used as a comparison, were removed when extracting the internal fixation. The positions where the cortical bones were removed were: under the swan-body and swan-neck of the SMC (Ref 1) and under the steel plate and opposite side to the steel plate (Ref 2).

3.2 The Making and Observation of the Light Microscopic Slices

The abovementioned cortical bones were made into light microscopic slices and observed with an Olympus PM-CBK-G biological microscope and photographed.

3.3 X-Ray Imaging

Images of bilateral bone shaft were taken 1 week after the removal of the internal fixation, and the conditions of the cortical bone at the broken and the normal parts were compared. Exposure factors were: voltage 50 kV, time 0.1 s, electric current 125 mA, and tube distance 100 cm.

3.4 The Quantitative Scanning Computed Tomography of the Cortical Bone

One week after removal of the internal fixation, cross-section scanning with a scanning thickness of 2 mm was carried out by a Somaton CT machine (Siemens Corp., Germany) on the cortical area of 1 cm distance to the center of the broken end. Four small cortical areas were selected randomly at the region of the cortical bone under the plate (the SMC group:cortex under swan-body and under swan-neck of SMC; the stainless steel plate group:cortex under plate and opposite side to plate) after scanning and imaging, and the average CT value of the cortical bone in each small area was automatically calculated by a computer. The average CT value and standard deviation of the cortical bone under the plate were calculated. Meanwhile, the CT value of the corresponding normal bone opposite to the fracture part at the same level was measured as a normal control. To reduce individual variation, the CT value in each group was converted to the percentage form of the normal value.

3.5 Main Observatory Indexes

(1) Histological observations on the osteocyte morphous and Haversian canal were carried out under the light microscope. (2) The healing condition of the fractures and state of stress-shielding were observed from the x-ray images. (3) The CT values of the cortical bones were measured by CT machine.

3.6 The Designer, Performer, and Evaluator

All three roles were performed by the author of this article, and nonblind evaluation was used.

3.7 Statistical Analysis

Under the direction of faculty in the teaching and research office of statistics at the Second Military Medical University, the author conducted statistical analyses with the aid of SPSS 12.0 software. A difference of $P < 0.05$ was considered as statistically significant.

4. Results

4.1 Histological Observation

The histological slices from all cases observed under the light microscope were of lamellar bone structure. Compared with the normal substantia ossea, the manifestations under the connective part and the compressive part of SMC group were not significantly different (see Fig. 2a, b); the lamellar bone had regular architecture, the bone cells in the bone matrix were in good shape, and there was no apparent osteoporosis or bone absorption. In the stainless steel plate group, under the steel plate there was obvious osteoporosis and bone absorption was observed. There were also large or small absorption cavities, expanded Haversian canals, shrinking osteocytes, and enlarged bone lacuna in the lamellar bone (see Fig. 3a). However, the lamellar bone opposite to the steel plate had less severe osteoporosis, less absorption cavities, and less shrinking osteocytes in both percentage and number compared to the substantia ossea under the plate (see Fig. 3b).

4.2 X-Ray Imaging Observation

The fracture in all the cases had achieved healing, and the fracture line disappeared. In the SMC group, there was no external callus under the connective part or the compressive part, and the thickness and density of the cortical bone were

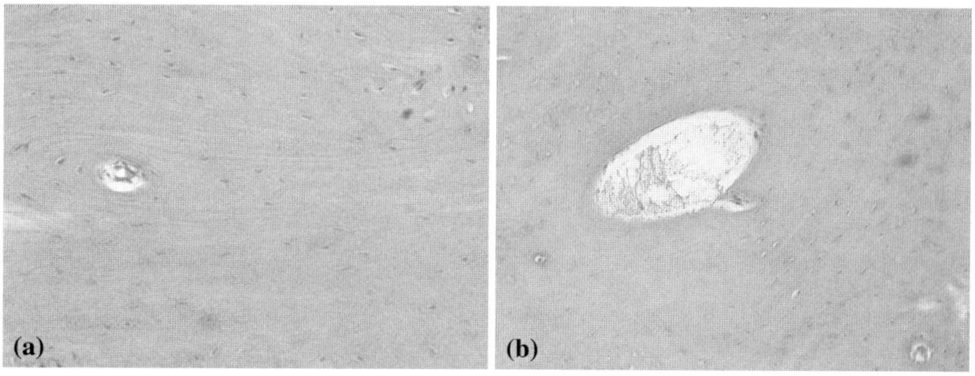

Fig. 2 (a) Under the body part of SMC, the lamellar bone architecture of the cortical bone was regular, the bone cells in the bone matrix were in good shape, and there was no bone absorption cavity. HE × 200. (b) Under the compressive part of SMC, the bone substance was similar to that of the normal cortical bone in shape, the lamellar bone architecture was regular, and the bone cells in the bone matrix were in good shape. HE × 200

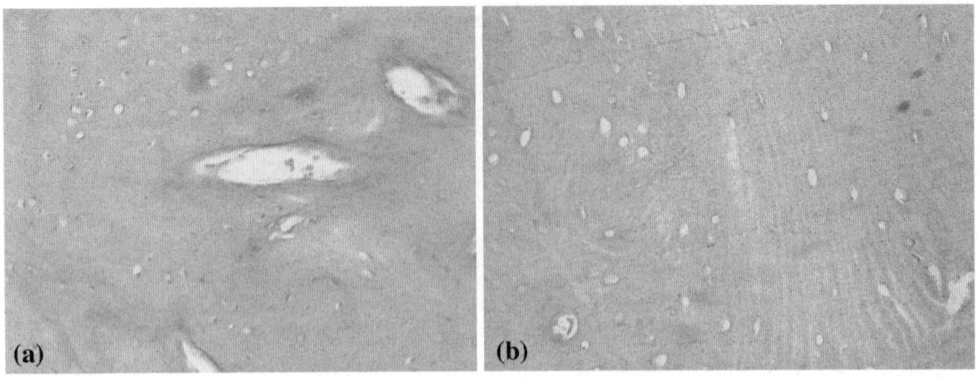

Fig. 3 (a) Under the steel plate, obvious osteoporosis and absorption were observed, and there were large or small absorption cavities in the architecture of the lamellar bone, expanded Haversian canals, atrophied bone cells and enlarged bone lacunas. HE × 200. (b) In the cortical bone opposite to the steel plate, the osteoporosis of the lamellar bone was apparently less severe than that of the bone segment under the plate, and the percentage and the number of the absorption cavities and atrophied bone cells were also less than those of the bone segment under the plate. HE × 200

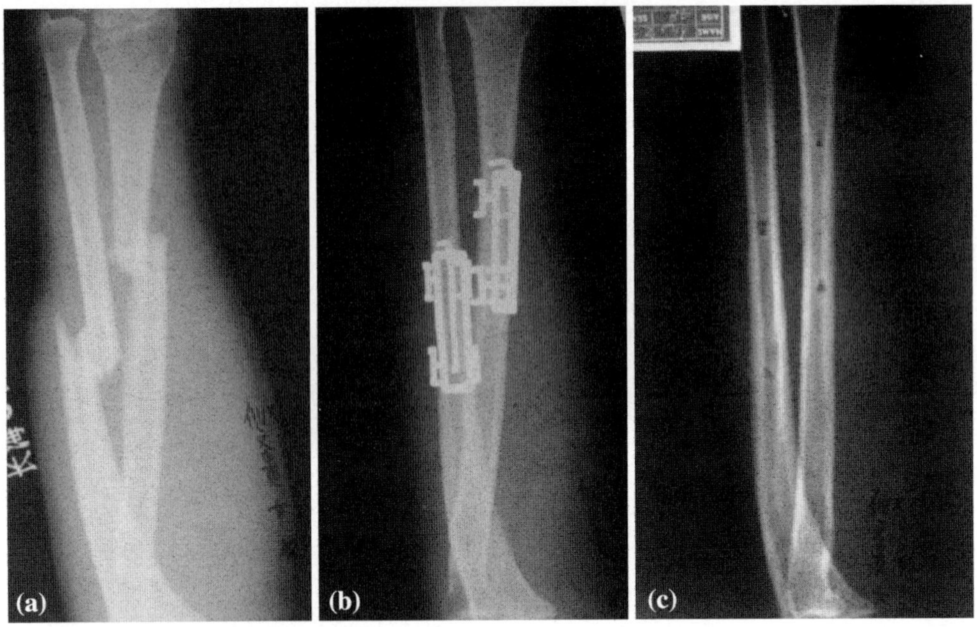

Fig. 4 (a) Fresh fracture of both the ulna and the radius shaft. (b) The healing conditions of the fracture of the ulna and the radius shaft after fixation of SMC, and there was no formation of external callus. (c) 11 months after the operation, the SMC was removed, and the thickness and density of the cortical bone were similar to those of the normal bone substance

similar to those of the normal bone; the cortical bone was in good shape, and its medullary cavity structure and gray scale had no significant differences compared with those of the normal side, indicating good molding of the bone structures (see Fig. 4a-c). In the stainless steel plate group, under the steel plate there was occasionally a small quantity of external callus, the cortical thickness became thinner, and there were pressure marks of deossification in the lateral side of the cortical bone after the removal of the steel plate. The cortical bone opposite to the steel plate had higher thickness and density than those of the cortical bone beneath the steel plate, the bone densities of the two parts were both slightly lower than that of the normal bone, the medullary cavity was recanalized on the whole. In addition, in some cases the density of the medullary cavity was higher than that of the opposite side (see Fig. 5a-c).

4.3 The Measurement of the CT Values of the Cortical Bone

There were no significant differences in the CT values between the connective part and the compressive part of the SMC group ($92.9 \pm 6.8\%$ vs. $93.3 \pm 5.6\%$, $P > 0.05$; see Fig. 6a). In the stainless plate group, the CT value of the cortical bone under the steel plate was lower than that of the bone opposite to the steel plate, indicating there were significant differences ($85.3 \pm 5.3\%$ vs. $91.2 \pm 4.9\%$, $P < 0.05$; see Fig. 6b). Comparison between groups showed that the CT values of the two parts in the stainless steel plate group were both lower than the CT values of the two parts in the SMC group ($P < 0.05$ or $P < 0.01$); the CT value of the cortical bone under the plate was the lowest, the CT value opposite to the steel plate was the lower, while those under the connective

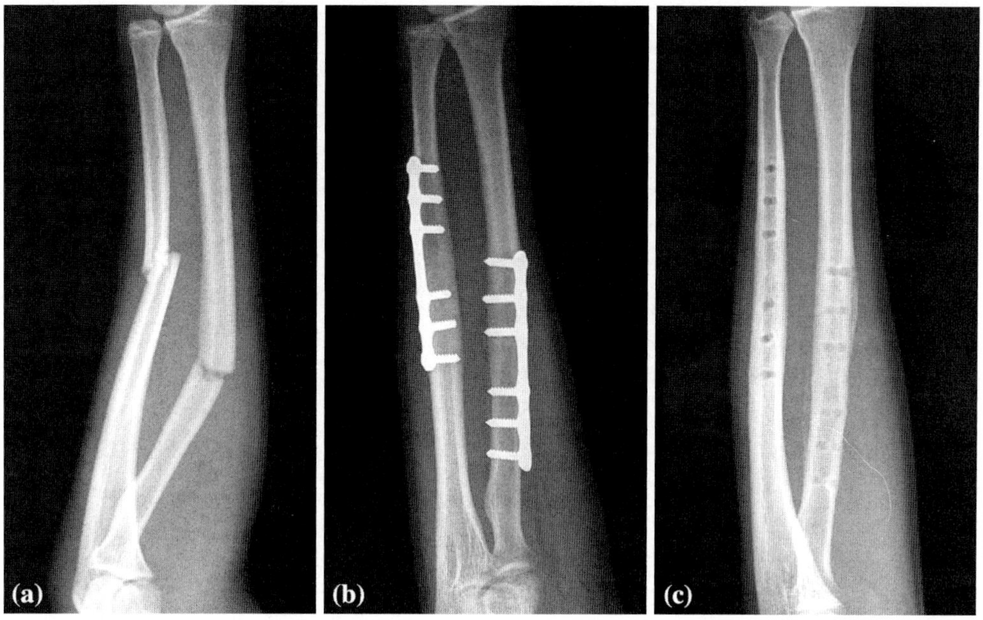

Fig. 5 (a) Fresh fracture of both the ulna and the radius shaft. (b) The healing conditions of the fracture of the ulna and the radius shaft after fixation of stainless steel plate. There was a small quantity of external callus. (c) 13 months after the operation, the steel plate was removed. The cortical bone absorption under the plate became thinner, and the thickness and density of the cortical bone opposite to the steel plate were higher than those of the cortical bone under the plate

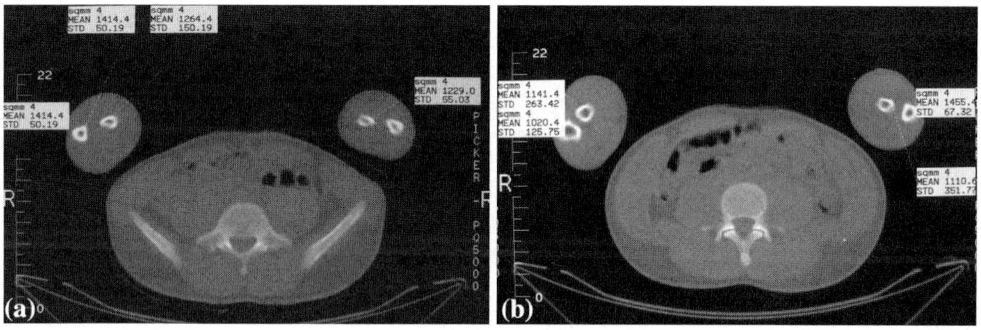

Fig. 6 (a) After the removal of SMC, CT scan was performed on the area with the original fractured ends of the ulna and the radius shaft as the center, and the average CT value of the specified area was measured and compared with that of the opposite normal bone shaft at the same level. (b) After the removal of the steel plate, CT scan was performed on the area with the original fractured ends of the ulna and the radius shaft as the center

part and the compressive part of SMC were at the same high level.

5. Discussion

5.1 The Development of Internal Fixation and Stress-Shielding Effect

Bone fracture refers to the consecutive destruction of the bones due to mechanical high force. Its union is a very complicated biological process influenced by many factors, especially a favorable local mechanical environment that is considered as one of the prerequisites to successful union. The importance of the stress environment to the fracture union has been generally recognized (Ref 14). During the treatment of internal fixation for bone fractures, the influence of bone plate on the blood supply of the fractured site should be reduced into the minimum. In addition, stress stimulation should be exerted on the fracture segment in order to promote bone-healing at the same time as maintaining the stability of the fractured site.

As early as in 1886, Hansmann (Ref 15) from Germany reported the method of treating bone fractures with a plate-screw system. In the past 30 years, the AO School, with Muller et al. (Ref 16) in Switzerland as the leader, established the principle of firm internal fixation that was widely accepted. The firm internal fixation plate can provide stable mechanical environment for the fractured site, ensuring its primary healing. The correct application of firm internal fixation is also conducive to the early exercise of the affected limb, avoiding the occurrence of such complications as myatrophy and anchylosis. However, a large amount of laboratory and clinical experiments have shown that at the advanced period of bone-healing—due to the stress-shielding effect of the bone plate on the fractured part—the local cortical bone lacked physical stimulation and the plate was hence unfavorable for the reconstruction of the bony callus, leading to local osteoporosis and thinning of the cortical bone. Re-fracture was liable to occur after the removal of the internal fixator.

As is known, when two or more materials with a different moduli of elasticity constitute a mechanical system, a redistribution of force will occur. Consequently, the material with greater modulus of elasticity will support more force, and the material with lesser modulus of elasticity will support less force; the strain will be reduced accordingly. This is called the stress-shielding effect. Due to the comparatively large difference between the modulus of elasticity of the material of the bone plate and the bone tissue, the influence of the firm compressive bone plate on the fixed part of bone is considerable. Cochran et al. (Ref 17) applied an AO compressive steel plate with four holes to fix the femur of dogs and found that after fixation, the average compressive strain of the fixed segment was decreased by 45% (especially that of the cortical bone under the bone plate, which was decreased by 84%). According to our current study and the studies of many other scholars, after firm fixation, the local cortical bone became thinner, some absorption cavities appeared, and the medullary cavity of bones became increasingly larger. In histology, there were large or small absorption cavities, expanded Haversian canals, shrinking osteocytes and enlarged bone lacuna in the bone architecture. The more conspicuous the abovementioned changes, the poorer the mechanical property of the bone will be. Due to the lack of stress stimulation in the fractured site, the cell function (controlled by mechanical environment) will be abnormal, such as a decrease of proliferation of the osteoblast and a delayed appearance of osteoblast, leading to local osteoporosis.

Therefore, the correct application of internal fixation and the reduction of the stress-shielding effect are important factors for improving the quality of bone-healing (Ref 18). At the end of the 1970s, many scholars put forth the concept of nonfirm internal fixation. Animal experiments have confirmed that the low-rigidity bone plate made of compound materials can significantly reduce the osteoporosis caused by the stress-shielding effect. However, this plate was weak in stabilizing the fixed site and had a comparatively high rate of delayed union and nonunion of the fracture. Subsequently, people began to use such absorbable biodegradation materials as polylactic acid and polyglycolic acid to make the bone plate, but the degradation speed was hard to control, and the initial mechanical intensity was also hard to meet the requirements of internal fixation of the long bone fracture.

The ideal bone plate should provide stable mechanical environment at the initial stage of bone-healing, and then decrease gradually, rather than abruptly, its stress-shielding effect, thus ensuring a gradual increase of the stress stimulation to the physiological level at the fracture site, hastening the process of bone-healing and molding and preventing the occurrence of local osteoporosis and re-fracture. To meet these ideal requirements, Zhang et al. invented the Nitinol SMC based on the shape memory effect of Nitinol. It is composed of a connective part (swan-body), two pairs of a horizontal encircling part (swan-wings), and a pair of a longitudinal compressive part (swan-necks). The swan-wings resemble a circle when holding the bone and provide point contact instead of close contact with the uneven bone surface. The swan-necks, driven by the body temperature and assisted by the shape memory effect, exert a persistent and dynamic longitudinal load effect on the fracture segment. The combination of the three parts of SMC constitutes the 3D fixation of the long bone shaft. The combination of this unique metal material with the geometric structure is, from the angle of biomechanics, conducive to the prevention of osteoporosis and osteanabrosis (Ref 7, 8).

5.2 The Relation Between Bone Tissue and Stress

Some studies (Ref 19) have indicated that changes of the bone structure are closely related to the mechanical properties of the bone. The mechanical environment of the fractured site impacts greatly on the process of bone-healing (Ref 20). Studies on biochemistry and molecular biology have revealed that under proper stress stimulation, the expression level of the related genes will be changed and the cytokine that regulates bone-healing will be produced (Ref 21). The primary callus reaction in the early stage of bone fracture is the basic reaction of bone tissues to injury, and the subsequent bone callus bridging process is closely related to the mechanical factors (Ref 22). In the early stage of bone-healing, the longitudinal compressive stress drives the differentiation of osteoblast and fibroblast into bone tissues, which is beneficial to the bone-healing. The sheering and the torsion of load will produce sheer stress force, which drives the fibroblast into fibrous tissues and redistributes the stress within the bone, leading to the over-concentration of layer interface stress of the fractured ends. Thus, it is unfavorable for the healing of fracture (Ref 23) and may directly destruct the newly born capillaries and bone callus. However, in the advanced stage of union, the various stresses are all capable of the reconstruction of bone callus, and the increase of sheer stress can promote the differentiation of osteoblast, resulting in more osteoid deposition and the mineralization of bones (Ref 24). If the stress is too small, the mechanical induction of tissue differentiation will be decreased, often resulting in delayed union and nonunion of the bone. If the stress is too large, the reactive surface absorption will take place in the bone-bone interface and the bone-internal fixation interface of the intravital bone, leading to osteanabrosis. When the stress surpasses a specific critical point, further differentiation or union may both stop.

Bone tissue may be functionally adaptive to stress, which ensures its growth in certain directions under the influence or direct action of stress. This is because the changes of stress in the bone tissue can induce the corresponding changes of the electric field, while the changes of the electric field will in turn stimulate the production of the bone tissue. The reaction and the functional adaptation of the bone tissue to the applied load are mainly due to reaction of the bone cell, ossein and bone extracellular fluid to stress or strain caused by the load, constituting a complicated auto-feedback regulatory system within the body. The manifestations are that the bone formation increases at the compressed side due to the stress effect, while bone absorption accelerates at the tension side. However, under certain compressive stress, the re-aggregation and re-absorption of bone tissue are mutually balanced. Additionally, if the compressive stress increases, the bone formation strengthens, and if the compressive stress decreases, the bone re-absorption strengthens (Ref 25, 26).

5.3 The Mechanism of Low Stress-Shielding Effect of SMC

This study has confirmed that the SMC has low or almost no stress-shielding effect on the fixed segment of bone. By comparison of the x-ray images, CT values, and histological observation results between the SMC group and the steel plate group, it was found that there was osteoporosis of the cortical bone under the plate in the steel plate group, while there was hardly any bone absorption and osteoporosis in the SMC group's cortical bone, suggesting the low stress-shielding effect of SMC on the fractured ends.

The SMC's effective mechanism may lie in the unique property of the nickel-titanium memory alloy and special structure of SMC. The elasticity modulus of the human bone is 7-30 GPa. Compared with stainless steel (183.6 GPa), the elasticity modulus of the nickel-titanium memory alloy (61.7 GPa) is similar to that of the human bone (Ref 27). Hence, the stress-shielding effect that existed in the utilization of medical metal materials is greatly reduced by the utilization of nickel-titanium memory alloy, and the weakening and absorption of bones close to the inserted materials are avoided.

The special structure of SMC forms a unique biomechanical environment at the local fracture site, which is very important for the speed and quality of the bone-healing. (1) The 3D multipoint fixation between the SMC and the fracture segment. The interaction between the standard inner diameter of SMC and the nonstandard outer diameter of the bone shaft forms the 3D multipoint fixation that spreads the pressure produced by the shape memory (Ref 28) and thus avoids the compression of local cortical bone due to the over-concentration of stress. Therefore, there is almost no absorption and atrophy of the bone substance. (2) The persistent longitudinal compressive stress effect on the fracture segment. The longitudinal compressive parts (swan-necks) continuously exert their force value on the anatomical fracture segment. Some studies (Ref 29) found that in the early stage of bone-healing, the compressive stress produced by the longitudinal load can drive the differentiation of osteoblasts and fibroblasts into bones, while in the middle and advanced stages, compressive stress is important to the molding of the fracture. The holistic 3D memory fixation of SMC on the fracture site and the dynamic memory compressive stress are more stable than the screw system and more customized to the bone mechanical behavior. This fixation mode may have the physiological guidance force value which stimulates the bone formation, and is thus bound to influence the process of bone-healing and the molding of bone tissues. (3) The integrative design of SMC forms the holistic effect of the fracture fixation. The device is capable of molding, and the original shape can be resumed, driven by the body temperature after placing into the body. It holds the bone in axial direction and exerts pressure on the fracture segment. These features combine into one and are mutually restricted, forming holistic 3D fixation and dynamic compressive effect that are conducive to the bone-healing and the reconstruction of the bone substance, and can alleviate or avoid the loss of bone substance and osteoporosis due to the stress-shielding effect (Ref 7, 8).

Although the stainless plate has a sliding compressive design, the screw can also provide compressive force for the fracture segment when sliding at the compressive hole. However, this compressive force can only be exerted passively and statically once and is bound to be weakened or disappear following the absorption of the fracture segment. More importantly, due to the apparent difference of the stress environment of the stainless steel plate and the SMC fixation approaches, there are obvious differences in the reconstruction activities of the bone tissue and the degree of the osteoporosis between these two approaches. Wu et al. (Ref 30) reported the influence of a Nitinol encircling connector and bone plate on the reconstruction of bones, which showed that at the advanced stage of the internal fixation of Nitinol encircling connector, there was less bone absorption, the degree of osteoporosis is less severe, and the speed of its formation is slower. The current experiment confirmed this conclusion.

6. Summary

In summary, after fixation of the long bone fracture of adult upper limb by SMC and achieving bone-healing, the internal fixator was removed and x-ray imaging observation, CT scanning of the cortical bone and histological observation were conducted. The results confirmed that the SMC has lesser stress-shielding effect than that of the stainless steel plate, is superior in the period of bone molding, and is helpful for the bone-healing. It is believed that the lesser stress-shielding effect of SMC is related to the special property of Nitinol and the unique geometric structure of SMC, suggesting that new modes of treatment for bone fracture may appear with the improvement of material property and architectonic development of internal fixator.

References

1. W.J. Buehler and F.E. Wang, A Summery of Recent Research on Nitinol Alloys and Their Potential Application in Ocean Engineering, *Ocean Eng.*, 1968, **1**, p 105–120
2. Y. Oshida and S. Miyazaki, Corrosion and Biocompatibility of Shape Memory Alloys, *Corrosion Eng.*, 1991, **40**, p 1009–1025
3. J.O. Sanders, A.E. Sanders, R. More, and R.B. Ashman, A Preliminary Investigation of Shape Memory Alloys in the Surgical Correction of Scoliosis, *Spine*, 1993, **18**, p 1640–1646
4. A. Kapanen, J. Ryhanen, A. Danilov, and J. Tuukkanen, Effect of Nickel-Titanium Shape Memory Metal Alloy on Bone Formation, *Biomaterials*, 2001, **22**, p 2475–2480
5. M. Assad, N. Lemieux, and C.H. Rivard, Comparative In Vitro Biocompatibility of Nickel-Titanium, Pure Nickel, Pure Titanium, and Stainless Steel: Genotoxicity and Atomic Absorption Evaluation, *Biomed. Mater. Eng.*, 1999, **9**, p 1–12
6. C.C. Zhang, S.G. Xu, J.L. Wang, B.Q. Yu, Q.G. Wang, Q.L. Zhang, Y.T. Liang, and R. Wang, Design and Clinical Application of Swan-like Memory Compressive Connector for Upper Limb Diaphysis, *Acad. J. Sec. Mil. Med. Univ.*, 2001, **22**, p 939–942 (in Chinese)
7. S.G. Xu, C.C. Zhang, J.C. Su, X.H. Liu, Y.N. Wu, J.G. Wu, Z.J. Xue, and Z.Q. Ding, Three-Dimensional Element Analysis of Swan-like Memory Compressive Connector for Treating Fractures and Nonunion of Humerus, *Acad. J. Sec. Mil. Med. Univ.*, 2001, **22**, p 943–945 (in Chinese)
8. S.G. Xu, C.C. Zhang, J.C. Su, W.M. Zeng, S.D. Gu, R.H. Fang, and L.C. Zhang, Biomechanical Study of Swan-Like Memory Compressive Connector for Treating Fractures and Nonunions of Humerus, *Acad. J. Sec. Mil. Med. Univ.*, 2001, **22**, p 946–948 (in Chinese)
9. J.G. Hazenberg, D. Taylor, and T. Clive Lee, Mechanisms of Short Crack Growth at Constant Stress in Bone, *Biomaterials*, 2006, **27**, p 2114–2122
10. S. Tadano and T. Okoshi, Residual Stress in Bone Structure and Tissue of Rabbit's Tibiofibula, *Biomed. Mater. Eng.*, 2006, **16**, p 11–21
11. J. Lin, P.W. Shen, and S.M. Hou, Complications of Locked Nailing in Humeral Shaft Fractures, *J. Trauma*, 2003, **54**, p 943–949
12. C. Bergot, A.M. Laval Jeantet, and K. Hutchinson, A Comparison of Spinal Quantitative Computed Tomography with Dual Energy X-ray Absorptiometry in European Women with Vertebral and Nonvertebral Fractures, *Calcif. Tissue. Int.*, 2001, **68**, p 74–82

13. J.A. MacNeil and S.K. Boyd, Accuracy of High-Resolution Peripheral Quantitative Computed Tomography for Measurement of Bone Quality, *Med. Eng. Phys.*, 2007, **29**, p 1096–1105
14. A.D. Bakker, J. Klein-Nulend, E. Tanck, I.C. Heyligers, G.H. Albers, P. Lips, and E.H. Burger, Different Responsiveness to Mechanical Stress of Bone Cells from Osteoporotic Versus Osteoarthritic Donors, *Osteoporos. Int.*, 2006, **17**, p 827–833
15. C. Hansmann, Eine neue Methode der Fixirung der Fragmente bei omplicierten Frakturen, *Verh. Dtsch. Ges. Chir.*, 1886, **15**, p 134–136
16. M.E. Muller, M. Allgöwer, and H. Willenegger, *Manual der Osteosynthese: AO-Technik*, Springer, Berlin, Heidelberg, 1969
17. G.V.B. Cochran, V.R. Palmieri, and R.E. Zickel, Aramid-Epoxy Composite Internal Fixation Plates: A Pilot Study, *Clin. Biomech.*, 1994, **9**, p 315–322
18. A.P. Datir, Stress-Related Bone Injuries with Emphasis on MRI, *Clin. Radiol.*, 2007, **62**, p 828–836
19. D. Taylor and T.C. Lee, Microdamage and Mechanical Behavior: Predicting Failure and Remodeling in Compact Bone, *J. Anat.*, 2003, **203**, p 203–211
20. O. Akkus and C.M. Rimnac, Cortical Bone Tissue Resists Fatigue Fracture by Deceleration and Arrest of Microcrack Growth, *J. Biomech.*, 2001, **34**, p 757–764
21. D.R. Epari, F. Kandziora, and G.N. Duda, Stress Shielding in Box and Cylinder Cervical Interbody Fusion Cage Designs, *Spine*, 2005, **30**, p 908–914
22. S. Nomura and T. Takano-Yamamoto, Molecular Events Caused by Mechanical Stress in Bone, *Matrix Biol.*, 2000, **19**, p 91–96
23. T.A. Einhorn and J.M. Lane, Significant Advances Have Been Made in the Way Surgeons Treat Fractures, *Clin. Orthop. Relat. Res.*, 1998, **355**(Suppl), p S2–S3
24. K.J. Jepsen, D.T. Davy, and D.J. Krzypow, The Role of the Lamellar Interface During Torsional Yielding of Human Cortical Bone, *J. Biomech.*, 1999, **32**, p 303–310
25. J.P. Lai, G. Wei, X.B. Li, and M. Xie, Research and Design of Bone Plate with Low Stress-shielding Effect, *Orthop. Biomech. Mater. Clin. Study*, 2007, **4**, p 51–52 (in Chinese)
26. W.Z. Xu, Y.J. Wang, X.Z. Xu, L.M. Wang, and Y.J. Wang, Treatment of Femoral Fractures with Stress-Controlled Auto-Compression Locking Nails, *Chin. J. Orthop. Trauma*, 2007, **9**, p 670–674 (in Chinese)
27. J. Rubin, X. Fan, D.M. Biskobing, W.R. Taylor, and C.T. Rubin, Osteoclastogenesis is Repressed by Mechanical Strain in an in Vitro Model, *J. Orthop. Res.*, 1999, **17**, p 639–645
28. Y. Wang, J. Tang, Z.D. Cui, S.L. Zhu, and X.J. Yang, Properties and Bioactive Surface Engineering on Nickel-Titanium Shape Memory Alloy, *Heat. Treat. Metals*, 2004, **29**, p 40–45 (in Chinese)
29. Y.Z. Zhang, Q.X. Wang, J.S. Pan, C.P. Zhang, L.L. Zhao, and Z.H. Song, The Design and Biomechanical Stability of the Automatic Physiological Compression Plate, *Chin. J. Orthop. Trauma*, 2005, **7**, p 66–68 (in Chinese)
30. X.T. Wu, K.R. Dai, and X.S. Zu, et al., The Effect of Internal Fixation of Nitinol Encircling Fixator and Steel Plate on the Healing and Reconstruction of Fracture, *Chin. J. Surg.*, 1998, **33**, p 481–484 (in Chinese)

Table 1 Patients' characteristics

Patient	Age	Sex	Time since event, years	Paresis	Spasticity score Flexors (extensors)	ROM
A	62	M	6	Left sensory-motor hemisyndrome	Ashworth 3 (3)	Full passive No active
B	64	F	9	Left sensory-motor hemisyndrome	Ashworth 2 (0)	Full passive No active

ROM, range of motion

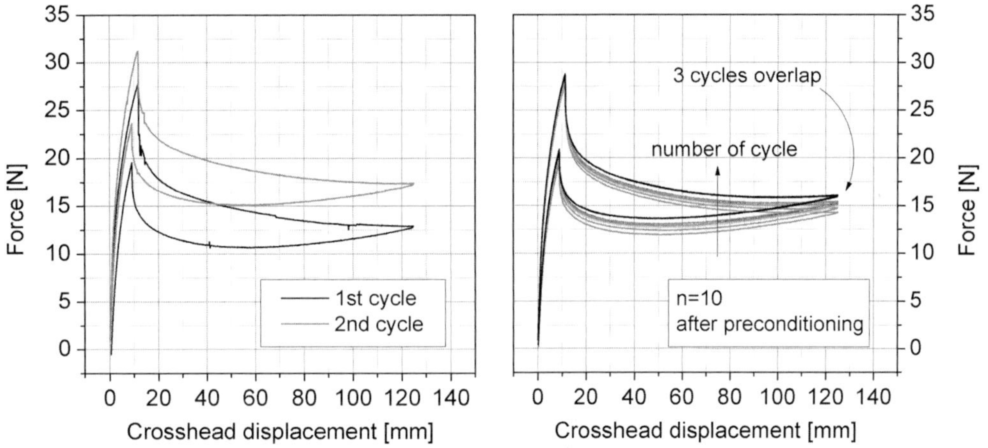

Fig. 5 Wire bending results. Left: first two deformation cycles on the as-thermally treated sample; right: deformation cycling on the same sample (after manual preconditioning cycles)

sample appears to be more stable; in fact, even if slight further stiffening occurs it can be noticed that the 8th, 9th and 10th cycles are completely overlapping (Fig. 5, right).

4.2 Mechanical Characterization of EDGES

Figure 6 reports the results of the bending tests conducted on the assembled orthosis. The corrective torque versus angle characteristic of EDGES was nonlinear and hysteretic in both trials. Sufficient corrective push is ensured also at lower angles. The orthosis produces a moment increasing with angle in the 20-80° range, after which it appears to reach a plateau at around 0.95 and 1.9 Nm when equipped with 2 and 4 wires, respectively.

4.3 Pre-Clinical Assessment

Variations of clinical conditions during the follow-up period were recorded and results are as in Table 2. It can be noticed that the general trend, even in these very chronic patients, is definitely toward an improvement of elbow posture, which is already present at 24 h and becomes more evident after 1 week. The decrease in spasticity obtained after 1 week of EDGES application is indeed a very interesting datum. All improvements appear to regress within 1 week of discontinuation. Minor skin rash was the only observed complication.

4.4 Optoelectronic Measurements

Optoelectronic measurements show very nicely the time course of elbow angle during the proposed clinical trials (Fig. 7). The quantitative analysis of movement confirms the general trend toward an increased elbow extension during therapy. Furthermore, it is very interesting to observe that, while EDGES allows for wide movement of the elbow and does not impede flexing synergies during standing-up and walking, it is capable of reducing the amplitude of this pathological movement.

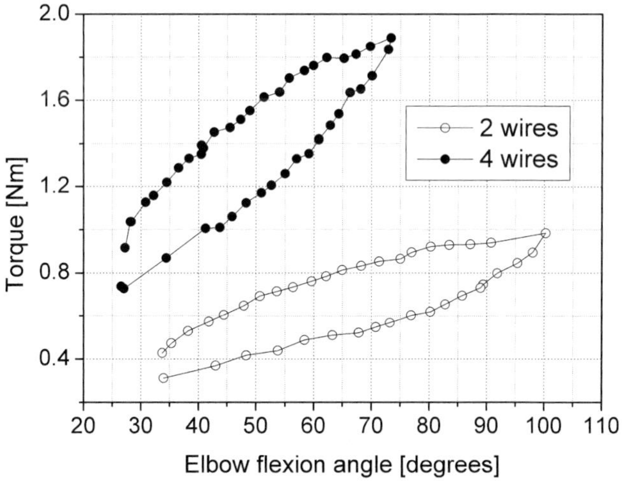

Fig. 6 EDGES bending results. Data are shown by the dots. Lines are guides for the eye. Both tests confirm that the orthosis has a non-linear hysteretic torque-angle behavior

Table 2 Clinical status as from manual assessment during therapy follow-up: resting elbow angle during different motion tasks and Ashworth spasticity score (without EDGES)

Patient	Time	Sitting, °	Standing, °	After walking, °	After relaxing, °	Ashworth
A	T_0	60	50	65	50	3
	3 h	45	45	60	50	3
	24 h	45	50	60	50	3
	1 week	35	40	55	40	2
	1 week after	50	60	65	60	3
B	T_0	70	110	100	60	2
	3 h	30	60	70	50	1+
	24 h	30	30	40	30	1
	1 week	25	30	40	20	1
	1 week after	65	80	95	60	2

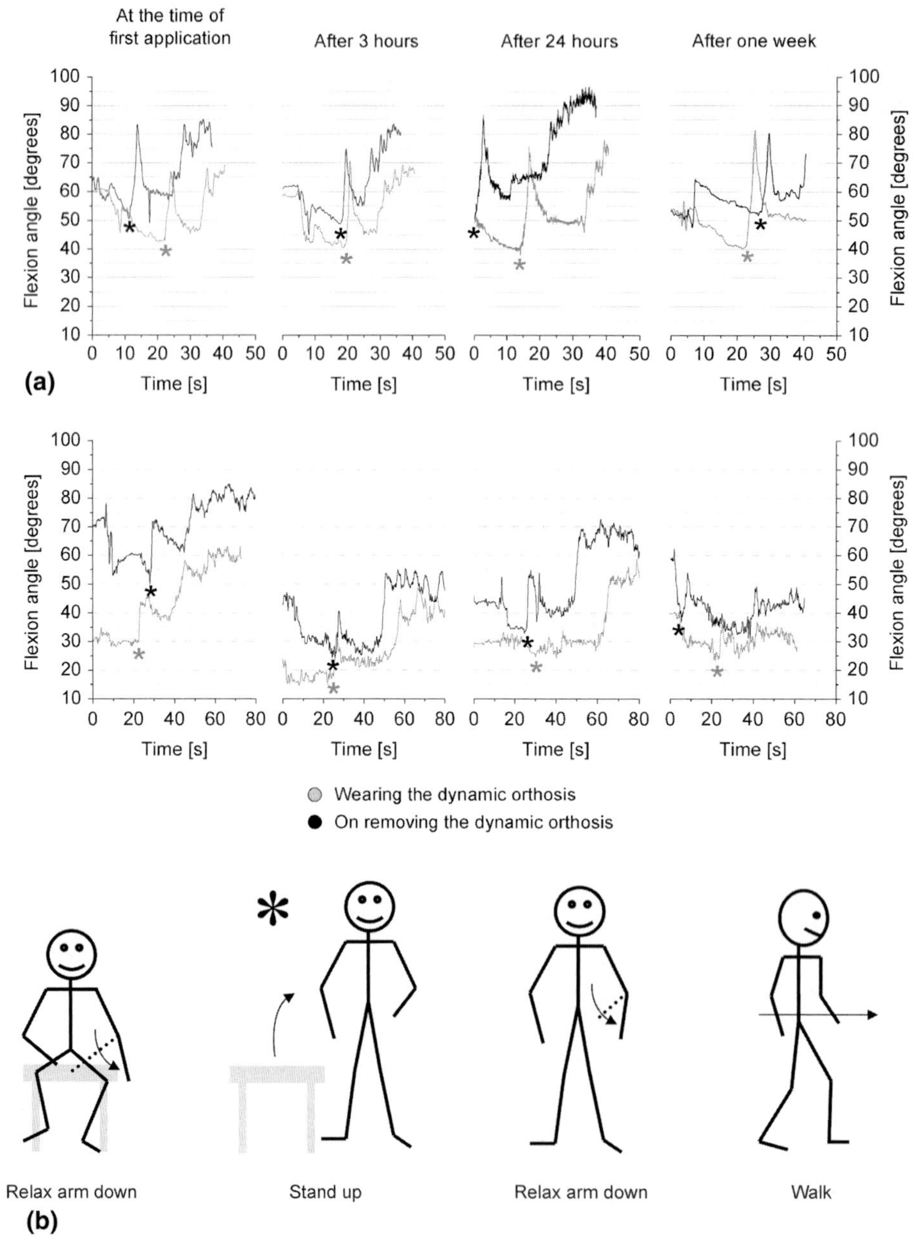

Fig. 7 Clinical tests on EDGES. Optoelectronic acquisition of patients' arm movement during the motion sequence sit – relax down the arm – stand up – relax down the arm – start walking. Stars mark the "stand up" command and are followed by a sudden and pronounced flexion of the elbow (flexor synergies)

Table 3 Outcomes of Gracies' questionnaire for orthosis acceptability

Question	A (24 h)	B (24 h)	A (1 week)	B (1 week)
How long did you wear EDGES today?	24 h	24 h	12 h	18 h
How did your arm feel while wearing EDGES?	8/10	8/10	7/10	8/10
Did you notice any changes when wearing EDGES?				
When walking	Better	Better	Better	Better
When toileting	No change	No change	No change	No change
When eating	No change	No change	No change	No change
When dressing	No change	No change	No change	No change
In ease of movement (muscle tone)	No change	No change	Better	No change
In the confidence in your arm	Better	Better	Better	Better
Would you wear EDGES for a few weeks if recommended?	Yes	Yes	Yes	Yes
Comments…	N/A	N/A	Cannot put on brace by myself. Some skin rash	Cannot sit putting the arm on the lap

4.5 Acceptability

The results of the acceptability questionnaires taken on the first and last days of therapy are presented in Table 3.

Patients' acceptation of EDGES was generally good after both 24 h and 1 week of therapy. Willingness to continue wearing it is preserved and only minor problems were observed, among which some light skin rash in the regions covered by the shells (far from the NiTi elements), which required no particular medication and that could possibly be avoided by a better lining than the present one. A particularly good result is in terms of improved ease of movement and confidence for the patient. It is interesting to note that both patients reported walking better when they are wearing this elbow splint, possibly suggesting that improved confidence and decreased flexor synergies bring about greater stability and equilibrium. Further comments related to the patients' inability to put on the brace by themselves or to rest the elbow on their laps when wearing EDGES.

5. Discussion

5.1 Choice of Material and Thermo-Mechanical History

This study aimed at demonstrating the possibility of using off-the-shelf basic pseudoelastic NiTi in orthotics. For this reason a commercial wire with an easily available formulation was utilized. The thermo-mechanical treatment was selected according to well-known principles of NiTi technology such as: first, good mechanical characteristics, working stability and fatigue life at higher loads can be obtained by an appropriate combination of cold working and annealing below re-crystallization temperature (Ref 7, 8); second, mechanical preconditioning can be employed to stabilize the material behavior for a specific application (Ref 8). According to literature (Ref 7, 9, 10) suitably fine and homogeneously dispersed Ti_3Ni_4 precipitates can be obtained by ageing at temperatures around 673 K for 1 h. That precipitation was obtained with this treatment and was confirmed by DSC analysis, which shows two-stage transformation to parent phase (Fig. 2, top left).

Although the above-mentioned technological criteria were applied, the present batch of material responded only partially well: in particular, static mechanical properties were actually very good, but the cycling behavior appeared rather poor, especially in the (rather severe) tensile tests.

5.2 Results of Tensile Tests on the NiTi Wire

The plateau stresses measured during tensile tests, although apparently high given the presented T_r and A_f, are consistent with the chosen ageing process and the underlying working history. The comprehensive 2005 review by Otsuka and Ren (Ref 7) duly shows specific evidence of how the combined effects of cold-working and precipitation ageing can powerfully increase strength in Ni-rich NiTi, even to the extent of allowing for large recovery stresses strictly below A_f. Even elsewhere it is reported (e.g. Ref 11) that in the intermediate temperature range (673-773 K) for ageing of cold-worked high-Ni NiTi, the main effect of residual work-hardening (and Ni-rich precipitates) on DSC-measured transformation temperatures is a strong decrease in M_s and M_f, while A_s and A_f tend to remain stable. On the other hand, lower-end intermediate treatment temperatures (e.g. 673 K rather than 773 K, the latter inducing too much softening) are typically employed to provide improved shape recovery forces (Ref 12). These concepts put together mean that the heights of the plateaux cannot be straightforwardly linked to the position of A_f alone. In other terms, dynamic explanations considering nonlinear interactions among the solid phases must be sought, as the calorimetric (viz. a Clausius-Clapeyron approach) cannot take into account microstructural rearrangements due to applied mechanical strains. This deviation from basic assumptions is also observable in the data presented within the cited review by Otsuka and Ren, and comes even more expected with our Ni-richer alloy ($Ti\text{-}Ni_{50.7}$ rather than their $Ti\text{-}Ni_{50.6}$). It is also likely that these effects might occur even more strikingly with a fine-grained and inclusion-rich matrix (as in the present case, see Fig. 8), where dislocation disentanglement and movement is further impeded. The phenomenology of pseudoelasticity in the present instance is probably more appropriately analyzed by observing that, from a thermodynamic perspective, a certain amount of work hardening and precipitation causes M_s to shift toward lower values and peaks to spread, thus increasing the energy requirement for inducing detwinned martensite from austenite. At the same time, the same factors have direct mechanical effects in the form of hardening, through defect formation and hindrance to material undergoing phase transitions. The large difference in transformation enthalpies between

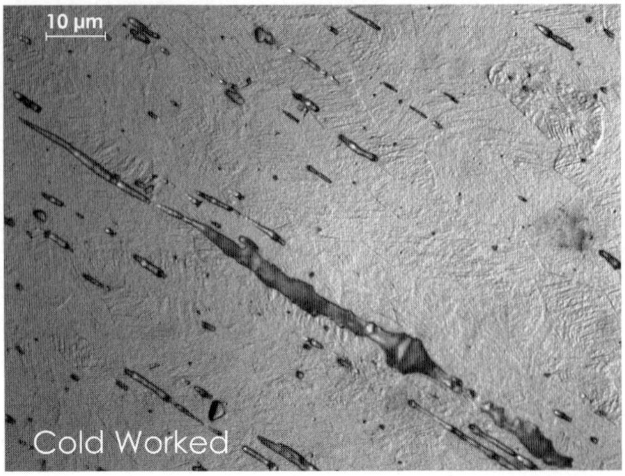

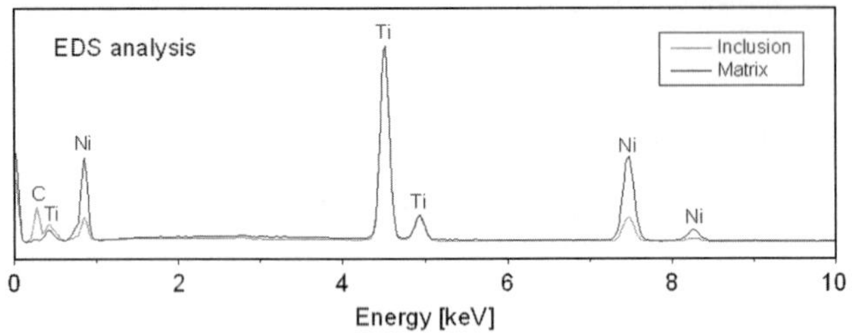

Fig. 8 Longitudinal 1000× micrographs for the cold worked (as-received) NiTi wire. The sample was polished and etched in a 10% HNO_3, 5% HF water solution for 20 s. Large defects are visible (mainly TiC, from EDS analysis, see bottom graphs). Grain size is in the order of 20 μm

the solution treated and aged samples (approximately 14 and 8 J/g, respectively) demonstrates that in the aged specimen some material either does not transform at all by sheer temperature changes or transforms so gradually that its DSC signal lies outside the main peaks but cannot be appreciated. Gradual transformation is also evidenced by the characteristic curved pattern of early shape recovery in tensile tests, signaling an important contribution of mechano-elastic terms in transformational energy.

5.3 Tensile Cycling and Microstructural Issues

In the light of the mechanical characterization results, it appears that cycling stability, particularly in tensile tests, is worse than generally expected, even considering that the chosen cycling protocol is rather heavy. Indeed, the fact of not heating the sample over A_f after every cycle produces an accumulation of strains within the matrix, so that finally nominal end-strains of 4% do not correspond to real strains of that amount. Thence a certain stress-hardening and structural rearrangements with slanting and lowering of the plateaux would be normal, probably not to the observed extent in "good" material. The excessively detrimental effect of cycling in the present test suggested that some investigation of the microstructural and chemical characteristics should be undertaken: light and scanning electron microscopy and the use of EDS elemental analysis demonstrated that the as-received material was fine grained (~20 μm), definitely textured along the drawing direction, and affected by abundant and large TiC inclusions across the whole sample, as clearly visible in Fig. 8. This was felt to explain the below-average cycling attributes and called for the choice of more refined materials for this type of applications in the future.

5.4 Results of Bending Tests on the NiTi Wire

Considering the wire bending tests, it can be noted that the major effect in the first cycles is a relative increase in measured stress values. This depends on the typical stiffening (very palpably perceived on manual deformation) correlated with the progressive increase in Young's modulus brought about by the strain-induced straightening of the R-phase notch. As a result of this, the whole curve is rotated counter-clockwise and the measured stresses tend to be higher. After the 50-cycle manual conditioning, the material shows a decrease in strength which can be associated with the decline in stress values produced by repeated deformation. Of course, the intensity of such decline is not comparable with the results of the tensile tests, as the severity of strain cycling is much reduced in the bending set-up. The final cycles are affected by another stiffening phase occurring in the transition between a manual two-way bending regime and a machine-controlled one-way one, probably implying, among other things, a slight shift of the principal inflection point and a variation of the loading path and speed. These minor adjustments should nonetheless be accompanied by a tuning of the characteristics of the newly involved material grains. After that, stabilization does occur as the test is conducted at fixed displacements, which means that wire curvature and fiber *real* strains remain close to constant. The use of EDGES on patients is more similar to the bending (rather than the tensile) testing procedure, so the presented results suggest fair working stability in spite of the tensile cycling curves.

5.5 Working Principles of the EDGES Orthosis

The purpose of including pseudoelastic elements in a positioning brace is to provide a corrective torque across an angle as wide as possible such that it is not negligible even when the arm is almost fully extended. This principle is very innovative with respect to traditional orthotics, which is based on fixing the position of the limb and await tissue remodeling by relaxation. In the present work, a stable corrective force (and not a stable position) drives muscle stretching. This force ought to be sufficiently strong to produce tissue lengthening by *viscoelastoplastic creep* and not as strong as to impede arm movements.

The values of corrective torque produced by EDGES, as measured from the bending tests (0.95 and 1.9 Nm), are considerably lower than maximal muscular flexor torques of around 50-75 Nm (Ref 13, 14), and somewhat less than the maximal passive-reflex torques, which are reported to be in the range 0.5-6 Nm (Ref 15, 16). These measurements confirm that EDGES can stretch muscles while not constraining elbow position during voluntary movements or involuntary jerks.

5.6 Pre-Clinical Considerations

The present study is far from a conclusive clinical validation of this EDGES orthosis. However, it yielded some interesting pre-clinical considerations worth taking into account in view of future developments. The clinical procedures carried out to assess the dynamic characteristics of EDGES when worn by spastic subjects showed that even in this unoptimized version it can produce corrective loads in a stable manner for at least a week and that it does not constrain the elbow in a fixed position. This is clearly visible from the strong oscillations in arm position detected and shown in Fig. 7. Flexor synergies, which are a very common feature of post-stroke syndromes, cause involuntary arm flexion during some types of body movements (standing up, walking…), as can be evinced also from the manual measures presented in Table 2. Constraining devices can therefore be ill-tolerated in these circumstances. Acceptability of EDGES was, on the other hand, very good, suggesting that longer orthotic treatments with this orthosis can be attempted. The ratings reported in the questionnaire of Table 3 were high and the daily application long, demonstrating that the patients felt no pain or discomfort during the therapy. Decrease in spasticity could be a wonderful result but it needs strong further evidence to be confirmed, as spasticity attenuation has never been conclusively proved in connection with orthotic treatments (Ref 17). Relapse of ill-posture and spasticity after treatment discontinuation was absolutely expected, as 1 week is not a sufficiently long time to obtain plastic remodeling of human muscular and nervous tissues. Stabilized benefits could only be attended, should treatment be prolonged by several weeks. Reversibility of patients' conditions in the present clinical experiment suggests that EDGES is not harmful or produces any permanent damage in a week's application.

The only observed complication was some light skin rash in patient A, which was considered a minor problem as it disappeared in a matter of hours without any need for medications. To fix this, a softer lining will be added to the next generation of EDGES.

Concerning the additional remarks in the questionnaires: first, it is believed that the willingness to be able to put on the orthosis without assistance, although suggesting that the patient was at ease with EDGES and confident with its working principles, actually points out a generalized problem with wearable aids, i.e. that they may be difficult to handle for impaired subjects. In the case of EDGES, there is still some marginal opportunity to improve on this, e.g. by implementing customized shells and connection bars, so that increased surface coherence between the orthosis and the body may guide easier fitting. Second, the question of whether allowing patients to rest the arm in their laps is moot, as this very attitude is reported to be among the leading causes for early development of ill-postures in paretic subjects (Ref 4). Most surely, a device provoking a certain degree of arm extension can make sitting positions less comfortable, but this question could probably be re-addressed by suggesting that the arm be rested at a more extended, rather than flexed angle, for instance on a table or portable tray.

6. Conclusions

A pseudoelastic positioning brace was designed and built based on commercial NiTi wires. A thermal treatment at 673 K 1 h + WQ to produce abundant precipitation and only limited material softening provided the wires with sufficient strength for the application. Orthosis characterization showed that it can generate corrective torques of 0.95 and 1.9 Nm when equipped with 2 and 4 wires, respectively. Clinical outcomes show very fast improvement of patients' posture within 24 h and further progress at 1 week. Spasticity appears to be decreasing during the first week, but this observation needs strong additional evidence to be confirmed. Patients' satisfaction was very high, as was acceptability at both 24 h and 1 week of treatment. This suggests that pseudoelastic NiTi is a suitable material for constructing elbow positioning braces that work on the principle of tissue lengthening by *viscoelastoplastic creep*. Other types of orthoses could be significantly enhanced with the use of pseudoelastic NiTi elements, whose quality must be controlled particularly in order to ensure good performance stability. As a conclusive consideration, however, it must be stressed that, in spite of the great informative value of our experience so far, true clinical validation of this type of pseudoelastic treatment will have to be sought in the future by longer application trials and randomized comparison studies with larger patient populations.

References

1. J. Mackay and G. Mensah, *Atlas of Heart Disease and Stroke*, World Health Organization, Geneva, 2004, p 50
2. E. Lundström, A. Terént, and J. Borg, Prevalence of Disabling Spasticity 1 Year After First-Ever Stroke, *Eur. J. Neurol.*, 2008, **15**(6), p 533–539
3. D.K. Sommerfeld, E.U. Eek, A.K. Svensson, L.W. Holmqvist, and M.H. von Arbin, Spasticity After Stroke: Its Occurrence and Association with Motor Impairments and Activity Limitations, *Stroke*, 2004, **35**(1), p 134–139
4. J.-M. Gracies, Pathophysiology of Spastic Paresis I: Paresis and Soft Tissue Changes, *Muscle Nerve*, 2005, **31**, p 535–551
5. S. Pittaccio, S. Viscuso, M. Caimmi, G. Gasperini, S. Pirovano, S. Besseghini, and F. Molteni, First Assessment of a New Dynamic Brace for Elbow Muscle Relaxation Based on Smart Materials, *Proceedings of the 4th World Congress of the International Society Physical and Rehabilitation Medicine*, U.W. Rah, Ed., June 10–14, 2007 (Seoul), Organizing Committee of ISPRM, 2007, p 110
6. J.-M. Gracies, J.E. Marosszeky, R. Renton, J. Sandanam, S.C. Gandevia, and D. Burke, Short-Term Effects of Dynamic Lycra Splints on Upper Limb in Hemiplegic Patients, *Arch. Phys. Med. Rehabil.*, 2000, **81**, p 1547–1555
7. K. Otsuka and X. Ren, Physical Metallurgy of Ti-Ni-Based Shape Memory Alloys, *Prog. Mater. Sci.*, 2005, **50**, p 511–678
8. S. Miyazaki, Thermal and Stress Cycling Effects and Fatigue Properties of Ni-Ti Alloys, *Engineering Aspects of Shape Memory Alloys*, T.W. Duerig et al., Eds. (London), Butterworth-Heinemann, 1990, p 401–407
9. S. Miyazaki and K. Otsuka, Development of Shape Memory Alloys, *ISIJ Int.*, 1989, **29**(5), p 353–377
10. S. Miyazaki, T. Imai, Y. Igo, and K. Otsuka, Effect of Cyclic Deformation on the Pseudoelasticity Characteristics of Ti-Ni Alloys, *Metall. Trans. A*, 1986, **17**(1), p 115–120
11. T. Todoroki and H. Tamura, Effect of Heat Treatment After Cold Working on the Phase Transformation on TiNi Alloy, *Trans. JIM*, 1987, **28**(2), p 83–94
12. Y. Suzuki, Shape Memory Processing, *Shape Memory Alloys*, English edition, H. Funakubo, Ed., OPA (Amsterdam), B.V., 1987, p 194–200
13. T.S. Buchanan, Evidence that Maximum Muscle Stress is Not a Constant: Differences in Specific Tension in Elbow Flexors and Extensors, *Med. Eng. Phys.*, 1995, **17**(7), p 529–536
14. A.A. Amis, D. Dowson, and V. Wright, Analysis of Elbow Forces Due to High-Speed Forearm Movements, *J. Biomech.*, 1980, **13**(10), p 825–831
15. B.D. Schmit and W.Z. Rymer, Identification of Static and Dynamic Components of Reflex Sensitivity in Spastic Elbow Flexors Using a Muscle Activation Model, *Ann. Biomed. Eng.*, 2001, **29**(4), p 330–339
16. P.H. McCrea, J.J. Eng, and A.J. Hodgson, Linear Spring-Damper Model of the Hypertonic Elbow: Reliability and Validity, *J. Neurosci. Methods*, 2003, **128**(1–2), p 121–128
17. N.A. Lannin, I. Novak, and A. Cusick, A Systematic Review of Upper Extremity Casting for Children and Adults with Central Nervous System Motor Disorders, *Clin. Rehabil.*, 2007, **21**(11), p 963–976

JMEPEG (2009) 18:814–817
DOI: 10.1007/s11665-009-9451-5

1059-9495/$19.00

Determination of Transformation Properties of Thin Medical Grade Ni-Ti Wire by High-Resolution Bend and Free Recovery Testing

A. Undisz, M. Fink, and M. Rettenmayr

(Submitted September 17, 2008; in revised form April 6, 2009)

Thin medical grade Ni-Ti wires, pseudo-elastic at ambient temperature, were annealed to set the final shape and the properties of the material. In order to investigate the influence of the annealing, high-resolution bend and free recovery tests were carried out. With the newly constructed device, the reverse deformation (free recovery) of the bent wires was recorded in detail. Wire segments with a weight of <1 mg are sufficient to attain reliable results. The high resolution of the testing device is based on a precise temperature control and a video monitoring system equipped with a microscope lens. In addition to the determination of the transformation temperatures, which is possible with common bend and free recovery tests, the high resolution allows for the identification and quantification of the reverse deformation stages attributed to the martensite-to-austenite transformation and the occurrence of the R-phase. The variations of the transformation properties attained by evaluating the different types of reverse-deformation curves are presented and discussed.

Keywords advanced characterization, biomaterials, heat treating

1. Introduction

The application of thin Ni-Ti wire for medical devices and implants requires specific and reliable information about material properties as, e.g., the transformation temperatures (Ref 1, 2). Short-time annealing that is necessary to set the shape of medical implants and devices has significant influence on the transformation temperatures (Ref 3). In order to warrant the functionality of an implant device, the transformation properties including the temperature range of the transformation and the corresponding reverse deformation degree are also of critical importance. Since the loading and straining conditions affect the transformation behavior and the transformation temperatures (Ref 4), the testing method should reproduce the loading and straining conditions of the material during its application (Ref 5). In surgical applications, the major deformation of pseudo-elastic Ni-Ti implants occurs when they are collapsed to small sizes and positioned through small catheters inside the patient. There, they further unfold and, thereby, fulfill their vessel-supporting or -occluding purpose. Similar straining conditions, such as during transcathetral surgeries, which result in bending of the wires, are generated during the bend and free recovery test. Since the type of deformation of the wires during this test is similar to that occurring in specific medical applications of interest, the determination of transformation temperatures and transformation characteristics is assumed to be of higher significance than when applying other methods without deformation or mechanical loading.

Results of bend and free recovery tests available in the literature occasionally exhibit low accuracy due to imprecise temperature control during the testing (Ref 6), or due to correlation of temperature and reverse deformation by only a small set of data points (Ref 7).

The experimental setup for bend and free recovery tests according to the standard test procedure ASTM F2082-03 (Ref 8) is based on contact methods to measure the reverse deformation of the material, but a non-contact method is suggested. Sometimes, in the production of medical implants, a thin wire with a diameter of <150 μm is required. Use of thin wire results in comparably small forces generated during the free recovery. In this case, the measurement of the reverse deformation with contact to the specimen is prone to deviations of the measured transformation temperatures and the transformation characteristics. In order to attain a reliable characterization of the transformation properties of thin wires, a new device has been constructed (Ref 3). The device combines a non-contact measurement technique of the reverse deformation, a precise temperature control by combining heating and cooling system, a temperature sensor right next to the specimen, and a detailed recording of the reverse deformation of the wire using a camera system equipped with a microscope lens. Using this device, not only the influence of annealing on the transformation temperatures (Ref 3), but also the effects on the transformation properties and the occurrence of the R-phase can be determined.

This article is an invited paper selected from presentations at Shape Memory and Superelastic Technologies 2008, held September 21-25, 2008, in Stresa, Italy, and has been expanded from the original presentation.

A. Undisz, M. Fink, and **M. Rettenmayr**, Institute of Materials Science and Technology, Metallic Materials, Friedrich-Schiller-University, Loebdergraben 32, 07743 Jena, Germany. Contact e-mail: Andreas.Undisz@uni-jena.de.

2. Experimental

Medical grade pseudo-elastic Ni-Ti wire with a diameter of ~150 μm and a Ni content of 50.8 at.% complying with the requirements of the standard SE508 was used. The manufacturing process is proprietary to the producer, and no details of annealing temperature and period of the straight annealing process were provided. However, it is clear from previous investigations that a certain defect density resulting from the drawing process remains stored inside the material. This was confirmed by annealing experiments that resulted in recrystallization under certain conditions (Ref 9). Similar to the application for medical implants, the thin Ni-Ti wires were neither solution-treated nor recrystallized before the heat treatments applied in the present study.

In order to investigate the influence of short-time annealing, long wires were treated in a muffle furnace at temperatures ranging from 450-610 °C for a period from 2 min to 30 min to simulate a shape-setting treatment, and to modify the materials surface with respect to biocompatibility aspects (Ref 10). After the annealing, pieces of 8 mm length (< 1 mg) were cut off the straight wires to be subjected to bend and free recovery tests.

Since the transformation properties of Ni-Ti are sensitive to mechanical loading, a non-contact optical method was chosen for the experimental setup (Fig. 1). The horizontal resolution of the video camera was 1024 pixels, and paired with the microscope lens, the identification of slightest differences in the wire position during their reverse deformation could be detected. A cooling/heating system that allows for controlling the temperature with an accuracy of ±0.1 °C in temperature ranging from −55 °C to +45 °C has been built and attached to the specimen chamber that was filled with ethanol (Fig. 1). According to the specifications of the material, the "as received" wire is assumed to be fully martensitic at −55 °C. The dimension of the specimen holder inside the measurement chamber was adapted according to the wire diameter to attain a defined strain from 2-2.5% of the external wire segment after bending as required in (Ref 8). After cooling the measurement chamber, the wire was placed on the specimen holder inside the ethanol. Due to the small amount of material, the temperature of the wire assimilates almost instantaneously to the temperature of the ethanol inside the measurement chamber. Afterward, the wire was bent until the desired curvature was obtained. The heating rate was set to 0.6 °C/min. The temperature of the ethanol bath was measured close to the specimen by a Ni/CrNi thermocouple, and pictures of the wire were taken by the video monitoring system, continuously. The constant slow heating rate ensured a uniform temperature inside the measurement chamber during the testing, and it allowed for the assimilation of the temperature of the wire to that of the ethanol.

3. Results and Discussion

Figure 2 shows a sequence illustrating the reverse deformation of a thin Ni-Ti wire during the free recovery. The pictures were recorded by the internal camera during heating of the ethanol bath. Figure 2(a) represents the bent wire at −55 °C and Fig. 2(d) was taken at 45 °C. Due to the microscope lens, the outline of the 150 μm thin wire is clearly visible after adjusting the focus. In the upper left corner, the thermocouple for measuring and recording the temperature is visible in a distance of ~2 mm from the wire.

Evaluating the reverse deformation of the differently annealed wires leads in general to 3 different types of reverse-deformation curves (Fig. 3). Due to the slow heating rate, each curve is defined by a set of more than 1000 data points. Type-1 curve (Fig. 3a) shows a single S-shape with definite beginning and end of the reverse deformation and a slope between these characteristic points. The curve is clearly attributed to a martensite-to-austenite transformation without R-phase. The transformation temperatures A_s (austenite-start) and A_f (austenite-finish) can be determined by tangent method (Ref 8).

Type-2 reverse-deformation curve (Fig. 3b) shows a double S-shape. The major portion of the reverse deformation occurs during the first S-shaped stage. The second stage has only a minor contribution to the total reverse deformation. This is attributed to a transformation path martensite-to-R-phase-to-austenite. The transformation from R-phase-to-austenite can only cause a deformation of ~10% of the deformation the transformation from martensite-to-austenite can. This is based on the differences in the crystal structures of the phases (Ref 11). The alteration of the dimensions of the unit cells of the different crystal structures from martensite-to-austenite or martensite-to-R-phase, respectively, are significantly larger than that from R-phase-to-austenite. Therefore, the detailed identification of the occurrence of R-phase and the accompanying small reverse deformation in the curves is difficult by bend and free recovery testing with lower resolution. If the resolution is not sufficient, the second S-shape of the reverse-deformation curve cannot be identified. However, using the presented

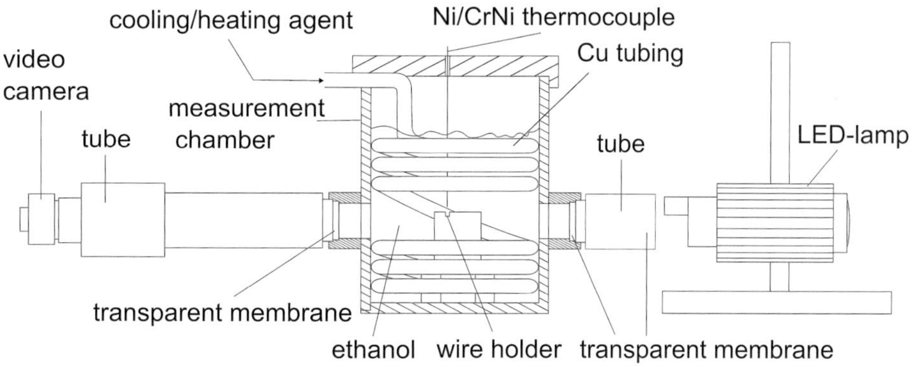

Fig. 1 Sketch of the high-resolution bend and free recovery test device

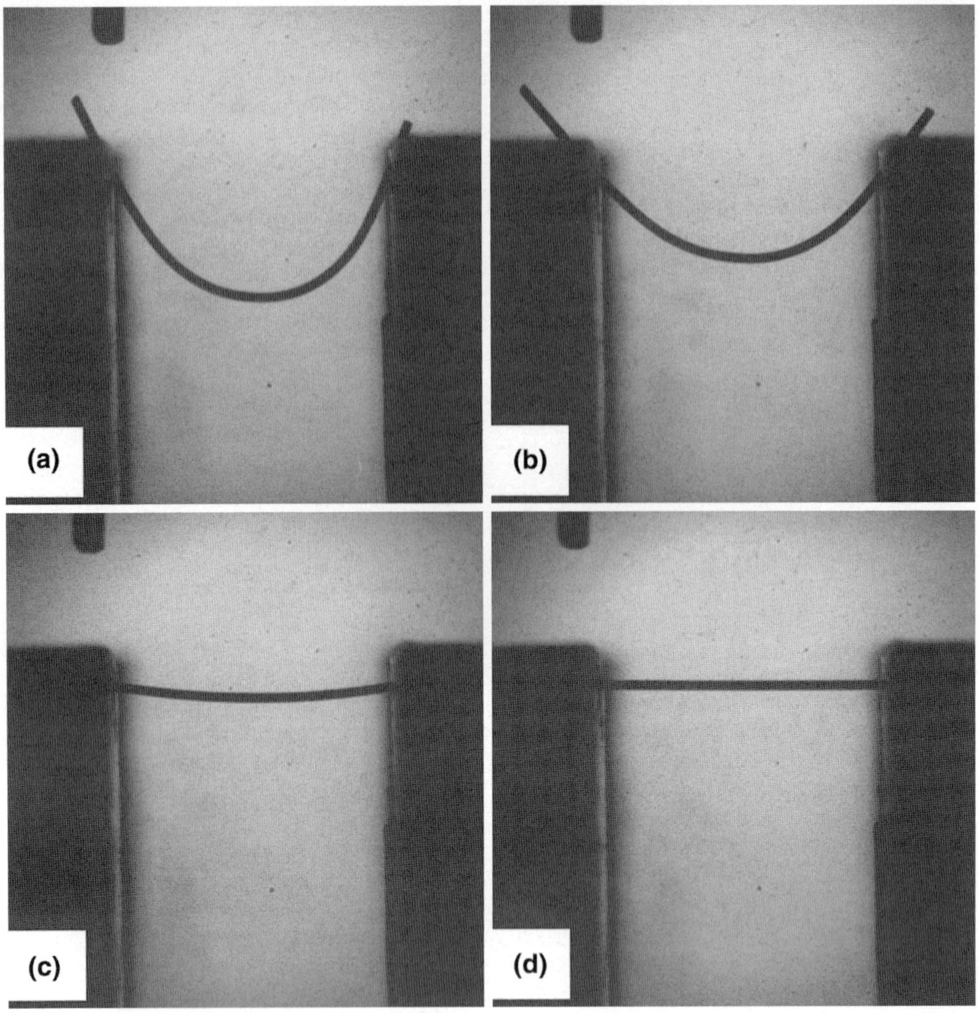

Fig. 2 Sequence illustrating the reverse deformation of a wire inside the measurement chamber from −55 °C (a) to +45 °C (d); in the upper left corner, the thermocouple is visible in a distance of ~2 mm from the wire

device, the R-phase-to-austenite transformation is clearly measurable, and in accordance to the evaluation of the transformation temperatures for Type-1 curves, in addition to the A_s and A_f temperatures, the R'_s and the R'_f temperatures can be attained from the curves of Type 2.

Type-3 (Fig. 3c) reverse-deformation curves are more complex. The curve does not show a double S-shape, but two different almost linear regimes. This indicates that the R-phase occurs in between the martensite-to-austenite transformation. However, the contribution of the second nearly linear stage to the total reverse deformation is ~30% and, therefore, too large to be caused solely by the R-phase-to-austenite transformation. Type-3 curves can be explained by the simultaneous occurrence and the overlap of both, the martensite-to-austenite and the martensite-to-R-phase-to-austenite transformations, possibly at different locations. Such a behavior has been demonstrated and described as multi-stage transformation in the literature (Ref 12). Unfortunately, the identification of all the characteristic transformation temperatures cannot be carried out using the tangent method. Only the R'_s and the A_f temperatures can be determined evaluating Type-3 curves.

The different types of reverse-deformation curves show a smooth transition depending on the prior annealing parameters. Type-3 curves were measured for the "as received" Ni-Ti wires. Though the exact conditions of the production process of the wires are not provided by the manufacturer, the occurrence of the R-phase can be attributed to the stored dislocations and the high defect density inside the material (Ref 12).

The smooth transition of the initial Type-3 curves to Type-1 or Type-2 curves seems to depend on the annealing temperature, mainly. An important temperature range for annealing has been identified to be ~520 °C. Annealing above 520 °C leads to Type-1, while annealing below leads to Type-2 reverse-deformation curves. The variation of annealing time does not cause a significant change of the type of the reverse-deformation curves, but with increasing annealing time, the characteristics of the curves become more articulated and clearer.

The origin of this behavior is complex and probably based on recovery, recrystallization, and precipitation processes that can occur separately or in combination during the annealing of the pre-deformed alloy. The results suggest that at temperatures above 520 °C, the defect density, which probably causes the occurrence of the R-phase, is reduced effectively. Therefore, a transformation path of martensite-to-austenite emerges. At temperatures below 520 °C, the reduction of the defect density during annealing is uncertain, but it was shown that annealing

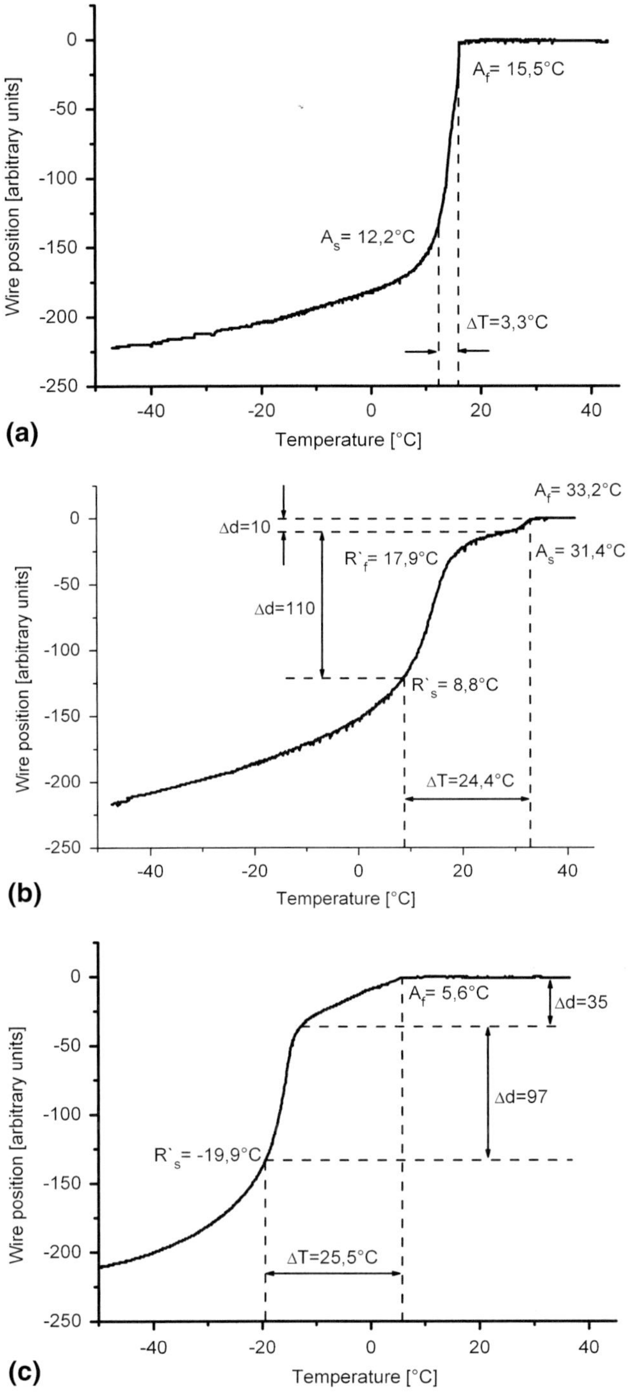

Fig. 3 Three types of reverse-deformation curves attained by bend and free recovery corresponding to the (a) reverse phase transformation from martensite-to-austenite, (b) martensite-to-R-phase-to-austenite transformation, (c) simultaneous martensite-to-austenite and martensite-to-R-phase-to-austenite transformations

in this temperature range leads to the formation and growth of Ni_4Ti_3 precipitates that promote the formation of the R-phase (Ref 12).

4. Summary

The influence of short-time annealing on the transformation properties of thin Ni-Ti wire was investigated by high-resolution bend and free recovery tests. Three different types of reverse-deformation curves related to a martensite-to-austenite, a martensite-to-R-phase-to-austenite and an overlapped martensite-to-R-phase-to-austenite and martensite-to-austenite transformation were identified. The major aspect for the change of the transformation characteristics appears to be the annealing temperature. Short-time annealing below and above ~520 °C causes two different types of reverse-deformation curves.

The contribution of each stage of the different transformation paths to the total reverse-deformation of the investigated wires can be quantified using the newly proposed device. Based on the results, the optimization of the materials behavior, e.g., the transformation path and the reverse-deformation degree can be attained and set with respect to medical application.

References

1. T.W. Duerig, A.R. Pelton, and D. Stockel, The Utility of Superelasticity in Medicine, *Biomed. Mater. Eng.*, 1996, **6**, p 255
2. A.R. Pelton, J. DiCello, and S. Miyazaki, Optimisation of Processing and Properties of Medical Grade Nitinol Wire, *Minim. Invasive Ther. Allied Technol.*, 2000, **9**, p 107
3. A. Undisz, M. Fink, and M. Rettenmayr, Response of Austenite Finish Temperature and Phase Transformation Characteristics of Thin Medical-grade Ni-Ti Wire to Short-time Annealing, *Scripta Mater.*, 2008, **59**, p 979
4. G.B. Stachowiak and P.G. McCormick, Shape Memory Behavior Associated with the R and Martensitic Transformations in a NiTi Alloy, *Acta Metall.*, 1988, **36**, p 291
5. J.D. Harrison, Measurable Changes Concomitant with the Shape Memory Effect Transformation, *Engineering Aspects of Shape Memory Alloys*, T.W. Duerig, K.N. Melton, D. Stöckel, and C.M. Wayman, Ed., Butterworth-Heinemann Ltd., London, 1990, p 106
6. M. Bartning and J. Simpson, Design and Performance of a Non-Contact Af Tester, *International Conference on Shape Memory and Superelastic Technologies*, A.R. Pelton, Ed. (Pacific Grove, CA), SMST Society, 2003, p 53
7. Y.N. Liu, X. Chen, and P.G. McCormick, Effect of Low Temperature Ageing on the Transformation Behaviour of near-equiatomic NiTi, *J. Mater. Sci.*, 1997, **32**, p 5979
8. ASTM, *Annual Book of ASTM Standards*, ASTM International, West Conshohocken, 2003
9. A. Undisz, S. Flauder, and M. Rettenmayr, Metallographische Präparation und Gefüge von Ni-Ti Legierungen mit Formgedächtnis/Pseudoelastizität. 42. Metallographietagung. Jena, Germany, 2008, *Sonderbände der praktischen Metallographie*, M. Rettenmayr and F. Mücklich, Ed., DGM Informationsgesellschaft, Frankfurt am Main
10. A. Undisz, F. Schrempel, W. Wesch, and M. Rettenmayr, In Situ Observation of Surface Oxide Layers on Medical Grade Ni-Ti Alloy During Straining, *J. Biomed. Mater. Res. Part A*, 2009, **88A**, p 1000
11. K. Otsuka, C.M. Wayman, T. Saburi, T. Tadaki, T. Maki, Y. Suzuki, J. Van Humbeeck, R. Stalmans, K. Uchino, M. Irie, K.N. Melton, I. Ohkata, and S. Miyazaki, *Shape Memory Materials*, Cambridge University Press, Cambridge, 1998
12. M.C. Carroll, C. Somsen, and G. Eggeler, Multiple-step Martensitic Transformations in Ni-Rich NiTi Shape Memory Alloys, *Scripta Mater.*, 2004, **50**, p 187

JMEPEG (2009) 18:818–823
DOI: 10.1007/s11665-009-9400-3

1059-9495/$19.00

Superelasticity of NiTi Ring-Shaped Springs Induced by Aging for Cranioplasty Applications

Henryk Z. Morawiec, Zdzisław H. Lekston, Kazimierz F. Kobus, Marek C. Węgrzyn, and Jan T. Drugacz

(Submitted September 11, 2008; in revised form January 16, 2009)

This paper concerns the application of titanium-nickel rings in modeling the cranium. After being fixed to the osseous margins, the ring's expansion at the same time broadens and shortens the cranium vault. The rings formed from a straight superelastic wire, flattened to an ellipse, do not show the presence of a typical force plateau but rather a pseudoelastic loop during loading-unloading in the relationship between the force and the deflection. Based on the idea that superelasticity in more complex shape-springs may be induced by the precipitation hardening process, the further studies were carried out on alloys with higher nickel contents (51.06 at.% Ni). The rings that had been formed were welded and aged at an optimal temperature and time. The improved superelastic behavior during compression and unloading the rings was obtained by introducing small deformation by drawing the quenched wires before forming the rings and aging. Very positive clinical reshaping by long-term distraction with the superelastic ring-shaped springs was achieved in young children under one year and a less spectacular effect was observed in the group of older children.

Keywords biomaterials, heat treating, intermetallics

1. Introduction

An alternative method for cranial modeling by surgery is the distraction of the cranium vault bone with the aid of steel springs, proposed by Lauritzen et al. (Ref 1). High efficacy of the spring-mediated cranioplasty was confirmed by Guimares-Ferreira et al. (Ref 2). As it is well-known steel springs lose force during their expansion. This explains why the application of superelastic NiTi springs with their force plateau for the long-term distraction of the mandibular in pigs has been confirmed (Ref 3). Based on these results, further research into superelastic springs and rings for cranioplasty has been undertaken (Ref 4). The basic aim of bone elongation was to achieve a continuous and constant force in a wide range of deformations using superelastic springs and rings. The very plateau stress-strain curves exhibited by superelastic NiTi alloys can be used advantageously to sustain constant forces or minimize their variations over a large range of deformations. It concerns the loading and unloading plateau connected with stress hysteresis.

This article is an invited paper selected from presentations at Shape Memory and Superelastic Technologies 2008, held September 21-25, 2008, in Stresa, Italy, and has been expanded from the original presentation.

Henryk Z. Morawiec and **Zdzisław H. Lekston**, Institute of Materials Science, University of Silesia, Bankowa 12, Katowice 40-007, Poland; **Kazimierz F. Kobus** and **Marek C. Węgrzyn**, Department of Plastic Surgery, Specialistic Medical Center, Polanica Zdrój, Poland; and **Jan T. Drugacz**, Clinic of Cranio-Maxillofacial Surgery, Medical University of Silesia, Katowice, Poland. Contact e-mail: henryk.morawiec@us.edu.pl.

This study presents the process of forming the superelastic rings from NiTi wire, which have been used for experimental elongation in the clinical research into modeling the cranial vault in children suffering from craniostosis.

2. Materials and Methods

The studies were carried out on three commercially available NiTi alloys in the form of straight wires. Their chemical composition and characteristics are presented in Table 1. The chemical analysis was carried out using the energy dispersive spectroscopy (EDS) method by the scanning electron microscope (SEM) JSM-6480. At the first stage of the studies, superelastic wires of 0.8, 1.0, and 1.2 mm diameter provided by SMATEC were used.

The rings were formed by bending a straight wire to a round form and then locating it in the holder which allows laser beam welding. To ensure a safe connection, both ends of the wire were overlap welded. The welding was carried out in a pure argon atmosphere using a Nd-YAG Dental Laser Welder type 2002 S with the power of 30 W. The characteristic temperatures and the course of the martensitic transformation were studied with the use of a differential scanning calorimetry (DSC) method and by the Perkin-Elmer DSC-7 instrument. The courses were registered at the cooling and heating speed of 10° per minute.

The aim of the second stage of the studies was to obtain superelastic behavior of rings during their deformation by flattening them to an ellipse. The alloys (nos. 2 and 3) marked NT-10 are destined to form complex shapes and induce superelasticity by aging. To obtain the best superelastic behavior, the optimization of the aging parameters was necessary. The aging was carried out in the temperature range 300-500 °C for different time holding. Before their implantation, the rings had undergone the process of steam passivity in an autoclave at 134 °C for 30 min.

Table 1 **Chemical composition and characteristics of NiTi alloys used in the studies**

Alloy	Ni, at.%	Ti, at.%	Al, at.%	Si, at.%	M_s, °C	A_f, °C	State of the wires	Delivering company
1	50.8 ± 0.27	49.0 ± 0.40	···	···	n.m. −15	n.m. +11	As received after quenching 800 °C/1 h	SMATEC
2	51.02 ± 0.36	48.71 ± 0.36	0.12	0.14	−80	−30	As received after quenching	AMT (NT-10)
					−65	−35	800 °C/1 h	
					−30	+15	After aging 500 °C/15 min	
3	51.06	48.9	···	···	n.m.	n.m.	As received after quenching 800 °C/1 h	SMATEC
					−32	+11	After aging 500 °C/30 min	

n.m., Not measurable

The structure of the obtained TiO_2 layer, about 4 nm thick, was amorphous, which was confirmed by a high-resolution electron microscopy (HREM) examination. The precipitation process was also controlled by a transmission electron microscopy (TEM) method in a JEM 3010.

The tensile tests of these wires were done on the Instron machine. The three-point bending tests in loading-unloading experiments were carried out on a specially computerized device equipped with a Hottinger force converter, a Pelton linear variable differential transformer, and a digital temperature indicator.

3. Results and Discussion

The idea of this work was to use rings instead of springs for cranioplasty, which by flattening to an ellipse would be connected to the skull and during long-term expanding would result in the elongation of the cranial bones perpendicularly to the sagittal direction and its contraction in a frontal plane, causing the craniofacial reshaping.

A flattened ring in the form of an ellipse is implanted onto the skull and the forces exerted on it are schematically shown in Fig. 1.

The problem was to achieve superelastic behavior of the rings, which means the presence of the broad force plateau on the force-deformation curve during force releasing. The rings formed from the superelastic wire of alloy 1 were welded with the use of a laser beam.

Their elastic characteristic during loading-unloading, which was obtained when the ring was flattened to an ellipse, does not show the presence of a typical force plateau but rather a linear relationship between the force and the deflection with a slope to the deflection axis as shown in Fig. 2.

In order to obtain typical superelastic properties of the ring ⇔ ellipse deflection, another method of superelastic induction for the rings was worked out. Based on the idea that superelasticity in more complex-shaped springs may be induced by the precipitating process, two alloys with higher nickel contents (alloys 2 and 3) were chosen.

The sequence of the martensitic transformation for the alloys 2 and 3 are shown on the DSC curves in Fig. 3.

The alloy 2 in as quenched state shows the direct B2 ⇔ B19′ and the reverse transformation, whereas after aging the transformation occurs through the R-phase, B2 ⇔ R ⇔ B19′. The preliminary slight deformation through drawing the wire of the parent B2-phase ($\varepsilon \sim 10\%$) and aging alloy 3 at 500 °C shows the same sequence through the R-phase. Both wires of alloys 2 and 3 after aging at 500 °C are at room temperature in the parent phase, as can be seen on the heating DSC curves in Fig. 3.

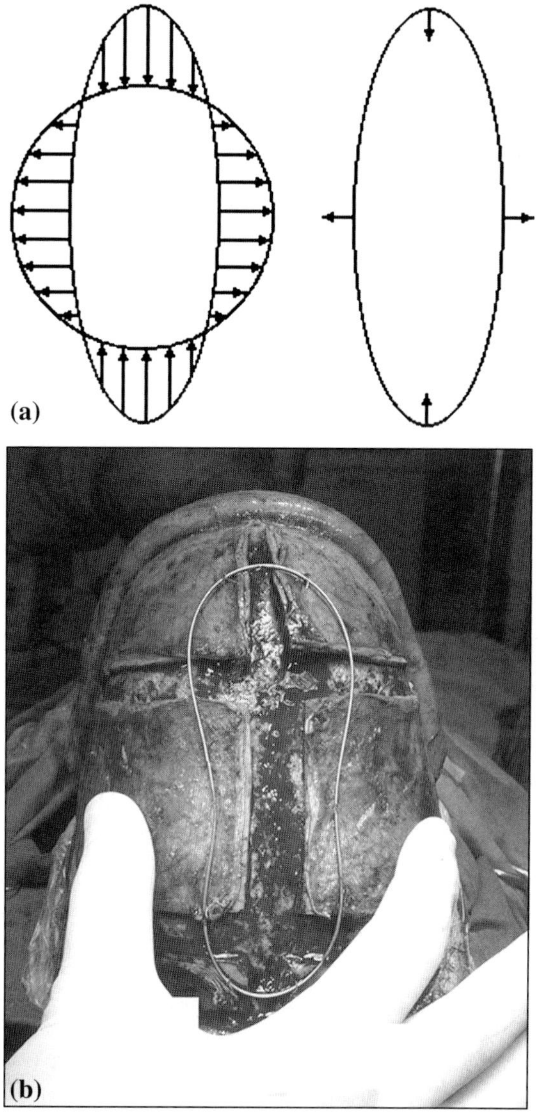

Fig. 1 Sketch of the forces exerted on the skull in a flattened ring (a) and its fixation on the cranium vault (b)

Aging of a straight wire of alloy 2 in the quenched state has shown the possibility of improving the elastic behavior of this wire, but still the amount of permanent strain is 2.5%. Considerable improvement of superelastic behavior has been

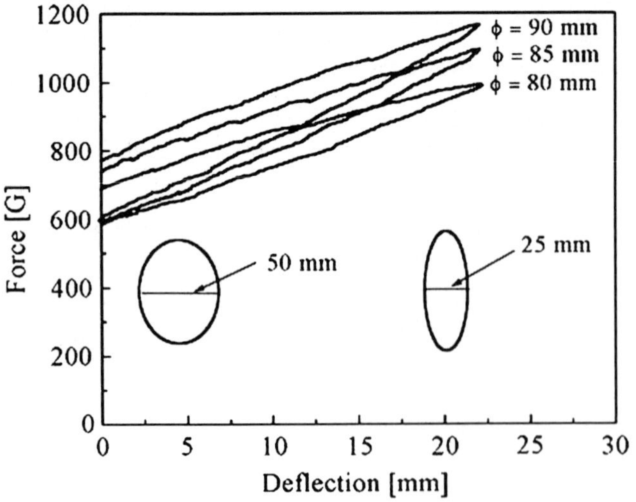

Fig. 2 Force vs. displacement for flattened rings formed from the straight superelastic wire (alloy no 1)

achieved by aging of the wires preliminary deformed by drawing. The three-point bending test for this state of alloy 3 after aging at 450 °C for 15 min shown in Fig. 4 exhibits a typical superelastic curve with a very low (about 0.1%) permanent strain.

The tree cycles of loading-unloading curves shown on this plot confirm the stability of the superelastic deformation.

The favorable affect of small deformation by drawing wires of alloy 3 on induced superelasticity is also presented in Fig. 5, which shows the effect of temperature aging on the hardening force-deflection curves by a three-point bending test. The hardening curves of the wires deformed by drawing show a more regular plateau and hysteresis force. As it can be seen, the higher temperature aging increases the force needed to start stress-induced martensite transformation, but the curve for the wire aged at 500 °C shows also the increase of the permanent strain up to 2.5%.

The rings formed were aged at the optimal temperature and time (500 °C/15′). The aging of rings results in hardening the parent phase by the precipitation of the coherent Ni_4Ti_3 phase. As a consequence, during deformation the parent phase exhibits superelastic behavior with a clear force plateau shown in Fig. 6 for rings of different diameters. As it can be seen, the level of force plateau lowers when the ring diameter increases.

The rings with the smallest diameter (D = 30 mm) show the highest force of the plateau, but simultaneously a large residual strain. This residual strain is caused by the dislocations which appear at the interface between the parent and the martensite phases due to the stress-induced martensite.

The smaller the ring's diameter the larger the pre-deformation during its formation and the higher the stress induced. It means that a relatively low deformation before the aging is favorable for the precipitation process. Identification of the precipitated particles was done using the TEM method. Figure 7 shows the disc shape coherent precipitates and their diffraction pattern.

The particles cause the matrix lattice distortions which achieved the maximum values perpendicular to the axis of the particles. In case of coherent Ni_4Ti_3 particles, this maximum appears in the $<111>$ direction of the matrix and are equal to

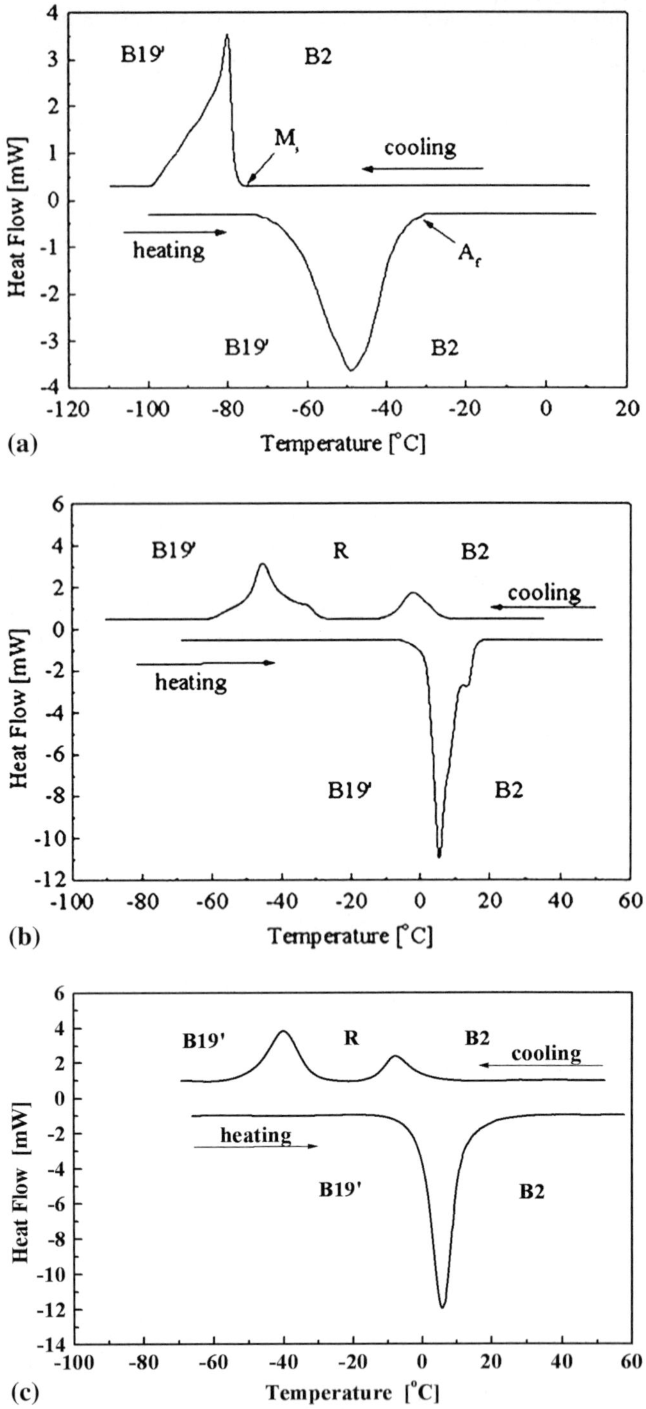

Fig. 3 DSC cooling and heating curves for alloy no 2 and 3 (a) for the delivery state of alloy 2, (b) after aging at 500 °C/15′, and (c) alloy no 3 deformed and aged at 500 °C/15′

the lattice misfit of both phases $\delta = (d_{111p} - d_{111B2})/d_{111B2} = 2.9\%$, they act as a tensile tension in this direction. Perpendicular to the $<111>$ direction: [110] and [112] directions, the lattice misfit is 1.4% and acts as a compression stress. As a result, the stress distribution around the particle is inhomogeneous but exhibits symmetry in respect to $<111>$ direction (Ref 5). The field stress distribution around the particles determined by Tirry and Schryvers has shown its lowering with a distance increase of the precipitates.

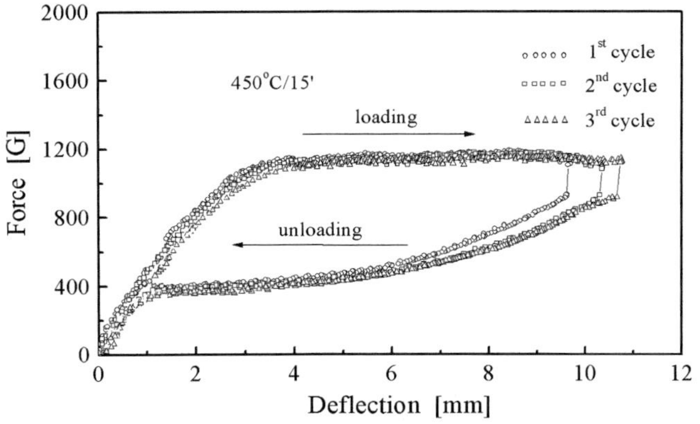

Fig. 4 Superelastic hysteresis of the three-point bending test for pre-strained wires of alloy 3 aged at 450 °C

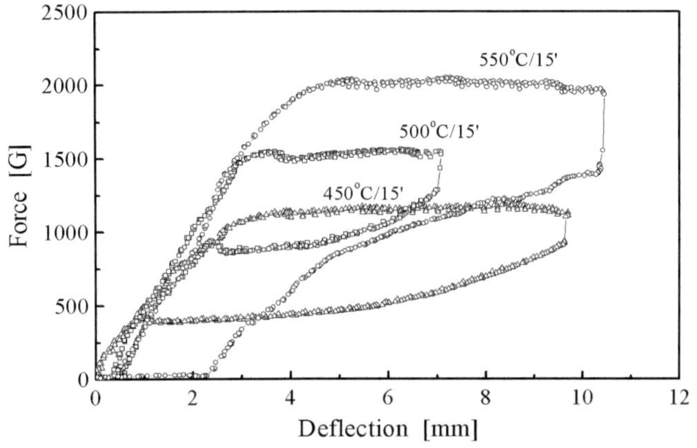

Fig. 5 Effect of temperature aging of alloy 3 on the superelastic force-deflection curves

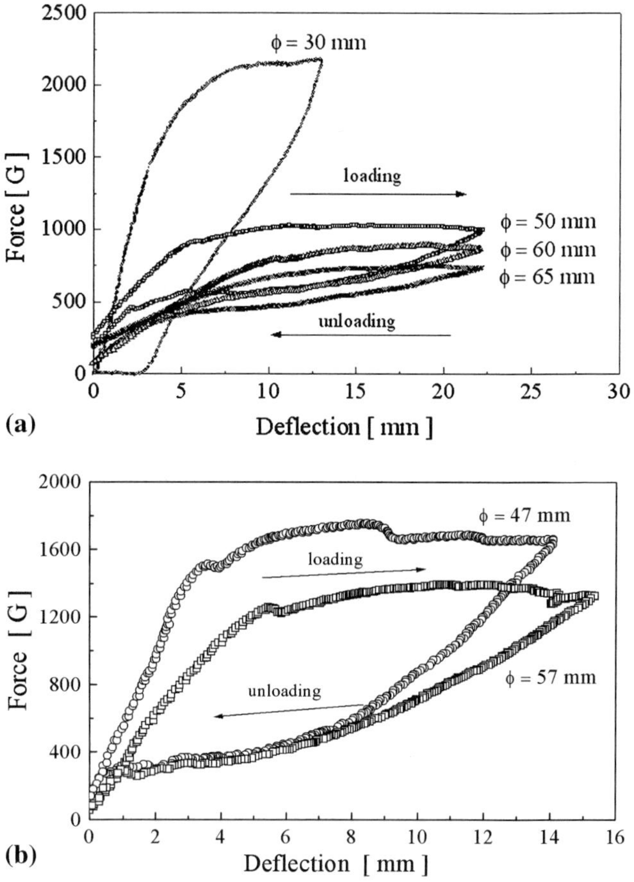

Fig. 6 Superelastic behavior of ring-shaped springs while deflecting to an elliptic shape and their reversion to the previous ring shape for (a) alloy 2 and (b) alloy 3

In accordance with the results obtained by Chumlyakov et al. (Ref 6), the superelastic effect is induced only under the following conditions: Ni_4Ti_3 particles are coherent with the B2 matrix and have the size of 50-100 nm. Coherent particles of Ni_4Ti_3 are the source of internal stresses, sites of the preferable martensite nucleation, and therefore assist lowering of the stress plateau. On the other hand, the particles do not undergo martensitic transformation and after the martensite is induced they become the source of the reverse stress fields and assist to store the elastic energy (Ref 7).

Clinical craniofacial reshaping by long-term distraction with the use of superelastic rings was carried out in the clinic of Plastic Surgery in Polanica (Ref 8). After the sparing excision of the cranium vault sutures in the shape of the letter "H," the compressed ring is given an oval shape in the sagittal axis and in this form it is fixed to the osseous margins. In four older children, the base parietal, frontal, and occipital bones were additionally resected approximately 2 cm. Ring expansion led to the expansion of the cranium vault bones perpendicularly in the sagittal direction, and to their compression in the anterior-posterior plane, which resulted in a desired cranium broadening and shortening. In these procedures, the rings with 90-100 mm diameters were used. The average operation time in case of the excision of the cranial sutures and rings placement was up to 100 min. The average hospitalization time was 7 days. After a 3-month period of distraction, the children were qualified for the removal of the rings. The ring was cut and removed through a small 0.5-cm incision on the welded site. The positive results of the operations carried out with the use of the superelastic rings can be seen in Fig. 8.

No early or late complications were observed during the implantation of the rings as well as during their removal. The change of cranium shape was approved as the criterion for the treatment results, since the objective measurement of the cranium capacity is still not very precise and the CT-3D tomography only in order to confirm clinical improvement appeared to be unadvisable. Better results were found in the cranium broadening than shortening, and mainly observed in younger patients under 1 year and less spectacular results in the group of older children. In every case, the shape of the head was subject to noticeable improvement. Classic procedures require extensive preparation and total reshaping of the calvaria. Large blood loss, long-term dissection, and the risk of infection or bone resorption pose a real threat that might be avoided in our method.

4. Conclusions

- The method of forming superelastic rings was worked out using the precipitation hardening of the rings previously

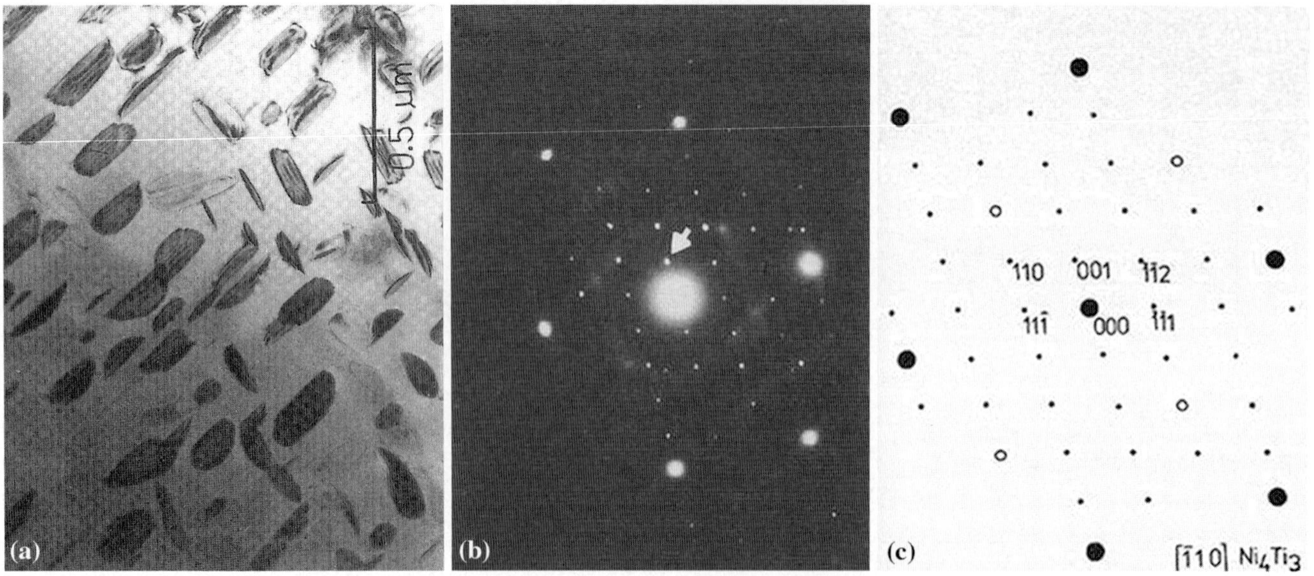

Fig. 7 Particles of Ni_4Ti_3 phase in the alloy aged at 500 °C for 1 h (a) bright-field image, (b) diffraction pattern from particle and (c) indexed electron diffraction pattern

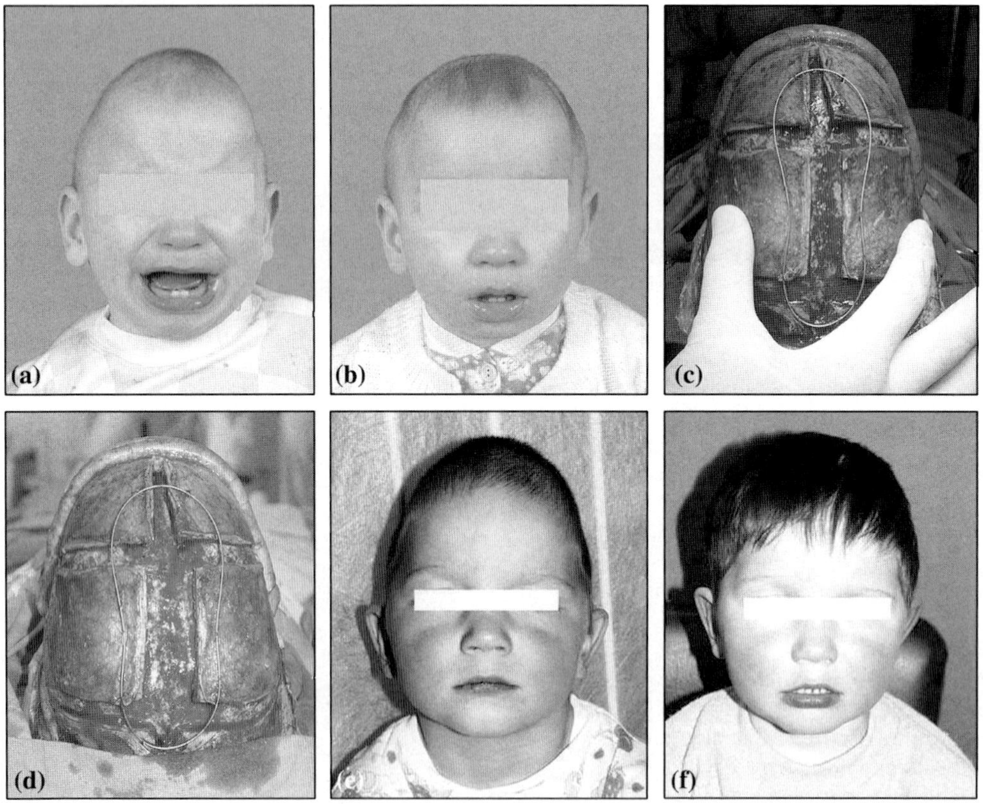

Fig. 8 Treatment of scaphocephaly with the use of superelastic ring-shaped springs

formed from wires of the NiTi alloy containing 51 at.% Ni.

- Superelastic rings deformed in the bending act with a constant force in the desired displacement range.
- Wires deformed by drawing and aged in the form of a ring improve the superelastic behavior of a ring-shaped spring.
- Clinical research confirmed the possibility of applying superelastic rings in cranioplasty.

Acknowledgment

Grateful acknowledgment is made for financial support by the Polish Ministry of Education and Science within the frame of the project PBR-4/rjp3/06.

References

1. C. Lauritzen, Y. Sugawara, O. Kocabalkan, and R. Olsson, Spring Mediated Dynamic Craniofacial Reshaping, *Scand. J. Plast. Reconstr. Hand Surg.*, 1998, **32**, p 331–338
2. J. Guimaraes-Ferreira, F. Gewalli, L. David, R. Olsson, H. Friede, and C.G.K. Lauritzen, Spring-Mediated Cranioplasty Compared with the Modified Pi-Plasty For Sagittal Synostosis, *Scand. J. Plast. Reconstr. Hand. Surg.*, 2003, **37**, p 208–215
3. Z.H. Lekston, J.T. Drugacz, and H. Morawiec, Application of Superelastic NiTi Wires for Mandibular Distraction, *Mat. Sci. Eng.*, 2004, **A378**, p 537–541
4. H.Z. Morawiec, Z.H. Lekston, K.F. Kobus, M.C. Węgrzyn, and J.T. Drugacz, Superelastic NiTi Springs for Corrective Skull Operations in Children with Craniosynostosis, *J. Mater. Sci: Mater. Med.*, 2007, **18**, p 1791–1798
5. W. Tirry and D. Schryvers, Quantitative Determination of Strain Fields Around Ni_4Ti_3 Precipitates in NiTi, *Acta Mater.*, 2005, **53**, p 1041–1049
6. Yu.I. Chumlyakov, I.V. Kireeva, E.I. Litviniva, and A.G. Lisyuk, Superelasticity During the Elastic Twinning, Slip and Martensitic Transformation, *Proceedings of the Second International Conference on Shape Memory and Superelastic Technologies*, 1st ed., A. Pelton, D. Hodgson, S. Russel, and T. Duerig, Eds., March 2–6, 1997 (Pacific Grove), International Organization on SMST, 1997, p 29–34
7. Yu.I. Chumlyakov, I.V. Kireeva, A.G. Lysyuk, and Y.L. Zuev, Superelasticity and Shape-Memory Effects in Ti-Ni Single Crystal, *Proceedings of the Second International Conference on Shape Memory and Superelastic Technologies*, 1st ed., A. Pelton, D. Hodgson, S. Russel, and T. Duerig, Eds., March 2–6, 1997 (Pacific Grove), International Organization on SMST, 1997, p 7-12
8. K.F. Kobus, M.C. Węgrzyn, Z.H. Lekston, H.Z. Morawiec, and J.T. Drugacz, Modeling of Scaphocephaly Using Superelastic Titanium-Nickel Rings: A Preliminary Report, *J. Craniofac. Surg.*, 2007, **18**, p 504–510

JMEPEG (2009) 18:824–830
DOI: 10.1007/s11665-009-9405-y

1059-9495/$19.00

SHADE: A Shape-Memory-Activated Device Promoting Ankle Dorsiflexion

S. Pittaccio, S. Viscuso, M. Rossini, L. Magoni, S. Pirovano, E. Villa, S. Besseghini, and F. Molteni

(Submitted September 12, 2008; in revised form January 13, 2009)

Acute post-stroke rehabilitation protocols include passive mobilization as a means to prevent contractures. A device (SHADE) that provides repetitive passive motion to a flaccid ankle by using shape memory alloy actuators could be of great help in providing this treatment. A suitable actuator was designed as a cartridge of approximately 150 × 20 × 15 mm, containing 2.5 m of 0.25 mm diameter NiTi wire. This actuator was activated by Joule's effect employing a 7 s current input at 0.7 A, which provided 10 N through 76 mm displacement. Cooling and reset by natural convection took 30 s. A prototype of SHADE was assembled with two thermoplastic shells hinged together at the ankle and strapped on the shin and foot. Two actuators were fixed on the upper shell while an inextensible thread connected each NiTi wire to the foot shell. The passive ankle motion (passive range of motion, PROM) generated by SHADE was evaluated optoelectronically on three flaccid patients (58 ± 5 years old); acceptability was assessed by a questionnaire presented to further three flaccid patients (44 ± 11.5 years old) who used SHADE for 5 days, 30 min a day. SHADE was well accepted by all patients, produced good PROM, and caused no pain. The results prove that suitable limb mobilization can be produced by SMA actuators.

Keywords actuator, orthosis, rehabilitation, shape memory alloys

1. Introduction

Ankle rehabilitation techniques are very commonly present in the clinical practice, as they are strongly indicated in the treatment of several different neuromuscular diseases. For instance, the distal lower extremities are very often affected by stroke events. Considering the social impact of stroke alone (annually, 15 million people worldwide suffer a stroke and 5 million of these are left permanently disabled) (Ref 1), it is easy to appreciate the importance of developing new strategies in this field. After stroke, a rehabilitation program should be performed as soon as the patient's conditions are sufficiently good. Moving the limbs passively can improve circulation, maintain joint flexibility and normal muscle tone, while the patient is unable to exercise actively (Ref 2, 3). Contracture prevention due to passive stretching in the acute phase after CNS injury is also reported (Ref 4). The motor impairment due to paresis after stroke is greatly aggravated by the muscle and joint contracture and the changes in muscle contractile properties caused by immobilization (Ref 5). AHA/ASA endorsed guidelines for 2005 (Ref 6), besides stressing the need for a multidisciplinary approach in post-stroke care, also indicate that a strong difference in the final treatment outcomes derives from the sheer time spent on physical rehabilitation. Favorable effects on activities of daily living are particularly evident if therapy input is augmented by at least 16 h within the first 6 months relative to standard practice (Ref 7).

Passive mobilization (Ref 8-10) may also contrast deafferentation and learned non-use, i.e. directly address impairment of the cortical pathways involved in active limb control. The way this happens is not clear. There is evidence, however, that functional recovery occurs while changes in sensorimotor activation take place in the cerebral cortex.

A simple, user-friendly and safe device for use in the first few weeks after the acute event, during or following manual treatment by a therapist, could lead to great benefits for the patient by preventing the insurgence of severe sequelae and would help limit health service costs. In order to be a reliable device for rehabilitation purposes, this orthosis should comply with a number of clinical and functional requirements, i.e. it should be capable of promoting cyclic mobilization of the ankle joint across at least a range of −5 to 10° (negative being toward plantarflexion), have suitable repetition times in a physiological range, be thermally and electrically safe for the patient, lightweight and compact to wear.

Such a device was designed as a prototype named SHADE. It mounts shape memory alloy (SMA) actuators, which are inherently compliant, among others, with the limited-weight and compactness requirements of wearable orthoses. To try and build such a device was an invaluable opportunity to study the possible applications of SMAs in rehabilitation medicine.

This article is an invited paper selected from presentations at Shape Memory and Superelastic Technologies 2008, held September 21-25, 2008, in Stresa, Italy, and has been expanded from the original presentation.

S. Pittaccio, S. Viscuso, E. Villa, and **S. Besseghini**, CNR Institute for Energetics and Interphases at Lecco, Lecco, Italy; and **M. Rossini, L. Magoni, S. Pirovano,** and **F. Molteni**, Ospedale Valduce Centro di Riabilitazione Villa Beretta, Costamasnaga, Italy. Contact e-mail: s.pittaccio@ieni.cnr.it.

2. Material and Characterization

2.1 Orthosis Design

A prototype of SHADE (Fig. 1) was assembled with two thermoplastic shells modeled on a prototype human lower limb of average size. These are hinged together at the ankle and strapped by Velcro® bands on the frontal aspect of the shin and on the foot. Two cartridge actuators are fixed on the front of the upper shell with inextensible threads connecting each NiTi wire to the foot shell. The fixation points for the actuators (on the shin) and the distal ends (foot ends) of the inextensible threads were optimized by a factorial experiment and principal component analysis scheme. This yielded the triplet of distances $a = 11.5$ cm, $b = 5.5$ cm, $c = 9.5$ cm for the fixtures with respect to the orthosis joint (Fig. 1). This configuration provides a theoretical maximum angular displacement of 28.67° with maximum force per actuator of around 10 N with a foot weighing 12-17 N. A dedicated software home-written in LabView (National Instruments, Austin, TX) controls an array of electronic relays (NI9481—National Instruments, Austin, TX) that switch on and off a dc-generator providing sufficient power to get shape recovery. All the electric connections and the thermal shielding are suitable to guarantee patients' safety.

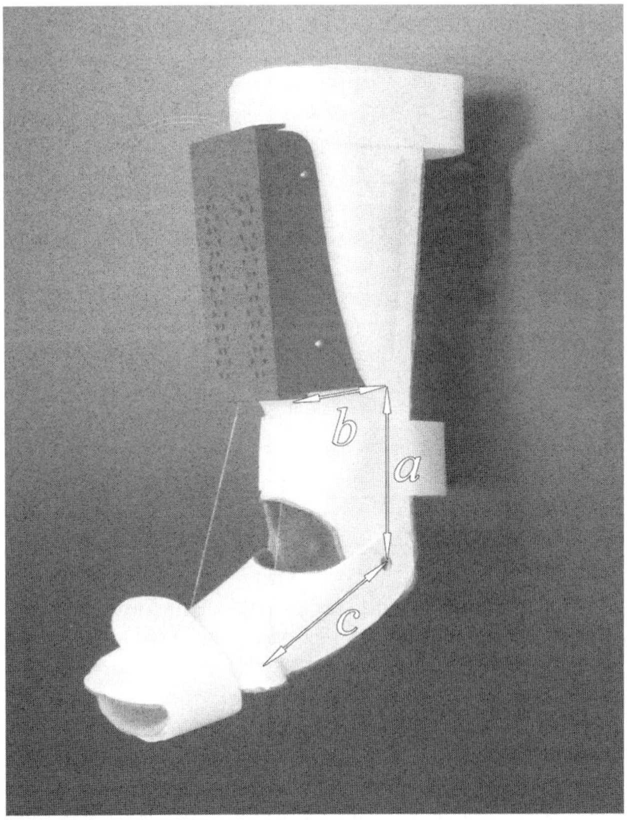

Fig. 1 Prototype of SHADE showing the two thermoplastic shells hinged at the ankle joint. The Velcro® straps around the calf and foot are visible; the cartridge actuators are enclosed in the vented box; the power wires (from the top) and the inextensible threads (from the bottom) are clearly exposed. Label a shows the distance of the thread exit from the ankle along the tibial shell, b shows the same distance along the orthogonal direction; and c shows the radial distance of the foot insertion of the thread from the ankle

2.2 Actuator Design

Choosing a lever arm of around 9.5 cm, to produce an angular movement of say 25°, the NiTi wire was taken 250 cm long (with a maximum expected deformation well beneath 4%, to guarantee cycling stability). The actuator was designed as a $150 \times 20 \times 15$ mm insulated aluminium cartridge wherein the necessary reference length of NiTi wire is led back-and-forth between two arrays of ten mini-pulleys. This makes it possible to confine a long wire within a limited space (Fig. 2). An end of the wire is connected to the electric contact and to an inextensible thread that transmits the force to the foot, while the other one is fixed to the housing and makes the second electric contact. A pseudoelastic spring (40 turns, 2 mm inner coil diameter, 300 μm wire) is connected to the moving end of the NiTi wire in order to keep it taut. Heating is provided by Joule's effect, while cooling is left to natural convection.

2.3 Shape Memory Material

A shape memory NiTi 250 μm diameter commercial wire stabilized for actuation was mounted in the two cartridges to ensure lasting performance for SHADE and industrial reproducibility.

Differential scanning calorimetry (on DSC 220 SSC/5200—Seiko Instruments, Tokyo, Japan) was carried out on a sample of wire, showing $A_f = 351$ K and $M_f = 274$ K (Fig. 3).

Tensile tests were conducted using an MTS 2/M thermomechanical test machine (MTS Systems, Eden Prairie, MN) equipped with a 2 kN load cell. The material was deformed up to an engineering strain of 5%, at 365, 380, and 390 K. The pseudoelastic plateau values varied as a function of temperature with a $\Delta\sigma/\Delta T$ ratio of 8.402 MPa/K (Fig. 3).

Activation tests were conducted on the same test machine, maintaining the crosshead fixed and injecting different currents (0.65, 0.7, 0.75, and 0.8 A) in the wire at a constant strain of 4% for a set period of 13 s. Results are shown in Fig. 4.

2.4 Power Dimensioning

The choice and the settings of a dc-generator to activate SHADE depend on both the application and the material properties. Actuators mount a 250 cm long 250 μm diameter wire, which implies an electric resistance ranging 45-54 Ω during transformation from fully austenitic to fully martensitic. Activation tests on the wire were analyzed to decide what current value is more appropriate to provide a maximal force of at least twice an estimated working load of 10 N per wire, minimizing current expenditure and heating time to reach 10 N (Fig. 4). The current selected was 0.7 A, corresponding to a voltage of 35 V, considering an average resistance of 49.5 Ω.

The temperature to achieve full transformation depends on the load applied. With the estimated working load of 10 N, the calculated $\Delta\sigma/\Delta T$ ratio and a DSC A_f of 351 K, this means 375 K. By applying 35 V to the ends of the cartridge actuator suspending a 10 N weight, it was demonstrated that this temperature is reached in around 7 s.

2.5 Technical Characterization

The cartridge was tested by applying to it increasing loads of 0.8, 0.9, 1.0, 1.1, and 1.2 kg (respectively 7.85, 8.83, 9.81, 10.79, and 11.76 N) and measuring the displacement achieved with a current injection of 0.7 A for 7 s. Cooling and reset by natural convection took around 30 s.

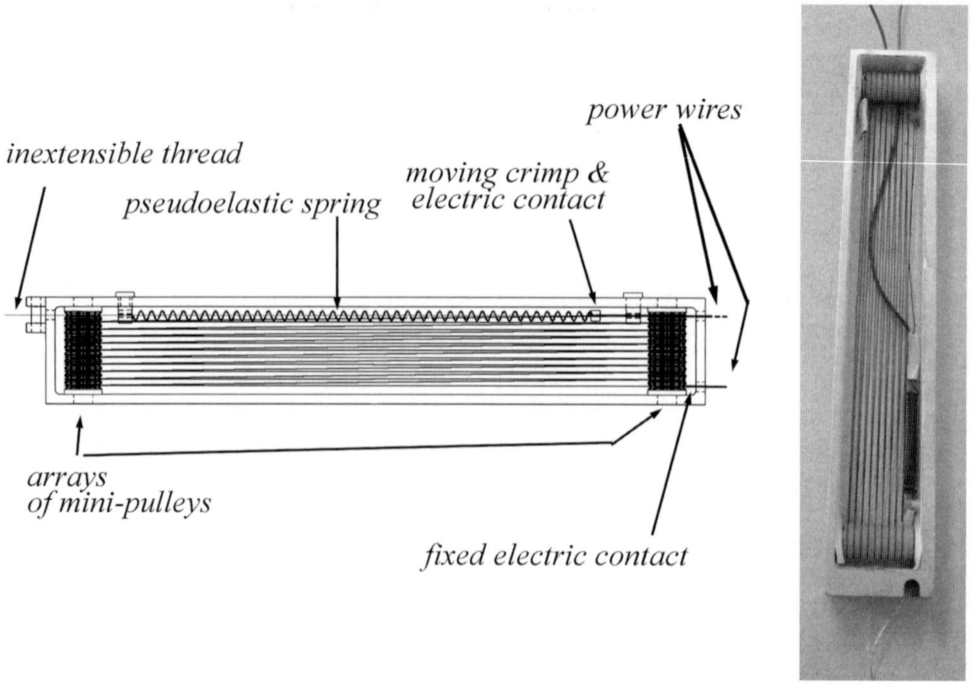

Fig. 2 Drawing and photograph of the cartridge actuator. SMA wire is pulled back-and-forth between two arrays of mini-pulleys. A pseudoelastic spring keeps the wire taut. Electric connections are made to both ends of the wire, so that the actuator can be worked by Joule's effect. While one end is fixed, the other end is connected to an inextensible thread that transfers force and displacement outside the cartridge

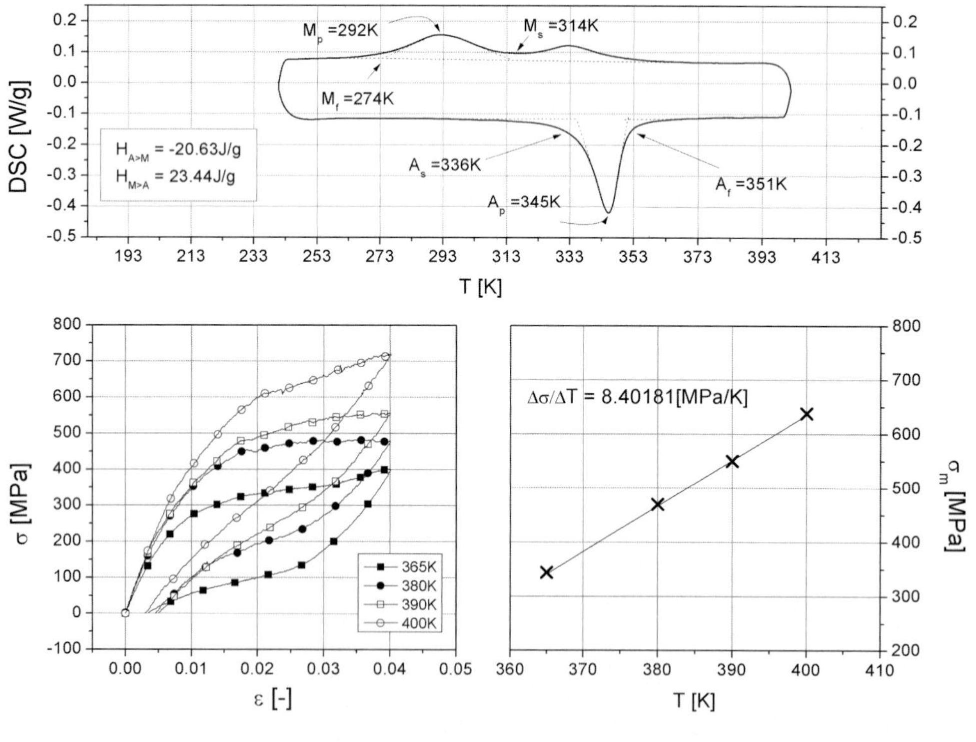

Fig. 3 SMA wire static characteristics. The top graph shows the results of DSC analysis on a sample of commercial wire, as received. The bottom left graph displays stress-strain curves for different working temperatures. The bottom right graph shows linear fitting of the plateau values (as measured near 2.5% strain) vs. working temperature. Clausius-Clapeyron constant is derived from the fit

Similarly the assembled SHADE orthosis was tested with a sequence of increasing loads (7.85, 8.83, 9.81, and 10.79 N) fixed on the foot shell at a distance of 15 cm from the ankle, while measuring the angular displacement by means of electrogoniometer SIM-HES-E.G 042 (Signo Motus, Messina, Italy). The current pattern was the same as that of the cartridge tests, with a 7 s step of 1.4 A (0.7 A per cartridge) and 30 s allowed for cooling.

3. Subjects and Clinical Methods

3.1 Patient Selection

A first group of 3 patients (A1, A2, A3) was enrolled to assess the efficacy of SHADE in producing ankle dorsiflexion in accordance with the clinical requirements and to prove short-term tolerability. Later on, acceptability of passive mobilization was evaluated on a second group of 3 inpatients (B1, B2, B3) who were treated with SHADE 30 min a day for 5 consecutive days, alongside standard physical therapy. The selection criteria for both groups were flaccid or slightly spastic (≤1 on the Ashworth Scale) hemiparesis involving the ankle as a consequence of first stroke, no major skin or articular pathologies, no severe cognitive impairment, able to answer questions. The patients' characteristics are presented in Table 1.

Patients were treated according to ethical principles, as approved by the hospital ethical committee, and gave informed consent.

3.2 Clinical Protocol

Patients underwent general clinical assessment prior to the start of the treatment. The first group used SHADE in single trials to measure the instantaneous range of motion by means of eight optoelectronic cameras sampling at 100 Hz (Elite Gait Eliclinic—BTS, Garbagnate Milanese, Italy). Six IR reflective markers were positioned on the *condyli tibiales*, the *malleoli*, *metatarsi I* and *V*, identifying the ankle angle as the one extending between the central axes of shin and foot. The additional measurement D and segment length L allowed adjusting the angle baseline (Fig. 5). The patients, wearing SHADE, were asked to try and lift the tip of the foot without any cues about to what extent or at what speed the task had to be carried out; irrespective of the results of this active trial, SHADE was then used to elicit passive dorsiflexion twice.

The second group of patients used SHADE 30 min for 5 days, subsequently to the usual physiotherapeutic manoeuvres. Acceptability of SHADE was investigated by observation and interviewing the patients.

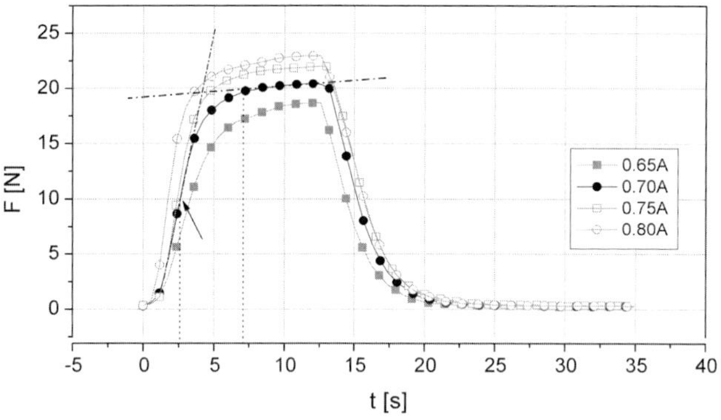

Fig. 4 SMA wire dynamic characteristics. This graph shows the wire force response to a 13 s square current pulse. The wire was kept at a constant strain of 4% while current was injected. The black line with filled circles relates to the current value selected for the application (0.7 A): rise time to 10 N approximates the one for 0.75 A (2.5 s) although current expenditure is less; transformation offset occurs near 4 s; power is sufficient to reach a force of 20 N in around 7 s

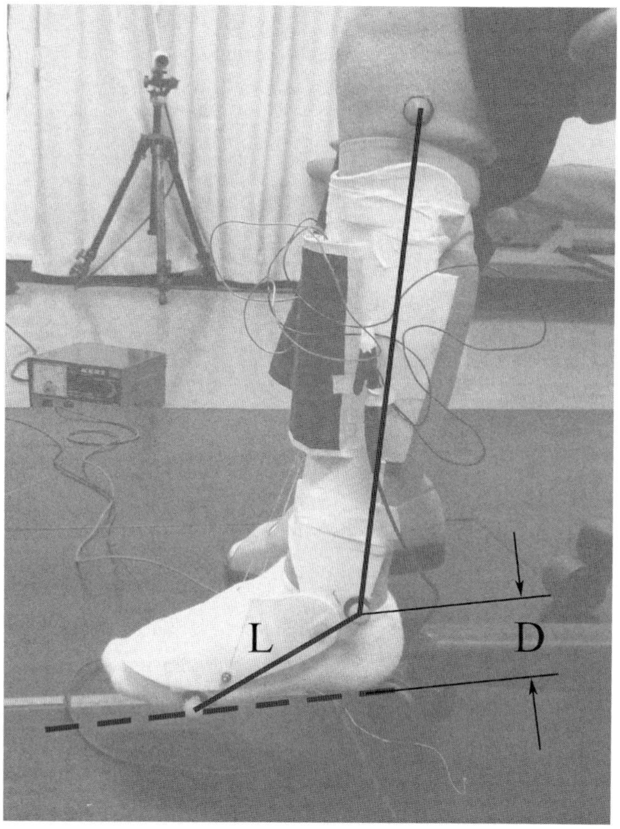

Fig. 5 IR-reflective markers for optoelectronic movement tracking were positioned on the *condyli tibiales*, the *malleoli*, *metatarsi I* and *V*, identifying the ankle angle as the one extending between the central axes of shin and foot. The additional manual measurement D and the segment lengths L acquired during tracking allowed adjusting of the angle baseline

Table 1 Patients' characteristics (PROM = passive range of motion; AROM = active range of motion)

Patient	Age	Sex	Diagnosis	Range of motion	Treatment
A1	63	F	Left hemiplegia post stroke (2003)	Full PROM, full AROM	Single trial
A2	53	M	Left hemiplegia post stroke (2004)	Full PROM, no AROM	Single trial
A3	58	M	Left hemiplegia post stroke (2007)	Full PROM, no AROM	Single trial
B1	48	M	Right hemiplegia post stroke (2007)	Full PROM, no AROM	30 min for 5 days
B2	53	M	Right hemiplegia post stroke (2007)	Full PROM, no AROM	30 min for 5 days
B3	31	F	Right hemiplegia post stroke (2007)	Full PROM, limited AROM	30 min for 5 days

4. Results

4.1 Technical Characterization of the Cartridge Actuator

The results of the characterization of a single actuator are shown in Fig. 6. Displacement peaks for a static load of about 8.83 N, with the free end of the wire moving by around 8 cm (3.2% strain, not considering localized deformation around the pulleys). Note that between-weight differences are all below 10%. Work capability continues to rise for heavier loads but appears to come to a maximum at around 12 N. These data suggest that the actuator should be able to lift the estimated load of 10 N across sufficiently large displacements.

4.2 Technical Characterization of SHADE

The results of the tests on SHADE are shown in Fig. 7. Angular displacement is very stable with weight (26.25° ± 1.5°, mean ± SD), while curves tend to drift toward lower angles (decreasing dorsiflexion) on increasing weights.

4.3 Clinical Outcomes

Optoelectronic movement analysis of the patients in group A revealed that SHADE is capable of producing (and even exceeding) the minimum required angle span (Fig. 8). In particular, with patient A3 it elicited motion from −4.5 to +19°, adding up to a total angular displacement of 23.5°, not too far from the theoretical limit. Patient A1 is particularly interesting, as she was able to produce active movements of her own. As can be appreciated from the graphs of her acquisitions, the speed of passive dorsiflexion (SHADE activation) closely matches the one of natural spontaneous motion. The reset speeds and angles are not matched, however.

Acceptability every day and after 5 consecutive days of exercise with SHADE was very good, as no patients complained of any pain (patient B1 only reported slight friction nuisance), while all of them said they would be willing to continue therapy with SHADE if recommended to do so.

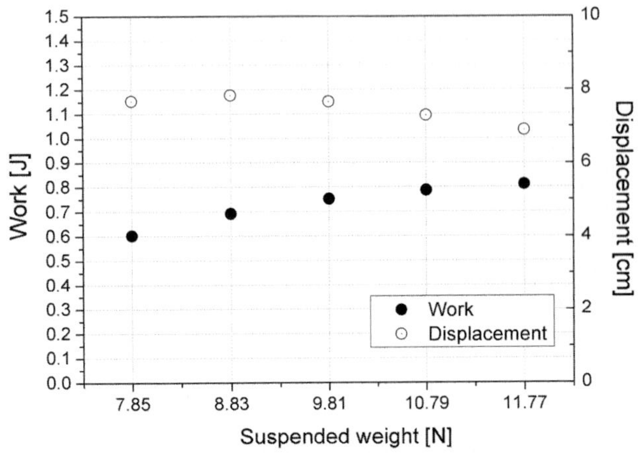

Fig. 6 Results of cartridge actuator characterization. Maximum displacement occurs for loading at around 0.9 kg (8.83 N). Maximum work output is achieved for loads above 1.2 kg (11.77 N)

5. Discussion

5.1 Actuator Design

In the light of the a priori choice to employ and study the application of shape memory materials to this case, the functional and clinical requirements set out in the introduction could be rephrased in terms of material and technical characteristics. In particular, the SMA actuators, whatever their shape, ought to overcome collectively loads in excess of 150 Ncm, provide angular motion across over 15° and have relatively short cooling times. Those issues have tread-offs in terms of

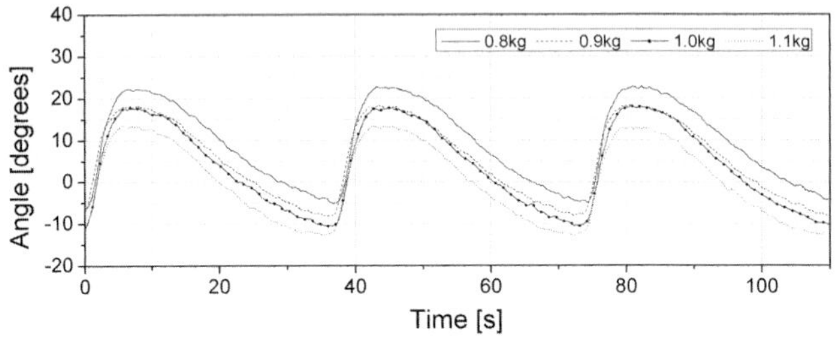

Fig. 7 Results of SHADE characterization. Angular range is stable with different loads, while maximum dorsiflexion angle decreases with load. Between-cycle repeatability is very good

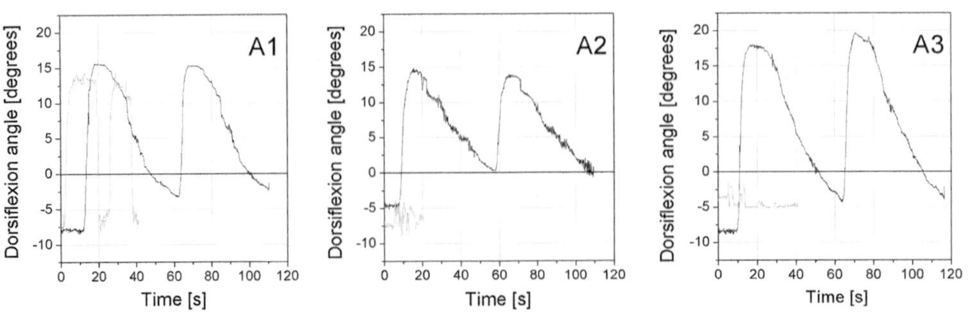

Fig. 8 Results of optoelectronic tracking of ankle motion during dorsiflexion mechanically produced by SHADE. Black lines represent passive exercise by SHADE; grey lines relate to the patients' attempts to carry out active ankle motion. The range of motion produced by SHADE is patient-dependent and adapts to the specific requirements. For patient A1 it is possible to appreciate that dorsiflexion speed (positive slope) is similar to the one spontaneously chosen by the patient during natural movement. Plantarflexion (negative slope) is limited to 30 s by the input current waveform. Natural convection allows resetting of the initial position to an extent dependent on the patient and on the selected plantarflexion time

actuator cross-dimensions (force versus cooling time) and actuator longitudinal dimensions (compactness versus maximal displacement, i.e. actuator recoverable strains times actuator length).

First, considering ankle biomechanics, it can be realized that, when using linear actuators, depending on what lever arm is utilized with respect to the joint pivot to convert linear motion into angular rotation, either fairly long displacements or large forces are needed to produce a sufficient foot movement. With a little lever arm it would be possible to achieve optimal compactness, but very large forces would then be required. Should these be obtained by significantly enlarging the diameter of the wire within the actuator, cooling time would rise dramatically; alternatively, if more actuators were employed, any advantage of compactness would be annuled. Therefore, a solution implying comparatively large lever arms was selected.

This posed the question of how to minimize overall actuator size while guaranteeing sufficient linear motion at reasonable strains. Although the actuator could theoretically mount either springs or wires, wires were ultimately selected. In fact, springs provide large displacements but are much less strong than the wire that is coiled to make them. Increasing the diameter of the spring wire to make the coiling stronger, on the other hand, is not practical, because it tends to increase both cooling times and material fragility due to excessive coiling strains. So the only solution with these constraints appeared to be an actuator mounting straight wires. The implementation with the use of mini-pulleys has several advantages but relies upon careful component design in order to minimize both friction and unwanted wire hardening, carrying along fragility and loss of shape memory properties. Due to the impossibility to eliminate these problems completely, it was paramount to allow for some safety ranges in the choice of construction elements both along the force and displacement dimensions. In particular, the choice of actuation force and electrical current was made considering a weight of 20 N per cartridge, which corresponded to twice the expected working load. This corrected for uncertainties in dimensioning and also took into account the possible abrupt failure of one actuator, so that the remaining one could temporarily take over.

5.2 Orthosis Performance

Technical tests were conceived so that the weights and lever arm produced plantarflexion moments in the range of those typical of foot weights (e.g. 15 N foot weight × 10 cm lever arm). These tests demonstrated that SHADE can produce a suitable range of motion (even exceeding the clinical requirements). Furthermore, the best match for the −5°/+10° range was achieved in the 1.0 kg test, most closely corresponding to the estimated foot weight. It is important to notice that also with the other weights (producing moments 80-110% of the reference one) the clinical requirements are always met.

Also, the clinical tests showed that SHADE can produce dorsiflexion beyond the minimum requirements. The dorsiflexion speed, which was implicitly chosen when deciding on the actuation current value, resulted appropriate to generate a natural movement. The reset speed is, on the other hand, much lower than the spontaneous one. This depends on the cooling rate, which is rather low, as convection is strongly hindered by the actuator housing walls (albeit they are vented). It was decided that an adjustable reset time of 30 s would be allowed during cyclic clinical working. This time is long enough to reach plantarflexion almost completely by means of the sheer foot weight without any forced convection. The chosen reset time, even though it is unnatural, could be suitable for physical rehabilitation because it allows the ankle to remain stretched for as long as 70% of each cycle, which is hoped to avert the development of contractures.

It was noted that the generated range of motion self-adapted to every patient's characteristics, i.e. it never happened that patients felt any pain due to an excessive pulling of their feet beyond the maximal individual passive range of motion (PROM): this is thanks to the existence of a detwinning plateau limiting the peak available force in SMA.

6. Conclusions

The actuators passed the technical tests and were able to produce, with good stability, the expected forces and displacements. Once put on, the orthosis and used in a clinical setting, they provided sufficient propulsion to elicit dorsiflexion up to almost +20°. It is believed that the use of such a device could be beneficial particularly in the early phases of post-stroke care, when it is vital to stop the insurgence of contractures and spasticity.

The main problem with the use of SMA in this context appears to be dealing with lengthy and hardly controllable cooling rates. This problem could well be lived with, if it was proven that long reset periods could provide extra time for therapeutic stretch. An important future development will be in the direction of including electromyographic control on SHADE activation, so that this orthosis can be used also as an aid and movement booster during early active workouts.

References

1. J. Mackay and G. Mensah, eds., *Atlas of Heart Disease and Stroke*, World Health Organization, Geneva, 2004
2. L.M. Brass, Stroke, Chap. 18, *The Yale University School of Medicine Heart Book*, 1st ed., B.L. Zaret, M. Moser, and L.S. Cohen, Eds., William Morrow & Co, 1992, p 215–234
3. NINDS (National Institute of Neurological Disorders and Stroke), Post-Stroke Rehabilitation Fact-Sheet, NIH Publication number: 02-4846, Office of Communications and Public Liaison—National Institute of Neurological Disorders and Stroke—National Institutes of Health, Bethesda, MD, 2002
4. L. Harvey and R. Herbert, Muscle Stretching for Treatment and Prevention of Contractures in People with Spinal Cord Injury, *Spinal Cord*, 2002, **40**, p 1–9
5. J.-M. Gracies, Pathophysiology of Spastic Paresis I: Paresis and Soft Tissue Changes, *Muscle Nerve*, 2005, **31**, p 535–551
6. P.W. Duncan, R. Zorowitz, B. Bates, J.Y. Choi, J.J. Glasberg, G.D. Graham, R.C. Katz, K. Lamberty, and D. Reker, Management of Adult Stroke Rehabilitation Care. A Clinical Practice Guideline, *Stroke*, 2005, **36**, p e100–e143
7. G. Kwakkel, R. van Peppen, R.C. Wagenaar, S. Wood Dauphinee, C. Richards, A. Ashburn, K. Miller, N. Lincoln, C. Partridge, I. Wellwood, and P. Langhorne, Effects of Augmented Exercise Therapy Time After Stroke: A Meta-Analysis, *Stroke*, 2004, **35**, p 2529–2536
8. S. Masiero, A. Celia, G. Rosati, and M. Armani, Robotic-Assisted Rehabilitation of the Upper Limb After Acute Stroke, *Arch. Phys. Med. Rehabil.*, 2007, **88**, p 142–149
9. S. Pittaccio, S. Viscuso, M. Rossini, L. Magoni, S. Pirovano, S. Besseghini, and F. Molteni, SHADE: An Active Orthosis for

Safeguarding Ankle Motion Skills in Dropfoot Patients, *Proceedings of the 4th World Congress of the International Society Physical and Rehabilitation Medicine*, U.W. Rah, Ed., June 10–14, 2007 (Seoul), Organizing Committee of ISPRM, 2007, p 360

10. S. Pittaccio, F. Zappasodi, S. Viscuso, F. Mastrolilli, M. Ercolani, F. Passarelli, F. Molteni, P.M. Rossini, and F. Tecchio, Cortical Correlate of Passive Mobilisation of the Ankle Joint by means of the SHADE Orthosis, *Clin. Neurophysiol.*, 2008, **119**(Suppl 1), p S20

JMEPEG (2009) 18:831–835
DOI: 10.1007/s11665-009-9402-1

1059-9495/$19.00

Design Considerations for Nitinol Bone Staples

Scott M. Russell

(*Submitted September 12, 2008; in revised form February 2, 2009*)

The use of Nitinol in orthopedic staples has a long, successful history beginning with their first commercial introduction in the early 1980s. Nitinol bone staples are generally inserted through a process of being chilled, opened, and inserted into predrilled holes. Upon insertion, they recover their preprogrammed shape either through springing back after removal of a constraint (using superelasticity) or thermal triggering (using thermal shape memory). In general, Nitinol bone staples fall into one of three categories: (1) those which are superelastic at room temperature, (2) those which recover their shape upon heating to body temperature, or (3) those which recover their shape upon heating above body temperature with the application of an external heat source. These three different design approaches—room temperature superelastic, body temperature activated, and heat activated—each have different performance characteristics. A version of the heat activated staple that uses a controlled heat source appears to have the best combination of clinical forces and procedural control.

Keywords biomaterials, medical devices, Nitinol, orthopedic

1. Introduction

Devices for fracture fixation, including bone plates and staples, were one of the very first applications of Nitinol in medicine. One of the first recorded patents for a Nitinol medical device was for a bone plate in 1974 (Ref 1). Bone staples soon followed, with the first commercial Nitinol staples released to the Chinese market in 1981, to the European market in 1990, and to the U.S. market in the mid-1990s (Ref 2).

Nitinol staples are part of a broader category of metallic compression staples used today for fracture fixation, arthrodesis and osteotomies. Bone staples are especially useful for treatment of bones in the hand, foot and ankle, and metallic bone staples of one form or another have been used in the U.S. since 1906 (Ref 3). The advantages of using staples for these procedures include good approximation of the bone fragments, dynamic compression of the fracture surfaces, avoidance of infection issues associated with external fixation, and reduced operating time as compared to bone screws (Ref 4). Application of bone staples is a four-step process: (1) temporarily fixate the two pieces of bone to maintain proper alignment prior to fixation, (2) predrill holes to accept the bone staple, (3) insert and seat the staple, and (4) trigger recovery of the staple's preprogrammed shape either by removal of a constraint or by application of heat (Ref 4).

This article is an invited paper selected from presentations at Shape Memory and Superelastic Technologies 2008, held September 21-25, 2008, in Stresa, Italy, and has been expanded from the original presentation.

Scott M. Russell, Benchmark Nitinol Device Technologies, LLC, San Jose, CA. Contact e-mail: scott@benchmarknitinol.com.

1.1 Types of Nitinol Bone Staples

With a typical Nitinol bone staple, holes are predrilled, an "open" Nitinol staple is inserted into the holes, and the staple recovers either superelastically or via shape memory to pull the fractured bones together and apply a compressive force to the fracture surfaces. Different staple designers and manufacturers have executed this basic approach in a number of different ways.

1.1.1 Room Temperature Superelastic (SE). In this form of Nitinol bone staple, the austenite finish (A_f) transformation temperature is near or less than room temperature and the staple must be held open by some type of device prior to deployment. This type of staple will attempt to spring closed at any temperature at or above room temperature as soon as the constraint is removed.

1.1.2 Body Temperature Activated (BT). This type of Nitinol staple has an A_f temperature below body temperature but above room temperature. Shape recovery is triggered by the thermal shape memory effect as the staple reaches body temperature. It is important to maintain the temperature of the staple sufficiently below body temperature during insertion to prevent premature deployment. This can be accomplished through a combination of external cooling and sufficiently rapid insertion.

1.1.3 Heat Activated (HA). This category of Nitinol staple has an austenite start (A_s) temperature near or slightly above body temperature but with an A_f temperature low enough to allow deployment without the application of excessive heat. Approximately 60 °C is generally recognized as the temperature at which tissue damage can occur, so the activation temperature for the staple must ideally be kept below this temperature (Ref 5). This type of staple can be activated in one of two ways: by simple application of a heating element such as a cautery device (Ref 6) or by using a specially designed external heat source that allows the amount and duration of the applied heat to be precisely controlled depending upon the degree of shape recovery desired (Ref 7). This latter type of staple (a subset of the HA category) will be referred to in subsequent discussion as "controlled heat activated."

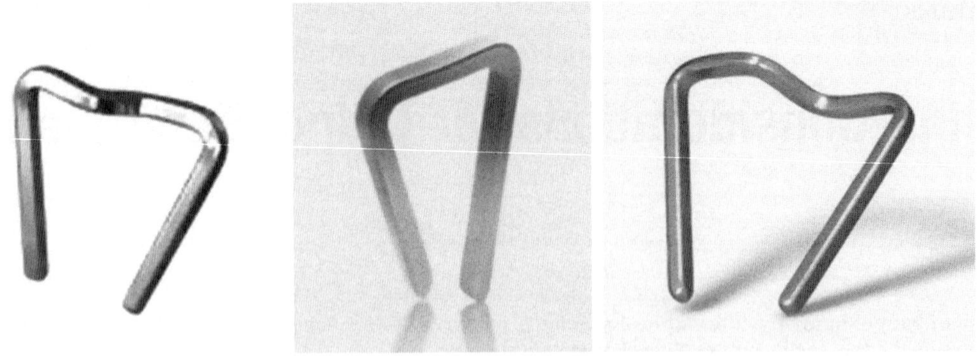

Fig. 1 Examples of commercially available Nitinol bone staples: (left to right) the heat activated OSStaple™ from BioMedical Enterprises, Inc., the room temperature superelastic EasyClip™ staple from Memometal, Inc., and the body temperature BioPro® Memory Staple

Table 1 Examples of commercially available Nitinol bone staples

Trade name	Company	Deployment	Indications
Memory™ Staple	Depuy	BT activated	Foot, ankle
Stimulink™	Intellifuse	Heat activated (Intellifuser™)	Head and face, soft tissue, hand and foot, bone fragments, lower limbs
Memoclip®	Memometal	Heat activated	Hand and foot
EasyClip®	Memometal	RT superelastic	Hand and foot
MEMODYN™	Telos Medical/ BioResearch Innovations	Heat activated (cautery)	Foot, ankle, hand, wrist
OSStaple™, BOSS™, OSSArc™, OSSpine™, OSSAnchor™ Grip	BioMedical Enterprises	Controlled heat activated (OSSforce™)	Head, face, sternum, pelvis, cervical spine, hand and foot, upper and lower extremities
BioPro® Memory Staple	BioPro	BT activated	Foot, ankle, hand, wrist

Staples of all three types are commercially available. An example of each type of staple is shown in Fig. 1. A list of some of the commercially available Nitinol bone staples, along with their type of deployment mechanism, is shown in Table 1.

2. Clinical Experience

The two most important factors that govern bone healing are blood supply and the biomechanical conditions, especially the stability of the fracture (Ref 8). A strong staple that reduces relative motion of the two parts of the fracture is important for stability. A key advantage of Nitinol staples over conventional metallic staples is the ability to apply a continuous compressive force to the fracture surfaces. Not only does this ensure stability of the fracture site, but such compressive forces are also known to speed healing (Ref 9). Optimization of such forces is important, however, as excessive forces can cause pressure necrosis and prevent healing while insufficient forces can slow healing and may even allow the fracture to reoccur (Ref 9).

A review of the literature (Ref 4, 8, 10-15) indicates a generally favorable opinion among physicians regarding Nitinol bone staples. A typical physician comment on Nitinol bone staples is, "The shape memory designed into the Nitinol staples studied shows the ability to produce an actively dynamic compressive force … The benefits of this type of device can provide unlimited options for repair of osteotomies, arthrodesis and fractures (Ref 14)."

3. Benchtop Measurement of Compression Forces

While Nitinol shape memory staples appear to be well-received in the marketplace, the different staple types have different procedural characteristics and compressive forces. To examine the compressive forces exerted by the different types of staples, an experimental study was conducted on the three different types of Nitinol bone staples: room temperature superelastic (SE), body temperature activated (BT), and heat activated (HA). In each case, the staples were subjected to a simulated deployment and the final forces were measured at body temperature. Similar staple designs and sizes were used in each case to eliminate differences in design as a factor in the results. In addition, all of the Nitinol materials were thermo-mechanically processed in a similar manner, with only the final aging heat treatment adjusted to create the different final transformation temperature characteristics. Interestingly, unlike most other medical device made from Nitinol, many commercial Nitinol bone staples are made from heavily annealed and aged Nitinol with very little residual cold work. This results in a lower strength material in general, but the lower strength is compensated for by thicker material cross-sections to achieve the desired compressive forces.

3.1 Experimental Protocol

Two different geometries of rectangular wire-based Nitinol staples were studied for the SE, BT, and HA staples: (1) 20 mm back, 20 mm legs, and a 2 × 3 mm wire cross-section (2020)

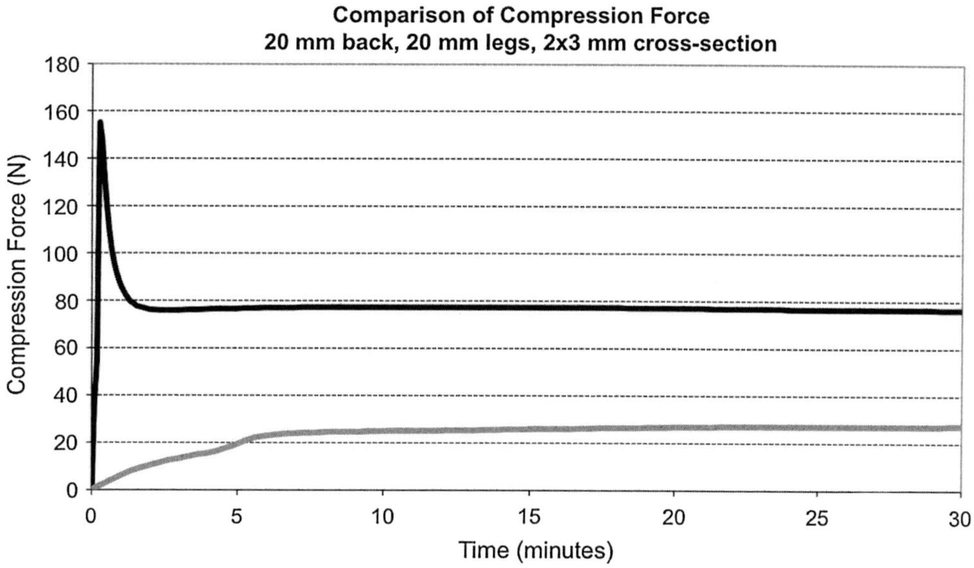

Fig. 2 Examples of compression forces as a function of time for 2020 BT (bottom, gray line) and HA (top, black line) staples. The SE staples showed similar recovery curves to the BT staple shown here

Table 2 Experimental procedures for measuring compressive forces exerted by three different types of Nitinol bone staples

SE	BT	HA
1. Pry open the staple mechanically at room temperature. 2. Insert the staple into the force-measuring jig. 3. Insert the jig into a temperature chamber at 37 °C and close the chamber door. 4. Record the compression force for 30 min.	1. Place the force-measuring jig in the 37 °C temperature chamber for 1 h to allow it to equilibrate. 2. Quickly insert the staple into the force-measuring jig using chilled forceps and close the chamber door. 3. Record the compression force for 30 min.	1. Place the force-measuring jig in the 37 °C temperature chamber for 1 h to allow it to equilibrate. 2. Insert the staple into the force-measuring jig. 3. Begin force data collection. 4. Heat the test staple to 55 °C using an electrical heating unit. 5. Close the temperature chamber door and record the compression force for 30 min.

and (2) 15 mm back, 12 mm legs, and a 2 × 2 mm wire cross-section (1512). The same staple design was used for all three staple types to enable direct comparison of the recovery mechanism independent of other possible differences in staple design. A staple design with an S-shaped back was chosen for the experiments as this is a common shape for commercial BT and HA staples; however, no SE staple is currently sold with an S-shaped back.

The starting material for all three staple types was a typical commercial-grade Nitinol alloy with 50.8 at.% Ni (balance Ti). The material was cold drawn to the appropriate rectangular shapes with approximately 30 to 40% final cold work. The staples were then heat treated by a commercial staple manufacturer (BME, Inc., San Antonio, TX) using their proprietary in-house process to set the staple shape, followed by aging between 450 and 550 °C to achieve the final desired transformation temperature characteristics. The final transformation temperatures of the staples were measured via bend and free recovery per ASTM F2082-06 (Ref 16). The measured transformation temperatures for the three different staple types were as follows: SE (A_s = 15 °C, A_f = 25 °C), BT (A_s = 30 °C, A_f = 37 °C), and HA (A_s = 38 °C, A_f = 43 °C). Each

Table 3 Force exerted by the three different staple types—SE, BT, and HA

Staple	Steady-state force, N, at 37 °C after deployment	Force magnitude scaled to BT staple
2020-2 × 3 SE	61.6	2.24×
2020-2 × 3 BT	27.5	1×
2020-2 × 3 HA	75.7	2.75×
1512-2 × 2 SE	50.3	2.14×
1512-2 × 2 BT	23.4	1×
1512-2 × 2 HA	30.5	1.30×

type of staple was tested per a specific experimental protocol as shown in Table 2.

3.2 Experimental Results

The measured steady-state forces exerted by the three different staple types are shown in Table 3. Figure 2 shows an example of compression forces as a function of time measured for the same size BT and HA 2020 staples. The spike in forces for the HA staple corresponds to the application of the heat source. Since the BT staple did not have a separate external heat source applied, it does not show a similar spike, but instead shows a gradual recovery to its final steady-state force as it reaches its final temperature of 37 °C. The shape of the SE recovery curves was similar to that of the BT curve as shown here.

The recovery forces were very sensitive to the precision of the span of the force-measuring jig. This was especially pronounced in the SE staples where the staple was seen to lose as much as 85% of its force if the jig was improperly set up with a span only 0.5 mm smaller than intended. In real-world clinical practice, this emphasizes the need for precision in the location of the predrilled holes to achieve predictable compressive forces.

4. Discussion

The well-known Clausius-Clapeyron equation determines the temperature-stress relationship of the Nitinol martensitic transformation:

$$\frac{d\sigma}{dT} = \frac{-\Delta H}{\varepsilon_0 T}$$

where σ is the stress, T is the temperature, H is the enthalpy of transformation, and ε_0 is the transformation strain.

Note that change in stress is inversely proportional to the change in temperature. In practice, one can use this relationship to estimate the relative differences between Nitinol recovery forces for devices with different transformation temperatures. For example, for superelastic devices, the larger the difference between ambient temperature and the A_f temperature, the higher the superelastic recovery force. Therefore, for devices recovering at body temperature, one would expect the recovery force of a "colder" staple (i.e., the SE staple) to be higher than that of a "warmer" staple (i.e., the BT staple) since the difference between the ambient temperature (37 °C) and the transformation temperature is greater for the superelastic staple. Indeed, this is shown to be the case for the SE and BT staples. However, it is more difficult to estimate the steady state recovery force for the HA staples as they involve the application of external heat to drive the temperature to a higher point (and conceivably drive the transformation to greater completion), followed by a reduction in temperature to the steady state 37 °C. Surprisingly, we see that after this activation process, these high-temperature HA staples can exhibit reasonably high steady-state compressive forces, sometimes exceeding even those of the SE staples. In spite of the fact that they are at a temperature below their A_s temperature when they cool back down to body temperature after being triggered at the initial high temperatures, the HA staples maintain a reasonably high holding force that is sustainable over time.

As expected, the lowest forces are consistently seen in the BT staples. Their forces take about 5 min to achieve their maximum level, and the final force level is low partly due to the stress-temperature relationship described above and partly due to the fact that a Nitinol material with its A_f temperature close to its use temperature will have difficulty reaching its total force potential based on the fact that as it transforms, it encounters greater resistance, which in turn effectively raises its A_f temperature. This continues until a force-transformation temperature equilibrium is reached at 37 °C. HA staples overcome this limitation by driving the staple beyond their inherent temperature limitations and, as a result, they are able to achieve their true compressive force potential to a greater degree.

One commercially available HA staple, the OSStaple™, has an additional benefit of using a special controller (the OSSforce™), which allows the amount of heat input and duration applied to the staple to be precisely controlled. This type of staple is referred to as "controlled heat activated." As a result of this additional control, the shape recovery (and resultant forces) of the staple can be more precisely controlled. This allows the physician to tune the implant for a particular application. For example, if the procedure calls for multiple staples to improve rotational or bending angle stability, each staple can be individually tuned at an appropriate force level to optimize the final force balance and fixation characteristics. This could also allow the physician to adjust for any imprecision in the location of the predrilled holes.

4.1 Procedural Factors

Other factors that influence staple choice include activation control, force control, ability to stabilize the fracture, and ease-of-use. A qualitative comparison of various characteristics of the different types of staples is given in Table 4. Overall, the HA staples achieve the best balance of high fixation forces and procedural control. The controlled heat activated subset of this larger category staples has all the benefits of HA staples with increased control of the recovery process and subsequent compressive forces.

5. Future Directions

Most staples today are rather simple in design, being made from rectangular Nitinol wire with a mostly constant cross-section. Recent advances in manufacturing techniques have resulted in more finely engineered designs, incorporating

Table 4 Qualitative comparison of different staple types

Characteristic	SE	BT	HA	Controlled heat activated
A_f range, °C	0-25	25-35	45-55	45-55
Activation control	++++	++	++++	++++
Force control	++	++	++	++++
Ease-of-use	+++	++	+++	+++
Fracture stability	++++	++	++++	++++
Compression force	High	Low	Medium/High	Optimizable

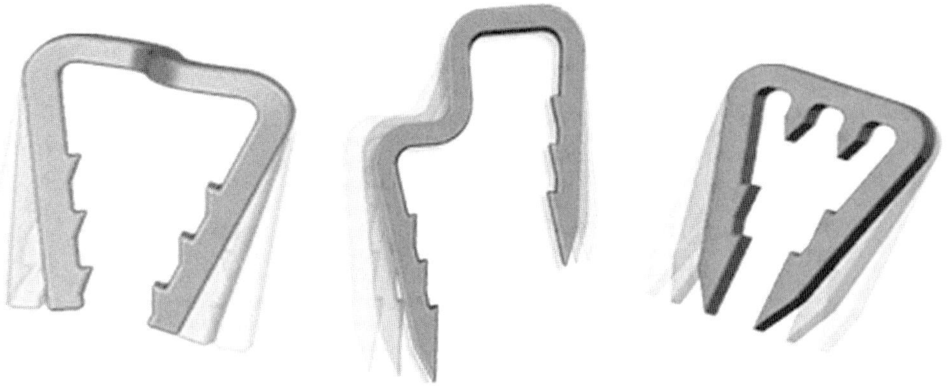

Fig. 3 The Barbed OSStaple™ (BOSS) from BioMedical Enterprises, Inc., is an example of a next generation staple, incorporating recent advances in manufacturing technology and staple design

design elements such as barbs and three-dimensional features to enhance staple security, prevent rotation of the fixation site, and allow tailoring of the compression forces at different parts of the staple. An example of such advanced staples is the Barbed OSStaple™ (BOSS) shown in Fig. 3. More manufacturers are likely to adopt similar sophisticated designs in the future.

References

1. A. Johnson and F. Alicandri, Thermoconstrictive Surgical Appliance, U.S. Patent 3,786,806, 2 Jan 1974
2. J. Ryhänen, "Biocompatibility Evaluation of Nickel-Titanium Shape Memory Metal Alloy," Dissertation, Department of Surgery, University of Oulu, Oulu, Finland, 1999
3. W.L. Bargar, N.A. Sharkey, and H.A. Paul, et al., Efficacy of Bone Staples for Fixation, *J. Orthop. Trauma*, 1987, **1**(4), p 326–330
4. T.M. Mereau, and T.C. Ford, Nitinol Compression Staples for Bone Fixation in Foot Surgery, *J. Am. Podiatr. Med. Assoc.*, 2006, **96**(2), p 102–106
5. J.D. Bronzino, *The Biomedical Engineering Handbook*, 3rd ed., CRC Press, Boca Raton, FL, 2006
6. Telos Medical Memodyn™ Compression Staple Product Information Available at http://www.telosmedical.com/memodyn.htm
7. BioMedical Enterprises, Inc. OSSforce™ Implant Controller Product Information Available at http://www.bme-tx.com/Product/OSSforce.html
8. N. Shibuya, et al., A Compression Force Comparison Study Among Three Staple Fixation Systems, *J. Foot Ankle Surg.*, 2007, **46**(1), p 7–15
9. A.K. Forsythe and C.J. Green, Bone Fracture Compression Device and Method of Usage, U.S. Patent 3,866,607, 18 Feb 1975
10. R.K. Choudhary, et al., First Metatarsophalangeal Joint Arthrodesis: A New Technique of Internal Fixation by Using Memory Compression Staples, *J. Foot Ankle Surg.*, 2004, **43**(5), p 312–317
11. J.T. Braun, et al., The Efficacy and Integrity of Shape Memory Alloy Staples and Bone Anchors with Ligament Tethers in the Fusionless Treatment of Experimental Scoliosis, *J. Bone Joint Surg.*, 2005, **87A**(9), p 2038–2051
12. J.J.G. Malal, G. Hegde, and R.D. Ferdinand, Tarsal Joint Fusion Using Memory Compression Staples—A Study of 10 Cases, *J. Foot Ankle Surg.*, 2006, **45**(2), p 113–117
13. P.J. Ronchetti and S.M. Topper, Lunocapitate Fusion Using the OSStaple Compression Staple, *Tech. Hand Up. Extrem. Surg.*, 2006, **10**(4), p 231–234
14. T.J. Chang and B.D. Overley, "An In Vitro Comparative Study of Screw and Nitinol Staple Compression: A Model Showing Active 'Dynamic' Compression," *Presented at the American College of Foot & Ankle Surgeons 65th Annual Scientific Conference*, Orlando, FL, March 2007
15. P. Fernández-de-Retana, D. Poggio, and J.P. Ortega, Technical Tip: First Metatarsophalangeal Arthrodesis with 20-mm Memory Compression Staples, *Foot Ankle Int.*, 2008, **29**(6), p 613–615
16. "Standard Test Method for Determination of Transformation Temperature of Nickel-Titanium Shape Memory Alloys by Bend and Free Recovery," ASTM F2082-06, ASTM International, West Conshohocken, PA, 2006

JMEPEG (2009) 18:836–842
DOI: 10.1007/s11665-009-9406-x

1059-9495/$19.00

Tensile Testing of Nitinol Tubes and Wires with Higher Strain Rates

Rainer Johe

(Submitted September 26, 2008; in revised form January 16, 2009)

During tensile testing of superelastic Nitinol material, the specimen temperature increases as result of the exothermic Austenite-to-Martensite phase transformation. The increase in specimen temperature has great influence on the stress-strain response—in particular, upper and lower plateau values—and limits the strain rate of the tensile test, so that for larger specimen dimension, the strain rate has to be reduced. A special setup of the tensile testing equipment has been developed using a fan to improve the heat exchange between the specimen and the ambiance to allow much higher strain rates as well as even gradient of the upper and lower plateau. It could be shown that the strain rate of the first loading and unloading cycle could be two to four times higher as recommended in ASTM F 2516-07 without any negative impact on the determined values. The needed time for tensile testing of Nitinol products could be reduced considerably. The improved heat exchange gives a better comparability and reproducibility of the tensile test data.

Keywords advanced characterization, biomaterials, mechanical testing

1. Introduction

Tensile testing of Nitinol is a curious affair. Common and well-known tensile testing methods as described, e.g. in ASTM E8, DIN EN 10002-1, or ISO 6892 are not adequate and do not relate to the pseudo-elastic properties and shape memory behavior.

For this reason, a new testing standard has been established; the procedure is defined in the ASTM F 2516 (Ref 5), which is the "Standard Test Method for Tension Testing of Nickel-Titanium Superelastic Materials" such as wire, strip or tubing in the superelastic condition.

Because the thermal exchange in conjunction with the strain rate has a huge impact on the gradient of the tensile test curve and the results derived, a comparative study with various testing speeds and setup modes on tubing and wire was done.

2. Testing Procedure and Technical Background

Starting tensile testing (see Fig. 1), Nitinol shows normal linear elastic behavior; that means increasing extension of the tensile test specimen leads to an increasing tensile stress response.

This article is an invited paper selected from presentations at Shape Memory and Superelastic Technologies 2008, held September 21-25, 2008, in Stresa, Italy, and has been expanded from the original presentation.

Rainer Johe, EUROFLEX GmbH, Kaiser-Friedrich-Str. 7, 75172 Pforzheim, Germany. Contact e-mail: r.johe@rau-pforzheim.de.

At a certain stress level, the characteristic, stress-induced Austenite-to-Martensite phase transformation of Nitinol begins: typically in the range of about 400 to 600 MPa depending on the Af-temperature of the specimen.

Due to mechanical stress, single lattice areas of Austenite begin to shear and change their shape to Martensite. The effect is that gradually the whole specimen starts elongating at a constant load. The stress-strain curve shows typically the so-called upper (loading) plateau. The Austenite-to-Martensite transformation is completed (end of plateau) at a total elongation of about 7%.

To see the reversible phase transformation back from Martensite-to-Austenite—which occurs due to the superelastic effect of Nitinol—the test specimen gets unloaded within the plateau after a total elongation of 6%. After passing through an unloading hysteresis of about 300 MPa, the backward Martensite-to-Austenite phase transformation starts forming the unloading or lower plateau because of the reversible lattice movement of Nitinol.

Ideally, the superelastic sample returns to the starting point without any residual strain.

There are two main effects which have remarkable influence on the gradient of the tensile test curves of Nitinol as follows:

1. The level of the upper and lower plateau depends on the test temperature that means increasing test temperature leads to increasing plateau levels (typically 5-6 MPa/K) (see Fig. 2) and
2. The stress-induced Austenite-to-Martensite transformation is an exothermic phase transformation (Martensite-to-Austenite is endothermic).

The exothermic Austenite-to-Martensite transformation leads to an increase in the specimen temperature during the loading cycle, and in consequence, to an ongoing increase of the plateau stresses with increasing elongation. On the other hand, during unloading, the specimen temperature decreases because of the endothermic Martensite-to-Austenite transformation which leads to a permanent decrease of the lower plateau stresses.

Consequently, it is essential to control the temperature of the tensile test specimen within close limits.

One option is to get an adequate thermal exchange between the specimen and the static ambient air by reducing the tensile test speed (mainly during the first cycle). The limited strain rate allows the heat to transfer out of the specimen.

A second option is to modify the tensile test equipment to improve the thermal exchange between the specimen and the circumfluent medium, e.g. water or stirred air.

The maximum crosshead speed (strain rates) according to ASTM F 2516-07 depending on the specimen dimensions are listed in the following table:

d, diameter or thickness (mm) for round tubing: d = OD − ID = 2× wall	Maximum crosshead speed in mm/min per mm of initial length: First cycle (load to 6% strain and unload)	Second cycle (load to failure)
$d \leq 0.2$	0.08 (8%/min)	0.8
$0.2 < d \leq 0.5$	0.04 (4%/min)	0.4
$0.5 < d \leq 2.5$	0.02 (2%/min)	0.2
$d > 2.5$	0.01 (1%/min)	0.1

The strain rate for the second cycle to failure is ten times higher than for the first cycle

3. Equipment and Samples

The tensile testing equipment has been modified, which means a ventilator has been integrated providing a constant forced flow of ambient air (no cooling) against the test specimen to transfer the exothermic heat out of the specimen during the loading cycle and to improve the thermal exchange so that the specimen can pick up heat during the unloading cycle.

So the dimensional range of 0.5-2.5 mm is typical size for the most common Nitinol products, hence sample material of superelastic Nitinol wires in that range have been chosen for investigation.

Samples of Nitinol tubing were taken with the range of $d = 0.5$ that means tubings right below and above this are chosen to see the influence of change in testing speed.

The tensile tests were performed with different sizes and various strain rates during the first cycle. For the second cycle of load-to-failure, the recommended strain rate as per ASTM F 2516-07 was used (Ref 5).

4. Results

4.1 Nitinol Wire with 0.6 and 2.4 mm

Lower and upper end of the range ($0.5 < d \leq 2.5$; strain rate, first cycle 2%/min)

Left side: stress-strain-curve in static ambient air (see Fig. 3 and 5)
Right side: stress-strain-curve in forced air flow (see Fig. 4 and 6)

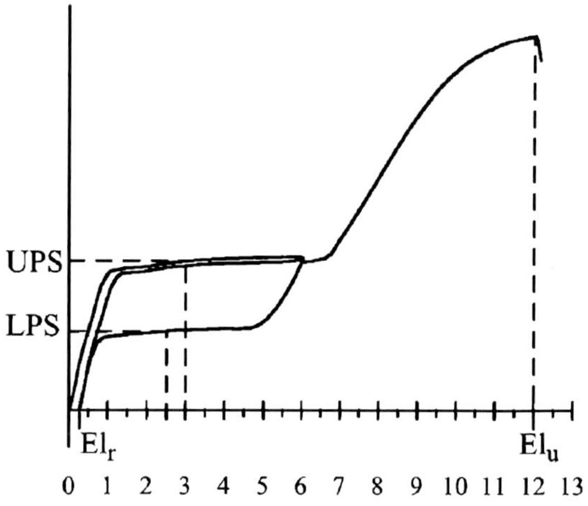

Fig. 1 Schematic stress-strain curve of superelastic Nitinol; picture taken from ASTM F 2516-07. Testing procedure: to see the superelastic behavior of Nitinol, a tensile test specimen gets pulled to the upper reversal point of 6% total elongation, unloaded to 7 MPa, and reloaded until failure

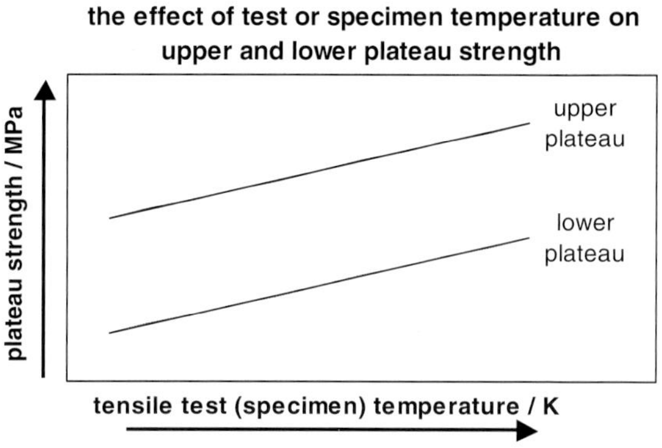

Fig. 2 Principal schematic dependence of upper and lower plateau stresses on the tensile test or specimen temperature. Upper plateau stresses (UPS) as well as lower plateau stresses (LPS) is increasing with increasing test or specimen temperature

4.2 Nitinol Tubing with 0.34 and 0.50 mm Wall

Mid of the range ($d = 0.68$ and 1.0; strain rate, first cycle 2%/min)

Left side: stress-strain curve in static ambient air (see Fig. 7 and 9)
Right side: stress-strain curve in forced air flow (see Fig. 8 and 10)

In Fig. 3, 5, 7, and 9, it can be seen that depending on the specimen dimension the stress-strain curve shows more or less a slope during the first loading cycle.

In case a fan is used—see Fig. 4, 6, 8, and 10—the slope of the stress-strain curve is much smaller than in static ambient air. That means, the forced air flow improves the heat exchange between specimen and ambient air.

UPS is about 30 MPa lower and LPS is about 30 MPa higher when measured using a fan than the respective values when measured without fan (Diagram 1 and 2). Therefore, the hysteresis between upper and lower plateau gets more pressed together in case of forced air flow.

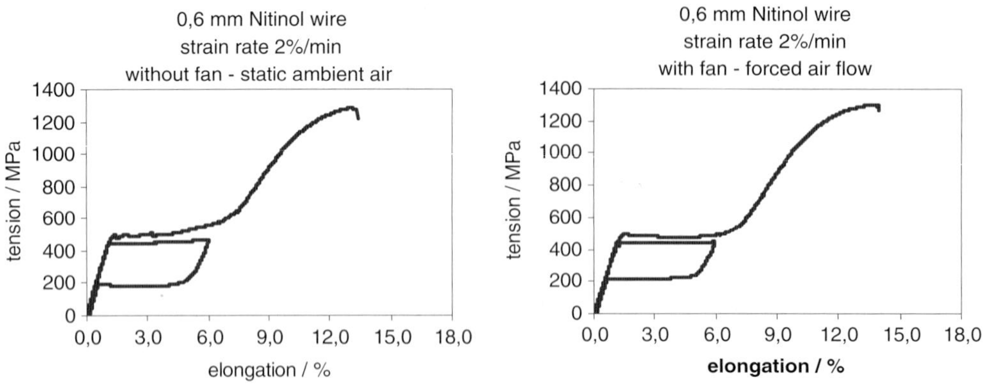

Fig. 3-4 Tensile test curves on a 0.6-mm Nitinol wire with the ASTM F 2516 recommended strain rate of 2%/min; *left*: without/*right*: with fan

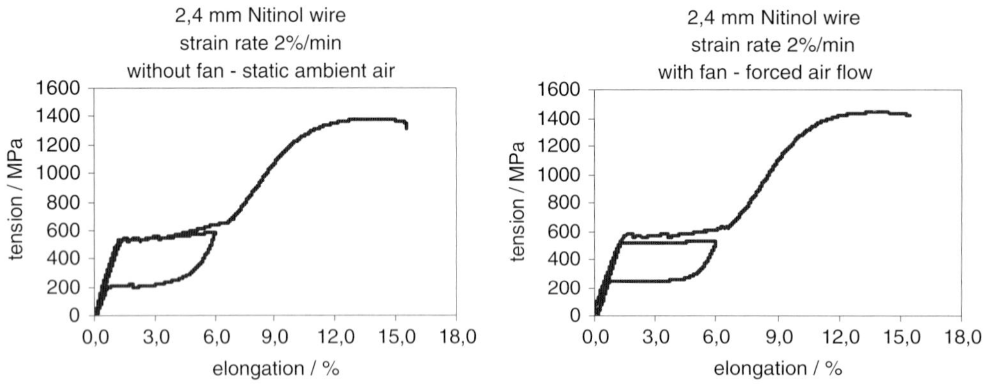

Fig. 5-6 Tensile test curves on a 2.4-mm Nitinol wire with the ASTM F 2516 recommended strain rate of 2%/min; *left*: without/*right*: with fan

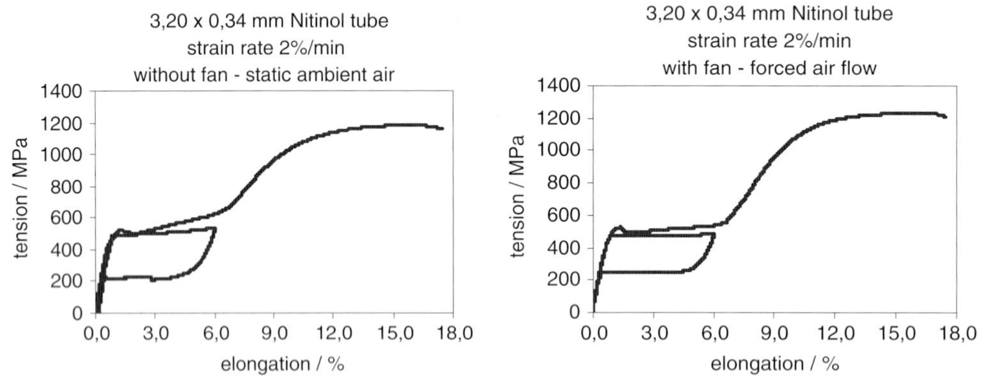

Fig. 7-8 Tensile test curves done on a 3.20×0.34 mm^2 Nitinol tube with the ASTM F 2516 recommended strain rate of 2%/min ($2\times$ wall = 0.68); *left*: without/*right*: with fan

Residual elongation (Permanent Set), UTS, and elongation are not influenced.

4.3 Nitinol Wire and Tubing—Tensile Test with Higher Strain Rates

In the following, only the first cycle is shown to see better the influence of various strain rates and the thermal exchange between specimen and ambient air to the slope of the stress-strain curve.

Left side: stress-strain curve in static ambient air (see Fig. 11, 13 and 15)

Right side: stress-strain curve in forced air flow (see Fig. 12, 14 and 16)

In Fig. 11, 13, and 15, it can be seen that—in case of static ambient air—the gradient of the upper plateau increases with increasing testing speed. The unloading plateau becomes more and more imprecise and the hysteresis between upper and lower plateau get wider with higher strain rates.

Whereas using a fan to force the air flow—see Fig. 12, 14, and 16—the upper plateau is still even and linear due to the improved thermal exchange between specimen and ambient air (Diagram 3 and 4).

Figures 17, 18, 19, and 20 show the heightened effect on 2.4-mm wire, which means—in case of static ambient air—the gradient of the upper plateau increases with increasing testing speed. On the right side, the gradient of the upper plateau is much more linear and even more so than on the left side because of the improved thermal exchange between specimen and ambient air.

Furthermore, there is a positive effect to the unloading plateau and the hysteresis between upper and lower plateau.

Figures 21 and 23 show that the increase in strain rate because of the thinner wall leads to a higher slope in the stress-strain curve of the 3.00 × 0.235 mm tubing.

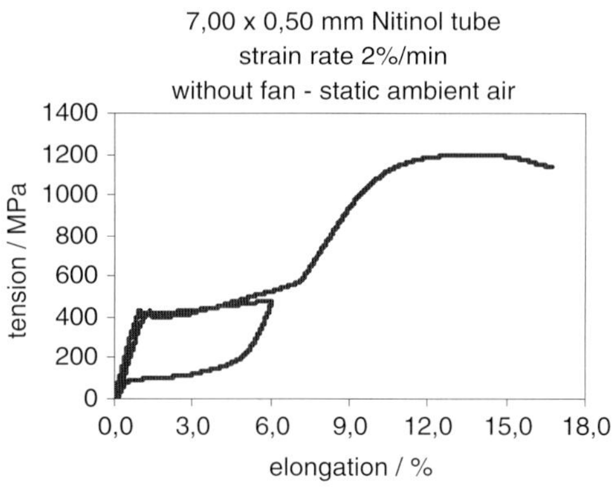

Fig. 9-10 Tensile test curves done on a 7.00 × 0.50 mm^2 Nitinol tube with the ASTM F 2516 recommended strain rate of 2%/min (2× wall = 1.0); *left*: without/*right*: with fan

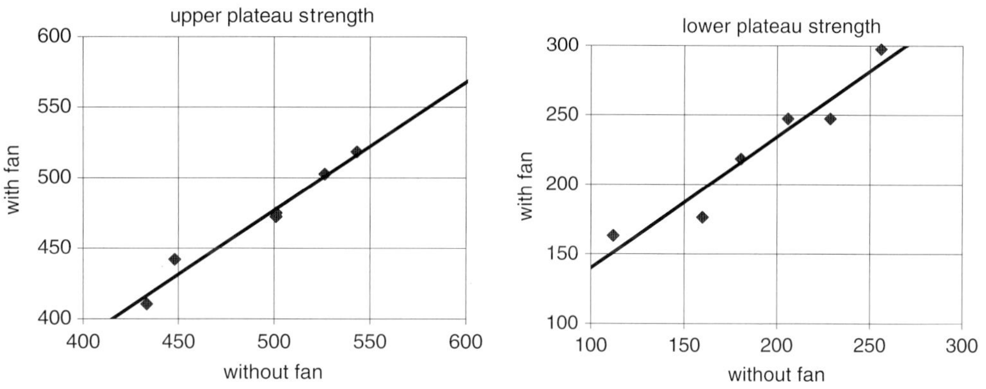

Diagram 1-2 Influence of forced air flow on UPS and LPS

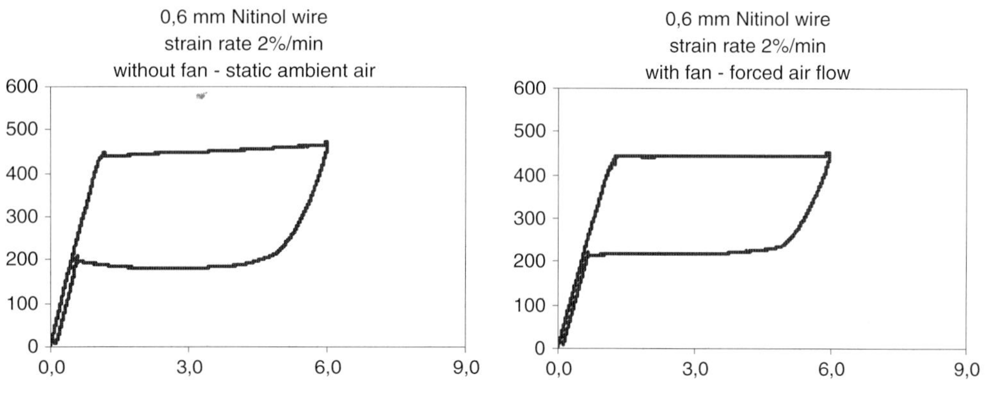

Fig. 11-12 Tensile test curves of a 0.6-mm Nitinol wire with the ASTM F 2516 recommended strain rate of 2%/min; *left*: without/*right*: with fan

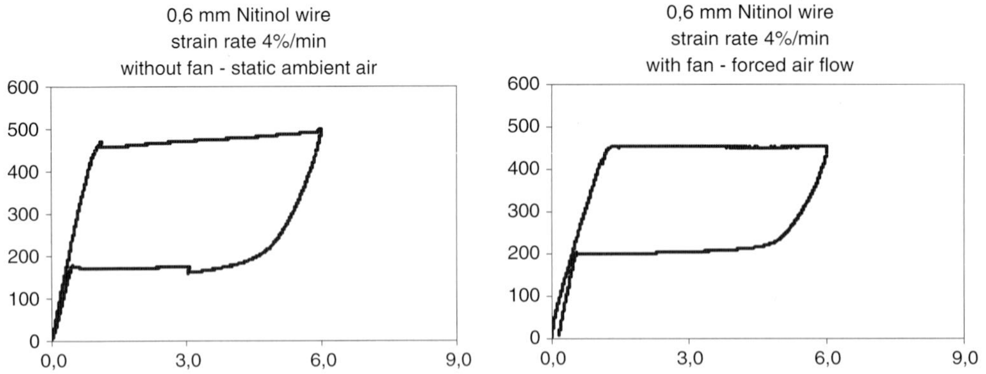

Fig. 13-14 Tensile test curves of a 0.6-mm Nitinol wire with two times higher strain rate, i.e. of 4%/min; *left*: without/*right*: with fan

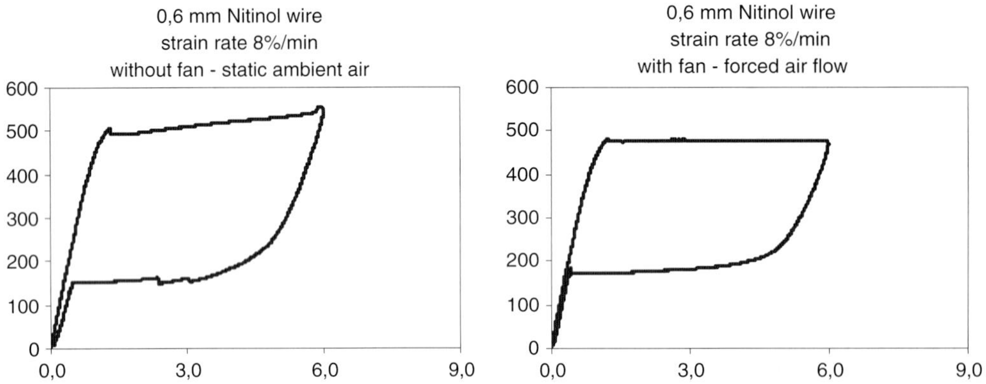

Fig. 15-16 Tensile test curves of a 0.6-mm Nitinol wire with four times higher strain rate, i.e. of 4%/min; *left*: without/*right*: with fan

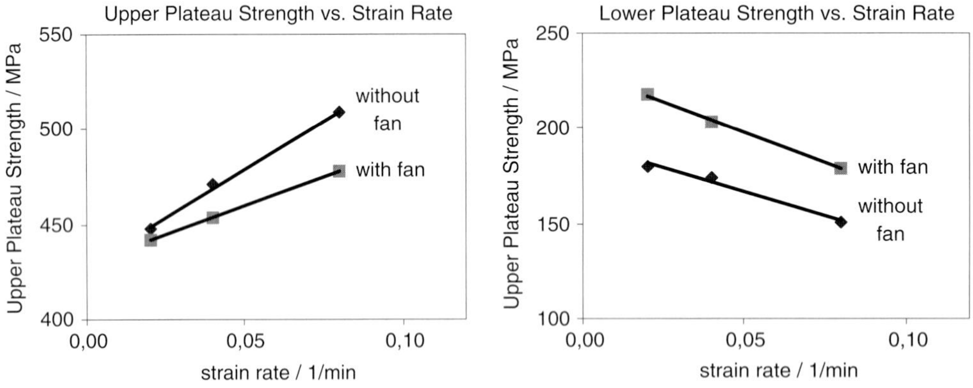

Diagram 3-4 Increase of UPS and decrease in LPS due to higher strain rates; effect of better heat exchange between specimen and ambient air using a fan

With the use of a fan—Fig. 22 and 24—the same material could be tested four times faster without increasing slope of the stress-strain curve. The upper plateau is linear and even because of the improved thermal exchange between specimen and ambient air.

Furthermore, there is a positive effect to the unloading plateau and the hysteresis between upper and lower plateau.

5. Conclusion

The exothermic Austenite-to-Martensite phase transformation during the loading cycle leads to an increase of the tensile test specimen temperature which has a huge influence on the stress-strain response. The exothermic heat depends on the

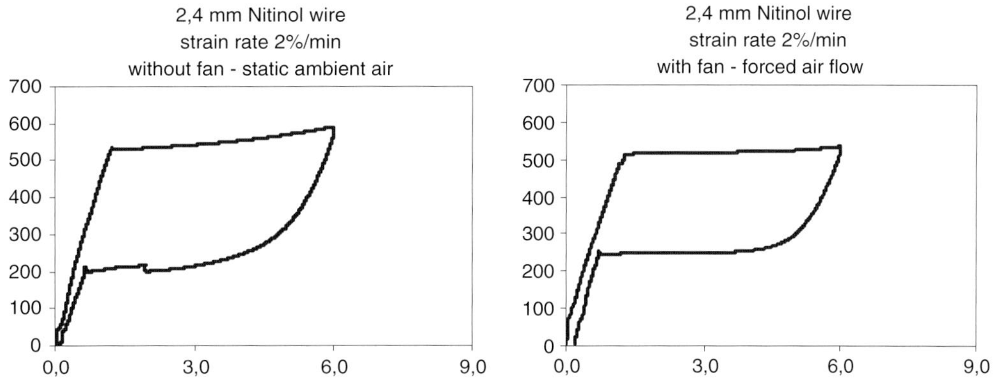

Fig. 17-18 Tensile test curves of a 2.4-mm Nitinol wire with the ASTM F 2516 recommended strain rate of 2%/min; *left*: without/*right*: with fan

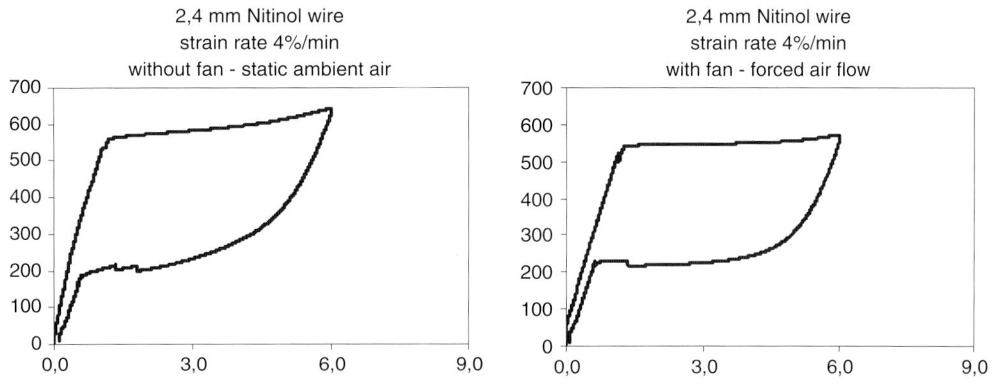

Fig. 19-20 Tensile test curves of a 0.60-mm Nitinol wire with two times higher strain rate of 4%/min; *left*: without/*right*: with fan

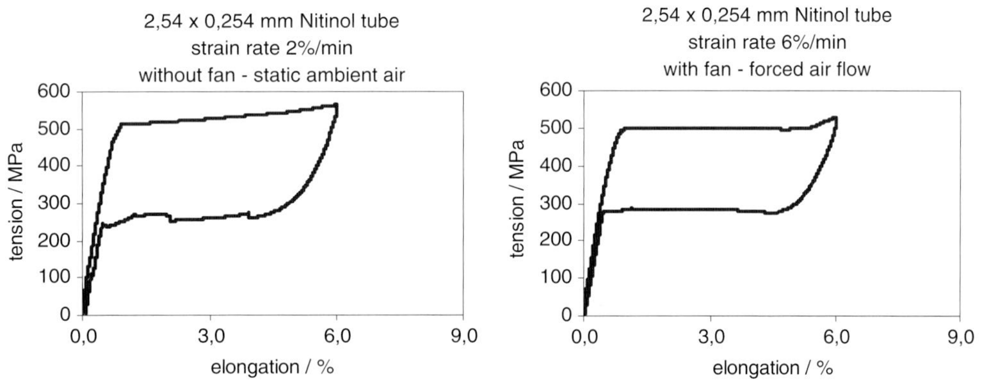

Fig. 21-22 Tensile test curves of a 2.54 × 0.254 mm^2 Nitinol tubing ($d > 0.5$). *Left side*: with the recommended strain rate of 2%/min without fan. *Right side*: Three times higher strain rate of 6%/min with fan

geometry of the tested specimen. In ASTM F 2516, the strain rate is limited depending on the wire diameter or the tube-wall thickness to allow the heat to transfer out of the specimen. For tubing, the defined steps are relatively large that means small differences in wall thickness could lead to different slopes in the stress-strain curves.

The use of a fan or any other equipment to improve the heat exchange between the specimen and the ambiance will reduce the slope effect on upper and lower plateau strength and will lead in consequence to more reproducibility of the data derived.

Improved heat exchange leads to lower UPS and higher LPS values because of the reduced gradient in the loading and unloading plateau.

Improved heat exchange allows higher strain rates during the first cycle without negative impact on UPS, LPS, Residual Elongation (Permanent Set), UTS, and elongation.

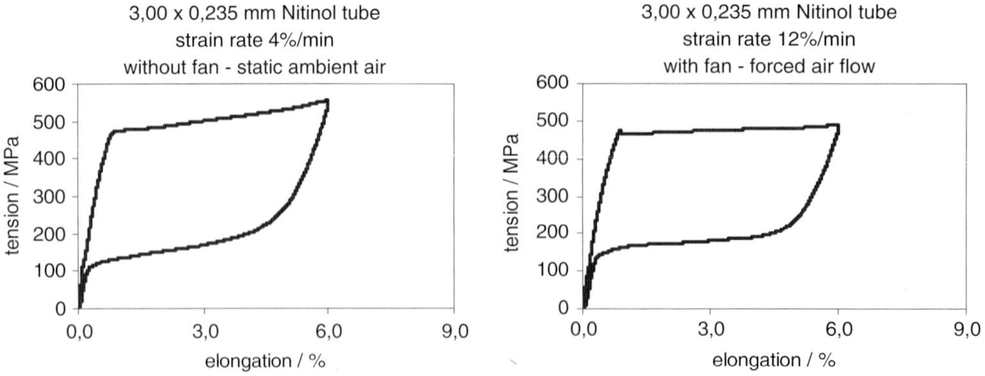

Fig. 23-24 Tensile test curves of a 3.00×0.235 mm^2 Nitinol tubing ($d < 0.5$). *Left side*: with the recommended strain rate of 4%/min without fan. *Right side*: Three times higher strain rate of 12%/min with fan

In the case of tubing, the impact of improved heat exchange is higher than that for wire because of the higher surface-to-volume ratio of tubing, which could allow three to four times higher strain rates during the first cycle.

The time needed for tensile testing could be reduced by about one-half.

The limitation of strain rates depending on rough dimensional groupings seems to be insufficient for all types of samplings. This could be a point for discussion in the next review of ASTM F 2516.

References

1. ASTM F 2004 Test Method for Transformation Temperature of Nickel-Titanium Alloys by Thermal Analysis
2. ASTM F 2005 Terminology for Nickel-Titanium Shape Memory Alloys
3. ASTM F 2063 Specification for Wrought Nickel-Titanium Shape Memory Alloys for Medical Devices and Surgical Implants
4. ASTM F 2082 Test Method for Determination of Transformation Temperature of Nickel-Titanium Shape Memory Alloys by Bend and Free Recovery
5. ASTM F 2516 Test Method for Tension Testing of Nickel-Titanium Superelastic Materials

JMEPEG (2009) 18:843–847
DOI: 10.1007/s11665-009-9489-4

1059-9495/$19.00

The Mechanical and Thermal Behaviors of Heat-Treated Ni-Rich NiTi Orthodontic Archwires

S.M. Seyyed Aghamiri, M. Nili Ahmadabadi, Sh. Raygan, I. Haririan, and M.S. Ahmad Akhondi

(Submitted September 16, 2008; in revised form April 21, 2009)

The near equiatomic nickel-titanium alloy is an outstanding intermetallic compound exhibiting distinctive properties associated with characteristic thermal and stress-induced martensitic transformations. The process of producing orthodontic wires has been modified to obtain the optimal shape memory behaviors. Phase transformation temperatures and load-deflection characteristics of this binary alloy are very significant variables in the performance of this alloy and can be manipulated by different thermomechanical treatments via inducing precipitation or dislocation networks in the matrix. In this study, one brand of commercial heat-activated nickel-titanium archwire (3 M Unitek) was selected and solution treated. Then, the wires annealed at 400 °C for 10, 30, and 60 min. Thermal transformation temperatures were determined using differential scanning calorimeter. It was showed that these temperatures increased with increasing the time of heat treatment and multistage transformation occurred as the result of inhomogeneities. In order to evaluate mechanical parameters of heat-treated archwires, they were placed on an archform fixture simulating maxillary dentition and load-deflection curves were obtained by three-point bending test at 37 °C. The results compared to as-received archwires and the best superelasticity was observed after 30 min aging.

Keywords biomaterials, heat treating, mechanical testing

1. Introduction

The shape memory nickel-titanium alloys (Nitinol) have been used in wide applications in medical fields due to their unique blend of mechanical properties and biocompatibility. Application of this alloy in orthodontics has been initiated from early 1970s (Ref 1, 2).

When a superelastic NiTi wire is deflected over a long distance, it eventually springs back to nearly its original position by reverse transformation from monoclinic to B2 structure while exerting light continuous forces (Ref 3-6).

Recently, a new generation of NiTi wires, i.e., shape memory wires (heat-activated NiTi wires) have been employed which exhibits clinically useful shape memory as well as superelasticity and higher transition temperature range (TTR) (Ref 7).

This article is an invited paper selected from presentations at Shape Memory and Superelastic Technologies 2008, held September 21-25, 2008, in Stresa, Italy, and has been expanded from the original presentation.

S.M. Seyyed Aghamiri, M. Nili Ahmadabadi, and **Sh. Raygan,** School of Metallurgy and Materials Engineering, University of Tehran, Tehran, Iran; **I. Haririan**, School of Pharmacy, Tehran University of Medical Science, Tehran, Iran; and **M.S. Ahmad Akhondi**, School of Dentistry, Tehran University of Medical Sciences, Tehran, Iran. Contact e-mails: sms.aghamiri@gmail.com and nili@ut.ac.ir.

Nickel amount, oxygen content, heat treatment, the amount of cold working, and deformation temperature are important factors that can have influence on the mechanical properties and superelasticity of NiTi alloy (Ref 1, 8). Since NiTi orthodontic wires generally contain over 50% Ni in their composition (Ni-rich), aging treatment will strongly influence thermal transformation behavior and stress deformation.

Many researchers have studied transformation behavior of Ni-rich NiTi alloys aged at various temperatures for different times (Ref 9-21). It has been confirmed that aging treatment introduces Ti_3Ni_4 metastable precipitates which change the transformation path to B2 $\rightarrow$ R $\rightarrow$ B19′. Age hardening of the alloy by such a precipitation also affects shape memory effect and superelasticity greatly (Ref 8, 22). Ni-rich NiTi alloys exhibit superelasticity in a wide temperature range as the critical stress for slip increases when high density of small precipitates is available in the B2 matrix.

In recent years, the effect of aging of this alloy on thermal multistage transformations observed in DSC curves has been a controversial topic. Three different explanations including local stress inhomogeneity, local composition inhomogeneity, and the most important theory based on grain boundary effect have been exhibited. TEM observations have explained sequential peaks of DSC curves of low nickel supersaturated Ni-rich alloys by heterogeneous precipitation between grain boundaries and grain interiors.

The purpose of this work was to study the effect of aging treatment at 400 °C for short times without cold work on the force-deflection characteristics and thermal transformation behaviors of solution-treated nickel-titanium orthodontic archwires. In the research, force-deflection characteristics of the aged archwires were compared to as-received archwires using a special three-point bending fixture.

2. Experimental Procedures

In this study, shape memory NiTi orthodontic archwires (Nitinol Heat-Activated, 3 M Unitek, Monrovia, CA, USA) with diameter of 0.41 mm were used as the Ni-rich alloy material.

In order to investigate the effect of heat treatments, the wires were first enclosed in evacuated quartz capsule and solution treated at 1000 °C for 90 min followed by water quenching. Then aging of solution-treated wires were performed at 400 °C for different times of 10, 30, and 60 min in air atmosphere.

The thermal transformation characteristics were determined using differential scanning calorimeter (Mettler-822^{e} DSC instrument). The samples were cut from the end part of heat-treated wires and cleaned in acetone containing ultrasonic cleaner. The complete cooling and heating cycles were done as follows.

The samples were quickly heated to 120 °C and held for 3 min following by cooling to −140 at a rate of 10 °C/min and held for 3 min before heating to 120 °C at the same rate. The transformation temperatures were specified by extrapolated tangential method. In the case of multistage overlapped peaks, only sharper peaks were considered. The characteristic temperatures were R_s, R_f, M_s, M_f and R'_s, R'_f, A_s, A_f as R phase, martensite and reverse R phase, austenite start and finish transformation temperatures, respectively. Some of the characteristic temperatures were missed due to overlapping of the peaks.

The force-deflection properties of as-received and heat-treated wires were studied by three-point bending (TPB) test (loading-unloading) in the simulated clinical conditions with a Zwick universal testing machine fitted with a 2.5 kN compression load cell and a crosshead speed of 1.0 mm/min in a dentition simulated fixture (Ref 23, 24) (Fig. 1).

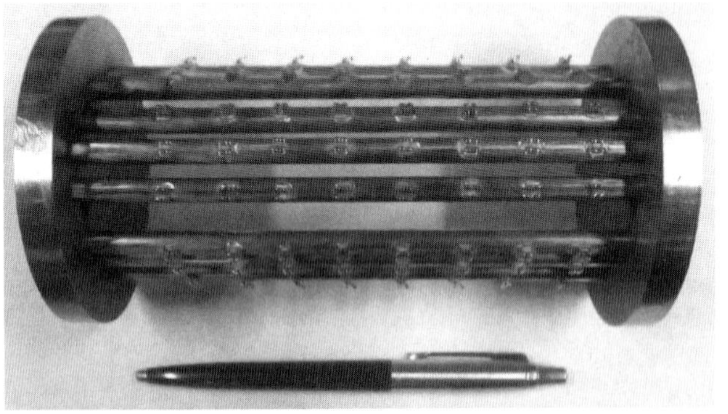

Fig. 1 Arch-form fixture simulating maxillary dentition for TPB tests (Ref 23, 24)

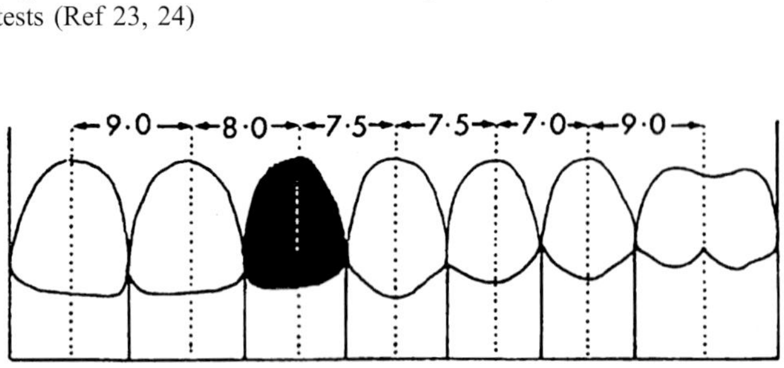

Fig. 2 Used Interbracket distance (mm) on fixture and location of irregular tooth (black tooth) (Ref 6)

By means of this fixture, the three-point bending test simulates wire force on the teeth in the oral configuration. The rods representing teeth No. 2 & 5 (left) and No. 3 & 5 (right) removed for dynamic three-point bending test. The wires were placed on the fixture through the brackets. The used inter-bracket distance on the fixture was selected according to Wilkinson et al.'s work (Fig. 2) (Ref 6). The tests were carried out on tooth No. 3 as irregular tooth to 4 mm deflection. The wires were tested in a water bath at 37 °C. The temperature control was done during each test with a K-type thermocouple in the range of accuracy of ±2 °C. Each bending test was repeated three times with the same new wire.

3. Results and Discussion

3.1 DSC and TPB Curves of As-Received Wires

Figure 3 shows the DSC curve of as-received orthodontic wire. As shown in the cooling cycle, two completely separated and distinct peaks can be seen. The peaks attributed to A → R and R → M transformations occur at 10 and −55 °C, respectively. In heating, only one sharp peak can be observed with the latent heat of −18 J/g, implying one stage transformation of M → A. The start and final transformation temperatures of the wires are shown in Table 1.

According to literature (Ref 22), two microscopic factors affect on the R phase formation prior to martensitic transformation: (1) introduction of dislocation network, (2) formation of Ti_3Ni_4 precipitates. In as much as combined cold work and aging treatments are a part of the process of orthodontic wire production, it is expected that the both factors induce internal stresses which consequently result in R phase transformation in DSC curve of NiTi wire.

TPB test of the wire at 37 °C in Fig. 4 demonstrates pseudoelastic behavior of the orthodontic wire with very limited residual strain. It should be noted the force level around 3 mm deflection in the unloading plateau drops due to slippage of wire through the brackets during loading as supported by Parvizi et al. (Ref 25).

According to Table 1, the A_f temperature of the as-received wire is about 19 °C. Thus, the deformation of wire has been done at about 20° above A_f temperature. So, the stress transformation behavior is reversible by unloading due to the higher critical stress for slip against the stress required to induce martensite.

Regarding the TPB curve of as-received wire (Fig. 4), the necessary force to move the austenite/martensite interface

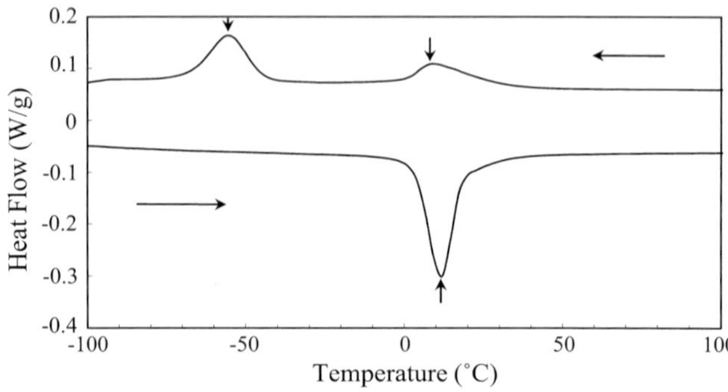

Fig. 3 DSC curve of as-received wire

Table 1 Transformation temperatures of as-received and heat-treated wires

	Temperature							
Condition	R_s	R_f	M_s	M_f	R'_s	R'_f	A_s	A_f
As-received wire	29	0	−44	−68	···	···	3	19
Solution-treated wire	···	···	−19	−26	···	···	−16	0
Aged wire at 400 °C/10 min	34	−20	−29	−60	−25	8	···	36
Aged wire at 400 °C/30 min	35	−14	−28	−54	−47	···	···	40
Aged wire at 400 °C/60 min	46	···	···	−13	−3	···	···	46

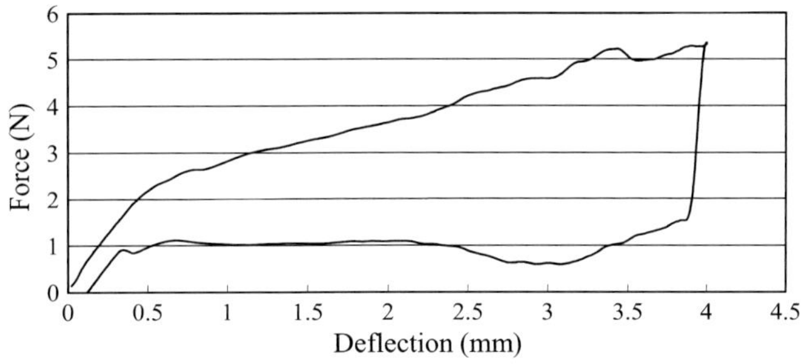

Fig. 4 TPB curve of as-received wire

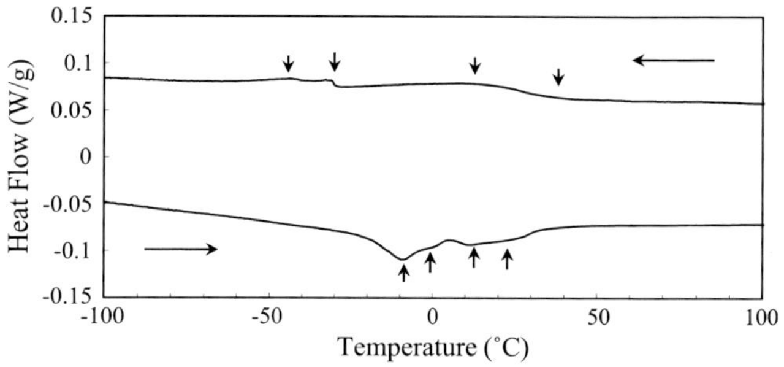

Fig. 6 DSC curve of solution treated and aged wire at 400 °C for 10 min

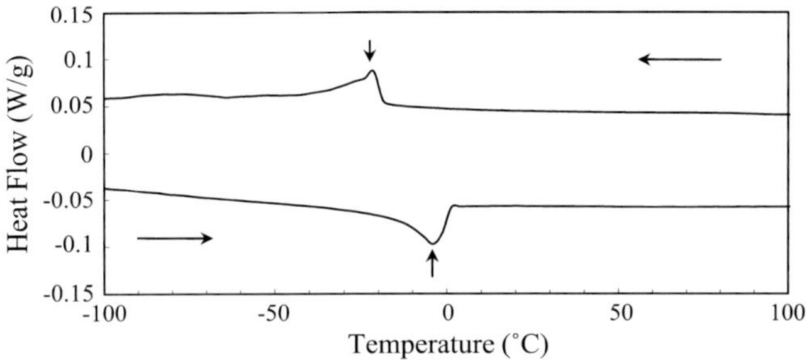

Fig. 5 DSC curve of solution-treated wire

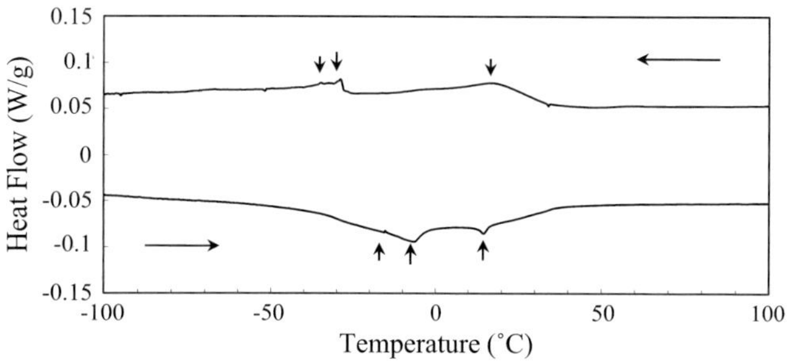

Fig. 7 DSC curve of solution treated and aged wire at 400 °C for 30 min

during loading increases progressively as the deflection increases. However, unloading force level is almost constant at about 1 N during a wide range of deflection. This characteristic is a clinical advantage so as to accomplish optimal tooth movement (Ref 1).

3.2 DSC Curves of Heat-Treated Wires

DSC results of solution-treated and subsequent aged wires at 400 °C for 10, 30, and 60 min have been depicted in Fig. 5-8.

The solution-treated wire exhibits direct A↔M transformation upon both cooling and heating cycles with thermal hysteresis of 18 °C, suggesting recovery of defects in the microstructure and solution of precipitates in the matrix.

It can be possible to clarify the composition of NiTi alloy in solution-treated state by the following relation (Ref 26):

$$M_s = (60 - (\text{at.\%Ni} - 50.0) \times 180)\,^\circ\text{C}$$

So the approximate composition of this orthodontic wire is about 50.5 at.% Ni. Considering the excess nickel content in composition, nucleation of metastable Ti_3Ni_4 particles can be expected during aging at appropriate temperatures.

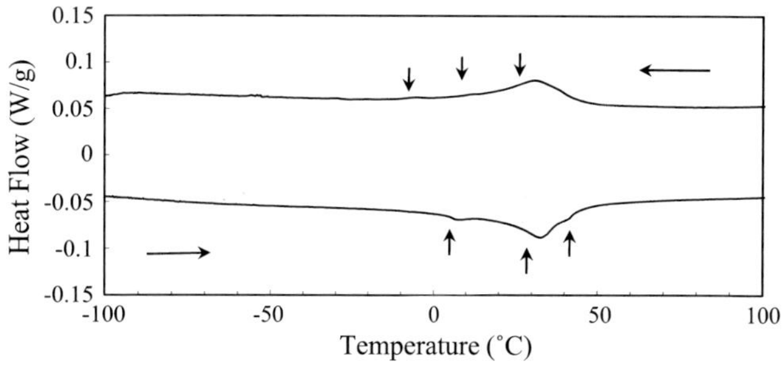

Fig. 8 DSC curve of solution treated and aged wire at 400 °C for 60 min

Figures 6-8 show that after aging treatment, A↔R and R↔M transformations take place as multistage transformations with determined transformation temperatures in Table 1.

The results indicate that transformation temperatures shift to higher temperatures with aging time. In addition, one-stage martensitic transformation in solution-treated wire turns to multistage transformation in a wide range of temperatures. It can be expected that due to dissolution of matrix and formation of coherent particles of Ti_3Ni_4, R phase transformation takes place in the way that its peak have partially overlapped on forward and reverse martensitic transformations.

Two points should be considered by comparing the DSC curves of aged wires: (1) The transformation temperatures particularly $R \leftrightarrow M$ transformation increases with increasing aging time, (2) the number of $R \leftrightarrow M$ transformation peaks reduces during both cooling and heating cycles with increasing aging time.

The changes in transformation temperatures with longer aging time can be related to increasing the density and size of Ni-rich Ti_3Ni_4 precipitates which result in depletion of Ni content of matrix. Therefore, the transformation temperatures move toward higher amounts. However, conversely $A \leftrightarrow R$ transformation peaks are not greatly affected by the evolution of precipitates due to low transformation strain.

As a clinical point of view, since the reverse martensitic transformation should be completed about 37 °C to have superelastic properties, it could be seen in Fig. 7 and 8 that 30 min aging has risen A_f temperature a little higher than 37 °C while 60 min aging has increased the transformation temperatures inappropriately, so that 60 min aged wire has a partial martensitic structure at 37 °C.

On the number of transformation peaks, some studies have expressed various stress fields around the Ti_3Ni_4 particles (Ref 9-13). On the other hand, many previous studies have reported that heterogeneous precipitation reaction happens preferentially around the grain boundaries and the grain interior regions are free of precipitates (Ref 13-21). These inhomogeneities bring about multistage transformations in DSC curves of aged samples in the present study. However, it seems that the inhomogeneties decrease as aging period increases and the precipitation is developed in more regions. It is well understood in the DSC curves that transformation features have changed from four stages in aged wires at 400 °C for 10 min to three stages after 30 and 60 min aging. In general, formation of sharp peaks of transformation under 37 °C with lower inhomogeneities is an advantage for engineering design of orthodontic wires which result in lessening residual martensite.

3.3 TPB Curves of Heat-Treated Wires

In Fig. 9-12, stress-induced transformation loop of heat-treated wires at different conditions can be observed.

It is obvious from the results that after unloading, only a small amount of deformation of solution-treated wire is reversible. This phenomenon reveals the low resistance of matrix to slip. Therefore, the plastic deformation of austenite by slip occurs in advance of the formation of stress-induced martensite by twinning mechanism.

It seems that after 10 min aging, localized precipitation leads to increase in magnitude of transformation strain recovery. After 30 min, more coherent particles precipitate and disperse in such a way that the strength of the matrix increases. According to DSC data in Table 1, although A_f temperature is about 40 °C, the superelasticity of the wire is revealed completely probably due to dynamic mechanical effects. The resistance of particles to the reverse martensitic transformation causes a considerable hysteresis, about 5.5 N, between loading and unloading plateaus.

The size of Ti_3Ni_4 precipitates is supposed to increase with increasing aging time up to 60 min (Ref 22). This increase may result in a decrease in the strength of the austenite matrix and the plastic deformation of the austenite may probably cause residual deformation in the matrix. Thus, the stress level of the unloading plateau shifts to a higher level and the stress hysteresis is reduced considerably. In addition, by increasing the size of precipitates and depletion of Ni content of the

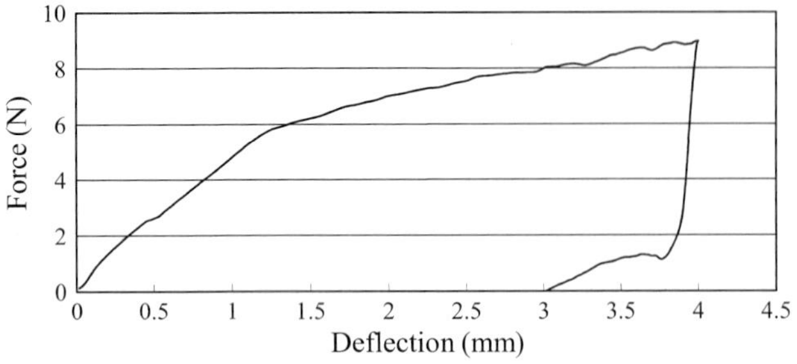

Fig. 9 TPB curve of solution-treated wire

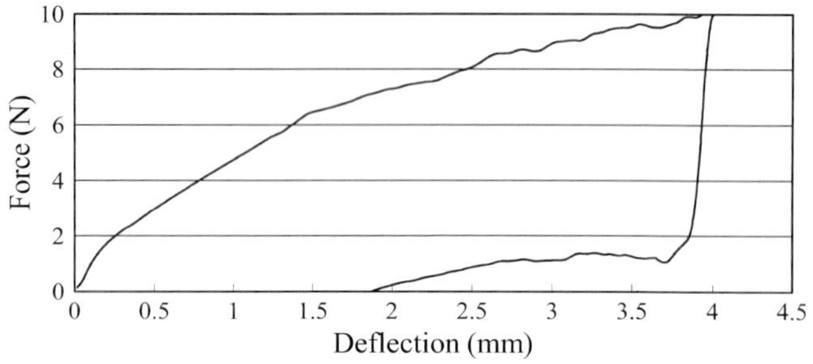

Fig. 10 TPB curve of solution treated and aged wire at 400 °C for 10 min

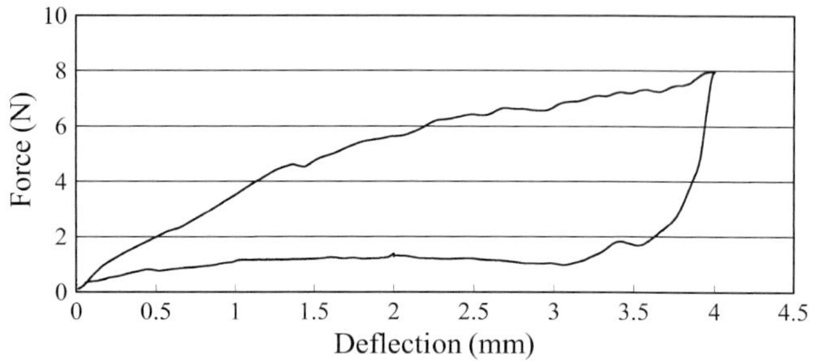

Fig. 11 TPB curve of solution treated and aged wire at 400 °C for 30 min

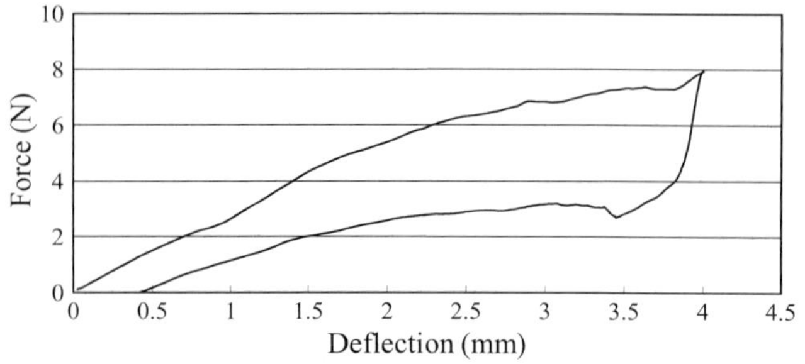

Fig. 12 TPB curve of solution treated and aged wire at 400 °C for 60 min

matrix, according to Table 1, the transformation temperatures have risen up and some remaining martensite in the matrix has caused residual deformation of the wire during TPB test.

4. Conclusions

1. Multistage martensitic transformations can be observed in DSC peaks for 10, 30, and 60 min aging of wires at 400 °C while there are quite sharp peaks in as-received wires.
2. Transformation temperatures shift to higher temperatures with increasing time of aging.
3. Strain recovery improved with increasing time of aging up to 30 min and TPB curve showed complete superelasticity after 30 min aging, but after 60 min aging, strain recovery decreased.
4. The superelasticity of the wire after 30 min aging was similar to as-received wire although the stress hysteresis of the aged wire is higher than for the others.
5. In terms of unloading plateau and residual deformation, 30 min aged wire shows the best result among heat-treated wires for orthodontic application but with quietly high loading plateau in comparison to as-received wire.

References

1. R. Sachdeva and S. Miyazaki, Superelastic Ni-Ti Alloys in Orthodontics, *Engineering Aspects of Shape Memory Alloys*, T.W. Duering, K.N. Melton, D. Stöckel, and C.M. Wayman, Ed., Butterworth-Heinemann, London, 1990, p 452-457
2. V.G. Dorodeiko, V.V. Rubanik, V.V. Rubanik Jr., and S.N. Miliukina, Heat Treatment of TiNi Wire Used for Interauterine Contraceptives, *Mater. Sci. Eng. A*, 2007, **481-482**, p 616–619
3. C. Gioka and T. Eliades, Superelasticity of Nickel-Titanium Orthodontic Archwires: Metallurgical Structure and Clinical Importance, *Hell. Orthod. Rev.*, 2002, **5**, p 111–127
4. D.C. Mallory, J.D. English, J.M. Powers, W.M. Brantley, and H.I. Bussa, Force-Deflection Comparison of Superelastic Nickel-Titanium Archwires, *Am. J. Orthod. Dentofacial. Orthop.*, 2004, **126**, p 110–112
5. B.K. Ruckerand and R.P. Kusy, Elastic Properties of Alternative Versus Single Stranded Leveling Archwires, *Am. J. Orthod. Dentofacial. Orthop.*, 2002, **122**(5), p 28–41
6. P.D. Wilkinson, P.S. Dysart, J.A.A. Hood, and G.P. Herbison, Load-Deflection Characteristics of Superelastic Nickel-Titanium Orthodontic Wires, *Am. J. Orthod. Dentofacial. Orthop.*, 2002, **121**, p 483–495
7. M. Iijima, H. Ohno, I. Kawashima, K. Endo, and I. Mizoguchi, Mechanical Behavior at Different Temperatures and Stresses for Superelastic Nickel-Titanium Orthodontic Wires Having Different Transformation Temperatures, *Dent. Mater.*, 2002, **18**, p 88–93
8. A.R. Pelton, J. DiCello, and S. Miyazaki, Optimisation of Processing and Properties of Medical Grade Nitinol Wire, *Min. Invas. Ther. Allied. Technol.*, 2000, **9**(1), p 107–118
9. L. Bataillard, J.-E. Bidaux, and R. Gotthardt, Interaction Between Microstructure and Multiple-Step Transformation in Binary NiTi Alloys Using In Situ Transmission Electron Microscopy Observations, *Phil. Mag. A*, 1998, **78**(2), p 327–344
10. J.I. Kim, Y. Liu, and S. Miyazaki, Ageing-Induced Two-Stage R-Phase Transformation in Ti-50.9 at.% Ni, *Acta. Mater.*, 2004, **52**, p 487–499
11. M. Peltonen, T. Lindroos, and M. Kallio, Effect of Ageing on Transformation Kinetics and Internal Friction of Ni-rich Ni-Ti Alloys, *J. Alloys Comp.*, 2007, **460**(1–2), p 237–245
12. J. Michutta, Ch. Somsen, A. Yawny, A. Dlouhy, and G. Eggeler, Elementary Martensitic Transformation Processes in Ni-Rich NiTi Single Crystals with Ni_4Ti_3 Precipitates, *Acta. Mater.*, 2006, **54**, p 3525–3542
13. L.J. Chiang, C.H. Li, Y.F. Hsu, and W.H. Wang, Age-Induced Four-Stage Transformation in Ni-Rich NiTi Shape Memory Alloys, *J. Alloys Comp.*, 2008, **458**, p 231–237
14. Z.G. Wang, X.T. Zu, X.D. Feng, H.Q. Mo, and J.M. Zhou, Calorimetric Study of Multiple-Step Transformation in TiNi Shape Memory Alloy with Partial Transformation Cycle, *Mater. Lett.*, 2004, **58**, p 3141–3144
15. Y. Zheng, F. Jiang, L. Li, H. Yang, and Y. Liu, Effect of Ageing Treatment on the Transformation Behaviour of Ti-509 at% Ni alloy, *Acta. Mater.*, 2008, **58**, p 736–745
16. P. Filip and K. Mazanec, On Precipitation Kinetics in TiNi Shape Memory Alloys, *Scri. Metall.*, 2001, **45**, p 701–707
17. Y. Zhou, J. Zhang, G. Fan, X. Ding, J. Sun, X. Ren, and K. Otsuka, Origin of 2-Stage R-Phase Transformation in Low-Temperature Aged Ni-rich Ti-Ni Alloys, *Acta. Mater.*, 2005, **53**, p 5365–5377
18. G. Fan, W. Chen, S. Yang, J. Zhu, X. Ren, and K. Otsuka, Origin of Abnormal Multi-Stage Martensitic Transformation Behavior in Aged Ni-Rich Ti-Ni Shape Memory Alloys, *Acta. Mater.*, 2004, **52**, p 4351–4362
19. G. Fan, Y. Zhou, W. Chen, S. Yang, X. Ren, and K. Otsuka, Precipitation Kinetics of Ti_3Ni_4 in Polycrystalline Ni-Rich TiNi Alloys and Its Relation to Abnormal Multi-Stage Transformation Behavior, *Mater. Sci. Eng. A*, 2006, **438-440**, p 622–626
20. O. Bojda, G. Eggeler, and A. Dlouhy, Precipitation of Ni_4Ti_3-Variants in a polycrystalline Ni-Rich NiTi shape memory alloy, *Scri. Mater.*, 2005, **53**, p 99–104
21. J.K. Allafi, G. Eggeler, W.W. Schmahl, and D. Sheptyakov, Quantitative Phase Analysis in Microstructures Which Display Multiple Step Martensitic Transformations in Ni-Rich NiTi Shape Memory Alloys, *Mater. Sci. Eng. A*, 2006, **438-440**, p 593–596
22. K. Otsuka and X. Ren, Physical metallurgy of Ti-Ni-Based Shape Memory Alloys, *Prog. Mater. Sci.*, 2005, **50**, p 511–678
23. M. Nili-Ahmadabadi, T. Shahhoseini, M. Haj-Fathalian, M.H. Parsa, H. Ghadirian, T. Hoseinzadeh-Nik, Cyclic Load-Deflection Characteristics of NiTi Orthodontic Archwires at Different Temperatures, *Proceedings of the 8th Asia-Pacific Conference on Materials Processing*, Guilin-Guanzhan, 2008
24. H. Ghadirian, "The Evaluation of Superealasticity of Nickel-Titanium Orthodontic Archwires in the Simulated Oral Environment," Ph.D. Thesis, Tehran University of Medical Sciences, 2008
25. F. Parvizi and W.P. Rock, The Load/Deflection Characteristics of Thermally Activated Orthodontic Archwires, *Eur. J. Orthod.*, 2003, **25**, p 417–421
26. W. Tang, Thermodynamic Study of the Low-Temperature Phase B19′ and the Martensitic Transformation in Near-Equiatomic Ti-Ni Shape Memory Alloys, *Metall. Trans. A*, 1997, **28**, p 537–544